STATISTICAL DATA ANALYSIS
AND INFERENCE

MONDAY

August 21

at 20.00 pm

Aula des Jeunes Rives

INDIAN CLASSICAL DANCES

KUCHIPUDI AND BHARATA NATYAM

by

TEJASWINI

supported by her disciples

ADITI BASAK and SMITA BASAK

CLASSICAL DANCES OF INDIA

There are six classical dance styles in India, each belonging to a different region, all having their origin in antiquity. These dance forms are based on the "Natya Sashtra" or the "Science of Dramaturgy" written in the 4th century B.C.

There are three aspects to Indian Classical Dancing - Nritta (pure dance) where the movements do not convey any meaning but create aesthetic pleasure, Nritya (expressive dance) where the meaning of the song is conveyed through hand gestures and facial expressions and Natya (dramatic element).

The dancing is religious in character using themes from Indian mythology and shows a high degree of sophistication and stylization. The art of dancing has been preserved intact by the patronage of kings and nobles and in the temples as a form of worship.

PROGRAM

This evening's recital will feature dances in two different classical dance styles of South India, Kuchipudi and Bharata Natyam. The numbers will be preceeded by an English commentary and demonstration.

1. **PRAYER** Kuchipudi

   This is an invocatory piece where the dancer pays obeisance to the Gods for a successful program and welcomes the audience.

2. **MANDODHARI SABDHAM** Kuchipudi

   This immortal piece is dedicated to Sri Krishnadevarayya, the famous king of Andhra in South India who ruled in the 16th century.

   A dancer enters the court of Sri Krishnadevarayya, sings the glory of the great patron King and then enacts the story of the Frog Maiden, Mandodhari to entertain him.

   According to legend Mandodhari is cursed to become a frog and lives in a pond, awaiting the arrival of Ravana the demon King of Lanka to release her. The court dancers protrays this story. The sequence of the story is as follows:

   A frog living in a pond amidst swimming fish and flying birds is transformed into a beautiful maiden, Mandodhari. The demon king Ravana spots her, falls in love with her, and carries her away triumphantly to his kingdom.

3. **JATISWARAM** Bharata Natyam & Kuchipudi

This is a pure dance where the emphasis is on rhythmic footwork. The dance will be performed simultaneously in the two dance styles - Bharata Natyam and Kuchipudi to show the similarities and differences in technique between the two styles.

4. **THILLANA** Kuchipudi

A scintillating dance number with sculpturesque poses and intricate rythmic patterns set to a fast tempo.

INTERMISSION

5. **PANDATTAM** Bharata Natyam

The dance portrays the beautiful princes Vasantavalli in the company of her friends playing with a ball. Several movements with the imaginary ball are depicted in the dance.

6. **KRISHNA SABDHAM** Kuchipudi

This dance is about the romantic God Krishna, set in an amorous mood. Satyabhama, the consort of Krishna, tries to win her Lord's affection by using her charm, smearing him with sandlewood paste, sprinkling him with rose water, making a garland of flowers, and giving him beetle leaves to chew. She sings and dances praising his beauty and valor. Finally Krishna succumbs to her charms.

7. **THARANGAM** Kuchipudi

Tharangam is the highlight of a Kuchipudi dance recital. The dance begins with a description of Lord Krishna. In the final segment, the dancer executes rhythmic footwork standing on the rim of a brass plate while balancing a pot of water on the head.

8. **MANGALAM**

The performance will end with a Mangalam, the final salutations to the Gods and audience.

ABOUT THE ARTISTS

TEJASWINI

Tejaswini is one of the foremost exponents of Kuchipudi and Bharata Natyam, two Indian Classical Dances having studied under such masters as Narasimhachari and Vempati Chinna Satyam. She started dancing at the age of six and since then has won the acclaim of the public and press across the world in India, Ceylon, Europe, Canada, and U.S.A. She has represented India at festivals held in Poland, France and Jamaica. In Ceylon she was awarded the title of "Nritya Deepika" - the Light of Dancing. The Telugu Association of Northern America honored her for her outstanding contribution to the art at their 6th meeting held at St. Louis.

Tejaswini's repertoire includes pure dance and dances involving mime. Tejaswini is noted for her precise, crisp movements combined with grace and elegance. Her rendering of the Kuchipudi style radiates energy and joyous abandon. The same dynamism is reflected in her delivery of abhinaya or expression. She conveys a wide range of moods and emotions with sincere feeling and ease.

Tejaswini has been the director of a school of Indian Classical Dancing, "Natya" in Buffalo since 1977 and several of her students have completed their arangetrams. Her production of the dance drama "Annamacharya" with her students was highly acclaimed in all the cities in which it was presented.

Tejaswini's academic stance lends the same lustre projected in her dancing. She received her Ph.D. from the Pennsylvania State University and is now an Associate Professor at the New York State University College at Buffalo. Her mental acuity is reflected in her style of presentation and meaningful expression of the rich repertoire of the Bharata Natyam and Kuchipudi styles. Her performances are entertaining both to the expert connoisseur of the art and the novice because of the commentary and demonstration that accompanies her dancing.

Tejaswini is a dedicated dancer who is working hard to keep alive the rich and ancient tradition of Indian dancing and to propogate the art by training the younger generation of Indian children in the U.S.A.

ADITI and SMITA BASAK

Aditi and Smita Basak have studied dance under Tejaswini for several years. They completed their dance graduation in 1985 and since then have been performing regularly individually and with their teacher. Aditi Basak is currently studying pre - medecine in U.S.A. while Smita Basak is in the 9th grade. Both sisters while being dedicated to Indian Classical dancing also perform Jazz, Tap dancing and Ballet.

STATISTICAL DATA ANALYSIS AND INFERENCE

edited by

Yadolah DODGE
University of Neuchâtel
Switzerland

1989

NORTH-HOLLAND
AMSTERDAM • NEW YORK • OXFORD • TOKYO

ISBN: 0 444 88029 1

ELSEVIER SCIENCE PUBLISHERS B.V.
Sara Burgerhartstraat 25
P.O. Box 211, 1000 AE Amsterdam, The Netherlands

Distributors for the United States and Canada:
ELSEVIER SCIENCE PUBLISHING COMPANY, INC.
655 Avenue of the Americas
New York, N.Y. 10010, U.S.A.

PRINTED IN THE NETHERLANDS

To my dear sons
Ali and Arash

C.R. Rao 1989

PREFACE

This volume contains invited papers presented at the International conference on Recent Developments in Statistical Data Analysis and Inference, held in Neuchâtel, Switzerland from August 21-24, 1989, in honor of C. Radhakrishna Rao, one of the most eminent statisticians of our time and a distinguished scientist.

Recently, the Times of India (December 31, 1988) after lengthy consultations with scientists in different disciplines chose C.R. Rao to be among the top ten greatest contributors to Indian Science.

C.R. Rao's contribution to the development of the modern theory of statistics has been so immense and has had such worldwide recognition that there is no need for comment in this preface. However, for those who might not be fully aware of the important role he played in the development of statistics in general and in particular in the growth of the Indian Statistical Institute (ISI), and in promoting statistical education and training in South East Asian countries, a brief summary of his biography is presented at the beginning of this volume.

Up to now he has been author or co-author of eleven books and over 280 research papers published in standard journals. He made many pioneering contributions to several areas of statistical theory and applications: Estimation, Multivariate Analysis, Design of Experiments and Combinatorics, Characterization of Probability Distributions, General Inverses of Singular Matrices. Most of the results obtained by him have been incorporated in standard text books on statistics.

The career of C.R. Rao has never been limited to the field of statistics only, as his other academic activities at the ISI attracted young enthousiastic scientists from all other horizons to the Institute. As Director of the Research and Training School at the Indian Statistical Institute, Dr. C.R. Rao has developed a variety of educational programs in the theory and applications of statistics at various levels. He has introduced undergraduate, graduate and Ph.D. programs and produced qualified statisticians who are working in almost all parts of the world. Under his leadership, the ISI has won international recognition as an outstanding center for statistical training and research.

This volume brings together a wide range of topics and perspectives in the field of statistics. It contains invited papers that are grouped for the ease of orientation in seven parts: (1) Prologue, which consists of C.R. Rao's contributions to statistics and three papers on linear models, orthogonal arrays and geometrical aspects of data analysis that Rao had special interest in at the early stage of his academic career, (2) Inference and Test of Hypotheses, (3) Parameter and Variance Estimation, (4) Distributions, Weighted Distributions and Characterization, (5) Linear Models and Matrices, (6) Multivariate Analysis and (7) Special Topics in Data Analysis.

As the editor of this volume, (and also the Conference Organizer) I would like to express my sincere appreciation to the authors, whose work and interest made this volume possible; to my assistants Sylvie Gonano, Sandrine König, Nicole Rebetez, Jean-Pierre Renfer and Séverine Pfaff; and to Linol Pradelier and Hendrik Oldenkott for their computer assistance; and to my secretary Sandra Rieder for her efficiency in handling all the correspondence; to Dr. Farhad Mehran, Chief Bureau of Statistics of the International Labour Office for his support and encouragement during the preparation of the conference; to Drs. Gerard Wanrooy and Ms. Inez van der Heide for handling this volume at Elsevier Science Publishers B.V. (North-Holland).

Finally, I wish to express my special thanks to the Swiss National Science Foundation; the Swiss Federal Statistical Office, and in particular its director, Mr. Carlo Malaguerra, and to Philip Morris Europe, especially Dr. M.C. Bourlas, director of Research and Development whose contributions made this conference possible.

University of Neuchâtel
Switzerland
August 1989

Yadolah Dodge

TABLE OF CONTENTS

PROLOGUE

Life is not void,
There is kindness, apple and faith,
Oh yes!
So long as there are poppies, one should live

Sohrab Sepehri
Persian Poet (1928-1980)

STATISTICAL DATA ANALYSIS AND INFERENCE
Y. Dodge (Editor)

C.RADHAKRISHNA RAO'S CONTRIBUTIONS TO STATISTICS

Yadolah Dodge
University of Neuchâtel
Switzerland

In January of 1944, at the age of 24, while giving a course on estimation to master students of Calcutta Uni University, C.R. Rao proved in his class a result on the asymptotic inequality first obtained by R.A. Fisher regarding the lower bound for the variance of an estimator for large samples. When a student asked, "why don't you prove it for finite samples ?", Rao went back home, worked all night and the next day proved what is now known as the Cramer-Rao inequality for finite samples. In what follows an attempt has been made to summarize a life of a dedicated scientist, whose many pioneering contributions to the field of statistics have found place in standard books on statistics.

1. BIOGRAPHY

Calyampudi Radhakrishna Rao was born on September 10, 1920, in Hadagali, Karnataka State, India. He received an M.A. with a first class and first rank in Mathematics from Andhra University in 1940. At that time there were not many job opportunities for mathematicians, so he tried to get a job in the military service. Being too young, he was disqualified for the job and therefore left for Calcutta, where he get to know the Indian Statistical Institute (ISI) founded a decade or so earlier by P.C. Mahalanobis. He was admitted

C.R.Rao 15 years old

to the ISI, and finished a one year training course in statistics. When Calcutta University started to offer a formal degree in statistics, Mahalanobis advised him to join the Calcutta University where he was awarded an M.A. degree in statistics with a first class, first rank and a gold medal in 1943. This is how began a long and fruitfull career of a scientist whose many pioneering contributions to the field of statistics have found place in standard books on statistics.

After completing his master's degree in 1943, he was asked to teach at the ISI and at Calcutta University. One day, while giving a course on estimation about the large sample properties of maximum likelihood, he proved of Fisher's asymptotic inequality that the asymptotic variance is not smaller than one divided by the information. Then a student asked, "why don't you prove it for finite samples ?" he went back home, worked all night and the next day he came up with the inequality, which is now known as the Cramer-Rao inequality. This result along with many other results such as the Rao-Blackwell theorem are in the same paper entitled "Information and accuracy attainable in the estimation of statistical parameters", published in the Bulletin of Calcutta Mathematical Society, 37, 1945, 81-91.

During this time, Mahalanobis asked him to work on anthropometric data which had been collected by Dr. D.N. Majumdar, an anthropologist in the state of Uttar Pradesh during the 1941 Indian population census. The analysis of this data set (5000 observations on 20 characters measured on each individual) was the first experience of C.R. Rao. The result of this report was published in book form jointly with Mahalanobis and Majumdar in 1950 entitled "Anthropometric Survey of the United Provinces' 1941: A statistical study" by the Asia Publishing House.

After getting Ph.D. from Cambridge in 1948

Marriage ceremony in Madras with Bhargavi, September 9, 1948

Upon a request from the anthropological museum of Cambridge, Mahalanobis sent C.R. Rao there to help the museum in analysis of data relating to skeletal material discovered and brought back to the museum by a British expedition from Jebel Moya in North Africa. He was employed by the museum for a couple of years. At the same time he joined King's College and registered for his Ph.D. degree under Sir R.A. Fisher. The late Fisher had a laboratory where he was breeding mice for mapping their chromosomes. There Rao worked on the linkage of four genes of a chromosome to find the distance between them. During the day he was working with the Cambridge museum on the anthropology project and in the evenings he was working on the linkage project for R.A. Fisher.

After two years C.R. Rao wrote the report on the skeletal material jointly with R.A. Mukherji and J.C. Trevor. The result was published as a book in 1955 by Cambridge University Press entitled "Ancient Inhabitants of Jebel Moya".

In 1948 C.R. Rao received his Ph.D. degree from Cambridge University under R.A. Fisher, with Wishard as the examiner. The thesis entitled "Statistical problems of biological classification" was based on the theoretical work he had done in connection with the anthropological project for the museum with many additional results. In that he expanded Fisher's discriminant function for classification within two categories to a method for many categories. He also developed a test for additional information for selection of a character.

He joined the ISI in 1944 as statistician and became professor in 1949, and the director of the Research and Training School in 1964. On the basis of his published work in statistics, he received in 1965 a Sc.D. from the Cambridge University. C.R. Rao has received many honorary degrees and awards during his career. Thus far he has been awarded, D.Sc. from

With Mahalanobis's in 1967

Andhra University, Waltair, India in 1967; D.Sc. from Leningrad University, Leningrad, U.S.S.R. in 1970; D.Litt. from Dehli University, Dehli, India in 1973; D.Sc. from Athens University, Athens, Greece in 1976; D.Sc. from Osmania University, Hyderabad, India in 1977; D.Sc. from the Ohio State University, Columbus, Ohio, U.S.A. in 1979; Professor Honorario from Universidad Nacional de San Marcos, Lima, Peru in 1982; D.Sc. from University of Philippines, Manila, Philippines in 1983; Ph.D. from University of Tampere, Finland in 1985; He also received Guy Medal in Silver of the Royal Statistical Society, U.K., for contributions to statistics in 1965; Megnadh Saha Medal of the Indian National Science

Receiving Bhatanagar Prize from Javaharlal Nehru in 1963

Receiving the Padma Bushan Award from President of India Zakir Hussain in 1968

Academy for outstanding contributions to science in 1969; Jagdish Chandra Bose Gold Medal and cash award by the Bose Institute for outstanding contributions to science in 1979; Silver plate bearing the monogram of the Andhra Pradesh Academy of Sciences for "outstanding

With Indira Gandhi in 1967

contributions to the growth and development of science in the field of Statistics and to the wellveing of the Society" in 1984; Padma Bhushan, one of the civilian awards of a high order of the Government of India, for services in the field of statistical education and research in 1968; Honored by the Andhra Pradish Akademi of Sciences at the Silver Jubilee Anniversary in 1988.

2. CONTRIBUTIONS TO STATISTICS

C.R. Rao is the author (or co-author) of eleven books and over 280 research papers published in standard journals. He made many pioneering contributions to several areas of statistical theory and applications - Estimation, Multivariate Analysis, Design of Experiments and Combinatorics, Characterization of Probability Distributions, Generalized Inverse of Singular Matrices, etc. Most of the results obtained by him have been incorporated in standard text books on statistics.

A brief description of Rao's contributions, concepts and the results associated with his name are as follows:

- **Estimation Theory**

  Fisher-Rao theorem, Cramer-Rao inequality, Rao-Blackwellization, second order efficiency, MINQUE (minimum norm quadratic estimation) in the Gauss-Markoff model, Scheffe- Lehmann-Rao theorem, unified theory of M-estimation.

Rao-Blackwell, Calcutta, 1976

With Harold Cramer in 1978

Rao (1945) established that the variance of an unbiased estimator of a parameter θ is not less than $1/I(\theta)$, where $I(\theta)$ is Fisher information on θ contained in a sample, which is known as Cramer-Rao inequality. It was also shown that given any unbiased estimator, one can construct a function of sufficient statistics which is unbiased and has possibly a smaller variance than that of the given estimator, a process known as Rao-Blackwellization. Using the Fisher information matrix, a quadratic differential Riemannian metric (known as Rao metric with the associated Rao distance) is introduced in the space of probability distributions. Rao's early paper (1945) dealt with the application of differential geometry in statistics, which is being actively pursued at present.

Fisher defined the efficiency of an estimator as its asymptotic variance. But there exists a wide class of estimators having the same asymptotic variance or efficiency (called the first order). To distinguish among them, the concept of second order efficiency was introduced by Rao (1961). One of the results in this study known as Fisher-Rao theorem provides an expression for maximal second order efficiency. Under some conditions, it is shown that the maximum likelihood estimator of a parameter in a multinomial distribution has the highest second order efficiency.

With Sir R.A.Fisher in 1956

Rao (1971) introduced a new principle known as MINQUE for the estimation of variance components in linear models. Explicit expressions were obtained for MINQUE's under additional conditions, such as unbiasedness, invariance and non-negativeness of an estimator.

A unified theory of linear estimation in the general Gauss-Markoff model without any assumptions on the design and error covariance matrices is developed. Three approaches were suggested (1973): generalized inverse partitioned matrix method, unified least squares thory and geometric method through a generalized projection operator.

Score and Lagrangian multiplier tests based on large samples were developed by Rao (1947) to test simple and composite hypotheses under very general conditions. Recent studies show that these tests have better local optimum properties compared to other competing tests, such as likelihood ratio and Wald's tests.

- **Multivariate Analysis**

  Familial correlations, Rao's U test (for the hypothesis of additional information due to some variables), canonical factor analysis, Linked Cross Sectional (LCS) survey in growth studies, graphical representation of populations by canonical coordinates to study clustering and interrelationships, F approxomation to Wilk's criterion in MANOVA, Rao's covariance structure, Rao's quadratic entropy, ANODIV (analysis of deversity), Rao's score and Lagrangian Multiplier tests.

In multivariate analysis. one usually faces the problem of extracting information from a large number of measurements $x_1, x_2, \ldots$ made on each sample unit or individual. Not all measurements carry independant information, and paradoxical situations arise where procedures based on a subset of measurements are more efficient than those based on the whole set. A question that arises then is whether the information in a given subset of measurements $(x_1, \ldots, x_p)$ is the same as that in the whole set $(x_1, \ldots, x_p, \ldots, x_{p+q})$ for answering a particular problem. In such a case, the set of measurements $x_{p+1}, \ldots, x_{p+q}$ becomes redundant. A test for this hypothesis known as Rao's U test was developed by Rao (1948).

The use of canonical variates (coordinates) in the representation of higher dimensional data in a lower dimensional space was discussed by Rao (1948). This is a useful device in studying clustering and inter-relationships among individuals or populations.

Using general diversity measures applicable to both qualitative and quantitative data, analysis of diversity was developed as a generalization af analysis of variance. The limitations of some of the well-known entropy functions (such as Shannon's) as measures of diversity were discussed. A new measure referred to as Rao's quadratic entropy was introduced and its usefulness in analysis of diversity demonstrated (1982, 1984). Further work on these lines using the concept of cross-entropy is being pursued.

- **Characterization Problems**

  Kagan-Linnik-Rao theorem, Rao's damage model, Rao-Rubin theorem, weighted distributions.

A number of characterizations of the normal, Poisson, Cauchy, stable, exponential, geometric and other distributions were obtained based on the properties of sample statistics. The results were presented in the book by Kagan, Linnik and Rao (1972). An interesting result which has led to considerable research by a number of workers is the characterization

With Bose of Bose-Einstein Statistics at Calcutta in 1973

of the Poisson distribution through what is called Rao's damage model and the Rao-Rubin condition (1965). This study also led to an extension of Deny's theorem on the solution of an integrated Cauchy functional equation and its application to numerous characterizations of the exponential, geometric and Pareto distributions (1982). A general class of distributions known as weighted distributions were introduced by Rao (1965), which have received several practical applications.

- **Combinatorics**

  Orthogonal arrays, Hamming-Rao bounds, cyclic generation or hyperplanes in a finite geometry, intra- and inter-group balanced designs.

Combinatorial arrangements known as orthogonal arrays were introduced in Rao (1949) for use in design of experiments. These arrangements are being used widely in multifactorial experiments to determine optimum combinations of factors in problems of production. These arrangements have applications in coding theory, a result of practical interest being the Hamming-Rao bound associated with orthogonal array.

- **Matrix Algebra**

  g-inverse of singular matrices and applications to statistics. Explicit representation of projection operators.

The problem of defining a generalized inverse (g-inverse) of a singular square or rectangular matrix, or a general linear operator, was studied by Rao (1967, 1973). It was shown that different classes of g-inverses can be defined for use in different problems, such as solving consistent equations, finding minimum norm solutions, computing least squares solutions of inconsistent equations, etc. An explicit expression for the projection operator onto the space spanned by the column vectors of a matrix X is given in terms of a generalized inverse, $X(X'X)^{-}X'$, which is now widely used in the discussion of linear models. A geometric approach leading to very general results was discussed by Rao and Yanoi (1985). G-inverses have received wide applications.

The concept of a projection operator on a subspace along a disjoint but not necessarily a complementary subspace was developed and used in providing a unified theory of linear estimation of parameters in a general Gauss-Markoff model. Explicit espressions were obtained for such projection operators.

- **Differential Geometric Methods in Statistics**

  Rao's distance, quadratic differential matrices.

Below is the list of C.R. Rao's published books:

1. **Linear Statistical Inference and Its Applications** First Edition (1965 John Wiley, New York), Second Edition (1973 John Wiley, New York), Taiwan (English) Edition (1976 Taiwan), Interscience (English) Edition (1975 Wiley Eastern, India), Russian Translation (1969 USSR Academy of Sciences, Moscow), German Translation (1972 Springer Verlag), Japanese Translation (1977 Tokyo Inc., Tokyo), Czech Translation (1978 Academia Publishers, Prague), Polish Translation (1982 Panstowowe Wydawnictwo Naukowe, Warsaw), Chinese Translation (1987 Academia Sinica Press Beijing)

2. **Advanced Statistical Methods in Biometric Research** First Eddition (1952 John Wiley, New York), Reprint (1971 Haffner, New York), Reprint (1974 Haffner, New York)

3. **Generalized Inverse of Matrices and Its Applications** (with S.K. Mitra) First Edition (1971, John Wiley, New York), Japanese Translation (1973 Tokyo)

4. **Characterization Problems of Mathematical Statistics** (with A. Kagan and Yu.V. Linnik) Russian Edition (1972 USSR Academy of Sciences, Moscow), English Translation (1973 John Wiley, New York)

5. **Computers and the Future of Human Society**. Andhra University, Waltair and Statistical Publishing Society, Calcutta, 1970

6. **Ancient Inhabitants of Jebel Moya** (with R.K. Mukherji and J.C. Trevor), Cambridge University Press, 1955
7. **Anthropometric Survey of the United Provinces, 1941. A Statistical Study** (with P.C. Mahalanobis and D.N. Majumdar), Asia Publishing House, 1950
8. **Race Elements of Bengal, a Quantitative Study** (with D.N. Majumdar), Asia Publishing House, 1958
9. **Formulae and Tables for Statistical Work** (with A. Matthai and S.K. Mitra). First Edition (1966 Statistical Publishing Society, Calcutta), Second Edition (with B. Ramamurti as additional author) (1975 Statistical Publishing Society, Calcutta)
10. **Estimation of Variance Components and its Applications** (with J. Kleffe). North-Holland, 1988
11. **Statistics and Truth: Putting Chance to Work** First Edition (1989 The Council of Scientific and Industrial Research, India), Reprint (1989 International Publishing House, U.S.A.)

Rao's book, **Linear Statistical Inference and its Applications**, appears in the list of 78 'Heavily Cited Works in Mathematics (during the period 1961-1972), Part I, Pure Mathematics' published in **Current Contents**, Vol. 47, November 1973. Recently, this book has been indentified as one of the most cited items in its field, according to the data from the **Science Citation Index** and the **Social Science Citation Index**. A brief commentary of this work is published in **This Week Citation Classics** section of **Current Contents**, CC/Number 12, March 24, 1980.

His book **Advanced Statistical Methods in Biometric Research**, appears in the list of 72 Heavily Cited Works in Mathematics (during the period 1961-1972), Part2, Applied Mathematics, published in **Current Contents**, Vol. 48, Nov. 28, 1973.

From the beginning of his statistical career, Dr. Rao has been interested in statistical applications. While working at the Anthropological Museum in Cambridge, U.K., he developed multivariate statistical methods for applications in anthropometric studies. He is, as mentioned above, the co-author of three books devoted to statistical analysis of anthropometric data collected on a large number of individuals.

3. PROFESSIONNAL ACTIVITIES

As Director of the Research and Training School at the Indian Statistical Institute, Dr. Rao has developed a variety of educational programs in the theory and applications of statistics at various levels. He has introduced undergraduate, graduate and Ph.D. programs and produced qualified statisticians who are working in almost all parts of the world. Under his leadership, the ISI has won international recognition as an outstanding Center for Statistical Training and Research. He has been a member of the Statistical Education committee of the International Statistical Institute for the last 30 years. He helped the institute in setting up statistcal training centers in many parts of the world, including a major center in Calcutta, called ISEC (International Statistical Education Center), to train statisticians from the Southeast Asian region, of which he is currently the chairman of the Board of Directors. Dr. Rao was the Chairman of the UN Committee which examined the demand for statistical personnel in Asian countries and recommended the establishment of an institute for statistical development of Southeast Asia. On the basis of these recommendations the Asian Statistical Institute (ASI) was established in Tokyo to provide training in statistics to statisticians working in the government and industrial organizations.

With Fisher and Yates in 1951

Dr. Rao served on governmental advisory committees and review committees of several academic institutes and international organisations. He is the principal investigator of several projects sponsored by the Air Force and Navy of the U.S. He was a member of the panel on small area statistics of population and income of the U.S. National Academy of Sciences. During the last six years he was a member of the visiting committee of the Board of Overseers of the Harvard University.

Dr. Rao was President of the International Statistical Institute, the Institute of Mathematical Statistics, the Biometric Society and the Forum for Interdisciplinary Mathematics. He has been the President of the Indian Econometric Society for five years and continues to be its Chairman.

He is the editor of two important journals, **Sankhyā** and **Multivatiate Analysis** and is on the Editorial Boards of numerous Statistical Journals.

With Kolmogorov at Calcutta in 1962

He organized several conferences and symposia in different areas of statistical theory and applications.

Dr. Rao provided statistical advice to numerous scientists and organizations in various statistical prjects involving design of experiments, sample surveys, quality control and medical diagnosis.

He was invited by several countries (Japan, U.S.S.R., Iran, Australia, Canada, U.K., Philippines, Venezuela, Peru and U.S.A.) to advise on the development of statistical courses at the universities and setting up organizations and institutes for statistical research, training and dissemination of knowledge. C.R. Rao has been elected a member of International Statistical Institute in 1951; he is a Fellow of Indian National Science Academy (1953); Fellow of Institute of Mathemetical Statistics, U.S.A. (1958); Fellow of Andhra Pradesh Academy of Sciences, U.S.A. (1965); Fellow of Royal Society (FRS), U.K. (1967); Honorary Fellow of Royal Statistical Society U.K. (1969); Fellow of American Statistical Association, U.S.A. and Fellow of International Econometry Society, U.S.A. (1972); Fellow of Indian Academy of Science, Honorary Member of Indian Society of Human Genetics, Honorary Fellow of King's College, Cambridge, U.K. (1974). This life fellowship is limited to only eleven persons at any time. Now he has the privilege of walking on the lawns of the King's College. He is also Foreign Honorary Member of American Academy of Arts and Science, U.S.A. (1975); Honorary Member of International Statistics Institute, The Netherlands (there are only 8 such members out of a total world membership of 1400), Founder Fellow of The Third World Academy of Sciences, Trieste Italy (1983); Fellow of The Indian Society for Medical Statistics, Honorary Fellow of Calcutta Statistical Association (1985); Honorary Life Member of Biometric Society (1986) (Only ten persons have been elected to this position since the society was founded (1948); Fellow of The National Academy of Sciences, India.

With J.Neyman at Calcutta in 1972

C.R. Rao was the Secretary and Director of Indian Statistical Institute from 1972 to 1976. He was appointed the Jawaharlal Nehru professor in 1976. In 1979, he was appointed a university professor in the Department of Mathematics and Statistics at the University of Pittsburgh. In 1988, he was appointed as National Professor by the government of India in recognition of his outstanding contributions to science. There are, at present, only three National Professors in India.

During The First International Conference on L_1-Norm at Neuchâtel, Switzerland in 1987

Currently, he is Eberly professor of statistics at the University of Pennsylvania, Adjunct professor at the University of Pittsburgh, U.S.A., and the National professor of India. He now plans to divide his time between these three positions.

4. OUTSIDE THE WORLD OF STATISTICS

C.R. Rao's interests vary widely from gardening to cooking, runing through photography, Indian classical dance and music. He developed a new dance academy in Delhi and was its president during which the academy brought back to value some neglected areas of dancing. He put her daughter in dance school and she currently runs a dance school in Buffalo, U.S.A.. She has given numerous performances in U.S.A., Europe and Asia.

C.R. Rao has a very special interest in gardening and grows exotic plants in his backyard. During my visit to the Indian statistical Institute in summer of 1976, while he was still the director of the ISI, I assisted to a quite memorable event. He called upon everyone to plant a tree and by the end of the day, we had planted about 100 trees, according to his plan. Of course, he used to observe the plants whenever he had time; through his observations and actual experiments made in Calcutta with T.A. Davis he discovered the right and left spiral phenomenon between different varieties of plants. The discovery of this interesting phenomenon lead to a series of articles (over 50) by T.A. Davis on palm trees and other plants.

C.R.Rao, Indian Statistical Institute, 1976

C.R. Rao's efforts in developing the field of statistics goes beyond anyone's imagination.

To see how this man is attached to his field and how he sees the future, I would like to bring the following quotation taken from Morris DeGroot in "Conversation with C.R. Rao".

DeGroot's question: "Are there other things that you like to do when you are not doing statistics ?".

Rao replies: "As I grow old I find that there's not much time to do many things. But in the next two or three years I may be visiting the third world countries and helping them in developing research in statistics. This is an assignment that the Third World Academy of Science, of which I am founder fellow, is trying to give me. I have not replied to their letter yet because I am not sure how my health will be in the coming years. But this is something which I would like to do because I know the third world countries are not developing basic research. They are dependent on foreign countries for all technological improvements, which is not good in the long run. That's the reason why we founded the Third World Academy. In addition, I may have some assignments in India which would keep me busy and occupied for a long time to come".

A good exemple for all of us.

References

[1] DeGroot M.(1987). **A Conversation with C.R.Rao**. Statistical Science. Vol.2, No.1, 53-57.

[2] Rao C.R.(1945). **Information and accuracy attainable in the estimation of statistical parameters**. Bull. Calc. Math. Soc., 37, 81-91.

[3] Rao C.R.(1961). **Asymptotic efficiency and limiting information**. Proc. Fourth Berkeley Symposium on Mathematical Statistics and Probability, Vol.1, 531-46.

[4] Rao C.R.(1971). **Estimation of variance components - Minque theory**. J. Multivariate Anal, 1, 445-56.

[5] Rao C.R.(1947). **Large sample tests of statistical hypotheses concerning several parameters with applications to problems of estimation**. Proc. Camb. Phil. Soc., 44, 50-57.

[6] Kagan A. Linnik Yu V and Rao C.R.(1972). **Characterization problems of mathematical statistics**. John Wiley, New York.

[7] Rao C.R.(1965). **On discrete distributions arising out of methods of ascertainement**. Sankhya A., 27, 311-24.

[8] Lau Ka-Sing and Rao C.R.(1982). **Integrated Cauchy functional equation and characterizations of the exponential law.** Sankhya A., 44, 72-90.

[9] Rao C.R.(1967). **Calculus of generalized inverses of matrices.** Part 1 - general theory, Sankhya A., 29, 317-50.

[10] Rao C.R.(1973). **Linear statistical inference and its application.** John Wiley, New York.

[11] Rao C.R. and Yanoi H.(1985). **Generalized inverse of linear transformations: A geometric approach.** Linear Algebra and Its Applications, 66, 87-98.

[12] Rao C.R.(1948). **Tests of significance in multivariate analysis.** Biometrika, 35, 58-79.

[13] Rao C.R.(1948). **The utilization of multiple measurements in problems of biological classification.** J. R. Stat. Soc., B 10, 159-203.

[14] Rao C.R.(1982). **Diversity and dissimilarity coefficients - a unified approach.** Theor. Pop. Biol., 21, 24-43.

[15] Rao C.R.(1984). **Convexity properties of entropy functions and analysis of diversity.** In Inequalities in Statistics and Probability, IMS Lecture Notes, 5, 68-77.

[16] Rao C.R.(1949). **On a class of arrangements.** Edin. Math. Proc., 8, 119-25.

STATISTICAL DATA ANALYSIS AND INFERENCE
Y. Dodge (Editor)

NONLINEAR ESTIMATION IN LINEAR MODELS

Frank Hampel

ETH Zürich, CH-8092 Zürich, Switzerland

Professor C.R. Rao has made great contributions to linear and quadratic estimation in linear models. The classical methods will remain a beautiful and important standard; but they can be extended to incorporate explicitly phenomena such as outliers and gross errors, thereby further narrowing the gap between mathematics and the problems of practical data analysis. It is now frequently possible to discuss realistically the efficiency losses of classical methods if the model, as usual, is only an approximation, and hence also to talk about the information losses within a wider range of models caused by restriction to a sufficient statistic.

The classical linear estimation methods (which in practice were always only conditional, given that no outliers were rejected) can be replaced by stable nonlinear methods, which are as close as reasonably possible to the classical ones under the ideal parametric model. These stable estimators can be viewed as nonlinear functionals which in some important cases have been shown to be even Fréchet-differentiable, while most maximum likelihood estimators are not, as had been found out by Kallianpur and Rao. There exists also a sharper version of the asymptotic Cramér-Rao bound for such functionals under an additional bound on the local instability. It seems interesting to have a new look also at the differential geometric view of statistics.

1. BRIEF OVERVIEW

The paper discusses some topics related to work by Professor Rao. In an extensive theoretical part, it shows some reasons for slightly deviating from strictly linear estimation methods. It then discusses numerically the efficiency losses of linear methods under high-quality data. Some examples of structured data are reanalyzed. Finally, the concepts of Fréchet-differentiability, Cramér-Rao bound, and differential geometry in statistics are briefly reconsidered. (Bayes and ridge regression methods are not treated in this paper.)

2. WHY NONLINEAR METHODS?

Professor Rao has beautifully demonstrated the elegance of linear statistical methods, both for linear models and in multivariate statistics (Rao [1],[2]). The mathematics speaks for itself, and in rereading Rao [2], I am again and again impressed by its mathematical breadth, depth, and intrinsic beauty, and by the broad horizon of its author. Most of the statistical methods discussed will remain, either as canonical or at least as a standard by which to gauge new developments. Another valuable feature is that the methods are applied to many data sets, which are usually given and analyzed in the concise form of sufficient statistics (such as means and sums of squares).

But here we have an aspect which merits more discussion. The sufficient statistics are only sufficient given an exact parametric model, and the information suppressed in going from the data to sufficient statistics is precisely the information about possible deviations from the model. In other words, once we have lost the original data, we are unable to check the model we are using. This is regrettable, for most models are only more or less approximative, and we ought to know how good the approximation is. There are famous examples of drastic changes of efficiency of classical methods by tiny changes of the model, such as the continuation (and clarification) by Tukey and Huber of the dispute between Fisher and Eddington about scale estimation (cf., e.g., Huber [3], p.1ff.). This implies that even if, for example, the normal model is a "good" approximation, it might still pay to use other methods than the linear and quadratic methods utilizing the sufficient statistics in this case.

It is also well known that gross errors, often resulting in big outliers, can occur; but they do not show up in the usual sufficient statistics, hence some kind of data cleaning procedure or analysis of residuals is always a necessary addendum to the standard analysis, except perhaps for high-quality data under full control. If any outlier is found and rejected, we have a nonlinear procedure for the original data set, although it may be conditionally linear on the remaining data, given that they are not rejected. It may also be viewed as linear in the (proper) subregion of the sample space in which none of the data are rejected. From a mathematical viewpoint, there is the question whether (potential) rejection of outliers is to be incorporated in the formal (overall) statistical procedure, or whether it is to be swept aside and left as a "dirty" task to the applied statistician, without any control about whether this task does not completely mess up the nice efficiency properties of the formal method under ideal circumstances, or even whether it achieves its minimum aim at all of discovering all distant outliers. I believe that now, in the computer age and with more advanced statistical theory than some decades ago, the time has come for the former approach.

There is a third reason, besides efficiency gain and besides prevention of disastrous results due to outliers, for looking at the data not only through the limiting optics of sufficient statistics. We may learn more about the data than from a mere routine analysis, namely by discovering unsuspected features, which may well lead to valuable progress in the subject matter science from which the data come and which may even be more valuable than the answer to the question originally asked. Some statistical examples are in Daniel and Wood [4] and Daniel [5], but the history of science contains many examples of unsuspected discoveries by open minds, though usually with only very little or very elementary statistics being used. (The question arises: is there a conflict between using statistics and thinking?) Of course, it depends on the research environment whether new hints and clues are acceptable at all and can be followed up, and in any case a balance has to be found between the mainstream of research and the new side trickles, some of which may become the mainstreams of tomorrow.

The general type of learning from data considered here may be formally considered as selection of a new, unsuspected model out of a huge class of conceivable models (such as: this subset of data, characterized by something perhaps only potentially measurable, behaves in this completely different way) or as a general kind of pattern recognition. Besides using the data, it also has to use background knowledge from the subject matter area, imagination, self-criticism, and intuition. Altogether, it leads to a highly nonlinear analysis of the data, even if the "default" analysis valid in a large part of the sample space is linear. At present, I believe it is not possible to do such an analysis formally and integrate it (at least partly) into mathematics, despite some attempts and first claims in the area of artificial intelligence, for example about the discovery of scientific laws. These results seem too superficial and restricted. Moreover, we still do not know how to measure the "simplicity" or "quality" of a scientific law in a sufficiently general way, apart from the sheer (also computational) complexity of the model selection task. But on the technical side of it, we do have some new tools, such as high-speed computers and general criteria for model fitting such as Akaike's information criterion; and perhaps it is a question of time until at least some partial, formal aspects of the task can be dealt with routinely on computers and treated by a mathematical theory. Even then, we shall still need the (in general non-unique) interpretation of the results in their broader context. The situation may develop similarly to the one in outlier rejection: the formal side of it can now be understood numerically and mathematically and improved upon practically to an extent which was hardly imaginable to most statisticians some decades ago, despite some ingenious work of early pioneers, such as Newcomb, Gosset, and Jeffreys. But still we need to try to integrate the outliers in their context as far as possible, and this remains (largely) an informal task.

The first two problems of linear methods mentioned can be dealt with by the same broad class of mildly nonlinear methods (which contains rejection rules among their worst members) keeping as much as possible the nice properties of linear methods under ideal circumstances, but, contrary to those, behaving stably both under smaller and under larger deviations from the ideal model. These methods also help with the third problem mentioned by exposing deviating substructures more easily and more clearly than do linear methods.

At this point we may ask whether there are any reasons left in favor of linear methods. There are. One is their simplicity, with all its ramifications, like an extensive and precise mathematical theory. Another reason is that least squares methods provide the best unbiased estimators if the errors are i.i.d. normal. This reason is connected with the first one, for in some sense the normal error distribution is the simplest probability distribution on the full real line. A probability text for physicists says that the normal distribution plays the same role in probability theory as does the straight line in geometry. This analogy becomes literally true if we consider the log density derivative (location score function) which is indeed linear for the normal. (This f'/f may actually be considered to be a more basic quantity than either density or cumulative distribution, for characterizing a probability distribution.) The fact that normal theory is so broad while nothing comparable exists for any other probability distribution is intimately connected with the simplicity of the normal. – A third reason for linear methods, again connected with simplicity, is that there are simple probabilistic rules for expectations and variances which can be used also in "nonparametric" situations, that is, when nothing except the existence of some moments is known about the underlying probability distribution.

But there is one argument often cited as justification for least squares, and for a "second-order theory" free of any relation to the normal distribution, which is dangerously misleading. This is the so-called Gauss-Markov theorem which says that least squares are optimal among all linear unbiased estimators independently of the underlying distribution (as long as some moments exist). The point, stressed already by Fisher [6], p.355 and before, in a superficially different context, is that outside a rather small neighborhood of the normal all linear methods are rather inefficient. It does not seem very convincing to concentrate on the best member of an altogether bad class, but this has been done in those cases in which any ties to even just approximate normality have been vigorously denied. Besides approximate normality, there remains again only simplicity of the methods (and not general efficiency) as a justifying reason. Simplicity of computations was once very important, but is not so anymore in the computer age. Simplicity of theory is still of great value, although in practical problems exact theoretical solutions for confidence intervals and tests are less needed than formerly due to the availability of jackknife, bootstrap, and manageable randomization tests.

In this connection the famous insensitivity of t-test and model I analysis of variance against deviations from normality has to be mentioned. This insensitivity is indeed a great asset of the methods; but what is often overlooked is that it is only a (reasonably good) insensitivity with respect to the level of the tests, and not at all with respect to the power. In other words, the standard errors of least squares estimators derived from second-order theory are correct, but they and the confidence intervals and acceptance regions are far too large compared with those of other, better estimators if the data are not rather nearly normally distributed.

An example for the slow spread of statistical knowledge even among statisticians is the great sensitivity of the level of χ^2- and F-tests for variances to slight deviations from normality, which is now known for about 60 years, but has been ignored even in recent statistics texts.

There is a kind of pseudologic found especially in applied statistics books which runs, for example, as follows. Before applying the 2-sample t-test, one has to make a χ^2-test for normality to ensure normality. Then one has to make an F-test for variances to ensure equality of variances. Then one is allowed to use the t-test. Much can be said about this argument (including the previous remarks about the F-test), but the main point is that the preliminary tests can at best prove <u>approximate</u> normality and equality of variances. There is a tacit assumption (or extrapolation) involved that the difference between approximate and exact properties is irrelevant. Sometimes this is true, but it has to be justified, and in other cases it is unfortunately wrong. While during the first decades of this century it was a great mathematical achievement to derive exact properties of statistical methods under exact model assumptions, it is now possible to study their properties (in good approximation) under full <u>neighborhoods</u> of parametric models. Since tests can at most prove (statistically) that the true distribution is in such a neighborhood (while they can only falsify the exact model), consideration of the behavior of statistics in full neighborhoods is necessary to change the above logical argument to a valid one.

Historically, there has been a shift of emphasis in statistical research over the decades. In the twenties and thirties, the basic structure of modern statistics was developed and its main mathematical problems were solved either then or soon thereafter; but at that time, this development was only possible at the price of leaving some subtle gaps in its connection to real applications. Great applied statisticians were always aware of these gaps, but both concepts and mathematical tools to fill them were still lacking. It is quite revealing to read the exchange of letters between Gosset and Fisher about this topic, where

Gosset implored Fisher urgently to do something about deviations from normality. Later, theoretical statistics became more and more mathematical, while losing more and more contact with the actual applied problems in their full breadth. Roughly during the last two decades, there was a revival of interest in "data analysis", including discussion of problems like that of "specification" (of a parametric model) explicitly ignored by Fisher ([6], p. 314) as part of theory, though stressed as important for applications. During about the same time the standard mathematical methodology developed by Fisher and others could be enlarged to include detailed knowledge about many of the gaps that, first out of necessity and later out of habit, had been ignored.

3. EFFICIENCY LOSSES OF LEAST SQUARES UNDER HIGH-QUALITY REAL DATA

Opinions seem presently divided about the importance of the efficiency losses of least squares due to the deviations from normality occurring in real data. There are some good reasons for this. It makes a difference whether we consider high-quality data or ordinary, routine data often including gross errors (and in the latter case, whether we consider least squares after rejection of outliers, or least squares in their strict mathematical sense). It also makes a difference whether we are willing to tolerate avoidable efficiency losses of, say, 2-5%, or 10-20%, or 50-90%. (This choice depends on circumstances, among other things; and amazing as it may sound, sometimes good reasons can be given for the last choice.) But there seem to be still other reasons, namely unawareness of some long-documented facts, and a psychological tendency to "save" the "good old methods" and sweep disturbing evidence under the rug.

Many statisticians do not seem to care much about the efficiency losses of least squares (although they may take very serious the 12% efficiency loss of the mean deviation against the standard deviation, or the 5% efficiency loss of the Hodges-Lehmann estimator and Wilcoxon test against mean and t-test under strict normality; they may even complain about the 0% asymptotic efficiency loss of the Fisher-Yates-Terry-Hoeffding-van der Waerden test against the t-test under strict normality, stating, mathematically absolutely correct: "There is a loss of information involved" by going from the data to the ranks). An example which is probably typical for the casual attitude towards deviations from normality is the brief note in McCullagh and Nelder [7], p. 36: "The assumption of Normality ... need not be relied upon in large samples: the central limit theorem offers protection from all but the most extreme distributional deviations but there is some loss of efficiency (Cox and Hinkley, 1968). The theory of least squares can be developed using only [the] first- and second-moment assumptions ...". The last sentence has already been commented upon above. The paper cited (Cox and Hinkley [8]) is psychologically very interesting. It gives, among other things, a table with the asymptotic efficiency of least squares under t-distributions with various degrees of freedom (starting with $\nu = 3$ d.f.), and then concludes in the Discussion: "For the Type VII distribution, the loss of efficiency is less than 10 per cent provided that λ_2 $[= (\nu+1)/2]$ is greater than about 4; $\lambda_2 \simeq 4$ is the value given by Jeffreys (1961, Section 5.7) for the distribution of observational errors in several series which he analyzed." It also states for which log gamma distributions the loss of efficiency is less than 10%, without any reference to real data. The casual reader may well take this to mean that "usually" the efficiency losses are less than 10% (or at most about 10%). He then has the authority of Cox and of Jeffreys backing him up, or so it seems.

But what does Jeffreys [9] really write in §5.7? First, he considers explicitly high-quality data, and not routine data with their gross errors (whose rejection still causes avoidable efficiency losses, see below).

Secondly, he carefully distinguishes between "ordinary" high-quality data under heterogeneous circumstances intended to measure something, and high-quality data under conditions made artificially as homogeneous as possible, in order to check empirically the theory of measurement. He also gives a most interesting discussion of the serial correlation of the data in connection with their longtailedness, which we cannot go into here. He shows that and why one gets for "ordinary" high-quality data a lower number of degrees of freedom (3.5 and 4.4 in his examples) than under homogeneous conditions. He continues (loc. cit., p. 316): "It seems that for uniform conditions, if independence of the errors can be attained, and if there is a single value of m $[= \lambda_2]$ suitable for such conditions, it is likely to be between 3 and 5. Such a departure from the normal law is serious ..." We have the complication of dependence (the stronger the serial dependence, the shorter-tailed are the data), but for fictive independent data, Jeffreys gives the range of ν from 5 to 9. Now $\nu = 7$ $[\lambda_2 = 4]$ lies in the middle between 5 and 9, but the efficiency loss under t_7 is not in the corresponding middle, the efficiencies under t_5, t_7, t_9 being 80%, 89%, 93%. (The formula is known to be $1-6/(\nu(\nu+1))$.) Hence the efficiency losses, according to Jeffreys (notabene for independent, homogeneous high-quality data), are likely to be between 7% and 20%, rather than below 10%.

This is not a final answer, however, both because of serial dependence (whose correlation with the longtailedness is still rather mysterious) and because of inhomogeneity, which counteract each other. The 7 homogeneous series studied by Jeffreys are fitted by symmetric Pearson curves starting with $t_{7.7}$ and going twice beyond the normal to Type II. But the two "realistic" series from astronomy (of about 4500 and 5000 observations), with $\nu = 4.4$ and 3.5 (and unknown correlations), lead to efficiencies of 75% and 62% ! This is the more realistic range of efficiencies of least squares estimators, still under high-quality data.

We can complement the numbers given in Jeffreys (loc. cit.) by some unpublished investigations by J.-D. Tacier. Tacier analyzed 8 large high-quality data sets from astronomy, geodesy, geophysics, metrology, and physics, including the 2885.5 determinations of the velocity of light by Michelson et al. [10] and the 8688 astronomical observations cited in Romanowski and Green [11]. The number of degrees of freedom of the t-distributions fitted ranges from 1.5 (!) to 8.4, with corresponding asymptotic efficiencies of the arithmetic mean from 0 to 92%. The Michelson data lead to ν about 3.8(?), while 70 means of groups of 12 of these data give $\nu = 1.5$. The set of 8688 data gives $\nu = 7.4$ (or 90% efficiency); the other values are $\nu = 4.4$; 4.6; 5.2; and 6.0. Thus efficiencies of the mean between 70% and 90% are common, and much lower values can occur.

The previous discussions consider only the arithmetic mean or, more generally, the linear estimators in the least squares method, as opposed to the quadratic estimate of scale. The asymptotic efficiency of the empirical variance and standard deviation under t_ν is $1 - 12/(\nu(\nu-1))$ (Fisher [6]) which becomes 0 for $\nu \leq 4$. For integer ν from 5 to 9 the efficiencies are 40%, 60%, 71%, 79%, 83% . Hence the losses in scale estimation are still considerably larger than for location.

In retrospect, the use of t_3 as an example for the possible distribution of high-quality data in the large Monte Carlo study in Andrews et al. [12] may have been a bit on the extreme side, though still within a realistic range. It should have been complemented by a t with a more "typical" number of degrees of freedom, such as t_4 or t_5. One might also consider inclusion of t_7 in addition, although we would probably learn little beyond what could be obtained from interpolation between t_3, t_5, and normal. Already t_3 gives a lot of quantitative and qualitative information on the behavior of estimators alternative to the mean under high-quality data.

One class of estimators which many applied statisticians are using is "the mean after rejection of outliers". The problem is that the remaining data after rejection are not normal either, and the resulting avoidable efficiency losses of the overall procedure are also known in some important cases from the Monte Carlo studies discussed in Relles and Rogers [13] and in Hampel [14]. The former investigates subjective rejection, the latter the most important classical rejection rules. If we would have strictly normal data and distant outliers, then some of the "objective" rules are very good (though even then some are mediocre to disastrous). But if we model our data by t_3 with or without outliers, then the avoidable efficiency losses of all rejection rules, including subjective rejection, are at least about 15-20%. While some "rejection rules" cannot even reject distant outliers, the more difficult problem is the "doubtful outliers" and the deviations from normality in the "flanks" of the distribution, about 2 to 3 or 4 standard deviations away, which occur also with high-quality data. All rejection rules can be arbitrarily bad under unfortunate, unrealistic, but theoretically possible circumstances. In usual situations, the mean used after one of the best rejection rules has been applied, avoids at least the disastrous efficiency losses possible for the "unprotected mean", but losses around 15-20% are still quite realistic.

A few words should perhaps be said to the paper by Stigler [15] which seems to have reinforced the belief that the mean loses only a few percent efficiency under high-quality data. There is valuable material in the paper (also pointing to a t with less than 5 degrees of freedom as possible model, corresponding to less than 80% efficiency of the mean), but unfortunately the main analysis by Stigler is basically wrong and misleading since it ignores the systematic errors in the physical constants. The point was brought out in the discussion by Pratt; in the rejoinder, Stigler claimed to have refuted Pratt's criticism by a theorem, but in its application he mixed up "increasing" and "nondecreasing". As a consequence, his counterargument was not valid, and the data show that his claim was also wrong. A much more informative analysis is the one by Eisenhart in the discussion (loc. cit.) which can be extended to other data sets with similar results.

4. SOME EXAMPLES OF STRUCTURED DATA

Structured data, as in the analysis of variance, provide much richer information and many more choices for modelling and interpretation than unstructured data. It may therefore be worthwhile to have a brief second look at a few of the examples in Rao [1] and Rao [2].

There is a 3×3 analysis of variance with measurements on skulls by 3 observers in 3 strata in Rao [1], p. 95 and Rao [2], p. 211. The residuals from a main effects model look reasonable, so one is led to the standard analysis as given in the books. By looking at the data, there is one small observation one might make, in a purely descriptive or "exploratory" sense: the first observer has both a smaller scatter of the residuals than the others and a (comparatively) widely different mean. By itself this means nothing, and there is nothing "significant" about it; <u>only</u> in the context of more background knowledge, here mainly about the observers, can it conceivably become an interesting hint reinforcing or weakening other hints or pieces of evidence (Is the first observer different from the others? Is he more careful? Does he – or the others – have a systematic error?). But we should not miss such clues, although we may forget them completely once they are shown to be irrelevant in the full context. – (The example, whose entries have different weights, can also be used to compare the proper analysis for the unbalanced data with a simplified "orthogonal" analysis ignoring the weights, at least in a first round.)

Another example with a much richer structure is the pig data in Rao [1], p.121f, and Rao [2], p.246. They give the growth rates of 30 pigs in 5 pens and of 2 sexes under 3 diets in a balanced design, with initial weight as concomitant variable. The data go back to Wishart [16] and Woodman et al. [17] and are also used in Scheffé [18], p.217, where still another intermediate reference is cited. The standard analysis of covariance with food $\times$ sex interaction gives residuals which are all between -0.60 and +0.43, except for two male pigs in the same pen with residuals -1.14 and +0.92 (cf. Hampel [14], p. 105). These residuals clearly stick out, for example in a stem-and-leaf display, but they are formally not significant in outlier tests (except with one very bad rule). A realistic P-value from outlier tests is somewhat below or around 10% . Hence no clearcut formal conclusion is possible, but the residuals are large enough to merit special consideration. The simplest explanation would be an inadvertent interchange of the food types (or some other part of the data). Indeed, when treatments A and B for the males in the second pen are interchanged, the new residuals become 0.07 and -0.25, and all residuals look rather satisfactory, lying between -0.59 and +0.41. Mixing up parts of data in a highly structured design is a common source of gross errors, and the position of the two absolutely largest residuals, differing only in one factor, points in this direction. On the other hand, a check of the literature on the data set given above revealed that the same set of data has been faithfully copied throughout all publications. Hence, if an interchange took place, it must have been before the first publication (or rather, before the data were put into final form). This is apriori the most likely time, but without more knowledge about the data acquisition and its reliability or perhaps access to original data sheets it seems hopeless to speculate more about this possibility.

Another interpretation offering itself is that two pigs actually behaved visibly different from the rest, one growing markedly slower (which would be explainable by some disease or a similar reason), and the other growing markedly faster, given the conditions they were in. If so, this would mean that the pigs were not homogeneous, but that there could be (a varying number of) stragglers on both sides. Their existence might have implications for the interpretation of and generalization from the results. Here it would be very helpful to know more about the results of other, similar data sets.

Wishart [16] gives the coefficient of the (orthogonal) quadratic growth term h in addition to the linear term g. This former term does not contain much information, but in addition to a slightly improved picture also for h by interchanging the diets of the above-mentioned pigs, there is a slight hint that interchange of diets for the first two pigs in the fourth pen might also simplify the data structure. Their g values and initial weights are almost identical, and nothing could be discovered about them by looking only at g. Both interchanges taken together clearly improve the significance of effects on h; but we have to remember that there is a selection effect in picking the largest, outlying residuals and looking for simple reinterpretations to decrease them. Alternatively, there are more stragglers among the pigs themselves which do not follow the simpler behavior shown by a large majority.

In any actual data analysis, it is important to discover such deviating substructures as early as possible and either to verify their unusual structure or to correct whatever might have gone wrong. Post festum, it is usually impossible to find out what really happened. We are reminded of the polite restraint by C. Daniel [5], p.40, when in reference to another famous data set he remarks: "If ... it could be done tactfully, we would ask the experimenters whether by any chance the reported 312 ... could have been misrecorded by 100. But this can hardly be done, even tactfully, for ancient data in another field."

It is on the one hand understandable and elegant, on the other hand regrettable that many data sets in Rao [2] are only given in condensed form. It may well be true that the main conclusions usually do not differ much after a refined residual analysis, but one can either improve or fortify them and often can learn new additional things. For example, it is well-known that the famous Iris data by Fisher (Rao [1], p.248, Rao [2], p.480) also contain (at least) one outlier (cf., e.g., Chambers et al. [19], p.148). It is an uncomfortable feeling for a data analyst not to know what is hidden behind a sufficient statistic, and how well the assumed model really fits.

5. SOME THEORETICAL DEVELOPMENTS

In a pioneering paper, Kallianpur and Rao [20] proposed to consider Fréchet-differentiable Fisher-consistent statistical functionals as estimators. They did not pursue this further (cf. also Kallianpur [21]) because most commonly used maximum likelihood estimators and other classical estimators are not Fréchet-differentiable. The reason is that they are too sensitive to outliers and hence not continuous in any reasonable sense; so they can only obey weaker concepts of differentiability. This looked like the premature end of an attempt to build a beautiful mathematical theory in statistics.

But recently (Clarke [22],[23]) Fréchet-differentiability could indeed be shown for a large and interesting class of estimators. These are so-called M-estimators with bounded ψ-functions and some additional smoothness conditions. They are defined as the solution θ of $\int \psi(x,\theta)dF_n(x) = 0$, where F_n is the empirical cumulative, and $\psi(.,.)$ characterizes the estimator. With $\psi = f'/f$ they include maximum likelihood estimators. They are also known under the name of "estimating equations" or "estimating functions", and their beginnings go back to the last century. Boundedness (and continuity) of the ψ–function is essential for Fréchet-differentiability, but otherwise Clarke has shown this property to hold for the maximum likelihood estimator for the Cauchy distribution with its non-monotone ψ–function, for vector-valued ψ–functions, and for so-called Huber-estimators for which the derivative of the ψ–function has jumps.

Putting a bound on the supremum norm of the Fréchet derivative, the latter one of which is proportional to the ψ–function, we obtain a sharper lower bound for the asymptotic variance of Fisher-consistent estimators (cf., e.g., Hampel et al. [24], p. 117). It seems more difficult to get exact results not only for the asymptotic Cramér-Rao bound, but also for finite n. One can get the opinion that the variance of an estimator is not the most natural quantity to consider, except for normally distributed estimators, but rather the "intrinsic accuracy" or Fisher-information of the distribution of the estimator, as has already been hinted at by Fisher. In cases of asymptotic normality, the asymptotic information and the asymptotic variance are again equivalent.

The bound on the ψ–function of M-estimators has the practical effect that it bounds the influence of small parts or small distortions of the data, such as a few outliers or a change from normal to t_3. It thus exposes the outliers better by not following them too much, and it makes irrelevant the precise form of the parametric model, say, in a full (weak topology) neighborhood of the normal including (but not restricted to) all t–distributions up to t_3. Within the neighborhood, the exact true model can never be specified anyway. But a totally different question is the (discontinuous) model choice, for example, whether to include an interaction, a quadratic term, or a new variable or not, or whether to allow for systematic heteroscedasticity of a simple specific form. Each such model should be tried and tested, as in classical statistics, only it is tested together with a small neighborhood of indistinguishable or immaterially differing potential true models. These tests are made easier and sharper by being less susceptible to features irrelevant for the model choice. (There seems to be a misunderstanding in the third paragraph of Part 4 and in the summary of Rao [25]; the two procedures are not alternatives, but complement each other.)

The methodology hinted at and described more fully in Huber [3] and Hampel et al. [24] fits beautifully into the framework of differential geometry applied to statistics (cf., e.g., Amari [26] or Amari [27]), for which Rao [28] was also one of the early pioneers. Given a parametric model within the space of all probability distributions, the functional we may want to use as estimator is fixed on the model to be equal to the parameter. Outside the model it is completely arbitrary and can be chosen according to some criteria. One is the information loss at the parametric model. But when we minimize this and try to introduce a topology into the space of probability distributions such as the weak topology or even the one given by the total variation metric, we may end up with hopelessly intertwined fiber bundles, as in the case of the expectation (arithmetic mean), which is nowhere continuous and not everywhere defined in any full neighborhood. The point is that at the price of a small loss of efficiency at the model, we can choose consistent functionals with beautiful topological and metric properties – and even with Fréchet differentiability.

REFERENCES

[1] Rao, C. R. (1952). *Advanced Statistical Methods in Biometric Research.* Wiley, New York.

[2] Rao, C. R. (1965). *Linear Statistical Inference and Its Applications.* Wiley, New York.

[3] Huber, P. J. (1981). *Robust Statistics.* Wiley, New York.

[4] Daniel, C., and Wood, F. S. ((1971)1980). *Fitting Equations to Data.* Wiley, New York.

[5] Daniel, C. (1976). *Applications of Statistics to Industrial Experimentation.* Wiley, New York.

[6] Fisher, R. A. (1922). On the mathematical foundations of theoretical statistics. Philos. Trans. Roy. Soc., London, Ser. A, **222**, 309-368.

[7] McCullagh, P., and Nelder, J. A. (1983). *Generalized Linear Models.* Chapman and Hall, London.

[8] Cox, D. R., and Hinkley, D. V. (1968). A note on the efficiency of least-squares estimates. J. R. Statist. Soc. B, **30**, 284-289.

[9] Jeffreys, H. (1961). *Theory of Probability.* 3rd edition. Clarendon Press, Oxford.

[10] Michelson, A. A., Pease, F. G., and Pearson, F. (1935). Measurement of the velocity of light in a partial vacuum. Contributions from the Mount Wilson Observatory, Carnegie Institution of Washington, Vol. XXII, No. 522, 259-294.

[11] Romanowski, M., and Green, E. (1965). Practical applications of the modified normal distribution. Bull. Géodésique **76**, 1-20.

[12] Andrews, D. F., Bickel, P. J., Hampel, F. R., Huber, P. J., Rogers, W. H., and Tukey, J. W. (1972). *Robust Estimates of Location. Survey and Advances.* Princeton University Press, Princeton, N.J.

[13] Relles, D. A., and Rogers, W. H. (1977). Statisticians are fairly robust estimators of location. J. Amer. Statist. Assoc. **72**, 107-111.

[14] Hampel, F. R. (1985). On the breakdown points of the mean combined with some rejection rules. Technometrics **27**, 95-107.

[15] Stigler, S. M. (1977). Do robust estimators work on real data? Ann. Statist. **6**, 1055-1098.

[16] Wishart, J. (1938). Growth-rate determinations in nutrition studies with the bacon pig, and their analysis. Biometrika **30**, 16-28.

[17] Woodman, H. E., Evans, R. E., Callow, E. H., and Wishart, J. (1936). The nutrition of the bacon pig. I. The influence of high levels of protein intake on growth, conformation and quality in the bacon pig. The Journal of Agricultural Science **26**, 546-619.

[18] Scheffé, H. (1959). *The Analysis of Variance.* Wiley, New York.

[19] Chambers, J. M., Cleveland, W. S., Kleiner, B., and Tukey, P. A. (1983). *Graphical Methods for Data Analysis.* Wadsworth, Belmont, Cal.

[20] Kallianpur, G., and Rao, C. R. (1955). On Fisher's lower bound to asymptotic variance of a consistent estimate. Sankhyā A **15**, 331-342.

[21] Kallianpur, G. (1963). Von Mises functions and maximum likelihood estimation. Sankhyā A **23**, 149-158.

[22] Clarke, B. R. (1983). Uniqueness and Fréchet differentiability of functional solutions to maximum likelihood type equations. Ann. Statist. **11**, 1196-1205.

[23] Clarke, B. R. (1986). Nonsmooth analysis and Fréchet differentiability of M-functionals. Prob. Th. Rel. Fields **73**, 197-209.

[24] Hampel, F. R., Ronchetti, E. M., Rousseeuw, P. J., and Stahel, W. A. (1986). *Robust Statistics: The Approach Based on Influence Functions.* Wiley, New York.

[25] Rao, C. R. (1987). Strategies of data analysis. Invited Paper 26.2, 46th Session ISI, Tokyo.

[26] Amari, S. (1985). *Differential Geometrical Methods in Statistics.* Springer Lecture Notes in Statistics, Vol. 28.

[27] Amari, S. (1987). Differential geometry in statistical inference. Invited Paper 6.1, 46th Session ISI, Tokyo.

[28] Rao, C. R. (1945). Information and accuracy attainable in the estimation of statistical parameters. Bull. Calcutta Math. Soc. **37**, 81-91.

STATISTICAL DATA ANALYSIS AND INFERENCE
Y. Dodge (Editor)

ON THE MAXIMUM NUMBER OF FACTORS IN TWO CONSTRUCTION METHODS FOR ORTHOGONAL ARRAYS*

A. S. Hedayat and J. Stufken*

Department of Mathematics, Statistics, and Computer Science, University of Illinois at Chicago, Chicago, Illinois 60680, U.S.A., and Department of Statistics, Iowa State University, Ames, Iowa 50011, USA.

ABSTRACT: An important problem in studying orthogonal arrays is the determination of the maximum number of factors that can be accommodated in such an array. A slightly easier and more tractable problem is to determine the maximum number of factors that can be accommodated in an orthogonal array if consideration is restricted to arrays that can be constructed through a particular method. This paper describes two such methods of construction and shows that the maximum numbers of factors under these two methods are the same.

1. INTRODUCTION.

Nowadays every statistician is aware of the crucial role of factorial designs and fractional factorial designs in experiments designed to study simultaneously the effect of various factors on one or more variables of interest. The study of these designs and of their applications has a long history and originates with the founding fathers of design of experiments, Fisher and Yates. For more detailed references see Raktoe, Hedayat and Federer [11].

One class of fractional factorial designs that has received a lot of attention in the literature, under various names, is the class of designs now widely known as orthogonal arrays. These designs were introduced and first studied by C.R. Rao in his papers listed as [12], [13], and [14]. Over the years many prominent researchers have been motivated by these papers and have contributed to reach the current level of knowledge on properties, existence and construction of orthogonal arrays. In addition, various other fractions that in some sense relax the combinatorial structure required by an orthogonal array, possibly at the expense of some desirable statistical properties, have resulted from Rao's initial papers.

There remain, however, an innumerable number of unanswered questions related to orthogonal arrays. Some of these have been brought back into prominence by the heightened interest in product quality control, where orthogonal arrays play an important role.

*Research is supported by Grant AFOSR-89-0221

This will hopefully be an incentive for some bright, young researchers to tackle some of the outstanding problems. A unified presentation of the available results on orthogonal arrays can be anticipated in the forthcoming monograph by Hedayat and Stufken [8].

The current paper discusses and compares two methods for the construction of orthogonal arrays. In particular, it is shown that the maximum numbers of factors that can be accommodated are the same for the two methods.

2. THE CONSTRUCTION METHODS.

For completeness we will define an orthogonal array in this section. We will then proceed to give the two methods of construction, one of which is due to Bose and Bush [2] and the other, although not presented with orthogonal arrays in mind, to Kempthorne [10]. The latter method is also presented in great detail for the special case with all factors at two levels by Box and Hunter [3] and [4].

A purely mathematical definition of an orthogonal array may be given as follows.

DEFINITION 2.1. *A $k \times N$ array based on s symbols, say $0, 1, \ldots, s-1$, is said to be an orthogonal array of strength t and index λ if for any $t \times N$ subarray each of the s^t vectors based on $0, 1, \ldots, s-1$ occurs λ times as a column.*

Obviously, it must hold that $N = \lambda s^t$. Such an array will be denoted by $OA(N, k, s, t)$, from which the index λ can be obtained by the preceding equation. For a statistician, the k rows correspond to k factors each of which can be at s different levels. The N columns correspond to the N level or treatment combinations at which a measurement of the variable of interest will be obtained. The strength t of an array has also a statistical implication. This is illustrated in the following theorem, which assumes the use of an orthogonal polynomial model in relating the k factors with the response variable. We will make that assumption throughout this paper, and will also assume that $t \geq 2$. The concept of orthogonal polynomial models and a proof of this theorem can be found in Raktoe, Hedayat and Federer [11].

THEOREM 2.1. *In an orthogonal array of strength t we have (i) if $t = 2u - 1$, then the general mean, main effects and interactions of at most $u - 1$ factors are estimable provided that interactions of at least u+1 factors are negligible, and (ii) if $t = 2u$, then the general mean, main effects and interactions of at most u factors are estimable provided that higher order interactions are negligible.*

One of the more interesting problems related to orthogonal arrays is that of the determination of the maximum number of factors, i.e. the maximum value of k, for given values of N, s and t. A solution, possibly for specific values of N, s and t, consists essentially of two steps. One is to obtain an $OA(N, k^*, s, t)$ with the claimed maximum number of factors k^*, the other is to show that an $OA(N, k, s, t)$ cannot exist if $k > k^*$. It suffices to show the latter for $k = k^* + 1$. To get some idea about the value of k^* it is useful to have a sharp upper bound for the maximum number of factors in an orthogonal array. The most general upper bound, applicable for any values of N, s and t, was obtained by Rao [13]. The maximum number of factors is implicitly given in the inequalities in the following theorem.

THEOREM 2.2. *In an orthogonal array, $OA(N, k, s, t)$, the following inequalities hold:*

(i)
$$N \geq 1 + \binom{k}{1}(s-1) + \ldots + \binom{k}{u}(s-1)^u \quad \text{if } t = 2u,$$

and

(ii)
$$N \geq 1 + \binom{k}{1}(s-1) + \ldots + \binom{k}{u}(s-1)^u + \binom{k-1}{u}(s-1)^{u+1} \quad \text{if } t = 2u+1.$$

Various authors, for example Blum, Schatz and Seiden [1], Bose and Bush [2] and Bush [5], have improved this upper bound for k under the assumption of specific values for one or more of N, s and t or specific relations between these design parameters. That the bounds by Bush [5] for arrays of index unity are indeed uniformly better than those from Theorem 2.2. was shown in Hedayat and Stufken [9].

In other research, authors have opted to answer a slightly less general question. Instead of determining k^* for given N, s and t, the problem of determining the maximum number of factors that can be accommodated in an orthogonal array that is constructed through a particular method is considered. This paper describes two construction methods for which this problem has been considered.

We will show that the maximum numbers of factors corresponding to these two constructions are the same. Since both methods require s to be a power of a prime, we will assume from hereon that this is indeed the case. The symbols $0, 1, \ldots, s-1$ will denote the elements of the finite Galois field $GF(s)$, where 0 denotes the zero element and $1, \ldots, s-1$ represent $x^0 = 1, \ldots x^{s-2}$, respectively, for a primitive element x.

The first method, due to Bose and Bush [2], starts with a $k \times n$ array A based on the elements of $GF(s)$. If A has the property that any t rows are linearly independent over $GF(s)$, then the matrix product (computed in $GF(s)$) AT, where T is an $n \times s^n$ array with its columns consisting of all possible n-tuples based on the elements of $GF(s)$, is an $OA(s^n, k, s, t)$ of index s^{n-t}.

This follows by observing that the restriction of AT to any t rows is obtained as A^*T, where A^* is the $t \times n$ array obtained as the restriction of A to those t rows. Since A^* has rank t, every possible t-tuple appears s^{n-t} times as a column in A^*T.

The second method was developed by Kempthorne [10], although the basic tools for it were also known by Fisher [7] and Finney [6]. With k factors at s levels, let $(\alpha_1, \alpha_2, \ldots, \alpha_k)_j$, with $\alpha_i \in GF(s)$, $i = 1, \ldots, k$, and $j \in GF(s)$ denote those treatment combinations $x_1 x_2 \ldots x_k$ for which

$$\alpha_1 x_1 + \alpha_2 x_2 + \ldots + \alpha_k x_k = j.$$

There are s^{k-1} treatment combinations that satisfy this equation, provided that $(\alpha_1, \alpha_2, \ldots, \alpha_k) \neq \underline{0}$. We denote this $1/s$ replicate formed by these treatment combinations by $I = (\alpha_1, \alpha_2, \ldots, \alpha_k)_j$, or for brevity by $I = (\underline{\alpha})_j$, where α is the row vector $(\alpha_1, \alpha_2, \ldots, \alpha_k)$.

If $\underline{\alpha}^{(1)}, \ldots, \underline{\alpha}^{(p)}$ are p linearly independent vectors over $GF(s)$, with

$$\underline{\alpha}^{(m)} = (\alpha_1^{(m)}, \alpha_2^{(m)}, \ldots, \alpha_k^{(m)})$$

say, then the p equations

$$\begin{aligned} \alpha_1^{(1)} x_1 + \alpha_2^{(1)} x_2 + \quad \ldots \quad + \alpha_k^{(1)} x_k &= j_1 \\ &\vdots \\ \alpha_1^{(p)} x_1 + \alpha_2^{(p)} x_2 + \quad \ldots \quad + \alpha_k^{(p)} x_k &= j_p \end{aligned}$$

induce a $1/s^p$ replicate, which we denote by $I = (\underline{\alpha}^{(1)})_{j_1} = \ldots = (\underline{\alpha}^{(p)})_{j_p}$. Any treatment combination in this fraction is also a solution of

$$\begin{aligned} &(\beta_1 \alpha_1^{(1)} + \ldots + \beta_p \alpha_1^{(p)}) x_1 + (\beta_1 \alpha_2^{(1)} + \ldots + \beta_p \alpha_2^{(p)}) x_2 + \cdots \\ &+ (\beta_1 \alpha_k^{(1)} + \ldots + \beta_p \alpha_k^{(p)}) x_k = \beta_1 j_1 + \ldots + \beta_p j_p, \end{aligned} \tag{2.1}$$

where $\underline{\beta} = (\beta_1, \beta_2, \ldots, \beta_p)$ is any vector based on elements from $GF(s)$. There are s^{p-1} such nonzero vectors $\underline{\beta}$, each inducing an equation through (2.1) that has to be satisfied. With $\gamma_i(\underline{\beta}) = \beta_1 \alpha_i^{(1)} + \ldots + \beta_p \alpha_i^{(p)}$ and $\underline{\gamma}(\underline{\beta}) = (\gamma_1(\underline{\beta}), \ldots, \gamma_k(\underline{\beta}))$ let $w(\underline{\gamma}(\underline{\beta}))$ denote the number of nonzero entries in $\underline{\gamma}(\underline{\beta})$. Let $w = \min_{\beta \neq 0} w(\underline{\gamma}(\underline{\beta}))$. Then the $1/s^p$ replicate obtained by the p equations is an $OA(s^{k-p}, k, s, w-1)$ of index $s^{k-p-w+1}$.

An alternative way, which will be easier for our purpose, to formulate this construction is as follows. Let B be any $p \times k$ matrix of rank p over $GF(s)$. Let $w(B) = \min\{w(\underline{\gamma}) : \underline{\gamma} \in$

$\mathcal{R}(B)$, $\underline{\gamma} \neq \underline{0}\}$, where $\mathcal{R}(B)$ denotes the row space of B and $w(\underline{\gamma})$, as before, the number of nonzero entries in $\underline{\gamma}$. With $\underline{\delta}$ as any $p \times 1$ vector based on $GF(s)$, it is easy to see that the s^{k-p} solutions to the equations

$$B\underline{x} = \underline{\delta} \tag{2.2}$$

form an orthogonal array $OA(s^{k-p}, k, s, w-1)$ of index $s^{k-p-w+1}$, where $w = w(B)$.

The relation between the two formulations is quite clear. The independent vectors $\underline{\alpha}^{(1)}, \ldots, \underline{\alpha}^{(p)}$ in the first formulation correspond to the p rows of B in the second formulation.

That we obtain orthogonal arrays through this method, as claimed, can be seen as follows. Select any $w-1$ rows in the array and select one of the s^{w-1} possible $(w-1)$-tuples. In our second formulation of the method this corresponds to selecting a value for each of $w-1$ selected x_i's. The equations in (2.2) reduce then to

$$B^*\underline{x}^* = \underline{\delta}^*, \tag{2.3}$$

where $\underline{x}^*$ is the vector of the remaining $k-w+1$ variables and B^* is obtained from B by deleting the appropriate $w-1$ columns.

Since for $\underline{y} \neq \underline{0}$, $w(\underline{y}'B^*) \geq w(\underline{y}'B) - (w-1) \geq 1$ by the definition of w, we see that B^* has rank p. Thus the equations in (2.3) have $s^{k-p-w+1}$ solutions, which is also equal to the number of solutions to (2.2) with $w-1$ of the x_i's fixed. Consequently, the second method gives an orthogonal array of strength $w-1$ and index $s^{k-p-w+1}$.

We conclude this section with a small example for each of the two methods.

EXAMPLE 2.1. For the first method with $s = 3$, $k = 4$ and $n = 2$ let

$$A = \begin{bmatrix} 1 & 0 \\ 0 & 1 \\ 1 & 1 \\ 1 & 2 \end{bmatrix}.$$

Any two rows are linearly independent, and hence

$$AT = \begin{bmatrix} 0 & 0 & 0 & 1 & 1 & 1 & 2 & 2 & 2 \\ 0 & 1 & 2 & 0 & 1 & 2 & 0 & 1 & 2 \\ 0 & 1 & 2 & 1 & 2 & 0 & 2 & 0 & 1 \\ 0 & 2 & 1 & 1 & 0 & 2 & 2 & 1 & 0 \end{bmatrix}$$

is an $OA(9, 4, 3, 2)$.

EXAMPLE 2.2. For the second method with $s = 3$, $k = 4$ and $p = 2$, let $\underline{\alpha}^{(1)} = (1,1,1,0)$ and $\underline{\alpha}^{(2)} = (0,1,2,1)$. Then $w(\underline{\gamma}(\underline{\beta})) = 3$ for $\underline{\beta} \in \{(0,1), (0,2), (1,0), (1,1), (1,2), (2,0), (2,1), (2,2)\}$. Hence the $1/3^2$ replicate generated by the two independent equations

$$\begin{aligned} x_1 + x_2 + x_3 &= j_1 \\ x_2 + 2x_3 + x_4 &= j_2 \end{aligned}$$

is an $OA(9,4,3,2)$, no matter what j_1 and j_2 are. For example, with $j_1 = j_2 = 0$ the resulting orthogonal array is as follows:

$$\begin{bmatrix} 0 & 0 & 0 & 1 & 1 & 1 & 2 & 2 & 2 \\ 0 & 1 & 2 & 0 & 1 & 2 & 0 & 1 & 2 \\ 0 & 2 & 1 & 2 & 1 & 0 & 1 & 0 & 2 \\ 0 & 1 & 2 & 2 & 0 & 1 & 1 & 2 & 0 \end{bmatrix}$$

The array B in the alternative formulation of the second construction corresponds to

$$B = \begin{bmatrix} 1 & 1 & 1 & 0 \\ 0 & 1 & 2 & 1 \end{bmatrix}.$$

The nonzero row vectors in its row space are $(0,1,2,1)$, $(0,2,1,2)$, $(1,1,1,0)$, $(1,2,0,1)$, $(1,0,2,2)$, $(2,2,2,0)$, $(2,0,1,1)$ and $(2,1,0,2)$, resulting in $w(B) = 3$.

3. THE MAXIMUM NUMBER OF FACTORS.

In order to obtain orthogonal arrays with the same number of treatment combinations and the same strength for the two methods in the previous section the relations $n = k - p$ and $w = t + 1$ must hold, where n, k, p and t are as previously introduced. In order to prove our claim that the maximum numbers of factors that can be accommodated under these two constructions are the same, it suffices now to show the following result.

THEOREM 3.1. *The largest value of k for which a $k \times (k-p)$ array A exists with any t rows linearly independent over $GF(s)$ equals the largest value of k for which a $p \times k$ array B of rank p exists with $w(B) \geq t + 1$.*

The proof consists of showing how such arrays A and B with the same value of k can be obtained from each other. First, assume that we have a $p \times k$ array B as in Theorem 3.1. Since the rank of B equals p, after a possible permutation of the columns and taking appropriate linear combinations of the rows we can convert B into

$$B^* = [C \;\; I_p],$$

where I_p denotes the $p \times p$ identity matrix. We have that $w(B^*) = w(B) \geq t + 1$. Now define an array A by

$$A = \begin{bmatrix} I_{k-p} \\ C \end{bmatrix}. \tag{3.1}$$

For $u \leq t$, any linear combination of u rows from C, with all u coefficients different from zero, contains at least $t+1-u$ nonzero elements since $w(B^*) \geq t+1$. But that implies that at least $u+(t+1-u)=t+1$ rows in A are needed to write $\underline{0}$ as a nontrivial linear combination of the rows in A. Consequently, any t rows in A are linearly independent.

For the converse, assume that we have a $k \times (k-p)$ array A for which any t rows are linearly independent. The rank of such an array could be less than $k-p$, so that we may not be able to convert it to a form as in (3.1). If the rank of A, say r, is less than $k-p$, remove any $k-p-r$ rows from A such that the remaining array has still rank r. Now add the same number of rows in such a way that the new array has rank $k-p$. Obviously, all these steps are possible and, moreover, result in an array for which any t rows are linearly independent. But this array can be converted to an array of the form

$$A^* = \begin{bmatrix} I_{k-p} \\ C \end{bmatrix} \tag{3.2}$$

after a possible permutation of the rows and taking appropriate linear combinations of the columns. The array A^* still has the property that any t rows are linearly independent. (Note that we can also obtain a form as in (3.2) by taking appropriate linear combinations of the rows. Then, however, it may no longer be true that any t rows of the resulting array are linearly independent). Now define an array B as

$$B = [C \quad I_p].$$

Obviously, the rank of B equals p. For a linear combination of the rows in B, say $\underline{x}B$ for a nonzero row vector $\underline{x}$, let $u = w(\underline{x})$. If $u > t$, then obviously $w(\underline{x}B) \geq t+1$. If $u \leq t$, then by the property of A^* $w(\underline{x}C) \geq t+1-u$. Hence, also for this case $w(\underline{x}B) = w(\underline{x}C)+u \geq t+1$. Thus we obtain that $w(B) \geq t+1$, which concludes the proof of Theorem 3.1.

As a closing remark, it would be interesting to know whether there are orthogonal arrays with design parameters N, s and t as in the two methods considered here and with a larger number of factors than can be obtained through these two methods. We are not aware of any examples of such orthogonal arrays, nor have we seen a proof of their nonexistence.

REFERENCES.

[1] Blum, J.R., Schatz, J.A. and Seiden, E., On 2-level orthogonal arrays of odd index, *J. Combin. Theory* **9** (1970), 239-243.

[2] BOSE, R.C. AND BUSH, K.A., Orthogonal arrays of strength two and three, *Ann. Math. Statist.* **23** (1952), 508-524.

[3] BOX, G.E.P. AND HUNTER, J.S., The 2^{k-p} fractional factorial designs, Part I, *Technometrics* **3** (1961), 311-351.

[4] BOX, G.E.P. AND HUNTER, J.S., The 2^{k-p} fractional factorial designs, Part II, *Technometrics* **3** (1961), 449-458.

[5] BUSH, K.A., Orthogonal arrays of index unity, *Ann. Math. Statist.* **23** (1952), 426-434.

[6] FINNEY, D.J., The Fractional replication of factorial arrangements, *Ann. Eugen.* **12** (1945), 291-301.

[7] FISHER, R.A., The theory of confounding in factorial experiments in relation to the theory of groups, *Ann. Eugen.* **11** (1942), 341-353.

[8] HEDAYAT, A. AND STUFKEN, J., Orthogonal arrays: The current status, to appear.

[9] HEDAYAT, A. AND STUFKEN, J., On the maximum number of constraints in orthogonal arrays, *Ann. Statist.* (1989), in press.

[10] KEMPTHORNE, O., A simple approach to confounding and fractional replication in factorial experiments, *Biometrika* **34** (1947), 252-272.

[11] RAKTOE, B.L., HEDAYAT, A. AND FEDERER, W.T., Factorial Designs, Wiley, New York, 1981.

[12] RAO, C.R., Hypercubes of strength '*d*' leading to confounded designs in factorial experiments, *Bull. Calc. Math. Soc.* **38** (1946), 67-78.

[13] RAO, C.R., On a class of arrangements, *Proc. Edinburgh Math. Soc.* (2) **8** (1947), 119-125.

[14] RAO, C.R., Factorial experiments derivable from combinatorial arrangements of arrays, *J. Roy. Statist. Soc.* **9** (1947), 128-139.

STATISTICAL DATA ANALYSIS AND INFERENCE
Y. Dodge (Editor)

SOME GEOMETRICAL ASPECTS OF DATA ANALYSIS AND STATISTICS

Josep M. Oller

Department of Statistics, University of Barcelona
Avda. Diagonal 645, 08028 Barcelona, Spain

In this paper we discuss some desirable properties that a distance between probability spaces must satisfy, and from these considerations we introduce the *information metric* in several different approaches. Once this information metric is introduced, its applications to data analysis and statistics are explored. For instance, the classic analysis of variance methods may be obtained through this geometrical approach, and a geometrical interpretation of likelihood estimation may be given in terms of this metric, by defining in a natural way a distance between the individuals of a statistical population. Finally we consider several geometric properties of a general class of elliptic probability distributions, exhibiting the differential metric, the sectional curvatures, the Ricci tensor, the geodesic equations and, in some cases, the evaluation of the Riemannian distance, called the *Rao distance*, for this family of probability distributions. Some statistical hypothesis tests are also discussed.

1. INTRODUCTION

Many geometrical tools, in particular the distance concept, are widely used in data analysis and applied research in order to study the similarity or dissimilarity between physical objects, usually, in a wide sense, populations. See, for instance, Mahalanobis *et al.* [1] and Rao [2] in anthropology, Cavalli-Sforza and Edwards [3], Nei [4] and Prevosti *et al.* [5] in genetics, Agresti and Agresti [6] in sociology, Hadjisavvas [7] and Provost and Vallee [8] in physics, among many others. In all these cases the distance used is considered as a measure of the information available about the differences between the objects compared. Frequently the distance is used to obtain graphical outputs, through taxonomical methods or multidimensional scaling techniques, and with the aid of these graphic representations to set up the relationship among the objects studied.

The distance concept is very suitable for this purpose, since it is an intuitive notion, closely related with our physical space perception and, at the same time, a well developed concept from a mathematical point of view. There are many distances and related indexes used in data analysis and a natural question arises: what is the most suitable distance to use in a particular problem? From a methodological point of view it would seem to be reasonable, in order to define a distance, to pay attention, mainly, to the formal properties of the underlying observation process rather than taking account of the physical nature of the objects studied. The distance is a property neither of physical objects nor of an observer: it is a property of the observation process.

From these considerations, a necessary first step is to associate an adequate mathematical object to every physical object to be compared. Usually, suitable where d_1 and d_2 are the distance between the original statistical models and the transformed ones respectively. In other words, any measurable function induces a transformation between statistical models which must be distance decreasing.

Furthermore, let P_T be the probability measure induced by T, and additionally let us assume that,

ii) $\forall b \in \mathcal{A}$ $\exists$ an $\mathcal{A}'$-mesurable map η_b : $\eta_b\colon \mathcal{X}' \to \mathbb{R}$ defined as

$$P(\, b \cap T^{-1}(c)\,) = \int_c \eta_b(T)\, dP_T \qquad \forall c \in \mathcal{A}' \;,\; \forall P \in S_{(\mathcal{X},\, \mathcal{A})}$$

in this case is reasonable that

$$d_1(P_1, P_2) = d_2(P_1' \,,\, P_2'\,) = d_2(P_1\, T^{-1}, P_2\, T^{-1})$$

in other words, if T is a *sufficient statistic* then the distance has to be invariant. In that case we may also call the map T an *admissible transformation*. Notice that η_b is only independent of P, *while* $P \in S_{(\mathcal{X},\, \mathcal{A})}$. Therefore the admissibility concept is dependent on the statistical model considered. The statistics T carry all the relevant information for discrimination purposes from $(\mathcal{X}, \mathcal{A})$ into $(\mathcal{X}', \mathcal{A}')$.

A useful approximation property, closely related to those previously mentioned, would be

iii) Let $\mathcal{A}_n$, $n \in \mathbb{N}$, be a sequence of σ-algebras satisfying :

$$\mathcal{A}_1 \subset \mathcal{A}_2 \subset \;\dots\; \subset \mathcal{A}_n \subset \;\dots$$

let $\mathcal{A}$ be the σ-algebra generated by $\bigcup_{n\in\mathbb{N}} \mathcal{A}_n$

and let P_n and Q_n be the restriction of P and Q on $\mathcal{A}_n$, then

$$\lim_{n\to\infty} d(P_n, Q_n) = d(P, Q)$$

Another convenient property is concerned with the stochastic independence concept. Let C_{S_1} and C_{S_2} be two statistical models built on $(\mathcal{X}_1, \mathcal{A}_1)$ and $(\mathcal{X}_2, \mathcal{A}_2)$ respectively, and consider the joint statistical model

$$C_S = \{\, (\mathcal{X}_1{\times}\mathcal{X}_2,\, \mathcal{A}_1{\otimes}\mathcal{A}_2,\, P_1{\times}P_2) \;:\; P_1 \in S_1,\, P_2 \in S_2 \,\}$$

that is the model built on stochastic independence assumptions. In this case

iv) $$d^2(P_1{\times}P_2, Q_1{\times}Q_2) = d_1^2(P_1, Q_1) + d_2^2(P_2, Q_2)$$

This condition establishes a relationship between orthogonality and stochastic dependence, which is particularly interesting in data analysis graphical outputs.

Hereafter we are going to restrict our problem to a very common and useful case: the parametric case. The statistical model is often defined through an n-dimensional C^∞ real and connected manifold $(M, \wp_M)$, where $M \subset \Pi$, and $\wp_M$ is a maximal C^∞ atlas on mathematical objects for our purpose are probability spaces, and then the problem of introducing a distance between physical objects is translated into the problem of introducing a distance between probability spaces. It is possible to simplify the problem slightly, without any practical loss of generality, by considering that all the probability spaces we want to compare are built on the same measurable space $(\mathcal{X}, \mathcal{A})$, where $\mathcal{X}$ is the sample space, and $\mathcal{A}$ is a σ-algebra of subsets of $\mathcal{X}$, and they only differ from the probability measure P, that is from the expectation degree of the different events corresponding to the observation process. Therefore our problem is now reduced to introducing a distance between probability measures.

On the other hand distances between probability measures and other geometrical structures are widely used in statistics, for a better understanding and also for the solution of many kinds of statistical problems: testing statistical hypotheses, discriminant analysis, regression models among others, see for instance Amari [9], [10], Bhattacharyya [11], Efron [12], Fisher [13], Mahalanobis [14], Matusita [15], and Rao [16], [17].

Therefore, the problem we have considered is a common problem to data analysis and mathematical statistics, although the purpose and philosophy is sometimes different.

Before we consider some desirable properties that a distance between probability measures must satisfy, let us remark that we sometimes have some *a priori information* about these measures not contained in the data to be analyzed. This situation enables us to restrict our analysis to a certain subset of probability measures. In the case where no a priori information is available this subset would be the set of all probability measures on $(\mathcal{X}, \mathcal{A})$.

2. SOME DESIRABLE PROPERTIES OF DISTANCES

Given a measurable space $(\mathcal{X}, \mathcal{A})$ let Π or $\Pi_{(\mathcal{X}, \mathcal{A})}$ be the set of all probability measures on $(\mathcal{X}, \mathcal{A})$. Let S or $S_{(\mathcal{X}, \mathcal{A})}$ be a subset of Π, $S \subset \Pi$. We may define a *statistical model* as a family of probability spaces

$$C_S = \{ (\mathcal{X}, \mathcal{A}, P) \ : \ P \in S \}$$

In this section we intend to discuss some desirable properties that a distance between probability spaces, which belong to a certain statistical model, must satisfy. First of all we pay attention to a general invariance property. Let $(\mathcal{X}', \mathcal{A}')$ be another measurable space, and let T be a measurable map from $(\mathcal{X}, \mathcal{A})$ into $(\mathcal{X}', \mathcal{A}')$. Given a statistical model C_S on $(\mathcal{X}, \mathcal{A})$, the transformation T induces, in a natural way, a transformed statistical model $C_{S'}$ on $(\mathcal{X}', \mathcal{A}')$

$$C_{S'} = \{ (\mathcal{X}', \mathcal{A}', P') \ : \ P' = P\,T^{-1} \quad P \in S_{(\mathcal{X}, \mathcal{A})} \}$$

Then, it is convenient to require that

i) $\qquad d_1(P_1, P_2) \geq d_2(P_1', P_2') = d_2(P_1\,T^{-1}, P_2\,T^{-1})$

M. For the sake of simplicity we may consider on M the topology induced through the atlas by the usual $\mathbb{R}^n$ topology. Moreover, if the probability measures of our family M are dominated by a common reference measure μ, taking into account the Radon-Nikodym theorem we are able to represent our measures manifold $(M, \wp_M)$ by a manifold of measurable functions, or, more precisely, equivalent classes of measurable functions, $(D^\lambda, \wp_{D^\lambda})$ where $D^\lambda \subset \mathscr{I}$, through the map

$$\Phi_\lambda : M \to D^\lambda \quad , \quad \Phi_\lambda(P) = \lambda(\frac{dP}{d\mu})$$

where λ is a strictly monotonous real function, $\mathscr{I}$ stands for the set of all measurable functions on $(\mathcal{X}, \mathcal{A})$ and the $\wp_{D^\lambda}$ is obtained by considering all the local charts of the form

$$(\Phi_\lambda(U), \theta \circ \Phi_\lambda^{-1}) \qquad \forall\, (U, \theta) \in \wp_M$$

We may call $(D^\lambda, \wp_{D^\lambda})$ the λ-representation of our statistical model. In a customary notation we denote by $\lambda(p(x|\theta))$ a point $q \in U \subset D^\lambda$ with coordinates $\theta = \theta(q)$. Particularly interesting are the following cases:

* $\lambda(x) = x$, in which case D^λ is a density function manifold and since $D^\lambda \subset L^1(\mu)$ we may call $D^1 \equiv D^\lambda$ the 1-representation of the statistical model.

* $\lambda(x) = 2\sqrt{x}$, in which case we have $D^\lambda \subset L^2(\mu)$ and we shall call $D^2 \equiv D^\lambda$ the 2-representation of the statistical model.

* $\lambda(x) = \log x$, in which case we shall call $D^l \equiv D^\lambda$ the l-representation of our statistical model.

Therefore, in the present case, the problem of introducing a distance between statistical models is reduced to the problem of defining a distance in a suitable function manifold. For the sake of simplicity, we shall restrict our study to function manifolds which satisfy some adequate regularity conditions such as for every local chart (U, θ) and given any point $q \in U \subset D^\lambda$ of coordinates $\theta = \theta(q)$ the functions in x , $\partial\lambda(p(x|\theta))/\partial\theta_i \quad i = 1, \dots, n$ are linearly independent, and belong to a convenient $L^\alpha(p(\cdot|\theta)\, d\mu)$. Also, the partial derivatives $\partial/\partial\theta_i$ and the integration with respect to the measure $p(\cdot|\theta)\, d\mu$ can always be interchanged.

Whatever *regular* statistical model we consider, any statistical distance must satisfy two further properties:

v) The distance has to be invariant under coordinate system transformations.

vi) The distance must be invariant under reference measure changes.

At this point we may remark that it would be interesting to relate the distance to the information concept, and in particular to the *information gain* concept or *Kullback divergence*, Kullback and Leibler [18], since we want the distance between two probability spaces to be a measure of the information that we have about the differences between them.

It is also convenient that the distance obtained should be related to classical statistical quantities. Although this requirement is not essential, it is useful from a practical point of view.

A simple and natural approach to this problem is obtained by considering the square of the distance between two neighboring points, and through this, we may introduce a Riemannian structure on the previously mentioned function manifolds. Standard results on Riemannian geometry may be found in Hicks [19] or Spivak [20].

3. THE RAO DISTANCE CONSTRUCTION

Let us try to define suitable Riemannian structures on the statistical manifolds introduced in the previous section. We are going to consider several approaches.

Let P and Q be two probability measures dominated by μ : $P, Q \ll \mu$, and let $p = dP/d\mu$ and $q = dQ/d\mu$ be their corresponding density functions.

Assume also that ϕ is a smooth function on $\mathbb{R}^+ \times \mathbb{R}^+ \to \mathbb{R}$ and consider the ϕ *-divergence* measure defined as

$$Ð_\phi(p, q) = \int_\chi \phi(p(x), q(x))\, d\mu(x)$$

assume that ϕ satisfies $\phi(x, y) \geq 0$ and $\phi(x, x) = 0 \quad x, y \in \mathbb{R}^+$. For instance, an easy and useful way to define such divergences is by considering

$$\phi(x, y) = (\psi(x) - \psi(y))^2 \qquad \text{or} \qquad \phi(x, y) = (\psi(x) - \psi(y))^2\, x$$

for a suitable smooth function ψ. Let us require that this measure has to be invariant under admissible reference measure changes. Let ν be a positive measure such that $\mu \ll \nu$ and $h = d\mu/d\nu \in L^1(\nu)$, the usual Radon-Nykodim derivative. If we change the reference measure from μ to ν, we induce a natural change on the densities

$$p = \frac{dP}{d\mu} \to \bar{p} = \frac{dP}{d\nu} = \frac{dP}{d\mu}\frac{d\mu}{d\nu} = ph \quad \text{and} \quad q = \frac{dQ}{d\mu} \to \bar{q} = \frac{dQ}{d\nu} = \frac{dQ}{d\mu}\frac{d\mu}{d\nu} = qh$$

therefore we need

$$\int_\chi \phi(p, q)\, d\mu = \int_\chi \phi(ph, qh)\, d\nu$$

for any densities p and q, considered on an arbitrary measurable space $(\chi, \mathcal{U})$. Then the function ϕ has to satisfy

$$\phi(xz, yz) = z\, \phi(x, y) \qquad \forall x, y, z \geq 0$$

and therefore

$$\phi = x\frac{\partial \phi}{\partial x} + y\frac{\partial \phi}{\partial y}$$

the general solution of which is given by

$$\Psi(\phi(x,y)/x,\ \phi(x,y)/y) = 0$$

where Ψ is an arbitrary function. Obviously there are many solutions, every one of them is a *globally invariant* ϕ-divergence. For instance if we let $\Psi(u,v) = uv - (u/v + v/u - 2)\ \log^2 (u/v)$, we obtain

$$\mathrm{Ð}_\phi(p,q) = \int_\chi p \log\frac{p}{q}\, d\mu + \int_\chi q \log\frac{q}{p}\, d\mu$$

which is the usual symmetric Kullback divergence, and if we let $\Psi(u,v) = (1 - \sqrt{u/v}\,)^2 - u$, we obtain

$$\mathrm{Ð}_\phi(p,q) = \int_\chi (\sqrt{p} - \sqrt{q}\,)^2 d\mu$$

which is the square of the Hellinger or Matusita distance, the ordinary distance in the Hilbert space $L^2(\mu)$ between the square root densities $\sqrt{p}$ and $\sqrt{q}$.

Let us now consider the parametric case. Let $(D^1, \wp_{D^1})$ be the 1-representation manifold of a statistical model, satisfying some adequate regularity conditions. Consider now two neighboring points with coordinates θ and $\theta + \Delta\theta$. In such a case it seems to be reasonable to define the squared distance between these points, Δs^2, as

$$\Delta s^2 = \kappa\, \mathrm{Ð}_\phi(p(\cdot\,|\theta), p(\cdot\,|\theta + \Delta\theta))$$

where $\kappa > 0$. Since $\phi(x,x) = D_2\,\phi(x,x) = 0$, the first order approximation is zero and the second order approximation results in

$$\Delta s^2 = \frac{\kappa}{2}\int_\chi D_{22}\,\phi(p(x|\theta), p(x|\theta))\,(\Delta p(x|\theta))^2\, d\mu(x) =$$

$$= \sum_{i,j=1}^{n}\left(\frac{\kappa}{2}\int_\chi D_{22}\,\phi(p(x|\theta),\ p(x|\theta))\,\left(\frac{\partial p(x|\theta)}{\partial\theta_i}\frac{\partial p(x|\theta)}{\partial\theta_j}\right) d\mu(x)\right)\Delta\theta_i\,\Delta\theta_j$$

and taking into account that

$$g_{ij}(\theta) = \frac{\kappa}{2}\int_\chi D_{22}\,\phi(p(x|\theta), p(x|\theta))\,\frac{\partial p(x|\theta)}{\partial\theta_i}\frac{\partial p(x|\theta)}{\partial\theta_j}\, d\mu(x)$$

$$i,j = 1, \ldots, n$$

are the components of a second order covariant symmetric tensor field and positive definite on the manifold $(D^1, \wp_{D^1})$, which may be regarded as the metric tensor field of this manifold, rendering $(D^1, \wp_{D^1})$ a Riemannian manifold structure. The corresponding

Riemannian metric is obviously invariant under any coordinate system change, and we have to determine ϕ in order to obtain invariance under reference measure changes. If we change the reference measure from μ to ν , $\mu \ll \nu$, as before, we need

$$\int_{\chi} D_{22}\,\phi(p,p)\,\frac{\partial p}{\partial\theta_i}\,\frac{\partial p}{\partial\theta_j}\,d\mu = \int_{\chi} D_{22}\,\phi(ph,ph)\,h\,\frac{\partial p}{\partial\theta_i}\,h\,\frac{\partial p}{\partial\theta_j}\,d\nu$$

$$i,j = 1,\dots,n$$

independently of the considered parametric family of probability measures, on an arbitrary measurable space $(\chi, \mathcal{a})$, and for all admissible reference measure changes.

If we let $\xi(x) \equiv D_{22}\,\phi(x,x)$, this requirement is satisfied if and only if

$$\xi(xy)\,y = \xi(x) \qquad x, y \geq 0$$

the solution of which is

$$\xi(x) = \frac{\gamma}{x} \qquad x > 0 \quad , \quad \xi(0) = 0$$

for an arbitrary positive constant γ . Therefore,

$$g_{ij}(\theta) = \frac{\kappa\gamma}{2}\int_{\chi}\frac{1}{p(x|\theta)}\,\frac{\partial p(x|\theta)}{\partial\theta_i}\,\frac{\partial p(x|\theta)}{\partial\theta_j}\,d\mu(x) =$$

$$= \frac{\kappa\gamma}{2}\int_{\chi}\frac{\partial\log p(x|\theta)}{\partial\theta_i}\,\frac{\partial\log p(x|\theta)}{\partial\theta_j}\,p(x|\theta)\,d\mu(x) = \frac{\kappa\gamma}{2}\,E_\theta\Big(\frac{\partial\log p(x|\theta)}{\partial\theta_i}\,\frac{\partial\log p(x|\theta)}{\partial\theta_j}\Big)$$

$$i,j = 1,\dots,n$$

which is, up to a proportionality constant, the Fisher information matrix. The same basic results would be obtained through any λ-representation of our statistical model.

Notice that every ϕ-divergence *locally invariant* defines, up to a proportionality constant, the same Riemannian metric. Taking for simplicity $\kappa\gamma/2 = 1$, we obtain the classical Riemannian metric on a parametric family of probability measures introduced by C. R. Rao [16],

$$ds^2 = \sum_{i,j=1}^{n} E_\theta\Big(\frac{\partial\log p(x|\theta)}{\partial\theta_i}\,\frac{\partial\log p(x|\theta)}{\partial\theta_j}\Big)\,d\theta_i\,d\theta_j$$

It is well known that the Fisher information matrix is invariant under sufficient-statistic transformations, so the corresponding Riemannian distance, called the *Rao distance*, remains unchanged, satisfying property ii of section 2, and since Kullback divergence satisfies property iii, it follows that ds^2 also does. Furthermore, if we have two stochastically independent parametric statistical models, it is also well know that

$$ds^2 = ds_1^2 + ds_2^2$$

and under suitable regularity conditions the Riemannian distance satisfies property iv of section 2.

Let us remark that any globally invariant ϕ-divergence is locally invariant, therefore the Riemannian distance obtained is the distance induced by the $L^2(\mu)$ Hilbert norm, on the 2-representation manifold of our statistical model. Moreover, the quadratic differential form defined before, ds^2 , is closely related to the information concept, since it is proportional, up to second order, to the information gain.

On the other hand if we define the information associated with a result $x \in \chi$ to be: $I(x|\theta) = \log(p(x|\theta))$ and consider two neighboring points with coordinates θ and $\theta + \Delta\theta$, then we have

$$\Delta s^2 = E_\theta(\Delta I^2) = \int_\chi (\, \Delta I(x|\theta)\,)^2\, p(x|\theta)\, d\mu(x)$$

Taking into account these considerations, it is called the *information metric*, the obtained Riemannian metric. Notice that any of these above-mentioned properties may be used to introduce the information metric on a regular statistical model.

Further results on the information metric may be found in Atkinson and Mitchell [21], Burbea [22], Burbea and Rao [23], [24], Campbell [25], [26], Mitchell [27], Oller and Cuadras [28], Oller [29], [30], Rao [31] and Skovgaard [32].

Let us now consider another approach to the same problem. Let $(D^\lambda, \wp_{D^\lambda})$ be the λ-representation function manifold associated with the statistical model. Let (θ, U) be a local chart, and let q be a point on this manifold, $q \in U \subset D^\lambda$. Let f be a C^∞ function in an open neighbourghood containing the point q. Now define the function Ψ as $\Psi = f \circ \theta^{-1}$, then Ψ is a differentiable function on an open $\mathbb{R}^n$ set containing the point $\theta(q) = \upsilon$. It is clear that

$$(\partial_i)_q f = D_i\,(f \circ \theta^{-1})(\theta(q)) = D\Psi(\upsilon)(e_i) \qquad i = 1, \dots, n$$

where $(\partial_i)_q$ $i = 1, \dots, n$ is the usual basis of the tangent space D_q^λ corresponding to the local chart (θ, U) , and e_i $i = 1, \dots, n$ is the canonical basis in $\mathbb{R}^n$. Furthermore, let us consider the map $\theta^{-1}: W \to L^2(\,\phi(\,p(\,\cdot\,|\upsilon))\, d\mu\,)$ such that $\theta^{-1}(\beta) = \lambda(\,p(\,\cdot\,|\beta)\,)$ where W is a $\mathbb{R}^n$ open set containing υ, and ϕ is a suitable μ-almost everywhere positive function. Under adequate regularity conditions, the map θ^{-1} is also differentiable at υ, and the differential is given by

$$D\theta^{-1}(\upsilon)\,(h) = \sum_{i=1}^{n} \lambda'(p(\,\cdot\,|\upsilon))\, \frac{\partial p(\,\cdot\,|\upsilon)}{\partial \theta_i}\, h_i$$

Therefore, there is a linear map H_f uniquely defined on the vector space

$$E_\upsilon^\lambda = \, < \lambda'(p(\,\cdot\,|\upsilon))\, \frac{\partial p(\,\cdot\,|\upsilon)}{\partial \theta_1}, \dots, \lambda'(p(\,\cdot\,|\upsilon))\, \frac{\partial p(\,\cdot\,|\upsilon)}{\partial \theta_n} >$$

defined by

$$H_f[\,\lambda'(p(\cdot\,|\upsilon))\,\frac{\partial p(\cdot\,|\upsilon)}{\partial\theta_i}\,] = D\Psi(\upsilon)(e_i) \qquad i=1,\dots,n$$

Notice that H_f may be interpreted as the Fréchet differential of any differentiable extension of f in an open set of $L^2(\,\phi(\,p(\cdot\,|\upsilon))\,d\mu\,)$ restricted to the subspace E_υ^λ . Taking into account the previous definition we have,

$$(\partial_i)_q f = H_f[\,\lambda'(p(\cdot\,|\upsilon))\,\frac{\partial p(\cdot\,|\upsilon)}{\partial\theta_i}\,]$$

or, in general, given any vector of D_q^λ , $Z_q = z_1\,(\partial_1)_q + \cdots + z_n\,(\partial_n)_q$ we have

$$Z_q f = H_f[\,\sum_{i=1}^{n} z_i\;\lambda'(p(\cdot\,|\upsilon))\,\frac{\partial p(\cdot\,|\upsilon)}{\partial\theta_i}\,]$$

that is, the vectors of the tangent space are reasonably represented by elements of $E_\upsilon^\lambda \subset L^2(\,\phi(\,p(\cdot\,|\upsilon))\,d\mu\,)$,

$$Z_q \to \Xi_\lambda(\,Z_q\,) = \sum_{i=1}^{n} z_i\;\lambda'(p(\cdot\,|\upsilon))\,\frac{\partial p(\cdot\,|\upsilon)}{\partial\theta_i}$$

then the tangent space at q , D_q^λ , may be represented by E_υ^λ .

Thus, we may define an inner product on D_q^λ by

$$g_{ij}(\upsilon) = \;<(\partial_i)_q\,,(\partial_j)_q>\;\equiv\;<\Xi_\lambda((\partial_i)_q)\,,\Xi_\lambda((\partial_j)_q)>\;=$$

$$= \int_\chi \lambda'(\,p(x|\upsilon))^2\,\frac{\partial p(x|\upsilon)}{\partial\theta_i}\,\frac{\partial p(x|\upsilon)}{\partial\theta_j}\,\phi(\,p(x|\upsilon))\,d\mu(x) \qquad i,j=1,\dots,n$$

If we require, as before, that this inner product has to be invariant under reference measure changes, for any parametric statistical model, built on an arbitrary measure space, we need

$$\lambda'(xy)^2\,y^2\,\phi(xy) = \lambda'(x)^2\,\phi(xy)\,y \qquad x,y\geq 0$$

which results in

$$\phi(y) = \frac{\kappa}{\lambda'(y)^2\,y} \qquad y>0 \quad , \qquad \phi(0)=0$$

where κ is an arbitrary positive constant. Therefore

$$g_{ij}(\upsilon) = \kappa\int_\chi \frac{1}{p(x|\upsilon)}\,\frac{\partial p(x|\upsilon)}{\partial\theta_i}\,\frac{\partial p(x|\upsilon)}{\partial\theta_j}\,d\mu(x) = \kappa\;E_\upsilon(\;\frac{\partial\log p(x|\upsilon)}{\partial\theta_i}\,\frac{\partial\log p(x|\upsilon)}{\partial\theta_j}\;)$$

$$i,\;j=1,\dots,n$$

Notice that this inner product is independent of λ, and can be used to define an inner product on the corresponding tangent space of the measure manifold

$(M, \wp_M)$, $M_{\Phi_\lambda^{-1}(q)}$. In this sense, *all* the statistical model functional representations are equivalent.

Observe also that if for simplicity we let $\kappa = 1$ we obtain the information metric as before.

Finally , notice that the inner product on E_υ^2 , is given by

$$< Z, W > = \int_\chi ZW \, d\mu \qquad \forall \, Z, W \in E_\upsilon^2$$

and the inner product on E_υ^l is the covariance, given by

$$< Z, W > = \int_\chi ZW p \, d\mu = \text{cov}(Z, W) \qquad \forall \, Z, W \in E_\upsilon^l$$

4. A DISTANCE BETWEEN STATISTICAL INDIVIDUALS

In this section we intend to show that a reasonable distance between statistical individuals may be obtained with the aid of the Riemannian structure introduced in the previous section. Additional material on this subject may be found in Miñarro and Oller [33].

Let us consider the l-representation manifold of a statistical model. Given a point of this manifold, coordinated by θ , $q = \log p(\cdot \,|\theta) \in D_q^l$ we may also consider the l-representation of the tangent space at q which we have discussed in the previous section, with the inner product defined by the covariance. We shall call any $x \in \chi$ a *statistical individual.* There is a natural way to represent the statistical individuals as linear forms on E_θ^l, through the map

$$\zeta : \chi \to E_\theta^{l\,*} \quad , \quad \zeta(x) = Y^* \;:\; Y^*(Z) = Z(x) \quad \forall Z \in E_\theta^l$$

that is, in component notation,

$$x \to \frac{\partial \log p(x|\theta)}{\partial \theta_1} e_1^* + \cdots + \frac{\partial \log p(x|\theta)}{\partial \theta_n} e_n^*$$

where

$$e_i^*\left(\frac{\partial \log p(x|\theta)}{\partial \theta_j}\right) = \delta_{ij} \qquad i, j = 1, \ldots , n$$

δ_{ij} being the Kronecker deltas. Through this representation, we may define a pseudodistance in χ by

$$d_\chi(u, v) = d_{E_\theta^{l\,*}}(\, \zeta(u), \zeta(v) \,) \qquad \forall u, v \in \chi$$

A proper distance may be obtained by defining an equivalence relation on χ as $u \sim v \;\Leftrightarrow\; d_\chi(u, v) = 0$, and extending the pseudodistance d_χ to the quotient set $\chi/\sim$.

Notice that the defined distance is *not* an intrinsic distance in the sense of Rao [34]. It is a distance between individuals *in* a concrete statistical population. Let us also remark that we could represent through the differential of the diffeomorphism Φ_{λ}^{-1}, $\Phi_{\lambda^*}^{-1}$, any statistical individual in a concrete statistical population as a linear form on $M_{\Phi_{\lambda}^{-1}(q)}$, with $\lambda(x) = \log(x)$. Therefore we look on any statistical individual as a first order covariant tensor field on the considered statistical manifold.

Introducing the following notation,

$$\partial_\theta \log p(x|\theta) \equiv \left(\frac{\partial \log p(x|\theta)}{\partial \theta_1}, \ldots, \frac{\partial \log p(x|\theta)}{\partial \theta_n} \right)'$$

and letting $G^{-1}(\theta)$ stand for the Fisher information matrix inverse, the distance between the individuals x and $\bar{x}$ is given by

$$d_\chi^2(\bar{x}, x) = < (\zeta(\bar{x}) - \zeta(x)), (\zeta(\bar{x}) - \zeta(x)) >_{E_\theta^{l^*}} =$$

$$= (\partial_\theta \log p(\bar{x}|\theta) - \partial_\theta \log p(x|\theta))' G^{-1}(\theta) (\partial_\theta \log p(\bar{x}|\theta) - \partial_\theta \log p(x|\theta))$$

Observe that this distance is preserved under sufficient-statistic transformations.

Another possible contruction of this individuals distance is based on the representation of the statistical individuals as scalar functions on D^λ

$$x \to f_x \ : \ q \to f_x(q) = q(x) = \lambda(p(x|\theta(q))) \qquad \forall q \in D^\lambda$$

and then each f_x would be represented by

$$Y^* \in D_q^{\lambda^*} \ : \ Y^*(Z) = Z f_x \qquad \forall Z \in D_q^\lambda$$

Therefore the components of Y^*, with respect the dual basis of $\partial_1, \ldots, \partial_n$, are given by $(\partial \lambda(p(x|\theta(q)))/\partial\theta_1, \ldots, \partial \lambda(p(x|\theta(q)))/\partial\theta_n)$ and we finally obtain the same basic results as before, provided that we require invariance under the reference measure changes.

As an application, given a sample, a statistical individual, we may define an estimation density procedure, by requiring that the expected square distance between our actual sample and any other possible sample from the population be minimal. In other words, given $\bar{x}$, we estimate the coordinates of the density minimizing the scalar function

$$\mathscr{H}_{\bar{x}}(\theta) = E_\theta(d_\chi^2(\bar{x}, x)) = \int_\chi d_\chi^2(\bar{x}, x)\, p(x|\theta)\, d\mu(x)$$

Computing this quantity, we have

$$\mathscr{H}_{\bar{x}}(\theta) = E_\theta\Big((\partial_\theta \log p(\bar{x}|\theta) - \partial_\theta \log p(x|\theta))' G^{-1}(\theta) (\partial_\theta \log p(\bar{x}|\theta) - \partial_\theta \log p(x|\theta)) \Big) =$$

$$= \partial_\theta \log p(\bar{x}|\theta)' G^{-1}(\theta) \partial_\theta \log p(\bar{x}|\theta) - 2\, \partial_\theta \log p(\bar{x}|\theta)' G^{-1}(\theta) E_\theta(\partial_\theta \log p(x|\theta)) +$$

$$+ E_\theta\Big(\partial_\theta \log p(x|\theta)' G^{-1}(\theta) \partial_\theta \log p(x|\theta) \Big)$$

since $E_\theta(\,\partial_\theta \log p(x|\theta)\,) = 0$ and $\partial_\theta \log p(\overline{x}|\theta)$ is constant, it turns out that

$$\mathcal{H}_{\overline{x}}\,(\theta) = E_\theta(\,d_\chi^2(\overline{x}, x)\,) = \partial_\theta \log p(\overline{x}|\theta)'\,G^{-1}(\theta)\,\partial_\theta \log p(\overline{x}|\theta) +$$

$$+\,E_\theta\Big(\,\partial_\theta \log p(x|\theta)'\,G^{-1}(\theta)\,\partial_\theta \log p(x|\theta)\,\Big)$$

but, taking into account that

$$E_\theta(\,\partial_\theta \log p(x|\theta)\,\partial_\theta \log p(x|\theta)'\,) = G(\theta)$$

it follows that

$$E_\theta\Big(\,\partial_\theta \log p(x|\theta)'\,G^{-1}(\theta)\,\partial_\theta \log p(x|\theta)\,\Big) = E_\theta\Big(\,\mathrm{tr}\,[\,G^{-1}(\theta)\,\partial_\theta \log p(x|\theta)\,\partial_\theta \log p(x|\theta)'\,]\,\Big) =$$

$$= \mathrm{tr}\,\Big(\,G^{-1}(\theta)\,E_\theta(\,\partial_\theta \log p(x|\theta)\,\partial_\theta \log p(x|\theta)'\,)\Big) = \mathrm{tr}\,(\,G^{-1}(\theta)\,G(\theta)\,) = \mathrm{tr}\,(\,I\,) = n$$

and finally we obtain

$$\mathcal{H}_{\overline{x}}\,(\theta) = E_\theta(\,d_\chi^2(\overline{x}, x)\,) = \partial_\theta \log p(\overline{x}|\theta)'\,G^{-1}(\theta)\,\partial_\theta \log p(\overline{x}|\theta) + n$$

Therefore the minimum value of $\mathcal{H}_{\overline{x}}\,(\theta)$ is attained when θ satisfies

$$\partial_\theta \log p(\overline{x}|\theta) = 0$$

which are the *likelihood equations*. Thus, the consistent roots of the likelihood equations define the minimum average squared distance estimator.

Some applications of the previously defined distance to discrimination and classification problems are given in Cuadras [35].

5. TESTING STATISTICAL HYPOTHESES THROUGH RAO DISTANCE

Another possible application of Rao distance is concerned with testing parametric statistical hypotheses. Given a parametric statistical model let us now consider a hypothesis testing problem defined by

$$H_0:\ g(\,\lambda(\,p(\,\cdot\,|\theta)))= 0 \qquad\qquad H_1:\ g(\,\lambda(\,p(\,\cdot\,|\theta)))\neq 0$$

for a smooth function g, or in the case where we have a global chart (θ, D^λ) with $\theta(D^\lambda) = \Theta$, where Θ is called a *parameter space*, we may consider the following coordinate-dependent version of the test

$$H_0:\ g(\theta) = 0 \qquad\qquad H_1:\ g(\theta) \neq 0$$

The null hypothesis defines a submanifold $\mathcal{M} \subset D^\lambda$, and, given a sample, we can define the distance from the likelihood estimation of θ , θ^* , to $\mathcal{M}$, as

$$\mathcal{R}_{\mathcal{M}}(\theta^*) \equiv \inf\{\,\rho(\theta, \theta^*)\ \ \theta \in \omega\}$$

where ρ is the Rao distance, the Riemannian distance on $(D^\lambda, \wp_{D^\lambda})$, expressed as a function of the coordinates, and $\omega = \theta(\mathcal{M})$. Then, given a sample $\overline{x}$, we are able to define a statistical test for solving the previous hypothesis testing problem, whose critical region is

$$W_\varepsilon \equiv \{ \bar{x} \in \mathcal{X} : \mathcal{R}_m(\theta^*(\bar{x})) > A_\varepsilon \}$$

where ε is the significance level of the test and the constant A_ε is chosen in such a way that

$$\text{Prob}\,(\bar{x} \in W_\varepsilon \mid H_0) = \varepsilon$$

As an example, see Burbea and Oller [36], consider a univariate linear normal model, defined by its density function as

$$p(Y \mid \beta, \sigma) = \pi^{-n/2}\,\sigma^{-n}\,\exp\{-\frac{1}{2\sigma^2}(Y - X\beta)'\,(Y - X\beta)\}$$

where Y is a $n\times 1$ random vector, X is a $n\times m$ constant matrix of rank $m \le n$, β is an $m\times 1$ parameter vector and σ is a positive scalar. The Rao distance between two univariate linear normal models is given by

$$\rho_{12} = \sqrt{2n}\,\log\frac{1+\Delta_{12}}{1-\Delta_{12}} = 2\sqrt{2n}\,\tanh^{-1}\Delta_{12}$$

where Δ_{12} is the *Möbius distance* given by

$$\Delta_{12} = \sqrt{\frac{(\beta_1-\beta_2)'(\beta_1-\beta_2) + 2n\,(\sigma_1-\sigma_2)^2}{(\beta_1-\beta_2)'(\beta_1-\beta_2) + 2n\,(\sigma_1+\sigma_2)^2}}$$

Consider now the linear hypothesis test given by

$$H_0 : H\beta = 0 \qquad\qquad H_1 : H\beta \neq 0$$

where H is an $r\times m$ constant matrix of rank r.

Let β^* and σ^* be the maximum-likelihood estimation of β and σ obtained from a sample of size 1, Y, given by

$$\beta^* = (X'X)^{-1}X'Y \qquad (\sigma^*)^2 = Y'\,(I - X(X'X)^{-1}X')\,Y\,/\,(n-m)$$

and let us define the function

$$f(\beta, \beta^*, \sigma, \sigma^*) = \frac{8\sigma\sigma^*}{\|\beta - \beta^*\|^2 + 2n\,(\sigma + \sigma^*)^2}$$

where $\|\beta - \beta^*\|$ stands for the usual euclidean norm. Then, it is easy to see that if we let

$$\Upsilon_H(\beta^*, \sigma^*) = \sup\{f(\beta, \beta^*, \sigma, \sigma^*) \;:\; \beta \in \mathbb{R}^m \;,\; H\beta = 0 \;,\; \sigma > 0\}$$

and taking into account that the functions $y \to \tanh^{-1} y$ and $y \to \sqrt{y}$ are strictly monotone increasing functions on $[0, \infty)$ it follows that the critical region may be expressed as

$$W_\varepsilon \equiv \{\, \bar{x} \in \chi \;:\; \Upsilon_H(\beta^*, \sigma^*) < B_\varepsilon \,\}$$

In order to maximize function $f(\beta, \beta^*, \sigma, \sigma^*)$ let us observe that fixing the norm $\|\beta - \beta^*\|$ this function attains its maximum at

$$\hat{\sigma} = \sqrt{\frac{\|\beta - \beta^*\|^2}{2n} + (\sigma^*)^2}$$

with maximal value

$$g(\|\beta - \beta^*\|) = f(\beta, \beta^*, \hat{\sigma}, \sigma^*) = \frac{2\sigma^*}{\sqrt{\dfrac{\|\beta - \beta^*\|^2}{2n} + (\sigma^*)^2} + \sigma^*}$$

On the other hand g is a monotone decreasing function of $\|\beta - \beta^*\|$, therefore in order to maximize $f(\beta, \beta^*, \sigma, \sigma^*)$ under the constraint $H\beta = 0$ it is sufficient to minimize $\|\beta - \beta^*\|$ under this constraint, and then substitute the resulting quantity in $g(\|\beta - \beta^*\|)$. Accordingly, if this minimum is attained at $\beta = \beta_0$ the previous critical region may be expressed as

$$W_\varepsilon = \{\, y \in \mathbb{R}^n : F \equiv \frac{\|\beta_0 - \beta^*\|^2 / r}{n\,(\sigma^*)^2 / (n - m)} > K_\varepsilon \,\}$$

where F is the classical statistic used in variance analysis, see Rao [37], such that, under the null hypothesis, it is distributed as a Fisher-Snedecor distribution with r and $n- m$ degrees of freedom, and K_ε is a constant chosen depending on the significance level of the test.

Therefore, the critical region defined for the present statistical test is the same as the critical region obtained through the standard analysis of variance methods, equivalent to *likelihood ratio test*. In this case, that is in order to test linear hypotheses in a univariate linear normal model, the likelihood ratio test and the *Rao distance test* are equivalent hypothesis tests.

It is also possible to use the Rao distance in order to compare two independent populations, $\theta_1, \theta_2 \in \Theta$ Consider the statistical hypothesis test given by

$$H_0 : \theta_1 = \theta_2 \qquad\qquad H_1 : \theta_1 \neq \theta_2$$

equivalent to

$$H_0 : \rho(\theta_1, \theta_2) = 0 \qquad\qquad H_1 : \rho(\theta_1, \theta_2) > 0$$

then, an obvious critical region for this test may be defined through the inequality $\rho^* > K$. Here ρ^* is the maximum likelihood estimation of ρ, and K is an appropiate constant which depends on the significance level of the test.

6. THE GEOMETRY OF ELLIPTIC PROBABILITY DISTRIBUTIONS

As an example we may expound some geometric results concerning the parametric family of multivariate elliptic probability distributions. Further results may be found in Burbea and Oller [38]. The density function of the previously mentioned family is given by

$$p(X \mid \mu, \Sigma) = \frac{\Gamma(n/2)}{\pi^{n/2}} \, |\Sigma|^{-1} \, F\left((X-\mu)' \, \Sigma^{-1} \, (X-\mu) \right)$$

where X is a $n{\times}1$ random vector , μ and Σ are the parameters, μ is a $n{\times}1$ vector and Σ is a $n{\times}n$ symmetric and strictly positive-definite matrix. F is a function $F: \mathbb{R}^+ \to \mathbb{R}^+$ satisfying

$$\int_0^\infty t^{n/2-1} F(t) \, dt = 1$$

Under certain additional conditions on F , such that

$$a = \frac{4}{n} \int_0^\infty t^{n/2} (\mathscr{L}F)^2(t) \, F(t) \, dt < \infty$$

$$b = \frac{4}{n(n+2)} \int_0^\infty t^{n/2+1} (\mathscr{L}F)^2(t) \, F(t) \, dt < \infty$$

where $\mathscr{L}F \equiv \frac{F'}{F} = (\log F)'$. Let us observe that in the particular case

$$F(t) = \frac{1}{2^{n/2} \, \Gamma(n/2)} \, e^{-t/2} \qquad t \geq 0$$

we have the *multivariate normal* distribution and in this case $a = 1$, and $b = 1$.

The information metric is given by

$$ds^2 = a \, d\mu' \, \Sigma^{-1} \, d\mu + \frac{b}{2} \operatorname{tr} (\Sigma^{-1} d\Sigma \, \Sigma^{-1} d\Sigma) + \frac{(b-1)}{4} \operatorname{tr}^2 (\Sigma^{-1} d\Sigma)$$

Notice that all the effects on the metric of F are summarized by a and b .

Let us denote the tangent vectors as

$$X = (X_\mu \, , X_\Sigma)$$

where X_μ and X_Σ are the components of the vector X in μ-direction and Σ-direction respectively.

The Riemann-Christoffel tensor field is given by

$$K(X, Y, Z, W) = \kappa\Big((X_\mu{}'\Sigma^{-1} Z_\mu)(Y_\mu{}'\Sigma^{-1} W_\mu) - (X_\mu{}'\Sigma^{-1} W_\mu)(Y_\mu{}'\Sigma^{-1} Z_\mu) \Big) +$$

$$+ \frac{a}{4} (Y_\mu{}'\Sigma^{-1} X_\Sigma - X_\mu{}'\Sigma^{-1} Y_\Sigma)\Sigma^{-1} (W_\Sigma \Sigma^{-1} Z_\mu - Z_\Sigma \Sigma^{-1} W_\mu) +$$

$$+ \frac{a}{4} X_\mu{}'\Sigma^{-1} (W_\Sigma \Sigma^{-1} Z_\Sigma - Z_\Sigma \Sigma^{-1} W_\Sigma)\Sigma^{-1} Y_\mu +$$

$$+ \frac{a}{4} Z_\mu{}'\Sigma^{-1} (Y_\Sigma \Sigma^{-1} X_\Sigma - X_\Sigma \Sigma^{-1} Y_\Sigma)\Sigma^{-1} W_\mu +$$

$$+ \frac{b}{4} \operatorname{tr} \Big(\Sigma^{-1} X_\Sigma \Sigma^{-1} Y_\Sigma \Sigma^{-1} (Z_\Sigma \Sigma^{-1} W_\Sigma - W_\Sigma \Sigma^{-1} Z_\Sigma) \Big)$$

where

$$\kappa = \frac{a^2}{4b} (1 + \frac{2(b-1)}{2b + (b-1)n})$$

observe that for multivariate normal distribution we have $\kappa = 1/4$.

On the other hand

$$< X, Y > = a X_\mu{}'\Sigma^{-1} Y_\mu + \frac{b}{2} \operatorname{tr} (\Sigma^{-1} X_\Sigma \Sigma^{-1} Y_\Sigma) + \frac{(b-1)}{4} \operatorname{tr} (\Sigma^{-1} X_\Sigma) \operatorname{tr} (\Sigma^{-1} Y_\Sigma)$$

Therefore, the surface element is given by

$$S(X, Y) = < X, X > < Y, Y > - < X, Y >^2 =$$

$$= \Big(a X_\mu{}'\Sigma^{-1} X_\mu + \frac{b}{2} \operatorname{tr} (\Sigma^{-1} X_\Sigma \Sigma^{-1} X_\Sigma) + \frac{(b-1)}{4} \operatorname{tr}^2 (\Sigma^{-1} X_\Sigma) \Big) \cdot$$

$$\cdot \Big(a Y_\mu{}'\Sigma^{-1} Y_\mu + \frac{b}{2} \operatorname{tr} (\Sigma^{-1} Y_\Sigma \Sigma^{-1} Y_\Sigma) + \frac{(b-1)}{4} \operatorname{tr}^2 (\Sigma^{-1} Y_\Sigma) \Big) -$$

$$- \Big(a X_\mu{}'\Sigma^{-1} Y_\mu + \frac{b}{2} \operatorname{tr} (\Sigma^{-1} X_\Sigma \Sigma^{-1} Y_\Sigma) + \frac{(b-1)}{4} \operatorname{tr} (\Sigma^{-1} X_\Sigma) \operatorname{tr} (\Sigma^{-1} Y_\Sigma) \Big)^2$$

and the sectional curvature $\overline{K}$ is obtained by replacing the previous formulas in

$$\overline{K}(X, Y) = \frac{K(X, Y, X, Y)}{S(X, Y)}$$

for instance, if we let $X_\Sigma = Y_\Sigma = 0$ we obtain

$$\overline{K}(X, Y) = \frac{\kappa}{a^2} = \frac{1}{4b} (1 + \frac{2(b-1)}{2b + (b-1)n})$$

The Ricci tensor is given by

$$\operatorname{Ric}(X, Y) = a\tau X_\mu{}'\Sigma^{-1} Y_\mu + \frac{n+1}{4} \operatorname{tr} (\Sigma^{-1} X_\Sigma \Sigma^{-1} Y_\Sigma) - \frac{1}{4} \operatorname{tr} (\Sigma^{-1} X_\Sigma) \operatorname{tr} (\Sigma^{-1} Y_\Sigma)$$

where $\tau = 1/(2b + n(b-1))$. For the multivariate normal case we have $\tau = 1/2$.

The geodesic equations are given by

$$\ddot{\mu} - \dot{\Sigma}\,\Sigma^{-1}\,\dot{\mu} = 0$$

$$\ddot{\Sigma} + \frac{a}{b}\,\dot{\mu}\dot{\mu}' + \frac{a\kappa}{2}\,\Sigma\,(\,\dot{\mu}'\,\Sigma^{-1}\,\dot{\mu}\,) \;-\dot{\Sigma}\,\Sigma^{-1}\,\dot{\Sigma} = 0$$

where $\dot{\mu} = d\mu/ds$, $\ddot{\mu} = d^2\mu/ds^2$, $\dot{\Sigma} = d\,\Sigma/ds$ and $\ddot{\Sigma} = d^2\,\Sigma/ds^2$. The Rao distance induced on the submanifold obtained by letting μ be constant is given by

$$\rho_{12} = \sqrt{\frac{b}{2}\sum_{i=1}^{n}\log^2\lambda_i \;+\; \frac{b-1}{4}\,(\sum_{i=1}^{n}\log\lambda_i\,)^2}$$

where λ_i are the eigenvalues of Σ_2 with respect Σ_1 . In the multivariate normal case, since $b = 1$ we recover the result obtained by Burbea [22] and Jensen, see Atkinson and Mitchell [21].

ACKNOWLEDGEMENTS

I should like to thank Professor Jacob Burbea, who helped me very much in the preparation of this paper, for his comments and suggestions.

REFERENCES

[1] Mahalanobis, P.C., Majumdar, D.N. and Rao, C.R., Sankhyà 9 (1949) 90.
[2] Rao, C.R., J. Roy. Statist. Soc. B 10 (1948) 159.
[3] Cavalli-Sforza, L.L. and Edwards, A.W.F., Amer. J. Hum. Genet. 19 (1967) 233.
[4] Nei, M., Japan J. Human Genet. 23 (1978) 341.
[5] Prevosti, A., Ocaña, J. and Alonso, G., Theoret. Appl. Genetics 45 (1975) 231.
[6] Agresti, A. and Agresti, B.F., Statistical Analysis of qualitative variation, in: K.F. Schussler (ed.), Social Methodology, 1978 pp 204-237.
[7] Hadjisavvas, N., Ann. Inst. Henri Poncaré XXXV 4 (1981) 287.
[8] Provost, J. P. and Vallee, G., Commun. Math. Phys. 76 (1980) 289.
[9] Amari, S., Biometrika 69 (1982) 1.
[10] Amari, S., Differential-Geometrical Methods in Statistics. Lecture notes in Statistics 28. (Springer-Verlag, New York, 1985).
[11] Bhattacharyya, A., Sankhyà 7 (1946) 401.
[12] Efron, B., Ann. Statist. 3 (1975) 1189.
[13] Fisher, R.A., Proc. Cambridge Phil. Soc. 22 (1925) 700.
[14] Mahalanobis, P.C., Proc. Nat. Inst. Sci. India 2 (1936) 49.
[15] Matusita, K., Ann. Inst. Statist. Math. 16 (1964) 305.
[16] Rao, C.R., Bull. Calcutta Math. Soc. 37 (1945) 81.
[17] Rao, C.R., Rao's Axiomatization of Diversity Measures, in: Kotz, S., Johnson, N.L. and Campbell, B.R. (eds.), Encyclopedia of Statistical Sciences (Wiley, New York, 1986) Vol. 7 pp. 614-617.
[18] Kullback, S. and Leibler, R. A., Ann. Math. Statist. 22 (1951) 79.
[19] Hicks, N.J., Notes on Differential Geometry, (Van Nostrand, Princeton, 1965).
[20] Spivak, M., A Comprehensive Introduction to Differential Geometry, (Publish or Perish Inc., Berkeley, CA. 1979).
[21] Atkinson, C. and Mitchell, A.F.S., Sankhyà 43 (1981) 345.
[22] Burbea, J., Expo. Math. 4 (1986) 347.
[23] Burbea, J. and Rao, C.R., J. Multivariate Anal. 12 (1982) 575.
[24] Burbea, J. and Rao, C.R., Probability Math. Statist. 3 (1984) 241.
[25] Campbell, L.L., Inform. Sci. 35 (1985) 199.
[26] Campbell, L.L., Proceed. Amer. Math. Soc. 98 (1) (1986) 135.

[27] Mitchell, A.F.S., Intern. Statist. Review 56 (1) (1988) 1.
[28] Oller, J.M. and Cuadras, C.M., Sankhyà 47 (1985) 75.
[29] Oller, J.M., Sobre la Curvatura Riemanniana asociada a Distribuciones de Series de Potencias Generalizadas Multivariantes, in: Pla, J. (ed.), Homenatge a Francesc d'A. Sales (Edicions Universitat de Barcelona, Barcelona, 1985).
[30] Oller, J.M., Sankhyà 49 (1987) 17.
[31] Rao, C.R., Differential metrics in Probability Spaces, in: S. S. Gupta (ed.) Differential Geometry in Statistical Inference, (IMS Lecture Notes - Monograph series 10, Hayward, California, 1987) pp. 217-240.
[32] Skovgaard, L. T., Scand. J. Statist. 11 (1984) 211.
[33] Miñarro, A. and Oller, J.M., A Distance between Statistical Individuals based on the Information Metric. Manuscript. University of Barcelona, Spain.
[34] Rao, C.R., Theoret. Pop. Biol. 21 (1982) 24.
[35] Cuadras, C. M., Distance Analysis in Discrimination and Classification using both Continuous and Categorical Variables, this volume.
[36] Burbea, J. and Oller, J.M., Statistics and Decisions 6 (1988) 209.
[37] Rao, C. R., Linear Statistical Inference and its Applications, (Wiley, New York, 1973).
[38] Burbea, J. and Oller, J.M., The Information Metric for Multivariate Elliptic Probability Distributions. In preparation.

Part One
INFERENCE AND TEST OF HYPOTHESES

Said one—"Folks of a surly Tapster tell,
And daub his Visage with the Smoke of Hell;
They talk of some strict Testing of us—Pish!
He's a Good Fellow, and 'twill all be well.—"

Khayyam Naishapuri-Rubáiyát:
Persian Astronomer, Mathematician
and Poet (12th Century)

STATISTICAL DATA ANALYSIS AND INFERENCE
Y. Dodge (Editor)

ON BAYES AND EMPIRICAL BAYES TWO-STAGE ALLOCATION PROCEDURES FOR SELECTION PROBLEMS*

Shanti S. GUPTA and TaChen LIANG

Department of Statistics, Purdue University, West Lafayette, IN 47907, USA
Department of Mathematics, Wayne State University, Detroit, MI 48202, USA

ABSTRACT: We investigate the problem of deriving two-stage allocation procedures for selecting the best normal population. If the prior distribution is assumed to be known, an exact Bayes two-stage allocation procedure is obtained. If the prior distribution depends on some unknown parameter, an adaptive two-stage allocation procedure is proposed. Using the empirical Bayes formulation, we prove that the proposed adaptive two-stage allocation procedure has some asymptotic optimality property.

1. INTRODUCTION

Suppose that an experimenter (a customer) wishes to purchase M items of some product. We assume that these items are supplied by k different manufacturers (suppliers), say, $\pi_1, \ldots, \pi_k$. At first, the experimenter carries out an inspection on m items of the product from each of the k suppliers in order to obtain data for determining their quality. Then, based on the resulting information, he allocates the remaining $M - km$ items to the k suppliers, say, $N_1, \ldots, N_k$, respectively, where N_i, $i = 1, \ldots, k$, are nonnegative integers such that $\sum_{i=1}^{k} N_i = M - km$. Let θ_i denote a measure of the quality of the product from the ith manufacturer π_i. Let $\theta_{[1]} \leq \ldots \leq \theta_{[k]}$ be the ordered values of the parameters $\theta_1, \ldots, \theta_k$. It is assumed that the exact pairing between the ordered and the unordered parameters is unknown. The supplier π_i with $\theta_i = \theta_{[k]}$ is called the *best*. Of course, the experimenter would ideally like to allocate (purchase) the remaining $M - km$ items from the best supplier. Thus, the experimenter is faced with the so-called two-stage allocation and selection problem.

For the two-stage allocation problem described above, we define the loss function:

$$L(\underset{\sim}{\theta}; m, N_1, \ldots N_k) = m \sum_{i=1}^{k} (\theta_{[k]} - \theta_i) + \sum_{i=1}^{k} N_i(\theta_{[k]} - \theta_i), \tag{1.1}$$

where $\underset{\sim}{\theta} = (\theta_1, \ldots, \theta_k)$, $0 \leq m \leq [\frac{M}{k}]$, $0 \leq N_i \leq M - km$, $i = 1, \ldots, k$, $\sum_{i=1}^{k} N_i = M - km$,

* This research was supported in part by the Office of Naval Research Contract N00014-88-K-0170 and NSF Grants DMS–8606964, DMS-8702620 at Purdue University.

and $[y]$ denotes the largest integer not greater than y. Note that the first summation in (1.1) is the loss due to the choice of the common initial number of items to be supplied by each of the k manufacturers, and the second summation in (1.1) is the loss due to the allocation made at the second stage. Our goal here is to derive optimal two-stage allocation procedures with respect to the loss function (1.1). We study the problem for normal populations, say $\pi_1, \ldots, \pi_k$, with unknown means $\theta_1, \ldots, \theta_k$, and a common known variance σ^2. The unknown means $\theta_1, \ldots, \theta_k$ are assumed to be independent and identically distributed (iid) with a normal prior distribution $N(\theta_0, \tau^2)$, where the parameter τ^2 may be either known or unknown.

We note that Somerville [6, 7] studied a two-stage minimax allocation procedure for the normal distribution model with a different loss function. However, since the loss function considered by Somerville [6] is not bounded, the minimax solution does not exist (see Ofosu [3] for a comment). Ofosu [4] also studied a two-stage allocation procedure via a Bayesian approach (see Gupta and Panchapakesan [2]).

2. NORMAL MODEL

Let $\pi_1, \ldots, \pi_k$ be normal populations with unknown means $\theta_1, \ldots, \theta_k$, and a common known variance σ^2. The unknown means $\theta_1, \ldots, \theta_k$ are assumed to be iid with a normal prior distribution $N(\theta_0,\ \tau^2)$. In this section, we assume that τ^2 is known. Also, for simplicity, we assume that $M = kN$ for some positive integer N.

2.1. Bayes Allocation Procedure for a Fixed m

First, we take m, $0 \leq m \leq N$, random observations from each of the k populations. Let $\overline{X}_i$ denote the sample mean of the m random observations taken from population π_i and let $\overline{x}_i$ denote the associated observed value, $i = 1, \ldots, k$. At the second stage, based on the observed values $\underset{\sim}{\overline{x}} = (\overline{x}_1, \ldots, \overline{x}_k)$, allocate $N_i(\underset{\sim}{\overline{x}})$ random observations to population π_i, $i = 1, \ldots, k$, where $N_1(\underset{\sim}{\overline{x}}), \ldots, N_k(\underset{\sim}{\overline{x}})$ are nonnegative integers such that $\sum_{i=1}^{k} N_i(\underset{\sim}{\overline{x}}) = k(N-m)$. Let $\overline{Y}_i$ denote the sample mean of the $N_i(\underset{\sim}{\overline{x}})$ random observations taken from the population π_i at the second stage, and let $\overline{y}_i$ be the associated observed value, $i = 1, \ldots, k$. Also, let $\underset{\sim}{\overline{y}} = (\overline{y}_1, \ldots, \overline{y}_k)$. Note that when either $m = 0$ or $m = N$, the above allocation procedure is reduced to a one-stage allocation procedure.

At stage two, given $\underset{\sim}{\overline{X}} = \underset{\sim}{\overline{x}}$ and $\underset{\sim}{\overline{Y}} = \underset{\sim}{\overline{y}}$, respectively, the posterior expected loss is:

$$\begin{aligned} r_m(\underset{\sim}{\overline{x}}, \underset{\sim}{\overline{y}}) &= E[L(\underset{\sim}{\theta}; m,\ N_1(\underset{\sim}{\overline{x}}), \ldots, N_k(\underset{\sim}{\overline{x}}))|\underset{\sim}{\overline{X}} = \underset{\sim}{\overline{x}},\ \underset{\sim}{\overline{Y}} = \underset{\sim}{\overline{y}}] \\ &= kNE[\theta_{[k]}|\underset{\sim}{\overline{X}} = \underset{\sim}{\overline{x}},\ \underset{\sim}{\overline{Y}} = \underset{\sim}{\overline{y}}] - \sum_{j=1}^{k}(m + N_j(\underset{\sim}{\overline{x}}))E[\theta_j|\underset{\sim}{\overline{X}} = \underset{\sim}{\overline{x}},\ \underset{\sim}{\overline{Y}} = \underset{\sim}{\overline{y}}]. \end{aligned} \tag{2.1}$$

Therefore, at stage one, given $\underset{\sim}{\overline{X}} = \underset{\sim}{\overline{x}}$, the posterior expected loss is given by

$$
\begin{aligned}
r_m(\overline{\underset{\sim}{x}}) &= E[r_m(\overline{\underset{\sim}{X}},\overline{\underset{\sim}{Y}})|\overline{\underset{\sim}{X}}=\overline{\underset{\sim}{x}}] \\
&= kNE[\theta_{[k]}|\overline{\underset{\sim}{X}}=\overline{\underset{\sim}{x}}] - \sum_{j=1}^{k}(m+N_j(\overline{\underset{\sim}{x}}))E[\theta_j|\overline{\underset{\sim}{X}}=\overline{\underset{\sim}{x}}] \\
&= kNE[\theta_{[k]}|\overline{\underset{\sim}{X}}=\overline{\underset{\sim}{x}}] - \sum_{j=1}^{k}(m+N_j(\overline{\underset{\sim}{x}}))\frac{\theta_0\sigma^2+m\tau^2\overline{x}_j}{\sigma^2+m\tau^2} \\
&= kNE[\theta_{[k]}|\overline{\underset{\sim}{X}}=\overline{\underset{\sim}{x}}] - m\sum_{j=1}^{k}\frac{\theta_0\sigma^2+m\tau^2\overline{x}_j}{\sigma^2+m\tau^2} \quad \sum_{j=1}^{k}N_j(\overline{\underset{\sim}{x}})\frac{\theta_0\sigma^2+m\tau^2\overline{x}_j}{\sigma^2+m\tau^2}.
\end{aligned}
\tag{2.2}
$$

For each observed $\overline{\underset{\sim}{X}}=\overline{\underset{\sim}{x}}$, let $A(\overline{\underset{\sim}{x}})=\{i|\overline{x}_i=\max_{1\le j\le k}\overline{x}_j\}$. Then, for a fixed m, the Bayes allocation at the second stage is to choose the nonnegative integers $N_1(\overline{\underset{\sim}{x}}),\dots N_k(\overline{\underset{\sim}{x}})$ such that $\sum_{i\in A(\overline{\underset{\sim}{x}})} N_i(\overline{\underset{\sim}{x}})=k(N-m)$. Then, conditional on m and the observed value $\overline{\underset{\sim}{X}}=\overline{\underset{\sim}{x}}$, the minimum posterior expected loss is:

$$
\begin{aligned}
r_m^B(\overline{\underset{\sim}{x}}) = kNE[\theta_{[k]}|\overline{\underset{\sim}{X}}=\overline{\underset{\sim}{x}}] &- m\sum_{j=1}^{k}\frac{\theta_0\sigma^2+m\tau^2\overline{x}_j}{\sigma^2+m\tau^2} \\
&- \frac{k(N-m)[\theta_0\sigma^2+m\tau^2\max_{1\le i\le k}\overline{x}_i]}{\sigma^2+m\tau^2},
\end{aligned}
\tag{2.3}
$$

and the minimum Bayes risk for a fixed m is:

$$
\begin{aligned}
r_m^B &= E[r_m^B(\overline{\underset{\sim}{X}})] \\
&= kNE[\theta_{[k]}] - m\sum_{j=1}^{k}\frac{\theta_0\sigma^2+m\tau^2E[\overline{X}_j]}{\sigma^2+m\tau^2} \\
&\quad - \frac{k(N-m)\{\theta_0\sigma^2+m\tau^2E[\max_{1\le j\le k}\overline{X}_j]\}}{\sigma^2+m\tau^2}.
\end{aligned}
\tag{2.4}
$$

Note that under the statistical model, $\overline{X}_1,\dots,\overline{X}_k$ are iid and have a marginal normal distribution with mean θ_0 and variance $\frac{\sigma^2}{m}+\tau^2$. Thus, $E[\max_{1\le i\le k}\overline{X}_i]=\theta_0+\sqrt{\frac{\sigma^2}{m}+\tau^2}E[\max_{1\le j\le k}Z_j]$ $=\theta_0+\sqrt{\frac{\sigma^2}{m}+\tau^2}\ \alpha$, where $Z_1,\dots,Z_k$ are iid $N(0,1)$ and $\alpha=E[\max_{1\le j\le k}Z_j]$. Also, $E[\theta_{[k]}]=\theta_0+\tau\alpha$. Hence, we have

$$
r_m^B = k\tau\alpha\{N-\frac{(N-m)\sqrt{m}\ \tau}{\sqrt{\sigma^2+m\tau^2}}\}. \tag{2.5}
$$

Note that the minimum Bayes risk r_m^B does not depend on the parameter θ_0.

2.2. Optimal Initial Sample Size

Next, we want to find an integer, say m_B, $0\le m_B\le N$ such that $r_{m_B}^B\le r_m^B$ for all integers m in $[0,N]$. We call such an integer m_B as an optimal initial sample size. When m_B is determined, a Bayes two-stage allocation procedure, say P_B, is given as follows:

First, take m_B random observations from each of the k populations. Compute the observed sample mean $\overline{x}_i$, $i = 1, \ldots, k$. Then, take $k(N - m_B)$ random observations from the population which yields the largest sample mean value.

Note that finding an integer m in $[0, N]$ to minimize the Bayes risk r_m^B is equivalent to finding an integer m in $[0, N]$ to maximize $(N-m)\sqrt{m}/\sqrt{\sigma^2 + m\tau^2}$ [see (2.5)]. In general, we assume m to be a variable and for each fixed $\tau^2 > 0$, let

$$H_{\tau^2}(m) = \frac{(N-m)^2 m}{\sigma^2 + m\tau^2} \tag{2.6}$$

be a function defined on the interval $[0, N]$. Then, the first derivative of the function $H_{\tau^2}(m)$ with respect to m is

$$H'_{\tau^2}(m) = \frac{(m-N)[(3m-N)\sigma^2 + 2m^2\tau^2]}{(\sigma^2 + m\tau^2)^2},$$

which is nonpositive if $\frac{N}{3} \le m \le N$. That is, the function $H_{\tau^2}(m)$ is nonincreasing in m for m in the interval $[\frac{N}{3},\ N]$. Thus, to find a number m in the interval $[0, N]$ to maximize the function $H_{\tau^2}(m)$, it suffices to consider those m in the subinterval $[0,\ \frac{N}{3}]$. Let

$$G(m) = (m-N)[(3m-N)\sigma^2 + 2m^2\tau^2],\ m \in [0,\ \frac{N}{3}].$$

Then,

$$G'(m) = (3m - 2N)(2\sigma^2 + 2m\tau^2) < 0, \quad \text{for all } m \in [0,\ \frac{N}{3}].$$

In other words, $G(m)$ is a decreasing function of m for $m \in [0,\ \frac{N}{3}]$. Also, note that $G(0) > 0$, $G(\frac{N}{3}) < 0$. Thus, there exists a unique number in $(0,\ \frac{N}{3})$, say m^*, such that $G(m^*) = 0$. Hence, $H'_{\tau^2}(m) > 0$ for all $m \in [0, m^*)$; $H'_{\tau^2}(m) < 0$ for all $m \in (m^*,\ \frac{N}{3})$, and $H'_{\tau^2}(m^*) = 0$. This implies that the function $H_{\tau^2}(m)$ achieves its maximum at $m = m^*$. Note that m^* is the positive solution of the equation $(3m - N)\sigma^2 + 2m^2\tau^2 = 0$. That is,

$$\begin{aligned} m^* &= (-3\sigma^2 + \sqrt{8N\tau^2\sigma^2 + 9\sigma^4})/(4\tau^2) \\ &= 2N\sigma/[\sqrt{8N\tau^2 + 9\sigma^2} + 3\sigma]. \end{aligned} \tag{2.7}$$

Let

$$m_B = \begin{cases} [m^*] & \text{if } H_{\tau^2}([m^*]) \ge H_{\tau^2}([m^*]+1), \\ [m^*]+1 & \text{if } H_{\tau^2}([m^*]) < H_{\tau^2}([m^*]+1). \end{cases} \tag{2.8}$$

Therefore, the minimum Bayes risk, denoted by r^B, of the Bayes two-stage allocation procedure is:

$$r^B = k\tau\alpha\{N - \frac{(N-m_B)\sqrt{m_B}\,\tau}{\sqrt{\sigma^2 + m_B\tau^2}}\}. \tag{2.9}$$

Remarks 2.1

a) For fixed N and σ^2, the optimal initial sample size m_B can be viewed as a function of the parameter τ^2, and hence is denoted by $m_B(\tau^2)$. From (2.6), (2.7), (2.8), one can see that

$$1 \leq m_B(\tau^2) \leq [\frac{N}{3}] + 1 \text{ for any } \tau^2 > 0.$$

Furthermore, we have the following results:

$$\lim_{\tau^2 \to \infty} m_B(\tau^2) = 1 \text{ and}$$

$$\lim_{\tau^2 \to 0} m_B(\tau^2) = \begin{cases} [\frac{N}{3}] & \text{if } N \equiv 0 \text{ or } 1 \pmod 3, \\ [\frac{N}{3}] + 1 & \text{if } N \equiv 2 \pmod 3). \end{cases}$$

b) From (2.7), m^* is a decreasing function of the parameter τ^2. Thus, from (2.8), one may expect that $m_B(\tau^2)$ is nonincreasing in τ^2. Actually, we have the following results:

$$\begin{cases} \text{If} \quad \tau_1^2 > \tau_2^2, \text{ then } m_B(\tau_1^2) \leq m_B(\tau_2^2) \\ \text{If} \quad m_B(\tau_1^2) < m_B(\tau_2^2), \text{ then } \tau_1^2 > \tau_2^2, \end{cases} \tag{2.10}$$

which can be obtained directly from the following lemma.

Lemma 2.1. Let $H_{\tau^2}(m) = \frac{(N-m)^2 m}{\sigma^2 + m\tau^2}$, $1 \leq m \leq [\frac{N}{3}] + 1$ and $\tau^2 > 0$. If $H_{\tau_2^2}(m) \geq H_{\tau_2^2}(m+1)$, then $H_{\tau_1^2}(m) > H_{\tau_1^2}(m+1)$ for all $\tau_1^2 > \tau_2^2$.

Proof: By the given condition,

$$\begin{aligned} 0 &\leq H_{\tau_2^2}(m) - H_{\tau_2^2}(m+1) \\ &= \frac{(N-m)^2 m[\sigma^2 + (m+1)\tau_2^2] - (\sigma^2 + m\tau_2^2)(N-m-1)^2(m+1)}{(\sigma^2 + m\tau_2^2)[\sigma^2 + (m+1)\tau_2^2]}. \end{aligned}$$

Let

$$h(\tau^2) = (N-m)^2 m[\sigma^2 + (m+1)\tau^2] - (\sigma^2 + m\tau^2)(N-m-1)^2(m+1).$$

Hence, $h(\tau_2^2) \geq 0$. Also, the first derivative of $h(\tau^2)$ with respect to τ^2 is

$$h'(\tau^2) = \frac{d\,h(\tau^2)}{d\tau^2} = m(m+1)[2(N-m) - 1] > 0 \text{ for all } 1 \leq m \leq [\frac{N}{3}] + 1,$$

which implies $h(\tau^2)$ is an increasing function of τ^2. Thus, $h(\tau_1^2) > h(\tau_2^2) \geq 0$ since $\tau_1^2 > \tau_2^2$. Therefore, we have $H_{\tau_1^2}(m) > H_{\tau_1^2}(m+1)$.

3. An Adaptive Two-Stage Allocation Procedure

In this section, we still assume the normal model except that the value of the parameter τ^2 is unknown. Thus, the Bayes two-stage allocation procedure derived in Section 2 can not be applied in this situation. To overcome this difficulty, we propose an adaptive two-stage allocation procedure via the empirical Bayes approach.

We now consider the following situation. Suppose that one is confronted repeatedly and independently with a sequence of the allocation problems as described in Section 1. We can then use the past observations at hand to construct an estimator for the unknown parameter τ^2. This estimator is then applied to form an adaptive two-stage allocation procedure for the next allocation problem. Suppose now, we are at time $t = n + 1$. We have already had n past observations at hand. We let m_j denote the adaptive optimal initial sample size taken at stage one at time $t = j$, $j = 1, \ldots, n$. The determination of m_j will be described later. From Remark 2.1 a), $1 \le m_j \le [\frac{N}{3}] + 1$. That is, we take at least one observation from each of the k populations at each time $j = 1, \ldots, n$. We let X_{ij} denote an observation taken from population π_i at time j, $j = 1, \ldots, n$. Then, under the normal model, X_{ij} has a marginal normal distribution with mean θ_0 and variance $\sigma^2 + \tau^2$. Also, following the usual empirical Bayes formulation (for example, see Robbins [5] or Gupta and Liang [1]), we can assume that X_{ij}, $j = 1, \ldots, n$; $i = 1, \ldots, k$, are independently distributed. In the following, we only consider the case when the parameter θ_0 is unknown. Let

$$\begin{cases} \overline{X}(n) = \frac{1}{kn} \sum_{i=1}^{k} \sum_{j=1}^{n} X_{ij}, \\ S^2(n) = \frac{1}{kn-1} \sum_{i=1}^{k} \sum_{j=1}^{n} (X_{ij} - \overline{X}(n))^2. \end{cases} \tag{3.1}$$

Then, $(kn-1)S^2(n)/(\sigma^2 + \tau^2)$ has a χ^2-distribution with $kn - 1$ degrees of freedom. Since τ^2 is positive, we suggest using

$$\tau^2_{n+1} = (S^2(n) - \sigma^2)^+ \tag{3.2}$$

to estimate the unknown parameter τ^2, where $y^+ = \max(0, y)$. When $\tau^2_{n+1} > 0$, we define m_{n+1}, the adaptive optimal initial sample size at time $t = n + 1$, to be an integer in the interval $[0, N]$ which maximizes the function $H_{\tau^2_{n+1}}(m) = \frac{(N-m)^2 m}{\sigma^2 + m\tau^2_{n+1}}$ among all the integers in the interval $[0, N]$. From Remark 2.1 a), $1 \le m_{n+1} \le [\frac{N}{3}] + 1$. When $\tau^2_{n+1} = 0$, we let $m_{n+1} = [\frac{N}{3}]$ (or $[\frac{N}{3}] + 1$) if $H_{\tau^2_{n+1}}([\frac{N}{3}]) \ge (<) H_{\tau^2_{n+1}}([\frac{N}{3}] + 1)$. Note that when $n = 0$, i.e. there is no past observation available, we arbitrarily choose an integer m_1 in the interval $[1, [\frac{N}{3}] + 1]$ as the initial sample size.

We then propose an adaptive two-stage allocation procedure, say P_{n+1}, at $t = n + 1$ as follows:

At time $t = n+1$, first take m_{n+1} observations from each of the k populations. Compute the observed sample mean $\overline{x}_i$ based on the m_{n+1} observations taken from population π_i, $i = 1, \ldots k$. Then, take $k(N - m_{n+1})$ random observations from the population which yields the largest sample mean value.

We denote the conditional Bayes risk given m_{n+1} and the Bayes risk of the adaptive two-stage allocation procedure P_{n+1} by $r_{n+1}(m_{n+1})$ and r_{n+1}, respectively. That is,

$$\begin{cases} r_{n+1}(m_{n+1}) = k\tau\alpha\{N - (N - m_{n+1})\sqrt{m_{n+1}}\ \tau/\sqrt{\sigma^2 + m_{n+1}\tau^2}\}, \\ r_{n+1} = E[r_{n+1}(m_{n+1})]; \end{cases} \tag{3.3}$$

where the expectation E is taken with respecpt to m_{n+1} or the probability space generated by $(X_{ij},\ j = 1, \ldots, n,\ i = 1, \ldots, k)$.

Note that $r_{n+1}(m_{n+1}) - r^B \ge 0$ since r^B is the minimum Bayes risk, and therefore $r_{n+1} - r^B \ge 0$. The two differences $r_{n+1}(m_{n+1}) - r^B$ and $r_{n+1} - r^B$ are always used as measures of the performance of the proposed two-stage allocation procedure P_{n+1}.

Definition 3.1

a) The sequence of adaptive two-stage allocation procedures $\{P_{n+1}\}$ is said to be asymptotically optimal in probability of order $\{\alpha_n\}$ if for any $\varepsilon > 0$, $P\{r_{n+1}(m_{n+1}) - r^B \geq \varepsilon\} \leq 0(\alpha_n)$ as $n \to \infty$ where $\{\alpha_n\}$ is a sequence of positive numbers such that $\lim_{n\to\infty} \alpha_n = 0$.

b) The sequence of adaptive two-stage allocation procedures $\{P_{n+1}\}$ is said to be asymptotically optimal of order $\{\beta_n\}$ if $r_{n+1} - r^B \leq 0(\beta_n)$ as $n \to \infty$ where $\{\beta_n\}$ is a sequence of positive numbers such that $\lim_{n\to\infty} \beta_n = 0$.

In the following, we will investigate some asymptotically optimal properties of the proposed adaptive two-stage allocation procedures $\{P_{n+1}\}$.

Let $I = \{m | m$ is an integer in $[1, [\frac{N}{3}] + 1]$ such that $H_{\tau^2}(m_B) - H_{\tau^2}(m) \neq 0\}$, and let $c = \min\{H_{\tau^2}(m_B) - H_{\tau^2}(m) | m \in I\}$. Then, by the definitions of m_B and the set I, $c > 0$.

Lemma 3.1.

a) Suppose that $m_{n+1} \in I$ and $m_{n+1} < m_B$. Then

$$c \leq H_{\tau^2}(m_B) - H_{\tau^2}(m_{n+1}) \leq d^{-1}(\tau_{n+1}^2 - \tau^2)$$

where $d^{-1} = N^4/(16\sigma^4)$.

b) Suppose that $m_{n+1} \in I$ and $m_{n+1} > m_B$. Then,

$$c \leq H_{\tau^2}(m_B) - H_{\tau^2}(m_{n+1}) \leq d^{-1}(\tau^2 - \tau_{n+1}^2).$$

Proof:

a) By Lemma 2.1, as $m_{n+1} \in I$ and $m_B > m_{n+1}$, we have $\tau^2 < \tau_{n+1}^2$. Thus, on the event that $m_{n+1} \in I$ and $\tau^2 < \tau_{n+1}^2$, we have

$$\begin{aligned} c &\leq H_{\tau^2}(m_B) - H_{\tau^2}(m_{n+1}) \\ &= \frac{(N-m_B)^2 m_B}{\sigma^2 + m_B\tau^2} - \frac{(N-m_{n+1})^2 m_{n+1}}{\sigma^2 + m_{n+1}\tau^2} \\ &= [\frac{(N-m_B)^2 m_B}{\sigma^2 + m_B\tau^2} - \frac{(N-m_B)^2 m_B}{\sigma^2 + m_B\tau_{n+1}^2}] \\ &\quad + [\frac{(N-m_B)^2 m_B}{\sigma^2 + m_B\tau_{n+1}^2} - \frac{(N-m_{n+1})^2 m_{n+1}}{\sigma^2 + m_{n+1}\tau_{n+1}^2}] \\ &\quad + [\frac{(N-m_{n+1})^2 m_{n+1}}{\sigma^2 + m_{n+1}\tau_{n+1}^2} - \frac{(N-m_{n+1})^2 m_{n+1}}{\sigma^2 + m_{n+1}\tau^2}]. \end{aligned} \tag{3.4}$$

In (3.4), $(N-m_B)^2 m_B/(\sigma^2 + m_B\tau_{n+1}^2) - (N-m_{n+1})^2 m_{n+1}/(\sigma^2 + m_{n+1}\tau_{n+1}^2) \leq 0$ which is obtained by the definition of m_{n+1}, and $(N-m_{n+1})^2 m_{n+1}/(\sigma^2 + m_{n+1}\tau_{n+1}^2) - (N-m_{n+1})^2 m_{n+1}/(\sigma^2 + m_{n+1}\tau^2) < 0$ by noting that $\tau^2 < \tau_{n+1}^2$. Thus, we obtain

$$\begin{aligned} c &\leq H_{\tau^2}(m_B) - H_{\tau^2}(m_{n+1}) \\ &\leq \frac{(N-m_B)^2 m_B}{\sigma^2 + m_B\tau^2} - \frac{(N-m_B)^2 m_B}{\sigma^2 + m_B\tau_{n+1}^2} \\ &= \frac{(N-m_B)^2 m_B^2(\tau_{n+1}^2 - \tau^2)}{(\sigma^2 + m_B\tau^2)(\sigma^2 + m_B\tau_{n+1}^2)} \\ &\leq \frac{N^4}{16\sigma^4}(\tau_{n+1}^2 - \tau^2) \\ &= d^{-1}(\tau_{n+1}^2 - \tau^2) \end{aligned} \tag{3.5}$$

which completes the proof of part a).

b) By Lemma 2.1 again, as $m_{n+1} \in I$ and $m_B < m_{n+1}$, we have $\tau^2 > \tau^2_{n+1}$. Thus, under the event that $m_{n+1} \in I$ and $\tau^2 > \tau^2_{n+1}$, we have

$$\begin{aligned} c &\le H_{\tau^2}(m_B) - H_{\tau^2}(m_{n+1}) \\ &= \frac{(N-m_B)^2 m_B}{\sigma^2 + m_B\tau^2} - \frac{(N-m_{n+1})^2 m_{n+1}}{\sigma^2 + m_{n+1}\tau^2} \\ &= \Big[\frac{(N-m_B)^2 m_B}{\sigma^2 + m_B\tau^2} - \frac{(N-m_B)^2 m_B}{\sigma^2 + m_B\tau^2_{n+1}}\Big] + \Big[\frac{(N-m_B)^2 m_B}{\sigma^2 + m_B\tau^2_{n+1}} - \frac{(N-m_{n+1})^2 m_{n+1}}{\sigma^2 + m_{n+1}\tau^2_{n+1}}\Big] \\ &\quad + \Big[\frac{(N-m_{n+1})^2 m_{n+1}}{\sigma^2 + m_{n+1}\tau^2_{n+1}} - \frac{(N-m_{n+1})^2 m_{n+1}}{\sigma^2 + m_{n+1}\tau^2}\Big], \end{aligned} \tag{3.6}$$

where $(N-m_B)^2 m_B/(\sigma^2 + m_B\tau^2) - (N-m_B)^2 m_B/(\sigma^2 + m_B\tau^2_{n+1}) < 0$ since $\tau^2 > \tau^2_{n+1}$ and $(N-m_B)^2 m_B/(\sigma^2 + m_B\tau^2_{n+1}) - (N-m_{n+1})^2 m_{n+1}/(\sigma^2 + m_{n+1}\tau^2_{n+1}) \le 0$, by the definition of m_{n+1}. Therefore,

$$\begin{aligned} c &\le H_{\tau^2}(m_B) - H_{\tau^2}(m_{n+1}) \\ &\le \frac{(N-m_{n+1})^2 m_{n+1}}{\sigma^2 + m_{n+1}\tau^2_{n+1}} - \frac{(N-m_{n+1})^2 m_{n+1}}{\sigma^2 + m_{n+1}\tau^2} \\ &= \frac{(N-m_{n+1})^2 m^2_{n+1}(\tau^2 - \tau^2_{n+1})}{(\sigma^2 + m_{n+1}\tau^2_{n+1})(\sigma^2 + m_{n+1}\tau^2)} \\ &\le d^{-1}(\tau^2 - \tau^2_{n+1}). \end{aligned} \tag{3.7}$$

Lemma 3.2.

a) $P\{r_{n+1}(m_{n+1}) > r_B\} \le P\{|\tau^2_{n+1} - \tau^2| \ge dc\}$.

b) $r_{n+1} - r_B \le k\alpha\tau^2[H_{\tau^2}(m_B)]^{\frac{1}{2}} P\{|\tau^2_{n+1} - \tau^2| \ge dc\}$.

Proof:

a)

$$\begin{aligned} &P\{r_{n+1}(m_{n+1}) > r_B\} \\ &= P\{H_{\tau^2}(m_B) - H_{\tau^2}(m_{n+1}) > 0,\ m_{n+1} \in I\} \\ &= P\{H_{\tau^2}(m_B) - H_{\tau^2}(m_{n+1}) \ge c,\ m_{n+1} \in I\} \\ &\qquad \text{(by the definition of the set } I) \\ &= P\{H_{\tau^2}(m_B) - H_{\tau^2}(m_{n+1}) \ge c,\ m_{n+1} \in I,\ m_B < m_{n+1}\} \\ &\qquad + P\{H_{\tau^2}(m_B) - H_{\tau^2}(m_{n+1}) \ge c,\ m_{n+1} \in I,\ m_B > m_{n+1}\} \\ &\le P\{\tau^2 - \tau^2_{n+1} \ge dc\} + P\{\tau^2_{n+1} - \tau^2 \ge dc\} \\ &\qquad \text{(by Lemma 3.1)} \\ &= P\{|\tau^2_{n+1} - \tau^2| \ge dc\}. \end{aligned}$$

b)

$$\begin{aligned}
&r_{n+1} - r_B \\
&= E[r_{n+1}(m_{n+1}) - r_B] \\
&- E[k\alpha\tau^2[(H_{\tau^2}(m_B))^{\frac{1}{2}} \quad (H_{\tau^2}(m_{n+1}))^{\frac{1}{2}}]] \\
&\le k\alpha\tau^2[H_{\tau^2}(m_B)]^{\frac{1}{2}} P\{H_{\tau^2}(m_B) - H_{\tau^2}(m_{n+1}) > 0\} \\
&= k\alpha\tau^2[H_{\tau^2}(m_B)]^{\frac{1}{2}} P\{H_{\tau^2}(m_B) - H_{\tau^2}(m_{n+1}) \ge c\} \\
&\le k\alpha\tau^2[H_{\tau^2}(m_B)]^{\frac{1}{2}} P\{|\tau^2_{n+1} - \tau^2| \ge dc\}
\end{aligned}$$

where the last equality is obtained from the definition of the constant c, and the last inequality is obtained from the proof of part a) of this lemma.

From Lemma 3.2, in order to investigate the asymptotic behavior of $P\{r_{n+1}(m_{n+1}) > r_B\}$ and $r_{n+1} - r_B$, it suffices to study the asymptotic behavior of the probability $P\{|\tau^2_{n+1} - \tau^2| \ge dc\}$.

Lemma 3.3. Let $\{\tau^2_{n+1}\}_{n=1}^{\infty}$ be a sequence of estimators defined in (3.2). Then, τ^2_{n+1} converges to τ^2 in probability. Furthermore, for any $\varepsilon > 0$, we have $P\{|\tau^2_{n+1} - \tau^2| \ge \varepsilon\} \le 0(n^{-1})$ as $n \to \infty$.

Proof: First note that $Y \equiv \frac{(kn-1)S^2(n)}{\sigma^2+\tau^2}$ follows a χ^2-distribution with $(kn-1)$ degrees of freedom. By the definition of τ^2_{n+1} given in (3.2), letting $\varepsilon_1 = \varepsilon/(\sigma^2 + \tau^2)$, we have

$$\begin{aligned}
&P\{|\tau^2_{n+1} - \tau^2| \ge \varepsilon\} \\
&= P\{\tau^2_{n+1} \ge \tau^2 + \varepsilon\} + P\{\tau^2_{n+1} \le \tau^2 - \varepsilon\} \\
&\le P\{S^2(n) \ge \tau^2 + \sigma^2 + \varepsilon\} + P\{S^2(n) \le \tau^2 + \sigma^2 - \varepsilon\} \\
&= P\{Y \ge (kn-1)(1+\varepsilon_1)\} + P\{Y \le (kn-1)(1-\varepsilon_1)\} \\
&= P\Big\{|\frac{Y-(kn-1)}{\sqrt{2(kn-1)}}| \ge \sqrt{\frac{kn-1}{2}}\varepsilon_1\Big\} \\
&\le \frac{2}{(kn-1)\varepsilon_1^2}
\end{aligned}$$

which can be obtained by Chebyschev's inequality. Hence we obtain that

$$P\{|\tau^2_{n+1} - \tau^2| \ge \varepsilon\} \le 0(n^{-1}) \text{ as } n \to \infty.$$

From Lemmas 3.2 and 3.3, we conclude the following theorem.

Theorem 3.1. The sequence of adaptive two-stage allocation procedures $\{P_{n+1}\}$ is asymptotically optimal in probability of order $\{n^{-1}\}$ and asymptotically optimal of order $\{n^{-1}\}$. That is,

$$P\{r_{n+1}(m_{n+1}) - r^B \ge \varepsilon\} \le 0(n^{-1}) \text{ as } n \to \infty \text{ for any } \varepsilon > 0,$$

and

$$r_{n+1} - r^B \le 0(n^{-1}) \text{ as } n \to \infty.$$

REFERENCES

[1] Gupta, S. S. and Liang, T. (1987). On some Bayes and empirical Bayes selection procedures. In Probability and Bayesian Statistics (Ed. R. Viertl), Plenum Publishing Corporation, 233–246.

[2] Gupta, S. S. and Panchapakesan, S. (1979). Multiple Decision Procedures. New York, John Wiley.

[3] Ofosu, J. B. (1974). Optimum sample size for a problem in choosing the population with the largest mean: some comments on Somerville's paper. Journal of American Statistical Association **69**, 270.

[4] Ofosu, J. B. (1975). A Bayesian procedure for selecting the largest of k normal means. Calcutta Statistical Association Bulletin **24**, 43–51.

[5] Robbins, H. (1983). Some thoughts on empirical Bayes estimation. Annals of Statistics **11**, 713–723.

[6] Somerville, P. N. (1970). Optimum sample size for a problem in choosing the population with the largest mean. Journal of American Statistical Association **65**, 763–775.

[7] Somerville, P. N. (1974). Rejoinder to Ofosu's comments. Journal of American Statistical Association **69**, 270–271.

STATISTICAL DATA ANALYSIS AND INFERENCE
Y. Dodge (Editor)
Elsevier Science Publishers B.V. (North-Holland), 1989

MULTI-SAMPLE FUNCTIONAL STATISTICAL DATA ANALYSIS

Emanuel Parzen

Department of Statistics
Texas A&M University
College Station, Texas 77843-3143

ABSTRACT. This paper discusses a functional approach to the problem of comparison of multi-samples (two samples or c samples, where $c \geq 2$). The data consists of c random samples whose probability distributions are to be tested for equality. A diversity of statistics to test equality of c samples are presented in a unified framework with the aim of helping the researcher choose the optimal procedures which provide greatest insight about how the samples differ in their distributions. Concepts discussed are: sample distribution functions; ranks; mid-distribution function; two-sample t test and nonparametric Wilcoxon test; multi-sample analysis of variance and Kruskal Wallis test; Anderson Darling and Cramer von Mises tests; components and linear rank statistics; comparison distribution and comparison density functions, especially for discrete distributions; components with orthogonal polynomial score functions; chi-square tests and their components.

1. INTRODUCTION.

We assume that we are observing a variable Y in c cases or samples (corresponding to c treatments or c populations). The samples can be regarded as the value of c variables $Y_1, \ldots, Y_c$ with respective true distribution functions $F_1(y), \ldots, F_c(y)$ and quantile functions $Q_1(u), \ldots, Q_c(u)$. We call $Y_1, \ldots, Y_c$ the conditioned variables (the value of Y in different populations).

The general problem of comparison of conditioned random variables is to model how their distribution functions vary with the value of the conditioning variable $k = 1, \ldots, c$, and in particular to test the hypothesis of homogeneity of distributions:

$$H_0 : F_1 = \ldots = F_c = F$$

The distribution F to which all the others are equal is considered to be the unconditional distribution of Y (which is estimated by the sample distribution of Y in the pooled sample).

2. DATA.

The data consists of c random samples

$$Y_k(j), j = 1, \ldots, n_k$$

for $k = 1, \ldots, c$. The pooled sample, of size $N = n_1 + \ldots + n_c$, represents observations of the pooled (or unconditional) variable Y. The c samples are assumed to be independent of each other.

3. SAMPLE DISTRIBUTION FUNCTIONS.

The sample distribution functions of the samples are defined (for $-\infty < y < \infty$) by

Research Supported by the U.S. Army Research Office

$$F_k\tilde{}(y) = \text{ fraction } \leq y \text{ among } Y_k(.).$$

The unconditional or pooled sample distribution of Y is denoted

$$F\hat{}(y) = \text{ fraction } \leq y \text{ among } Y_k(.), k = 1, \ldots, c.$$

We use ^ to denote a smoother distribution to which we are comparing a more raw distribution which is denoted by a ˜. An expectation (mean) computed from a sample is denoted $E\tilde{}$.

4. RANKS, MID-RANKS, AND MID-DISTRIBUTION FUNCTION.

Nonparametric statistics use ranks of the observations in the pooled sample; let

$R_k(t)$ denote the rank in the pooled sample of $Y_k(t)$.

One can define $R_k(t) = NF\hat{}(Y_k(t))$.

In defining linear rank statistics one transforms the rank to a number in the open unit interval, usually $R_k(t)/(N+1)$. We recommend $(R_k(t) - .5)/N$. These concepts assume all observations are distinct, and treat ties by using average ranks. We recommend an approach which we call the "mid-rank transform" which transforms $Y_k(t)$ to $P\hat{}(Y_k(t))$, defining the mid-distribution function of the pooled sample Y by

$$P\hat{}(y) = F\hat{}(y) - .5p\hat{}(y).$$

We call

$$p\hat{}(y) = \text{ fraction equal to } y \text{ among pooled sample}$$

the pooled sample probability mass function.

5. SAMPLE MEANS AND VARIANCES.

When the random variables are assumed to be normal the test statistics are based on the sample means (for $k = 1, \ldots, c$)

$$Y_k^- = E\tilde{}[Y_k] = (1/n_k) \sum_{t=1}^{n_k} Y_k(t).$$

We interpret Y_k^- as the sample conditional mean of Y given that it comes from the kth population. The unconditional sample mean of Y is

$$Y^- = E\tilde{}[Y] = p_{.1}Y_1^- + \ldots + p_{.c}Y_c^-,$$

defining

$$p_{.k} = n_k/N$$

to be the fraction of the pooled sample in the kth sample; we interpret it as the empirical probability that an observation comes from the kth sample.

The unconditional and conditional variances are denoted

$$\text{VAR}\tilde{}[Y] = (1/N)\sum_{k=1}^{c}\sum_{j=1}^{n_k}\{Y_k(j) - Y^-\}^2$$

$$\text{VAR}\tilde{}[Y_k] = (1/n_k)\sum_{j=1}^{n_k}\{Y_k(j) - Y_k^-\}^2$$

Note that our divisor is the sample size N or n_k rather than $N - c$ or $n_k - 1$. The latter then arise as factors used to define F statistics.

We define the pooled variance to be the mean conditional variance:

$$\sigma\hat{}^2 = \sum_{k=1}^{c} p_{.k}\ \text{VAR}\tilde{}[Y_k]$$

6. TWO SAMPLE NORMAL T TEST.

In the two sample case the statistic to test H_0 is usually stated in a form equivalent to

$$T = \{Y_1^- - Y_2^-\}/\sigma\hat{}\{(N/(N-2))((1/n_1) + (1/n_2))\}^{.5}$$

We believe that one obtains maximum insight (and analogies and extensions) by expressing T in the form which compares Y_1^- with Y^-:

$$T = \{(N-2)p_{.1}/(1-p_{.1})\}^{.5}\{Y_1^- - Y^-\}/\sigma\hat{}$$

The exact distribution of T is $t(N-2)$, t-distribution with $N-2$ degrees of freedom.

7. TWO-SAMPLE NONPARAMETRIC WILCOXON TEST.

To define the popular Wilcoxon non-parametric statistic to test H_0 we define W_k to be the sum of the n_k ranks of the Y_k values; its mean and variance are given by

$$E[W_k] = n_k(N+1)/2, \ \text{VAR}[W_k] = n_1 n_2(N+1)/12$$

The usual definition of the Wilcoxon test statistic is

$$T_k = \{W_k - E[W_k]\}/\{\text{VAR}[W_k]\}^{.5}.$$

The approach we describe in this paper yields as the definition of the nonparametric Wilcoxon test statistic (which can be verified to approximately equal the above definition of T_1, up to a factor $\{1-(1/N)^2\}^{.5}$)

$$T_1 = \{12(N-1)p_{.1}/(1-p_{.1})\}^{.5}(R_1^- - .5\},$$

defining

$$R_1^- = (1/n_1)\sum_{t=1}^{n_1}(R_1(t) - .5)/N$$
$$= (W_1/n_1 N) - (1/2N)$$

One reason we prefer this form of expressing non-parametric statistics is because of its relation to mid-ranks;

$$R_k^{-} = E^{\sim}[P^{\wedge}(Y_k)]$$

One should notice the analogy between our expressions for the parametric test statistic T and the nonparametric test statistic T_1; the former has an exact $t(N-2)$ distribution and the latter has asymptotic distribution Normal$\{0,1\}$.

8. TEST OF EQUALITY OF c SAMPLES NORMAL CASE.

The homogeneity of c samples is tested in the parametric normal case by the analysis of variance which starts with a fundamental identity which in our notation is written

$$\mathrm{VAR}^{\sim}[Y] = \sum_{k=1}^{c} p_{.k}\{Y_k^{-} - Y^{-}\}^2 + \sigma^{\wedge 2}$$

The F test of the one-way analysis of variance can be expressed as the statistic

$$\begin{aligned} T^2 &= \sum_{k=1}^{c} p_{.k}|T_k|^2, \\ &= \sum_{k=1}^{c}(1-p_{.k})|TF_k|^2, \end{aligned}$$

defining

$$\begin{aligned} T_k &= (N-c)^{.5}\{Y_k^{-} - Y^{-}\}/\sigma^{\wedge} \\ TF_k &= \{(N-c)p_{.k}/(1-p_{.k})\}^{.5}\{Y_k^{-} - Y^{-}\}/\sigma^{\wedge} \end{aligned}$$

The asymptotic distribution of $T^2/(c-1)$ and TF_k^2 are $F(c-1, N-c)$ and $F(1, N-c)$ respectively.

9. TEST OF EQUALITY OF c SAMPLES NONPARAMETRIC KRUSKAL-WALLIS TEST.

The Kruskal-Wallis nonparametric test of homogeneity of c samples can be shown to be

$$\begin{aligned} TKW^2 &= \sum_{k=1}^{c}(1-p_{.k})|TKW_k|^2. \\ TKW_k &= \{12(N-1)p_{.k}/(1-p_{.k})\}^{.5}\{R_k^{-} - .5\} \end{aligned}$$

The asymptotic distributions of TKW^2 and TKW_k^2 are chi-squared with $c-1$ and 1 degrees of freedom respectively.

10. COMPONENTS.

We have represented the analysis of variance test statistic T^2 and the Kruskal-Wallis test statistic TKW^2 as weighted sums of squares of statistics TF_k and TKW_k respectively which we call components, since their values should be explicitly calculated to indicate the source of the significance (if any) of the overall statistics. Other test statistics that can be defined can be shown to correspond to other definitions of components.

11. ANDERSON DARLING AND CRAMER VON MISES TEST STATISTICS.

Important among the many test statistics which have been defined to test the equality of distributions are the Anderson-Darling and Cramer-von Mises test statistics. They will be introduced below in terms of representations as weighted sums of squares of suitable components.

12. COMPARISON DISTRIBUTION FUNCTIONS AND COMPARISON DENSITTY FUNCTIONS.

We now introduce the key concepts which enable us to unify and choose between the diverse statistics available for comparing several samples. To compare two continuous distributions $F(.)$ and $H(.)$, where H is a true or smooth and F is a model or raw, we define the comparison distribution function

$$D(u) = D(u; H, F) = F(H^{-1}(u))$$

with comparison density

$$d(u) = d(u; H, F) = D'(u) = f(H^{-1}(u))/h(H^{-1}(u)).$$

Under $H_0 : H = F$, $D(u) = u$ and $d(u) = 1$. Thus testing H_0 is equivalent to testing $D(u)$ for uniformity.

Sample distribution functions are discrete. The most novel part of this paper is that we propose to form an estimator $D\tilde{}(u)$ from estimators $H\tilde{}(.)$ and $F\tilde{}(.)$ by using a general definition of $D(.)$ for two discrete distributions $H(.)$ and $F(.)$ with respective probability mass functions p_H and p_F satisfying the condition that the values at which p_H are positive include all the values at which p_F are positive.

13. COMPARISON OF DISCRETE DISTRIBUTIONS.

To compare two discrete distributions we define first $d(u)$ and then $D(u)$ as follows:

$$\begin{aligned} d(u) &= d(u; H, F) = p_F(H^{-1}(u))/p_H(H^{-1}(u)), \\ D(u) &= \int_0^1 d(t)dt. \end{aligned}$$

We apply this definition to the discrete sample distributions $F\hat{}$ and $F\tilde{}_k$ to obtain

$$d_k\tilde{}(u) = d(u; F\hat{}, F_k\tilde{})$$

and its integral $D_k\tilde{}(u)$.

We obtain the following definition of $d_k\tilde{}(u)$ for the c sample testing problem with all values distinct:

$$\begin{aligned} d_k\tilde{}(u) &= N/n_k \text{ if } (R_k(j) - 1)/N < u < R_k(j)/N, j = 1, \ldots, n_k, \\ &= 0, \qquad \text{otherwise.} \end{aligned}$$

A component, with score function $J(u)$, is a linear functional

$$T_k\tilde{}(J) = \int_0^1 J(u) d_k\tilde{}(u) du$$

It equals

$$(1/n_k) \sum_{j=1}^{n_k} N \int_{(R_k(j)-1)/N}^{R_k(j)/N} J(u) du$$

which can be approximated by $E\tilde{}[J(P\hat{}(Y_k))]$.

14. LINEAR RANK STATISTICS.

The concept of a linear rank statistic to compare the equality of c samples does not have a universally accepted definition. One possible definition is

$$T_k\tilde{}(J) = (1/n_k) \sum_{j=1}^{n_k} J((R_k(j) - .5)/N)$$

However we choose the definition of a linear rank statistic as a linear functional of $d_k\tilde{}(u)$, which we call a component; it is approximately equal to the above formula.

We define

$$T_k\tilde{}(J) = ((N-1) \text{ VAR}[J(U)] p_{.k}/(1-p_{.k}))^{.5} \int_0^1 J(u)\{d_k\tilde{}(u) - 1\} du \qquad (!)$$

where U is Uniform$\{0,1\}$, $E[J(U)] = \int_0^1 J(u)du$,

$$\text{VAR}[J(U)] = \int_0^1 \{J(u) - E[J(U)]\}^2 du.$$

Note that the integral in the definition of $T_k\tilde{}(J)$ equals

$$\int_0^1 J(u) d\{D_k\tilde{}(u) - u\}.$$

The components of the Kruskal-Wallis nonparametric test statistic TKW^2 for testing the equality of c means have score function $J(u) = u - .5$ satisfying

$$E[J(U)] = .5, \text{ VAR}[J(U)] = 1/12.$$

The components of F test statistic T^2 have score function

$$J(u) = \{Q\hat{}(u) - Y\bar{}\}/\sigma\hat{}$$

where $Q\hat{}(u)$ is sample quantile function of the pooled sample Y.

15. GENERAL DISTANCE MEASURES.

General measures of the distance of $D\tilde{}(u)$ from u and of $d\tilde{}(u)$ from 1 are provided by the integrals from 0 to 1 of

$$\{d\tilde{}(u) - 1\}^2, \quad \{D\tilde{}(u) - u\}^2, \quad \{D\tilde{}(u) - u\}^2/u(1-u), \quad \{d\hat{}(u) - 1\}^2$$

where $d\hat{}(u)$ is a smooth version of $d\tilde{}(u)$. We will see that these measures can be decomposed into components which may provide more insight; recall basic components are linear functionals defined by (!)

$$T\tilde{}(J) = \int_0^1 J(u) d\tilde{}(u) du.$$

If $\phi_i(u)$, $i = 0, 1, 2, \ldots$, are complete orthonormal functions with $\phi_0 = 1$, then H_0 can be tested by diagnosing the rate of increase (as a function of $m = 1, 2, \ldots$) of

$$\int_0^1 \{d_m(u) - 1\}^2 du = \sum_{i=1}^{m} |T\tilde{}(\phi_i|^2$$

which measure the distance from 1 of the approximating smooth densities

$$d_m(u) = \sum_{i=1}^{m} T\tilde{}(\phi_i)\phi_i(u).$$

16. ORTHOGONAL POLYNOMIAL COMPONENTS.

Let $p_i(x)$ be Legendre polynomials on (-1,1):

$$\begin{aligned} p_1(x) &= x \\ p_2(x) &= (3x^2 - 1)/2, \\ p_3(x) &= (5x^3 - 3x)/2, \\ p_4(x) &= 35x^4 - 30x^2 + 3. \end{aligned}$$

Define Legendre polynomial score functions

$$\phi L_i(u) = (2i + 1)^{.5} p_i(2u - 1).$$

One can show that an Anderson-Darling type statistic, denoted $AD(D\tilde{})$, can be represented

$$\begin{aligned} AD(D\tilde{}) &= \int_0^1 \{\{D\tilde{}(u) - u\}^2 / u(1-u)\} du \\ &= \sum_{i=1}^{\infty} |T\tilde{}(\phi L_i)|^2 / (i(i+1)) \end{aligned}$$

Define cosine score functions by

$$\phi C_i(u) = 2^{.5} cos(i\pi u).$$

One can show that a Cramer–von Mises type statistic, denoted $CM(D\tilde{})$, can be represented

$$\begin{aligned} CM(D\tilde{}) &= \int_0^1 \{D\tilde{}(u) - u\}^2 du \\ &= \sum_{i=1}^{\infty} |T\tilde{}(\phi C_i)|^2 / (i\pi)^2 \end{aligned}$$

In addition to Legendre polynomial and cosine components we consider Hermite polynomial components corresponding to Hermite polynomial score functions

$$\phi H_i(u) = (i!)^{-.5} H_i(\Phi^{-1}(u))$$

where $H_i(x)$ are the Hermite polynomials:

$$\begin{aligned} H_1(x) &= x, \\ H_2(x) &= x^2 - 1, \\ H_3(x) &= x^3 - 3x, \\ H_4(x) &= x^4 - 6x^2 + 3. \end{aligned}$$

17. QUARTILE COMPONENTS AND CHI-SQUARE.

Quartile diagnostics of the null hypothesis H_0 are provided by components with quartile "square wave" score functions

$$\begin{aligned} SQ_1(u) &= -2^{.5}, \quad && 0 < u < .25, \\ &= 0, && .25 < u < .75, \\ &= 2^{.5}, && .75 < u < 1; \\ SQ_2(u) &= 1, && 0 < u < .25, \\ &= -1, && .25 < u < .75, \\ &= 1, && .75 < u < 1; \\ SQ_3(u) &= 0 && \text{if } 0 < u < .25 \text{ or } .75 < u < 1, \\ &= -2^{.5}, && .25 < u < .5, \\ &= 2^{.5}, && .5 < u < .75. \end{aligned}$$

A chi-squared portmanteau statistic, which is chi-squared(3), is

$$\begin{aligned} CQ_k &= (N-1)p_{.k}/(1-p_{.k}) \sum_{i=1}^{3} |T^{\sim}(SQ_i)|^2 \\ &= (N-1)p_{.k}/(1-p_{.k}) \int_0^1 \{dQ_k(u) - 1\}^2 du \end{aligned}$$

defining the quartile density (for $i = 1, 2, 3, 4$)

$$dQ_k(u) = 4\{D_k^{\sim}(i(.25)) - D_k^{\sim}((i-1).25), (i-1).25 < u < i(.25)$$

A pooled portmanteau chi-squared statistic is

$$CQ = \sum_{k=1}^{c} (1 - p_{.k}) CQ_k$$

18. DIVERSE STATISTICS AVAILABLE TO TEST EQUALITY OF c SAMPLES.

The problem of statistical infereence is not that we don't have answers to a given question; usually we have too many answers and we don't know which one to choose. A unified framework may help determine optimum choices. To compare c samples we can compute the following functions and statistics:

1) comparison densities: $d_k\breve{}(u)$,
2) comparison distributions $D_k\breve{}(u)$,
3) quartile comparison density $dQ_k(u)$, quartile density chi-square

$$CQ_k = (N-1)p_{.k}/(1-p_{.k}) \int_0^1 \{dQ_k(u) - 1\}^2 du.$$

4) non-parametric regression smoothing of $d_k\breve{}(u)$ using a boundary Epanechnikov kernel, denoted $d_k\hat{}(u)$,
5) Legendre components and chi-squares up to order 4 are defined using definition (!) of $T_k\breve{}$:

$$\begin{aligned} TL_k(i) &= T_k\breve{}(\phi L_i) \\ CL_k(m) &= \sum_{i=1}^{m} |TL_k(i)|^2 \\ CL(m) &= \sum_{k=1}^{c} (1-p_{.k}) CL_k(m) \\ AD_k &= \sum_{i=1}^{\infty} |TL_k(i)|^2 / i(i+1) \\ AD &= \sum_{k=1}^{c} (1-p_{.k}) AD_k \end{aligned}$$

6) Cosine components and chi-squares up to order 4 are defined:

$$\begin{aligned} TC_k(i) &= T_k\breve{}(\phi C_i) \\ CC_k(m) &= \sum_{i=1}^{m} |TC_k(i)|^2 \\ CC(m) &= \sum_{k=1}^{c} (1-p_{.k}) CC_k(m) \\ CM_k &= \sum_{i=1}^{\infty} |TC_k(i)|^2 / (i\pi)^2 \\ CM &= \sum_{k=1}^{c} (1-p_{.k}) CM_k \end{aligned}$$

7) Hermite components and chi-squares up to order 4 are defined:

$$TH_k(i) = T_k\tilde{}(\phi H_i)$$

$$CH_k(m) = \sum_{i=1}^{m} |TH_k(i)|^2$$

$$CH(m) = \sum_{k=1}^{c} (1 - p_{.k}) CH_k(m)$$

8) density estimators $d_k\hat{}(u)$ computed from components up to order 4,

9) entropy measures with penalty terms which can be used to determine how many components to use in the above test statistics

19. EXAMPLES OF DATA ANALYSIS.

The interpretation of the diversity of statistics available is best illustrated by examples.

In order to compare our methods with others available we consider data analysed by Boos (1986) on ratio of assessed value to sale price of residential property in Fitchburg, Mass., 1979. The samples (denoted I, II, III, IV) represent dwellings in the categories single-family, two-family, three-family, four or more families. The sample sizes (54, 43, 31, 28) are proportions .346, .276, .199, .179 of the size 156 of the pooled sample. We compute Legendre, cosine, Hermite components up to order 4 of the 4 samples; they are asymptotically standard normal. We consider components greater than 2 (3) in absolute value to be significant (very significant).

Legendre, cosine, and Hermite components are very significant only for sample I, order 1 (-4.06, -4.22, -3.56 respectively). Legendre components are significant for sample IV, orders 1 and 2 (2.19, 2.31). Cosine components are significant for sample IV, orders I and II (2.36, 2.23) and sample III, order 1 (2.05). Hermite components are significant for sample IV, orders 2 and 3 (2.7 and -2.07).

Conclusions are that the four samples are not homogeneous (have the same distributions). Samples I and IV are significantly different from the pooled sample. Estimators of the comparison density show that sample I is more likely to have lower values than the pooled sample, and sample IV is more likely to have higher values. While all the statistical measures described above have been computed, the insights are provided by the linear rank statistics of orthogonal polynomials rather than by portmanteau statistics of Cramer-von Mises or Anderson-Darling type.

20. CONCLUSIONS. The goal of our recent research (see Parzen (1979), (1983)) on unifying statistical methods (especially using quantile function concepts) has been to help the development of both the theory and practice of statistical data analysis. Our ultimate aim is to make it easier to apply statistical methods by unifying them in ways that increase understanding, and thus enable researchers to more easily choose methods that provide greatest insight for their problem. We believe that if one can think of several ways of looking at a data analysis one should do so. However to relate and compare the answers, and thus arrive at a confident conclusion, a general framework seems to us to be required.

One of the motivations for this paper was to understand two-sample tests of the Anderson-Darling type; they are discussed by Pettitt (1976) and Scholz and Stephens (1987). This paper provides new formulas for these test statistics based on our new definition of sample comparison density functions. Asymptotic distribution theory for rank processes defined by Parzen (1983) is given by Aly, Csorgo, and Horvath (1987); an excellent review of theory for rank processes is given by Shorack and Wellner (1986).

However one can look at k sample Anderson-Darling statistics as a single number formed from combining many test statistics called components. The importance of components is also advocated by Boos (1986), Eubank, LaRiccia, and Rosenstein (1987) and Alexander (1989). Insight is greatly increased if instead of basing one's conclusions on the values of single test statistics, one looks at the components and also at graphs of the densities of which the components are linear functionals corresponding to various score functions. The question of which score functions to use can be answered by considering the tail behavior of the distributions that seem to fit the data.

REFERENCES

Alexander, William (1989) "Boundary kernel estimation of the two-sample comparison density function" Texas A&M Department of Statistics Ph.D. thesis.

Aly, E.A.A., M. Csorgo, and L. Horvath (1987) "P-P plots, rank processes, and Chernoff-Savage theorems" in *New Perspectives in Theoretical and Applied Statistics* (ed. M.L. Puri, J.P. Vilaplann, W. Wertz) New York: Wiley 135-156.

Boos, Dennis D. (1986) "Comparing k populations with linear rank statistics" *Journal of the American Statistical Association*, 81, 1018-1025.

Eubank, R.L., V.N. LaRiccia, R.B. Rosenstein (1987) "Test statistics derived as components of Pearson's Phi-squared distance measure" *Journal of the American Statistical Association*, 82, 816-825.

Parzen, E. (1979) "Nonparametric statistical data modeling" *Journal of the American Statistical Association*, 74, 105-131.

Parzen, E. (1983) "FunStat quantile approach to two-sample statistical data analysis" Texas A&M Institute of Statistics Technical Report A-21 April 1983.

Pettitt, A.N. (1976) "A two-sample Anderson Darling statistic" *Biometrika*, 63, 161-168.

Scholz, F.W, and M.A. Stephens (1987) "k-sample Anderson-Darling tests" *Journal of the American Statistical Association*, 82, 918-924.

Shorack, Galen and Jon Wellner (1986) *Empirical Processes With Applications to Statistics* New York: Wiley.

Score functions for location: SQ1 (square wave), phiL1 (light curve), phiC1 (dark curve).

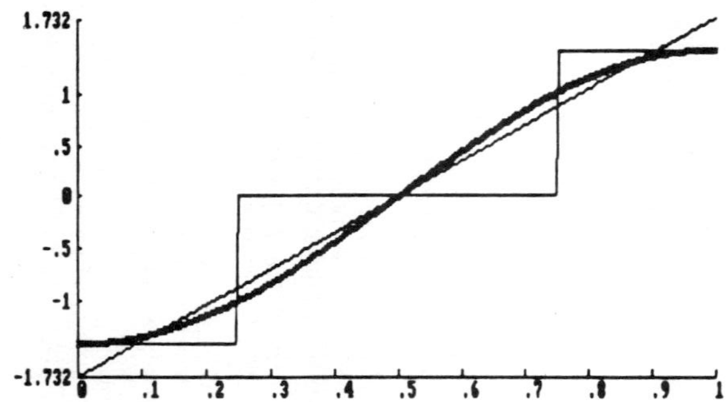

Score functions for skewness: SQ3 (square wave), phiL3 (light curve), phiC3 (dark curve).

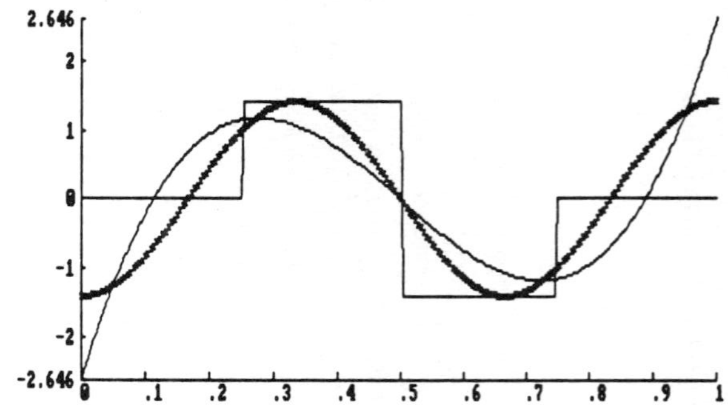

Score functions for scale: SQ2 (square wave), phiL2 (light curve), phiC2 (dark curve).

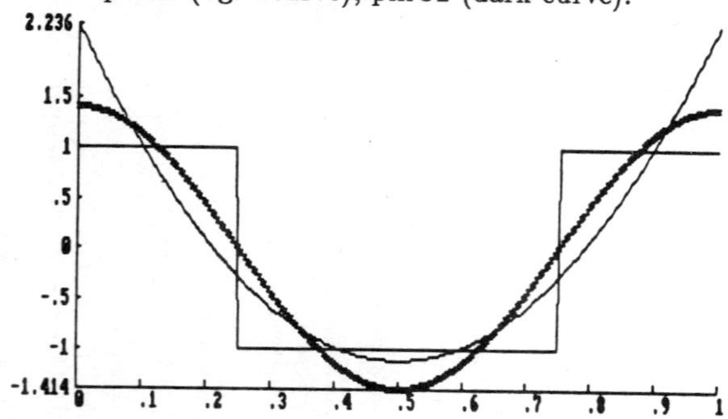

Score functions for kurtosis: phiL1 (light curve), phiC1 (dark curve).

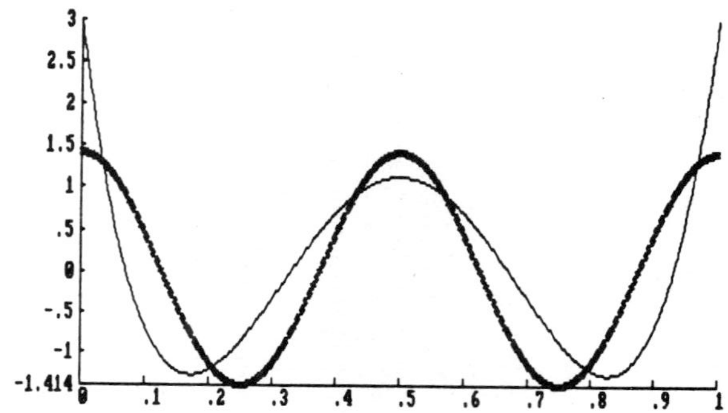

For samples I and IV, sample comparison distribution function $D^{\sim}(u)$

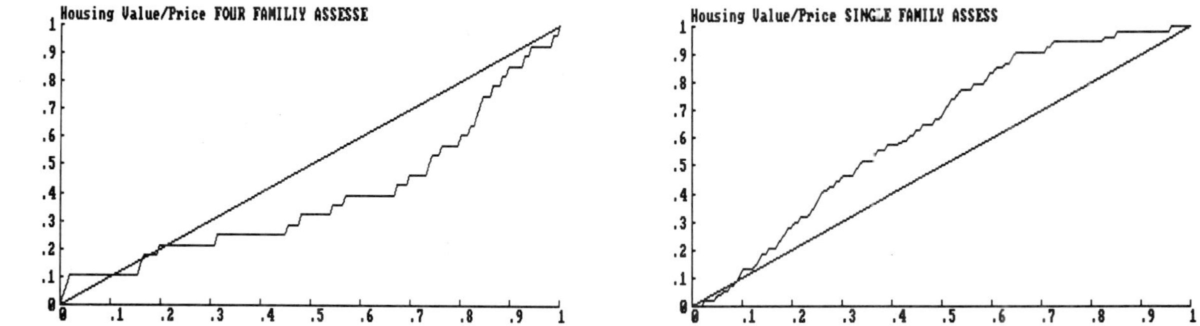

For samples I and IV, sample comparison density $d^{\sim}(u)$, sample quartile density $dQ^{\sim}(u)$ (square wave), nonparametric density estimator $d^{\wedge}(u)$

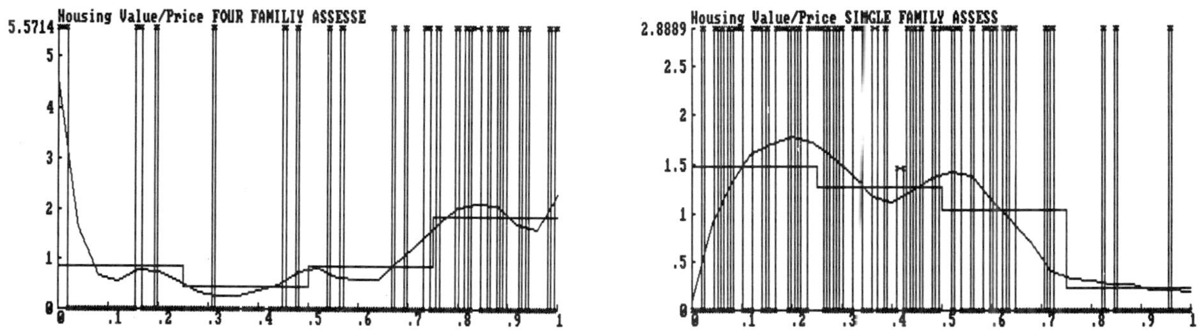

For samples I and IV, Legendre, cosine, and Hermite orthogonal polynomial estimator of order 4 of the comparison density, denoted $d_4(u)$, compared to sample quartile density $dQ\tilde{}(u)$.

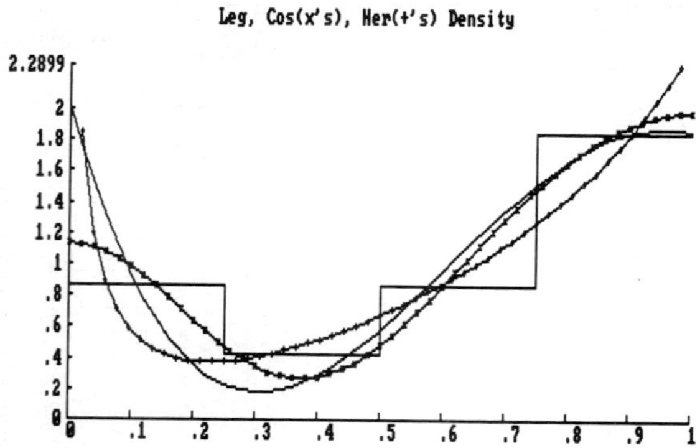

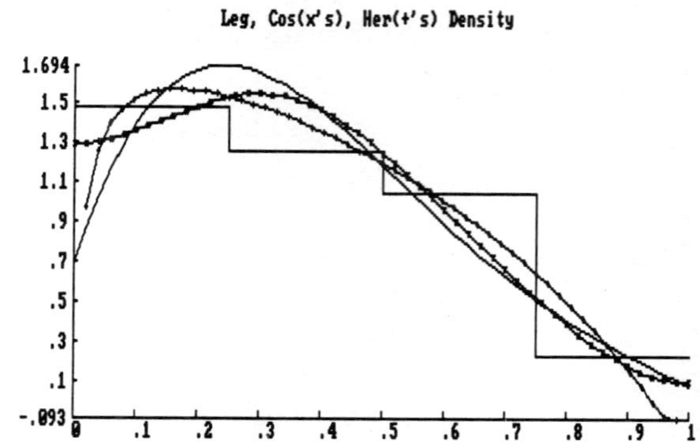

STATISTICAL DATA ANALYSIS AND INFERENCE
Y. Dodge (Editor)
Elsevier Science Publishers B.V. (North-Holland), 1989

DISCUSSING TRUTH OR FALSITY BY COMPUTING A Q-VALUE

Willem SCHAAFSMA, Jeroen TOLBOOM, and Bert VAN DER MEULEN

Department of Mathematics, University of Groningen,
P.O. Box 800, 9700 AV Groningen, The Netherlands

The Q–values to be introduced are meant to bridge part of the gap between the Bayesians, Fisherians and Neymanians.
Given a testing problem with hypothesis $H:\theta\in\Theta_H$ and alternative $A:\theta\in\Theta_A$, procedures $\alpha:\mathfrak{X}\to[0,1]$ are constructed for weighting the evidence. The weight $\alpha(x)$ is assigned to H while the complementary weight $\varphi(x)=1-\alpha(x)$ is assigned to A. Such weights can be used in solving decision–making problems provided that the loss function is constant on Θ_H and Θ_A, respectively.
The Bayesians suggest that the posterior probabilities of Θ_H and Θ_A should be used as the weights $\alpha(x)$ and $\varphi(x)$. For a restricted class of problems, Fisher introduced P–values. These can be regarded as weights in favour of H and became part of The Grammar of Science. In de Neyman–Pearson theory $\alpha(x)$ is constructed as the smallest value of the nominal level of significance for which H is rejected. In the present paper Q–values $\alpha:\mathfrak{X}\to[0,1]$ are constructed as appropriate estimators of the indicator function $I_H(\theta)$ of Θ_H. Various restrictions by unbiasedness, all different from the usual one, are considered.
In a number of situations classical P–values are recovered as optimal Q–values. Note that the posterior probability of Θ_H is obtained as the outcome, if the Q–value α is constructed by minimizing the Bayes risk. Asymptotic theory suggests that in some situations consensus will appear if the sample size is increased and the principles used for generating Q–values are sufficiently flexible. Practical implications are discussed.

1. INTRODUCTION

Discussing the truth or falsity of a statistical hypothesis $H:\theta\in\Theta_H$ goes beyond the making of one of the two statements suggested in the Neyman–Pearson theory. In fact Neyman and Pearson themselves already remarked that two–sided testing problems should be regarded as *three – decision* problems. The first author and his client, the physical–anthropologist Van Vark, were interested in the problem of *classifying with a doubtful region* discussed in Rao (1952). These inputs from Neyman and Pearson, and Rao led to Chapter 3 of Schaafsma (1966) where any testing problem was considered from the three–decision point of view of choosing one of the statements or "modalities"

a_0: H and A are neither rejected nor accepted
a_1: H is accepted and A is rejected
a_2: H is rejected and A is accepted

on the basis of the loss function

$$L(\theta,a_0)=a;\ L(\theta,a_1)=0;\ L(\theta,a_2)=b\ (\theta\in\Theta_H)$$
$$L(\theta,a_0)=a;\ L(\theta,a_1)=b;\ L(\theta,a_2)=0\ (\theta\in\Theta_A)$$

where $b(a)$ represents the loss resulting from an error of the first (second) kind. This formulation was in line with the two-decision formulation in Lehmann (1959) Section 1.5 where it was established that, under certain type I conditions, minimax risk and unbiased size-$a(a+b)^{-1}$ are equivalent.

For the three-decision formulation this equivalence is lost. The trivial procedure $d\equiv a_0$ is the unique minimax risk procedure. Hence alternative methodology had to be developed. The attention was focussed on clear but controversial optimum properties like minimax regret (D_L) and minimax regret SMR(D_L) where D_L is the class of three-decision procedures satisfying Lehmann's decision-theoretic concept of unbiasedness and SMR stands for the restriction to the class of somewhere minimum risk procedures. It was concluded in Schaafsma (1969) that loss-function formulations of multiple-decision problems lead to considerable methodological uncertainty if it comes to the point of proposing a particular procedure.

It is obvious from a practical point of view that a three-decision formulation is less refined than a formulation with, e.g., five modalities. This is in line with Savage (1971) who wrote that there are situations where yes, no, or maybe is simply not enough, and with Reichenbach (1949) who discussed the probabilities of the single case. The first author became interested in such "probabilities" when he and Van Vark were discussing the problem of making a sex-diagnosis on the bases of skull measurements. This led to the POSCON project. This name symbolizes that the aim of the project and of the corresponding computer program is to construct CONfidence intervals for POSterior probabilities. Note that Rao's problem of classifying with a doubtful region and the POSCON project belong to the area of making inferences about the truth or falsity of certain random events. We have seen that Rao's problem has an analogue in the theory of testing statistical hypotheses. The POSCON project should have a similar analogue. This will be presented in Section 2. A historical introduction is as follows.

In Fisher's "Statistical Methods" the testing of a statistical hypothesis meant something like (1) choosing the appropriate test statistic $t:\mathfrak{X}\to\mathbb{R}$, (2) determining the outcome $t(x)$, (3) computing the probability $\mathrm{P}=\mathrm{P}_0(T>t(x))$ of exceedance under the "null" hypothesis H, (4) making the interpretation that the deviation from H is not statistically significant if $\mathrm{P}>.05$ and highly significant if $\mathrm{P}<.01$. Such P-values became part of The Grammar of Science.

Neyman and Pearson started their research by focussing on the choice of test statistic. They introduced their likelihood-ratio principle in 1928. Trying to establish that this principle is fundamental they focussed on a two-decision formulation and on the restriction to the class of level-α tests where $\alpha = .05$ is in line with Fisher's recommendations. The mathematical beauty of the Neyman-Pearson theory and its decision-theoretic procedure-oriented extensions have had the effect that the testing of a statistical hypothesis H became more or less synonymous with the application of a level-α test. Fisher was not fond of this. The neo-Bayesians also concluded that much is lost if discussing truth or falsity is narrowed down to making an accept-reject statement. Fortunately the Neyman-Pearson theory allows an extension where Fisher's P-value is recovered by constructing the smallest value of α such that the hypothesis H is rejected in favour of the alternative A if the optimal Neyman-Pearson level-α test is applied to the data x. It is obvious that this Neyman-Pearson approach to the construction of P-values is somewhat circular. The literature shows that there is some controversy about P-values and posterior probabilities. The Bayesians claim that the definition of P-values is unnatural, and the objectivists claim that posterior probabilities are not compelling (because θ is not the outcome of a random variable). Recent publications, especially by Berger and coauthors, show that the discussion has not yet come to an end. Our contribution to the discussion is as follows.

2. Q-VALUES

It seems natural to discuss the truth or falsity of $H:\theta \in \Theta_H$ by constructing a procedure $\alpha:\mathcal{X} \rightarrow [0,1]$ for weighting the evidence. The weight $\alpha(x)$ is assigned to the hypothesis H and the weight $\varphi(x) = 1 - \alpha(x)$ to the alternative $A:\theta \in \Theta_A$.

In some situations discussing the truth or falsity of hypotheses is a goal in itself and it might suffice to communicate the weights $\alpha(x)$ and $\varphi(x)$. In some other situations the data x is not collected for the purpose of having a sophisticated discussion, but for taking a decision, i.e. choosing an element a from some action space $\mathcal{A}$, e.g. the space $\{a_0, a_1, a_2\}$ discussed at the beginning of Section 1. In this respect it may be useful to compute the weights $\alpha(x)$ and $\varphi(x)$, at least if the choice of action has to be based on a loss function L which is *constant* on Θ_H and Θ_A, respectively, i.e.

$$L(\theta,a) = w_{H,a}(\theta \in \Theta_H); \; L(\theta,a) = w_{A,a}(\theta \in \Theta_A).$$

If this restrictive assumption is not approximately satisfied then the weights $\alpha(x)$ and $\varphi(x)$ are not sufficiently informative to be used as the basis of the choice of a decision. However, if the loss function is as indicated then the weights $\alpha(x)$ and $\varphi(x)$ can be used by choosing the action $a = d(x)$ in such a way that (see Figure 1) the linear combination $w_{H,a}\alpha(x) + w_{A,a}\varphi(x)$ is as small as

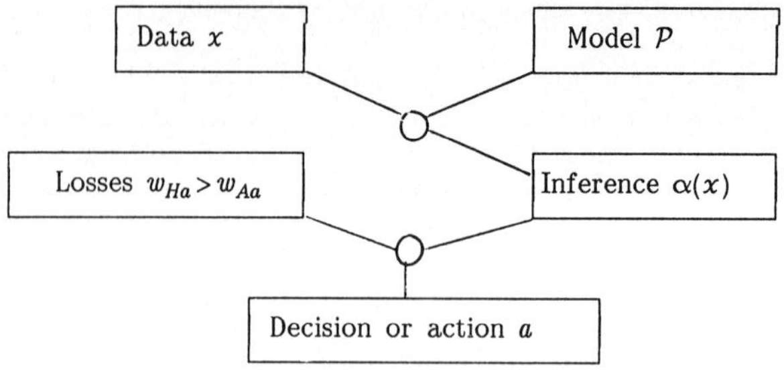

FIGURE 1. Two stages in discussing the choice of a decision.

possible as a function of a. The fundamental issue involved in this two-stage approach is whether the procedure $d:\mathcal{X}\to A$ thus constructed can compete with similar procedures based on a direct decision-theoretic approach. Note, however, that considerable methodological uncertainty is involved in the direct approach. It is natural to expect that this type of uncertainty will not be ameliorated if one focusses on the more fundamental problem of constructing a procedure $\alpha:\mathcal{X}\to[0,1]$ for weighting the evidence.

It is easy to make certain proposals. The Bayesians claim righteously that if θ were the outcome of a random variable T with known distribution τ and, conditionally given $T=\theta$, X has distribution P_θ as specified by the model, then one should use the posterior probability

$$\alpha(x) = \mathrm{P}(T \in \Theta_H | X = x).$$

We call this claim right because it complies with the rules of standard probability theory. Unfortunately we are usually not in the situation that θ is the outcome of a random variable T with known distribution τ. In such situations the symbol P has a metastatistical flavour and $\alpha(x)$ is not computable in principle. It is of some interest to consider the true value

$$\mathrm{P}_\theta(T \in \Theta_H | X = x)$$

of the posterior probability if T and θ are identified. This, however, is not much of a probability because it is equal to 1 if $\theta \in \Theta_H$ and equal to 0 if $\theta \in \Theta_A$. The argument suggests, however, that $\alpha(x)$ should be as close as possible to the indicator function $\mathrm{I}_H(\theta)$ of Θ_H or, in technical terms, $\alpha:\mathcal{X}\to[0,1]$ can be regarded or an *estimator* of $\mathrm{I}_H(\theta)$.

Fisher was fond of maximum likelihood. This suggests to consider procedure

$$\alpha_{\mathrm{ML}}(x) = \mathrm{I}_H(\hat{\theta}) = \begin{cases} 1 & \text{if } \hat{\theta}(x) \in \Theta_H \\ 0 & \text{if } \hat{\theta}(x) \in \Theta_A \end{cases}$$

where $\hat{\theta}: \mathcal{X} \to \Theta$ is the maximum likelihood estimator of θ. Neyman and Pearson were fond of the likelihood-ratio principle. A slight modification of this principle leads to the proposal.

$$\alpha_{\mathrm{MLR}}(x) = \sup_{\theta \in \Theta_H} p_\theta(x) / \{\sup_{\theta \in \Theta_H} p_\theta(x) + \sup_{\theta \in \Theta_A} p_\theta(x)\}$$

for estimating $\mathrm{I}_H(\theta)$. An advantage of α_{ML} and α_{MLR} is that they are well defined and simply to compute. This advantage may be lost if we consider Fisher's P value or its Neyman–Pearson analogue. These are only well defined in a rather restricted class of mathematically pleasant situations. (See Lehmann, 1959, 1986). We restrict the attention to such nice situations. Being well defined is not a complete justification for the P–value or its Neyman–Pearson analogue because, e.g., α_{ML} and α_{MLR} are also well defined. Fisher tried to give a justification by introducing the concept of fiducial probability. This was too mathematical to be really satisfactory. A decision–theoretic justification of P–values is as follows.

Consider the two–stage approach displayed in Figure 1 and suppose that $\#(A) = 2$. The intuitive background of this restriction is obvious: P–values are in line with the Neyman–Pearson theory which is based on the same restriction $A = \{a_H, a_A\}$. We assume that $a_{H(A)}$ is the right decision if $\theta \in \Theta_{H(A)}$ and that $w_{H,a(H)} = w_{A,a(A)} = 0$. (A generalization is possible but not elaborated upon for the sake of simplicity.) The result of Lehmann quoted in Section 1 implies that a direct approach on the basis of Wald's minimax risk principle leads to the optimal Neyman–Pearson level–$w_{A,a(H)}/(w_{H,a(A)} + w_{A,a(H)})$ test as the uniformly minimum risk procedure in the class of all minimax rules. Thus action a_A is chosen if and only if the hypothesis H is rejected in favour of the alternative A on the basis of the data x and at the above–mentioned nominal level of significance. This, however, is equivalent to the condition that the P–value $\alpha_{\mathrm{P}}(x)$ is smaller than the indicated level of significance. But this is equivalent to

$$w_{H,a(A)} \alpha_{\mathrm{P}}(x) < w_{A,a(H)}(1 - \alpha_{\mathrm{P}}(x)).$$

Hence, if $A = \{a_H, a_A\}$ consists of *two* decisions, then the optimal two–decision procedure $d: \mathcal{X} \to A$ corresponds to the two–stage approach of Figure 1 provided that $\alpha(x) = \alpha_{\mathrm{P}}(x)$ is the classical P–value.

This decision–theoretic justification of P–values is not compelling if we are interested in procedures $\alpha: \mathcal{X} \to [0,1]$ for their own sake, or for the purpose of solving problems with $\#(\mathcal{A})\geq 3$. On the other hand it is satisfying to know that P–values lead to optimal procedures if $\#(\mathcal{A})=2$. Note, however, that the optimum property of having uniformly minimum risk among the minimax procedures is nice but not compelling either because there are many situations where minimax risk procedures are not attractive from an overall point of view.

What we need is a criterion for comparing procedures $\alpha:\mathfrak{X}\rightarrow[0,1]$ for weighting the evidence. For that purpose we focus on the idea that the indicator function $I_H(\theta)$ of Θ_H should be estimated *with squared error loss.* Procedures $\alpha:\mathfrak{X}\rightarrow[0,1]$ characterized this way are called Q–values (Q resembling P and standing for Quadratic). Imposing the loss function

$$L(\theta,a)=(I_H(\theta)-a)^2 \quad (\theta\in\Theta,\ a\in[0,1])$$

has the effect that the performance of a Q–value α is characterized by the risk function

$$R_\alpha(\theta)=R(\theta,\alpha)=\begin{cases} E_\theta(1-\alpha(X))^2 & \text{if } \theta\in\Theta_H \\ E_\theta(\alpha(X))^2 & \text{if } \theta\in\Theta_A \end{cases}$$

We propose to judge procedures for weighting the evidence, e.g. α_{ML}, α_{MLR} or α_P, by means of their risk function. By doing so (see Figures 3 and 4) we give an impression of the methodological uncertainty involved in the ultimate choice of such procedure $\alpha:\mathfrak{X}\rightarrow[0,1]$. Similar uncertainties are experienced by the Bayesian when he contemplates about the choice of a prior distribution τ in situations where he knows almost nothing about θ. Such discussions of methodological uncertainty are not alien to Neyman's nondogmatic approach to statistics.

The loss function requires some explanation. Note that this loss function is *proper* in the sense that the Bayesian solution

$$\alpha_\tau(x)=P(T\in\Theta_H|X=x)$$

is obtained if one postulates a prior distribution τ and minimizes the Bayes risk

$$r(\tau,\alpha)=\int R(\theta,\alpha)d\tau(\theta).$$

(The family of all loss functions with this property of properness will be characterized elsewhere). We hope that the Bayesians and the objectivists will agree that it is interesting to study procedures $\alpha:\mathfrak{X}\rightarrow[0,1]$ for weighting the evidence from the Q–value point of view, i.e. by making comparisons on the basis of the risk functions indicated.

3. RESTRICTING THE CLASS OF Q-VALUES

We are interested in situations where the type-I assumptions are satisfied: (1) $E_\theta\varphi(X)$ is a continuous function of θ for each test function $\varphi:\mathfrak{X}\to[0,1]$, (2) the common boundary $\Theta_B=[\Theta_H]\cap[\Theta_A]$ is nonempty. It is obvious that $I_H(\theta)$ does not admit an unbiased estimator: $E_\theta\alpha(X)=I_H(\theta)$ cannot be satisfied because the left-hand side is a continuous function of θ whereas the right-hand side is discontinuous.

DEFINITION: The Q-value $\alpha:\mathfrak{X}\to[0,1]$ is said to be

(1) *Fisher unbiased* ($\alpha\in D_F$) if $\alpha(X)$ has the uniform distribution on $[0,1]$ for all $\theta\in\Theta_B$,

(2) *Reflection unbiased* ($\alpha\in D_R$) if $\theta\in\Theta_B$ implies that the distributions of $\alpha(X)$ and $\varphi(X)=1-\alpha(X)$ coincide,

(3) *Similarity unbiased* ($\alpha\in D_S$) if $E_\theta\alpha(X)=\frac{1}{2}$ for all $\theta\in\Theta_B$,

(4) *Lehmann unbiased* ($\alpha\in D_L$) if

$$R(\theta,\alpha)=E_\theta L(\theta,\alpha(X))\le E_\theta L(\theta',\alpha(X))\ \ (\theta,\theta'\in\Theta).$$

It is easy to see that (1) $\alpha\in D_F$ implies $R(\theta,\alpha)=3^{-1}$ if $\theta\in\Theta_B$, (2) the procedure $\alpha_0\equiv\frac{1}{2}$ is minimax in the sense that $\inf_\alpha\sup_\theta R(\theta,\alpha)=R(\theta,\alpha_0)=4^{-1}$ $(\theta\in\Theta)$ (3) $D_F\subset D_R\subset D_S$, (4) $\alpha\in D_S$ if and only if α is a continuous-risk procedure, i.e. $R(\theta,\alpha)$ is a continuous function of θ, (5) $\alpha\in D_L$ if and only if $\varphi=1-\alpha$ is unbiased size$-\frac{1}{2}$ for problem (H,A), (6) $D_L\subset D_S$. It is less easy to see that *classical P-values have uniformly minimum risk if the attention is restricted to the class $D_{F,I}=D_F\cap D_I$ of Q-values.* Here D_I is a suitably chosen class of invariant (or equivariant, see Section 4) estimators of $I_H(\theta)$. The restrictions by Fisher unbiasedness or reflection unbiasedness are not satisfactorily motivated if one adopts the Q-value point of view. If the distribution of X is discrete for $\theta\in\Theta_B$ then $D_F=\phi$. Thus our main focus is on D_S. In this respect the following result is crucial.

FUNDAMENTAL LEMMA. The convex optimization problem

$$\textit{Minimize}\ \int\{\alpha(x)\}^2\ g(x)d\mu(x)$$

$$\textit{Restrictions}\ \ 0\le\alpha(x)\le 1\ \ (x\in\mathfrak{X}),\ \int\alpha(x)\ f(x)\ d\mu(x)=\tfrac{1}{2}$$

has a solution of the form $ch\wedge 1$. Here $h=f/g$ $(f,g\ge 0$ is assumed). The constant c has to be determined such that $\int(ch\wedge 1)f\ d\mu=\frac{1}{2}$.

About the proof. The first author obtained the result by applying Hilbert-space methods, guessing that the solution is of the form $ch\wedge 1$, and adapting the proof of the Neyman-Pearson Fundamental Lemma. The result was obtained independently by Klein Haneveld (Operations Research).

4. $X \sim N(\theta,1)$; $H{:}\theta<0$ VERSUS $A{:}\theta>0$

Suppose that $I_H(\theta)=I_{(-\infty,0)}(\theta)$, or $I_{(-\infty,0)}(\theta)+\frac{1}{2}I_{\{0\}}(\theta)$, has to be estimated with squared error loss on the basis of the outcome x of an $N(\theta,1)$ variate X, where $\theta\in\Theta=\mathbb{R}-\{0\}$, or $\mathbb{R}$. The classical P–value is given by

$$\alpha_{\mathrm{P}}(x)=1-\Phi(x).$$

The maximum–likelihood estimator is given by

$$\alpha_{\mathrm{ML}}(x)=\begin{cases}1 & \text{if } x<0\\ 0 & \text{if } x>0\end{cases}.$$

Using the modified likelihood–ratio principle one obtains

$$\alpha_{\mathrm{MLR}}(x)=\begin{cases}\exp\,(\frac{1}{2}x^2)/\{\exp(\frac{1}{2}x^2)+1\} & \text{if } x<0\\ 1/\{\exp(\frac{1}{2}x^2)+1\} & \text{if } x>0\end{cases}$$

The problem is invariant under the transformations $x\to -x$, $\theta\to-\theta$, $a\to 1-a$. Let D_{I} denote the class of equivariant Q–values, i.e. those which satisfy $\alpha(x)+\alpha(-x)=1$ for all $x\in\mathbb{R}$. Restricting the attention to D , the envelope risk

$$R_{\mathrm{I}}^*(\theta)=\inf_{\alpha\in D_{\mathrm{I}}} R(\theta,\alpha)$$

can be obtained by constructing the class $C_{\mathrm{I}}=\{\alpha_\eta;\eta\in[0,\infty)\}$ of SMR (Somewhere Minimum Risk) (D_{I}) procedures. The Q–value α_η with minimum risk in η among all equivariant Q–values is obtained as the Bayes rule with respect to $\tau=\frac{1}{2}\delta_\eta+\frac{1}{2}\delta_{-\eta}$. We have

$$\alpha_\eta(x)=e^{-\eta x}/(e^{-\eta x}+e^{\eta x}).$$

The envelope risk $R_{\mathrm{I}}^*(\theta)=R(\theta,\alpha_\theta)=R(-\theta,\alpha_\theta)=R_{\mathrm{I}}^*(-\theta)$ is displayed as a function of $\theta\in[0,\infty)$ in Figure 2 (the solid line). It is of some interest to

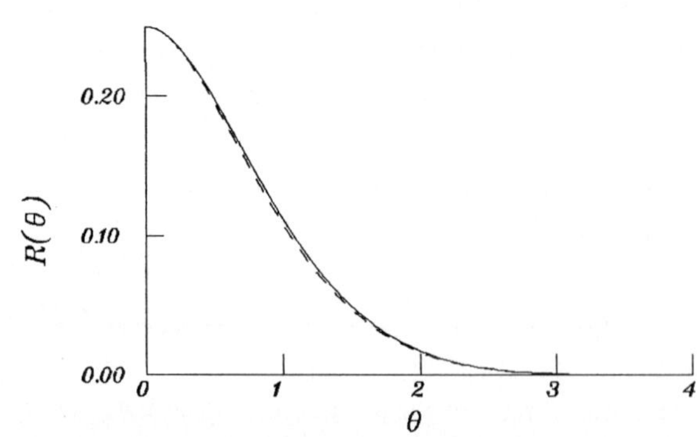

Figure 2. The envelop risks of Section 4 (solid line) and Section 5 (broken line).

note that $\int R_I^*(\theta)d\theta = .5088$. (Integration from $-\infty$ as to $+\infty$ by numerical methods.) Henceforth Q-values $\alpha \in D_I$ are characterized by their regret function (or shortcoming) $S(\theta,\alpha) = R(\theta,\alpha) - R_I^*(\theta)$.

We are interested in minimizing the area under the graph of the regret function and, of course, also in minimizing the maximum. As

$$\int_{-\infty}^{+\infty} S(\theta,\alpha)d\theta = \int_{-\infty}^{\infty} R(\theta,\alpha)d\theta - \int_{-\infty}^{\infty} R_I^*(\theta)d\theta,$$

the minimization of the area can equally well be carried out for the risk function. Note that

$$r(\lambda,\alpha) = \int_{-\infty}^{\infty} R(\theta,\alpha)d\theta - \int_{-\infty}^{\infty} [\{1-\alpha(x)\}^2\{1-\Phi(x)\} + \{\alpha(x)\}^2\Phi(x)]dx$$

is obviously as small as possible if the integrand is minimized for all x separately. By doing so, the classical P-value α_P is recovered. This is not surprising from the Bayesian viewpoint because using Lebesgue measure as an improper prior, one arrives at N $(x,1)$ as the corresponding posterior distribution of θ and $\alpha_P(x) = 1-\Phi(x)$ as the posterior probability of a negative outcome of θ. Note that α_P is admissible in the decision-theoretic sense while $r(\lambda,\alpha_P) = \pi^{-\frac{1}{2}}$, $r(\lambda,\alpha_{ML}) = 2^{\frac{1}{2}}\pi^{-\frac{1}{2}}$, $r(\lambda,\alpha_{MLR}) - r(\lambda,\alpha_P) = \int \{\alpha_{MLR}(x) - \alpha_P(x)\}^2 dx = 0.116$ (numerical integration). In terms of the regret function we obtain

$$s(\lambda,\alpha_P) = .055, \quad s(\lambda,\alpha_{ML}) = .29, \quad s(\lambda,\alpha_{MLR}) = .17$$

if $s(\lambda,\alpha)$ is defined as the area under the graph of the regret function.

The minimization of the maximum of the regret function (the minimax risk procedure $\alpha_0 \equiv \frac{1}{2}$ is useless) can be achieved numerically by constructing the least favorable distribution. This turns out to be of the form $\kappa\delta_{-\eta} + (1-2\kappa)\delta_0 + \kappa\delta_\eta = \tau_{\kappa,\eta}$ where $\kappa \in (0,\frac{1}{2})$ and $\eta > 0$. By maximizing the corresponding Bayes regret $s(\tau_{\kappa,\eta}\ \alpha_{\kappa,\eta})$ we obtained $\kappa^* = .2378$, $\eta^* = 1.5987$. It was verified numerically that the regret function of the corresponding Bayes estimator assumes its maximum value in $-\eta^*$, 0, and η^* (see Figure 3). Hence the minimax regret equivariant Q-value α_{MRI} has been found.

It is easier to minimize the maximum of the regret function if the attention is restricted to the class $C_I = \{\alpha_\eta; \eta \in [0,\infty)\}$ of SMR (D_I) procedures. The corresponding procedure α_{MRSMRI} is obtained for $\eta = .4240$ (see Figure 3).

The estimators α_{ML} and α_{MLR} were introduced on the basis of intuitive considerations. The following argument may also be of interest. If $|\theta|$ were known then we would use α_η with $\eta = |\theta|$. As we do not know $|\theta|$, we may replace θ in $|\theta|$ by its maximum likelihood estimate x. We obtain the *intuitive* Q-value

$$\alpha_{INT}(x) = \exp(-x|x|)/\{\exp(-x^2) + \exp(x^2)\}$$

(Note that $|X|$ is not the maximum likelihood estimator of $\eta=|\theta|$ if a reduction by invariance to $|X|$ or X^2 is applied. Note that α_{INT} and α_{MLR} have a similar functional form.) Yet another type of optimum property is obtained if one

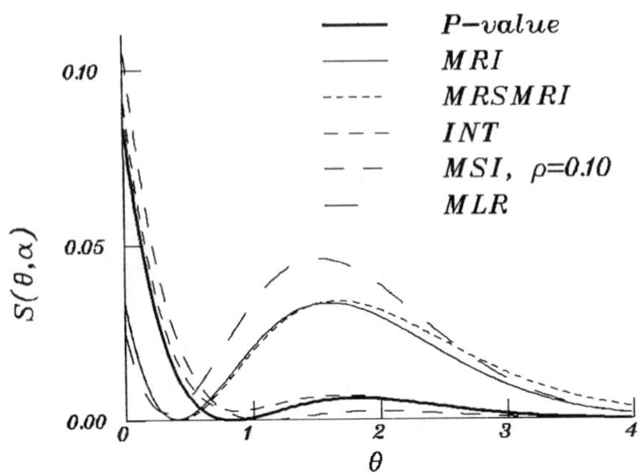

Figure 3. Regret functions of various Q–values for discussing whether θ is negative or positive.

considers the set $V_\rho(\alpha)=\{\theta;R(\theta,\alpha)>\rho\}$ of all parameter values where the risk is larger that some lower bound $\rho\in(0,4^{-1})$, e.g. $\rho=.10$. The *minimal set* (D_{I}) rule $\alpha_{\mathrm{MSI},\rho}$ is defined by minimizing $V_\rho(x)$, in the set-theoretic sense, among all Q–values in the class D_{I}. It can easily be established that this rule is of the form α_η where η is defined by $R^*_{\mathrm{I}}(\eta)=\rho$.

Conclusion. We conclude from Figure 3 that the mathematical statistician is confronted with considerable methodological uncertainty if he is asked to propose a Q–value. As θ is completely unknown in practice there is nothing else to rely on. Mathematical statisticians are supposed to be the experts in statistical uncertainty and that implies that they should carry the main responsibility for cutting this Gordian knot. The authors recommend α_{P} because this is suggested by Figure 3, at least to some extent, and because this is in line with (1) the Bayesians who use Lebesgue measure as uninformative improper prior and (2) the objectivists who recommend P–values.

5. X~N(θ,1); H:θ=0 VERSUS A:θ>0

Suppose that $\mathrm{I}_H(\theta)=\mathrm{I}_{\{0\}}(\theta)$ has to be estimated with squared error loss on the basis of the outcome x of an N(θ,1) variate X, where $\theta\in\Theta=[0,\infty)$. The classical P–value $\alpha_{\mathrm{P}}=1-\Phi$ is the same as in Section 4. This is not very satisfactory from the Q–value point of view because "the estimator should be affected if $(-\infty,0)$ is replaced by $\{0\}$". Note that $\alpha_P \in D_F \subset D_S$ and that α_P is uniformly best in the class D_F. This result is not very compelling because Fisher-unbiasedness is very restrictive and not well motivated from the Q-value point of view. The class D_S is much more attractive, especially if optimal (D_S) procedures happen to be Lehmann-unbiased, i.e. such that $E_0\alpha(X)=\frac{1}{2}$ and $E_\theta\alpha(X)<\frac{1}{2}$ for all $\theta>0$. It is obvious that some type of requirement is needed. We focus on the class D_S of continuous-risk procedures.

The corresponding envelope risk R_S^* can be obtained by constructing the class $C_S=\{\alpha_\eta;\eta\geq 0\}$ of SMR (D_S) procedures. The Q-value α_η with minimum risk in η is obtained by applying the Fundamental Lemma of Section 3. We have

$$\alpha_\eta(x)=ce^{-\eta x}\wedge 1$$

where c is such that $E_0\alpha_\eta(X)=\frac{1}{2}$. The envelope risk has been presented in Figure 2 (the broken line). As in Section 4, the performance of $\alpha\in D_S$ is characterized by the regret function $S(\theta,\alpha)=R(\theta,\alpha)-R_S^*(\theta)$. In Section 4 we were fascinated by α_P because it minimized the area under the risk as well as under the regret function. In the present situation these areas can be made smaller by leaving the class D_F. It seems reasonable to stay within the class D_S of continuous-risk procedures because the area can trivially be made equal to zero by using $\alpha\equiv 0$. The minimum area (D_S) procedure α_{MA} is obtained by applying the Fundamental Lemma of Section 3 to the following optimization problem.

$$\textit{Minimize}\ \int_0^\infty \int_{-\infty}^\infty \{\alpha(x)\}^2\ \varphi(x-\theta)dx\ d\theta=\int_{-\infty}^\infty \{\alpha(x)\}^2\ \Phi(x)dx$$

$$\textit{Restrictions}\ 0\leq\alpha(x)\leq 1,\ \int_{-\infty}^\infty \alpha(x)\ \varphi(x)dx=\tfrac{1}{2}.$$

The solution to this problem is given by

$$\alpha_{MA}(x)=\{c(\varphi/\Phi)\}\wedge 1$$

where c has to be determined such that $E_0\alpha_{MA}(X)=\frac{1}{2}$. We have

$$\int_0^\infty R(\theta,\alpha_P)d\theta=\tfrac{1}{2}\pi^{-\frac{1}{2}}=.2821>.2803=r(\lambda,\alpha_{MA}).$$

The difference becomes more spectacular if we subtract the area under the envelope risk $R_\mathcal{E}^*$ which is about .25 (see Figures 2 and 4). As in Section 4 we have constructed the minimax regret element α_{MRSMR} in the class C_S of SMR(D_S) procedures. The construction of the minimax regret (D_S) Q-value is beyond our capabilities. Regret functions were plotted in Figure 4.

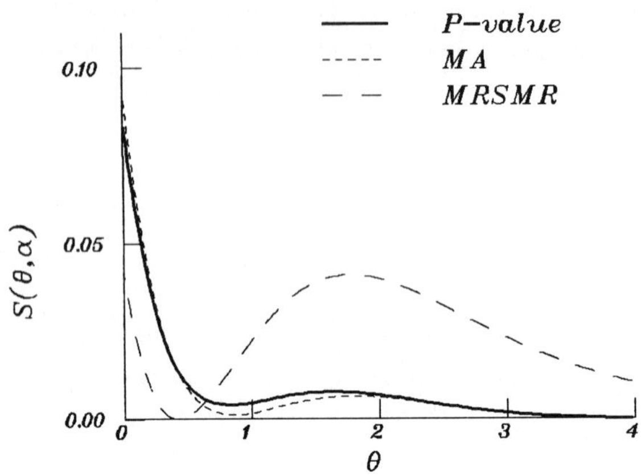

Figure 4. Regret functions of various Q–values for testing $\theta=0$ against $\theta>0$.

Conclusion. As in Section 4, the statistician is confronted with considerable methodological uncertainty if he has to recommend a procedure $\alpha:\mathbb{R}\to[0,1]$ for weighting the evidence. We recommend the Q–value α_{MA} though we will not object if somebody else recommends the classical P–value α_P (or some of the other procedures to be mentioned in the forthcoming POSCON book).

6. SOME ASYMPTOTIC THEORY

Constructing a Q–value is somewhat peculiar from an estimation point of view. The parametric function $I_H(\theta)$ to be estimated is locally constant if $\theta\notin\Theta_B$ and discontinuous for $\theta\in\Theta_B$. The usual requirement of unbiasedness makes no sense and the Q–values constructed are not such that $n^{\frac{1}{2}}(\alpha_n - I_H(\theta))$ has a nondegenerate limiting normal distribution. Nevertheless we would like to prove that the differences of opinion among the Bayesians and the objectivists tend to vanish in a very strong sense if the data accumulates. We, statisticians, accept the idea that Nature or The Data should dictate our opinions. There are some difficulties if the sample size is small because we have to choose a method to let the data speak. The methodological uncertainties involved should vanish if the sample size is increased. This, at least, is something we hope for. These hopes come true in situations similar to that of Section 4. The situation is less favorable if the truth or falsity of a point–null hypothesis has to be discussed. In the present section we try to give an impression of the general theory to be developed. We focus on the following extension of the problem of Section 4.

Let $X_1,X_2,\ldots$ be independent and identically distributed random variables, $X_i \sim N(\theta,1)$ as in Section 4, $\theta\in\Theta=\mathbb{R}-\{0\}$, $\Theta_H=(-\infty,0)$, $\Theta_A=(0,\infty)$. Suppose that the outcome of $X_1,\ldots,X_n$ is available and that the truth or falsity of $H:\theta<0$ has to be discussed on the basis of this outcome. Let $\alpha_n:\mathbb{R}^n\to[0,1]$ be the Q–value of choice. We are interested in asymptotic properties of the sequence $\{\alpha_n;n=1,2,\ldots\}$. It is in line with Section 4 to be interested in the asymptotic behaviour of the area $r_n(\lambda,\alpha_n)$ under the risk curve. It is also interesting, especially from a Bayesian point of view, to study the asymptotic behaviour of $r_n(\nu,\alpha_n)=\int R_n(\theta,\alpha_n)d\nu(\theta)$ where ν is some proper prior. Finally it is interesting to study the rate at which $R_n(\theta,\alpha_n)$ tends to 0 for $\theta\neq 0$ fixed.

The specific choice of $\alpha_n:\mathbb{R}^n\to[0,1]$ depends on the philosophy of the statistician. The objectivist will argue that $X=n^{-\frac{1}{2}}(X_1+\ldots+X_n)$ is a complete sufficient statistic if $(X_1,\ldots,X_n)$ is observed and that the problem of estimating $I_H(\theta)$ reduces to that of Section 4. (For the Bayesian the situation is different because his prior changes if θ is replaced by $\theta n^{\frac{1}{2}}$.) Hence it is reasonable for the objectivist to focus on sequences $\{\alpha_n;n=1,2,\ldots\}$ satisfying

$$\alpha_n(x_1,\ldots,x_n)=\alpha(n^{-\frac{1}{2}}(x_1+\ldots+x_n))$$

Theorem 1. If the above-mentioned condition is satisfied, then

$$\lim_{n\to\infty} r_n(\lambda,\alpha_n)n^{\frac{1}{2}}=r_1(\lambda,\alpha_1)\geq r(\lambda,1-\Phi)=\pi^{-\frac{1}{2}}$$

If the density f of a prior $d\nu=f(\theta)d\lambda$ is bounded, and continuous at 0, then

$$\lim_{n\to\infty} r_n(\nu,\alpha_n)n^{\frac{1}{2}}=f(0)\ r_1(\lambda,\alpha_1)\geq f(0)\ r_1(\lambda,1-\Phi)=\lim_{n\to\infty}\ r_n(\nu,\alpha_{\mathrm{P},n})n^{\frac{1}{2}}$$

where $\alpha_{\mathrm{P},n}(x_1,\ldots,x_n)=1-\Phi\{n^{-\frac{1}{2}}(x_1+\ldots+x_n)\}$.

The proof of this theorem follows from

$$n^{\frac{1}{2}}r_n(\nu,\alpha_n)=n^{\frac{1}{2}}\int R(\theta n^{\frac{1}{2}},\alpha_1)d\nu(\theta)=\int R(\theta,\alpha_1)\ f(\theta n^{-\frac{1}{2}})d\theta\to\int R(\theta,\alpha_1)f(0)d\theta$$

If we have prior information τ and choose the Bayes procedure $\alpha_{n,\tau}=d_{1,\tau_n}$ where τ_n is the distribution of $\theta n^{\frac{1}{2}}$ if θ has distribution τ, then it turns out that all Bayes procedures have the same asymptotic risk as that of the objectivist's procedure. We state this assertion for normal priors.

Theorem 2. If $\tau=N(\mu,\sigma^2)$ and $\nu=N(\eta,\gamma^2)$ are normal priors, then

$$\lim_{n\to\infty} r_n(\nu,\alpha_{n,\tau})n^{\frac{1}{2}}=\lim_{n\to\infty} r_n(\nu,\alpha_{n,\nu})n^{\frac{1}{2}}=\lim_{n\to\infty} r_n(\nu,\alpha_{\mathrm{P},n})n^{\frac{1}{2}}=(2\pi\gamma^2)^{-\frac{1}{2}}\ r_1(\lambda,1-\Phi)=(2^{\frac{1}{2}}\pi\gamma)^{-1}$$

The proof will be given in the forthcoming POSCON book.

With respect to the speed of convergence towards 0 of the risk function at a fixed alternative we state a result of Dehling–Schaafsma (1989) about the construction of Q–values if, e.g., $\Theta_H=\{0\}$ and $\Theta_A=\{\theta\}$ are simple. (They do not focus on classes of unbiased procedures.)

Theorem 3. $\lim_{n\to\infty} n^{-1}\log \inf_{\{\alpha_n;R_n(0,\alpha_n)=c\}} R_n(\theta,\alpha_n) = -\frac{1}{2}\theta^2$

Natural generalizations of this result contain the Kullback–Leibler number $-K(P_\theta,P_0)$ as the right-hand side.

7. PRACTICAL IMPLICATIONS?

One of our clients, the pediatrician R.Y. Tamminga, is involved in the care for children with cancer. One of his interests is the growth of these children during therapy and thereafter. Like many of his colleagues he is also interested in the growth of these children before the diagnosis was made. This information is of interest to study the pathogenesis of the disease. Unfortunately measurements at diagnosis are the first ones available for a patient. Thus Tamminga studies whether the distribution of , e.g., height at diagnosis differs from that of healthy children of the same age. For that purpose and for some other ones, he uses the growth curves presented by the pediatrician W.J. Gerver. The data in Table 1 has to be used to establish whether statistically significant differences exist between the two diagnostic categories indicated. The meaning of these figures is explained in a minute.

	leukaemia			Non–leukaemia						
Variable	n_1	m_1	s_1	n_2	m_2	s_2	s	t	P	Q
1. Height	44	.26	1.16	50	−.07	1.17	1.17	1.37	.17	.27
2. Arm span	23	.82	1.21	32	.31	1.18	1.19	1.59	.11	.20
3. Sitting h.	39	.22	1.20	40	−.07	1.10	1.15	1.12	.26	.38
4. Head circ.	45	.04	.89	51	−.42	1.13	1.02	2.20	.03	.06

Table 1. Data of R.Y. Tamminga (Academic Hospital Groningen)

The original scores for height, etc., were transformed into standard–deviation scores by using Gerver (1988). If a child is taken at random from the overall population and measured at an arbitrary age t during its development then the corresponding standard–deviation score follows the $N(0,1)$ distribution, at least approximately. Table 1 presents sample sizes (n), means (m), and estimated standard deviations ($s = \{(n-1)^{-1}\Sigma(x_i - m)^2\}^{\frac{1}{2}}$) for the two diagnostic categories indicated. Using the pooled estimate of the common standard deviation (s) the outcome t of Student's two-sample statistic is computed. The P-values are based on the exact distribution of Student's t. The Q-values are obtained by computing

$$\alpha_{MA}(t) = 2^{-\frac{1}{2}} \exp(-\tfrac{1}{2}\, t^2)$$

because $\alpha_{MA}: \mathbb{R} \rightarrow [0,1]$ is the minimum area procedure if the attention is restricted to the class D_S of continuous-risk procedures for estimating $I_{\{0\}}(\theta)$ on the basis of the outcome t of $T \sim N(\theta,1)$. The optimality of α_{MA} can be established by applying the Fundamental Lemma of Section 3. Should we confront Tamminga with these Q-values? Two different lines of thought are as follows.

(1) If the main aim of the analysis is to decide whether or not the results for the separate measurements are statistically significant at the level $\alpha = .05$, then one should base this decision on whether or not the P-value is smaller that $\alpha = .05$ because this is in line with the Neyman-Pearson theory.

(2) If the main aim is to discuss the truth or falsity of the hypothesis that no differences exist between the two groups if the indicated measurements are considered, then there exist reasons to prefer the Q-values.

The main idea is, however, that considerable methodological uncertainty is generated if one tries to consider procedures $\alpha: \mathfrak{X} \rightarrow [0,1]$ for weighting the evidence. We recommend Q-values like α_{MA} because (1) we consider it appropriate to minimize the area under the risk and regret function, (2) we like to require that $R(\theta,\alpha)$ is continuous as a function of θ, (3) we do not like to require that α should be Fisher-unbiased. We will not make objections, however, if somebody else recommends a P-value, a multiple decision approach, or a Bayesian approach (provided that τ is not too weird).

REFERENCES

[1] Dehling, H. and Schaafsma, W., Discussing truth or falsity of a statistical hypothesis, simple versus simple revisited, (1989), in preparation, Dept. Math. Groningen University.

[2] Gerver, W.J.M., Measurements of the body proportions in childeren, the Oosterwolde study (Thesis, University of Groningen, 1988).

[3] Lehmann, E.L., Testing Statistical Hypotheses (Wiley, New York, 1959, 1986).

[4] Rao, C.R., Advanced Statistical Methods in Biometric Research (Wiley, New York, 1952).

[5] Reichenbach, H., The Theory of Probability (Un. California Press, 1949).

[6] Savage, L.J., Elicitation of personal probabilities and expectations, Journ. Am. Stat. Ass. (1971) 783–801.

[7] Schaafsma, W., Hypothesis testing problems with the alternative restricted by a number of inequalities (Noordhoff, Groningen, 1966).

[8] Schaafsma, W., Minimax risk and unbiasedness for multiple decision problems of type I, Ann. Math. Statist. (1969), 1684–1720.

SUMMARY

The truth or falsity of a statistical hypothesis $H:\theta\in\Theta_H$ *can be discussed by estimating the indicator function* $\mathrm{I}_H(\theta)$ *of* Θ_H *with squared error loss. Classical* P–*values, posterior probabilities and other* Q–*values may be compared on the basis of their mean squared error. This may indicate that some procedure is more reasonable than another one. Uniformly best procedures, however, will not exist unless unreasonable restrictions are imposed.*

STATISTICAL DATA ANALYSIS AND INFERENCE
Y. Dodge (Editor)

LINEAR PIVOTALS AND THE BAYES- NON-BAYES COMPROMISE

G. A. Barnard, University of Essex, U. K.

Changes in attitudes to Bayesian assumptions during this century are reviewed. Attention is drawn to the fact that the "incompatibilities" between Fisherian inferences and Bayesian inferences are less than has often been thought. The properties of linear pivotals and their associated likelihoods are indicated as having special relevance to a Bayes- non-Bayes compromise.

The attainment of three score years and ten is an occasion for retrospect and prospect. In presenting some results of my own reflections, having passed that stage, I hope I may encourage our distinguished dedicatee to proffer more of his own. The need for such reflections has appeared more urgent since closer contact with areas of application of statistics, particularly in medicine, has impressed upon me the extent to which there is a lack of understanding of the statistician's roles.

The past twenty five years has seen a marked revival of interest in Bayesian approaches to statistical inference. Advocates of such approaches have sometimes exhibited almost missionary zeal in their revolt against some "classical" procedures. Emotive words such as "incoherent" have been given technical meanings in a way reminiscent of Neyman's use of "biased" to describe some very useful tests of significance such as the F test. While the eloquence and depth of thought of the pioneers of the current movement served very well to get rid of a degree of doctrinal rigidity that had come to infect a great deal of statistical practice, we have now surely had enough of sharp controversy and we should now turn our thoughts towards integration. The areas of application of statistical methods are now so wide as to call for a correspondingly wide variety of approaches whose relationships to each other are worth exploring further. I am thus pursuing a theme which has already been pressed by Box(1980) and by Good (1989).

It seems worth while to explore how the "classical" modes of statistical inference came to be regarded as so strongly "anti-Bayesian". Why were many leading figures of the early modern period so anxious to avoid any taint of "a priori" probabilities in their work? Were there any features special to their historical situation and their special interests which led them to take the positions they adopted?

Under the influence of Quetelet, Galton, Fechner, and others, the latter part of the XIX Century saw a great increase in the amount of data available concerning relative frequencies in existent populations, and a corresponding increase in the mathematical apparatus available for describing these frequencies. The rediscovery of Mendel's work at the turn of the century was the first occasion on which a scientific theory was propounded in which the results of repeatable experiments were described in terms of probability, interpretable as frequency in an indefinitely large population. The resulting concept, which we would now call "experimental probability" or, in Jeffreys' term, "chance", was in fact markedly different from the notion involved in what had come to be called the "equal distribution of ignorance"(EDOI). This was the idea that, confronted with a set of alternatives concerning which nothing was known to give any "sufficient reason" for preferring any one alternative to any other, one should assign equal probabilities to all alternatives. In Karl Pearson's work extending Galton's studies of correlations between relatives in human populations it was born in upon him that such correlations departed markedly from the uniform distribution between -1 and +1 which EDOI would seem to imply. Pearson's mistaken notion that Fisher's maximum likelihood method derived from EDOI led to the first serious break between these two great men.

R.A. Fisher exerted by far the greatest influence on the subject from the early 1920's until around 1940, when his dominance was challenged, particularly in the United States, by J. Neyman. Neyman gave up his original "Bayesian" views in the early 1930's, under the influence of E, S. Pearson who himself was quite strongly influenced by Fisher. Why did Fisher attack the Bayesian approach so strongly? He gives us clues in Chapter II of his "Statistical Methods and Scientific Inference". Of special importance, it seems to me, is the fact that in the midst of a review of the attacks by

Boole and Venn on the illegitimate use of the "equal distribution of ignorance" he gives an example of the application of statistical methods in the Mendelian theory. He considers a black mouse B which may be homozygous for coat colour, H, or heterozygous, H'. B is mated with a brown mouse and the result E is that all 7 offspring are black. From Mendel's laws we have $Pr\{E|H\}=1$, while $Pr\{E|H'\}=1/2^7$. If nothing is known about B's ancestry, EDOI would suggest that the prior probabilities of H and H' should each be taken as ½, so that, given E, the probability of H is 128/129. If, on the other hand, it is known that B was itself the offspring of a mating of a heterozygous black with a brown mouse, then the Mendelian laws would imply that H amd H' were *known to be* equally probable, and given E the probability of H would again be 128/129. The scientific attitude to the number 128/129 in the two cases would be very different. As Fisher says, "if knowledge of the origin of the mouse were lacking, no experimenter would feel he had warrant for arguing as if he knew that of whjich in fact he was ignorant."In the absence of knowledge B's provenance, on the data E alone, the evidence in favour of H would be measured by the *likelihood ratio* 128/1 in favour of H. Without further observational data this likelihood ratio would not be convertible to a probability.

To stress the difference implied here let us call black mice which are known to be offspring of a heterozygous black with a brown mouse, K-black mice; and we call black mice of unknown provenance, U-black mice. (K for Known, U for unknown). From a large collection of K-blacks we select those, the KE-blacks, which produced just 7 blacks when mated with brown mice, and similarly select the UE-blacks from a large collection of U-blacks. Now we mate each of these mice with a brown mouse. Mendel's theory asserts (assuming no differential viability etc.) that the probability of brown mice among the KE-black offspring is 1/258. It makes no such prediction concerning the U-blacks. This implies, it seems to me, that we must differentiate in some way between probabilities derived from EDOI and those derived from Medelian theory.

Fisher was a natural scientist who made major contributions to genetical problems such as the disentangling of the Rhesus blood groups and to geophysical problems connected with plate tectonics. But the reason why Dawkins, for example, judges him to have been the greatest biologist of the century, is Fisher's work establishing Darwin's theory of natural and sexual selection as completealy compatible with Mendelism. These two theories form the core of biology and are the basis of the tremendous developments we continue to see in that field. For the first quarter of this century Darwinism was the subject of bitter attack from many philosophers and theologians in addition to some scientists. Fisher did this purely scientific work while having to support a large family by solving problems in agriculture. The personal sacrifices involved were very great indeed. The Wordsworth quotation which regularly appeared on the masthead of "Nature" in those days: "To the solid ground of Nature/Trusts the mind that builds for aye" would have held a deep significance for him and he would have rejected the notion that the picture of the natural world which he was helping to build involved any irreducible subjective element. Probability, for him, was experimental probability, the concept involved in the Mendelian theory and in the later developed theory of quantum mechanics. Uncertainty of a kind necessarily attached to any scientific hypothesis, but this uncertainty was not measurable in terms of experimental probability. It could sometimes be measured, *objectively*, though in a relative sense, in terms of his concept of likelihood.

Fisher's attempts to interpret his concept of experimental probability in terms of relative frequency in an "hypothetical infinite population" were justifiably criticised by Jeffreys. In the post-Wittgenstein era Fisher might have argued that the meaning of his concept was to be found in the use made of it in quantum physics and in genetics. But the period around 1920, when Fisher's ideas were in their major formative phase, was one in which the many different forms of uncertainty which we now can perceive as distinguishable were all lumped together under the heading of "probability". In particular, Jeffreys was beginning his major work on the theory of scientific method in which he formulated his simplicity postulate in terms of a "probability" concept encompassing both experimental probability (which he later came to call "chance") and a notion closely related to Fisher's likelihood. Fisher was later able to return Jeffreys criticism with criticism of his own by equating Jeffreys' notion of probability with his experimental probability. The mutual criticisms were friendly and good humoured on both sides, as their correspondence shows. The geneticist had great respect for the geophysicist, and vice versa. Indeed, as Jeffreys said when interviewed by Lindley, when he and Fisher agreed (as, for example, over the Behrens-Fisher problem) they felt sure they were right; and when they disagreed, then each was doubtful. Contrary to a widespread impression, Fisher's note (1938) on a paper by Jeffreys clearly indicates that, so far as practical procedures were concerned, there were very few differences between them.

Egon Pearson was one of the first to welcome Fisher's advances in theoretical statistics, though filial respect for his father meant that his expression of his welcome was restrained. That he followed Fisher in his rejection of Bayesian modes of argument fitted well with his main area of statistical application, in industry, where in the early days of mass production problems of industrial quality control naturally laid emphasis on the operating characteristics of statistical procedures. His great collaborator Neyman was initially prepared to adopt the traditional prior probability approach, but under Egon's influence he eventually abandoned this position and the final form of the Neyman-Pearson theory concentrates wholly on operating characteristics. But whereas Pearson was wont to say that their theory should be thought of as only one way of enabling users of statistics to "get their thinking into gear" with their practice, the temperamentally more assertive Neyman argued, contrary to Fisher, Jeffreys and Egon Pearson, that there could be no such thing as "statistical *inference*", only *"inductive behaviour"* this latter necessarily involving an *act of will*.

Neyman belonged to the remarkable group of Polish mathematicians in the 1920's grouped around the journal Fundamenta Mathematicae whose studies in logic, point set theory, measure theory and functional analysis laid the basis for much of modern pure mathematics. The publication in 1933 of Kolmogoroff's "Grundlagen der Wahrscheinlichkeitsrechnung" was seen as a final canonisation of probability theory into a respectable branch of pure mathematics, cleansed of the vagueness and ambiguities that had thus far beset it. It may be conjectured that Neyman regarded this as of the utmost importance* and this led him to attempt an account of statisticalinference using only the concept of probability as defined by Kolmogoroff, eschewing notions such as Fisher's likelihood. It may be significant also that Kolmogoroff's axiomatisation takes as its initial notion that of unconditional probability, defining conditional probability in terms of unconditional; it was not until 1955 when Renyi produced his work on conditional probability spaces that the conditional nature of all probabilities was fully incorporated into the mathematical theory. Meanwhile perhaps the most serious weakness of the Neyman-Pearson theory, and the source of most of the strong contradictions between it and Bayesian approaches, was its neglect of conditionality. Although, of course, it is possible to develop a theory of conditional inference using Kolmogoroff's axiomatisation, it comes more naturally via Renyi's formulation.

---

*It is noticeable that "sets of measure zero" and other such measure-theoretic terminology appear in Neyman's papers after 1933, but not before.

It is a great pity that Kiefer's work on conditional inference from a Neymannian point of view was cut short by his untimely death; had he been able to continue it there might have eventually been a considerable convergence, at least in the sphere of most applications.

The Ramsey-de Finetti concept of personal probability, adapted so skilfully for statisticians by Good, Savage, Lindley and others, introduced a new element into the debate. Natural science, as understood by Karl Pearson, Fisher and Jeffreys had as its main function the proper understanding of natural phenomena, applications being valuable incidentals. But after the second world war the majority of scientists were employed by Governments and commercial or industrial organisation whose aims were not primarily the better understanding of natural phenomena but rather the study of how such phenomena could be used to human advantage. The terms we use to denote the two kinds of scientific activity -- pure and applied science-- still, unfortunately, often carry with them a connotation of a lower status for applications. Betterment of the human condition is a no less lofty aim, nor one easier of achievement, than mere understanding of the natural world. As one of the older generation Fisher appeared to express a Brahminical view in (1955) though he was careful to hedge his statements. The essential point is, that in applications of science we make decisions which express our best judgements to the date at which decisions need to be made. We cannot, as we can in pure science, content ourselves with a weighing up of evidence without coming to a definite conclusion. In making decisions personal probability assessments may play a role which would not be allowable in pure science. Of course the distinction between pure and applied science can never be made absolute, so there can be no absoluteness about the distinction we are drawing here.

All this leads to the conclusion that there is a role for personal probability assessments in much scientific work, so long as the personal assessments are clearly recognised as such and properly allowed for.

If a "Bayes- non-Bayes" compromise, as Good has called it, comes to be accepted by statisticians generally, we may reach a state analogous to that which exist in pure mathematics. For much of the first half of this century, workers in the foundations of mathematics were engaged in sometimes bitter disputes as between the "formalists", the "logicists", and the "intuitionists". For the first group mathematics consisted in setting up formal systems in which a recursive definition of a well formed formula (wff) was given, along with a recursive definition of a derivable wff. The only requirement was, that the system should not be trivial in the sense that not all wff's should be derivable wff's. For the intuitionists, on the other hand, the results of mathematics should be based on our intuitive perception of counting numbers and of the continuum, and mathematical proofs should specify a way in which anyone possessing such perceptions could construct an entity corresponding to the mathematical result. There is now a broad consensus among mathematicians that each point of view provides useful insights into a complex area of mental activity with, for example, the intuitionist point of view being specially relevant to the construction of computer programmes and the formalist approach facilitating the assimilation of a large mass of mathematical theory.

It seems to me, however, essential that statisticians should distinguish between the types of uncertainty involved in whatever assumptions they make, or whatever conclusions they draw. Over the past century we have built up a reputation for objectivity which we need to be careful not to imperil. Even if we insist that all probabilities must be regarded as personal, some probabilities are more broadly shared than others, and are less liable to revision in the light of relatively small amounts of further information. Judgements of exchangeability, although in principle no less personal than other probabilistic judgements, do appear to command more general assent, so that models involving assumptions that observations are independent and identically distributed, without specifying what the common distribution is, are frequently adopted by Bayesians and non-Bayesians alike. So long as we carefully note the differential status of the assumptions we make - and the use of terms like "experimental probability", "likelihood", and "confidence" (perhaps, even, fiducial probability!) can help in this - there should be no reason why the advice we present to our fellow researchers should appear to be confusing.

Linear Pivotals and their Likelihoods

To illustrate one way in which reflections such as those above can be translated into practice, I would like to draw attention to work in the theory of estimation in which David Sprott and his students Steve Chamberlin and Ramon Viveros have played the major part.

Suppose we are given an n-vector $\mathbf{x}$ of observations from a population whose shape is known to be given by a density function $\varphi(\mathbf{p})$, but whose location and scale parameters λ and σ are unknown. The fact that these parameters are parameters of location and scale is expressed by the fact that

$$\mathbf{p} = (\mathbf{x} - \lambda \mathbf{1})/\sigma$$

To obtain pivotals for λ and σ we transform $\mathbf{p}$ to (s, t, c) where

$$\mathbf{p} = s(t\mathbf{1} + \mathbf{c})$$

and $\mathbf{1}'\mathbf{c} = 0$ and $\mathbf{c}'\mathbf{c} = n(n-1)$. Then taking s, t, and the first $n-2$ components of $\mathbf{c}$ as the new independent variables, the Jacobian of the transformation is $\Delta(\mathbf{c})s^{n-1}$ where $\Delta(\mathbf{c})$ is expressible as a determinant involving only $\mathbf{c}$. Thus the joint density of $(s, t, \mathbf{c})$ is

$$\Delta(\mathbf{c})s^{n-1}\varphi(s(t\mathbf{1} + \mathbf{c}))$$

while in terms of observations and parameters

$$s = s_x/\sigma\sqrt{n}, \quad t = (x. - \lambda)\sqrt{n}/s_x, \text{ and } \mathbf{c} = (\mathbf{x} - x.\mathbf{1})\sqrt{n}/s_x$$

where s_x denotes the usual sample standard deviation and $x.$ denotes the sample mean. Thus s is a pivotal for σ, t is a pivotal for λ, while $\mathbf{c}$ involves the observations only, and so is ancillary. The joint density of (s, t), conditional on $\underline{c}_o$, the observed value of $\mathbf{c}$, is then

$$Ks^{n-1}\varphi[s(t\mathbf{1} + \underline{c}_o)].$$

where K is a normalising constant.

The pivotal t is linear in its unknown parameter, and so is a linear pivotal. And s is convertible to a linear pivotal by putting

$$z = \ln s = \ln s_x/\sqrt{n} - \zeta$$

where $\zeta = \ln\sigma$. Since $ds = \exp z.dz$, the joint conditional density of (t, z) is

$$\psi(t, z) = K\exp\{nz - \ln\varphi[e^z(t\mathbf{1} + \underline{c}_o)]\}.$$

We <u>define</u> the likelihood for the parameters (λ, ζ), given by the linear pivotals t, z, to be

$$L(\lambda, \zeta) = \psi(t_o, z_o),$$

where t_o, z_o denote, respectively, the results of substituting the observed values $x._o$, s_{xo} for $x.$ and s_x in the expressions for t and z.

It will be found that $L(\lambda, \sigma)$ is equivalent to the likelihood $L^*(\lambda, \sigma)$ based on the original observations, i.e. that the ratio L/L^* does not involve the parameters. Since this is true for any $\mathbf{x}$, no information is lost in reducing the observations to t_o, z_o, together with the joint density $\psi(t, z)$. We can conventionally write the "estimation statement":

$$\lambda = x._o - (s_{xo}/\sqrt{n})t$$

$$\zeta = \ln s_{xo}/\sqrt{n} - z,$$

$$\psi(t, z) = K\exp\{nz - \ln\varphi[e^z(t\mathbf{1} + \underline{c}_o)]\}.$$

it being understood that we are hereby asserting that the relevant density of (t, z) is given by the expression for ψ, and that the observed values of $x.$ and s_x are as given.

For example we take the sleep differences (mis)quoted by "Student" to illustrate his t test, where a sample of n=10 observations, assumed i.i.d. normal, gave a mean $x. = +1.58$, with $s_x^2=1.513$, $s_x/\sqrt{n} = .3890$, $\ln .3890=-.9442$. The estimation statement is then:

$$\lambda = 1.58 - .3989t,$$

$$\zeta = -.9442 - z,$$

$$\psi(t, z) = K\exp 5\{2z - e^{2z}(t^2+9)\}$$

the term corresponding to $10\ln\sqrt{2\pi}$ in $\ln\varphi$ having been absorbed in the normalising constant K. This statement has a number of interpretations:

(i)the likelihood interpretation: The model and the data imply that

$$\ln L(\lambda, \zeta) = 5\{2(-.9442-\zeta) - \{[(1.58-\lambda)/.3989]^2 + 9\}\exp 2(-.9442-\zeta)\}$$

from which we find the maximum likelihood estimates of (λ, ζ) to be $(1.58, \ln\sqrt{(1.513)})$, giving the usual maximum likelihood estimate $\hat{\sigma}= \sqrt{[(9/10)1.513]} = 1.1669$.

If we are prepared to assume a prior density $\pi(\lambda, \zeta)$ for (λ, ζ) we can infer a posterior probability density

$$P(\lambda, \zeta) = K^* \pi(\lambda, \zeta)\exp L(\lambda, \zeta),$$

where K^* is determined, if necessary, by the condition that P should integrate to 1. From P, given a loss function $\Lambda(D, D')$, such as the squared distance $\delta^2(D, D')$, measuring the cost of judging the point (λ, ζ) to be at D when in fact it is at D', we can determine the optimum estimate of (λ, ζ). If we join Jeffreys in taking $\pi(\lambda, \zeta)$ constant as an "ignorance" or "non-informative" prior we obtain results like those of Jeffreys and Box and Tiao, and these will agree numerically with those in the confidence interpretation.

(ii)the confidence interpretation: If we determine K by the condition that $\psi(t, z)$ should integrate to 1, then given a Γ between 0 and 1 we can find $h(\Gamma)$ such that

$$\int_{\psi > h} \psi(t, z)\,dt\,dz = \Gamma$$

and then $\{(\lambda, \zeta): \psi(t_o, z_o)\}$ is a "shortest" Γ-confidence set for (λ, ζ).

If we compute

$$\chi(t) = \int\psi(t,z)dz$$

we can use $\chi(t)$ in a similar way to obtain confidence intervals for λ, ignoring ζ. Similarly for confidence intervals for ζ, or, equivalently, for σ. The functions ψ and χ serve in this way to generate confidence distributions.

(iii)the "significance test"interpretation: If hypothetical values (λ_o, ζ_o) entail values for $\psi(t_o, z_o)$ which are improbably small, doubt is cast upon the corresponding hypothesis.

(iv)the fiducial interpretation: If nothing at all is otherwise known about (λ, ζ), since t and z reproduce the original likelihood function, they contain all the information available about the parameters. It would seem therefore that we would be justified in referring to the confidence distributions generated in (ii) as fiducial distributions.

In mathematics we use the symbol 2 to refer to an integer, a rational number, a real number, a Boolean algebra, and sometimes in other senses. Similar systematic ambiguities can be useful with estimation statements provided that, as with 2, we are prepared to make distinctions when these are necessary.

In the example given, the likelihood function given by the pair of linear pivotals is exactly equivalent to that of the original data. In other cases approximations may be needed. A remarkable case is treated by Sprott of an exponential survival time distribution with unknown mean θ. n items are observed for fixed periods of time and of these k are seen to fail at times t_j, $j=1,2,..,k$, while the remainder are censored at times T_i, $i=1,2,...,n-k$. The original relative likelihood function is

$$R(\theta) = (\hat{\theta}/\theta)^k \exp[k(1-\hat{\theta}/\theta]$$

where $\hat{\theta} = T/k$, $T = \Sigma t_j + \Sigma T_i$. Transforming the parameter to $\phi = \theta^{-1/3}$ gives a linear pivotal $\sqrt{I(\hat{\phi},k)}[\hat{\phi}-\phi]$, where $I(\hat{\phi},k)$ is the observed Fisher information. Taking this to be exactly standard normal, instead of only asymptotically so, gives a remarkably accurate reproduction of the original likelihood even when n is quite small. The use of such pivotals after parameter transformations is an area which continues to be worth exploring; sometimes a Student, or some other standard distribution for the pivotal gives a better approximation to the original likelihood than does the normal distribution.

Incoherence

The use of tests of significance has been attacked as involving integration over the sample space leading to incoherence in the technical sense that if the resulting P value is used as a basis for betting, a "sure loss" strategy can be devised. As Fisher pointed out, the Neyman-Pearson concept of the power function of a test of significance is a special case of a likelihood function based, not on the whole data, but merely on the knowledge whether or not the sample point has fallen into the critical region of the test in question. In so far as no account is taken of possible changes in the likelihood function as the sample point moves within the critical region, or within its complement, information is certainly lost and "sure loss" strategies become possible. But while the statistician should always be prepared to maximise the amount of information used in his inferences, there are many occasions where there is no point is striving one's hardest to do this, and a simple test of significance is adequate. To do this, when appropriate, could not properly be termed "incoherent", and I would like to suggest that a modification of an old word, "impalestric" (unathletic) would serve better to indicate what is being done.

References

Box, G. E. P. (1980) Sampling and Bayes' inference in scientific modelling and robustness. J. Roy. Stat. Soc. A **143** 383-404.

Box, G. E. P. and Tiao, G. C. (1973) Bayesian Inference in Statistical Analysis. Addison Wesley, Reading, Mass.

Good, I. J. (1987) Foundations and Philosophy of Probability and Statistics: an International Symposium. To appear in Journal of Statistical Planning and Inference.

Fisher, R. A. (1938): Comment on H. Jeffreys paper. Annals of Eugenics, **8**, 151 (Coll. pap. 159A)

-- (1955): Statistical Methods and Scientific Induction. J. Roy. Stat. Soc. B, **17**, 69-78.

Sprott, D. A. and Viveros, R. (1984). The interpretation of maximum likelihood estimation. Canad. J. Statist. **12**, 27-38.

STATISTICAL DATA ANALYSIS AND INFERENCE
Y. Dodge (Editor)

BAYES MODIFIED MINIMAX EXPERIMENTAL DESIGN

Ronald V. CANFIELD

Department of Mathematics and Statistics
Utah State University
Logan, UT 84322-3900 USA

Existing Bayes strategies for design have been known to be inadequate when the prior density does not describe a real random variable. The responses of the designs to conservative prior information are illogical in that the more conservative the prior density, the smaller the design sample size required. Thus prior ignorance is rewarded. A design strategy based upon a Bayesian modification of the minimax principle is developed for test of the form $H_0:\theta \in \Theta_0$ vs. $H_1:\theta \in \Theta_1$.

The noninformative prior for hypothesis testing is defined and the Bayesian modified minimax test with noninformational prior is compared with the classical minimax test.

1. INTRODUCTION

Determining sample size using prior information for the decision problem $H_0:\theta \in \Theta_0$ against the alternate $H_1:\theta \in \Theta_1$ is considered in this work. The term preposterior analysis is used in Bayesian literature for this activity

Let D represent a family of decision functions δ_n and $L(\theta,\delta_n(\underline{X}))$ the loss function for using decision $\delta_n(\underline{X})$ when the parameter has value θ where $\underline{X}$ represents a random sample of size n. The risk function is defined

$$R(\theta,\delta_n) = E_{\underline{X}}(L(\theta,\delta_n(\underline{X})) \tag{1}$$

and the Bayes risk is given by

$$r(\delta_n) = E_{\theta}(R(\theta,\delta_n)), \tag{2}$$

where $\pi(\theta)$ represents the prior density of θ. The appropriate Bayes decision function $\delta_n(\underline{X})$ (if it exists) minimizes (2). A theoretically satisfying approach for determining sample size is given by Berger (1985). This method includes sampling expense in the loss function. The optimal sample size for a test minimizes (2). This design strategy will be referred to as S1. In many applications opposing interests demand that individual protection be given to the two types of decision error. Under this obligation, sample cost is not a consideration and the Bayes risk (2) is separated into the two components

$$r(\delta) = \int_{\theta\varepsilon\Theta_1}\int_{\underline{x}:\delta_n(\underline{x})=a_0} L(\theta,a_0)f(\underline{x}|\theta)\pi(\theta)d\underline{x}\,d\theta$$

$$+\int_{\theta\varepsilon\Theta_0}\int_{\underline{x}:\delta_n(\underline{x})=a_1} L(\theta,a_1)f(\underline{x}|\theta)\pi(\theta)d\underline{x}\,d\theta, \qquad (3)$$

where a_0 represents the action accept H_0 and a_1 reject H_0. The first component on the right of (3) is referred to as consumer risk and the second, producer risk. Sample size is determined to satisfy acceptable constraints on the Bayes risks which affect the opposing interests (e.g.[4]. This design strategy will be referred to as S2.

Operational characteristics of the design strategies based on (2) were investigated in [3]. It was shown that although the strategies may have theoretical appeal, the designs are in general unreasonable in application. Characteristics of existing Bayes experimental design properties [3] are reviewed in the next section. A classical minimax design strategy is then developed for later comparison with the Bayes modified approach. The concept of a noninformative prior in hypothesis testing is discussed. It is shown that the definition as used in problems of estimation is not appropriate for hypothesis testing. A noninformative prior is defined. The Bayes modified minimax design strategy is developed and compared with the classical minimax approach. An application is discused in the final selection.

2. Bayes Risk Design

The relationship between information in a prior and the resulting experimental design (i.e., sample size) using the Bayes Risk strategy [2] is reviewed in this section. The term "prior information" as it applies to Bayesian analysis may have two meanings. It commonly refers to the collection of prior knowledge with respect to distributional parameters. However it can also be used in an information theoretic context. To avoid confusion in this paper, the term "prior knowledge" will be used to describe the collection of information which is used to determine the prior density, where as "prior information" describes the theoretical characteristics of the prior density.

The relationship between "information" and "design" is easily understood by considering the following example representing the applied Bayesian's dream with regard to prior knowledge, i.e., a random sample from the population to be tested.

Consider the problem of designing a test of the hypothesis $H_0:\theta\le\theta_0=0$ versus $H_1:\theta>0$ where θ is the mean of a normal random variables with known variance $\sigma^2=1$. Prior knowledge for designing the test is assumed to come from a prior random sample of size n_0. By suffiency it is only necessary to deal with the random variables $\bar{X}_0$ and $\bar{X}$ when using (2) where $\bar{X}_0$ is mean if the prior sample and $\bar{X}$ is the sample mean of the future data. The Bayes decision function for this test is

$$\delta(\bar{X}) = \begin{cases} a_0 \text{ (accept } H_0) \text{ if } \bar{X}<K \\ a_1 \text{ (reject } H_0) \text{ if } \bar{X}\ge K \end{cases} \tag{4}$$

and the loss function to be used is

$$L(\theta,\delta(\bar{X}))= \begin{cases} \theta \text{ if } \delta(\bar{X})=a_0 \text{ and } \theta>0 \\ -\theta \text{ if } \delta(\bar{X})=a_1 \text{ and } \theta\le 0 \\ 0 \text{ otherwise.} \end{cases} \tag{5}$$

Design constraints for this example are that consumer and producer risks in (3) must not exceed 0.01.

An appropriate prior density for this situation is the posterior density resulting when a noninformative prior is used with the prior sample, i.e.,

$$\Pi(\theta) = \sqrt{n_0}\,\exp(-n_0(\theta-\bar{X}_0)^2/2)/\sqrt{2\pi} \tag{6}$$

Note that information in the prior always increases as prior knowledge (i.e., sample size) increases. It follows from (4),

(5) and (6) that the consumer risk component of (3) is

$$r_c = \frac{\sqrt{nn_0}}{2\pi} \int_0^{\infty} \int_{-\infty}^{K} \theta \exp(-(n(\bar{X}-\theta)^2 + n_0(\theta-\bar{X}_0)^2)/2)\, d\bar{X} d\theta,$$

and the producer risk component is

$$r_p = -\frac{\sqrt{nn_0}}{2\pi} \int_{-\infty}^{0} \int_{K}^{\infty} \exp(-(n(\bar{X}-\theta)^2 + n_0(\theta-9\bar{X}_0)^2)/2)\, d\bar{X} d\theta.$$

The design sample size n is of course a function of the random variable $\bar{X}_0$. However it satisfies the purpose of this example to assume that $\bar{X}_0$= constant (e.g., the true mean of the population). Table 1 shows the design sample size for various n_0 assuming that $\bar{X}_0$= 0.1. The test critical value K is chosen to satisfy the risk constraints $r_c \le 0.01$ and $r_p \le 0.01$.

Table 1. Design Sample Sizes using Bayes Risk Criterion with $\bar{X}_0$= 0.1

Prior Sample Size	Design Sample Size
1	10
5	19
15	26
20	27
25	27
50	19
75	5
80	1

The most interesting feature of this example is the impact of ignorance on design. A prior sample of size n_0=1 requires only 10 additional experimental points while a prior sample of size 25 requires 27 additional points. The design sample size decreases for n_0>25 (approximately) and does not reach the value at n_0=1 until n_0>50. This response seems totally illogical, i.e., the risk constraints are more easily satisfied if there is little prior information (prior sample size).

This response of sample size to prior knowledge can be justified with the following argument. A noninformative prior (in the information theoretic sense) has a great deal of probability for θ at it's extremes where experimental data is decision efficient. Thus smaller sample sizes will satisfy risk constraints. As the prior becomes more informative it will reduce the density of θ in these data efficient regions and at the same time begin to concentrate θ density where data is less efficient for both H_0 and H_1. The natural response is an increase in design sample size. As the prior becomes even more informative a point will be reached where the density will likely concentrate in either the H_0 or the H_1 region, thus reducing the requirement for sample data.

Validity of this justification lies in the interpretation of θ. It is required that θ be a true random variable which occurs in any given interval according to probabilities dictated by its density function. Thus, for example, if the prior density is noninformative, θ will occur with greater frequency in the extremes of its range. This also suggests that since prior knowledge is used to estimate the prior density, more prior knowledge will result in a more accurate estimate which will not necessarily be more informative (in the information theoretic sense).

It is generally held that the prior density is an expression of prior knowledge or ignorance about the parameter θ and in most situations there is nothing random about θ (e.g., [1], [2]).Prior ignorance does not dictate a density to the parameter θ. The prior density has no more influence on the parameter θ than the probability of a head has on the result of flipping a coin after it has been flipped. The falicy of a random variable interpretation of θ in example 1 is clear. More prior knowledge (i.e. samples) always reduces the prior variance and hence increases information. In the limit θ is degenerate. These are not the properties of a true random variable.

3. MINIMAX EXPERIMENTAL DESIGN

Minimax theory plays an important role as a bench mark for Bayesian analysis. Intuitively, the Bayesian approach when there is no prior information and the minimax approach should yield similar results. Indeed there are many situations in which the two approaches give the same result [1]. The minimax principle is used in this section as a design principle for the hypothesis $H_0{:}\theta \in \Theta_0$ vs $H_1{:}\ \theta \in \Theta_1$ using a loss function. This will be used as a comparison with the Bayes modified approach.

Assuming existance, let $\theta^* \in \Theta_0 \cup \Theta_1$ be defined such that

$$R(\theta^*,\delta_n) = \max_{\theta\varepsilon\Theta_0\cup\Theta_1} R(\theta,\delta_n) \tag{7}$$

The following theorem is useful for determining the experimental design.

2.1 Theorem.

$$R(\theta^*,\delta_n) = \max\left[\max_{\theta\varepsilon\Theta_1} L(\theta,a_0)\, P_\theta(\delta_n(\underline{X} = a_0),\ \max_{\theta\varepsilon\Theta_0} L(\theta,a_1)\, P_\theta(\delta_n(\underline{X}) = a_1)\right] \tag{8}$$

Proof. By definition

$$R(\theta^*,\delta_n) = \max_{\theta\varepsilon\Theta_0\cup\Theta_1} E_{\underline{X}}(L(\theta,\delta(\underline{X}))$$

$$= \max_{\theta\varepsilon\Theta_0\cup\Theta_1}\left[\int_{\delta_n(\underline{x})=a_0} L(\theta,a_0)\, f_\theta(\underline{X})d\underline{X} + \int_{\delta_n(\underline{X})=a_1} L(\theta,a_1) f_\theta(\underline{X})d\underline{X}\right]$$

$$= \max_{\theta\varepsilon\Theta_0\cup\Theta_1}\left[L(\theta,a_0)P(\delta_n(\underline{X})=a_0) + L(\theta,a_1)P(\delta_n(\underline{X})=a_1)\right]$$

The result follows since $L(\theta,a_i) = 0$ if $\theta \varepsilon \Theta_i, i = 0,1$.

Experimental design for strategy S1 is determined by choosing δ_n to minimize $R(\theta^*,\delta_n)$, where the design sample size is the n of the minimizing δ_n.

Experimental design for the constrained risk strategy S2 is determined by the decision rule δ_n (and sample size n) to constrain the maximum consumer and producer risks in (3). Theorem 2.1 provides a convenient representation of these maximums where the first component on the right in (8) is the max consumer risk and the second component is the max producer risk.

3. THE NONINFORMATIVE PRIOR DENSITY

The concept of a noninformative prior density is important in Bayesian analysis as a point of comparison with classical methods. Such comparisons are useful to signal the possibility of a failure of the Bayesian approach to produce a practical design [1]. The noninformative prior as defined in [2] is used in the context of estimation. This definition is not appropriate for hypothesis testing as illustrated by the following example.

Let $F_0(y;\theta)$ and $F_1(y;\phi)$ be different families of cdf's with parameters θ and ϕ respectively. Suppose Y is a random variable with cdf either F_0 of F_1. It is desired to test the hypothesis that $Y \sim F_0$ against the alternate that $Y \sim F_1$. This hypothesis can be parameterized as follows. The cdf of the random variable Y is given by $F(y) = (1-\eta)\ F_0(y;\theta) + \eta\ F_1(y;\theta)$ where $\eta = 0$ or 1.The hypothesis can be given in terms of the parameter η as $H_0{:}\eta = 0$ vs. $H_1\eta = 1$. It is intuitively obvious that a noninformative prior for η allocates a probability of 0.5 to both $\eta = 0$ and $\eta = 1$.

More commonly, the hypothesis involves a portion of the parameter space of a random variable, e.g. $H{:}\theta \in \Theta_0$ vs $H_0{:}\theta \in \Theta_1$. In this case it is still reasonable to reparameterize using η as above. Thus in order for a prior density $\pi(\theta)$ to be noninformative with respect to π, it is necessary that

$$\int_{\Theta_0} \pi(\theta)d\theta = \int_{\Theta_1} \pi(\theta)d\theta = 0.5$$

It is convenient to use the representation of a prior given in [1],

$$\pi(\theta) = \begin{cases} \pi_0 g_0(\theta) & \text{if } \theta \in \Theta_0 \\ \pi_1\ g_1(\theta) & \text{if } \theta \in \Theta_1 \end{cases}$$

where $\pi_0 + \pi_1 = 1$, $\pi_i \geq 0$ and $g_i(\theta)$ is a proper density function, $i = 0,1$.

4. THE BAYES MODIFIED MINIMAX PRINCIPLE

4.1 Design Criteria

The primary criticism of procedures based on the minimax principle is that they tend to be too conservative. The function of prior information is to bring a more reasonable perception of the state of nature. As discussed previously, the Bayes risk (2) does not adequately represent the risks for hypothesis testing. In this section a modification of the risk $R(\theta,\delta_n)$ is defined. This prior modified risk is substituted for $R(\theta,\delta_n)$ in (2) to define the Bayes modified maximum risk for decision rule δ_n.

Definition 4.1. The prior modified maximum risk for $\delta_n \varepsilon D$ is

$$R_m(\theta,\delta_n) = \begin{cases} R(\theta_0^*,\delta_n) & \text{for } \theta \varepsilon \Theta_0 \\ R(\theta_1^*,\delta_n) & \text{for } \theta \varepsilon \Theta_1 \end{cases}$$

where θ_0^* is the θ value which maximizes $R(\theta,\delta_n)$ for $\theta \varepsilon \Theta_0$ and θ_1^* the value which maximizes $R(\theta,\delta_n)$ for $\theta \varepsilon \Theta_1$.

Definition 4.2. The Bayes modified maximum risk is defined

$$B(\delta_n) = E_\theta[R_m(\theta,\delta_n)].$$

Theorem 4.1

$$B(\delta_n) = \pi_0 L(\theta_0^*,a_1)\ P_{\theta_0^*}(\delta_n(\underline{X})=a_1)$$

$$+\ \pi_1 L(\theta_1^*,a_0) P_{\theta_1^*}(\delta_n(\underline{X})=a_1) \qquad (9)$$

The proof is similar to that of theorem 2.1.

The first term on the right of (9) is the Bayes modified maximum producer risk and the second term is the maximum consumer risk.

4.2 Comparison with Minimax Design

Consider first the Bayesian modified approach with a noninformative prior. It can be seen from definition 4.1 that a noninformative prior will have little influence in the determination of θ_0^* and θ_1^*. It follows that θ_0^* and θ_1^* will also maximize $R(\theta,\delta_n)$ in Θ_0 and Θ_1 respectively. It is evident that the only difference between the risk components of the Bayes and minimax approaches is the presence of $\pi_0 = \pi_1 = 0.5$ in $B(\delta_n)$, e.g., the Bayes modified consumer risk is one half the minimax consumer risk. Therefore given this scale factor for interpreting risk, the Bayes modified and minimax approaches are equivalent for strategy S2.

Minimization of $B(\delta_n)$ and $R(\theta^*,\delta_n)$ for design strategy S1 will not necessarily produce the same result. For the Bayes case δ_n is selected to minimize $B(\delta_n)$ which is a sum of consumer and producer risk component. The minimax strategy chooses δ_n to minimize $R(\theta^*,\delta_n)$ which is the maximum of consumer and producer risk components. Under certain symmetry conditions it can be shown that the two approachs will yield the same design for common one way hypotheses (e.g.,$H_0:\theta \le \theta_0$). These conditions are symmetric loss function and symmetric prior and random variable (or sufficient statistic) densities. If the departure from symmetry is not great, the two approaches should yield similar results.

4.3 Increasing Prior Information

Consider now the effect of increasing prior information on design. This discussion is limited to the case H_0: $\theta \le \theta_0$ vs $H_1:\theta > \theta_0$. It is assumed that the loss function $L(\theta,a_0)$ is monotone increasing for $\theta > \theta_0$, $L(\theta,a_1)$ is monotone decreasing for $\theta < \theta_0$ and $L(\theta,\delta_n(\underline{X})) = 0$ otherwise. It is also assumed that the prior is unimodel with mode at $\theta_m \ne \theta_0$. Without loss of generality assume $\theta_m < \theta_0$. Then $g_0(\theta)$ will have a mode at θ_m and the mode of $g_1(\theta)$ will occur at θ_0. It can be seen from definition 4.1 that the effect of the prior in the maximization operations which determine θ_0^* and $\theta_1^*$is a tendency to shift θ_0^* and θ_1^* toward the modes of $g_0(\theta)$ and $g_1(\theta)$ respectively. Thus θ_0^* tends toward θ_m where the risk is not zero. At the same time θ_1^*tends toward θ_0 where risk is zero.

This tendency reduces the consumer risk and permits choice of $\delta_n \varepsilon$ D to act more strongly to reduce the producer risk (at any given η value). Thus the risks are both reduced. For the constrained risk strategy S2, the sample size can then be reduced and still maintain the constraint. In the limit, the prior compresses all probability at θ_m, then $\theta_0^* = \theta_m$so that the consumer risk is zero. Then the δ_n can be chosen $\delta_n(\underline{X}) = a_0$for all $\underline{X}$ and the producer risk is zero. Thus no samples are required.

In summary, an increase in prior information has a tendancy to reduce the design sample size n.

5. Application

In most tests there is a natural family of decision functions which serves to limit the family D. For example if θ is the mean of a random variable and the hypothesis is $H_0:\theta\le\theta_0$ vs. $H_1:\theta>\theta_0$, the Bayes decision function is of the form $\delta_n(\underline{X}) = a_1$ if $\bar{X} > k$

for many distributions (e.g., normal and Bernoulli). The following application is motivated by the need to monitor a producer's compliance to certain design specifications.

Electronic systems with several million states and responsesd are difficult to manually test for faults prior to actual operation due to the large number of possibilities. It is also difficult to maintain such equipment if there exists no automatic system of isolating faults. Therefore, automatic test systems are used where the fault coverage is specified by contract. Statistical tests are used to monitor contractor compliance. Sampling expense motivates the desire for Bayes methods. It is clear that the standard Bayes risk approach is nonproductive.

The test design using the Bayes modified minimax approach is a straight forward application of the method developed in section 4. Although the method requires numerical optimization, it is possible to provide a few tables covering a wide range of prior distributions. For example, the prior probabilities π_0, and π_1, are not required for table construction.

A very serious impediment to the application of Bayes methods in general has been associated with specification of the prior density. A useful feature of Bayes modified minimax approach to experimental design is that very simple prior densities can be used effectively. For example, an informative prior density can be determined by choosing only the prior probabilities π_0 and π_1(the $g_0(\theta)$ and $g_1(\theta)$ can be noninformative). This is much simpler and easier to determine than choosing from a large family of densities (e.g. the gamma family). Thus it is more likely a group of "experts" can agree upon a prior.

5. CONCLUSIONS

The risk foundation for experimental design based on (2) seems to be inappropriate for most applications. The Bayes modified minimax approach has been suggested as an alternative. It is consistent with minimax designs when a noninformative prior is used. Designs based on the Bayes modified approach respond to "amount" if prior information in an intuitively reasonable manner.

REFERENCES

[1] Berger, J.O., Statistical Decision Theory and Bayesian Analysis (Springer-Verlag, NY, 1985).

[2] Box, G.E.P. and Tiao, G.C., Bayesian Inference in Statistical Analysis (Addison Wesley, Reading, Mass., 1973).

[3] Canfield, R.V., Bayesian Experimental Design: A Paradox, submitted to Technometrics.

[4] Martz, H.F. and Waller, R.A., Bayesian Reliability Analysis (John Wiley, NY, 1982)

STATISTICAL DATA ANALYSIS AND INFERENCE
Y. Dodge (Editor)

A SIMPLE CLASS OF TESTS LOCALLY BETTER THAN THE SCORE TEST

TAPAS K. CHANDRA

Division of Theoretical Statistics and Mathematics, Indian Statistical Institute, 203 Barrackpore Trunk Road, Calcutta 700 035, INDIA

Abstract. A simple example of unbiased test procedures for the standard two-sided testing problem is given, whose powers, under contiguous alternatives, are larger than that of the score test. The comparison is valid up to terms of order $o(n^{-1})$.

1. INTRODUCTION AND AN EXAMPLE

Let $\{X_n\}_{n\geq 1}$ be a sequence of iid random variables with common density $f(x;\Theta)$, $\Theta \in (H) \subset R^1$, (H) being an open set. Consider the testing problem

$$H_o : \Theta = \Theta_o \text{ vs. } H_1 : \Theta \neq \Theta_o . \qquad \dots (1)$$

We fix a sequence of contiguous alternatives $\Theta_n = \Theta_o + n^{-\frac{1}{2}}\delta$, $\delta \in R^1$. It is recently shown in Chandra and Joshi (1983), Chandra and Mukerjee (1984,1985), and Chandra and Samanta (1988), that there exists an asymptotically locally unbiased (up to $o(n^{-1})$) two-sided test based on Rao's statistic (see Rao (1965), page 349) whose power at Θ_n is larger that those of a fairly large family of asymptotically locally unbiased two-sided tests (provided that δ and the common size α of the test remain small enough). For the precise statement and other related references, the interested reader may consult the above references.

In this paper, we construct an example where there exists a family of asymptotically locally unbiased two-sided tests for the testing problem (1), some of which can have larger local powers than that of the test based on Rao's statistic. We next show that in this example the locally most powerful unbiased test (see Ferguson (1967), page 237) for (1) belongs to this family and is superior to Rao's test. It must be emphasized here that the Rao's test is much simpler than the locally most powerful unbiased test.

Example. Let X and Y be independent random variables following $N(\Theta;1)$ and $N(\Theta^2;1)$ respectively, $\Theta \in R^1$. Let $X_1, X_2, \dots$ and $Y_1, Y_2, \dots$ be independent copies of X and Y respectively. Let $C_{1,n}$ and $C_{2,n}$ be any two constants, which may depend on the size α of the tests, such that

$$0 < \liminf C_{i,n} \leq \limsup C_{i,n} < \infty \qquad i = 1,2.$$

Define two statistics, $T_{1,n}$ and $T_{2,n}$, as follows :

$$T_{1,n} = n^{\frac{1}{2}}\,\overline{X} + C_{1,n}\,\overline{Y} + C_{2,n}\,\overline{Y}^2,$$
$$T_{2,n} = n^{\frac{1}{2}}\,\overline{X} - C_{1,n}\,\overline{Y} - C_{2,n}\,\overline{Y}^2, \qquad \ldots (2)$$

where $\overline{X} = n^{-1}(X_1 + \ldots + X_n)$, $\overline{Y} = n^{-1}(Y_1 + \ldots + Y_n)$. We consider the testing problem (1) with $\Theta_o = 0$. Our suggested critical region is

$$\{T_{1,n} > z + t_{1,n}\} \cup \{T_{2,n} < - z + t_{2,n}\} \qquad \ldots (3)$$

for suitable $t_{1,n}$ and $t_{2,n}$. Here for $\alpha \in (0,1)$,

$$\int_z^\infty \phi(t)dt = \alpha/2\,, \qquad \phi(t) = (2\pi)^{-\frac{1}{2}} \exp(-t^2/2).$$

As in Chandra and Joshi (1983), we determine the cut-off points, up to $o(n^{-1})$, such that

$$P_{\Theta_o}(T_{1,n} > z + t_{1,n}) = \frac{1}{2}\,\alpha + n^{-\frac{1}{2}}\,d_1 + n^{-1}\,d_2\,,$$

$$P_{\Theta_o}(T_{2,n} <-z + t_{2,n}) = \frac{1}{2}\,\alpha - n^{-\frac{1}{2}}\,d_1 + n^{-1}\,d_2\,;$$

in this example,

$$d_1 = 0, \quad d_2 = z/2, \quad I = 1$$

(see Remark 4.1 of Chandra and Joshi (1983)). To determine $t_{i,n}$ and the power at Θ_n of (3), one needs the Edgeworth expansions (up to $o(n^{-1})$) of $T_{i,n}$ under Θ_n. It is easy to verify that

$$E_{\Theta_n}(T_{1,n}) = \delta + n^{\;1}(C_{1,n}\,\delta^2 + C_{2,n})$$

$$\text{var}\,\Theta_n(T_{1,n}) = 1 + n^{-1}\,C_{1,n}^2\,, \quad \mu_3,\,\Theta_n(T_{1,n}) = 0$$

$$\mu_4,\,\Theta_n(T_{1,n}) = 3 + n^{-1}\,6\,C_{1,n}^2\,.$$

Thus the third and fourth order approximate cumulants of $T_{1,n}$ are zero. Hence it is very easy to write down the asymptotic expansion of the power at Θ_n of test (3).

Using Equation (2.6) of Chandra and Joshi (1983), one gets

$$t_{1,n} = n^{-1}C_1', \quad C_1' = - d_2 + C_{2,n} + \frac{1}{2}\,z\,C_{1,n}^2\,, \qquad \ldots (4)$$

and

$$P_{\Theta_n}(T_{1,n} > z + n^{-1}\,C_1')$$

$$= \int_{z-\delta}^\infty \phi(t)dt + n^{-1}\phi(z-\delta)\,\{d_2 - \frac{\delta}{2}\,C_{1,n}^2 + \delta^2\,C_{1,n}\,\} + o(n^{-1}).$$

From this, it follows that

$$t_{2,n} = n^{-1} C_2' , \quad C_2' = - C_1' \qquad \ldots \qquad (5)$$

and

$$P_{\Theta_n}(T_{2,n} < - z + n^{-1} C_2')$$

$$= \int_{z+\delta}^{\infty} \phi(t)dt + n^{-1} \phi(z+\delta) \{d_2 + \frac{\delta}{2} C_{1,n}^2 + \delta C_{1,n}\} + o(n^{-1}).$$

Let P_{Θ_n} and $P_{\Theta_n}^{Rao}$ be respectively the power at Θ_n of the test with the critical region (3) (with $t_{i,n}$, $i = 1,2$ given by (4) and (5)), and the Rao's test. Note that when $C_{1,n} = C_{2,n} = 0$, $T_{1,n} = T_{2,n} =$ Rao's statistic. Hence

$$\lim_{n\to\infty} \{n(P_{\Theta_n} - P_{\Theta_n}^{Rao})\}$$

$$= \delta C_{1,n} \exp \{- \tfrac{1}{2}(z^2 + \delta^2)\} [- \frac{C_{1,n}}{2} (\exp(z\delta) - \exp(-z\delta))$$

$$+ \delta (\exp(z\delta) + \exp(-z\delta))]$$

$$= \delta^2 C_{1,n} \exp \{- \tfrac{1}{2}(z^2 + \delta^2)\} \sum_{r=0}^{\infty} \frac{(z\delta)^{2r}}{(2r+1)!} [4r + 2 - C_{1,n}z].$$

Clearly the test (3) is 'better' than Rao's test if

$$0 < C_{1,n} < 2z^{-1} \qquad (C_{2,n} \text{ is arbitrary}).$$

Since usually δ will be small, the most important term in the above expression is the coefficient of δ^2 which is

$$2 C_{1,n} - z C_{1,n}^2 \leq z^{-1} \qquad \text{for all } C_{1,n}$$

equality being true when $C_{1,n} = z^{-1}$. It may be noted that if $C_{1,n} = z^{-1}$, and α and δ are small, the above limit is expected to be small.

2. LOCALLY MOST POWERFUL TEST

We consider the same testing problem as discussed in the example of the last section. Let $L(\Theta)$ be the likelihood function when the observations are $(X_1, Y_1),\ldots, (X_n, Y_n)$. Let

$$Z_n = n^{-\frac{1}{2}} \frac{d}{d\Theta} \log L(\Theta)|_{\Theta = 0} ,$$

$$W_n = n^{-1} \frac{d^2}{d\Theta^2} \log L(\Theta)|_{\Theta = 0} .$$

Then the locally most powerful unbiased (LMPU) test is given by the critical region

$$Z_n^2 + W_n > k_{1,n} + 2k_{2,n} Z_n \qquad \ldots. \quad (6)$$

where $k_{1,n}$ and $k_{2,n}$ are to be chosen such that the size condition and the local unbiasedness condition are satisfied. Here one may note that the Fisher information at $\Theta = 0$ is 1, and the Efron curvature at $\Theta = 0$ is 2. It can be shown that as $n \to \infty$

$$k_{1,n} \to z^2 - 1 \text{ and } k_{2,n} \to 0 . \qquad \ldots. \quad (7)$$

In our example

$$Z_n = n^{\frac{1}{2}} \overline{X} , \quad W_n = 2\overline{Y} - 1 . \qquad \ldots. \quad (8)$$

It is well-known that

$$P_\Theta (M_n^c) = o(n^{-1}) ,$$

uniformly over compact subsets of δ , where

$$M_n = \{|n^{\frac{1}{2}} \overline{X}| > (3 \log n)^{\frac{1}{2}}, \ |n^{\frac{1}{2}} \overline{Y}| > (3 \log n)^{\frac{1}{2}} \}$$

and M_n^c is the complement of M_n. We first rewrite the region (6) as follows:

$$Z_n > k_{2,n} + A_n \{1 - 2\overline{Y}/A_n^2\}^{\frac{1}{2}} ,$$

or $\qquad \ldots. \quad (9)$

$$Z_n < k_{2,n} - A_n \{1 - 2\overline{Y}/A_n^2\}^{\frac{1}{2}} ;$$

here

$$A_n = (k_{1,n} + 1 + k_{2,n}^2)^{\frac{1}{2}} \to z . \qquad \ldots. \quad (10)$$

Now on M_n^c ,

$$A_n \{1 - 2\overline{Y}/A_n^2 \}^{\frac{1}{2}} = A_n \{1 - \overline{Y}/A_n^2 - \overline{Y}^2/(2A_n^4) + o(n^{-1})\}.$$

Thus the power, $P_{\Theta_n}^{LMPU}$, of the critical region (6) is the same, up to $o(n^{-1})$, as that of the following region C_n :

$$Z_n + (nA_n^2)^{\frac{1}{2}} (n^{\frac{1}{2}} \overline{Y}) + (2nA_n^3)^{-1} (n^{\frac{1}{2}} \overline{Y})^2 > k_{2,n} + A_n \text{ or } < k_{2,n} - A_n .$$

From Chandra and Joshi (1983, page 232), one now has

$$\begin{aligned} P_{\Theta_n}(C_n) = & \int_{A_n + k_{2,n} - \delta}^{\infty} \phi(t)dt + \int_{A_n - k_{2,n} + \delta}^{\infty} \phi(t)dt \\ & + n^{-1} \phi(A_n + k_{2,n} - \delta) (-\delta/(2A_n^2) + \delta^2/A_n) \\ & + n^{-1} \phi(A_n - k_{2,n} - \delta) (\delta/(2A_n^2) + \delta^2/A_n) \\ & + o(n^{-1}) \end{aligned}$$

$$= \int_{A_n + k_{2,n} - \delta}^{\infty} \phi(t)dt + \int_{A_n - k_{2,n} + \delta}^{\infty} \phi(t)dt$$

$$+ n^{-1}\phi(z - \delta)(- \delta/(2z^2) + \delta^2/z)$$

$$+ n^{-1}\phi(z + \delta)(\delta/(2z^2) + \delta^2/z)$$

$$+ o(n^{-1}),$$

by (7) and (10). By directly computing the derivative of $P_{\Theta_n}(C_n)$ with respect to δ at $\delta = 0$, one verifies that the local unbiasedness holds up to $o(n^{-1})$ iff

$$n(A_n - k_{2,n}) - (A_n + k_{2,n})) \to 0$$

$$\text{i.e. iff } n\, k_{2,n} \to 0.$$

The size condition is

$$\int_{A_n + k_{2,n}}^{\infty} \phi(t)dt + \int_{A_n - k_{2,n}}^{\infty} \phi(t)dt + o(n^{-1})$$

$$= \alpha + 2\, n^{-1}\phi(z)d_2$$

i.e., $n(A_n - z) \to -d_2$ It, therefore, follows that the LMPU test belongs to the family of tests considered in the previous section. In fact, to get the local power at Θ_n of the LMPU test, we need only put $C_1 = z^{-1}$, $C_2 = (2z^3)^{-1}$ in the expression for P_{Θ_n}.

REFERENCES

[1] Chandra, T. K. and Joshi, S. N. (1983), Comparison of the likelihood ratio, Rao's and Wald's tests and a conjecture of C. R. Rao Sankhyā, Ser A, 45, 226-246

[2] Chandra, T. K. and Mukherjee, R. (1984), On the optimality of Rao's statistics, Comm. in Statist A - Theory and Methods, 13, 1507-1515.

[3] Chandra, T.K. and Mukherjee, R. (1985) , Comparison of the likelihood ratio, Wald's and Rao's tests Sankhya, Ser A, 47, 271-284.

[4] Chandra, T. K. and Samanta, T. (1988), On the second order local comparison between perturbed maximum likelihood estimators and Rao's statistic as test statistics, J. Mult. Anal., 25, 201-222.

[5] Ferguson, T. S. (1967), Mathematical Statistics : A Decision Theoretic Approach. Academic Press, New York.

[6] Rao, C. R. (1965), Linear Statistical Inference and Its Applications. John Wiley, New York.

STATISTICAL DATA ANALYSIS AND INFERENCE
Y. Dodge (Editor)

COMPARISON OF TESTS IN THE PRESENCE OF A NUISANCE PARAMETER

Rahul MUKERJEE

Division of Theoretical Statistics and Mathematics, Indian Statistical Institute, 203 Barrackpore Trunk Road, Calcutta 700 035, INDIA

We consider a very large class of tests containing in particular the likelihood ratio, the conditional likelihood ratio, Rao's and Wald's tests. In the presence of a nuisance parameter and under contiguous alternatives the third-order power of tests within the class has been studied. A third order optimality property of Rao's test has been proved. The performance of the conditional likelihood ratio test vis-a-vis the usual likelihood ratio test has been investigated. No assumption has been made regarding curved exponentiality.

1. INTRODUCTION

Let $\{X_j\}$, $j \geq 1$, be a sequence of i.i.d. random variables with common density $f(x; \theta, m)$, where θ, the parameter of interest, and m, the nuisance parameter are both supposed to be one-dimensional. The study of tests for $H_o : \theta = \theta_o$ against the alternative $\theta \neq \theta_o$ has received considerable attention in recent years - see the very illuminating work by Cox and Reid [5] and the discussion thereon. Considering contiguous alternatives $\theta_o + dn^{-\frac{1}{2}}$, where n is the sample size, this paper presents a simple formula using which third-order powers of any two tests in a large family can easily be compared. In particular, the optimality of Rao's test is established in a very large class of tests. The performance of the conditional likelihood ratio test, proposed by Cox and Reid [5], vis-a-vis the usual likelihood ratio test has also been explored. In passing, it is observed that identity of power up to the first order implies that up to the second order. It may be noted that in comparing tests we are considering their locally unbiased (up to $o(n^{-1})$) versions. Chandra and Joshi [3], who initiated such studies following a suggestion of J. K. Ghosh, discuss why this may be a meaningful way of comparing tests for two-sided alternatives.

The techniques used in this paper are similar to those in Chandra and Joshi [3], Mukerjee and Chandra [9] and Mukerjee [8] who compared tests in the one-parameter case under the absence of any nuisance parameter. Kumon and Amari [7] developed an elegant differential geometric approach for comparison of tests in curved exponential families in the presence of nuisance parameters (see also Amari [1]). The derivations in the present work do not require any assumption regarding curved exponentiality. Also, compared to the differential geometric approach, the final results here can be presented in a rather elementary form which perhaps makes applications to particular situations simpler. There is some similarity between our Theorem 1 and Theorem 2 in Pfanzagl and Wefelmeyer [11]. However, as discussed in Mukerjee [8], our results do not seem to follow from theirs. For further literature references in this area, see Chandra and Joshi [3] and Kumon and Amari [7].

In this paper the nuisance parameter m has been supposed to be one-dimensional primarily for the sake of notational simplicity and the arguments used here should be applicable even when m is multi-dimensional. The supposition that Θ, the parameter of interest, is one-dimensional is, however, non-trivial. In particular, if Θ and m are both multi-dimensional then in general one cannot employ parametric orthogonality as used here.

2. NOTATION, PRELIMINARIES AND A BASIC RESULT

Let $X_1, X_2, \ldots,$ be i i.d. random variables each with p.d.f. $f(x; \Theta, m)$, where $\Theta \in R^1$, or an open subset thereof. Interest lies in the problem of testing $H_0 : \Theta = \Theta_0$. We consider contiguous alternatives of the form $\Theta_n = \Theta_0 + dn^{-\frac{1}{2}}$. Let m_0^* be the maximum likelihood estimator of the nuisance parameter m given $\Theta = \Theta_0$. For $i,j,i',j' = 0,1,2,\ldots,$ define

$$K_{ij}(\Theta,m) = E_{\Theta,m}[\partial^{i+j} \log f(X;\Theta,m)/\partial\Theta^i \partial m^j],$$

$$K_{ij.i'j'}(\Theta,m) = E_{\Theta,m}[\{\partial^{i+j}\log f(X;\Theta,m)/\partial\Theta^i\partial m^j\}\{\partial^{i'+j'}\log f(X;\Theta,m)/\partial\Theta^{i'}\partial m^{j'}\}],$$

$$H_{ij}(\Theta,m) = n^{-\frac{1}{2}} \sum_{t=1}^{n} [\partial^{i+j}\log f(X_t;\Theta,m)/\partial\Theta^i\partial m^j - K_{ij}(\Theta,m)],$$

$$K_{ij}^* = K_{ij}(\Theta_0,m_0^*), \quad H_{ij}^* = H_{ij}(\Theta_0,m_0^*), \quad K_{ij.i'j'}^* = K_{ij.i'j'}(\Theta_0,m_0^*),$$

$$L_{ij} = K_{ij}(\Theta_0,m), \quad L_{ij.i'j'} = K_{ij.i'j'}(\Theta_0,m),$$

$$S_{ij} = H_{ij}(\Theta_0,m), \quad V_{ij} = H_{ij}(\Theta_n,m)$$

Since Θ and m are both one-dimensional, we assume parametric orthogonality (vide Cox and Reid [5]), i.e., $K_{11}(\Theta,m) = 0$ for every Θ,m. Then under standard regularity conditions the per observation information matrix is given by diag (a_1,a_2) where $a_1 = -K_{20}(\Theta,m)$, $a_2 = -K_{02}(\Theta,m)$. Let

$$a_1^* = -K_{20}^*, \quad a_2^* = -K_{02}^*, \quad a_{10} = -L_{20}, \quad a_{20} = -L_{02}$$

We consider a family F of test procedures as described below. For contiguous alternatives $\Theta_n = \Theta_0 + d\, n^{-\frac{1}{2}}$ and for every test procedure in F, a set A_n with $P_{\Theta_n}(A_n) = 1 + o(n^{-1})$, uniformly over compact subsets of d, can be obtained (cf. Chandra and Joshi [3]) such that over A_n the test procedure is given by critical region of the form

$$W_n > z + n^{-\frac{1}{2}} b_1(m_o^*) + n^{-1} c_1(m_o^*) + o(n^{-1})$$
$$\text{or } W_n < -z + n^{-\frac{1}{2}} b_2(m_o^*) + n^{-1} c_2(m_o^*) + o(n^{-1}) , \tag{2.1}$$

where z is the upper $(\alpha/2)$-point of a normal deviate,

$$W_n = (a_1^*)^{-\frac{1}{2}} H_{10}^* + n^{-\frac{1}{2}} Q_1 + n^{-1} Q_2 ,$$
$$Q_v = g_v(Q_{v1}, Q_{v2}, \ldots, Q_{vr_v}) , \tag{2.2}$$
$$Q_{vs} = n^{-\frac{1}{2}} \sum_{t=1}^{n} [q_{vs}(X_t, m_o^*) - p_{vs}(\Theta_o, m_o^*)] \quad (1 \le s \le r_v,\ v = 1,2),$$

and $b_i(.)$, $c_i(.)$ $(i = 1,2)$ are smooth functions (i.e , admit sufficiently many derivatives), free from n, to be so chosen that the test procedure has size $\alpha + o(n^{-1})$ and is locally unbiased up to $o(n^{-1})$. In (2.2), the $g_v(.)$ are polynomials in $Q_{vs}(1 \le s \le r_v)$ the coefficients of various terms in $g_v(.)$ being smooth functions of m_o^* (see e.g, (2.3) below). Also the $q_{vs}(.)$ are smooth functions such that

$$E_{\Theta,m}[q_{vs}(X_j, m)] = p_{vs}(\Theta, m) \quad \forall\ \Theta,\ m,$$

$(1 \le s \le r_v;\ v = 1,2)$ which are assumed to exist.

It may be made explicit that Q_1, Q_2 and also the q_{vs}'s depend on the particular test procedure in F under consideration. The family F is very rich and includes the likelihood ratio, conditional likelihood ratio, Rao's and Wald's tests (Rao ([12], pp. 417-418), Cox and Reid [5]). In particular, following Chandra and Joshi [3] it can be seen that the expressions for Q_1 for these tests are given respectively by

$$Q_1^{LR} = \tfrac{1}{2}(a_1^*)^{-3/2} H_{10}^* H_{20}^* + \tfrac{1}{6} K_{30}^*(a_1^*)^{-5/2}(H_{10}^*)^2 ,$$
$$Q_1^{CLR} = Q_1^{LR} + \tfrac{1}{2} K_{12}^*(a_2^*)^{-1}(a_1^*)^{-\frac{1}{2}} , \tag{2.3}$$
$$Q_1^{RAO} = 0 ,$$
$$Q_1^{Wald} = (a_1^*)^{-3/2} H_{10}^* H_{20}^* - \tfrac{1}{2} K_{20.10}^*(a_1^*)^{-5/2}(H_{10}^*)^2 .$$

Let

$$Q_{vs}^* = n^{-\frac{1}{2}} \sum_{t=1}^{n} \{q_{vs}(X_t, m) - p_{vs}(\Theta_n, m)\},$$

$$Q_{vs}^{**} = n^{-\frac{1}{2}} \sum_{t=1}^{n} \{\partial q_{vs}(X_t, m)/\partial m - \bar{p}_{vs}(\Theta_n, m)\},$$

$$\bar{p}_{vs}(\Theta, m) = E_{\Theta,m}[\partial q_{vs}(X, m)/\partial m] \quad (1 \le s \le r_v;\ v = 1,2).$$

The following assumptions are made. Since we are working with polynomial statistics, these should not be hard to verify.

Assumption 1. (i) The joint characteristic function of $(V_{10}, V_{01}, Q_{v1}^{*}, \ldots, Q_{vr_v}^{*})'$, $v = 1,2$, under Θ_n, admits an expansion up to $o(n^{-\frac{1}{2}})$ and for that adequately many moments exist.

(ii) The joint characteristic function of $(V_{10}, V_{01}, V_{20}, V_{11}, V_{02}, Q_{v1}^{*}, \ldots, Q_{vr_v}^{*}, Q_{v1}^{**}, \ldots, Q_{vr_v}^{**})'$, $v = 1,2$, under Θ_n, admits an expansion up to $o(1)$ and for that adequately many moments exist.

Assumption 2. The following expectations exist and can be calculated up to the stated orders of approximation using the expansions for characteristic functions considered in Assumption 1 :

$$E_{\Theta_n}(Q_1) = C_1(d) + n^{-\frac{1}{2}} C_2(d) + o(n^{-\frac{1}{2}}) ,$$

$$E_{\Theta_n}(Q_2) = C_3(d) + O(n^{-\frac{1}{2}}), \quad E_{\Theta_n}(Q_1^2) = C_4(d) + O(n^{-\frac{1}{2}}) ,$$

$$E_{\Theta_n}(Q_1 S_{20}) = C_5(d) + O(n^{-\frac{1}{2}}), \quad E_{\Theta_n}(Q_1 S_{01}) = C_6(d) + O(n^{-\frac{1}{2}}) ,$$

$$E_{\Theta_n}(Q_1 S_{11} S_{01}) = C_7(d) + O(n^{-\frac{1}{2}}), \quad E_{\Theta_n}(Q_1 S_{01}^2) = C_8(d) + O(n^{-\frac{1}{2}}) ,$$

$$E_{\Theta_n}(Q_1 S_{01} S_{10}) = C_9(d) + O(n^{-\frac{1}{2}}) ,$$

where the $C_j(d)$ $(1 \leq j \leq 9)$ are free from n,

Define

$$C_{10}(d) = (a_{20} a_{10}^{\frac{1}{2}})^{-1} C_7(d) + \tfrac{1}{2}(a_{20}^2 a_{10}^{\frac{1}{2}})^{-1} L_{12} C_8(d) + \tfrac{1}{2}(a_{20} a_{10}^{3/2})^{-1}(L_{21} + L_{01.20}) C_9(d).$$

It may be noted that the $C_j(d)$'s $(1 \leq j \leq 10)$ may also involve m. Since each $g_v(.)$ is a polynomial, it follows from (2.2) that for fixed m, $C_j(d)$ is a polynomial in d for each j (cf. the proof of Lemma 1 in Mukerjee [8] ; see also section 3). For each j, let $C_j^{(r)}(d)$ be the rth partial derivative of $C_j(d)$ and let $J_r(.)$ be the Hermite polynomial of degree $r (r = 0,1,2,\ldots)$. Then as in Mukerjee [8], one obtains

$$\sum_r C_j^{(r)}(d)\, a_{10}^{-r/2}\, J_r(u - d a_{10}^{\frac{1}{2}})/r\,! = C_j^{*}(u) \qquad (1 \leq j \leq 10;\ u = z \text{ or } -z) \qquad (2.4)$$

where the $\{C_j^{*}(u)\}$ are free from d (and also from n) but may involve m. We are now in a position to state the following result :

Theorem 1. If $b_i(.)$, $c_i(.)$ $(i = 1,2)$ are chosen subject to conditions of size and local unbiasedness up to $o(n^{-1})$, then the (local) power function of a test procedure of the form (2.1) is given by

$$P = P_o + n^{-\frac{1}{2}} P_1 + n^{-1} P_2 + o(n^{-1}) ,$$

where P_o, P_1, P_2 are free from n and

(i) $P_o = \int_e^{\infty} \phi(y)dy + \int_{-\infty}^{e^*} \phi(y)dy ,$

$e = z - d\, a_{10}^{\frac{1}{2}}$, $e^* = -z - d\, a_{10}^{\frac{1}{2}}$, $\phi(y)$ being the standard normal density,

(ii) P_1 is free from Q_1, Q_2 and hence the same for all tests in the family F,

(iii) $P_2 = P_{20}(d,m) + \phi(e)\{-(2za_{10}^{\frac{1}{2}})^{-1}(w_{11}+w_{21}) + dw_{11} + d^2w_{12}\}$

$$+ \phi(e^*)\{(2za_{10}^{\frac{1}{2}})^{-1}(w_{11}+w_{21}) + dw_{21} + d^2w_{22}\},$$

with $P_{20}(d,m)$ free from Q_1, Q_2 and

$$w_{11} = \frac{1}{2}\, a_{10}^{\frac{1}{2}}\{(C_1^*(z))^2 - 2C_{10}^*(z) - C_4^*(z) - (a_{20}a_{10}^{\frac{1}{2}})^{-1}L_{12}C_1^*(z)\} ,$$

$$w_{21} = \frac{1}{2}\, a_{10}^{\frac{1}{2}}\{-(C_1^*(-z))^2 + 2C_{10}^*(-z) + C_4^*(-z) + (a_{20}a_{10}^{\frac{1}{2}})^{-1}L_{12}C_1^*(-z)\} ,$$

$$w_{12} = -\frac{1}{2}\, a_{10}^{-\frac{1}{2}}\, L_{10.20}\; z\, C_1^*(z) + \frac{1}{2}\, C_5^*(z) ,$$

$$w_{22} = -\frac{1}{2}\, a_{10}^{-\frac{1}{2}}\, L_{10.20}\; z\, C_1^*(-z) - \frac{1}{2}\, C_5^*(-z) .$$

The above result is analogous to Theorem 1 in Mukerjee [8] and can be proved along the same line but with much more algebra. The details are omitted here in order to save space but may be obtained from the author. The proof essentially makes use of parametric orthogonality, Assumptions 1 and 2 together with standard regularity conditions, the fact that

$$m_o^* = m + n^{-\frac{1}{2}}\, a_{20}^{-1}\, V_{01} + o(n^{-\frac{1}{2}}), \tag{2.5}$$

and Edgeworth expansions for the distributions of $W_n - n^{-\frac{1}{2}}b_i(m_o^*) - n^{-1}c_i(m_o^*)$, $i = 1,2$ (vide (2.1)), the validity of which is assumed. Analogously to (3.14) in Mukerjee [8], the expressions for $b_1(.)$, $b_2(.)$, under the conditions of size and local unbiasedness upto $o(n^{-1})$, turn out as

$$b_1(m) = C_1^*(z) + \frac{1}{6}(a_{10})^{-3/2}\, L_{10.10.10}\, z^2 - \frac{1}{2}(a_{20}a_{10}^{\frac{1}{2}})^{-1}\, L_{12} ,$$

$$b_2(m) = b_1(m) + C_1^*(-z) - C_1^*(z) ,$$

where

$$L_{10.10.10} = K_{10.10.10}(\Theta_o,m) , \quad \text{and}$$

$$K_{10.10.10}(\Theta,m) = E_{\Theta,m}[\{\partial \log f(X;\Theta,m)/\partial\Theta\}^3].$$

The expressions for $c_1(.)$, $c_2(.)$ can similarly be worked out.

Theorem 1 demonstrates that in the present setting identity of power up to the first order implies that up to the second order (cf. Pfanzagl [10], Bickel, Chibisov and van Zwet [2]; see also the discussion by Amari following Cox and Reid [5]). Also Theorem 1(iii) shows that P_2 does not depend on Q_2. It is possible to derive a further simplification of Theorem 1 along the line of Theorem 2 in Mukerjee [8]. This is omitted since, as seen in the next section, the comparisons in which we are interested can be made directly from Theorem 1.

3. PERFORMANCE OF THE CONDITIONAL LIKELIHOOD RATIO AND RAO'S TESTS

Consider now a subclass F_1 of F for which Q_1 is of the form

$$Q_1 = h_o(m_o^*) + h_1(m_o^*) H_{10}^* H_{20}^* + h_2(m_o^*)(H_{10}^*)^2 , \tag{3.1}$$

where $h_i(.)$, $i = 0,1,2$, are smooth functions. The class F_1 is also fairly large and by (2.3) includes the likelihood ratio, conditional likelihood ratio, Rao's and Wald's tests. By (2.5), parametric orthogonality and simple regularity conditions,

$$H_{10}^* = V_{10} + d\, a_{10} + O(n^{-\frac{1}{2}}) ,$$

$$H_{20}^* = V_{20} + d\, L_{10.20} - a_{20}^{-1} L_{01.20} V_{01} + O(n^{-\frac{1}{2}}) ,$$

so that by (3.1),

$$Q_1 = h_o(m) + h_1(m)(V_{10} + d\, a_{10})(V_{20} + d\, L_{10.20} - a_{20}^{-1} L_{01.20} V_{01})$$
$$+ h_2(m)(V_{10} + d\, a_{10})^2 + O(n^{-\frac{1}{2}}) . \tag{3.2}$$

Using parametric orthogonality again,

$$S_{10} = V_{10} + d\, a_{20} + O(n^{-\frac{1}{2}}), \quad S_{01} = V_{01} + O(n^{-\frac{1}{2}}) ,$$
$$S_{11} = V_{11} - d\, L_{21} + O(n^{-\frac{1}{2}}), \quad S_{20} = V_{20} + d\, L_{10.20} + O(n^{-\frac{1}{2}}). \tag{3.3}$$

Observe that under Θ_n, up to the first order of approximation, the joint distribution of $(V_{10}, V_{01}, V_{11}, V_{20})'$ is four variate normal (vide Assumption 1) with a null mean vector and a dispersion matrix

$$\begin{bmatrix} a_{10} & 0 & -L_{21} & L_{10.20} \\ 0 & a_{20} & -L_{12} & L_{01.20} \\ -L_{21} & -L_{12} & L_{11.11} & L_{20.11} \\ L_{10.20} & L_{01.20} & L_{20.11} & L_{20.20} - a_{10}^2 \end{bmatrix},$$

again employing parametric orthogonality and standard regularity conditions. Hence by (3.2), (3.3), after some algebra,

$$C_1(d) = h_o(m) + h(m)(1 + d^2a_{10}) ,$$

$$C_4(d) = h_o^2(m) + 2h_o(m)h(m)(1 + d^2a_{10}) + h_2(m)a_{10}(h(m) + h_1(m)L_{10.20})(3 + 6d^2a_{10} + d^4a_{10}^2) + h_1^2(m)[a_{10}D(1 + d^2a_{10}) + L_{10.20}^2(2 + 5d^2a_{10} + d^4a_{10}^2)] ,$$

$$C_5(d) = h_o(m)dL_{10.20} + h_1(m)[a_{10}Dd + dL_{10.20}^2(2 + d^2a_{10})] + h_2(m)da_{10}L_{10.20}(3 + d^2a_{10}) ,$$

$$C_7(d) = - h_o(m)L_{12} - h(m)L_{12}(1 + d^2a_{10}) ,$$

$$C_8(d) = - a_{20}C_7(d)/L_{12} ,$$

$$C_9(d) = 0 ,$$

$$C_{10}(d) = \frac{1}{2}(a_{20}a_{10}^{\frac{1}{2}})^{-1}C_7(d) ,$$

where the $C_j(d)$'s are as defined in the last section and

$$D = L_{20.20} - a_{10}^2 - a_{20}^{-1} L_{01.20}^2 , \quad h(m) = h_1(m)L_{10.20} + h_2(m)a_{10} .$$

Hence by (2.4), for $u = z$ or $-z$,

$$C_1^*(u) = h_o(m) + h(m)u^2 , \quad C_4^*(u) = \{C_1^*(u)\}^2 + u^2h_1^2(m)D^* ,$$

$$C_5^*(u) = a_{10}^{-\frac{1}{2}}[h_o(m)L_{10.20}u + h_1(m)D^*u + L_{10.20}h(m)u^3] ,$$

$$C_{10}^*(u) = - \frac{1}{2} L_{12}(a_{20}a_{10}^{\frac{1}{2}})^{-1} C_1^*(u) ,$$

$$w_{11} = - w_{21} = - \frac{1}{2} a_{10}^{\frac{1}{2}} z^2 h_1^2(m)D^* , \tag{3.4}$$

$$w_{12} = w_{22} = \frac{1}{2} a_{10}^{-\frac{1}{2}} z h_1(m)D^*,$$

where the w_{ij}'s are as in Theorem 1 and $D^* = a_{10}D - L_{10.20}^2$.

The quantity D^* may be interpreted in terms of a version of Efron's curvature (Efron [6]), appropriate in the present context, as follows. Let $G(m)$ be the 2×2 dispersion matrix at θ_o of the least squares linear regression residuals of $\partial \log f(X;\theta_o,m)/\partial\theta$ and $\partial^2 \log f(X;\theta_o,m)/\partial\theta^2$ on $\partial \log f(X;\theta_o,m)/\partial m$, and define a modified version of Efron's curvature at θ_o by $\nu(m) = |G(m)|^{\frac{1}{2}} a_{10}^{-3/2}$. Then using parametric orthogonality it can be seen that $D^* = \nu^2(m)a_{10}^3$. Hence by Theorem 1 and (3.4), proceeding as in Chandra and Mukerjee [4], Mukerjee and Chandra [9], for every test in the class F_1,

$$P_2 - P_{20}(d,m) = \phi(z)\nu^2(m)a_{10}^3 \exp(- \frac{1}{2} d^2a_{10}) \sum_{r=1}^{\infty} T_{1r}(z,d,m) \tag{3.5a}$$

$$= z\phi(z)\nu^2(m)a_{10}^{3/2} \sum_{r=1}^{\infty} T_{2r}(z,d,m), \tag{3.5b}$$

where

$$T_{1r}(z,d,m) = (z^2d^2a_{10})^r[(2r-1)(za_{10}^{3/2})^{-1}h_1(m) - z\,h_1^2(m)]/(2r-1)! \tag{3.6a}$$

$$T_{2r}(z,d,m) = (d^2a_{10})^r[(2r-1)J_{2r-2}(z)h_1(m) - z\,a_{10}^{3/2}\,J_{2r-1}(z)h_1^2(m)]/(2r-1)! \tag{3.6b}$$

The relations (3.5b) and (3.6b) present an extension of the findings in Chandra and Mukerjee [4] to the present set-up. Several observations follow from (3.4), (3.5a,b) and (3.6a,b).

(a) Let F_1^* be a subclass of F_1 containing those tests for which $h_1(.)$ is free from z. By (2.3), the likelihood ratio, conditional likelihood ratio, Rao's and Wald's tests belong to F_1^*. For Rao's test $h_1(m) = 0$. Hence denoting the expression for P_2 for Rao's test by P_2^{Rao}, it follows from (3.5b), (3.6b) that for every test in F_1^*,

$$P_2^{Rao} - P_2 = -z\phi(z)\nu^2(m)a_{10}^{3/2}\sum_{r=1}^{\infty}T_{2r}(z,d,m), \tag{3.7}$$

and that for every positive integer r' if one truncates the infinite series in (3.7) keeping the first r' terms then the resulting expression would be positive for sufficiently large z (i.e., for sufficiently small α) provided $\nu(m) \neq 0$ and $h_1(m) \neq 0$. In this sense Rao's test is optimal in F_1^* and hence superior to the likelihood ratio, conditional likelihood ratio and Wald's tests. In particular, by (3.6b), (3.7), for every test in F_1^*,

$$P_2^{Rao} - P_2 = z\phi(z)\nu^2(m)a_{10}^{5/2}(z^2a_{10}^{3/2}h_1^2(m) - h_1(m))d^2 + O(d^4)\ ,$$

and the coefficient of d^2 in the above is positive for large z provided $\nu(m) \neq 0$ and $h_1(m) \neq 0$.

(b) Rao's test, however, is not optimal in the entire class F_1 (or in the class F) in the above sense. Consider a test in F_1 for which $h_1(m) = (2z^2a_{10}^{3/2})^{-1}$, i.e., $h_1(m_o^*) = (2z^2a_1^{*3/2})^{-1}$. Let P_2^* denote the expression for P_2 for this test. They by (3.5a), (3.6a),

$$P_2^* - P_2^{Rao} = \frac{1}{2}\phi(z)\nu^2(m)z^{-3}\exp(-\frac{1}{2}d^2a_{10}) \times \sum_{r=1}^{\infty}(z^2d^2a_{10})^r(2r-\frac{3}{2})/(2r-1)!$$

which is positive (in fact all partial sums of the above are positive) provided $\nu(m) \neq 0$.

(c) It can be seen that in general there does not exist a test which maximizes P_2, uniformly in δ and m, in the class F_1. However, using (3.5b) ,

(3.6b), if one writes $P_2 - P_{20}(d,m) = Rd^2 + O(d^4)$, where

$$R = z\phi(z)\nu^2(m)a_{10}^{5/2}(h_1(m) - z^2a_{10}^{3/2}h_1^2(m)),$$

then the test considered in (b) above maximizes R.

(d) By Theorem 1 and (3.4), the third-order power of a test in F_1 does not involve $h_0(.)$. Hence by (2.3), (3.1), up to the third order of comparison , the conditional likelihood ratio test and the usual likelihood ratio test have identical (local) power functions.

REFERENCES

[1] Amari, S., Differential-Geometrical Methods in Statistics (Springer-Verlag, New York, 1985).

[2] Bickel, P.J., Chibisov, D.M. and van Zwet, W.R., On the efficiency of first and second order, Int. Statist. Rev. 49 (1981) 169-175.

[3] Chandra, T.K. and Joshi, S.N., Comparison of the likelihood ratio, Rao's and Wald's tests and a conjecture of C.R. Rao, Sankhyā A45 (1983) 226-246.

[4] Chandra, T.K. and Mukerjee, R., Comparison of the likelihood ratio, Wald's and Rao's tests, Sankhyā A47 (1985) 271-284.

[5] Cox, D.R. and Reid, N., Parameter orthogonality and approximate conditional inference (with discussion), J. Roy. Statist. Soc. B49 (1987) 1-39.

[6] Efron, B., Defining the curvature of a statistical problem (with applications to second order efficiency), Ann. Statist. 3(1975) 1189-1242.

[7] Kumon, M. and Amari, S., Differential geometry of testing hypothesis - a higher order asymptotic theory in multi-parameter curved exponential family (Preprint, Tokyo University, 1985).

[8] Mukerjee, R., Third-order comparison of unbiased tests - a simple formula for the power difference in the one-parameter case, Sankhyā A51 (1989), in print.

[9] Mukerjee, R. and Chandra, T.K., Comparison between the locally most powerful unbiased and Rao's tests, J. Multivariate Analysis 22 (1987) 94-105.

[10] Pfanzagl, J., First order efficiency implies second order efficiency, in : Jureckova, J., (ed), Contributions to Statistics : J. Hajek Memorial Volume (Academia ,Prague, 1979) pp. 167-196.

[11] Pfanzagl, J. and Wefelmeyer, W., An asymptotically complete class of tests, Z. Wahrsch. verw. Gebiete 45(1978) 49-72.

[12] Rao, C.R., Linear Statistical Inference and Its Applications, 2nd ed. (John Wiley, New York, 1973).

STATISTICAL DATA ANALYSIS AND INFERENCE
Y. Dodge (Editor)

TESTS FOR REDUNDANCY OF SOME VARIABLES IN MULTIVARIATE ANALYSIS

Yasunori FUJIKOSHI

Department of Mathematics, Faculty of Science, Hiroshima University, Hiroshima 730, Japan

The main object of the present paper is to review the problems of formulating and testing the hypothesis of redundancy of some variables in each of various multivariate situations, including multivariate regression model, multivariate inverse regression model, discriminant analysis, covariate discriminant analysis, multivariate growth curve model, canonical correlation analysis, principal component analysis. The hypothesis in each of multivariate situations are formulated from three different approaches. Two alternative tests as well as the LR test are given.

1. INTRODUCTION

In most multivariate situations, it is of interest to find out whether the addition of a new set of variables gives additional information for an aim of statistical analysis. For example, in the area of discriminant analysis, it is of interest to find out whether the addition of new variables contributes to the discrimination between the groups. In the area of canonical correlation analysis, it is of interest to find out whether the addition of variables to one or both sets of variables contributes to the degree of relationships between the sets of variables. An approach for such problems is to formulate and test the hypothesis of redundancy of some variables in each of multivariate situations.

Rao [26], [27], [30] discussed the redundancy of a given set of variables, for purposes of discrimination between two groups with the same covariance matrix. Since then, this idea has been extended to various multivariate situations including multiple discriminant analysis, covariate discriminant analysis, canonical correlation analysis, multivariate inverse regression model, multivariate linear model, multivariate growth curve model, principal component analysis. The main object of the present paper is to review the problems of formulating and testing the hypothesis of redundancy of some variables in each of various multivariate situations. We also attempt to characterize the hypothesis of redundancy from three different approaches and to propose two alternative tests as well as the LR (= likelihood ratio) test for the hypothesis of redundancy. Let $\underset{\sim}{y}$ be a vector of p random variables, and let $\underset{\sim}{y}$ partition as $\underset{\sim}{y} = (\underset{\sim}{y}_1', \underset{\sim}{y}_2')'$, $\underset{\sim}{y}_i$: $p_i \times 1$. Then we characterize the hypothesis of redundancy of $\underset{\sim}{y}_2$ or sufficiency of $\underset{\sim}{y}_1$ in each of various multivariate situations from the following three approaches:

(1) If we have the transformed variables, $\underset{\sim}{\gamma}_j'\underset{\sim}{y}$, j=1,...,m useful in an aim of statistical analysis, then it is natural to define the hypothesis of redundancy of $\underset{\sim}{y}_2$ as $\underset{\sim}{\gamma}_{2j} = 0$, j=1,...,m, where $\underset{\sim}{\gamma}_j = (\underset{\sim}{\gamma}_{1j}', \underset{\sim}{\gamma}_{2j}')'$, $\underset{\sim}{\gamma}_j$: $p_i \times 1$.

(2) In some situations it is possible to express the contribution of $\underset{\sim}{y}$ for an aim of statistical analysis as a criterion $\Delta_p^2(\underset{\sim}{y})$ such as Mahalanobis distance, the sum of squares of canonical correlations, etc. In the situations we can define the hypothesis of redundancy of $\underset{\sim}{y}_2$ as $\Delta_p^2(\underset{\sim}{y}) = \Delta_{p_1}^2(\underset{\sim}{y}_1)$.

(3) We examine the hypothesis of redundancy of $\underset{\sim}{y}_2$ in terms of the conditional distribution of $\underset{\sim}{y}_2$ given $\underset{\sim}{y}_1$ or the random vector $\underset{\sim}{y}_2^*$ obtained by substracting from $\underset{\sim}{y}_2$ its regression on $\underset{\sim}{y}_1$.

The third approach is useful in obtaining the LR criterion for the hypothesis of redundancy of $\underset{\sim}{y}_2$. We note that the hypotheses discussd in this paper are useful for the problem of selecting the best subset of variables. The likelihood ratio criterions are easily extended to a selection criterion, based on Akaike's information criterion (Akaike [1]). For these results, see, e.g., Fujikoshi [9], [10], [11], Fujikoshi and Nishii [12], Nishii [24]. For applications of these hypotheses to a simultaineous procedure, see McKay [20], [21], [22], [23], etc.

2. MULTIVARIATE REGRESSION MODEL

Let $\underset{\sim}{y} = (y_1,\ldots,y_p)'$ and $\underset{\sim}{x} = (x_1,\ldots,x_q)'$ be the vectors of p response variables and q explanatory variables, respectively. It is assumed that

$$\underset{\sim}{y} = \underset{\sim}{\alpha} + \beta'\underset{\sim}{x} + \underset{\sim}{\varepsilon} \tag{2.1}$$

where $\underset{\sim}{\alpha}$ is a vector of unknown parameters, β is a q×p matrix of unknown parameters, and $\underset{\sim}{\varepsilon}$ is an error vector having a multivariate normal distribution $N_p(\underset{\sim}{0}, \Sigma)$. Further, it is assumed that N independent observations $\underset{\sim}{y}_j$, $\underset{\sim}{x}_j$ (j=1,...,N) on $\underset{\sim}{y}$ amd $\underset{\sim}{x}$ are available. Then, letting $Y = [\underset{\sim}{y}_1,\ldots,\underset{\sim}{y}_N]'$ and $X = [\underset{\sim}{x}_1,\ldots,\underset{\sim}{x}_N]'$, we have a multivariate regression model

$$Y = \underset{\sim}{1}_N\underset{\sim}{\alpha}' + X\beta + \varepsilon, \tag{2.2}$$

where $\underset{\sim}{1}_N = (1,\ldots,1)'$, $\varepsilon = [\underset{\sim}{\varepsilon}_1,\ldots,\underset{\sim}{\varepsilon}_N]$, and $\underset{\sim}{\varepsilon}_1,\ldots,\underset{\sim}{\varepsilon}_N$ are independently and identically distributed as $N_p(\underset{\sim}{0}, \Sigma)$.

In order to formulate the redundancy of a subvector of $\underset{\sim}{x}$, we partition $\underset{\sim}{x}$ as $(\underset{\sim}{x}_1', \underset{\sim}{x}_2')'$, $\underset{\sim}{x}_i$: q_i×1. It is natural to define the redundancy of $\underset{\sim}{x}_2$ in the presence of $\underset{\sim}{x}_1$ for purposes of predicting $\underset{\sim}{y}$ from $\underset{\sim}{x}$ as

$$H: \beta_2 = 0, \tag{2.3}$$

where $\beta = [\beta_1' \ \beta_2']'$, β_i: q_i×p. This testing problem has been extensively discussed in a frame work of multivariate linear hypotheses (see, e.g., Anderson [2], Seber [32]). We shall see that some testing problems discussed in this paper can be reduced to the ones in a multivariate linear model, after a conditional set-up. Let $X = [X_1 \ X_2]$, X_i: $N \times q_i$, and let P_Ω and P_ω be the projection operators on $\mathbb{R}[\underset{\sim}{1}_N \ X]$ and $\mathbb{R}[\underset{\sim}{1}_N \ X_1]$, respectively, where $\mathbb{R}[A]$ is the space spanned by the column vectors of A. The LR test for H is based on

$$\Lambda = |S_e|/|S_e + S_h|, \tag{2.4}$$

where $S_e = Y'(I_N-P_\Omega)Y$ and $S_h = Y'(P_\Omega-P_\omega)Y$. The null distribution of Λ is the Wilks' Λ-distribution $\Lambda_p(q_2, N-q-1)$. We may use the following trace test statistics

$$T_0^2 = \text{tr } S_h S_e^{-1} \quad \text{and} \quad V = \text{tr } S_h(S_h+S_e)^{-1} \tag{2.5}$$

whose null distributions are denoted as T_0^2 (p, q_2, N-q-1) and V(p, q_2, N-q-1), respectively. For a summary of the distributional results on Λ, T_0^2 and V, see Siotani and et al. [34, p.247].

Let

$$\begin{aligned} S &= \sum_{j=1}^{N} \begin{pmatrix} \underset{\sim}{x}_j - \bar{\underset{\sim}{x}} \\ \underset{\sim}{y}_j - \bar{\underset{\sim}{y}} \end{pmatrix} \begin{pmatrix} \underset{\sim}{x}_j - \bar{\underset{\sim}{x}} \\ \underset{\sim}{y}_j - \bar{\underset{\sim}{y}} \end{pmatrix}' \\ &= \begin{pmatrix} S_{xx} & S_{xy} \\ S_{yx} & S_{yy} \end{pmatrix}, \end{aligned} \tag{2.6}$$

where $\bar{\underset{\sim}{x}} = N^{-1}\Sigma_{j=1}^{N}\underset{\sim}{x}_j$ and $\bar{\underset{\sim}{y}} = N^{-1}\Sigma_{j=1}^{N}\underset{\sim}{y}_j$. Further, we decompose S_{xx}, S_{xy} as

$$S_{xx} = \begin{pmatrix} S_{aa} & S_{ab} \\ S_{ba} & S_{bb} \end{pmatrix}, \quad S_{xy} = \begin{pmatrix} S_{ab} \\ S_{by} \end{pmatrix}, \tag{2.7}$$

corresponding to the partition $(\underset{\sim}{x}_1', \underset{\sim}{x}_2')'$ of $\underset{\sim}{x}$. Then S_e and S_h are expressed as

$$\begin{aligned} S_e &= S_{yy\cdot x} = S_{yy} - S_{yx}S_{xx}^{-1}S_{xy}, \\ S_h &= S_{yx}S_{xx}^{-1}S_{xy} - S_{ya}S_{aa}^{-1}S_{ay}. \end{aligned} \tag{2.8}$$

3. MULTIVARIATE INVERSE REGRESSION MODEL

We consider the situation where the vector $\underset{\sim}{y}$ of response variables in the model (2.1) is observed. but the vector $\underset{\sim}{x}$ of explantory variables is unknown. The problem of estimating $\underset{\sim}{x}$ from $\underset{\sim}{y}$ is called a multivariate inverse regression problem (Brown [3]). If $\underset{\sim}{\alpha}$, β and Σ are known, we may estimate $\underset{\sim}{x}$ by minimizing

$$(\underset{\sim}{y} - \underset{\sim}{\alpha} - \beta'\underset{\sim}{x})'\Sigma^{-1}(\underset{\sim}{y} - \underset{\sim}{\alpha} - \beta'\underset{\sim}{x}). \tag{3.1}$$

For the case p ≥ q, the solution is uniquely given by

$$\hat{x} = \Xi(y - \alpha), \tag{3.2}$$

where $\Xi = (\beta\Sigma^{-1}\beta')^{-1}\beta\Sigma^{-1}$. We partition y, α, Ξ and Σ as

$$y = \begin{pmatrix} y_1 \\ y_2 \end{pmatrix}, \quad \alpha = \begin{pmatrix} \alpha_1 \\ \alpha_2 \end{pmatrix}, \quad \Sigma = \begin{pmatrix} \Sigma_{11} & \Sigma_{12} \\ \Sigma_{21} & \Sigma_{22} \end{pmatrix}, \tag{3.3}$$

$$\Xi = [\Xi_1 \ \Xi_2], \quad y_i: p_i \times 1, \quad \alpha_i: p_i \times 1, \quad \Xi_i: q \times p_i, \quad \Sigma_{ii}: p_i \times p_i.$$

We define the redundancy of y_2 in the presense of y_1 for the multivariate inverse problem as

$$H: \Xi_2 = 0 \tag{3.4}$$

which was introduced by Fujikoshi and Nishii [12].

Theorem 3.1. If $p \geq q$, the following statements are equivalent.

[3.1] $\Xi_2 = 0$,

[3.2] $\mathrm{tr}\ \beta'\beta\Sigma^{-1} = \mathrm{tr}\ \beta_1'\beta_1\Sigma_{11}^{-1}$,

[3.3] $\eta = \beta_2 - \beta_1\Sigma_{11}^{-1}\Sigma_{12} = 0$.

Proof. It is easily seen that

$$\Xi = (\beta\Sigma^{-1}\beta')^{-1}[\beta_1\Sigma_{11}^{-1} - \eta\Sigma_{22\cdot 1}^{-1}\Sigma_{21}\Sigma_{11}^{-1}, \ \eta\Sigma_{11\cdot 2}^{-1}]$$

and

$$\mathrm{tr}\ \beta'\beta\Sigma^{-1} = \mathrm{tr}\ \beta_1'\beta_1\Sigma_{11}^{-1} + \mathrm{tr}\ \eta'\eta\Sigma_{22\cdot 1}^{-1}.$$

These give the desired result.

We note that the equivalence of [3.2] and [3.3] holds for any p and q. The statement [3.3] is the same as Rao's additional information hypothesis in a multivariate linear model. Rao [29], [31] has discussed the problem of testing "$\eta = 0$" in the model (2.2), based on a conditional model of Y_2 given Y_1, where $Y = [Y_1 \ Y_2]$ and Y_i; $N \times p_i$. The rows of Y_2 given Y_1 are independently normally distributed as $N_{p_2}(\cdot, \Sigma_{22\cdot 1})$ and

$$E[Y_2|Y_1] = X\eta + \underset{\sim}{1}_N\underset{\sim}{\alpha}_2' + (Y_1 - \underset{\sim}{1}_N\underset{\sim}{\alpha}_1')\Sigma_{11}^{-1}\Sigma_{12}. \tag{3.5}$$

Therefore, the LR test for "$\eta = 0$" or equivalntly "$\Xi_2 = 0$" is based on

$$\begin{aligned}\Lambda &= |S_{22\cdot x} - S_{21\cdot x}S_{11\cdot x}^{-1}S_{12\cdot x}|/|S_{22\cdot 1}| \\ &= \frac{|S_{yy\cdot x}|}{|S_{11\cdot x}|}\cdot\frac{|S_{11}|}{|S_{yy}|}\end{aligned} \tag{3.6}$$

whose null distribution is $\Lambda_{p_2}(q, N-q-1-p_1)$, where S is defined by (2.6), $S_{22\cdot 1} = S_{22} - S_{21}S_{11}^{-1}S_{12}$, and

$$S_{yy} = \begin{pmatrix} S_{11} & S_{12} \\ S_{21} & S_{22} \end{pmatrix}, \quad S_{yy\cdot x} = \begin{pmatrix} S_{11\cdot x} & S_{12\cdot x} \\ S_{21\cdot x} & S_{22\cdot x} \end{pmatrix},$$

$$S_{ii}: p_i \times p_i, \qquad S_{ii\cdot x}: p_i \times p_i.$$

The test statistics similar to the ones in (2.5) are also used.

4. DISCRIMINANT ANALYSIS

4.1. The case of two groups. Let $\underset{\sim}{x}$ be the vector of p random variables $x_1, \ldots, x_p$ measurable on the individuals of each of two populations π_1 and π_2. Consider the partition $\underset{\sim}{x} = (x_1', x_2')'$, $\underset{\sim}{x}_i: p_i \times 1$. It is assumed that for i = 1, 2,

$$\begin{aligned}E[\underset{\sim}{x}|\pi_i] &= \underset{\sim}{\mu}^{(i)} = \begin{pmatrix} \underset{\sim}{\mu}_1^{(i)} \\ \underset{\sim}{\mu}_2^{(i)} \end{pmatrix}, \\ \mathrm{Var}[\underset{\sim}{x}|\pi_i] &= \Sigma = \begin{pmatrix} \Sigma_{11} & \Sigma_{12} \\ \Sigma_{21} & \Sigma_{22} \end{pmatrix}.\end{aligned} \tag{4.1}$$

The vector of coefficients in the linear discriminant function is given by

$$\chi = \Sigma^{-1}(\underset{\sim}{\mu}^{(1)} - \mu^{(2)}) = (\chi_1', \chi_2')'. \tag{4.2}$$

Letting $\boldsymbol{\delta} = (\boldsymbol{\mu}^{(1)} - \boldsymbol{\mu}^{(2)}) = (\boldsymbol{\delta}_1', \boldsymbol{\delta}_2')'$, the square of Mahalanobis distance between π_1 and π_2 based on $\mathbf{x}$ is $\Delta_p^2 = \boldsymbol{\delta}'\Sigma^{-1}\boldsymbol{\delta}$ while that based on $\mathbf{x}_1$ is $\Delta_{p_1}^2 = \boldsymbol{\delta}_1'\Sigma_{11}^{-1}\boldsymbol{\delta}_1$. Rao [26], [27], [30] discussed the redundancy of $\mathbf{x}_2$ in the presence of $\mathbf{x}_1$ for purposes of discrimination, based on the approaches mentioned in Section 1. The hypothesis of redundancy of $\mathbf{x}_2$ may be defined as one of the following statements:

[4.1] $\boldsymbol{\gamma}_2 = \mathbf{0}$,

[4.2] $\Delta_p^2 = \Delta_{p_1}^2$,

[4.3] $\boldsymbol{\mu}_2^{(1)} - \Sigma_{21}\Sigma_{11}^{-1}\boldsymbol{\mu}_1^{(1)} = \boldsymbol{\mu}_2^{(2)} - \Sigma_{21}\Sigma_{11}^{-1}\boldsymbol{\mu}_1^{(2)}$.

For a proof of the equivalence of [4.1]~[4.3] and some other statements, see Rao [30].

Suppose that random samples of sizes N_i from π_i are available. Then, the LR test for the redundancy of $\mathbf{x}_2$ is obtained by using the conditional distribution of $\mathbf{x}_2$ given $\mathbf{x}_1$ and the statement [4.3]. Let D_p and D_{p_1} be the sample Mahalanobis distances based on $\mathbf{x}$ and $\mathbf{x}_1$, respectively. The LR test is to reject the hypothesis if

$$F = \frac{N-p-1}{p_2} \cdot \frac{D_p^2 - D_{p_1}^2}{n(N_1^{-1}+N_2^{-1}) + D_{p_1}^2} \tag{4.3}$$

is large, where $N = N_1 + N_2$ and $n = N-2$. The null distribution of F is the F-distribution with p_2 and $N-p-1$ degrees of freedom.

3.2. The case of multiple groups. Consider $k+1$ p-variate populations $\pi_i (i=1,\ldots,k+1)$ with means $\boldsymbol{\mu}^{(i)}$ and the same covariance matrix Σ. Suppose that $\boldsymbol{\mu}^{(i)}$ and Σ are unknown, and random samples of sizes N_i from π_i are available. Let Ω be the population between-groups covariance matrix difined by

$$\Omega = \sum_{i=1}^{k+1} g_i(\boldsymbol{\mu}^{(i)} - \bar{\boldsymbol{\mu}})(\boldsymbol{\mu}^{(i)} - \bar{\boldsymbol{\mu}})', \tag{4.4}$$

where $\{g_i\}$ is a set of positive numbers such that $\sum g_i = 1$, and $\bar{\boldsymbol{\mu}}$

$= \sum g_i \mu^{(i)}$. We may define g_i by $N_i/\sum N_i = N_i/N$. Discriminant analysis generally has two aims; One is to interpret the differences among k+1 populations in terms of only a few discriminant functions. The other is to assign the individuals under consideration to one of k+1 populations. For the first aim, we use the discriminant functions $\gamma_j' x$, $j=1,\ldots,m = \operatorname{rank}(\Omega)$ defined by the solutions of

$$\Omega\gamma_j = \theta_j \Sigma\gamma_j, \quad \gamma_i'\Sigma\gamma_j = \delta_{ij}, \tag{4.5}$$

where $\theta_1 \geq \cdots \geq \theta_m$ are the possibly nonzero roots of $\Sigma^{-1}\Omega$. and $\delta_{ij} = 1$ for $i = j$, 0 for $i \neq j$. In order to formulate the redundancy of x_2, we partion $\mu^{(i)}$, γ_j and Ω as in (4.1), i.e.

$$\mu^{(i)} = \begin{pmatrix} \mu_1^{(i)} \\ \mu_2^{(i)} \end{pmatrix}, \quad \gamma_j = \begin{pmatrix} \gamma_{1j} \\ \gamma_{2j} \end{pmatrix}, \quad \Omega = \begin{pmatrix} \Omega_{11} & \Omega_{12} \\ \Omega_{21} & \Omega_{22} \end{pmatrix}. \tag{4.6}$$

The statements [4.1]~[4.3] are extended to the case of multiple groups as follows:

Theorem 4.1. The following statements are equivalent.

[4.4] $\gamma_{2j} = 0$, $j=1,\ldots,m = \operatorname{rank}(\Omega)$,

[4.5] $\operatorname{tr} \Omega\Sigma^{-1} = \operatorname{tr} \Omega_{11}\Sigma_{11}^{-1}$,

[4.6] $\mu_2^{(1)} - \Sigma_{21}\Sigma_{11}^{-1}\mu_1^{(1)} = \cdots = \mu_2^{(k+1)} - \Sigma_{21}\Sigma_{11}^{-1}\mu_1^{(k+1)}$.

The statement [4.6] has been used in Rao [29], [31] etc. The staments [4.4] and [4.6] which are shown to be equivalent to [4.6] were introduced by Fujikoshi [8] and McKay [22], respectively.

When the random samples from Π_i are multivariate normal, the LR test is obtained as a special case of the one for redundancy in a multivariate linear model (see Section 6). Let W and B be the matrices of sums of squares and products due to within-groups and between-groups, respectively. We partion W, B and T = W + B as in (4.1). Then the LR test is based on

$$\Lambda = |W_{22\cdot 1}|/|T_{22\cdot 1}| = \frac{|W|}{|T|} \cdot \frac{|T_{11}|}{|W_{11}|} \tag{4.7}$$

whose null distribution is $\Lambda_{p_2}(k, N-(k+1)-p_1)$. As alternative test statistics, we propose

$$T_0^2 = tr(T_{22\cdot 1} - W_{22\cdot 1})W_{22\cdot 1}^{-1}, \text{ and}$$
$$V = tr(T_{22\cdot 1} - W_{22\cdot 1})T_{22\cdot 1}^{-1} \tag{4.8}$$

whose null distributions are $T_0^2(p_2, k, N-(k+1)-p_1)$ and $V(p_2, k, N-(k+1)-p_1)$, respectively.

5. COVARIATE DISCRIMINANT ANALYSIS

5.1. The hypothesis of redundancy. Cochran and Bliss [4] discussed the classification problem when some variables (called covariables) are known to be the same means in two groups π_1 and π_2. The discriminant analysis in this case is called a covariate discriminant analysis. Let $\underset{\sim}{y}_x$ and $\underset{\sim}{y}_z$ be the vectors of discriminate variables and of covariables, respectively, measurable the individuals of each of k+1 populations π_i (i=1,... ,k+1). It is assumed that for i=1,...,k+1,

$$E\left[\begin{pmatrix} \underset{\sim}{y}_x \\ \underset{\sim}{y}_z \end{pmatrix} | \pi_i\right] = \underset{\sim}{\eta}^{(i)} = \begin{pmatrix} \underset{\sim}{\mu}^{(i)} \\ \underset{\sim}{\xi} \end{pmatrix},$$
$$Var\left[\begin{pmatrix} \underset{\sim}{y}_x \\ \underset{\sim}{y}_z \end{pmatrix} | \pi_i\right] = \Sigma = \begin{pmatrix} \Sigma_{xx} & \Sigma_{xz} \\ \Sigma_{zx} & \Sigma_{zz} \end{pmatrix}. \tag{5.1}$$

The population between-groups covariance matrix in (4.4) is equal to

$$\Omega = \sum_{i=1}^{k+1} g_i(\underset{\sim}{\eta}^{(i)}-\bar{\underset{\sim}{\eta}})(\underset{\sim}{\eta}^{(i)}-\bar{\underset{\sim}{\eta}})' = \begin{pmatrix} \Omega_{xx} & 0 \\ 0 & 0 \end{pmatrix}, \tag{5.2}$$

where $\bar{\underset{\sim}{\eta}} = \Sigma g_i \underset{\sim}{\eta}^{(i)}$ and $\Omega_{xx} = \Sigma g_i(\underset{\sim}{\mu}^{(i)}-\bar{\underset{\sim}{\mu}})(\underset{\sim}{\mu}^{(i)}-\bar{\underset{\sim}{\mu}})'$. In order to formulate the redundancy of both the subvectors of $\underset{\sim}{y}_x$ and $\underset{\sim}{y}_z$, we partition $\underset{\sim}{y}$ and Σ as

$$
\underset{\sim}{y} = \begin{pmatrix} \underset{\sim}{y}_x \\ \\ \underset{\sim}{y}_z \end{pmatrix} = \begin{pmatrix} \underset{\sim}{y}_1 \\ \underset{\sim}{y}_2 \\ \underset{\sim}{y}_3 \\ \underset{\sim}{y}_4 \end{pmatrix}, \quad \Sigma = \begin{pmatrix} \Sigma_{11} & \Sigma_{12} & \Sigma_{13} & \Sigma_{14} \\ \Sigma_{21} & \Sigma_{22} & \Sigma_{23} & \Sigma_{23} \\ \Sigma_{31} & \Sigma_{32} & \Sigma_{33} & \Sigma_{34} \\ \Sigma_{41} & \Sigma_{42} & \Sigma_{43} & \Sigma_{44} \end{pmatrix}, \quad \begin{matrix} y_1 : p_i \times 1, \\ \Sigma_{ii} : p_i \times p_i . \end{matrix} \tag{5.3}
$$

Let $\underset{\sim}{y}_a = (\underset{\sim}{y}_1', \underset{\sim}{y}_3')'$ and $\underset{\sim}{y}_b = (\underset{\sim}{y}_2', \underset{\sim}{y}_4')'$. Similar notations are used for $\underset{\sim}{\gamma}_j$, Σ and Ω. For example,

$$
\underset{\sim}{\gamma}_{aj} = \begin{pmatrix} \underset{\sim}{\gamma}_{1j} \\ \underset{\sim}{\gamma}_{3j} \end{pmatrix}, \quad \Sigma_{ab} = \begin{pmatrix} \Sigma_{12} & \Sigma_{14} \\ \Sigma_{32} & \Sigma_{24} \end{pmatrix}. \tag{5.4}
$$

With the above notations, we define the redundancy of $\underset{\sim}{y}_b$ in the presence of $\underset{\sim}{y}_a$, as one of the following statements:

[5.1] $\underset{\sim}{\gamma}_{bj} = (\underset{\sim}{\gamma}_{2j}', \underset{\sim}{\gamma}_{4j}')' = \underset{\sim}{0}$,

[5.2] $\mathrm{tr}\ \Omega\Sigma^{-1} = \mathrm{tr}\ \Omega_{aa}\Sigma_{aa}^{-1}$,

[5.3] $\underset{\sim}{\mu}_b^{(1)} - \Sigma_{ba}\Sigma_{aa}^{-1}\underset{\sim}{\mu}_a^{(1)} = \cdots = \underset{\sim}{\mu}_b^{(k+1)} - \Sigma_{ba}\Sigma_{aa}^{-1}\underset{\sim}{\mu}_a^{(k+1)}$.

The equivalence of [5.1]~[5.3] follows from Theorem 4.1 since Theorem 4.1 holds for any $\underset{\sim}{\mu}^{(i)}$ and $\Sigma > 0$. For a direct proof, see Fujikoshi and Khatri [14].

5.2. Tests for redundancy. First we derive a canonical form for testing the redundancy of $\underset{\sim}{y}_b$ in the presence of $\underset{\sim}{y}_a$, based on random samples of sizes N_i from multivariate normal populations π_i $(i=1,\ldots,k+1)$. Let W, B and T = W+B be the matrices of sums of squares and products due to within-groups, between-groups and the total variation, respectively. Then, W and B are independently distributed as a central Wishart distribution $W_p(\Sigma, n)$ and a noncentral Wishart distribution $W_p(\Sigma, n; N\Omega)$, where $N = N_1 + \cdots + N_{k+1}$, $n = N-k-1$, and Ω is defined by (5.2) with $g_i = N_i/N$. Defining U, V and ν such that $B = U'U$, $W = V'V$ and $\nu'\nu = N\Omega$, we have the following canonical form: Let U: k×p and V: n×p be the random matrices whose rows are independently distributed as $N_p(\cdot, \Sigma)$ with

$$E\begin{pmatrix} U \\ V \end{pmatrix} = E\begin{pmatrix} U_1 & U_2 & U_3 & U_4 \\ V_1 & V_2 & V_3 & V_4 \end{pmatrix} = \begin{pmatrix} \nu_1 & \nu_2 & 0 & 0 \\ 0 & 0 & 0 & 0 \end{pmatrix}. \tag{5.5}$$

The hypothesis of [5.3] can be written as

$$H_o: [\nu_2 \ 0] - [\nu_1 \ 0]\Sigma_{aa}^{-1}\Sigma_{ab} = 0 \tag{5.6}$$

which can be decomposed as

$$H_{o1}: \nu_2 - \nu_1\Sigma_{11\cdot z}^{-1}\Sigma_{12\cdot z} = 0 \text{ and } H_{o2}: \nu_1\Sigma_{11\cdot 3}^{-1}\Sigma_{14\cdot 3} = 0. \tag{5.7}$$

Fujikoshi and Khatri [14] has obtained the LR test for (5.6) under (5.5). However, the test statistics in the general case is very complicated, and its null distribution is not known. Here we consider the LR test for redundancy of $\underset{\sim}{y}_2$ only, as an important special case. Using the same notations for U and V as in (5.3), we can write the problem of testing the redundancy of $\underset{\sim}{y}_2$ as follows: The elements of the canonical model in (5.4) are;

$$E\begin{pmatrix} U_1 & U_2 & U_z \\ V_1 & V_2 & V_z \end{pmatrix} = \begin{pmatrix} \nu_1 & \nu_2 & 0 \\ 0 & 0 & 0 \end{pmatrix}. \tag{5.8}$$

We wish to test

$$H_o: \nu_2^* = \nu_2 - \nu_1\Sigma_{11\cdot z}^{-1}\Sigma_{12\cdot z} = 0 \tag{5.9}$$

versus alternatives $\nu_2^* \neq 0$. Let

$$\begin{pmatrix} \beta_1 \\ \beta_z \end{pmatrix} = \begin{pmatrix} \Sigma_{11} & \Sigma_{12} \\ \Sigma_{z1} & \Sigma_{zz} \end{pmatrix}^{-1}\begin{pmatrix} \Sigma_{12} \\ \Sigma_{z2} \end{pmatrix}.$$

Then, using $\beta_1 = \Sigma_{11\cdot z}^{-1}\Sigma_{12\cdot z}$ we have

$$E[\begin{pmatrix} U_2 \\ V_2 \end{pmatrix} | \begin{pmatrix} U_1 & U_z \\ V_1 & V_z \end{pmatrix}] = \begin{pmatrix} \nu_2^* \\ 0 \end{pmatrix} + \begin{pmatrix} U_1 & U_z \\ V_1 & V_z \end{pmatrix}\begin{pmatrix} \beta_1 \\ \beta_2 \end{pmatrix}. \tag{5.10}$$

The testing problem (5.9) under the conditional model (5.10) is the one of testing a linear hypothesis in a multivariate linear model. Therefore, the LR test is based on

$$\Lambda = |W_{22\cdot 1z}|/|T_{22\cdot 1z}|, \tag{5.11}$$

where $W_{22\cdot 1z} = W_{22} - W_{2(1z)}W_{(1z)(1z)}^{-1}W_{(1z)2}$ and $T_{22\cdot 1z} = T_{22} - T_{2(1z)}T_{(1z)(1z)}^{-1}T_{(1z)2}$. The null disstribution of Λ is $\Lambda_{p_2}(k, n-(p-p_2))$.

6. MULTIVARIATE LINEAR MODEL

Consider a matrix Y: N×p whose rows are N independent observation vectors on a vector $\underset{\sim}{y}$ of p random variables. It is assumed that the rows of Y are independently distributed with common covariance matrix Σ and means

$$E[Y] = X\Xi, \tag{6.1}$$

where X is a known N×q matrix of rank q and Ξ is a q×p matrix of unknown parameters. Suppose we are interesting in departures from the nullity of the hypothesis H_o: $C\Xi = 0$, where C is a known s×q matrix of rank s. As a measure of the departures we consider

$$\Delta_p^2 = \text{tr}\ \Omega\Sigma^{-1}, \tag{6.2}$$

where $\Omega = (C\Xi)'D^{-1}C\Xi$, and D is a given positive definite matrix. If we take q = k+1, s = k and $C = [I_k\ -\underset{\sim}{1}_k]$ in multiple discriminant analysis, then Ω is equal to the population between-groups covariance matrix in (4.4). We partition $\underset{\sim}{y}$, Ξ, Σ and Ω as in (3.3). We define the redundancy of $\underset{\sim}{y}_2$ in the presence of $\underset{\sim}{y}_1$ for departures of the nullity of the hypothesis H_o as

$$\Delta_p^2 = \Delta_{p_1}^2\ (= \text{tr}\ \Omega_{11}\Sigma_{11}^{-1}). \tag{6.3}$$

Letting $\eta = \Xi_2 - \Xi_1\Sigma_{11}^{-1}\Sigma_{12}$, it is easily seen that

$$\begin{aligned}\Delta_p^2 &= \text{tr}\ C\Xi\Sigma^{-1}(C\Xi)'D^{-1} \\ &= \Delta_{p_1}^2 + \text{tr}\ C\eta\Sigma_{22\cdot 1}^{-1}(C\eta)'D^{-1}.\end{aligned} \tag{6.4}$$

This shows that (6.3) is equivalent to $C\eta = 0$. Further, let $\underset{\sim}{\gamma}_j$, j = 1,...,m = rank($\Omega$) be the characteristic vectors defined by

$$\Omega\chi_j = \theta_j \Sigma\chi_j, \tag{6.5}$$

where $\theta_1 \geq \cdots \geq \theta_m$ and $\chi_i'\Sigma\chi_j = \delta_{ij}$. We note that (6.3) is equivalent to

$$[-\Sigma_{21}\Sigma_{11}^{-1} \quad I_{p_2}]\Omega = 0. \tag{6.6}$$

Theorem 6.1. The following statements are equivalent.

[6.1] $\chi_{2j} = \underset{\sim}{0}$, $j = 1,\ldots,m = \text{rank}(\Omega)$,

[6.2] $\text{tr}\ \Omega\Sigma^{-1} = \text{tr}\ \Omega_{11}\Sigma_{11}^{-1}$,

[6.3] $C\eta = 0$, where $\eta = \Xi_2 - \Sigma_{21}\Sigma_{11}^{-1}\Xi_1$.

The comments for Theorem 4.1 may be also extended to this theorem.

Let Y partition as $[Y_1\ Y_2]$, Y_i: $N\times p_i$ and assume that the rows of Y are independently distributed as $N_p(\cdot, \Sigma)$. Then

$$E[Y_2|Y_1] = X\eta + Y_1\Sigma_{11}^{-1}\Sigma_{12}. \tag{6.7}$$

Let W and B be the matices of sums of squares and products due to error and due to the hypothesis H_o: $C\Xi = 0$, i.e.

$$\begin{aligned} W &= Y'\{I_N - X(X'X)^{-1}X'\}Y, \\ B &= Y'X(X'X)^{-1}C'\{C(X'X)^{-1}C'\}^{-1}C(X'X)^{-1}XY. \end{aligned} \tag{6.8}$$

Using the conditional set-up (6.7), Rao [29], [31] obtained the LR criterion λ for H: $C\eta = 0$ as

$$\lambda^{2/N} = \Lambda = |W_{22\cdot 1}|/|T_{22\cdot 1}|,$$

where T = W + B and W are partitioned as in (4.7). The null distribution of Λ is $\Lambda_{p_2}(s, N-q-p_1)$. The power of this test has been considered by Fujikoshi [7]. The test statistics as in (4.8) are also proposed for the hypothesis of redundancy of $\underset{\sim}{y}_2$.

7. MULTIVARIATE GROWTH CURVE MODEL

The multivariate growth curve model introduced by Potthoff and Roy [25] is defined by restricting $\Xi = \eta B$ in (6.1), where B is a k×p known matrix of rank k and η is a q×k matrix of unknown parameters, i.e.

$$E[Y] = X\eta B. \tag{7.1}$$

Rao [28], [29] gave the following reductions: Suppose Q_1 and Q_2 are p×k and p×(p-k) matrices, respectively, such that $Q = [Q_1\ Q_2]$ is a matrix of full rank and

$$BQ_1 = I_k, \quad BQ_2 = 0. \tag{7.2}$$

Letting $Y_x = YQ_1$ and $Y_z = YQ_2$, we have

$$E[Y_x\ Y_z] = [X\eta\ 0]. \tag{7.3}$$

For a notational convenience, we assume that the rows of $[Y_x\ Y_z]$ consist of the N independent observations on $\underset{\sim}{y}_x = (y_1,\dots,y_k)'$ and $\underset{\sim}{y}_z = (y_{k+1},\dots,y_p)'$, and the covariance matrix of $\underset{\sim}{y} = (\underset{\sim}{y}_1', \underset{\sim}{y}_2')'$ is Σ. We now consider the situation where we are interesting in departures from the nullity of H_o: $C\eta = 0$, where C is an s×q known matrix of rank s. Following the same lines as in Section 6, let us consider $\Delta_p^2 = \text{tr}\ \Omega\Sigma^{-1}$ with

$$\Omega = \{C[\eta\ 0]\}'D^{-1}\{C[\eta\ 0]\} = \begin{pmatrix} (C\eta)'D^{-1}C\eta & 0 \\ 0 & 0 \end{pmatrix}, \tag{7.4}$$

where D is a given positive definite matrix. Using the same notations as in Section 5.1 we define the redundancy of $\underset{\sim}{y}_b = (\underset{\sim}{y}_2', \underset{\sim}{y}_4')'$ as

$$\text{tr}\ \Omega\Sigma^{-1} = \text{tr}\ \Omega_{aa}\Sigma_{aa}^{-1} \tag{7.5}$$

which is equivalent to

$$H: C\{[\eta_2\ 0] - [\eta_1\ 0]\Sigma_{aa}^{-1}\Sigma_{ab}\} = 0. \tag{7.6}$$

It is easily seen that this testing problem can be reduced as the one in (5.6). So, the detail may be omitted.

8. CANONICAL CORRELATION ANALYSIS

8.1. Redundancy on one set of variables. Let $\underset{\sim}{y}_x$ and $\underset{\sim}{y}_z$ be two vectors of p (= $p_1 + p_2$) and q (= $p_3 + p_4$) random variables, respectively and a nonsingular covariance matrix

$$\Sigma = \begin{pmatrix} \Sigma_{xz} & \Sigma_{xz} \\ \Sigma_{zx} & \Sigma_{zz} \end{pmatrix}. \tag{8.1}$$

When we are concerned to summarize the relationships between $\underset{\sim}{y}_x$ and $\underset{\sim}{y}_z$ in terms of only a few transformed variables, the canonical correlation variables $(\underset{\sim}{\gamma}'_{xj}\underset{\sim}{y}_x,\ \underset{\sim}{\gamma}'_{zj}\underset{\sim}{y}_z)$, j = 1,...,m = rank($\Sigma_{xz}$) due to Hotelling [15] are used. The vectors $\underset{\sim}{\gamma}_{xj}$ and $\underset{\sim}{\gamma}_{zj}$ of coefficients in $\underset{\sim}{y}_x$ and $\underset{\sim}{y}_z$, respectively, are defined by

$$\begin{aligned} &\Sigma_{xz}\Sigma_{zz}^{-1}\Sigma_{zx}\underset{\sim}{\gamma}_{xj} = \rho_j^2\Sigma_{xx}\underset{\sim}{\gamma}_{xj}, \quad \underset{\sim}{\gamma}'_{xj}\Sigma_{xx}\underset{\sim}{\gamma}_{xj} = \delta_{ij}, \\ &\Sigma_{zx}\Sigma_{xx}^{-1}\Sigma_{xz}\underset{\sim}{\gamma}_{zj} = \rho_j^2\Sigma_{zz}\underset{\sim}{\gamma}_{zj}, \quad \underset{\sim}{\gamma}'_{zj}\Sigma_{zz}\underset{\sim}{\gamma}_{zj} = \delta_{ij}, \end{aligned} \tag{8.2}$$

where $\rho_1^2 \geq \cdots \geq \rho_m^2$ are the characteristic roots of $\Sigma_{xz}\Sigma_{zz}^{-1}\Sigma_{zx}\Sigma_{xx}^{-1}$. We use the notations for partitions of $\underset{\sim}{y} = (\underset{\sim}{y}'_x,\ \underset{\sim}{y}'_z)'$, $\underset{\sim}{\gamma} = (\underset{\sim}{\gamma}'_x,\ \underset{\sim}{\gamma}'_z)'$, $\underset{\sim}{\mu} = (\underset{\sim}{\mu}'_x,\ \underset{\sim}{\mu}'_z)'$ and Σ as in (5.3) and (5.4). On the other hand, the magnitude of relationships between $\underset{\sim}{y}_x$ and $\underset{\sim}{y}_z$ can be measured by

$$\begin{aligned} \Delta^2(\underset{\sim}{y}_x,\ \underset{\sim}{y}_z) &= \operatorname{tr} \Sigma_{xz}\Sigma_{zz}^{-1}\Sigma_{zx}\Sigma_{xx}^{-1} \\ &= \rho_1^2 + \cdots + \rho_m^2 . \end{aligned} \tag{8.3}$$

For simplicity, first we consider the redundancy of $\underset{\sim}{y}_2$ only, which is defined as one of the following statements:

[8.1] $\underset{\sim}{\gamma}_{2j} = \underset{\sim}{0}$, j=1,...,m,

[8.2] $\Delta^2(\underset{\sim}{y}_x,\ \underset{\sim}{y}_z) = \Delta^2(\underset{\sim}{y}_1,\ \underset{\sim}{y}_z)$,

[8.3] $\operatorname{Cov}(\underset{\sim}{y}_2 - \Sigma_{21}\Sigma_{11}^{-1}\underset{\sim}{y}_1,\ \underset{\sim}{y}_z) = \Sigma_{2z\cdot 1} = 0$,

[8.4] $B_{z2} = 0$, where $\Sigma_{zx}\Sigma_{xx}^{-1} = [B_{z1}\ B_{z2}]$.

Laha [18], Siotani [33] and McKay [21] proved the equivalence of the statements [8.2], [8.3] and [8.4]. The statement [8.1] was obtained by Fujikoshi [8].

Suppsoe that the distribution of $\underset{\sim}{y}$ is normal. Let S be the sample covariance matrix, based on a sample of size N = n+1. We partition S as in (5.3). The conditional distribution of $(\underset{\sim}{y}_2', \underset{\sim}{y}_z')'$ given $\underset{\sim}{y}_1$ is normal with means

$$E\left[\begin{pmatrix} \underset{\sim}{y}_2 \\ \underset{\sim}{y}_z \end{pmatrix} \Big| \underset{\sim}{y}_1\right] = \begin{pmatrix} \underset{\sim}{\mu}_{2x} \\ \underset{\sim}{\mu}_z \end{pmatrix} + \begin{pmatrix} B_{21} \\ B_{z1} \end{pmatrix} (\underset{\sim}{x}_1 - \underset{\sim}{\mu}_1), \tag{8.4}$$

and covariance matrix

$$\operatorname{var}\left[\begin{pmatrix} \underset{\sim}{y}_2 \\ \underset{\sim}{y}_z \end{pmatrix} \Big| \underset{\sim}{y}_1\right] = \begin{pmatrix} \Sigma_{22\cdot 1} & \Sigma_{2z\cdot 1} \\ \Sigma_{z2\cdot 1} & \Sigma_{2z\cdot 1} \end{pmatrix}, \tag{8.5}$$

where

$$\begin{pmatrix} B_{21} \\ B_{z1} \end{pmatrix} = \begin{pmatrix} \Sigma_{21} \\ \Sigma_{z1} \end{pmatrix} \Sigma_{11}^{-1}. \tag{8.6}$$

The hypothesis of redundancy of $\underset{\sim}{y}_2$ means that $\underset{\sim}{y}_2$ and $\underset{\sim}{y}_z$ are conditionally independent. Therefore, the LR test is based on

$$\begin{aligned} \Lambda &= \begin{vmatrix} S_{22\cdot 1} & S_{2z\cdot 1} \\ S_{z2\cdot 1} & S_{zz\cdot 1} \end{vmatrix} / \{|S_{22\cdot 1}||S_{zz\cdot 1}|\} \\ &= |S_{22\cdot 1z}|/|S_{22\cdot 1}| \end{aligned} \tag{8.7}$$

whose null distribution is $\Lambda_{p_2}(p_3+p_4, n-p+p_2)$ or $\Lambda_{p_3+p_4}(p_2, n-p_1-p_2)$. The test statistics which are analogous to (2.5) are

$$\begin{aligned} T_0^2 &= \operatorname{tr} S_{2z\cdot 1} S_{zz\cdot 1}^{-1} S_{z2\cdot 1} S_{22\cdot 1z}^{-1}, \quad \text{and} \\ V &= \operatorname{tr} S_{2z\cdot 1} S_{zz\cdot 1}^{-1} S_{z2\cdot 1} S_{22\cdot 1}^{-1} \end{aligned} \tag{8.8}$$

whose null distributions are $T_0^2(p_2, p_3+p_4, n-p+p_2)$ and $V(p_2, p_3+p_4, n-p+p_2)$, respectively.

8.2. Redundancy on two sets of variables. We consider the case of redundancy of some variables of both the two sets of variables, i.e., redundancy of $\underset{\sim}{y}_b = (\underset{\sim}{y}_2', \underset{\sim}{y}_4')'$. The extension of the statements [8.1]~[8.4] to this case was given by Fujikoshi [8].

Theorem 8.1. The following staments are equivalent.

[8.5] $\gamma_{bj} = (\gamma'_{2j}, \gamma'_{4j})' = \underset{\sim}{0}, \; j = 1, \ldots, m = \text{rank}(\Sigma_{xy}),$

[8.6] $\Delta^2(y_x, y_z) = \Delta^2(y_1, y_3),$

[8.7] $\Sigma_{2z\cdot 1} = 0, \; \Sigma_{4x\cdot 3} = 0 \; (\text{or } \Sigma_{41\cdot 3} = 0),$

[8.8] $B_{23} = 0, \quad B_{41} = 0, \quad \Sigma_{24\cdot 13} = 0,$

where

$$\begin{pmatrix} B_{21} & B_{23} \\ B_{41} & B_{43} \end{pmatrix} = \begin{pmatrix} \Sigma_{21} & \Sigma_{23} \\ \Sigma_{41} & \Sigma_{43} \end{pmatrix}\begin{pmatrix} \Sigma_{11} & \Sigma_{13} \\ \Sigma_{31} & \Sigma_{33} \end{pmatrix}^{-1},$$

$$\Sigma_{24\cdot 13} = \Sigma_{24} - (\Sigma_{21} \; \Sigma_{23})\begin{pmatrix} \Sigma_{11} & \Sigma_{13} \\ \Sigma_{31} & \Sigma_{33} \end{pmatrix}^{-1}\begin{pmatrix} \Sigma_{14} \\ \Sigma_{34} \end{pmatrix}. \tag{8.9}$$

In the problem of testing the hypothesis of redundancy of y_b it is assumed that y is multivariate normal. The conditional distribution of y_b given y_a is a (p_2+p_4)-variate normal distribution with means

$$E[\begin{pmatrix} y_2 \\ y_4 \end{pmatrix} | \begin{pmatrix} y_1 \\ y_3 \end{pmatrix}] = \begin{pmatrix} \mu_2 \\ \mu_4 \end{pmatrix} + \begin{pmatrix} B_{21} & B_{23} \\ B_{41} & B_{43} \end{pmatrix}\begin{pmatrix} y_1-\mu_1 \\ y_3-\mu_3 \end{pmatrix} \tag{8.10}$$

and

$$\text{Var}[\begin{pmatrix} y_2 \\ y_4 \end{pmatrix} | \begin{pmatrix} y_1 \\ y_3 \end{pmatrix}] = \begin{pmatrix} \Sigma_{22\cdot 13} & \Sigma_{24\cdot 13} \\ \Sigma_{42\cdot 13} & \Sigma_{44\cdot 13} \end{pmatrix}.$$

The statement [8.8] is useful in obtaining the LR criterion. It may be noted that the testing problem is fairly different from the one in the case of redundancy of one set of variables. It is shown that the LR test is based on

$$\begin{aligned} \Lambda &= \begin{pmatrix} S_{22\cdot 13} & S_{24\cdot 13} \\ S_{42\cdot 13} & S_{44\cdot 13} \end{pmatrix} / \{|S_{22\cdot 1}||S_{44\cdot 3}|\} \\ &= \Lambda_{(1)}\Lambda_{(2)}, \end{aligned} \tag{8.11}$$

where

$$\Lambda_{(1)} = \frac{|S_{22\cdot 13}|}{|S_{22\cdot 1}|}, \quad \Lambda_{(2)} = \frac{|S_{44\cdot x3}|}{|S_{44\cdot 3}|}. \tag{8.12}$$

We propose the following two test statistics as a competitor to the LR test:

$$T_o^2 = T_{o(1)}^2 + T_{o(2)}^2, \quad \text{and} \quad V = V_{(1)} + V_{(2)}, \tag{8.13}$$

where

$$T_{o(1)}^2 = \mathrm{tr}(S_{22\cdot 1} - S_{22\cdot 13})S_{22\cdot 13}^{-1}, \quad T_{o(2)}^2 = \mathrm{tr}(S_{44\cdot 3} - S_{44\cdot x3})S_{44\cdot x3}^{-1},$$

$$V_{(1)} = \mathrm{tr}(S_{22\cdot 1} - S_{22\cdot 13})S_{22\cdot 1}^{-1}, \quad V_{(2)} = \mathrm{tr}(S_{44\cdot 3} - S_{44\cdot x3})S_{44\cdot 3}^{-1}.$$

It is possible to decompose Λ as $\{|S_{22\cdot 1z}|/|S_{22\cdot 1}|\}\cdot\{|S_{44\cdot 13}|/|S_{44\cdot 3}|\} = \tilde{\Lambda}_{(1)}\tilde{\Lambda}_{(2)}$. The test statistics T_o^2 and V corresponding to this decomposition may be also proposed.

Theorem 8.2. Under the hypothesis H_o it holds that

(1) $\Lambda_{(1)}$ and $\Lambda_{(2)}$ are independently distributed as $\Lambda_{p_2}(p_3, n-p_1-p_3)$ and $\Lambda_{p_4}(p_1+p_2, n-p+p_4)$, respectively,

(2) $T_{o(1)}^2$ and $T_{o(2)}^2$ are independently distributed as $T_o^2(p_2, p_3, n-p_1-p_3)$ and $T_o^2(p_4, p_1+p_2, n-p+p_4)$, respectively.

(3) $V_{(1)}$ and $V_{(2)}$ are independently distributed as $V(p_2, p_3, n-p_1-p_3)$ and $V(p_4, p_1+p_2, n-p+p_4)$, respectively.

Proof. For a proof of (1), see Fujikoshi [8]. Without loss of generality, we may assume that S is distributed as $W_p(\Sigma, n)$. It has been essentially proved (Fujikoshi [8]) that

(i) $S_{22\cdot 13}$ and $S_{22\cdot 3} - S_{22\cdot 13}$ are independently distributed as $W_{p_2}(n-p_1-p_3, \Sigma_{22\cdot 13})$ and $W_{p_2}(p_3, \Sigma_{22\cdot 13})$, respectively

(ii) $S_{44\cdot x3}$ and $S_{44\cdot 3} - S_{44\cdot x3}$ are independently distributed as $W_{p_4}(n-p+p_4, \Sigma_{44\cdot x3})$ and $W_{p_4}(p_1+p_2, \Sigma_{44\cdot x3})$, respectively,

(iii) $\{S_{22\cdot 13}, S_{22\cdot 3} - S_{22\cdot 13}\}$ and $\{S_{44\cdot x3}, S_{44\cdot 3} - S_{44\cdot x3}\}$ are statistically independent.

The results (2) and (3) follows from (i)~(iii).

From Theorem 8.2 we can obtain asymptotic expansions of the null distributions of Λ, T_o^2 and V. Wijsman [37] and Fujikoshi et al. [13] have obtained asymptotic distributions of certain statistics useful in studying the effect of additional variables on the canonical correlations.

9. PRINCIPAL COMPONENT ANALYSIS

Let $\underset{\sim}{x}$ be a vector of p random variables with mean $\underset{\sim}{\mu}$ and covariance matrix Σ. It is assumed that Σ is positive definite. When we are concerned to explain the variation of $\underset{\sim}{x}$ in terms of only a few transformed variables, the principal components $y_j = \gamma_j'(\underset{\sim}{x}-\underset{\sim}{\mu})$, $j = 1,\ldots,p$ are used. Here the vectors γ_j of coefficients in $\underset{\sim}{x}$ are defined by

$$\Sigma\gamma_j = \lambda_j\gamma_j, \quad \gamma_i'\gamma_j = \delta_{ij}, \tag{9.1}$$

where $\lambda_1 \geq \cdots \geq \lambda_p > 0$ are the characteristic roots of Σ. In order to formulate the hypothesis of redundancy of variables for this purpose of principal component analysis, we partition $\underset{\sim}{x}$, γ_j and Σ into the first q and last p-q components as

$$\underset{\sim}{x} = \begin{pmatrix} \underset{\sim}{x}_1 \\ \underset{\sim}{x}_2 \end{pmatrix}, \quad \gamma_j = \begin{pmatrix} \gamma_{1j} \\ \gamma_{2j} \end{pmatrix}, \quad \Sigma = \begin{pmatrix} \Sigma_{11} & \Sigma_{12} \\ \Sigma_{21} & \Sigma_{22} \end{pmatrix}. \tag{9.2}$$

We note that it is impossible to define the hypothesis of redundancy of $\underset{\sim}{x}_2$ as "$\gamma_{2j} = \underset{\sim}{0}$ for all j". For, if $\gamma_{2j} = \underset{\sim}{0}$ for all j, we have

$$\Sigma_{11}\gamma_{1j} = \lambda_j\gamma_{1j}, \quad j = 1,\ldots,p,$$

and hence $\lambda_{q+1} = \cdots = \lambda_p = 0$ which contradicts to the nonsigularity of Σ. This means that such a hypothesis should be defined not for all the principal components, but for only a subset of principal components. Let v denote a subset of $m(\leq q)$ distinct integers between 1 and p such that $\lambda_i \neq \lambda_j$ for all $i \in v$ and $j \notin v$. The hypothesis of redundancy of $\underset{\sim}{x}_2$ for the m principal components y_j, $j \in v$ is defined (Flury [5], [6]) as

$$H_v(p,\ q):\ \gamma_{2j} = \underset{\sim}{0} \text{ for all } j \in v. \tag{9.3}$$

Of course $H_v(p,\ q)$ makes sense only if m, the number of

principal components analysised simultaneously, does not exceed q. Let w be the set of p - m indices not in v, and let

$$A = \begin{pmatrix} I_q \\ 0 \end{pmatrix}, \quad B = \begin{pmatrix} 0 \\ I_{p-q} \end{pmatrix}. \tag{9.4}$$

Further, let P_v and P_w be the projection operators on the spaces $R_v = \mathbb{R}[\chi_j, j\in v]$ and $R_w = \mathbb{R}[\chi_j, j\in w]$, respectively. Then it is easily seen that the following statements are equivalent:

[9.1] $\chi_{2j} = \underset{\sim}{0}$ for all $j \in v$,

[9.2] $\mathbb{R}[A] \supset R_v$,

[9.3] $\mathbb{R}[B] \subset R_w$,

[9.4] $P_w B = B$.

We now consider to test $H_v(p, q)$, based on a sample of size N = n+1 (> p) from a multivariate normal population. Tyler [35], [36] discussed the problem of testing the hypothesis [9.3] or [9.4] in a general situation and proposed a test statistic $T_n(B)$, based on the asymptotic distribution of $(I_p - \hat{P}_w)B$. Let $\ell_1 > \cdots > \ell_p > 0$ be the characteristic roots of the sample covariance matrix S and let $\underset{\sim}{h}_1, \ldots, \underset{\sim}{h}_p$ be the corresponding orthonormal characteristic vectors. We partition $\underset{\sim}{h}_j$ and S as in (9.2). Then the test statistic $T_n(B)$ can be reduced as

$$T_n = n \sum_{j\in v} \underset{\sim}{h}'_{2j} \{ \sum_{i\in w} \frac{\ell_i \ell_j}{(\ell_i - \ell_j)^2} \underset{\sim}{h}_{2i} \underset{\sim}{h}'_{2i} \}^{-1} \underset{\sim}{h}_{2j} \tag{9.5}$$

whose asymptotic null distribution is a chi-square distribution with m(p-q) degrees of freedom.

Flury [5], [6] discussed the test for redundancy of variables in the comparison of two covariance matrices and obtained the statistic (9.5) as its special case. For procedures for selecting variables in pricipal component analysis, see, e.g., Jolliffe [16], [17] McCabe [19], etc. However, the hypothesis $H_v(p, q)$ and its test are not discussed in these papers. Fujikoshi et al. [13] derived asymptotic distributions of increases in certain functions of the characteristic roots of the sample covariance matrix when a new set of variables is added.

References

[1] Akaike, H., Information theory and an extension of the maximum likelihood principle, in: B. N. Petrov and F. Czáki, eds., 2nd International Symposium on Information Theory (Akademiai Kiadó, Budapest, 1973) 267-281.

[2] Anderson, T. W., An Introduction to Multivariate Analysis, 2nd ed., Wiley, New York, 1984.

[3] Brown, P., Multivariate calibration, J. R. Ststist. Soc. B 44(1982) 287-321.

[4] Cochran, W. G. and Bliss, C. I., Discriminant functions with covariance, Ann. Math. Statist. 19(1948) 151-176.

[5] Flury, B. N., An asymptotic test for redundancy of variables in the comparison of two covariance matrix, Statist. Prob. Letters 4(1986) 123-126.

[6] Flury, B. N., Analysis of linear combinations with extreme ratios of variance, J. Amer. Statist. Assoc. 80(1985) 915-922.

[7] Fujikoshi, Y., The power of the likelihood ratio for additional information in a multivariate linear model, Ann. Inst. Statist. Math. 33(1981) 279-285.

[8] Fujikoshi, Y., A test for additional information in canonical correlation analysis, Ann. Inst. Statist. Math. 34(1982) 137-144.

[9] Fujikoshi, Y., A criterion for variable selection in multiple discriminant analysis, Hiroshima Math. J. 13(1983) 203-214.

[10] Fujikoshi, Y., Selection of variables in discriminant analysis and canonical correlation analysis, in: P. R. Krishnaiah, ed., Multivaiate Analysis- VI(Elsevier Science Publishers B. V., Amsterdam, 1985) 219-236.

[11] Fujikoshi, Y., Selection of variables in two-group discriminant analysis by error rate and Akaike's information criteria, J. Multivariate Anal. 17(1985) 27-37.

[12] Fujikoshi, Y. and Nishii, R., Selection of variables in a multivariate inverse regression, Hiroshima Math. J. 16 (1986) 269-277.

[13] Fujikoshi, Y., Krishnaiah, P. R. and J. Schmidhammer, Effect of additional variables in principal component analysis and canonical correlation analysis, in: A. K. Gupta, ed., Advances in Multivariate Statistical Analysis (D. Reidel Publishing Company, Dordrecht, 1987) 45-61.

[14] Fujikoshi, Y. and Khatri, C. G., A study of redundancy of some variables in covariate discriminant analysis, Submitted for Publicaitons.

[15] Hotelling, H., Relations between two sets of variates, Biometrika 28(1936) 321-377.

[16] Jolliffe, I. T., Discarding variables in a principal component analysis I: Artificial data, Appl. Statist. 21(1972) 160-173.

[17] Jolliffe, I. T., Discarding variables in a principal component analysis II: Real data, Appl. Statist. 22(1973) 21-31.

[18] Laha, R. G., On some problems in canonical correlations, Sankhyā 14(1954) 61-66.

[19] McCabe, G. P., Principal variables, Technometrics 26(1984) 137-144.

[20] McKay, R. J., Simultaneous procedures in discriminant analysis involving two groups, Technometrics 18(1976) 47-53.

[21] McKay, R. J., Variable selection in multivariate regression: An application of simultaneous test procedures, J. Roy. Statist. Soc. Ser. B 39(1977) 371-380.

[22] McKay, R. J., Simultaneous procedures for variable selction in multiple discriminant analysis, Biometrika 64(1977) 283-290.

[23] McKay, R. J., The adequacy of variable subsets in multivariate regression, Technometrics 21(1979) 475-479.

[24] Nishii, R., Criteria for selection of response variables and the asymptotic properties in a multivariate calibration, Ann. Inst. Statsit. Math. 38(1986) 319-329.

[25] Potthoff, R. R. and Roy, S. N., A generalized multivariate analysis of variance model useful specially for growth curve problems, Biometrika 51(1964) 313-326.

[26] Rao, C. R., tests with discriminant functions in multivariate analysis, Sankhyā 7(1946) 407-414.

[27] Rao, C. R., Tests of significance in multivariate analysis, Biometrika 35(1948) 58-79.

[28] Rao, C. R., The theory of least squares when parameters are stochastic and its application to the analysis of growth curves. Biometrika 52(1965) 447-458.

[29] Rao, C. R., Covariance adjustment and related problems in multivariate analysis, in: P. R. Krishnaiah, ed., Multivariate Analysis (Academic Press, New York, 1966) 87-103.

[30] Rao, C. R., Inference on discriminant function coefficients, in: R. C. Bose and Others, eds., Essays in Probability and Statistics (Univ. of North Carolina Press, Chapel Hill, N. C., 1970) 587-602.

[31] Rao, C. R., Linear Statistical Inference and Its Applications, Wiley, New York, 1973.

[32] Seber, G. A. F., Multivariate Observations, Wiley, New York, 1984.

[33] Siotani, M., Effect of the additional variates on the canonical correlation coefficients, Proc. Inst. Statist. Math. 5(1957) 52-57.

[34] Siotani, M., Hayakawa, T. and Fujikoshi, Y., Modern Multivaraite Statistical Analysis: A Graduate Course and Handbook, American Sciences Press, INC., Ohio, 1985.

[35] Tyler, D. E., Asymptotic inference for eigenvectors, Ann. Statist. 9(1981) 725-736.

[36] Tyler, D. E., A class of asymptotic tests for principal component vectors, Ann. Statist. 11(1983) 1243-1250.

[37] Wijsman, R. A., Asymptotic distribution of the increase of the largest canonical correlation when one of the vectors is augmented, J. Multivariate Anal. 18(1986) 169-177.

STATISTICAL DATA ANALYSIS AND INFERENCE
Y. Dodge (Editor)

MODEL CHOICE IN THE CONTEXT OF SIMULTANEOUS INFERENCE

Tomáš Havránek, Ondřej Soudský

Institute of Computer Science, Czechoslovak Academy of Sciences, Pod vodárenskou věží 2
182 07 Prague, Czechoslovakia

Methods of model choice from large finite sets of models are discussed from the statistical point of view, namely the question of generalized power of these methods is considered.

1. INTRODUCTION

In the last years some methods for model choice were suggested particularly in the field of multivariate analysis (see e.g. [1],[2]). In these methods, a search for a set of acceptable models is implemented. Perhaps it is better to speak more precisely about a set of models supported by a given data set. The methods are usually described more or less from the computational point of view and a great attention is devoted to questions of computer realizability of these methods as well as to concise representation of the sets of models considered. Moreover some computer programs mainly for the case of discrete multivariate data were written [3], [4], [5], [6]. After having some experience with these tools it is now time to reconsider these methods from the statistical point of view, mainly from the side of simultaneous inference. Some results are known showing namely that the probability that a "true" model will be covered can be in some cases computed. The size of the set of redundant models, i.e. models suggested by procedures considered but incomparable or simpler than the true model is, as a rule, unknown. The present paper is, in a degree, a continuation of [7].

2. MODEL SELECTION

Consider a finite family of models $\mathcal{F}$. Such a family can be for example formed by hierarchical log-linear models for multivariate categorial data or by subsets of a set of regressors in multiple linear regression. Such a family of models is usually ordered by a partial ordering $\leq$; if for $m_1, m_2 \in \mathcal{F}$, $m_1 \leq m_2$ (and $m_1 \neq m_2$), the model m_1 is considered to be simpler than m_2. Usually, the family $\mathcal{F}$ with a partial ordering forms a lattice $(\mathcal{F}, \leq)$, i.e. for each two models m_1 and $m_2 \in \mathcal{F}$ there is unique lowest upper bound in $\mathcal{F}$ ($m_1 \vee m_2$, the join of m_1 and m_2) and a unique greatest lower bound in $\mathcal{F}$ ($m_1 \wedge m_2$, the meet of m_1 and m_2). In terms of $\leq$, the set min(S) of minimal elements and the set of maximal elements max(S) can be defined for a set $S \subseteq \mathcal{F}$. If for each m_1, m_2 it is neither $m_1 \leq m_2$ nor $m_1 \geq m_2$, the set S is incomparable (then min(S) = max(S) = S).

Suppose now that additionally a decision rule d is given such that for a data set M, we can say that a model $m \in \mathcal{F}$ is rejected (d(m,M)=r) or accepted (d(m,M)=a); for accepted we shall use frequently "adequate" in the sequel.

For a given $(\mathcal{F}, \leq)$, d and M, let us state our goal as to find the set A(M) of minimal adequate models, A(M)=min{$m \in \mathcal{F}$; d(m,M)=a}, and dually, the set of maximal rejected models R(M)=max{$m \in \mathcal{F}$; d(m,M)=r}.

From the computational point of view we should add: to find A(M) and R(M) without fitting and evaluating all models or, better, with evaluating a small proportion of all models.

The underlying idea is that if a model is adequate, then all more complex models are "uninteresting" and similarly, if a model is rejected then all simpler models are as well "uninteresting". This view can be formalized into two rules:

(a) if for a model m_1, d(m,M)=a, then all models m_2 such that $m_1 \leq m_2$ are considered to be accepted.

(b) If for a model m_1, d(m,M)=r, then all models m_2 such that $m_2 \leq m_1$ are considered to be rejected.

If a model is not evaluated at M, i.e. d(m,M) is not actually computed, but it is considered to be accepted due to (a), it can be called *weakly accepted*. Similarly for *weakly rejected* models.

The above rules can be used in two linked directions: first, we can say that if one was able to find A(M) and R(M) then the status of all models in $\mathcal{F}$ is given as (weakly) accepted or (weakly) rejected; this is the *interpretational* point of view. Second, both rules can be used in a computational procedure for finding (or approximation) of A(M) and R(M); this is the *computational* point of view (see [2]).

Both rules were used in model selection procedures to reduce the number of models fitted, for example in [8] the rule (a) was used in the context of multiple regression, in [9] for a multiple comparison problem. The rule (b) was used in [10] for contingency tables and a similar rule was used in [11], [12] a [13] for the best subset selection procedures. It is clear, that a procedure based on both rules (a) and (b) could be far more effective in reducing the number of models to be fitted than a procedure using one rule only. Such procedure was proposed in [5] and [2] with application to contingency tables and multivariate regression and further applied to discriminant analysis in [14], to some other multivariate problems in [15] and to robust regression in [16] and [17].

3. COHERENCE AND SIGNIFICANCE

Clearly, in the general case, the following unwelcome situation can occur: $d(m_1,M) = r$ and for some $m_2 \le m_1$, $d(m_2,M) = a$. Then the model m_2 could be weakly rejected by the rule (b), but it is not rejected when evaluating $d(m_2,M)$ or, on the other hand, m_1 could be weakly accepted by the rule (a), but it is rejected when $d(m_1,M)$ is evaluated.

A decision rule d for which such situation can occur for no data M is called, according to [18], *coherent* (with respect to $(\mathcal{F},\le)$.

Let us now mention two simple facts:

(i) for a coherent decision rule, A(M) and R(M) determine due to (a) and (b) status of all models in $\mathcal{F}$ as accepted or rejected.

(ii) The algorithm of [2] using the rules (a) and (b) finds A(M), R(M) for each starting set $S\subseteq\mathcal{F}$ of incomparable models.

In [2], [14], [15], [16] and [17] coherent decision rules for various problems were proposed in the context of the algorithm of [2].

Consider now the case, when d(m,.) is a goodness-of-fit test on a significance level α. Then the probability of non rejecting the "true" hypothesis m by d(m,.) is greater or equal to $1-\alpha$. More precisely for each sample size n, $P(\{M\in\mathcal{M}_n;\ d(m,M)=a\}) \ge 1-\alpha$ under m; here $\mathcal{M}_n$ is the corresponding sample space.

Denote by $A^+(M)$ the set of all models (weakly) accepted from A(M) by (a). Consider now $(\mathcal{F},\le)$ and a coherent d being a goodness-of-fit test for each $m\in\mathcal{F}$ on a significance level less or equal α. Then the following simple but important fact holds: for each sample size n,

$$P(\{M \in \mathcal{M}_n;\ m \in A^+(M)\}) \geq 1 - \alpha$$

Proof: clearly, due to the coherence, $\{M \in \mathcal{M}_n;\ d(m,M)=a\} = \{M \in \mathcal{M}_n;\ m \in A^+(M)\}$. Hence $A^+(M)$ can be regarded as a $1-\alpha$ *confidence set* for the true model. The above property can be called the *confidence property*.

Consider now, as an example, the case of model selection in linear regression $y=\mu+\beta_1 x_1+\ldots+\beta_p x_p+\varepsilon$ with p regressor variables. Models are specified by sets of indexes; for example, m = {1,2,4} means that the variables x_1, x_2 and x_4 are present in the regression equation. The hypothesis to be tested for each model m is

$$H_o : \beta_j = 0 \quad \text{for } i \notin m .$$

The usual F-statistic for testing this hypothesis is

$$F_m = \frac{(SS_F - SS_m)/(p-q)}{SS_F/(n-p-1)}$$

where SS_m and SS_F are the residual sums of squares for m and the full model respectively, n is the sample size and q is the number of regressors under m.

In [2], following [19], as coherent decision rule was suggested to reject m if

$$F_m \geq c^1_\alpha(q)\ , \text{ where}$$

$$c^1_\alpha(q) = (p/(p-q))\ F_{1-\alpha}(p, n-p-1).$$

Since $c^1_\alpha(q) \geq c_\alpha(q)=F_{1-\alpha}(p-q, n-p-1)$, the above rule is a conservative test for H_0 on the significance level less or equal to α. The coherent rules considered in [14], [15], [16] and [17] are of the similar nature.

The present approach seems to have a serious flaw: tests are likely "too conservative"; note that for the confidence property of $A^+(M)$, the significance level of individual coherent test could be equal to α. What are individual significance levels α' for the above tests can be seen for typical values n=100 and p=10 in Table 1.

It is clear, that in this particular but typical case, individual test based on the coherent rule are very conservative. (Compare with discussion in [19] and [20]). One can expect that this indicates a considerable loss of power, i.e. under a given $m_0(\beta) > m$,

$$\gamma^1 = P(F_m \geq c_\alpha^1(q))$$

should be considerably smaller than $\gamma = P(F_m \geq c_\alpha(q))$.

Table 1

q	$c_\alpha(q)$	$c_\alpha^1 = \frac{p}{p-q} c_\alpha(0)$	α'
9	3.94986	19.3885	0.00003
8	3.11456	9.69425	0.00016
7	2.71002	6.46283	0.00052
6	2.47470	4.84713	0.00139
5	2.31698	3.87770	0.00317
4	2.20227	3.23142	0.00642
3	2.11466	2.76979	0.01188
2	2.04438	2.42357	0.02037
1	1.98691	2.15428	0.03279
0	1.93885		

In Table 2, for n=100, p=10 and α=0.05, the non-centrality parameters for γ=0.90, 0.75 and 0.50 are presented and for these non-centrality parameters, the corresponding values of γ^1 are computed. The computational methods of [21] were used. (As the alternative models m_0, the models with q_0=q+1 were considered).

Table 2

q	$\lambda_{0.90}$	$\lambda_{0.75}$	$\lambda_{0.5}$	$\gamma^1_{0.9}$	$\gamma^1_{0.75}$	$\gamma^1_{0.5}$
9	10.74	7.096	3.927	0.145	0.050	0.011
8	13.14	8.927	5.158	0.272	0.115	0.031
7	14.82	10.22	6.027	0.390	0.190	0.061
6	16.25	11.31	6.770	0.500	0.275	0.102
5	17.52	12.29	7.435	0.599	0.363	0.153
4	18.69	13.18	8.044	0.685	0.452	0.213
3	19.77	14.01	8.615	0.758	0.537	0.281
2	20.79	14.80	9.150	0.816	0.617	0.353
1	21.76	15.55	9.660	0.863	0.688	0.426
0	22.70	16.26	10.15			

A global view for $q = 5$ to the power γ and γ^1 as a function of the non centrality parameter λ is displayed in Fig.1.

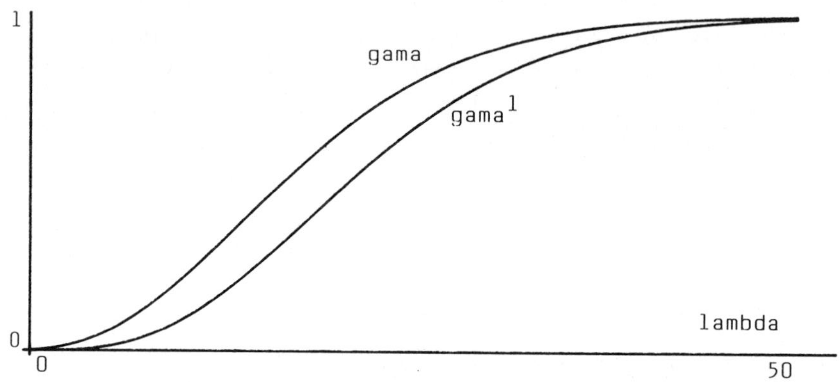

FIGURE 1

4. SIZE AND POWER

The above examples indicate that the size of the set of adequate $A^+(M)$ models can be far greater then necessary. The analogy with confidence intervals is straightforward: for $(-\infty, +\infty)$ is clearly the confidence level greater or equal than $1-\alpha$, but the interval brings no useful information.

In general, we can ask how close $A^+ = A^+(M)$ is to the true model m_0. Clearly the cone $(m_0)^+=\{m\in\mathcal{F}; m_0\leq m\}$ is from this point of view uninteresting.

The set in question is then the random set $A^+-(m_0)^+$ and its "probabilistic size" depends not only on the model m_0 (structure), but on particular values of the parameters of m_0 (i.e. on $m_0(\theta)$).

We can expect that, roughly speaking, if for an individual test the α-levels α_m increases to overall α, then the "size" of $A^+-(m_0)^+$ decreases. But a formal statement in this direction seems to be hard to formulate and prove.

At first glance, much simpler is to consider the "numerical" size $s=\#(A^{+}-(m_0)^{+})$ that is a discrete random variable with distribution depending not only on m_0, but on values of parameters of m_0 as well. Generally, max $s(m_0) = \#\mathcal{F} - \#(m_0)^{+}$ and for the regression case, max $s(m_0) = 2^{p}-2^{p-q}$. Note that $s=0$ is equivalent to $A^{+}=(m_0)^{+}$, hence evaluating of $P_{m_0}(s>0)$ and $E_{m_0}s$ could be of interest. Similarly a reduced numerical size can be defined as $rs=(A^{+}\cap(m_0)^{-})$, where $(m_0)^{-}=(m\in F;\ m\leq m_0)$. In this case "concurrent" incomparable models are not considered; max $rs(m_0) = \#(m_0)^{-}$ and for the regression case, max $rs(m_0) = 2^{q-1}$.

Again, the evaluating of $P_{m_0}(s>0)$ and $E_{m_0}s$ in particular cases seems to be hardly tractable. A way is to provide computer simulation experiments using some programs realizing the model search algorithms, but these experiments could be expensive.

What is more important, that nor $P_{m_0}(s>0)$ neither $E_{m_o}s$ give a global quality of the procedure. One way to obtain such global view is to use the Bayesian approach and evaluate

$$S = \sum_{m\in\mathcal{F}} \left(\int E_{m(\theta)}s\ f_m(\theta)d\theta\right)p(m)$$

for some reasonable priors $(f_m(\theta);m\in\mathcal{F})$ and $p(m)$.

Another possible approach is the local one: consider for a given $m_0(\theta_0)$ a model $m_1 < m_0$ and evaluate

$$\beta_{m_0,m_1}(\theta_0)=P_{m_0(\theta_0)}(m_1(\hat{\theta}_1) \in A^{+})\ ;$$

both $m_1(\hat{\theta}_1)$ and A^{+} are random $(m_1(\hat{\theta}_1)\in A^{+}$ is the random event $(M\in\mathcal{M}_n;\ m_1(\hat{\theta}_1))\in A^{+}(M)))$. Table 2 refers to such values for the regression case. If the decision rules for individual models are tests on the α_m-level, it is possible to ask whether under $\alpha_{m_0}\nearrow\alpha$ $P_{m_0(\theta_0)}(m_1(\hat{\theta}_1) \in A^{+})$ decreases.

If such decision rule is coherent and monotonous (i.e. $d_{\alpha_1}(m,M)=r$ implies $d_{\alpha_2}(m,M)=r$ for $\alpha_1 < \alpha_2$) then the answer is positive.

It seems worth to concentrate the attention to immediate *predecessors* of m_0, i.e. to models $m\in B(m_0) = \max\ (m;\ m<m_0)$. Note that

$$P_{m_0(\theta_0)}(\exists m\in B(m_0),\ m\in A^{+})=P_{m_0(\theta_0)}(rs>0).$$

To summarize, one can define

$$\beta_{m_0}(\theta_0)= \max_{m\in B(m_0)} \beta_{m_0,m}(\theta_0)$$

and

$$\beta = \max_{m_0} \sup_{\theta_0} \beta_{m_0}(\theta_0) .$$

But, in general, if the decision rules d(m,.) are (goodness-of-fit) tests on significance levels less or equal to α, one can expect that $\beta \geq 1-\alpha$. Consider the regression example, when $m \in B(m_0)$ are models with one additional regressor missing, say $i \in m_0$ and $i \notin m_i$, $m_i \in B(m_0)$. Then by limiting $\beta_i \to 0$ one can see that

$$\sup_{\theta} \beta_{m_0}(\theta) \geq 1 - \alpha.$$

The Bayesian approach can be more informative, considering

$$\beta = \max_{m_0 \in \mathcal{F}} \int \beta_{m_0}(\theta)\, f(\theta)\, d\theta$$

with some appropriate prior.

In all particular cases the evaluating of necessary probabilities seems to be hard. Known are up to now only some simulation results mainly for the non-coherent case of likelihood ratio χ^2-tests in contingency tables. In [22] the case of 2x2x2 contingency tables is considered with model m_0 = (AB, BC), i.e. conditional independence of A and B, and m_1=(A,B,C) both with equiprobable marginals. The second model is characterized in fact by p_{ijk}=0.125, the first one by various values of $\theta_0=(p_{111},p_{112})$. Percentages of "acceptance" of the simpler model m_1 under the true model m_0 in 4000 simulations for each sample size are given in Table 3.

Table 3

n	θ_0 0.1125	0.100	0.0625
10			49.5
25			23.7
50			6.1
100		45.2	0.07
200		20.2	0.0
300	≥50.0	8.8	
500	38.0	1.3	
1000	14.6	0.0	
2000	1.3		
3000	0.0		

From these results as well as from a more detailed study of similar cases [23] a high dependence on sample sizes and θ_0 values is apparent and no clear summarization is possible.

Consider now some relations concerning the decision rules. First, let $d_\alpha(m,M) = a$ iff $t(m,M)<c_\alpha$ for some overall α (clearly $t(m,.) \geq c_\alpha$ could correspond to some individual level $\alpha_m \leq \alpha$). If now a $c_{\alpha'}>c_\alpha$ is used then naturally $\beta_{m'}< \beta_m$. This is a particular case, if the coherence is present, of the relation

$$A^+ \supset (A')^+$$

cf. the notion of resolution in [18]. More generally, if d' is *sharper* than d (d'⊨d, [15]), i.e. $\forall(m,M)$, $d'(m,M)=a \Rightarrow d(m,M)=a$, then under coherence, $A^+ \supset (A')^+$ (supposing that for some data $A \neq A'$).

Under such deterministic relation between A^+ and $(A')^+$, clearly the global sensitivity linked with d' is better: $\forall(m_0,\theta_0)$, $s(d) > s(d')$, as well the local sensitivity is better for d', i.e.

$$\beta_{m'} < \beta_m .$$

under mild probabilistic assumptions excluding equality.

Hence, the deterministic relations between decision rules are to be studied [1] and under preserving the coherence property, the rules (and procedures) with minimal A^+ for each $M \in \mathcal{M}_m$ have to be chosen.

4. DISCUSSION

In our opinion, the way of increasing α to obtain more powerful coherent simultaneous test procedures as discussed, for example in [20] and [19] is not applicable here; to have the confidence level equal e.g. to 0.59 (cf. [19], p.472) is unacceptable.

An open question is whether, generally, it is reasonable to insists on coherence principle. Apart from non-uniqueness of the obtained solution, non-coherent procedures give results covering the true model more closely. A necessary condition is that the procedures are only mildly non-coherent as declared for likelihood ratio χ^2-test based procedures in [5]. An attempt to evaluate the approximate coherence in this case is contained in [24].

More generally, important is to note that non-coherence for a rule d with individual significance levels $\alpha_m \leq \alpha$ can break the confidence property only if from $d(m,M)=r$ the rejection is deduced for $m_1 \le m$. Using of the rule (a) only has no effect to this property.

Consider now the following case: For $(\mathcal{F},\le)$, let d^c be a coherent decision rule with $\alpha_m^c \le \alpha$ (and for the majority of models with $\alpha_m^c < \alpha$) and let d^n be a non coherent rule with $\alpha_m^n = \alpha$ such that $d^n \models d^c$.

Apply the following two-step procedure:

(1) find $A_d c(M)$ and $R_d c(M)$ by the general algorithm (see [2], [15]), using both rules (a) and (b).

(2) use $A_d c(M)$ as the starting set and apply d_n and the rule (a) only.

For the step (2) it is necessary to distinguish between accepted ($d(m,M)=a$ computed) and weakly accepted models: as a result we obtain the set of minimal accepted models $A_d n|_d c(M)$, $A_d^+ n|_d c(M)$ having the confidence property. Clearly, $A_d^+ n|_d c(M) \subseteq A_d^+ c(M)$ and, as a rule, $A_d^+ n|_d c(M) \ne A_d^+ c(M)$ and $\#A_d^+ n|_d c(M) \ll \#A_d^+ c(M)$.

The step (2) should be explained in more detail from the computational point of view. For any $S \subseteq \mathcal{F}$ denote $S_n^a = \{m \in S;\ d^n(m,M)=a\}$, $S_n^r = \{m \in S;\ d^n(m,M)=r\}$. Now the second step can be decomposed as follows:

Input: $S := A_d c(M)$, $A := R := \emptyset$.

(i) $A := S_n^a$, $R := S_n^r$, if $R=\emptyset$ then stop.

(ii) put $S := \min(A_d^+ c(M) - (A^+ \cup R))$, if $S=\emptyset$ then stop.

(iii) evaluate S, put $A := \min(A \cup S_n^a)$, $R := R \cup S_n^r$ and go to (ii).

Output: A, R.

Note that A *represents* the set A^+, R *contains all models* from $A_d^+ c(M)$ explicitly tested and rejected.

The minimal elements are in (ii) considered since evaluating and possibly accepting of them gives the greatest gain in reducing the subset $N \subseteq A_d^+ c(M)$ of not yet decided models w.r.t. d_n. How to find these minimal elements without checking explicitly all sets involved? First, rewrite S as $\min\{m \in \mathcal{F};\ \exists m_a \in A_d c,\ m_a \le m\ \&\ \forall m_a \in A,\ \neg(m_a \le m)\ \&\ m \notin R\}$. Let $D \subseteq \mathcal{F}$, denote by $C(D)$ the set of immediate successors of D, $C(D) = \min\{m \in \mathcal{F}-D;\ \exists m' \in D,\ m' < m\}$. Note that $\min(\{m \in \mathcal{F};\ m \notin A^+\} \cap \{m \in \mathcal{F}-R;\ \exists m_r \in R,\ m_r < m\}) = \{m \in \mathcal{F};\ \forall m_a \in A,\ \neg(m_a \le m)\} \cap C(R)$. Hence we can put S equal to successors of R that are not in A^+. Using this procedure, the situation $m_1 < m_2 < m_3$ with $m_1, m_3 \in R$ and $m_2 \notin R$ cannot occur, hence we can represent R by $\max(R)$ and use $S := C(\max(R)) \cap \{m \in \mathcal{F};\ \forall m_a \in A,\ \neg(m_a \le m)\}$. Using this representation and, in particular cases, the natural hierarchy of models linked with the ordering $\leq$, the step (2) becomes to be tractable. Clearly its complexity depends heavily on the size of A_d^+c. As *output* we obtain A (and additionally max(R), A and R defining the status of all models in A_d^+c w.r.t. d^n). $A^+ \subseteq A_d^+c$ but A^+ "covers" the true model more tightly than A_d^+c and satisfies the confidence property.

We can conclude that, in the present context, it will be necessary not only to study more deeply procedures mentioned here but additionally to focus attention to multistage procedures perhaps inspired by sequentially rejective test procedures, cf. the discussion in [25]. Last developments in the field of the power of multiple test procedures [27] are to be considered as well. We hope that in the near future new results could appear giving new insight into the topic considered.

REFERENCES

[1] Hájek, P. and Havránek, T., Mechanizing hypothesis formation - mathematical foundations of a theory (Springer-Verlag, Heidelberg, 1978).

[2] Edwards, D. and Havránek, T., A fast model selection procedure, J. Amer. Statist. Assoc. (1987), 205-213.

[3] Hájek, P., Havránek, T. and Chytil, M., The GUHA method (in Czech, Academia, Prague, 1983).

[4] Havránek, T., The present state of the GUHA software, Int. J. Man-Machine Studies 15(1981), 253-264.

[5] Edwards, D. and Havránek, T., A fast procedure for model search in multidimensional contingency tables, Biometrika 72 (1985), 339-351.

[6] Badsberg, J.H., A guide to COCO (a program for model search among hierarchical models on large complete contingency tables), (SRU, Copenhagen, 1988).

[7] Havránek, T., Model search in large model families, in.: Proceedings of the First World Congress of the Bernoulli society, (VNU Press, Utrecht, 1987).

[8] Cox, D.R. and Snell, E.J., The choice of variable in observational studies, Applied Statistic 23 (1974), 51-59.

[9] Cox, D.R. and Spjotvoll, E., On partitioning means into groups, Scandinavian Journal of Statistics 9 (1982), 147-152.

[10] Havránek, T., A procedure for model search in multidimensional contingency tables, Biometrics 40 (1984), 95-100.

[11] Beale, E.M.L., Kendall, M.G. and Mann, D.W., The discarding of variables in multivariate analysis, Biometrika 54 (1967), 357-365.

[12] Lamotte, L.R. and Hocking, R.R., Computational efficiency in the selection of regression variables, Technometrics 12 (1970), 83-93.

[13] Furnival, G.M. and Wilson, R.W.J., Regression by leaps and bounds, Technometrics 16 (1974), 499-511.

[14] Havránek, T. and Edwards, D., On model choice in multivariate statistical problems, in: Zítek, F.(ed.),

Proceedings of DIANA II Conference (Mathematical Institute, Czechoslovak Academy of Sciences, Prague 1987), 161-174.

[15] Havránek, T., On a general algorithm for model choice in multivariate analysis, Statistics 19 (1988), 465-475.

[16] Antoch, J., Algorithmic development in variable selection procedures, in: DeAntoni, F., Lauro, N. and Rizzi, A (eds.), COMPSTAT 86 (Physica-Verlag, Heidelberg, 1986), 83-90.

[17] Antoch, J., Variable selection in linear model based on trimmed least squares estimation, in: Dodge, Y.(ed.), Statistical Data Analysis Based on the L_1-Norm and Related Methods (Elsevier, Amsterdam, (1987), 231-245.

[18] Gabriel, K.R., Simultaneous test procedures - some theory of multiple comparisons, Annals of Mathematical Statistics 40 (1969), 224-250.

[19] Aitkin, M.A., Simultaneous inference and the choice of variable subsets in multiple regression, Technometrics 16 (1974), 221-227.

[20] Gabriel,K.R., A procedure for testing the homogeneity of all sets of means in analysis of variance, Biometrics 20 (1964), 459-477.

[21] Narula, S.C. and Weistroffer, Computation of probability and non-centrality parameter of a non-central F-distribution, Commun. Statist.- Computa. Simula 15 (1986), 871-878.

[22] Králová, D. and Havránek, T., On power of model search procedures, in: European Congress on Simulation (Academia, Prague,1987), vol.A., 267-270.

[23] Oler, J., Noncentrality parameters in chi-squared goodness-of-fit analyses with application to log-linear procedures, J. Amer. Statist. Assoc. 80 (1985), 181-189.

[24] Havránek, T. and Pokorný, D., On the GUHA approach to model search in connection to generalized linear models, in: Gilchrist, R., Francis, B. and Whittaker, J., Generalized Linear Models (Lecture Notes in Statistics, vol.32, Springer-Verlag, Berlin, 1985), 82-92.

[25] Hommel, G., Multiple test procedures for arbitrary dependence structures, Metrika

[26] Bauer, P., On the assesment of the performance procedures of multiple test procedures, Biom. J. 29 (1987), 894-906.

STATISTICAL DATA ANALYSIS AND INFERENCE
Y. Dodge (Editor)

FIRST ZEROS OF EMPIRICAL CHARACTERISTIC FUNCTIONS AND EXTREME VALUES OF GAUSSIAN PROCESSES

Jürg HÜSLER

Dept. of math. Statistics, University of Bern, Sidlerstr.5, CH-3012 Bern, Switzerland

Abstract: Testing the symmetry of an underlying distribution can be based on the empirical characterstic function. A crucial role of such tests is played by the first zero of the empirical characteristic function. We discuss in this paper the behaviour of such zeros depending on the underlying distribution function. In some cases, the first zero of the empirical characteristic function is related to the extreme value or extreme crossing events of a nonstationary Gaussian process. Recent results on such extreme values are used to derive the weak limit laws of the first zeros.

1. INTRODUCTION

In recent papers statistical procedures were developed based on the behaviour of the empirical characteristic function. They includes tests e.g. for the symmetry (Csörgő and Heathcote (1987)) and tests for normality (Welsh (1984)) as well as estimation procedures (Csörgő (1983), Heathcote and Welsh (1983), Welsh and Nicholls (1987)). It was mentioned by Welsh (1986) that these procedures depend on a working interval in the neighbourhood of the origin, where the real part of the characteristic function does not vanish. The size of this interval is hence determined by zeros of the characteristic function. He discussed the estimation of these zeros .

Let the random variables $X_i, i \geq 1$, be independent and identically distributed as X with distribution function F and with real part of the characteristic function

$$u(t) = E(\cos tX).$$

The corresponding empirical real part is denoted by

$$U_n(t) = n^{-1} \sum_{j=1}^{n} \cos(tX_j),$$

the first zero of this function $U_n(t)$ by

$$R_n = \inf\{t > 0 : U_n(t) = 0\},$$

and the first zero of the theoretical function $u(t)$ by

$$r_0 = \inf\{t > 0 : u(t) = 0\}$$

where $r_0\,(\leq \infty)$.

Welsh (1986) proved that for most cases with $r_0 < \infty$, $R_n \to r_0$ a. s. as $n \to \infty$. Hüsler and Heathcote (1988) investigated the limit distributions of R_n for cases with $r_0 < \infty$ and $r_0 = \infty$. The treatment of these two cases use different methods. If $r_0 < \infty$, many of the limit distributions of R_n are normal which can be shown by a straightforward proof. If $r_0 = \infty$, the evaluation of the asymptotic behaviour of R_n is more involved since $R_n \to \infty$. We note here that the limit distribution of R_n is in some cases a Gumbel or a part of a normal distribution. Since $U_n(t)$ is a stochastic process which converges, suitably normalized, to a nonstationary Gaussian process, the distribution of R_n is asymptotically related to the distribution of extreme crossings of a nonstationary Gaussian process. This relation is discussed in the next section. The recent work of Hüsler (1988) on such extreme crossings is strongly used to derive the limit distributions of R_n.

In this paper we review some of these asymptotic results on R_n and present theorems, which are slightly more general than corresponding ones in the paper of Hüsler and Heathcote (1988).

2. THE RELATION TO EXTREME CROSSINGS

Note the basic equivalence of the two events $\{R_n > r_n\}$ and $\{U_n(t) > 0, t \le r_n\}$. Under the mild condition $E|X|^\gamma < \infty$ for some $\gamma > 0$, $\{\sqrt{n}(U_n(t) - u(t)), t > 0\}$ converges in distribution to a continous Gaussian process $\{Y(t), t > 0\}$ with zero mean and covariance function

$$\sigma(t,s) = \frac{1}{2}[u(t-s) + u(t+s) - 2u(t)u(s)]$$

(Csörgő (1981)). Hence $\{R_n > r_n\}$ is also equivalent to

$$\{n^{1/2}[U_n(t) - u(t)] > -n^{1/2}u(t), t \le r_n\}.$$

If $\limsup_{n\to\infty} r_n < \infty$, as in the case of a finite first zero, the stated weak convergence implies that

$$P\{R_n > r_n\} \sim P\{Y(t) > -n^{1/2}u(t), t \le r_n\} = P\{Y(t) < n^{1/2}u(t), t \le r_n\} \tag{1}$$

In some cases $\min_{t \le r_n} \sqrt{n}u(t) \to \infty$ as $n \to \infty$. The event in the last probability is thus an extreme crossing event for a nonstationary Gaussian process. To evaluate such a probability for these particular processes we use the following result which is a particular case of Theorem 4.1 of Hüsler (1988). This is mainly used for $r_0 = \infty$, where an additional argument is used to insure that the two probabilities in (1) are also equivalent for $r_n \to \infty$. It is made possible by the strong approximation of the standardized process $U_n(t)$ by Gaussian processes which are in distribution equivalent to $Y(t)$. The strong approximation of empirical characteristic processes is discussed in Csörgő (1981).

We showed that the limiting Gaussian process $Y(t)$ respectively the standardized process $Y^*(t) = Y(t)/\sigma(t)$ is locally stationary (for a definition see Berman (1974)) where $\sigma(t) = \sigma(t,t)$. This is true if the following conditions hold: We assume that $u'(t)$ exists for $t > 0$ and that

$$u(t) \to 0,\ u'(t) \to 0 \text{ as } t \to \infty \tag{2}$$

and in addition either i)

$$u''(t) \text{ exists for } t > 0 \text{ with } u''(t) \to 0 \text{ as } t \to \infty \text{ and } EX^2 < \infty, \tag{3}$$

or ii)

$$1 - u(t) = c|t|^\alpha + o(|t|^\alpha) \text{ as } t \to 0 \text{ with } c > 0,\ 0 < \alpha \le 2, \text{ together with (3) if } \alpha \ge 1. \tag{4}$$

We say that $Y^*(\cdot)$ is locally stationary with index 2 in case i) , i. e. $1-u(t)$ is regularly varying with index 2 as $t \to 0$, and with index α in case ii). For such processes we have the following approximation of the extreme crossing probabilities where we assume that $u(t)$ is sufficiently smooth for $t \to \infty$ (for details see Hüsler (1988) and Hüsler and Heathcote (1988)). A sufficient condition for the smoothness used in the following is for instance $u'(t)/u(t) = O(1)$ as $t \to \infty$.

Theorem 1: *Let $Y^*(t), t > 0$, be a locally stationary Gaussian process with index α, $0 < \alpha \le 2$, by assuming the above conditions (2) and (3) or (2) and (4). If $r_n \to \infty$ such that $\min_{t \le r_n} \sqrt{n}u(t) \to \infty$, then as $n \to \infty$*

$$P\{Y(t) < n^{1/2}u(t), T < t \le r_n\} \to \exp(-J)$$

where $T > 0$ and

$$J = \lim_{n\to\infty} H_\alpha C^{1/\alpha}(\infty) \int_T^{r_n} \psi(\sqrt{n}u(t))\, dt$$

with $C(\infty) = -u''(0)/2$ if $\alpha = 2$, and $= c$, if $\alpha < 2$, H_α a constant depending on α only (see e.g. Leadbetter et al. (1983)) and $\psi(x) = \exp(-x^2/2)x^{2/\alpha-1}/\sqrt{2\pi}$.

In some cases, we can approximate $\sqrt{n}u(t)$ by a constant value u_n which simplifies the above result and where the smoothness conditions is naturally satisfied.

Corollary 2: *Let X and $u(\cdot)$ be such that $Y^*(\cdot)$ satisfies the condition of Theorem 1. If $r_n \to \infty$ and $u_n \to \infty$, then*

$$P\{Y(t) < u_n, t \le r_n\} \to \exp(-J)$$

with

$$J = \lim_{n\to\infty} H_\alpha C^{1/\alpha}(\infty) r_n \psi(u_n).$$

With these results we may discuss now the asymptotic behaviour of the empirical first zero R_n.

3. ASYMPTOTIC DISTRIBUTIONS OF R_n

a) First we consider the case of a finite zero $r_0 < \infty$. Basically, the behaviour of $u(t)$ in a neighbourhood of r_0 plays the dominant role. For r_0 may be an isolated zero or not; also $u(t)$ may take negative values in a right neighbourhood of r_0. This last case is the most interesting case, for which Welsh (1986) showed that $R_n \to r_0$ a. s. as $n \to \infty$. The most natural situation is given if

$u'(r_0)$ exists and is negative. We showed (Hüsler and Heathcote (1988)) for this usual case that with a suitable normalization R_n has a normal limit distribution. Slightly more general we state

Theorem 3: *Assume that $E|X|^\alpha < \infty$ for some $\alpha > 0$ and that $u(t)$ is strictly decreasing in $(r_0 - \delta, r_0 + \delta)$ for some $\delta > 0$ with $u(2r_0) > -1$. Let $r_n = r_n(z)$ be such that*

$$\sqrt{n}u(r_n)/\sigma(r_n) \to -z \text{ as } n \to \infty$$

for all $z \in \mathbb{R}$. Then

$$P\{R_n \le r_n(z)\} \to \Phi(z) \text{ as } n \to \infty, z \in \mathbb{R}.$$

Remark: Such normalization $r_n(z)$ exists by assumptions, since $\sigma(r_n) = \sigma(r_0) + o(1) > 0$. Hence we may select $r_n(z)$ as $r_n(z) = u^{-1}(z\sigma(r_0)/\sqrt{n})$ where $u^{-1}(\cdot)$ is an inverse function of $u(\cdot)$ in $(r_0 - \delta, r_0 + \delta)$.

Assuming the existence of the first derivative $u'(r_0) < 0$, we find easily the normalization

$$r_n(z) = r_0 + z\sigma(r_0)/(|u'(r_0)|\sqrt{n})$$

for $z \in \mathbb{R}$.

The proof follows the lines of Theorem 3.1 of Hüsler and Heathcote (1988) and is thus omitted. As basic hint we mention that we showed that

$$P\{R_n \le r_n(z)\} = P\{Y^*(t) < n^{1/2}u(t)/\sigma(t), t \le r_n\} + o(1)$$

$$= P\{Y^*(r_n) < n^{1/2}u(r_n)/\sigma(r_n)\} + o(1)$$

This result is only partially true if r_0 is not an isolated zero or if $u(t)$ is not negative in $(r_0, r_0 + \delta)$. The result holds then only for all $z \le 0$.

b) We consider now some results in the case $r_0 = \infty$. It is obvious that the occurence of a first zero R_n before some $y_n \ll r_n$ tends to 0 for a suitably chosen $y_n \to \infty$ and mild conditions on X. This can be shown (see Hüsler and Heathcote 1988)) by using e.g. Fernique's lemma (see Marcus (1981)). Thus the behaviour of an exceedance of $Y(\cdot)$ above $\sqrt{n}u(t)$ has to be considered only in the interval (y_n, r_n) . The discussion is therefore based on the behaviour of $u(t)$ as $t \to \infty$. Since we discussed in Section 2 sufficient conditions such that $Y^*(\cdot)$ is locally stationary, we state the following results assuming that this is true for some index $\alpha, 0 < \alpha \le 2$.

Theorem 4: *Assume that $u(t) = L(t)t^{-p}$ for some $p > 0$ as $t \to \infty, r_0 = \infty, L(t)$ a slowly varying function such that $|L(tz)/L(t) - 1| \le \delta/\log t$ as $t \to \infty$, for $z \in (1 - \epsilon, 1]$ and $\delta > 0, \delta = \delta(\epsilon) \to 0$ as $\epsilon \to 0$. Let X and $u(\cdot)$ be such that $Y^*(\cdot)$ is locally stationary with index $\alpha, 0 < \alpha \le 2$, by assuming the conditions (2) and (3) or (2) and (4). Let*

$$r_n = (2pL^2((\frac{n}{\log n})^{1/2p})(\frac{n}{\log n})^{1/2p}\Big(1 + (\frac{3}{2} - \frac{1}{\alpha} + \frac{1}{2p})\frac{\log_2 n}{\log n} -$$

$$-\frac{1}{p}\frac{L^2((n/\log n)^{1/2p})}{\log n} + \frac{z + A}{\log n}\Big)$$

with

$$A = \frac{1}{\alpha}\log C(\infty) - \frac{1}{2}\log(2\pi) + \frac{1}{2p}\log 2 + (\frac{1}{2} + \frac{1}{2p} - \frac{1}{\alpha})\log p.$$

Then i)

$$P\{Y(t) \le \sqrt{n}u(t), T \le t \le r_n\} \to \exp(-e^{-z})$$

for all $z \in I\!R$, *as* $n \to \infty$ *and any* $T > 0$.

ii) If in case $\alpha < 2$: $p > (\alpha+2)/\alpha$ *or in case* $\alpha = 2$: $p > (\beta+2)/\beta$, *for some* $\beta \ge 2$ *with* $E|X|^\beta < \infty$, *then*

$$P\{R_n \le r_n\} \to 1 - \exp(-e^{-z}) \text{ for all } z \in I\!R.$$

The first part of the proof follows the lines of Theorem 4.1 in Hüsler and Heathcote (1988) which assumes slightly more restrictive that $L(t) \to a$, as $t \to \infty$, $a > 0$. The proof is mainly based on Theorem 1 and Corollary 2. The second statement is then implied by the strong approximation argument. An example for such a case is given by the Exponential or Laplace r.v.'s.

In the last case we deal with distributions including the stable laws. We assume that $u(t) = \exp(-\gamma|t|^\beta + o(|t|^\beta))$ as $t \to \infty, \gamma > 0$.

Theorem 5: *Let* X *be such that* $u(t) = \exp(-\gamma|t|^\beta + g(t))$ *as* $t \to \infty, \gamma > 0, 1 < \beta$, $g(t) = o(|t|^\beta)$ *and that* $Y^*(\cdot)$ *is locally stationary with index* $\alpha, 0 < \alpha \le 2$, *by assuming the conditions* (2) *and* (3) *or* (2) *and* (4). *If* $g(r_n) - g((\log n/2\gamma)^{1/\beta}) \to 0$ *with*

$$r_n = \Big[\frac{\log n}{2\gamma}\Big(1 - \frac{2\log z/\sqrt{2} + 2g((\log n/2\gamma)^{1/\beta})}{\log n}\Big)\Big]^{1/\beta}, \quad z > 0,$$

as $n \to \infty$, *then we get for* $z > 0$

$$P\{Y(t) \le \sqrt{n}u(t), 0 \le t \le r_n\} \to \Phi(z)$$

and

$$P\{R_n \le r_n\} \to \Phi(-z) \text{ as } n \to \infty.$$

The results extends Theorem 4.3 and Corollary 4.5 of Hüsler and Heathcote (1988) by assuming the more general form of $u(t)$ for $t \to \infty$ with the additional term $g(t)$. The behaviour of $u(t)$ as $t \to 0$ is also more general, such that Y^* is locally stationary with some index α, not depending on β. Again the same method is used for the proof of this slightly more general theorem, by using Corollary 2 for the event of no exceedance of $Y(t)$ above $\sqrt{n}u(t)$ in $(0, y_n)$ with $y_n < r_n$ suitably chosen, by using a direct argument for no exceedance in (y_n, r_n) and the strong approximation argument again.

For $z < 0$, thus larger r_n values, only bounds on the tail of the distribution can be found in general. If in the above case $\beta \le 1$, then the used proof has to be drastically changed since the boundary $\sqrt{n}u(t)$ has a rather different behaviour in the neighbourhood of r_n. We proved a weaker limit result.

Theorem 6: *Assume that X such that $u(t) = \exp(-\gamma|t|^\beta)$ with $0 < \beta \leq 1$, $\gamma > 0$ and such that $Y^*(\cdot)$ is locally stationary with index α, $0 < \alpha \leq 2$. Then as $n \to \infty$*

$$R_n(2\gamma/\log n)^{1/\beta} \to 1 \text{ i. p.} .$$

The theoretical analysis clarifies rather well the simulated results on R_n in Csörgő and Heathcote (1987) for absolutely continous distributions. Results for discrete distribution can be rather different, as found for instance for the Bernoulli distribution in Hüsler and Heathcote (1988).

REFERENCES

[1] Berman, S. M. (1974) Sojourns and extremes of Gaussian processes. Ann. Probab. 2, 999-1026 (Correction (1980) 8, 999).

[2] Csörgő, S. (1981) Limit behaviour of the empirical characteristic function. Ann. Probab. 9, 130-144.

[3] Csörgő, S. and Heathcote, C. R. (1987) Testing for symmetry. Biometrika 74, 177-184.

[4] Heathcote, C. R. and Welsh, A. H. (1983) The robust estimation of autoregressive processes by functional least squares. J. Appl. Probab. 20, 737-753.

[5] Hüsler, J. (1988) Extreme values and high boundary crossings of locally stationary Gaussian processes. To appear.

[6] Hüsler, J. and Heathcote, C. R. (1988) Estimating the first zero of an empirical characteristic function. To appear.

[7] Leadbetter, M. R. , Lindgren, G. and Rootzén, H. (1983) Extremes and related properties of random sequences and processes. Series in Statistics, Springer, New York.

[8] Marcus, M. B. (1981) Weak convergence of the empirical characteristic function. Ann. Probab. 9, 194-201.

[9] Welsh, A. H. (1984) A note on scale estimates based on the empirical characteristic function and their application to test for normality. Statist. Probab. Letters 2, 345-348.

[10] Welsh, A. H. (1986) Implementing empirical characteristic function procedures. Statist. Probab. Letters 4, 65-67.

[11] Welsh, A. H. and Nicholls, D. F. (1987) Robust estimation of regression models with dependent regressor: The functional least squares approach. Econometric Theory. To appear.

Part Two
PARAMETER AND VARIANCE ESTIMATION

Ah, make the most of what we yet may spend,
Before we too into the Dust descend;
Dust into Dust, and under Dust, to Lie,
Sans Wine, sans Song, sans Singer, and—sans Ende!

Khayyam Naishapuri-Rubáiyát

STATISTICAL DATA ANALYSIS AND INFERENCE
Y. Dodge (Editor)

COMPARISON OF EXPERIMENTS WITH WEIGHTED DISTRIBUTIONS

M. J. BAYARRI AND M. H. DE GROOT

Department of Statistics and Operations Research
University of Valencia
Av. Dr. Moliner 50
46100 Burjasot, Valencia, Spain
and
Department of Statistics
Carnegie Mellon University
Pittsburgh, PA 15213, U.S.A.

In this paper the standard statistical experiment in which a random sample is drawn from some distribution involving an unknown parameter θ is compared with experiments in which samples are obtained from different weighted versions of that distribution, so that the underlying probability of entering the sample is multiplied by some weight function. This comparison is carried out by means of the concepts of sufficiency and pairwise sufficiency as developed in Blackwell's theory of the comparison of statistical experiments. Many examples are presented which illustrate the wide variety of effects that weight functions can have on the information about θ.

1 Introduction

In many situations, experimenters are not able to draw a random sample from the population in which they are interested, and statistical models that incorporate the nonrandomness or bias in their observations must be developed. In this paper we will discuss problems in which the sampling mechanism distorts the probabilities with which possible values in the population enter the sample by multiplying them by some weight function.

Consider a random variable X that is distributed over a certain population according to the density $g(x \mid \theta)$ and suppose that is it desired to make inferences about the unknown value of the parameter θ $(\theta \epsilon \Omega)$. The usual statistical analysis assumes that a random sample from $g(x \mid \theta)$ can be obtained. Suppose, however, that in fact the probabilty or density that a particular observation x will actually be obtained is proportional to $w(x)g(x \mid \theta)$, where $w(x)$ is a nonnegative weight function that may itself depend on some parameters whose values are unknown. Thus the observed sample is actually a random sample from the weighted density

$$f(x \mid \theta) = \frac{w(x)g(x \mid \theta)}{E_\theta[w(X)]} \, . \tag{1.1}$$

The expectation in the denominator is simply the normalizing factor that makes the density $f(x \mid \theta)$ integrate to 1.

The concept of a weighted distribution such as (1.1) can be traced to Fisher (1934), although these models were first formulated in a unified way by Rao (1965). Good surveys on this topic are Patil (1984) and Rao (1985). A recent application is Iyengar and Greenhouse (1988).

Our main goal in this paper is to compare the information about θ that is obtained by taking observations from the weighted density (1.1) for various weight functions. In particular, we will often be interested in comparing the information in a random sample from a specified weighted density $f(x \mid \theta)$ with that in an unweighted random sample from the underlying density $g(x \mid \theta)$. There are many different ways to measure and compare the information in a statistical experiment. Probably the most common one in statistics is based on Fisher information. Comparison of some weighted and unweighted distributions according to this criterion can be found in Bayarri and DeGroot (1987b) and Patil and Taillie (1987).

In spite of its widespread use in statistics, Fisher information does not have any clear-cut operational interpretation in statistical decision theory. Therefore, rather than study Fisher information in this paper, we will consider concepts of information based on the theory of the comparison of statistical experiments as developed originally by Blackwell (1951, 1953). This theory is very restrictive in the sense that it is relatively rare for one experiment to be regarded as more informative than another one. Nevertheless, when this does occur, every decision maker would prefer the more informative experiment, regardless of his or her prior distribution and utility function. Blackwell's method can be described as follows:

Let $\mathcal{E}_X = \{X, \mathcal{X}; g(\cdot \mid \theta), \theta \epsilon \Omega\}$ denote a statistical experiment in which a random variable or random vector X defined on some sample space $\mathcal{X}$ is to be observed, and the distribution $g(\cdot \mid \theta)$ of X depends on a parameter θ taking values in the parameter space Ω. Also, let $\mathcal{E}_Y = \{Y, \mathcal{Y}; f(\cdot \mid \theta), \theta \epsilon \Omega\}$ denote another statistical experiment with the same parameter space Ω. Then the experiment $\mathcal{E}_X$ is said to be *sufficient* for the experiment $\mathcal{E}_Y$ (denoted $\mathcal{E}_X \succeq \mathcal{E}_Y$) if there exists a stochastic transformation of X to a random variable $Z(X)$ such that, for each $\theta \epsilon \Omega$, the random variable $Z(X)$ and Y have identical distributions. A stochastic transformation from X to Z is defined to be a function $h(z \mid x)$ defined on $\mathcal{Y} \times \mathcal{X}$ such that for each $x \epsilon \mathcal{X}, h(z \mid x)$ is a density on $\mathcal{Y}$ with respect to some fixed σ-finite measure.

An important property of the relationship $\mathcal{E}_X \succeq \mathcal{E}_Y$ should be noted. Let $\mathcal{E}_X^n$ denote the experiment in which $\mathcal{E}_X$ is independently performed n times so that a random sample $X_1, \ldots, X_n$ is obtained, and let $\mathcal{E}_Y^n$ be defined analogously. Then $\mathcal{E}_X \succeq \mathcal{E}_Y$ implies $\mathcal{E}_X^n \succeq \mathcal{E}_Y^n$ for every value of n.

The relation $\succeq$ introduces a partial order among all experiments $\mathcal{E}$ with parameter space Ω. A somewhat less restrictive partial order that still makes use of the basic concept of sufficiency is based on the notion of pairwise sufficiency. An experiment $\mathcal{E}_X$ is said to be *pairwise sufficient* for the experiment $\mathcal{E}_Y$ (denoted $\mathcal{E}_X \succeq_2 \mathcal{E}_Y$) if for every pair of values $\theta_1, \theta_2 \epsilon \Omega$, $\mathcal{E}_X$ is sufficient for $\mathcal{E}_Y$ when the parameter space is restricted to contain just the two values θ_1 and θ_2. Clearly, if $\mathcal{E}_X \succeq \mathcal{E}_Y$ then $\mathcal{E}_X \succeq_2 \mathcal{E}_Y$. However, the converse does not necessarily hold.

If θ is a k-dimensional vector and Ω is an open subset of R^k, we shall let $I_X(\theta)$ and $I_Y(\theta)$ denote the $k \times k$ Fisher information matrices for the experiments $\mathcal{E}_X$ and $\mathcal{E}_Y$, respectively, under the standard regularity conditions. We shall use the notation $\mathcal{E}_X \succeq_F \mathcal{E}_Y$ whenever $I_X(\theta) - I_Y(\theta)$ is nonnegative definite for all $\theta \epsilon \Omega$. The relationship $\mathcal{E}_X \succeq \mathcal{E}_Y$ implies a similar ordering in terms of Fisher information; i.e., if $\mathcal{E}_X \succeq \mathcal{E}_Y$ then $\mathcal{E}_X \succeq_F \mathcal{E}_Y$. However, the converse does not necessarily hold. Moreover, since the Fisher information can be obtained from the Kullback -Leibler information by considering pairs of values of θ that are arbitrarily close to each other, it can be shown that if $\mathcal{E}_X \succeq_2 \mathcal{E}_Y$ then $\mathcal{E}_X \succeq_F \mathcal{E}_Y$. Some of these relations and other properties of the comparison of experiments are described in Stein (1951), Stone (1961), Kullback (1968, pages 26-28), Torgersen (1970, 1972, 1976), Hansen and Torgersen (1974), and Goel and DeGroot (1979).

In Section 2, we introduce a special class of weighted distributions known as selection models, and study pairwise sufficiency in different selection models based on underlying normal distributions. In Section 3, selection is applied to binomial and Poisson distributions, and we compare the information in the resulting experiments. In Section 4, we consider different weight functions applied to standard statistical distributions in order to illustrate the wide variety of effects that weighting can have even in common situations. Finally, in Section 5, we study the size-biased weight function $w(x) = x$, and show that its effect on the three standard families of discrete distributions, binomial, Poisson, and negative binomial, is different for each family.

2 Selection Models for Normal Distributions

In many problems in practice, observations are obtained only from certain selected portions of the population, either because experimental conditions make it impossible to obtain data from the whole population or because the experimenter chooses to restrict the observations in this way. In such problems, the density of each observation in the sample has the form

$$f(x \mid \theta) = \frac{g(x \mid \theta)}{\Pr(X \epsilon S \mid \theta)} \quad \text{for} \quad x \epsilon S , \tag{2.1}$$

where $g(x \mid \theta)$ is the underlying density over the whole population and S is the subset of the population to which observations are restricted. It is assumed that $\Pr(X\epsilon S \mid \theta) > 0$ for all $\theta\epsilon\Omega$. It can be seen that (2.1) is a special case of (1.1) in which the weight function $w(x)$ is the indicator function of the selection set S,

$$w(x) = \begin{cases} 1 & \text{for} \quad x\epsilon S , \\ 0 & \text{for} \quad x \epsilon S^c . \end{cases} \tag{2.2}$$

The model (2.1) is called a *selection model* or a *truncation model*, and a random sample from (2.1) is called a *selection sample.* Selection samples occur frequently in practice, and several examples were given in Bayarri and DeGroot (1987a, 1988) together with a Bayesian analysis of these models. The name "selection models" in this context is due to Fraser (1952, 1966), although the term "selection" was used in a more general setting by Tukey (1949).

It should be emphasized that a selection sample is *not* necessarily less informative about θ than an unrestricted random sample. In this section we will present examples pertaining to normal distributions with various selection sets S in which the selection sample is more informative than the unrestricted sample, as well as examples in which the converse is true.

We begin by considering an example in which the experiment based on a selection sample is pairwise sufficient for the analogous experiment based on an unrestricted random sample. Consider a population in which X has a normal distribution with a known mean, which we take to be 0, and an unknown precision θ. (The precision of a normal distribution is the reciprocal of its variance.) For $i = 1, 2$, let $\mathcal{E}_i$ denote the experiment in which a single observation X_i is obtained by restricting X to the tails $X \geq \tau_i$ and $X \leq -\tau_i$, where $\tau_i > 0$ is given and fixed. (It should be noted that this experiment is equivalent to one in which X is restricted to the upper tail only.) Since $Y_i = X_i^2$ is a sufficient statistic for θ in this problem, we can regard the experiment $\mathcal{E}_i$ as consisting of the observation of Y_i lying in the selection set $S = \{Y_i \geq \tau_i^2\}$. The density of Y_i in this experiment $\mathcal{E}_i$ is

$$f_i(u \mid \theta) = \frac{1}{\sqrt{2\pi}} \left(\frac{\theta}{y}\right)^{1/2} \frac{e^{-\frac{1}{2}\theta y}}{1 - G(\tau_i^2 \theta)} \quad \text{for} \quad y \geq \tau_i^2 , \tag{2.3}$$

where G is the d.f. of the χ^2-distribution with one degree of freedom.

We assume that $0 \leq \tau_1 < \tau_2 < \infty$, and we will carry out a comparison of the experiments $\mathcal{E}_1$ and $\mathcal{E}_2$. It will be shown that the experiment $\mathcal{E}_2$ with the larger truncation point τ_2 is pairwise sufficient for the experiment $\mathcal{E}_1$. Since an unrestricted random observation can be obtained by taking $\tau_1 = 0$, it follows that every experiment in which a selection sample of this type is obtained is pairwise sufficient for the original unrestricted experiment.

Consider the experiments $\mathcal{E}_1$ and $\mathcal{E}_2$, and suppose that the parameter space contains just two values θ_0 and θ_1. For any experiment $\mathcal{E}$ and any value of α $(0 < \alpha < 1)$, let $\beta(\alpha \mid \mathcal{E})$ denote the probability of a type 2 error when the likelihood-ratio test for distinguishing between θ_0 and θ_1 is carried out with the specified probability α of a type 1 error. It was shown by Torgersen (1970, 1976) that for any two experiments $\mathcal{E}$ and $\mathcal{E}^*$ with the same parameter space $\Omega = \{\theta_0, \theta_1\}, \mathcal{E} \succeq \mathcal{E}^*$ if and only if $\beta(\alpha \mid \mathcal{E}) \leq \beta(\alpha \mid \mathcal{E}^*)$ for all values of α $(0 < \alpha < 1)$. Hence, in the problem we are now considering, the desired conclusion that $\mathcal{E}_2 \succeq_2 \mathcal{E}_1$ can be obtained by showing that, for any pair of values θ_0 and θ_1, the function $\beta(\alpha \mid \mathcal{E}_2)$ is not larger than the function $\beta(\alpha \mid \mathcal{E}_1)$ over the interval $0 < \alpha < 1$.

Without loss of generality we will assume that $\theta_0 = 1$ and $\theta_1 < 1$. With this choice of θ_1 it can be seen that for testing

$$\begin{aligned} H_0 &: \ \theta = \theta_0 = 1 , \\ H_1 &: \ \theta = \theta_1 , \end{aligned} \tag{2.4}$$

the likelikhood-ratio test at level α based on the experiment $\mathcal{E}_i$ rejects H_0 when $Y_i \geq c_i(\alpha)$, where

$$\Pr[Y_i \geq c_i(\alpha) \mid \theta = 1] = \alpha . \tag{2.5}$$

Thus,

$$c_i(\alpha) = G^{-1}[1 - \alpha + \alpha G(\tau_i^2)] . \tag{2.6}$$

For notational convenience, let $\beta_i(\alpha) = \beta(\alpha \mid \mathcal{E}_i)$ for $i = 1, 2$. It follows from standard statistical theory that both $\beta_1(\alpha)$ and $\beta_2(\alpha)$ are continuous, convex, decreasing functions of α such that $\beta_1(0) = \beta_2(0) = 1$ and $\beta_1(1) = \beta_2(1) = 0$. Therefore, in order to show that

$\beta_2(\alpha) \leq \beta_1(\alpha)$ for $0 < \alpha < 1$, it is sufficient to show that there exists a value α^* in the interval $0 < \alpha^* < 1$ such that $\beta_2'(\alpha) < \beta_1'(\alpha)$ for $\alpha < \alpha^*$ and $\beta_2'(\alpha) > \beta_1'(\alpha)$ for $\alpha > \alpha^*$. For any given value of α, let $L_i(\alpha)$ denote the likelihood ratio for the observation Y_i evaluated at $Y_i = c_i(\alpha)$.

Since $\beta_i'(\alpha) = -L_i(\alpha)$, it will be true that $\mathcal{E}_2 \succeq_2 \mathcal{E}_1$ if there exists a value α^* such that

$$\begin{aligned} \frac{L_1(\alpha)}{L_2(\alpha)} &< 1 \quad \text{for} \quad 0 < \alpha < \alpha^* , \\ \frac{L_1(\alpha)}{L_2(\alpha)} &> 1 \quad \text{for} \quad \alpha^* < \alpha < 1 . \end{aligned} \tag{2.7}$$

In this problem

$$L_i(\alpha) = \frac{f_i[y = c_i(\alpha) \mid \theta = \theta_1]}{f_i[y = c_i(\alpha) \mid \theta = 1]} \tag{2.8}$$

which, from (2.3), is found to be

$$L_i(\alpha) = \theta_1^{1/2} \frac{1 - G(\tau_i^2)}{1 - G(\theta_1 \tau_i^2)} \exp[\frac{1}{2}(1 - \theta_1) c_i(\alpha)] . \tag{2.9}$$

Hence,

$$\frac{L_1(\alpha)}{L_2(\alpha)} = A \exp\{\frac{1}{2}(1 - \theta_1)[c_1(\alpha) - c_2(\alpha]\} , \tag{2.10}$$

where A is a constant not involving α. In order to establish the relations (2.7), it is sufficient to show that (2.10) is an increasing function of α. In turn, since $\theta_1 < 1$ it is sufficient to show that $D(\alpha) = c_1(\alpha) - c_2(\alpha)$ is an increasing function of α; that is, that

$$c_1'(\alpha) > c_2'(\alpha) \quad \text{for} \quad 0 < \alpha < 1 . \tag{2.11}$$

From (2.6),

$$c_i'(\alpha) = -\frac{1 - G(\tau_i^2)}{g\{G^{-1}[1 - \alpha + \alpha G(\tau_i^2)]\}} , \tag{2.12}$$

where g denotes the p.d.f. corresponding to G. Since $\tau_1 < \tau_2$, we can establish (2.11) by showing for each fixed α, that the right-hand side of (2.12) is a decreasing function of τ_i. If we let $b = 1 - G(\tau_i^2)$, then this is equivalent to showing that $b/g[G^{-1}(1 - \alpha b)]$ is a decreasing function of b for each fixed α. Finally, if we multiply this expression by α and let $u = G^{-1}(1 - \alpha b)$, then this is equivalent to showing that

$$[1 - G(u)]/g(u) \tag{2.13}$$

is an increasing function of u. But (2.13) is simply the reciprocal of the hazard-rate function of the χ^2 distribution with one degree of freedom, which is known to be decreasing. Therefore, $\mathcal{E}_2 \succeq_2 \mathcal{E}_1$.

It follows from this result that $\mathcal{E}_2^n \succeq_2 \mathcal{E}_1^n$, and therefore that a selection sample from the upper tail of a normal distribution with known mean and unknown precision is pairwise sufficient for a selection sample of the same size from the upper tail with a smaller truncation point. Moreover, a selection sample from the upper tail is pairwise sufficient for a random sample from the entire distribution.

It is striking that the reverse relationships hold when the mean is unknown and the precision is known. To see this, assume now that X has a normal distribution with unknown mean θ and known precision. For $i = 1, 2$, let $\mathcal{E}_i$ denote the experiment in which a single observation Y_i is obtained by restricting X to the upper tail $X \geq \tau_i$, where $-\infty < \tau_1 < \tau_2 < \infty$. Here it can be shown, by a proof completely analogous to the one just given (Bayarri and DeGroot, 1987b) that $\mathcal{E}_1^n \succeq_2 \mathcal{E}_2^n$. In particular, by letting $\tau_1 \to -\infty$, it follows that a random sample from the entire normal distribution is pairwise sufficient for a selection sample from the upper tail.

We do not know whether the relation $\succeq_2$ can be replaced by the stronger relation $\succeq$ in these problems.

If X has a normal distribution with unknown mean θ and known precision, and the observation Y is obtained by restricting X to the two tails $X \leq \tau_1$ and $X \geq \tau_2$, then it can be shown that neither the observation of X nor of Y is pairwise sufficient for the other. This is shown by noting that neither $\mathcal{E}_x \succeq_F \mathcal{E}_Y$ nor $\mathcal{E}_Y \succeq_F \mathcal{E}_x$ (Bayarri and DeGroot, 1987b).

3 Binomial and Poisson Selection

In this section we will compare the binomial and Poisson distributions with their zero-truncated versions in which the zero class is missing and observations are restricted to be positive. Selection models of this type have been widely studied in the statistical literature (David and Johnson, 1952; Irwin, 1959; Cohen, 1960; Dahiya and Gross, 1973; Sanathanan, 1977; Blumenthal and Sanathanan, 1980; and Blumenthal, 1981).

Let $\mathcal{E}_x$ denote the experiment in which an unrestricted observation X is taken from one of these two families, and let $\mathcal{E}_Y$ denote the experiment in which an observation Y is taken from the corresponding zero-truncated version. Then a straightforward calculation shows that $\mathcal{E}_X \succeq_F \mathcal{E}_Y$ (Bayarri and DeGroot, 1987b). In fact, it follows from a general theoretical development of sufficiency that $\mathcal{E}_X \succeq_2 \mathcal{E}_Y$. (We learned these results in independent personal communcations from Prem K. Goel and Erik N. Torgersen.) In this section we will show that in spite of these relations, it is *not* true that $\mathcal{E}_X \succeq \mathcal{E}_Y$.

Suppose first that X has a binomial distribution with parameters n and θ, so its p.f. is given by

$$g(x \mid \theta) = \binom{n}{x} \theta^x \bar{\theta}^{n-x} \quad \text{for} \quad x = 0, 1, \ldots, n \,, \tag{3.1}$$

where θ is unknown $(0 < \theta < 1)$ and $\bar{\theta} = 1 - \theta$. Then Y has the following p.f.:

$$f(y \mid \theta) = \frac{1}{1 - \bar{\theta}^n} \binom{n}{y} \theta^y \bar{\theta}^{n-y} \quad \text{for} \quad y = 1, \ldots, n \,. \tag{3.2}$$

In order for the relation $\mathcal{E}_X \succeq \mathcal{E}_Y$ to hold, there must exist a stochastic transformation $h(y \mid x)$ such that for all values of θ in the interval $0 < \theta < 1$,

$$\sum_{x=0}^{n} h(y \mid x) g(x \mid \theta) = f(y \mid \theta) \quad \text{for} \quad y = 1, \ldots, n \,. \tag{3.3}$$

The stochastic transformation $h(y \mid x)$ must be a nonnegative function such that

$$\sum_{y=1}^{n} h(y \mid x) = 1 \quad \text{for} \quad x = 0, 1, \ldots, n \,. \tag{3.4}$$

Since for any given value of y and any possible stochastic transformation h, the left-hand side of (3.3) must be a polynomial in θ whereas the right-hand side, as given by (3.2), is not a polynomial, it follows immediately that (3.3) cannot be satisfied for all values of θ. Hence, it cannot be true that $\mathcal{E}_X \succeq \mathcal{E}_Y$. Of course, it also cannot be true that $\mathcal{E}_Y \succeq \mathcal{E}_X$ since $\mathcal{E}_X \succeq_F \mathcal{E}_Y$ and the experiments are not equivalent.

Next, consider the truncated Poisson distribution with the zero class missing. Let X have a Poisson distribution with mean θ, so its p.f. is given by

$$g(x \mid \theta) = \frac{e^{-\theta} \theta^x}{x!} \quad \text{for} \quad x = 0, 1, 2, \ldots \,, \tag{3.5}$$

where θ is unknown $(\theta > 0)$. Then Y has p.f.

$$f(y \mid \theta) = \frac{e^{-\theta} \theta^y}{y!(1 - e^{-\theta})} \quad \text{for} \quad y = 1, 2, \ldots \,. \tag{3.6}$$

In order for the relation $\mathcal{E}_X \succeq \mathcal{E}_Y$ to hold, there must exist a stochastic transformation $h(y \mid x)$ defined for $x = 0, 1, 2, \ldots$, and $y = 1, 2, \ldots$ such that for all values of $\theta > 0$,

$$\sum_{x=0}^{\infty} \frac{h(y \mid x) e^{-\theta} \theta^x}{x!} = \frac{e^{-\theta} \theta^y}{y!(1 - e^{-\theta})} \quad \text{for} \quad y = 1, 2, \ldots \,. \tag{3.7}$$

By expanding $1 - e^{-\theta}$ in a Taylor series, it is found that (3.7) is equivalent to

$$\left[\sum_{x=0}^{\infty} \frac{h(y \mid x)}{x!} \theta^x \right] \left[\sum_{n=1}^{\infty} \frac{(-1)^{n-1}}{n!} \theta^n \right] = \frac{\theta^y}{y!} \quad \text{for} \quad y = 1, 2, \ldots \,. \tag{3.8}$$

Next, by taking $y = 1$ and equating the coefficients of θ^k on both sides of (3.8) it is found successively that the only possible values of h are $h(1 \mid 0) = 1$, $h(1 \mid 1) = \frac{1}{2}, h(1 \mid 2) = \frac{1}{6}, h(1 \mid 3) = 0$, and $h(1 \mid 4) = -\frac{1}{30}$, which is inadmissible. Thus, no such stochastic transformation can exist and therefore it cannot be true that $\mathcal{E}_X \succeq \mathcal{E}_Y$.

4 Other Weight Functions

In this section we will present some results on sufficiency obtained when various weight functions are applied to standard statistical distributions. Throughout this section and the next one we shall let $\mathcal{E}_X$ denote an experiment in which an observation X is taken from some specified underlying density $g(x \mid \theta)$ and $\mathcal{E}_Y$ denote an experiment in which an observation Y is taken from the weighted density $f(x \mid \theta)$ given by (1.1) for some specified weight function $w(x)$. We begin with an exponential example in which an unweighted random sample is sufficient for a sample from the weighted distribution.

Theorem 4.1 Let $g(x \mid \theta)$ be the exponential density with parameter θ,

$$g(x \mid \theta) = \theta e^{-\theta x} \quad \text{for} \quad x > 0\,, \tag{4.1}$$

where θ is unknown ($\theta > 0$), and let

$$w(x) = e^{-ax} \quad \text{for} \quad x > 0\,, \tag{4.2}$$

where a is a fixed constant ($a > 0$). Then $\mathcal{E}_X \succeq \mathcal{E}_Y$.

Proof. It follows from (1.1) that Y has an exponential distribution with parameter $\theta + a$. To construct a stochastic transformation from X to Y, let U be a random variable independent of X having an exponential distribution with parameter a. Then the random variable $Z = \min\{X, U\}$ has the same distribution as Y for all values of the parameter θ. □

It is interesting to consider the effect of some other weight functions $w(x)$ for the exponential density (4.1). If $w(x) = x^a$ for some fixed $a > 0$, then it will be seen later in this section that $\mathcal{E}_Y \succeq \mathcal{E}_X$. If $w(x)$ represents selection from the upper tail of the exponential distribution, i.e.,

$$w(x) = \begin{cases} 1 & \text{for} \quad x \geq a, \\ 0 & \text{for} \quad x < a\,, \end{cases} \tag{4.3}$$

then the experiments $\mathcal{E}_X$ and $\mathcal{E}_Y$ are equivalent in the sense that each is sufficient for the other.

We next present a weight function for the normal distribution that results in an equivalent experiment.

Theorem 4.2. Let $g(x \mid \theta)$ be the normal density with unknown mean θ and unit variance. Let $w(x) = e^{ax}$ for some constant a $(-\infty < a < \infty)$. Then $\mathcal{E}_X \succeq \mathcal{E}_Y$ and $\mathcal{E}_Y \succeq \mathcal{E}_X$.

Proof. It follows from (1.1) that Y has a normal distribution with mean $\theta + a$ and unit variance. Hence, $X + a$ and Y have the same distribution for all values of θ. □

Up to this point we have presented no examples in which a sample from a weighted distribuiton is sufficient for (and not equivalent to) a sample from the unweighted underlying distribution. The next two theorems provide such examples.

Theorem 4.3. Let $g(x \mid \theta)$ be the p.f. of a Poisson distribution with unknown mean θ $(\theta > 0)$, and let $w(x) = a^x$ for some fixed constant a $(a > 0, a \neq 1)$. Then $\mathcal{E}_X \succeq \mathcal{E}_Y$ if $a < 1$ and $\mathcal{E}_Y \succeq \mathcal{E}_X$ if $a > 1$.

Proof. It follows from (1.1) that Y has a Poisson distribution with mean $a\theta$. Suppose first that $a < 1$. Let Y^* be a random variable such that the conditional distribution of Y^* given $X = x$ is binomial with parameters x and a. ($Y^* = 0$ with probability one given $X = 0$.) Then it can be shown that the marginal distribution of Y^* is the same as that of Y for all values of θ. Hence, $\mathcal{E}_X \succeq \mathcal{E}_Y$.

Next, suppose that $a > 1$. Let X^* be a random variable such that the conditional distribution of X^* given $Y = y$ is binomial with parameters y and $1/a$. ($X^* = 0$ with probability one given $Y = 0$.) Then it can be shown that the marginal distribution of X^* is the same as that of X for all values of θ. Hence, $\mathcal{E}_Y \succeq \mathcal{E}_X$. □

The next result is of interest because it shows the different effect that a given weight function can have on the information that is obtained about the different individual parameters in a multiparameter family.

Theorem 4.4 Let $g(x \mid \alpha, \beta)$ be the gamma density with parameters α and β,

$$g(x \mid \alpha, \beta) = \frac{\beta^\alpha}{\Gamma(\alpha)} x^{\alpha-1} e^{-\beta x} \quad \text{for} \quad x > 0\,, \tag{4.4}$$

where $\alpha > 0$ and $\beta > 0$, and let

$$w(x) = x^a \quad \text{for} \quad x > 0\,, \tag{4.5}$$

where a is a fixed constant $(a > 0)$. If the parameter α is unknown and β is known, then $\mathcal{E}_X \succeq \mathcal{E}_Y$. If the parameter β is unknown and α is known, then $\mathcal{E}_Y \succeq \mathcal{E}_X$.

Proof. It follows from (1.1) that Y has a gamma distribution with parameters $\alpha + a$ and β. Assume first that α is unknown and β is known. Let U be a random variable independent of X and having a gamma distribution with parameters a and β. Then $X + U$ has the same distribution as Y for all values of α. Hence, $\mathcal{E}_X \succeq \mathcal{E}_Y$.

Assume next that β is unknown and α is known. Let V be a random variable independent of Y and having a beta distribution with parameters α and a. Then VY has the same distribution as X for all values of β. Hence $\mathcal{E}_Y \succeq \mathcal{E}_X$. □

5 Size-Biased Weight Functions

One of the most frequently-occurring weight functions in the statistical literature is the *size-biased* weight function $w(x) = x$. A wide variety of applications in which size-biased weight functions arise are described by Patil (1984) and Rao (1985). We applied this weight function to the gamma distribution in Theorem 4.4. In this section, we will study the effect of size-biased weight functions on the "Big 3" families of discrete distributions — binomial, Poisson, and negative binomial. Surprisingly, it will be shown that the effect is quite different in each of the three families.

First, we will show that for the binomial distribution, a random sample is sufficient for a size-biased sample.

Theorem 5.1. Let $g(x \mid \theta)$ be the p.f. of a binomial distribution with parameters n and θ, where θ is unknown $(0 < \theta < 1)$, and let $w(x) = x$ for $x = 0, 1, \ldots, n$. Then $\mathcal{E}_X \succeq \mathcal{E}_Y$.

Proof. It follows from (1.1) that $Y - 1$ has a binomial distribution with parameters $n - 1$ and θ. It is clear on intuitive grounds that X is more informative than $Y - 1$. In fact, the stochastic transformation that establishes this result is as follows: Let $Z = Y - 1$. Then $h(z \mid x) = \binom{n-1}{z} / \binom{n}{x}$ for $z = x - 1$ and $z = x$. Therefore, $\mathcal{E}_X \succeq \mathcal{E}_Z$. Since the observation of $Z = Y - 1$ is obviously equivalent to the observation of Y, it follows that $\mathcal{E}_X \succeq \mathcal{E}_Y$. □

Next, we will show that for the Poisson distribution, random and size-biased samples are equally informative.

Theorem 5.2. Let $g(x \mid \theta)$ be the p.f. of a Poisson distribution with mean θ, where θ is unknown $(\theta > 0)$, and let $w(x) = x$ for $x = 0, 1, 2, \ldots$. Then $\mathcal{E}_X$ and $\mathcal{E}_Y$ are equivalent in the sense that $\mathcal{E}_X \succeq \mathcal{E}_Y$ and $\mathcal{E}_Y \succeq \mathcal{E}_X$.

Proof. This result follows immediately from the fact that $Y - 1$ has a Poisson distribution with mean θ. □

Finally, we show that for the negative binomial distribution the sufficiency relation is reversed from the one obtained for the binomial distribution.

Theorem 5.3. Let $g(x \mid \theta)$ be the p.f. of the negative binomial distribution with parameters r and θ,

$$g(x \mid \theta) = \binom{r+x-1}{x} \theta^r (1-\theta)^x \quad \text{for} \quad x = 0, 1, 2, \dots , \tag{5.1}$$

where r is a fixed positive constant and θ is unknown $(0 < \theta < 1)$. Let $w(x) = x$ for $x = 0, 1, 2, \dots$. Then $\mathcal{E}_Y \succeq \mathcal{E}_X$.

Proof. It follows from (1.1) that $Y-1$ has a negative binomial distribution with parameters $r+1$ and θ. Let $Z = Y - 1$ and define a stochastic transformation from Z to a new random variable X^* as follows:

$$h(x \mid z) = \binom{r+x-1}{x} \Big/ \binom{r+z}{z} \quad \text{for} \quad x = 0, 1, \dots, z . \tag{5.2}$$

Then the marginal distribution of X^* is the same as the distribution of X for all values of θ. Therefore, $\mathcal{E}_Z \succeq \mathcal{E}_X$. Since $\mathcal{E}_Y$ and $\mathcal{E}_Z$ are equivalent experiments, $\mathcal{E}_Y \succeq \mathcal{E}_X$. □

Acknowledgments

This research was supported in part by the National Science Foundation under grant DMS-8701770. We are indebted to Prem K. Goel for several helpful and friendly conversations on this topic.

References

Bayarri, M. J., and DeGroot, M. H. (1987a). Bayesian analysis of selection models. *The Statistician* **36**, 137-146.

Bayarri, M. J., and DeGroot, M. H. (1987b). Information in selection models. In *Probability and Bayesian Statistics* (R. Viertl, ed.), 39-51. New York: Plenum Press.

Bayarri, M. J., and DeGroot, M. H. (1988). A Bayesian view of weighted distributions and selection models. In *Accelerated Life Testing and Experts' Opinions in Reliability* (C. A. Clarotti and D. V. Lindley, eds.), 70-82. Amsterdam: North-Holland.

Blackwell, D. (1951). Comparison of experiments. *Proceedings of the Second Berkeley Symposium on Mathematical Statistics and Probability*, 93-102. Berkeley, California: University of California Press.

Blackwell, D. (1952). Equivalent comparison of experiments. *Annals of Mathematical Statistics* **24**, 265-272.

Blumenthal, S. (1981). A survey of estimating distributional parameters and sample sizes from truncated samples. In *Statistical Distributions in Scientific Work* (C. Taillie, G. P. Patil, and B. Baldessari, eds.), Vol. 5, 75-86. Dordrecht: Reidel.

Blumenthal, S., and Sanathanan, L. P. (1980). Estimation with truncated inverse binomial sampling. *Communications in Statistics* **A9**, 977-1017.

Cohen, A. G. (1960). Estimation in truncated Poisson distributions when zeros and some ones are missing. *Journal of the American Statistical Association* **55**, 342-348.

Dahiya, R. G., and Gross, A. J. (1973). Estimating the zero class from a truncated Poisson sample. *Journal of the American Statistical Association* **68**, 731-733.

David, F. N., and Johnson N. L. (1952). The truncated Poisson. *Biometrics* **8**, 275-285.

Fisher, R. A. (1934). The effect of methods of ascertainment upon the estimation of frequencies. *Annals of Eugenics* **6**, 13-25.

Fraser, D. A. S. (1952). Sufficient statistics and selection depending on the parameters. *Annals of Mathematical Statistics* **23**, 417-425.

Fraser, D. A. S. (1966). Sufficiency for selection models. *Sankhā* A **28**, 329-334.

Goel, P.K., and DeGroot, M. H. (1979). Comparison of experiments and information measures. *Annals of Statistics* **7**, 1066-1077.

Hansen, O. H., and Torgersen, E. N. (1974). Comparison of linear normal experiments. *Annals of Statistics* **2**, 367-373.

Irwin, J. O. (1959). On the estimation of the mean of a Poisson distribution from a sample with the zero class missing. *Biometrics* **15**, 324-326.

Iyengar, S., and Greenhouse, J. B. (1988). Selection models and the file drawer problem. *Statistical Science* **3**, 109-135.

Kullback, S. (1968). *Information Theory and Statistics.* New York: Dover.

Patil, G. P. (1984). Studies in statistical ecology involving weighted distributions. In *Statistics: Applications and New Directions*, 475-503. Calcutta: Indian Statistical Institute.

Patil, G. P., and Taillie, G. (1987). Weighted distributions and the effects of weight functions on Fisher information. Unpublished technical report. Department of Statistics, Pennsylvania State University.

Rao, C. R. (1965). On discrete distributions arising out of methods of ascertainment. In *Classical and Contagious Discrete Distributions* (G. P. Patil, ed.), 320-333. Calcutta: Statistical Publishing Society.

Rao, C. R. (1985). Weighted distributions arising out of methods of ascertainment: What population does a sample represent?. In *A Celebration of Statistics: The ISI Centenary Volume* (A. G. Atkinson and S. E. Fienberg, eds.), 543-569. New York: Springer-Verlag.

Sanathanan, L. P. (1977). Estimating the size of a truncated sample. *Journal of the American Statistical Association* **72**, 669-672.

Stein, C. (1951). Information and the comparison of experiments. Unpublished report. University of Chicago.

Stone, M. (1961). Non-equivalent comparisons of experiments and their use for experiments involving location parameters. *Annals of Mathematical Statistics* **32**, 326-332.

Torgersen, E. N. (1970). Comparison of experiments when the parameter space is finite. *Z. Wahrscheinlichkeitstheorie und Verw. Gebiete* **16**, 219-249.

Torgersen, E. N. (1972). Comparison of translation experiments. *Annals of Mathematical Statistics* **43**, 1383-1399.

Torgersen, E. N. (1976). Comparison of statistical experiments. *Scandinavian Journal of Statistics* **3**, 186-208.

Tukey, J. W. (1949). Sufficiency, truncation and selection. *Annals of Mathematical Statistics* **20**, 309-311.

STATISTICAL DATA ANALYSIS AND INFERENCE
Y. Dodge (Editor)

OPTIMAL ESTIMATION FOR WEIGHTED DISTRIBUTIONS: SEMI-PARAMETRIC MODELS

V.P. Godambe and M.B. Rajarshi

University of Waterloo, Waterloo, Ontario, Canada N2L 3G1 and University of Poona, Pune 411007, India

Utilizing the theory of estimating functions (Godambe, 1960, 1985), Godambe & Thompson (1989) obtained optimal estimation for generalized linear models or semi-parametric models. Their results are extended to cover response dependent sampling (Scott & Wild, 1986), including weighted distributions (Rao, 1965).

1. INTRODUCTION AND NOTATION

We consider a finite population $\mathcal{P}$ of N labelled individuals i, $\mathcal{P} = \{i : i = 1, 2, \ldots, N\}$. With each individual i is associated a random (response) variate y_i with distribution given by $f_i(y_i; \boldsymbol{\theta})$ which is completely specified upto an unknown parameter, possibly vector valued, $\boldsymbol{\theta}$. If $\boldsymbol{y} = (y_1, y_2, \ldots, y_N)$, the probability density of $\boldsymbol{y}$ is given by

$$Prob(\boldsymbol{y}; \boldsymbol{\theta}) = \prod_1^N f_i(y_i; \boldsymbol{\theta}). \tag{1}$$

One situation in which response dependent sampling or a weighted distribution arises is from a postulated experiment as follows.

EXPERIMENT 1. After the variate $\boldsymbol{y}$ is *realized* (to be distinguished from *observed*) from the distribution (1), for each individual $i \in \mathcal{P}$, nature performs a Bernoulli experiment with chance of success $w_i(y_i)$. Only if the experiment results in a 'success', (i, y_i) is included in the data

$$d = \{(i, y_i) : i \in s\} \quad , s \subseteq \mathcal{P}; \tag{2}$$

the subset s is also called a 'sample' in survey sampling.

REMARK 1. Throughout, we assume that the chance functions $w_i(y_i)$ in the Experiment 1 are independent of $\boldsymbol{\theta}$. However, many results, particularly of Section 3 and 4 are valid even without this assumption.

Now, denoting generally 'probability density' or 'probability' under $\boldsymbol{\theta}$ by $Prob(.;\boldsymbol{\theta})$, we have

$$Prob(d;\boldsymbol{\theta}) = \int Prob(s,\boldsymbol{y};\boldsymbol{\theta}) \prod_{i \notin s} dy_i.$$

If further $Prob(.|.)$ denotes the conditional probability density, from (1), we have

$$Prob(d;\boldsymbol{\theta}) = \int Prob(s|\boldsymbol{y};\boldsymbol{\theta}) \prod_{i=1}^{N} f_i(y_i;\boldsymbol{\theta}) \prod_{i \notin s} dy_i.$$

Now, since in the Experiment 1,

$$Prob(s|\boldsymbol{y};\boldsymbol{\theta}) = \prod_{i \in s} w_i(y_i) \prod_{i \notin s} (1 - w_i(y_i)),$$

we have

$$Prob(d;\boldsymbol{\theta}) = \prod_{i \in s} \{f_i(y_i;\boldsymbol{\theta}) w_i(y_i) / w_{i0}(\boldsymbol{\theta})\} \prod_{i \in s} w_{i0}(\boldsymbol{\theta}) \prod_{i \notin s} \{1 - w_{i0}(\boldsymbol{\theta})\} \tag{3}$$

where

$$w_{i0}(\boldsymbol{\theta}) = \int w_i(y_i) f_i(y_i;\boldsymbol{\theta}) dy_i, \ i = 1, 2, \ldots, N. \tag{4}$$

In the following section, we consider maximum likelihood (ml) estimation of the unknown parameters in (3) on the basis of the data d in (2).

REMARK 2. It is important to note that though our entire discussion is within the set-up of a finite population $\mathcal{P}$ above, we will, in general *not* study estimation of the parameters of $\mathcal{P}$, like functions $t(\boldsymbol{y})$. We restrict ourselves to the estimation of the parameter $\boldsymbol{\theta}$ of the distribution in (1), the only exception being estimation of the size of $\mathcal{P}$, $|\mathcal{P}| = N$.

2. ML ESTIMATION : PARAMETRIC MODEL

For the parametric model (1), the ml equations for $\boldsymbol{\theta}$, based on the data d are obtained as usual from (3). Denoting $\boldsymbol{\theta} = (\theta_1, \theta_2, \ldots, \theta_k)$, where θ_j's are scalars, we have, for $j = 1, 2, \ldots, k$,

$$\frac{\partial}{\partial \theta_j} \log Prob(d;\boldsymbol{\theta}) = \sum_{i \in s} \frac{\partial}{\partial \theta_j} \log\left(\frac{f_i w_i}{w_{i0}}\right) + \sum_{i \in s} \frac{1}{w_{i0}} \frac{\partial w_{i0}}{\partial \theta_j} - \sum_{i \notin s} \frac{1}{1 - w_{i0}} \frac{\partial w_{i0}}{\partial \theta_j} = 0. \tag{5}$$

An interesting special case is obtained when in (1) and (3), $f_i = f$, $w_i = w$ and $w_{i0} = w_0$. Denoting $|s|$ by n, we have, from (5)

$$\sum_{i\in s}\frac{\partial}{\partial\theta_j}\log\left(\frac{f(y_i;\boldsymbol{\theta})\,w(y_i)}{w_0(\boldsymbol{\theta})}\right)+\frac{n-w_0(\boldsymbol{\theta})\,N}{w_0(\boldsymbol{\theta})\,(1-w_0(\boldsymbol{\theta}))}\frac{\partial w_0}{\partial\theta_j}=0\;\;;\;j=1,2,\ldots,k. \tag{6}$$

It is important to note that the ml equations (6), even when the population size $|\mathcal{P}|=N$ is unknown, do *not* reduce to the equations

$$\sum_{i\in s}\frac{\partial}{\partial\theta_j}\log\left(\frac{f(y_i;\boldsymbol{\theta})\,w(y_i)}{w_0(\boldsymbol{\theta})}\right)=0\;;\;j=1,2,\ldots,k, \tag{7}$$

which are usually found in the literature on weighted distributions (cf. Rao, 1965). The ml estimate of N obtained from the distribution (3) of the *full data* is $\hat{N}=n$! On the other hand, the natural estimate of N, when θ is known, is given by $[n/w_0(\boldsymbol{\theta})]$. This natural estimate is the ml estimate but with respect to the marginal distribution of only a *part* data $|s|=n$ viz., $Prob(n|N,\boldsymbol{\theta})$, (Feldman & Fox, 1968). Thus, the conventional ml theory does *not* provide equations (7) and $n-w_0N=0$ for joint estimation of $\boldsymbol{\theta}$ and N. Note that n/w_0 is also the uniformly minimum variance unbiased estimator with respect to the distribution $Prob(n|N,\boldsymbol{\theta})$.

The weakness of ml estimation, described above, is corrected in the next section with an extension of ml estimation provided by the theory of estimating functions (Godambe & Thompson, 1989; hereafter, for brevity, just refered to as GT). This theory provides equation (6) and the equation $n-w_0(\boldsymbol{\theta})=0$ as jointly optimal for $\boldsymbol{\theta}$ and N, implying equation (7), which is in common use.

3. ESTIMATING FUNCTIONS

To modify or extend ml estimation discussed in Section 2, we first note that the 'Experiment 1' of Section 1 is statistically equivalent to the following Experiment 2. We note here that, in (3), (f_iw_i/w_{i0}) due to (4), is a probability density function.

EXPERIMENT 2. With each individual $i\in\mathcal{P}$, nature performs a Bernoulli experiment with the chance of 'success' $w_{i0}(\boldsymbol{\theta})$ given by (4). Only if the Bernoulli experiment results in a success, a value y_i is drawn from the distribution (f_iw_i/w_{i0}); (i,y_i) then becomes a part of statistician's data.

THEOREM 1. Experiment 2 above is logically equivalent to Experiment 1 of Section 1.

PROOF. A generic outcome in both, Experiment 1 and Experiment 2 is d of (2). Further, the probability density of d in both the experiments is given by (3) and (4).

Now the following Experiment 2 above, we define for each individual $i\in\mathcal{P}$ and a *specified*

j, random variates

$$h_i = \phi_i \frac{\partial}{\partial \theta_j} \log [f_i w_i / w_{i0}] \ ;$$

$$h_{i+N} = \phi_i - w_{i0} \ , \quad i = 1, 2, \ldots, N, \tag{8}$$

where ϕ_i equals 1, if the corresponding Bernoulli experiment results in a success, otherwise $\phi_i = 0$. Denoting the expectation in Experiment 2 by $\mathcal{E}_2$, we note that in (8), for $i, i' = 1, 2, \ldots, 2N$ and $i \neq i'$,

$$\mathcal{E}_2 (h_i) = 0 \ \ , \ \ \mathcal{E}_2 (h_i h_i') = 0 \ ; \tag{9}$$

That is, according to GT, the $2N$ estimating functions in (8) are mutually *orthogonal.* Hence, to estimate $\boldsymbol{\theta}$, we obtain the optimal combination g (upto a constant multiple) of $2N$ estimating functions in (8) as

$$g = \sum_{i=1}^{N} \left\{ h_i \frac{\mathcal{E}_2 (\partial h_i / \partial \theta_j)}{\mathcal{E}_2 (h_i^2)} + h_{i+N} \frac{\mathcal{E}_2 (\partial h_{i+N} / \partial \theta_j)}{\mathcal{E}_2 \left(h_{i+N}^2 \right)} \right\}$$

$$= \sum_{i=1}^{N} \phi_i \frac{\partial}{\partial \theta_j} \log \left(\frac{f_i w_i}{w_{i0}} \right) + \sum_{i=1}^{N} (\phi_i - w_{i0}) \frac{\partial w_{i0} / \partial \theta_j}{w_{i0} (1 - w_{i0})} \tag{10}$$

From (10), we get the optimal estimating equation $g = 0$, which is the same as the ml equation (5).

As a special case, when in (1) and (3), $f_i = f$, $w_i = w$ and $w_{i0} = w_0$, we have

$$g = \sum_{i \in s} \frac{\partial}{\partial \theta_j} \log \frac{f(y_i; \boldsymbol{\theta}) \, w(y_i)}{w_0(\boldsymbol{\theta})} + \frac{[n - w_0(\boldsymbol{\theta}) N] \, \partial w_0(\boldsymbol{\theta}) / \partial \theta_j}{w_0(\boldsymbol{\theta}) \, (1 - w_0(\boldsymbol{\theta}))} \tag{11}$$

as in (6). Here, (unlike when f_i, w_i vary for different i's), a case of the population size $|\mathcal{P}| = N$ being *unknown* arises naturally. When N is unknown, it is more logical to combine in (8) function h_{i+N} , $i = 1, 2, \ldots, N$ into single function

$$\sum_{i=1}^{N} h_{i+N} = \sum_{i=1}^{N} (\phi_i - w_0) = n - w_0 N \ , \tag{12}$$

which is orthogonal to h_i's; $i = 1, 2, \ldots, N$. Here, N is an integer-valued parameter. Hence, replacing the corresponding derivative by its appropriate modification (Godambe & Thompson, 1984), we get from (12), the optimum estimating function $\{(n - w_0 N) w_0 / [N w_0 (1 - w_0)]\}$. Hence, according to GT, equations $g = 0$ and $n - w_0 N = 0$ provide joint optimal estimation of $\boldsymbol{\theta}$ and N. Further, the commonly used equation (7) is a logical consequence of our theory.

It is important to note that for simplicity of presentation, in (8), we did not include estimating functions $\partial \log (f_i w_i / w_{i0}) / \partial \theta_j$ for all $j = 1, 2, \ldots, k$ to avoid derivatives $(\partial^2 / \partial \theta_j \, \partial \theta_{j'})$

$\log(f_i w_i)/w_{i0}$. Of course, even with inclusion of such functions in (8), the estimating functions (10), (11) and (12) would continue to be 'optimal'; only the derivation would be longer. This longer derivation was unnecessary here, in veiw of the development of a general semi-parametric model in the next section.

4. SEMI-PARAMETRIC MODELS

In (1), we made a full distributional assumption that y_i was distributed as $f_i(y_i;\boldsymbol{\theta})$. Following the developments of the theory and practice of generalized linear models (McCullagh & Nelder, 1983), we now replace the distributional assumption by assumptions of just the mean α, variance σ^2 and the skewness coefficeint γ: for $i = 1, 2, \ldots, N$,

$$\int y_i f_i dy_i = \alpha_i(\boldsymbol{\theta})$$

$$\int (y_i - \alpha_i)^2 f_i dy_i = \sigma_i^2(\boldsymbol{\theta})$$

$$\int (y_i - \alpha_i)^3 f_i dy_i = \gamma_i(\boldsymbol{\theta})\sigma_i^3(\boldsymbol{\theta}) \quad , \tag{13}$$

$\alpha_i, \sigma_i, \gamma_i$ begin specified functions of $\boldsymbol{\theta}$. Further assuming that $y_i > 0$, we let in Experiment 1, that is, in (3) and (4),

$$w_i(y_i) = k_i y_i, \text{ (and hence) } w_{i0}(\boldsymbol{\theta}) = k_i \alpha_i(\boldsymbol{\theta}) \ , \tag{14}$$

where k_i is a known constant and α_i is given by (13), $i = 1, 2, \ldots, N$. (This corresponds to the size-biased sampling in the literature.) Recalling that $\mathcal{E}_2$ denotes the expectation with respect to the Experiment 2 of Section 3, we have from (13), for $i = 1, 2, \ldots, N$,

$$\mathcal{E}_2(y_i) = \left(\sigma_i^2 + \alpha_i^2\right)/\alpha_i = m_i(\boldsymbol{\theta}) \text{ say,}$$

$$\mathcal{E}_2(y_i - m_i)^2 = \sigma_i^2 \left\{(\gamma_i \sigma_i/\alpha_i) + 1 - \left(\sigma_i^2/\alpha_i^2\right)\right\}. \tag{15}$$

Now, in (8), we take for $i = 1, 2, \ldots, N$

$$h_i = \phi_i(y_i - m_i) \quad \text{and} \quad h_{i+N} = \phi_i - w_{i0} \tag{16}$$

where m_i and w_{i0} are respectively given by (15) and (14). Note that the estimating function h_i, $i = 1, 2, \ldots, N$ in (16) satisfies (9) and hence are mutually *orthogonal.* There is some arbitrariness in the choice of the functions h_i but this arbitrariness is only apparent. This is explained in terms of some appealing properties by GT. One rather intuitive alternative, in view of (14), is to take h_i to be $\phi_i(y_i - \alpha_i)/y_i$ instead of $\phi_i(y_i - m_i)$, as in (16). This alternative, apart from its mathematical unmanageability, seems to be logically unsound. For the factor $(y_i - \alpha_i)$ in it is suggested by a possiblity of the basic distribution f_i in (1) belonging to an exponential family and the factor $(1/y_i)$ is suggested by (14). However, the exponentiality of f_i implies that the distribution $(f_i w_i / w_{i0})$ in Experiment 2 also belongs to an exponential family. Hence, our choice of $h_i = \phi_i(y_i - m_i)$. This is further supported by the investigation of Scott & Wild (1986).

The optimum combination of the estimating functions in (16), according to GT, is given by

$$g = \sum_{i=1}^{N} \phi_i(y_i - m_i) \frac{\partial m_i / \partial \theta_j}{\mathcal{E}_2 (y_i - m_i)^2} + \sum_{i=1}^{N} (\phi_i - w_{i0}) \frac{\partial w_{i0} / \partial \theta_j}{w_{i0}(1 - w_{i0})} \tag{17}$$

for $j = 1, 2, \ldots, k$, where m_i, $\mathcal{E}_2 (y_i - m_i)^2$ and w_{i0} are as in (14) and (15), and

$$\frac{\partial m_i}{\partial \theta_j} = -\left\{ \frac{1}{\alpha_i} \frac{\partial \sigma_i^2}{\partial \theta_j} - \left[\frac{\sigma_i^2}{\alpha_i^2} - 1 \right] \frac{\partial \alpha_i}{\partial \theta_j} \right\}$$

$i = 1, 2, \ldots, N$; $j = 1, 2, \ldots, k$. Thus, finally we have,

$$g = \sum_{i \in s} (y_i - m_i) \frac{\left\{ (\sigma_i / \alpha_i)^2 - 1 \right\} (\partial \alpha_i / \partial \theta_j) - (\partial \sigma_i^2 / \partial \theta_j) / \alpha_i}{\sigma_i^2 \left\{ (\gamma_i \sigma_i / \alpha_i) + 1 - (\sigma_i / \alpha_i)^2 \right\}}$$

$$+ \sum_{i \in s} (\partial \log w_{i0} / \partial \theta_j) + \sum_{i \notin s} \{ \partial \log (1 - w_{i0}) / \partial \theta_j \} , \tag{18}$$

for $j = 1, 2, \ldots, k$.

5. EXPONENTIAL FAMILIES

A simplification of the estimating function g of (18) occurs when the distribution f_i in (1) belongs to the exponential family of distributions,

$$f_i = \operatorname{Exp} \{ y_i A_i(\boldsymbol{\theta}) + B_i(\boldsymbol{\theta}) + C_i(y_i) \} ,$$

A_i, B_i, C_i begin functions of the indicated arguments (McCullagh & Nelder, 1983; Section 2.2). For then, as shown by GT

$$\gamma_i \sigma_i (\partial \alpha_i / \partial \theta_j) = \partial \sigma_i^2 / \partial \theta_j \tag{19}$$

for $i = 1, 2, \ldots, N$ and $j = 1, 2, \ldots, k$. Note that (19) may hold without f_i being a member of the exponential family of distributions. Under the assumption (19), the first factor of (18) reduces to $-\{(y_i - m_i)/\sigma_i^2\}(\partial\alpha_i/\partial\theta_j)$. But, because of (14) and (15),

$$-\frac{y_i - m_i}{\sigma_i^2}\frac{\partial\alpha_i}{\partial\theta_j} = -\frac{y_i - \alpha_i}{\sigma_i^2}\frac{\partial\alpha_i}{\partial\theta_j} + \frac{\partial \log w_{i0}}{\partial\theta_j} \tag{20}$$

Actually, the equality (20) is true not only for $w_i(y_i) = k_i y_i$ as in (14) but for *any* chance function $w_i(y_i)$ in Experiment 1. Since $w_i(y_i)$ are independent of $\boldsymbol{\theta}$, exponentiality of f_i, as said before, implies exponentiality of $(f_i w_i/w_{i0})$. Hence,

$$\frac{\partial \log f_i}{\partial\theta_j} - \frac{\partial \log (f_i w_i/w_{i0})}{\partial\theta_j}$$

$$= \frac{y_i - \alpha_i}{\sigma_i^2}\frac{\partial\alpha_i}{\partial\theta_j} - \frac{y_i - m_i}{\mathcal{E}_2(y_i - m_i)^2}\frac{\partial m_i}{\partial\theta_j} \; . \tag{21}$$

Now, in (21), the left hand side is independent of y_i, hence we must have for $i = 1, 2, \ldots, N$ and $j = 1, 2, \ldots, k$

$$\frac{\partial\alpha_i/\partial\theta_j}{\sigma_i^2} = \frac{\partial m_i/\partial\theta_j}{\mathcal{E}_2(y_i - m_i)^2}\;;$$

$$\alpha_i - m_i = -\frac{\sigma_i^2(\partial \log w_{i0}/\partial\theta_j)}{\partial\alpha_i/\partial\theta_j}\;, \tag{22}$$

and consequent direct simplification of the optimum estimating function g of (17). We can summarize the above results in the following

THEOREM 2. For any given chance function $w_i(y_i)$, independent of $\boldsymbol{\theta}$, if $\mathcal{E}_2(y_i - m_i) = 0$, $i = 1, 2, \ldots, N$, then the optimum estimating functions based on (16) are given by g of (17). Further, $g = 0$ are also ml equations if the underlying distributions f_i in (1) belong to some exponential families. Under this exponentiality, the function g in (17) is simplified through the relation (22).

6. AN ADDITIONAL PARAMETER

Here, we assume that the chance function w_i of the Experiment 1 depends on an additional parameter, λ say:

$$w_i = w_i(y_i; \lambda) \tag{23}$$

Of course, as before w_i continues to be independent of $\boldsymbol{\theta}$. For simplicity, let λ be real. It follows from GT that the estimating functions based on (16), which are jointly optimal for estimating $\boldsymbol{\theta}$ and λ, are given by the function g in (17) and

$$g_1 = \sum_{i=1}^{N} \phi_i (y_i - m_i) \frac{\partial m_i / \partial \lambda}{\mathcal{E}_2 (y_i - m_i)^2} + \sum_{i=1}^{N} (\phi_i - w_{i0}) \frac{\partial w_{i0} / \partial \lambda}{w_{i0} (1 - w_{i0})} . \tag{24}$$

Now we have a generalization of Theorem 2.

THEOREM 3. If, for a fixed λ and a fixed $\boldsymbol{\theta}$ *respectively*, the distributions $(f_i w_i / w_{i0})$ in Experiment 2 of Section 3 belong to some exponential families, the jointly optimum estimating equations $g = 0$ and $g_1 = 0$ given by (17) and (24) are also ml equations for the parameters $\boldsymbol{\theta}$ and λ.

The conditions for a fixed $\boldsymbol{\theta}$ in the above Theorem will evidently by satisfied if, in (23),

$$w_i = \text{Exp}\,\{y_i A_i(\lambda) + B_i(\lambda) + C_i(y_i)\}$$

where A_i, B_i and C_i are functions of the indicated arguments.

We note here that under a full distributional assumption as in (1), with $f_i = f$ and $w_i = w$, $i = 1, 2, \ldots, N$, the optimal estimating equations $g = 0$ and $n - w_0 N = 0$ obtained from (11) and (12), in the presence of an additional parameter λ mentioned above, need only an obvious notational modification: $\boldsymbol{\theta} = (\theta_1, \theta_2, \ldots, \theta_k, \lambda)$.

7. ILLUSTRATIONS

1. Aerial survey of Animal Populations

An aerial survey of animal populations attempts to record as many groups of animals as possible. The survey purports to estimate the total number of animals of a species in a given area. Let N be the unknown number of groups in a given quadrat and let y_i be the number of animals in the i-th group, $i = 1, 2, \ldots, N$. The entire group of animals is observed if any member of the group is observed. Cook & Martin (1974) assume that all animals in a group are observed independently of each other. If λ denotes the probability that an animal is observed, we have, as in section 6,

$$\begin{aligned} w_i &= w_i (y_i, \lambda) = w (y_i, \lambda) \\ &= 1 - (1 - \lambda)^{y_i} , \end{aligned}$$

$y_i = 1, 2, \ldots \ : \ i = 1, 2, \ldots, N$. Further, in terms of our notaiton in (1), Cook & Martin (1974) assume that for $i = 1, 2, \ldots, N$,

$$f_i (y_i, \boldsymbol{\theta}) = f (y_i, \theta) = e^{-\theta} \frac{\theta^{y_i - 1}}{(y_i - 1)!} \tag{25}$$

$y_i = 1, 2, \ldots, \ : \ \theta$ being a real parameter. Now in (4), for $i = 1, 2, \ldots, N$,

$$w_{i0} = w_0 = 1 - e^{-\theta\lambda}(1-\lambda) \quad . \tag{26}$$

From the 'data' obtained from an Aerial Moose Survey of North-East Minnesota (1970), Cook & Martin (1974) obtain ml estiamtes of $\theta, \lambda \ : \ \hat{\theta} = .84$ and $\hat{\lambda} = .89$. Now the data recorded the number (n) of groups observed. Hence, using the equations $g = 0$ and $n - w_0 N = 0$ obtained from (11) and (12), we, in addition to getting $\hat{\theta} = .84$ and $\hat{\lambda} = .89$, get an estimate of the unknown N, $\hat{N} = [n/\hat{w}_0] = 119$. In this connection, we refer to the last paragraph of Section 6.

2. Regression

Now, assuming in (1), $\boldsymbol{\theta} = \theta$, a scalar parameter, we follow the set-up of Section 4. Let in (13), $\alpha_i(\boldsymbol{\theta}) = \theta x_i$, x_i, $i = 1, 2, \ldots, N$ being some known covariates. Further, we assume that $\sigma_i^2(\theta)$ and $\gamma_i(\theta)$ in (13) to be independent of $\theta \ : \ \sigma_i^2 = \sigma^2$ known, $\gamma_i = 0$ say, $i = 1, 2, \ldots, N$. Now, from (14) and (15), we have

$$m_i(\theta) = \frac{\sigma^2 + (\theta x_i)^2}{\theta x_i} \ ,$$

$$\mathcal{E}_2 (y_i - m_i)^2 = \sigma^2 \left\{ 1 - \frac{\sigma^2}{(\theta x_i)^2} \right\} \ .$$

Hence, in (18),

$$g = -\sum_{i \in s} (y_i - m_i) \frac{x_i}{\sigma^2} + \frac{n}{\theta} + \sum_{i \notin s} \frac{k x_i}{1 - k\theta x_i} \ .$$

If, in the above, σ^2 is unknown, joint optimal estimation for θ and σ^2 can be obtained, as in GT, by including in (16) additional estimating functions based on the second moments of $\boldsymbol{y}$, which are orthogonal to h_i, $i = 1, 2, \ldots, 2N$.

ACKNOWLEDGEMENTS

Work of the first author (V.P. Godambe) was in part supported by UGC (India) and NSERC (Canada) grants.

REFERENCES

[1] Cook, R.D. and Martin, F.B., (1974) A model for quadrat sampling with 'visibility bias' *J. Ame. Statist. Assoc.* **69**, 345-349.

[2] Feldman, D. and Fox, M.,(1968) Estimation of the parameter n in the Binomial model *J. Ame. Statist. Assoc.* **63**, 150-158.

[3] Godambe, V.P., (1960) An optimum property of regular maximum likelihood estimation *Ann. Math. Statist.* **31**, 1208-1211.

[4] Godambe, V.P., (1985) The foundations of finite sample estimation in stochastic processes *Biometrika* **72**, 419-428.

[5] Godambe, V.P. and Thompson, M.E., (1984) Robust estimation through estimating equations *Biometrika* **71**, 115-125.

[6] Godambe, V.P. and Thompson, M.E., (1989) An extension of quasi-likelihood estimation (with discussion) *J. Statist. Plan. Infer.*, (to appear).

[7] McCullagh, P. and Nelder, J.A., (1983) *Generalized linear models*, London: Chapman and Hall.

[8] Rao, C.R., (1965) On discrete distributions arising out of methods of ascertainment in *Classical and Contagious Discrete Distributions*, (Ed. G.P. Patil) 320-332; Calcutta: Statistical Publishing Society.

[9] Scott, A.J. and Wild, C.J., (1986) Fitting logistic models under case control or choice-based sampling *J. Roy. Statist. Soc. B* **48**, 170-182.

STATISTICAL DATA ANALYSIS AND INFERENCE
Y. Dodge (Editor)

ON EFFICIENCY FOR QUASI-LIKELIHOOD AND COMPOSITE QUASI-LIKELIHOOD METHODS

C.C. HEYDE

Department of Statistics, IAS,
Australian National University,
GPO Box 4, Canberra, ACT 2601, Australia

Quasi-score estimating functions based on optimizing within different classes of estimating functions can be derived in many contexts and it is important to be able to compare them and to combine them. Two related general approaches to the problem of combination are outlined in this paper.

1. INTRODUCTION

For a Conference in honour of C.R. Rao it seems appropriate to discuss a topic concerning efficiency in estimation, a subject to which he has contributed so much. His work on what is now known as the Cramér–Rao inequality (Rao ([7]) and his series of important papers on asymptotic efficiency in the early 1960s, which are well summarized in his book Rao [8], should specifically be noted.

Our focus in this paper is on extensions of the form of likelihood type estimation often called *pseudolikelihood*, apparently first proposed by Besag [1]. The term *composite likelihood* has also been used (Lindsay [6]) and we shall adopt this form since it is descriptive of the method of construction.

The setting which we shall consider is one of general stochastic processes in which maximum likelihood is difficult or impossible to use; the context may be a semiparametric one, for example. However, we suppose that we are able to piece together a set of quasi-likelihood estimating functions, possibly based on conditional or marginal information. The problem is how to combine these most efficiently for estimation purposes. The case of composite likelihoods has been considered by Lindsay [6]. Here we take a more general approach in which specific parametric assumptions are unnecessary. For a discussion of the principles of quasi-likelihood appropriate for such a setting the reader is referred to the papers of Godambe and Heyde [2] and Heyde [5].

2. THE SETTING

Let $\{X_t, 0 \le t \le T\}$ be a sample in discrete or continuous time drawn from a process taking values in r–dimensional Euclidean space whose distribution involves a "parameter" θ taking values in an open subset Θ of p–dimensional Euclidean space. The setting may be parametric or non–parametric; θ could be, for example, the mean of stationary process. The true value of the "parameter" is θ_0 and this is to be estimated.

Suppose that the possible probability measures for $\{X_t\}$ are $\{P_\theta\}$ and that each $(\Omega, \mathcal{F}, P_\theta)$ is a complete probability space.

We shall confine attention to the class $\mathcal{G}$ of zero mean square integrable estimating functions $G_T = G_T(\{X_t, 0 \le t \le T\}, \theta)$ for which $EG_T(\theta) = 0$ for each P_θ. Here G_T is a vector of dimension p. Estimators $\hat{\theta}$ are found by solving the estimating equation $G_T(\hat{\theta}) = 0$.

We consider a class of $\mathcal{H} \in \mathcal{G}$ of estimating functions G_T which are a.s. differentiable with respect to the components of θ and such that

$$E\dot{G}_T = (E\partial G_{T,i}/\partial\theta_j)$$

and EG_TG_T' are nonsingular, the prime denoting transpose.

Our assessment of the performance of an estimating function G_T involves focusing on the quantity

$$\mathcal{E}(G_T) = (E\dot{G}_T)' (EG_TG_T')^{-1} (E\dot{G}_T)$$

which may be thought of as an information matrix (and is indeed the Fisher information matrix when $G_T = U_T$, the true score function). Efficiency can be measured through the information matrix and if $G_T^* \in \mathcal{H}_1 \subseteq \mathcal{H}$ satisfies

$$\mathcal{E}(G_T^*) - \mathcal{E}(G_T)$$

nonnegative definite for all $G_T \in \mathcal{H}_1$ we shall say that G_T^* is a *quasi–score estimating function* within $\mathcal{H}_1$. An estimator $\hat{\theta}$ obtained from $G_T^*(\hat{\theta}) = 0$ will be referred to as a *quasi–likelihood estimator*. For a comprehensive explanation of these concepts, which convey the advantages of great generality and flexibility, see Godambe and Heyde [2], Heyde [5].

Quasi-score estimating functions can usually be found in practice through the application of the following theorem (Heyde [4]).

Theorem 1. *Suppose that* $\mathcal{H}_1 \subseteq \mathcal{H}$ *is a convex set. Then* $G_T^* \in \mathcal{H}$ *is a quasi-score estimating function within* $\mathcal{H}_1$ *if and only if*

$$(E\dot{G}_T)^{-1} EG_T G_T^* = (E\dot{G}_T)^{-1} EG_T^* G_T^{*\prime}$$

for all $G_T \in \mathcal{H}_1$.

It should be noted that distinct subsets of $\mathcal{H}$ may produce distinct quasi-score estimating functions and that issues of comparison and possible combination are of considerable importance.

3. COMPOSITE QUASI-LIKELIHOODS

Now we shall suppose that the setting is such that k distinct estimating functions $G_{i,T}(\theta)$, $1 \le i \le k$, are available. Our first approach to the composition problem is to seek a quasi-score estimating function from within the set

$$\mathcal{K} = \left\{ \sum_{i=1}^{k} \alpha_{i,T}(\theta)\, G_{i,T}(\theta) \right\}$$

where the weighting matrices $\alpha_{i,T}$ are $p \times p$ constants.

Suppressing the dependence on T and θ for convenience, we write

$$H = \sum_{i=1}^{k} \alpha_i G_i = \boldsymbol{\alpha}'\mathbf{G}, \qquad H^* = \sum_{i=1}^{k} \alpha_i^* G_i = \boldsymbol{\alpha}^{*\prime}\mathbf{G}$$

where

$$\boldsymbol{\alpha}' = (\alpha_1 \vdots \cdots \vdots \alpha_k), \qquad \boldsymbol{\alpha}^{*\prime} = (\alpha_1^* \vdots \cdots \vdots \alpha_k^*),$$

$$\mathbf{G}' = (G_1' \vdots \cdots \vdots G_k'),$$

so that

$$E\dot{H} = \sum_{i=1}^{k} \alpha_i E\dot{G}_i = \boldsymbol{\alpha}' E\dot{\mathbf{G}},$$

$$EHH^{*\prime} = \sum_{i=1}^{k} \sum_{j=1}^{k} \alpha_i EG_i G_j' \alpha_j^{*\prime} = \boldsymbol{\alpha} E\mathbf{G}\mathbf{G}'\boldsymbol{\alpha}^{*\prime}.$$

Then, from Theorem 1 we see that H^* is a quasi-score estimating function within $\mathcal{K}$ if

$$\alpha^{*\prime} = (EGG')^{-1} E\dot{G} .$$

This provides a general solution although it may be less than straightforward to calculate α^* in practice. Note in particular that since each individual $G_i \in \mathcal{K}$, the formulation ensures that

$$\mathcal{E}(H^*) - \mathcal{E}(G_i)$$

is nonnegative definite, $1 \le i \le k$, so that composition is generally advantageous.

Various simplifications are sometimes possible. For example if the G_i are quasi-score (or true score) estimating functions then (as they are specified only up to a constant (matrix) multiplier) we can suppose that each G_i is normalized so that

$$E\dot{G}_i = EG_iG_i' , \quad 1 \le i \le k .$$

Circumstances under which equal weighting is obtained are especially important. Clearly this holds in the particular case when $EG_iG_j' = 0$, $i \ne j$, that is, the estimating functions are orthogonal.

There is also another approach to the problem of composition which is closely related but quite distinct and has been pioneered by Heyde [3]. This is applicable only to the case of martingale estimating functions but it is nevertheless of use in a wide variety of contexts. Many models admit a semimartingale description which naturally leads to martingale estimating functions and, furthermore, true score functions are martingales under mild conditions. It is principally spatial processes that are excluded.

The setting here is one of martingale estimating functions G_T, each having an associated family of martingales $\int_0^T a_t(\theta)\, dG_t(\theta)$ where $a_t(\theta)$ is a predictable matrix. It turns out that the quasi-score estimating function within this family is

$$\int_0^T (d\bar{G}_t)' \, (d\langle G\rangle_t)^+ \, dG_t .$$

For details see Godambe and Heyde [2], especially Section 5 and Heyde [3]. Here A^+ refers to the Moore–Penrose generalized inverse of a matrix A, $\langle M\rangle_T$ is the quadratic characteristic of a martingale M and $d\bar{M}_t = E(d\dot{M}_t \mid \mathcal{F}_{t-})$, the conditional expectation taken with respect to a standard filtration of past-history σ–fields.

For the composition problem we here suppose that distinct martingale estimating functions $G_{i,T}$, $1 \le i \le k$, are available, none being expressible as a stochastic integral with respect to one of the others. Then, we can combine the estimating functions into a $2p \times 1$ vector martingale

$$G_T = (G'_{1,T}, \ldots, G'_{k,T})'$$

and, via a closely related analogue of Theorem 1, the quasi-score estimating function within the family $\int_0^T \mathbf{b}_t \, dG_t$ for predictable $\mathbf{b}_t$ is

$$\int_0^T (d\bar{G}_t) \, (d\langle G \rangle_t)^+ \, dG_t \, .$$

This simplifies down to a tractable expression in the case $k = 2$ which has been discussed by Heyde [3] and the reader is referred to that paper for details and an example. Again it is basically only in the case of orthogonality of the G_i's that drastic simplification occurs and the composite quasi-score is the sum of the component quasi-scores.

REFERENCES

[1] Besag, J.E., The Statistician, 24 (1975), 179–195.
[2] Godambe, V.P. and Heyde, C.C., Int. Statist. Rev. 55 (1987), 231–244.
[3] Heyde, C.C., Stoch. Processes Applic. 25 (1987), 281–287.
[4] Heyde, C.C., Contemp. Math. 80 (1988), 241–247.
[5] Heyde, C.C., Quasi-likelihood and optimality for estimating functions: some current unifying themes, Bulletin of the ISI, Vol. LIII, 1989, to appear.
[6] Lindsay, B.G., Contemp. Math. 80 (1988), 221–239.
[7] Rao, C.R., Bull. Cal. Math. Soc. 37 (1945), 81–91.
[8] Rao, C.R., Linear Statistical Inference and its Applications (Wiley, New York, 1965).

STATISTICAL DATA ANALYSIS AND INFERENCE
Y. Dodge (Editor)

IMPROVED ESTIMATORS OF DISPERSION OF AN INVERSE GAUSSIAN DISTRIBUTION

Nabendu PAL

Department Of Mathematical Sciences
University of Arkansas
Fayetteville, Arkansas, U.S.A. 72701
and
Department of Mathematics and Statistics
University of Maryland Baltimore County
Baltimore, Maryland, U.S.A. 21228

Bimal K. SINHA

Department of Mathematics and Statistics
University of Maryland Baltimore County
Baltimore, Maryland, U.S.A. 21228

In this paper improved estimators of dispersion of an inverse gaussian distribution under quadratic and entropy losses are derived. Stein type testimators dominating the best equivariant estimators are obtained in both the cases.

1. INTRODUCTION

The importance of an inverse Gaussian distribution as a mathematical model for analyzing positively skewed data is quite well known in the literature [Banerjee and Bhattacharya(1976), Tweedie(1957a, 1957b), Folks and Chikkara(1978), Lancaster(1972), Whitmore(1979,1986)]. A univariate inverse Gaussian distribution which obviously belongs to a two parameter exponential family and which arises as a first passage time distribution of a barrier in a brownian process with a positive drift presents striking similarities with the normal distribution and provocating departures from it [see Folks and Chikkara(1978)].

The probability density function $f(x;\mu,\lambda)$ of an inverse Gaussian distribution $IG(\mu,\lambda)$ can be written in many equivalent forms using different reparametrization as

$$\begin{aligned}
f(x;\mu,\lambda) &= \sqrt{\lambda/2\pi x^3}e^{-\frac{\lambda}{2\mu^2 x}(x-\mu)^2} && (1.1)\\
f(x;\mu,\phi) &= \sqrt{\mu\phi/2\pi x^3}e^{-\frac{\phi x}{2\mu}+\phi-\frac{\mu\phi}{2x}} && (1.2)\\
f(x;\phi,\lambda) &= \sqrt{\lambda/2\pi x^3}e^{-\frac{\phi^2 x}{2\lambda}+\phi-\frac{\lambda}{2x}} && (1.3)\\
f(x;\alpha,\lambda) &= \sqrt{\lambda/2\pi x^3}e^{-\alpha\lambda x+\lambda\sqrt{2\alpha}-\frac{\lambda}{2x}} && (1.4)
\end{aligned}$$

where $\phi = \lambda/\mu$, $\alpha = \sqrt{2\mu}$, and $\mu, \lambda, \phi, \alpha$ are all positive. Here μ stands for the mean of the inverse Gaussian distribution and λ is known as the inverse measure of dispersion so that $\sigma^2 = 1/\lambda$ is the direct measure of dispersion.

Our object in this paper is to treat the problem of point estimation of σ^2 when μ is unknown on the basis of a random sample of size n from an inverse Gaussian distribution. Denoting the samples by $X_1, X_2, ..., X_n$, and writing $\bar{X} = \frac{1}{n}\sum_{i=1}^n X_i$, $U = \sum_{i=1}^n (X_i^{-1} - \bar{X}^{-1})$, it is well known that [Tweedie(1957a, 1957b), Folks and Chikkara(1978)] $\bar{X} \sim IG(\mu, n\lambda)$, $U \sim \sigma^2\chi^2_{n-1}(central)$, and $\bar{X}, U$ are independent. Ordinarily, σ^2 is estimated by $\hat{\sigma}^2 = \frac{1}{n}U$ (maximum likelihood estimator) or $\hat{\sigma}_1^2 = \frac{1}{n+1}U$ (best equivariant estimator under squared error loss) or $\hat{\sigma}_2^2 = \frac{1}{n-1}U$ (best equivariant estimator under entropy loss). It is evident that all the above estimators ignore $\bar{X}$. This situation is analogous to the estimation of variance of a normal population with unknown mean. Following Stein's(1964) ideas for a normal population, we derive below improved estimators of σ^2 under both squared error and entropy losses. Some ideas and techniques of Sinha(1976) and Sinha and Ghosh(1987) who generalized Stein's result from a univariate normal to the multivariate case are also used in the sequel.

The result for the squared loss is described in Section 3 while Section 4 deals with the entropy loss. In Section 2 we have stated and proved a result of somewhat independent interest, which is repeatedly used in Sections 3 and 4 to yield the desired dominance results. It may be noted that the form of the improved estimator of σ^2 under squared error loss appeared in a recent paper of Pandey and Malik(1988), although no dominance over $\hat{\sigma}_1^2$ is claimed. Moreover, Bravo and MacGibbon(1988) discussed a somewhat related but different problem of estimation of common σ^2 using samples from two independent inverse Gaussian distributions with different means.

2. SOME USEFUL RESULTS

It is well known that if $X \sim N(\mu, \sigma^2)$, then

(i) $(X-\mu)^2/\sigma^2 \sim \chi_1^2(central)$ and

(ii) $(X-\mu_0)^2/\sigma^2 \sim \chi_1^2(noncentral\ with\ noncentrality\ parameter(\mu-\mu_0)^2/\sigma^2)$.

While a result analogous to (i) holds for an inverse Gaussian distribution, namely, if $X \sim IG(\mu, \lambda)$, then the exponent $\lambda(X-\mu)^2/(\mu^2 X) \sim \chi_1^2(central)$ [Shuster(1968)], the result for the noncentral case is far from being true. In fact, if $X \sim IG(\mu, \lambda)$, then $V = \lambda(X - \mu_0)^2/(\mu_0^2 X)$ has a quite complicated distribution for $\mu_0 \neq \mu$. It can be shown that the moment generating function of V has the expression

$$M_V(t) = \frac{1}{\sqrt{1-2t}} exp\{-\frac{2t\lambda}{\mu_0} + \frac{\lambda}{\mu} - \frac{\lambda}{\mu}\sqrt{(1-2t)(1-2t\mu^2/\mu_0^2)}\},\ t < \frac{1}{2} \tag{2.1}$$

which can be used to derive the distribution of V. Among other rare departures, this is a point of departure from normal theory.

The following result which essentially provides an upper bound for $E(V^k)/E(V^{k+1})$ for some positive real k is crucial for our purpose. It may be noted that the proof for a similar result for a normal distribution is rather straightforward in view of the property (ii) mentioned above.

LEMMA 2.1 *Let*

$$I_k(A,B) = \int_0^\infty e^{-Ax-\frac{B}{x}} x^{-\frac{3}{2}}\{\frac{(x-1)^2}{x}\}^k dx,$$

where $k > 0, A > 0, B > 0$. *Define* $\psi_{k,k+1}(A,B) = \frac{I_k(A,B)}{I_{k+1}(A,B)}$. *Then*

$$\psi_{k,k+1}(A,B) \leq \psi_{k,k+1}(B,B), \quad \forall 0 < A < \infty. \tag{2.2}$$

Proof: Write $A = B + \delta$ so that

$$I_k(A,B) = \int_0^\infty e^{-B(x+\frac{1}{x})-\delta x} x^{-\frac{3}{2}} \{\frac{(x-1)^2}{x}\}^k dx$$

Case 1: $\delta \geq 0$. Write

$$\begin{aligned} I_k(A,B) &= \int_1^\infty e^{-B(x+\frac{1}{x})-\delta x} x^{-\frac{3}{2}} \{\frac{(x-1)^2}{x}\}^k dx \\ &\quad + \int_0^1 e^{-B(x+\frac{1}{x})-\delta x} x^{-\frac{3}{2}} \{\frac{(x-1)^2}{x}\}^k dx \\ &= \int_1^\infty e^{-B(x+\frac{1}{x})-\delta x} x^{-\frac{3}{2}} \{\frac{(x-1)^2}{x}\}^k dx \\ &\quad + \int_1^\infty e^{-B(x+\frac{1}{x})-\delta/x} x^{-\frac{1}{2}} \{\frac{(x-1)^2}{x}\}^k dx \end{aligned} \tag{2.3}$$

(making the transformation $x \to \frac{1}{x}$ over (0,1) in the second term of (2.3))

$$= \int_1^\infty e^{-B(x+\frac{1}{x})} (e^{-\delta x} + x e^{-\delta/x}) x^{-\frac{3}{2}} \{\frac{(x-1)^2}{x}\}^k dx \tag{2.4}$$

so that

$$\frac{I_{k+1}(A,B)}{I_k(A,B)} = E_\delta\{\frac{(Y-1)^2}{Y}\} \tag{2.5}$$

where the random variable Y has the pdf

$$Y \sim f_\delta(y) = \frac{e^{-B(y+\frac{1}{y})}(e^{-\delta y} + y e^{-\delta/y}) y^{-\frac{3}{2}} \{\frac{(y-1)^2}{y}\}^k}{\int_1^\infty e^{-B(y+\frac{1}{y})}(e^{-\delta y} + y e^{-\delta/y}) y^{-\frac{3}{2}} \{\frac{(y-1)^2}{y}\}^k dy}, \quad 1 < y < \infty. \tag{2.6}$$

We now show that $\pi(y) = \frac{f_\delta(y)}{f_0(y)} \uparrow y$ for any $\delta > 0$. Since $(y-1)^2/y \uparrow y$ for $y \geq 1$, it would then follow that

$$E_{\delta>0}\{\frac{(Y-1)^2}{Y}\} > E_{\delta=0}\{\frac{(Y-1)^2}{Y}\}$$

which implies (2.2). Clearly, $\pi(y) = c(e^{-\delta y} + y e^{-\delta/y})/(1+y)$ where c is a suitable positive constant. Note that

$$\begin{aligned} \pi'(y) &= c\{\frac{-\delta e^{-\delta y} + e^{-\delta/y} + \frac{\delta}{y} e^{-\delta/y}}{(1+y)} - \frac{e^{-\delta y} + y e^{-\delta/y}}{(1+y)^2}\} > 0 \\ &\leftrightarrow e^{-\delta/y}(1 + \delta + \frac{\delta}{y}) > e^{-\delta y}(1 + \delta + \delta y). \end{aligned} \tag{2.7}$$

To prove that (2.7) holds for all $y > 1$, let $\varphi(y) = (1 + \delta + \delta y)e^{-\delta y}$ so that the inequality (2.7) reads as

$$\varphi(1/y) > \varphi(y). \tag{2.8}$$

(2.8) then follows once we show that $\varphi(y) \downarrow y$, since $y > \frac{1}{y}$ for $y > 1$. But

$$\varphi'(y) = -\delta^2(1+y)e^{-\delta y} < 0,$$

proving (2.2).

Case 2: $\delta < 0$. Making the transformation $x \to \frac{1}{x}$ over $(1, \infty)$ in (2.3), we can write

$$I_k(A,B) = \int_0^1 e^{-B(x+\frac{1}{x})} x^{\frac{3}{2}} (e^{-\delta x} + x e^{-\delta/x}) \{\frac{(x-1)^2}{x}\}^k dx$$

so that

$$\frac{I_{k+1}(A,B)}{I_k(A,B)} = E_\delta\{\frac{(Y-1)^2}{Y}\} \tag{2.9}$$

where the random variable Y has the pdf

$$Y \sim f_\delta(y) = \frac{e^{-B(y+\frac{1}{y})}(e^{-\delta y} + y e^{-\delta/y}) y^{-\frac{3}{2}} \{\frac{(y-1)^2}{y}\}^k}{\int_0^1 e^{-B(y+\frac{1}{y})}(e^{-\delta y} + y e^{-\delta/y}) y^{-\frac{3}{2}} \{\frac{(y-1)^2}{y}\}^k dy}, \quad 0 < y < 1. \tag{2.10}$$

We now show that $\pi(y) = f_\delta(y)/f_0(y) \uparrow \frac{1}{y}$ for any $\delta < 0$. Since $(y-1)^2/y \uparrow \frac{1}{y}$ for $0 < y < 1$, it would then follow that

$$E_{\delta<0}\{\frac{(Y-1)^2}{Y}\} > E_{\delta=0}\{\frac{(Y-1)^2}{Y}\}$$

which implies (2.2). Let $\delta = -\delta^*$, $\delta^* > 0$. Clearly,

$$\pi(y) = \frac{f_{-\delta^*}(y)}{f_0(y)} = c\frac{e^{\delta^* y} + y e^{\delta^*/y}}{1+y}$$

for a suitable positive constant c. Again,

$$\pi'(y) < 0 \Leftrightarrow e^{\delta^* y}\{(1+y)\delta^* - 1\} < e^{\delta^*/y}\{(1+\frac{1}{y})\delta^* - 1\}. \tag{2.11}$$

Let $\varphi_*(y) = e^{\delta^* y}\{(1+y)\delta^* - 1\}$ so that the inequality (2.11) reads as

$$\varphi_*(y) < \varphi_*(1/y). \tag{2.12}$$

Since $y < \frac{1}{y}$ for $0 < y < 1$, (2.12) follows once we show that $\varphi_*(y) \uparrow y$. But $\varphi_*'(y) = (1+y)\delta^{*2} e^{\delta^* y} > 0$. Hence (2.12) follows and hence the fact that $\pi(y) \downarrow y$ i.e.,$\pi(y) \uparrow \frac{1}{y}$. This proves (2.2) for $\delta < 0$. This completes the proof of the lemma.

3. ESTIMATION UNDER SQUARED ERROR LOSS

In this section we consider the problem of estimation of σ^2 under the squared error loss function

$$L_1(\hat{\sigma^2}, \sigma^2) = (\hat{\sigma^2} - \sigma^2)^2 \tag{3.1}$$

Clearly, when μ is unknown, the best equivariant estimator of σ^2 (under location shift) is

$$\hat{\sigma_1^2} = \frac{1}{n+1} U \tag{3.2}$$

which, as noted before, ignores $\bar{X}$. On the other hand, if μ is assumed to be known, say $\mu = \mu_0$, then using the fact that $\sum_1^n (X_i - \mu_0)^2/(X_i \mu_0^2) \sim \sigma^2 \chi_n^2$, an admissible estimator of

σ^2 is given by $\frac{1}{n+2}\frac{1}{\mu_0^2}\sum_1^n(X_i-\mu_0)^2/X_i$. Note that this estimator, apart from $\frac{1}{n+2}$, can be expressed as

$$\begin{aligned}\frac{1}{\mu_0^2}\sum_1^n\frac{(X_i-\mu_0)^2}{X_i} &= \sum_1^n(X_i^{-1}-\bar{X}^{-1})+n\{\frac{(\bar{X}-\mu_0)^2}{\mu_0^2\bar{X}}\} \\ &= U\{1+n\frac{(\bar{X}-\mu_0)^2}{U\mu_0^2\bar{X}}\}. \end{aligned} \tag{3.3}$$

Motivated by (3.3), quite generally when μ is unknown, we consider estimators of σ^2 of the form

$$\tilde{\sigma}_1^2 = U\phi(\frac{V}{U}) \tag{3.4}$$

where $V = n\frac{(\bar{X}-\mu_0)^2}{\mu_0^2\bar{X}}$ for some positive μ_0, and ϕ is a real valued function over $(0,\infty)$. Clearly, V and U are statistically independent since $\bar{X}$ and U are so. Now the risk of $\tilde{\sigma}_1^2$ can be computed as

$$\begin{aligned}E(\tilde{\sigma}_1^2-\sigma^2)^2 &= E(U\phi(\frac{V}{U})-\sigma^2)^2 \\ &= E_W(E(U\phi(W)-\sigma^2)^2 \mid W)\end{aligned}$$

where $W=\frac{V}{U}$. The best choice of $\phi(w)$ for fixed $W=w$ is then obtained by minimizing $E((U\phi(W)-\sigma^2)^2 \mid W=w)$ with respect to ϕ resulting in

$$\begin{aligned}\phi_{opt}^1(w) &= \sigma^2\frac{E_{\mu,\sigma}(U\mid W=w)}{E_{\mu,\sigma}(U^2\mid W=w)} \\ &= \sigma^2 w\frac{E_{\mu,\sigma}(V\mid W=w)}{E_{\mu,\sigma}(V^2\mid W=w)}.\end{aligned} \tag{3.5}$$

We simplify (3.5) in the following lemma.

LEMMA 3.1 *Let U,V,W be defined as above. Then*

$$\frac{E_{\mu,\sigma}(V\mid W=w)}{E_{\mu,\sigma}(V^2\mid W=w)} = \frac{1}{n}\frac{E_{A,B}(V_*^k)}{E_{A,B}(V_*^{k+1})} \tag{3.6}$$

where $V_*=(X-1)^2/X$ and the random variable X has the inverse Gaussian pdf of the form (1.2) given by constant.$x^{-\frac{3}{2}}e^{-Ax-B/x}$ where $A=\frac{n\lambda}{2}(\frac{1}{\mu^2}+\frac{1}{w\mu_0^2})$, $B=\frac{n\lambda}{2}(1+\frac{1}{w})$ and $k=(n+1)/2$.

Proof. Since $\bar{X}\sim IG(\mu,n\lambda)$, $U\sim\sigma^2\chi_{n-1}^2(central)$, and $\bar{X},U$ are independent, the joint density of $(\bar{X},U)$ is

$$g(\bar{x},u) \propto \bar{x}^{-\frac{3}{2}}u^{\frac{n-1}{2}-1}e^{-\frac{n\phi^2\bar{x}}{2\lambda}-\frac{n\lambda}{2\bar{x}}-\frac{\lambda u}{2}}$$

where $\phi=\lambda/\mu$. Making the transformation $(\bar{X},U)\to(\bar{X},W)$ where $W=\frac{V}{U}$ and $V=n\frac{(\bar{X}-\mu_0)^2}{\mu_0^2\bar{X}}$, and noting that the Jacobian of the transformation is v/w^2, the joint probability density function of $(\bar{X},W)$ is given by

$$h(\bar{x},w) \propto v^{\frac{n-1}{2}}\bar{x}^{-\frac{3}{2}}w^{-\frac{n+1}{2}}e^{-\frac{n\phi^2\bar{x}}{2\lambda}-\frac{n\lambda}{2\bar{x}}-\frac{v\lambda}{2w}}.$$

Now

$$\frac{E(V \mid W = w)}{E(V^2 \mid W = w)} = \frac{\int_0^\infty \frac{v^{(n+1)/2}}{\bar{x}^{3/2}} e^{-\frac{n^2\phi^2\bar{x}}{2n\lambda} - \frac{n\lambda}{2\bar{x}} - \frac{\lambda v}{2w}} d\bar{x}}{\int_0^\infty \frac{v^{(n+3)/2}}{\bar{x}^{3/2}} e^{-\frac{n^2\phi^2\bar{x}}{2n\lambda} - \frac{n\lambda}{2\bar{x}} - \frac{\lambda v}{2w}} d\bar{x}}$$

$$= \frac{\int_0^\infty \frac{v^{(n+1)/2}}{\bar{x}^{3/2}} e^{-A\bar{x} - B/\bar{x}} d\bar{x}}{\int_0^\infty \frac{v^{(n+3)/2}}{\bar{x}^{3/2}} e^{-A\bar{x} - B/\bar{x}} d\bar{x}}$$

$$= \frac{1}{n} \frac{E_{A,B}(V_*^k)}{E_{A,B}(V_*^{k+1})}$$

where $V_* = (X-1)^2/X$, $k = (n+1)/2$, and X has the pdf $\propto x^{-3/2}e^{-Ax-B/x}$. This completes the proof of the lemma.

Using Lemma 2.1, it follows that

$$\frac{1}{n} \frac{E_{A,B}(V_*^k)}{E_{A,B}(V_*^{k+1})} = \frac{1}{n}\psi_{k,k+1}(A,B) \le \frac{1}{n}\psi_{k,k+1}(B,B)$$

$$= \frac{1}{n} \frac{\int_0^\infty \{\frac{(x-1)^2}{x}\}^k e^{-Bx-B/x} x^{-3/2} dx}{\int_0^\infty \{\frac{(x-1)^2}{x}\}^{k+1} e^{-Bx-B/x} x^{-3/2} dx}$$

$$= \frac{2B}{n} \frac{\int_0^\infty \{\frac{2B(x-1)^2}{x}\}^k x^{-3/2} e^{-B\frac{(x-1)^2}{x}} dx}{\int_0^\infty \{\frac{2B(x-1)^2}{x}\}^{k+1} x^{-3/2} e^{-B\frac{(x-1)^2}{x}} dx}$$

$$= \frac{2B}{n} \frac{E(\chi_1^2)^k}{E(\chi_1^2)^{k+1}} = \frac{\lambda}{n+2}(1 + \frac{1}{w}) \tag{3.7}$$

since $2B = n\lambda(1 + \frac{1}{w})$. In the above, we have used the fact that $2B(X-1)^2/X \sim \chi_1^2$ [see Shuster(1968)]. From (3.6) and (3.7), we get

$$\phi_{opt}^1(w) \le \frac{1}{n+2}(w+1). \tag{3.8}$$

We are now in a position to state the main result.

THEOREM 3.1 *Under the squared error loss function $L_1(\delta, \sigma^2)$, any estimator of the form $\tilde{\sigma}_1^2 = U\phi_1(W)$ is dominated by $\tilde{\tilde{\sigma}}_1^2 = U\tilde{\phi}_1(W)$ where*

$$\tilde{\phi}_1(W) = Min\{\phi_1(W), \frac{1}{n+2}(1+W)\}. \tag{3.9}$$

Proof. This is immediate since $\tilde{\phi}_1(W) \le \phi_1(W)$ and the loss function is convex.

In particular, if $\phi_1(W) = \frac{1}{n+1}$, defining $\tilde{\phi}_1(W) = Min\{\frac{1}{n+1}, \frac{1}{n+2}(1+W)\}$, we get the following corollary.

COROLLARY 3.1 *Under the squared error loss, the best equivariant estimator $\hat{\sigma}_1^2 = \frac{1}{n+1}U$ of σ^2 is dominated by the estimator $\hat{\hat{\sigma}}_1^2$ given as*

$$\hat{\hat{\sigma}}_1^2 = \begin{cases} \frac{1}{n+2}\sum_1^n (X_i - \mu_0)^2/(X_i\mu_0^2) & \text{if } \frac{n(\bar{X}-\mu_0)^2}{\bar{X}\mu_0^2}/U < \frac{1}{n+1} \\ \frac{1}{n+1}U & \text{otherwise.} \end{cases} \tag{3.10}$$

As in Stein(1964), the estimator $\hat{\hat{\sigma}}_1^2$ can be interpreted as a testimator. Here we use $\frac{1}{n+1}U$ if the hypothesis $\mu = \mu_0$ is rejected at a certain level, and $\frac{1}{n+2}\sum_1^n (X_i - \mu_0)^2/(X_i\mu_0^2)$, otherwise.

REMARK 3.1 *Instead of using $L_1(\delta, \sigma^2)$, we could use the invariant squared error loss function given by $L_1^*(\delta, \sigma^2) = (\frac{\delta}{\sigma^2} - 1)^2$. The same dominance result still holds for L_1^*. A different type of invariant loss function, namely, $L_1^{**} = (\frac{\sigma^2}{\delta} - 1)^2$ can also be used in this context and similar results can be derived.*

4. ESTIMATION UNDER ENTROPY LOSS

Under the entropy loss function

$$L_2(\delta,\sigma^2) = (\frac{\delta}{\sigma^2}) - log(\frac{\delta}{\sigma^2}) - 1, \tag{4.1}$$

the best equivariant estimator of σ^2 is

$$\hat{\sigma_2^2} = \frac{1}{n-1}U. \tag{4.2}$$

Proceeding as in the previous case and using the same notation as before, if we consider an estimator of the form (3.4), the optimal choice of ϕ, minimizing $\phi(w)E(\frac{1}{\sigma^2}U \mid W = w) - log\phi(w)$ for fixed $W = w$ is given by

$$\begin{aligned}\phi^2_{opt}(w) &= \frac{\sigma^2}{E(U \mid W = w)} \\ &= \frac{w\sigma^2}{E(V \mid W = w)}.\end{aligned} \tag{4.3}$$

This is simplified in the following lemma.

LEMMA 4.1 *Let A, B, k and V_* be defined as before. Then*

$$\frac{1}{E_{\mu,\sigma}(V \mid W = w)} = \frac{1}{n}\frac{E_{A,B}\{V_*^{k-1}\}}{E_{A,B}\{V_*^k\}}. \tag{4.4}$$

Proof. Similar to lemma 3.1.

Using Lemma 2.1 and Lemma 4.1,

$$\begin{aligned}\frac{1}{E_{\mu,\sigma}(V \mid W = w)} &= \frac{1}{n}\frac{I_{k-1}(A,B)}{I_k(A,B)} = \frac{1}{n}\psi_{k-1,k}(A,B) \\ &\leq \frac{1}{n}\psi_{k-1,k}(B,B) \\ &= \frac{2B}{n}\frac{E(\chi_1^2)^{k-1}}{E(\chi_1^2)^k} = \frac{\lambda}{n}(1+\frac{1}{w}).\end{aligned} \tag{4.5}$$

From (4.3), (4.4) and (4.5), we get

$$\phi^2_{opt}(w) \leq \frac{1}{n}(w+1). \tag{4.6}$$

The main result of this section is the following.

THEOREM 4.1 *Under the entropy loss function $L_2(\delta,\sigma^2)$, any estimator of the form $\tilde{\sigma_2^2} = U\phi_2(W)$ is dominated by $\tilde{\tilde{\sigma_2^2}} = U\tilde{\phi}_2(W)$ where*

$$\tilde{\phi}_2(W) = Min\{\phi_2(W), \frac{1}{n}(1+W)\}. \tag{4.7}$$

Proof. Immediate since $\tilde{\phi_2(W)} \leq \phi_2(W)$, and the loss function is convex.

In particular, if $\phi_2(W) = \frac{1}{n-1}$, defining $\tilde{\phi}_2(W) = Min\{\frac{1}{n-1}, \frac{1}{n}(1+W)\}$, we get the following corollary.

COROLLARY 4.1 *Under the entropy loss, the best equivariant estimator $\hat{\sigma_2^2} = \frac{1}{n-1}U$ is dominated by the estimator $\hat{\hat{\sigma_2^2}}$ given as*

$$\hat{\hat{\sigma_2^2}} = \begin{cases} \frac{1}{n}\sum_1^n(X_i-\mu_0)^2/(X_i\mu_0^2) & if \quad \frac{n(\bar{X}-\mu_0)^2}{\bar{X}\mu_0^2}/U < \frac{1}{n-1} \\ \frac{1}{n-1}U & otherwise. \end{cases} \tag{4.8}$$

REMARK 4.1 *A second type of entropy loss function, namely, $L_2^{**} = (\sigma^2/\delta) - log(\sigma^2/\delta) - 1$ can also be used and similar results can be obtained.*

5. ACKNOWLEDGEMENT

Our sincere thanks are due to Dr. Sanat K. Sarkar of Temple University for some helpful discussion in connection with the proof of Lemma 2.1.

6. REFERENCES

[1] Banerjee, A. and Bhattacharya, G. K., A Purchase incidence model with inverse Gaussian inter purchase times, Journal of the American Statistical Association, 71(1976), pp.823-829.

[2] Bravo, G. and MacGibbon, B., Improved estimation for the parameters of an inverse Gaussian distribution, Communications in Statistics, Theory and Methods, 17(12)(1988), pp.4285-4299.

[3] Folks, J. L. and Chhikara, R. S., The inverse Gaussian distribution and its statistical applications- A Review, Journal of the Royal Statistical Society, Series B, 40(1978), pp.263-289.

[4] Lancaster, A., A stochastic model for the duration of a strike, Journal of the Royal Statistical Society, Series A, 135(1972), pp.257-271.

[5] Pandey, B. N. and Malik, H. J., Some improved estimators for a measure of dispersion of an inverse Gaussian distribution, Communications in Statistics, Theory and Methods, 17(11)(1988), pp.3935-3949.

[6] Shuster, J., On the inverse Gaussian distribution function, Journal of the American Statistical Association, 63(1968), pp.1514-1516.

[7] Sinha, B. K., On improved estimators of the generalized variance, Journal of Multivariate Analysis, 6(1976), pp.617-625.

[8] Sinha, B. K. and Ghosh, M., Inadmissibility of the best equivariant estimators of the variance-covariance matrix, the precision matrix and the generalized variance under entropy losses, Statistics and Decisions, 5(1987), pp.201-227.

[9] Stein, C., Inadmissibility of the usual estimator for the variance of a normal distribution with unknown mean, Annals of the Institute of Statistical Mathematics, 16(1964), pp.155-160.

[10] Tweedie, M. C. K., Statistical properties of inverse Gaussian distributions I, The Annals of Mathematical Statistics, 28(1957a), pp.362-377.

[11] Tweedie, M. C. K., Statistical properties of inverse Gaussian distributions II, The Annals of Mathematical Statistics, 28(1957b), pp.696-705.

[12] Whitmore, G. A., An inverse Gaussian model for labor turnover, Journal of the Royal Statistical Society, Series A, 142(4)(1979), pp.468-479.

[13] Whitmore, G. A., Normal-Gamma mixtures of inverse Gaussian distribution, Scandanavian Journal of Statistics, 13(1986), pp.211-220.

STATISTICAL DATA ANALYSIS AND INFERENCE
Y. Dodge (Editor)

APPLICATIONS OF EDGEWORTH EXPANSIONS TO BOOTSTRAP - A REVIEW

Gutti Jogesh Babu

The Pennsylvania State University, Department of Statistics,
219 Pond Laboratory, University Park, PA 16802, USA

The bootstrap and other resampling plans have gained popularity among applied scientists in the past few years. In general, the standard bootstrap method is known to give as good an approximation as the normal theory for the sampling distribution of a statistic. The main emphasis of the review is on the results which show that the bootstrap approximation is better in some cases than the classical normal approximation. This was first observed in the case of sample mean in 1981 by Kesar Singh. The result has been extended by him and the present author to a wide class of statistics. These results mainly use the Edgeworth expansions. This article reviews recent results on the use of Edgeworth expansions in bootstrap and other related resampling procedures.

1. *Introduction.*

Efron's (1979) resampling method known as the bootstrap provides, in general, as good an estimate as that of the classical Gaussian approximation for the sampling distribution function of a statistic. Singh (1981) has shown, by deriving a two-term Edgeworth expansion (EE) of the student's t-statistic, that the bootstrap provides a better approximation than the Gaussian approximation. This is the basis for the subsequent investigations into the superiority of the bootstrap method, in obtaining confidence intervals, in computing coverage probabilities and estimating the sampling distributions. These results heavily depend on EE. Some of this work is reviewed in this paper.

Let $T(\underset{\sim}{Z},F)$ be a statistic based on a random sample $\underset{\sim}{Z} = (Z_1,\ldots,Z_n)$ from an unknown k-variate population F. Efron's bootstrap consists of approximating the sampling distribution of $T(\underset{\sim}{Z},F)$ by that of $T(\underset{\sim}{Z}^*,F_n)$, where $\underset{\sim}{Z}^* = (Z_1^*,\ldots,Z_n^*)$ is a random sample from the empirical distribution function F_n of X. The sampling distribution of $T(\underset{\sim}{Z},F)$ is needed to make inferences about the population parameters.

The bootstrap procedure can be generalized by using a general estimate G_n of F based on $\underset{\sim}{Z}$, in the place of F_n. A variation of this procedure called the Bayesian bootstrap is considered in Section 6.

Singh (1981) has shown that if F is univariate non-lattice distribution function and $E_F|Z_1|^3 < \infty$, then with probability 1,

$$\sup_x |P_F(T(\underset{\sim}{Z},F)\le x) - P_{F_n}(T(\underset{\sim}{Z}^*, F_n)\le x)| = o(n^{-1/2}) \tag{1.1}$$

holds in the special case

$$T(\underset{\sim}{Z},F) = \sqrt{n}(\bar{Z}-E_F(Z_1))/\sqrt{\mathrm{Var}(Z_1)},$$

where

$$\bar{Z} = \frac{1}{n}\sum_{i=1}^{n} Z_i.$$

The proof of (1.1) depends on an interesting observation which is exploited by Babu and Singh (1983, 1984) to establish (1.1) for a wide class of statistics. Bose (1988) used these techniques to establish (1.1) for the least-squares estimates in autoregressions.

To describe the main idea, let the distribution $D(\cdot,F)$ of $T(\underset{\sim}{Z},F)$ admit a two term EE,

$$D(x,F) = \int_{\infty}^{x}(1+n^{-1/2}p(y,F))\varphi(y)dy + o(n^{-1/2}), \tag{1.2}$$

where φ denotes the density of the standard normal distribution and $P(\cdot,F)$ denotes a polynomial whose coefficients depend upon F. If G_n is any estimate of F such that G_n is close to F in a suitable sense, then it is reasonable to expect

$$D(x,G_n) = P_{G_n}(T(\underset{\sim}{Z}^*,G_n)\le x)$$
$$= \int_{\infty}^{x}(1+n^{-1/2}p(y,G_n))\varphi(y)dy + o(n^{-1/2}), \tag{1.3}$$

where $\underset{\sim}{Z}^* = (Z_1^*,\ldots,Z_n^*)$ is a random sample of size n from G_n. Typically the coefficients of the polynomial $p(y,G_n)$ are close to those of $p(y,F)$. In such a case (1.2) and (1.3) imply

$$D(x,F) - D(x,G_n) = o(n^{-1/2}). \tag{1.4}$$

If the remainders in (1.2) and (1.3) are suitably integrable, we obtain approximations analogous to (1.4) for bias, standard error and similar other functions.

To obtain the accuracy of order $o(n^{-1/2})$ by bootstrap method the leading terms in the EEs of $D(x,F)$ and $D(x,G_n)$ should be matched. Otherwise the difference of the leading terms would give rise to an error which is typically of order $O(n^{-1/2})$. This point is further elaborated towards the end of Section 2. The studentized statistics are perhaps the most important examples of statistics with first order approximations totally free from the underlying population.

Crucial to the results of this type are the investigations of Bhattachyarya and Ghosh (1978), who established the validity of the formal EE under very general conditions. In particular, their results are applicable to statistics which can be written as smooth functions of multivariate means.

2. Approximation Along a Sample Path.

We start with some notations and definitions. A distribution function F on R^k is called strongly non-lattice (*s.n.l.*) if $|\int e^{it'x}dF(x)| < 1$ for all $t \neq 0$. Let $\|x\|$ denote the Euclidean norm of x,

$$\rho_s = \int \|x\|^s dF \text{ and } \rho_{s,n} = \int \|x\|^s dG_n.$$

We write $G_n \xrightarrow{(s)} F$ if G_n converges weakly to F and $\rho_{s,n} \longrightarrow \rho_s < \infty$. Let φ_U and Φ_U denote the density and the distribution function (d.f.) of a normal vector with mean zero and dispersion U. Let μ, μ_n be the mean vectors and Σ and Σ_n be the dispersion matrices of F and G_n. Σ is clearly positive definite if F is *s.n.l.* Let $\Sigma^{1/2}$ be the matrix such that $\Sigma^{1/2}\Sigma^{-1}(\Sigma^{1/2})' = I$, the identity matrix, where A' denotes transpose of A. For any real function f, any positive definite k×k matrix U, r > o and ε > o, let

$$M_r(f) = \sup_x (1+\|x\|)^{-r}|f(x)|$$

and

$$\omega(f,\varepsilon,U) = \int(\sup\{|f(y)-f(x)|:\|x-y\|\leq\varepsilon\})d\Phi_U(x).$$

Finally, let $Q_n(P,\cdot)$ denote the distribution of $\sqrt{n}(\bar{V}-E(V_1))$, where $V_1, \ldots, V_n$ are independent with common distribution P.

Generalizing Singh's (1981) result, Babu and Singh (1984) proved the following theorem.

Theorem 2.1. Let F *be s.n.l. and for some* $s \geq 3$, $G_n \xrightarrow{(s)} F$ *with probability 1. Let* $\bar{G}_n(x) =$

$G_n(x\Sigma_n^{1/2})$ *and* $\bar{F}(x) = F(x\Sigma^{1/2})$. *Then with probability 1*

$$|\int f(x)d(Q_n(\bar{G}_n,x) - Q_n(\bar{F},x))| \leq M_s(f) \cdot o(n^{-1/2}) + C\omega(f,\delta_n,I) \quad (2.1)$$

for all measurable f *with* $M_s(f) < \infty$, *where* C *is a positive constant and* $\delta_n = o(n^{-1/2})$.

The assumption on G_n holds in the case of $G_n = F_n$, whenever $\rho_s < \infty$. This theorem, together with the results of Bhattacharya and Ghosh (1978), establishes superiority of bootstrap over the normal approximation for smooth functions of multivariate means. Recall that $\underset{\sim}{Z}^* = (Z_1^*,\ldots,Z_n^*)$ is an i.i.d. sample from G_n. We have

Theorem 2.2. Let F *be s.n.l. and with probability* 1, $G_n \xrightarrow{(3)} F$. *Let* H *be thrice continuously differentiable in a neighborhood of* μ *and let* $\ell(x)$ *denote the vector of first order partial derivatives of* H *evaluated at* x. *Let* D_F *and* D_{G_n} *denote the distribution of*

$$T(\underset{\sim}{Z},F) = \sqrt{n}(H(\bar{Z})-H(\mu)) \text{ and } T(\underset{\sim}{Z}^*,G_n) = \sqrt{n}(H(\bar{Z}^*)-H(\mu_n)).$$

If $\ell(\mu) \neq o$, *then with probability* 1,

$$\sqrt{n}\,\sup_x|D_{G_n}(x\sigma_{G_n}) - D_F(x\sigma_F)| \longrightarrow o \tag{2.2}$$

as $n \longrightarrow \infty$, *where* $\sigma_F^2 = \ell(\mu)'\Sigma\ell(\mu)$ *and* $\sigma_{G_n}^2 = \ell(\mu_n)'\Sigma_n\ell(\mu_n)$.

Theorem 2.2 is established by approximating $T(\underset{\sim}{Z},F)$ by

$$t(\underset{\sim}{Z},F) = \ell(\mu)'(\sqrt{n}(\bar{Z}-\mu)) + \sqrt{n}(\bar{Z}-\mu)'M(\mu)(\bar{Z}-\mu), \tag{2.3}$$

where $M(\mu)$ denotes the matrix of partial derivatives of H at μ of order 2.

Example 2.3. For Student's t-statistic, the condition of strong non-latticeness is satisfied if the underlying population X has a continuous distribution J. This can be seen by expressing t-statistic as a function of the mean of $Z_i = (X_i,X_i^2)'$, where $X_1, X_2, \ldots, X_n$ are independent copies of X. If the distribution of X_1 on the line is continuous, then clearly $Z_1 = (X_1,X_1^2)'$ has a *s.n.l.* distribution on R^2. Thus if X has finite 6th moment, then with probability 1,

$$\sqrt{n}\,\sup_x|P_J(\sqrt{n}(\bar{X}-E_J(X_1))\le xs_n) - P_{J_n}(\sqrt{n}(\bar{X}^*-E_{J_n}(X_1^*)\le xs_n^*)| \longrightarrow o \tag{2.4}$$

as $n \longrightarrow \infty$, where $X_1^*, \ldots, X_n^*$ are i.i.d. from the empirical distribution J_n of $X_1, \ldots, X_n$,

$$ns_n^2 = \sum_{i=1}^n (X_i-\bar{X})^2 \text{ and } n(s_n^*)^2 = \sum_{i=1}^n (X_i^*-\bar{X}^*)^2. \tag{2.5}$$

If the *s.n.l.* assumption is dropped, we do not have the standard EE for multivariate means, and hence (2.1) does not hold in general. See for example (1.7) of Singh (1981), who has shown that the left hand side of (2.4) tends to a positive constant. However, we can get around this problem for statistics like ratio estimators $H(\bar{Z}) = \bar{Y}/\bar{X}$, where the distribution of $Z_i = (X_i,Y_i)'$ is F, the variables $X_1, \ldots, X_n$ are independent with a common lattice distribution and $Y_1, \ldots, Y_n$ are independent with a common continuous distribution. Let

$$\mu = E_F(Z_1) = (\mu_x,\mu_y)'.$$

If $\mu_x \neq o$, then the main term in the expansion of $H(\bar{Z}) - H(\mu)$ is

$$(\bar{Y}-\bar{X}(\mu_y/\mu_x))(1-(\bar{X}-\mu_x)\mu_x^{-1})\mu_x^{-1}. \tag{2.6}$$

The linear term in (2.6) is

$$\bar{W} = \frac{1}{n}\sum_{i=1}^{n} W_i,$$

where

$$W_i = \mu_x^{-1} Y_i - X_i(\mu_y/\mu_x^2) \tag{2.7}$$

has continuous distribution. Edgeworth expansions for such cases are studied in Babu and Singh (1989). We state here the result in the bivariate case which is useful in studying estimators like the ratio estimator considered above. Let η be a real valued symmetric random variable with finite third moment and whose characteristic function vanishes outside a compact interval. Let η be independent of $\underset{\sim}{Z}$ and let the dispersion of Z_1 be Σ. The next theorem, due to Babu and Singh (1989) states that by adding a negligible smooth variable to $\sqrt{n}(\bar{X}_1-\mu_x)$, one can still get the same EE as in the case of *s.n.l.* distributions.

Theorem 2.4. Suppose X_1 *has a lattice distribution and* Y_1 *has a continuous distribution. Let* $\rho_3 < \infty$ *and let* f *be a real valued measurable function* $\mathbb{R}^2$ *bounded by* 1. *Suppose* P_n *denotes the distribution of* $\sqrt{n}(\bar{Z}-\mu) + (\eta n^{-3/8}, 0)$ *and* R_n *denotes the formal two term* EE, *i.e.,*

$$R_n(x) = (1+p(x,F)n^{-1/2})\varphi_\Sigma,$$

where p *is a third degree polynomial, whose coefficients are continuous functions of moments of* Z_1 *of the orders not more than* 3. *Then*

$$\int fd(P_n-R_n) \le C\omega(f,\delta_n,\Sigma) + o(n^{-1/2}), \tag{2.8}$$

where C *is an absolute constant and* $\delta_n = o(n^{-1/2})$, *not depending upon* f.

Using the results of Bhattacharya and Ghosh (1978) we obtain

Corollary 2.5. Let (a,b) *be non-zero vector,* L *be a* 2×2 *matrix and let* $\mu = (o,o)'$. *Then under the conditions of Theorem* 2.4, *we have*

$$\sqrt{n}\sup_x |P(a\bar{X}+b\bar{Y})\sqrt{n} + a\eta n^{-3/8} + \sqrt{n}\,\bar{Z}'L\bar{Z}' < x(a,b)\Sigma(a,b)')$$
$$- \int_{-\infty}^{x}(1+n^{-1/2}r(y))\rho(y)dy| \longrightarrow o \tag{2.9}$$

as $n \longrightarrow \infty$, *where* r *is a polynomial of degree* ≤ 3.

If a = o, that is, if the lattice variable does not appear in the linear term as in the case of ratio estimator, then there is no need to add η. In particular both $\sqrt{n}(\bar{Y}/\bar{X}-(\mu_y/\mu_x))$ and $\sqrt{n}\,\bar{W}(2-(\bar{X}/\mu_x))$ have the same two term EE as in the case of *s.n.l.* (X_1,Y_1), where W_i is described in (2.7). As a consequence, under the existence of 3rd moment for Z_1, even when X_1 has lattice distribution we can establish that

$$\sqrt{n}|P_F(\sqrt{n}((\bar{Y}/\bar{X})-(\mu_y/\mu_x))\le zd_n)-P_{F_n}(\sqrt{n}((\bar{Y}^*/\bar{X}^*)-(\bar{Y}/\bar{X}))\le zd_n^*)|\longrightarrow o \qquad (2.10)$$

uniformly in z, where $(X_1^*,Y_1^*)'$, ..., $(X_n^*,Y_n^*)'$ are independent vectors with common distribution function F_n, and

$$d_n^2 = \frac{1}{n}\sum_{i=1}^{n}(\bar{X}Y_i-\bar{Y}X_i)^2(\bar{X})^{-4}$$

and

$$(d_n^*)^2 = \frac{1}{n}\sum_{i=1}^{n}(\bar{X}^*Y_i^*-\bar{Y}^*X_i^*)^2(\bar{X}^*)^{-4}$$

are sample variances of the linear terms in the expansions of the statistics considered in (2.10).

So far we have concentrated on studentized statistics, where bootstrap gives "total $n^{-1/2}$-term correction". For non-studentized statistics Liu and Singh (1987) show that bootstrap gives "partial $n^{-1/2}$-term correction". To be precise, under regularity conditions, we have with probability 1,

$$D_F(x) - D_{F_n}(x) = \Phi(x/\sigma_F) - \Phi(x/\sigma_{F_n}) + o(n^{-1/2}), \qquad (2.11)$$

uniformly in x, where σ_F, σ_{F_n}, D_F and D_{F_n} are defined in Theorem 2.2. As a consequence, under regularity conditions

$$\sqrt{n}(D_F(x)-D_{F_n}(x)) \text{ and } \sqrt{n}(D_F(x)-\Phi(x/\sigma_{F_n}))$$

converge in distribution respectively to $N(o,b^2)$ and $N(p(x,F)\varphi(x/\sigma_F),b^2)$, where p is the polynomial appearing in the two term EE of D_F and b is a positive number depending on x, σ_F and the asymptotic variance of $(\ell(\bar{Z})-\ell(\mu))'\Sigma\ell(\mu)$. So the amount of reduction in mean square error due to the bootstrap generally depends on the bias of $T(\bar{Z},F)$ and skewness of D_F. For details see Liu and Singh (1987), where similar comparisons are made also between the bootstrap p-values and the p-values given by the central limit theorem.

3. *The Bootstrap and Empirical Edgeworth Expansion.*

The results of Section 2 suggest that Empirical Edgeworth Expansions (EEE) and the bootstrap yield the same approximations up to the order $o(n^{-1/2})$. Here and in what follows by EEE we mean the formal EE, where the population moments are replaced by the corresponding sample moments. If that is the case, there is no need to do the computer intensive bootstrap analysis, rather one could analytically obtain the EEE. Going beyond two term EE, Bhattacharya and Qumsiyeh (1989) established that the bootstrap distribution of studentized statistics is more accurate than even the two term EEE.

Theorem 2 of Babu and Singh (1984) essentially states that the bootstrap estimate of the distribution of multivariate sample mean differs from EEE by only $o(n^{1-s/2})$ if $\rho_s < \infty$ for some $s \geq 3$. This result is extended by Bhattacharya (1987) to establish the same, under certain moment conditions, for a wide class of statistics which are smooth functions of the multivariate sample means. These results are further utilized by Bhattacharya and Qumsiyeh (1989) to obtain the next theorem.

Theorem 3.1. Let $\rho_8 < \infty$ and Cramér's condition

$$\limsup_{\|t\|\to\infty} |E_F(\exp(it'Z_1))| < 1 \tag{3.1}$$

hold. Suppose H is a real-valued function on $\mathbb{R}^k$ which is four times continuously differentiable in a neighborhood of $\mu = E_F(Z_1)$. Then

$$n(D_{F_n}(x\sigma_{F_n}) - D_F(x\sigma_F)) \longrightarrow N(o,\sigma^2(x)),$$

where D_F and D_{F_n} are defined in Theorem 2.2,

$$\sigma^2(x) = a(x)\varphi^2(x)$$

and a is a fourth degree polynomial.

Using Theorem 3.1, Bhattacharya and Qumsiyeh (1989), show that for Student's t-statistic, even in L_p-norm, the bootstrap estimate of the distribution is better than the two term EEE. To state the result, let $X_1, \ldots, X_n$ be independent random variables from a common distribution function J. Let $X_1^*, \ldots, X_n^*$ be i.i.d. from J_n. Let

$$T_n = \sqrt{n}(\bar{X}-\mu)/s_n \text{ and } T_n^* = \sqrt{n}(\bar{X}^*-\bar{X})s_n^*,$$

where s_n and s_n^* are defined in (2.5). For any random variable W, let $\|W\|_p = (E(|W|^p))^{1/p}$.

Theorem 3.2. Let $1 \leq p < \infty$. Suppose Cramér's condition (3.1) holds for $Z_1 = (X_1, X_1^2)'$ and $E_J|X_1|^{12p+14} < \infty$. Then

$$\lim_{n\to\infty} n\|P_{J_n}(T_n^* \leq x) - P_J(T_n \leq x)\|_p < \lim_{n\to\infty} n\|\psi_n(x) - P_J(T_n \leq x)\|_p, \tag{3.2}$$

for all points x such that $q(x) \neq o$, where ψ_n is the two term EEE and $q(x)\varphi(x)$ is the coefficient of $\frac{1}{n}$ in the three term EE. If $q(x) = o$, then equality holds in (3.2).

Remark 3.3. If $s_n = o$, then all the X_i and X_i^* are the same and in this case the expression in $\|\cdot\|_p$ in (3.2) may be given any value between o and 1. Further, $P(s_n=o)$ is exponentially small and it is zero if F is continuous.

It may be possible to prove (3.2) under fewer moment restrictions than assumed. The method of proof of Theorem 3.2 given in Bhattacharya and Qumsiyeh (1989) can be extended to general studentized statistics $\sqrt{n}(H(\bar{Z})-H(\mu))\sigma_n^{-1}$ considered earlier.

4. Bootstrap Confidence Intervals.

The results described in the previous sections either hold with probability one along a sample path or use expansions which hold along a sample path. As a consequence, it is not possible to deduce results on confidence intervals from these. Some kind of probability error estimates are needed for the remainders. For the simple case of population mean μ, the results of the earlier sections and the formal EEs suggest, in the univariate case, that for $0 < \alpha < 1$,

$$P_F(\mu<\bar{Z}+Z_\alpha s_n n^{-1/2}) = \alpha - (\mu_3/6\sigma^3\sqrt{n})\varphi(Z_\alpha)(1+2Z_\alpha^2) + o(n^{-1/2}) \quad (4.1)$$

$$P_F(\mu<\bar{Z}-b_{1-\alpha}n^{-1/2}) = \alpha - (\mu_3/2\sigma^3\sqrt{n})\varphi(Z_\alpha)Z_\alpha{}^2 + o(n^{-1/2}) \quad (4.2)$$

and

$$P_F(\mu<\bar{Z}-b_{1-\alpha}^*n^{-1/2}) = \alpha + o(n^{-1/2}), \quad (4.3)$$

where $\Phi(Z_\alpha) = \alpha$, $\sigma^2 = \text{var } Z_1$, μ_3 denotes the third central moment of Z_1 and $b_{1-\alpha}$ and $b_{1-\alpha}^*$ denote respectively, the $(1-\alpha)$th quantile of the bootstrap distribution of $\sqrt{n}(\bar{Z}-\mu)$ and Student's t-statistic. By comparing (4.1) and (4.2), we notice that the one sided confidence interval (4.1) given by the normal theory is better than (4.2) which is obtained using the non-studentized bootstrap, if $|Z_\alpha| > 1$. However, as expected the studentized bootstrap corrects for skewness and seems to yield better confidence interval (4.3). An improved version of (4.3) is obtained in Babu and Bose (1989), where the following general result is established.

Theorem 4.1. In addition to the conditions of Theorem 2.2, let $E_F\|Z_1\|^{12} < \infty$ *and the Cramer's condition hold for* Z_1. *Then uniformly in* $0 < \alpha < 1$, *we have*

$$D_F(t_\alpha\sigma_F) = \alpha + O(n^{-1}(\log n)^{3/2}), \quad (4.4)$$

where σ_F, σ_{F_n}, D_{F_n} *and* D_F *are defined in the statement of Theorem 2.2 and*

$$t_\alpha = \inf\{x:D_{F_n}(x\sigma_{F_n}) \geq \alpha\} \quad (4.5)$$

is the α*th quantile of the distribution of* $T(\underset{\sim}{Z}^*,F_n)/\sigma_{F_n}$.

The power of log n in (4.4) can be reduced by assuming additional smoothness conditions on H. A version of the theorem for studentized statistics is also established in Babu and Bose (1989). These results are established by first obtaining EE conditional on the sample path.

By considering additional terms in the conditional EEs, Hall (1986) obtains additional terms in (4.4). To state his result, let Z_i^o denote the vector obtained after adjoining $Z_i = (Z_{i1},\dots,Z_{ik})$, the set of products $Z_{ir}Z_{ij}$, $1 \le r \le j \le k$, which do not already appear in Z_i. Let F^o denote the distribution of Z_1^o, $\mu^o = E_{F^o}(Z_1^o)$ and $\underset{\sim}{Z}^o = (Z_1^o,\dots,Z_n^o)$. Clearly $T(\underset{\sim}{Z},F)/\sigma_{F_n}$ can be rewritten as $\sqrt{n}(K(\bar{\underset{\sim}{Z}}^o)-K(\mu^o))$ for some smooth K. Similarly, Z_i^1 is obtained by adjoining Z_i^o those triples $Z_{ir}Z_{im}Z_{ij}$ that do not already appear. Let Σ^* denote the sample dispersion of $\underset{\sim}{Z}^*$ and $(\sigma^*)^2 = \ell(\bar{Z}^*)'\Sigma^*\ell(\bar{Z}^*)$. Hall (1986) obtained the following theorem for studentized statistics.

Theorem 4.2. Suppose K *is four times continuously differentiable and that Cramér's condition holds for* Z_1^1. *If for some* $\varepsilon > o$, $E_{F^o}\|Z_1^o\|^{8+\varepsilon} < \infty$, *then*

$$D_F(t_\alpha^*\sigma_{F_n}) = \alpha + n^{-1}\psi(Z_\alpha)\varphi(Z_\alpha) + o(n^{-1}), \tag{4.6}$$

where

$$t_\alpha^* = \inf\{t : D_{F_n}(t\sigma^*) \ge \alpha\}$$

and ψ *is an odd polynomial of degree at most five.*

The expression for ψ in general is too complicated, see (5.25), (5.26) and (5.28) of Hall (1986). The moment condition above turns out in general to be $E_F\|Z_1\|^{16+\varepsilon} < \infty$ for the original variables Z_i. For the Student's t-statistic $\sqrt{n}(\bar{X}-\mu)(\frac{1}{n}\Sigma(X_i-\bar{X})^2)^{-1/2}$, ψ is given by

$$\psi(Z) = -(Z/6)(1+2Z^2)(\mu_4\sigma^{-4}-3-(3/2\sigma^3)\mu_3^2),$$

where σ^2 = var X_1 and μ_3 and μ_4 denote third and fourth central moments of X_1 respectively. Hall (1988) gives a detailed comparison of various types of bootstrap confidence intervals.

Another way to assess the accuracy of the bootstrap method that helps in getting good confidence intervals is to obtain probabilistic bounds for the deviation of the sampling distribution from the bootstrap distribution. Let H be a smooth function and K_n, K_n^* denote the distributions of $T(\underset{\sim}{Z},F)/\sigma_F$ and $T(\underset{\sim}{Z}^*,F_n)/\sigma_{F_n}$ respectively. Using EE, Babu and Bose (1986) show, under certain conditions on the moments and Cramér's condition for F, that

$$P(\sup_x|K_n(x)-K_n^*(x)| \ge \varepsilon_n n^{-1/2}) = O(n^{-1-\delta}),$$

where $\varepsilon_n \longrightarrow o$ is an appropriate sequence and $\delta > o$ depends on the moment assumptions. Some bounds for ε_n are also obtained.

In the next section we concentrate on the so called accelerated bias corrected bootstrap confidence intervals.

5. *Accelerated Bias Correction.*

Let $\hat{\theta} = \theta(F_n)$ be an estimation of $\theta = \theta(F)$ based on n independent variables $\underset{\sim}{Z} = (Z_1,\ldots,Z_n)$ with common distribution F. To improve the confidence intervals for $\theta(F)$, Efron (1982) introduced a bias-corrected bootstrap interval (BC). It is assumed that normality is achieved by *some* transformation g, say

$$g(\hat{\theta}) - g(\theta) \sim N(-v_o,1), \tag{5.1}$$

where v_o is the bias correction. Taking (5.1) literally gives confidence intervals for $g(\theta)$, which can be converted back to that of θ. The bootstrap method does this without the knowledge of g and gives confidence intervals that are *first-order correct*.

In searching for alternative methods for confidence intervals and coverage probabilities for moderate samples, Efron (1987) introduced the accelerated bias corrected (ABC) confidence interval. This method takes bias and skewness corrections into account.

This method, in addition, assumes the existence of an acceleration constant a_o such that

$$(g(\hat{\theta})-g(\theta))/(1+a_o g(\theta)) \sim N(-v_o,1). \tag{5.2}$$

A good approximation for v can be taken as $\Phi^{-1}(\hat{G}(\theta(F_n)))$, where $\hat{G}$ denotes the bootstrap distribution of $\theta(F_n)$. Hall (1988) argues that for statistics, which are smooth functions of sample means of multivariate statistics, a_o can be taken as 1/6th of the skewness factor of the first order approximation of $T(\underset{\sim}{Z},F)/\sigma_F$. Bickel (1988) heuristically argues that it also holds for the statistics $\hat{\theta} = \theta(F_n)$ admitting von Mises (stochastic) expansion around the parameter $\theta = \theta(F)$.

$$\theta(F_n) = \theta(F) + \int \theta_1(x,F)dF_n(x) + \frac{1}{2}\iint \theta_2(x,y,F)dF_n(x)dF_n(y) + e_n(F), \tag{5.3}$$

where

$$\int \theta_1(x,F)dF(x) = o,$$

θ_2 is symmetric,

$$\int \theta_2(x,y)F)dF(y) = o$$

for all x and the remainder $e_n(F) = o_p(n^{-1})$. Here θ_1, the first order compact (or Hadamard) derivative of θ is known as the influence function and θ_2 is the second order compact derivative of $\theta(\cdot)$.

Konishi (1989) formally constructs a transformation g and estimates the two correction factors v_o and a_o such that the normal approximation to the distribution of (5.2) is valid with a remainder of the order $o(n^{-1/2})$.

Suppose $\hat{\theta} = \theta(F_n)$ admits an Edgeworth expansion

$$P(\sqrt{n}(\hat{\theta}-\theta)<x\sigma(\theta))=\Phi(x) - \frac{1}{\sqrt{n}}\{b(\theta) + \tfrac{1}{6}k(\theta)(x^2-1)\}\varphi(x) + O(n^{-1}), \tag{5.4}$$

where the asymptotic variance of $\sqrt{n}(\theta(F_n)-\theta(F))$ is given by

$$(\sigma(\theta))^2 = \mathrm{var}(\theta_1(Z_1,F)) = \int \theta_1^2(x,F)dF(x) \tag{5.5}$$

and the bias and skewness factors are given by

$$b(\theta) = \frac{1}{2\sigma(\theta)} \int \theta_2(x,x,F)dF(x) \tag{5.6}$$

and

$$k(\theta) = \frac{1}{\sigma^3(\theta)} \{\int\theta_1^3(x,F)dF(x) + 3\int\int T_1(x,F)T_1(y,F)T_2(x,y,F)dF(x)dF(y)\}. \tag{5.7}$$

Suppose $\sigma(\theta)$ has a derivative $\sigma'(\theta)$ with respect to θ.

Theorem 5.1. If $3\sigma'(\theta) \neq k(\theta)$, *then* v_o, a_o *and the transformation* g *that satisfies*

$$P(g(\hat{\theta})-g(\theta)<(x-v_o)(1+a_o g(\theta))) = \Phi(x) + o(n^{-1/2}) \tag{5.8}$$

are given by

$$g(\hat{\theta}) = (\exp(a_o f(\hat{\theta}))-1)a_o^{-1},$$

$$v_o = -\frac{1}{\sqrt{n}}\{\hat{b}(\theta) - \tfrac{1}{6}\hat{k}(\theta)\},\ \textit{and}\ a_o = \frac{1}{\sqrt{n}}\{\hat{\sigma}'(\theta)- \tfrac{1}{3}\hat{k}(\theta)\},$$

where f *is a solution of the differential equation*

$$\sigma(\theta)f'(\theta) = \sqrt{n}$$

and $\hat{b}$, $\hat{k}$, $\hat{\sigma}'$ *are estimators of* $b(\theta)$, $k(\theta)$ *and* $\hat{\sigma}'(\theta)$ *respectively, with possible estimation errors of the order* $O_p(n^{-1/2})$.

If $3\sigma'(\theta) = k(\theta)$, *then* g *can be taken as* f.

The constants a_o and v_o obtained from Theorem 5.1, can be used to define

$$z[\alpha] = v_o + (v_o+z_\alpha)(1-a(v_o+z_\alpha))^{-1}.$$

Then the ABC confidence interval of level $1-2\alpha$ for θ is given by

$$[\hat{G}^{-1}(\Phi(z[\alpha])),\ \hat{G}^{-1}(\Phi(z[1-\alpha]))], \tag{5.9}$$

where $\hat{G}$ is the bootstrap distribution of $\hat{\theta}$. Further from (5.8) it can be argued that the coverage probability of the confidence interval (5.9) is $1-2\alpha + o(n^{-1/2})$.

To illustrate the result we consider the following

Example 5.2. Let $L(\theta)$ be the log-likelihood function of a sample of size n drawn from a distribution F depending on an unknown θ. Let $\hat{\theta}$ be the maximum likelihood estimator of θ. If C_2 is the variance of a score function, then it is known that

$$b(\theta) = \frac{\sqrt{n}}{6C_2^{3/2}}(C_1-2C_3),\ k(\theta) = \frac{\sqrt{n}}{C_2^{\ 3/2}}(C_1-C_3)$$

and

$$\sigma'(\theta) = -\frac{\sqrt{n}}{6C_2^{\ 3/2}}(C_3-2C_1)$$

where

$$C_1 = E(\frac{\partial^3}{\partial\theta^3}L(\theta)) \text{ and } C_3 = E((\frac{\partial}{\partial\theta}L(\theta))^3).$$

It follows that

$$v_o = a_o = \frac{1}{6}(C_3/C_2^{3/2}) = \frac{1}{6}\{\text{skewness of } \frac{\partial L(\theta)}{\partial\theta}\}.$$

This agrees with conclusions of Theorem 2 of Efron (1987).

6. *The Bayesian Bootstrap.*

Rubin (1981) suggested an alternative, smoother bootstrap called the Bayesian Bootstrap (BB). Contrary to the standard bootstrap, each Bayesian bootstrap replication generates a posterior probability for each Z_i, where the values that are not in the sample have zero posterior probability, as in the case of empirical distribution function F_n. In order to compute BB estimate of a parameter T(F), let $S_n = \{(x_1,\dots,x_n):x_i \geq o,\ \sum_{i=1}^{n} x_i = 1\}$ and let $\underset{\sim}{V} = (V_1,\dots,V_n)$ be a random vector independent of $\underset{\sim}{Z}$ such that $P(\underset{\sim}{V}\varepsilon S_n) = 1$. Let $\underset{\sim}{v} = (v_1\dots,v_n)$ be a realization of $\underset{\sim}{V}$ and $G_{\underset{\sim}{v}}$ denote the distribution which gives weight v_i to Z_i. The statistic $T(G_{\underset{\sim}{V}})$ is called the BB estimate of T(F). The distribution of $T(G_{\underset{\sim}{V}})$ given $\underset{\sim}{Z}$ is called the BB distribution of T(F). This is very much similar to the standard bootstrap. However, the resulting distributions are interpreted differently, since BB simulates the posterior distribution of the parameter T(F), whereas the bootstrap simulates the sampling distribution of a statistic estimating T(F). For example in the case of the population mean $(T(F)=\mu)$, $T(G_{\underset{\sim}{V}}) = \sum_{i=1}^{n} V_i Z_i$.

In view of their simple structure, Dirichlet distributions on S_n are preferred for BB. Making use of EEs, Tu (1987) obtained the following result for the univariate sample mean.

Theorem 6.1. *Let* $(V_1,\ldots,V_n)$ *be a random vector, independent of* $\underset{\sim}{Z}$, *having the Dirichlet distribution* $D(4,4,\ldots,4)$. *Let* F *be univariate non-lattice distribution function and* $E_F|Z_1|^3 < \infty$. *Then*

$$\sqrt{n}\ \sup_x|R_n(x) - P(\sqrt{n}(\bar{Z}-E(Z_1)) \le x\sigma)| \longrightarrow o,$$

with probability 1 *as* $n \longrightarrow \infty$, *where*

$$\sigma^2 = \mathrm{Var}(Z_1),\ D_n = \sum_{i=1}^{n} (Z_i-\bar{Z})V_i,\ \bar{D}_n = (\sum_{i=1}^{n} (Z_i-\bar{Z})^2 V_i)^{1/2}$$

and R_n *denotes the conditional distribution of* $D_n/\bar{D}_n$ *given* $\underset{\sim}{Z}$.

Zheng (1987) extended these results to $Z_1, \ldots, Z_n$ independent but not identically distributed. Following the ideas of Section 2, Theorem 6.1 can easily be extended to statistics which are smooth functions of multivariate means.

7. *Concluding Remarks.*

In Section 2, we considered statistics which can be approximated by a linear term and a quadratic term, i.e.,

$$H(\bar{Z}) - H(\mu) = \ell(\mu)'(\bar{Z}-\mu) + (\bar{Z}-\mu)'M(\mu)(\bar{Z}-\mu)$$
$$+ \text{ a remainder.}$$

If $\ell(\mu)$ is a null vector and $M(\mu)$ is non-null matrix then typically $n(H(\bar{Z})-H(\mu))$ is asymptotically distributed as a linear combination of chi-squares. The following example (see Babu (1984)) shows that the corresponding bootstrap statistic $n(H(\bar{Z}^*) - H(\bar{Z}))$ does not have the same asymptotic distribution.

Example 6.1. Let $H(x) = x^2$, Z_i, i.i.d. standard normal variables and $G_n = F_n$. Since $\sqrt{n}\ \bar{Z}$ has standard normal distribution, $n(H(\bar{Z})-H(0))$ has chi-square distribution with one degree of freedom. But with probability one,

$$\limsup_{n\to\infty} P_{F_n}(H(\bar{Z}^*)-H(\bar{Z})<u) \ge \frac{1}{2}$$

for all $u > o$ so $n(H(\bar{Z}^*)-H(\bar{Z}))$ cannot converge weakly to the chi-square distribution with one degree of freedom.

The main cause of the failure of the bootstrap here, is the possibility of $n\ell(\bar{Z})$ being very large, even when $\ell(\mu) = o$. However, it can be shown that the distribution of $n(H(\bar{Z})-H(\mu))$ is close to the distribution of the modified bootstrap statistic

$$T_1(\bar{Z}^*,F_n) = n(H(\bar{Z}^*)-H(\bar{Z})-\ell(\bar{Z})'(\bar{Z}^*-\bar{Z})). \tag{6.1}$$

For details see Babu (1984).

Similar situations arise in the case of statistics $\hat{\theta} = \theta(F_n)$ admitting von Mises expansion around $\theta = \theta(F)$. If $\theta_1(x,F)$ in (5.3) vanishes a.e.(F), then typically the asymptotic distribution of $n(\hat{\theta}-\theta)$ is expressible in terms of a (finite or infinite) mixture of independent chi-square random variables. In this case, too, a similar modification as in (6.1) is needed. For discussion on these statistics see Sen (1988).

It is worthwhile to investigate such statistics using EE and find out whether they are second order correct.

The standard bootstrap procedure cannot be applied directly for the errors-in-variables models. This involves recovering the residuals and then properly rescaling them. The procedure is given in detail for the model with known error variance ratio in Linder and Babu (1988). The bootstrap distribution of the slope is shown to approximate the sampling distribution as well as the one given by the normal theory. In this case, further investigations on EEs for weighted averages of random vectors are needed to establish results similar to Theorem 2.2.

In conclusion it may be mentioned that Babu and Singh (1985) contain EEs for sample means for samples from finite multivariate populations, under very mild conditions on the population.

References

[1] Babu, G. J. (1984). Bootstrapping statistics with linear combinations of chi-squares as weak limit. *Sankhya, Series A, 46,* 85-93.

[2] Babu, G. J. and Bose, A. (1986). Accuracy of the bootstrap approximation. Preprint.

[3] Babu, G. J. and Bose, A. (1989). Bootstrap confidence intervals. *Statistics and Probability Letters, 7,* 151-160.

[4] Babu, G. J. and Singh, K. (1983). Non-parametric inference on means using bootstrap. *Annals of Statistics, 11,* 999-1003.

[5] Babu, G. J. and Singh, K. (1984). On one term Edgeworth correction by Efron's bootstrap. *Sankhya, Series A, 46,* 219-232.

[6] Babu, G. J. and Singh, K. (1985). Edgeworth expansions for sampling without replacement from finite populations. *Journal of Multivariate Analysis, 17,* 261-278.

[7] Babu, G. J. and Singh, K. (1989). On Edgeworth expansions in the mixture cases. *Annals of Statistics, 17,* in press.

[8] Bhattacharya, R. N. (1987). Some aspects of Edgeworth expansions in statistics and probability. *New Perspectives in Theoretical and Applied Statistics* (M. L. Puri, J. P. Vilaplanou and W. Wertz, editors), 157-171, Wiley, New York.

[9] Bhattacharya, R. N. and Ghosh, J. K. (1978). On the validity of the formal Edgeworth expansion. *Annals of Statistics, 6,* 434-451.

[10] Bhattacharya, R. N. and Qumsiyeh, M. (1980). Second order and L^p-comparisons between the bootstrap and empirical Edgeworth expansion methodologies. *Annals of Statistics, 17,* in press.

[11] Bickel, P. J. (1988). Discussion on the paper by Peter Hall entitled 'Theoretical comparison of bootstrap confidence intervals.' *Annals of Statistics, 16,* 959-961.

[12] Bose, A. (1988). Edgeworth correction by bootstrap in autoregressions. *Annals of Statistics, 16,* 1709-1722.

[13] Efron, B. (1979). Bootstrap Methods: Another look at the jackknife. *Annals of Statistics, 7,* 1-26.

[14] Efron, B. (1982). *The jackknife, the bootstrap, and other resampling plans.* CBMS 38, SIAM-NSF.

[15] Efron, B. (1987). Better bootstrap confidence intervals. *Journal of the American Statistical Association, 82,* 171-185.

[16] Hall, P. (1986). On the bootstrap and confidence intervals. *Annals of Statistics, 14,* 1431-1452.

[17] Hall, P. (1988). Theoretical comparison of bootstrap confidence intervals. *Annals of Statistics, 16,* 927-953.

[18] Konishi, S. (1989). Normalizing transformations and nonparametric confidence intervals. Center for Multivariate Analysis, *Tech. Report No. 89-1,* The Pennsylvania State University.

[19] Linder, E. and Babu, G. J. (1988). Bootstrapping the linear functional relationship with known error variance ratio. Preprint.

[20] Liu, R. and Singh, K. (1987). On a partial correction by the bootstrap. *Annals of Statistics, 15,* 1713-1718.

[21] Rubin, D. B. (1981). The Bayesian bootstrap. *Annals of Statistics,* 9, 130-134.

[22] Sen, P. K. (1988). Functional approaches in resampling plans: A review of some recent developments. *Sankhya, Series A, 50,* in press.

[23] Singh, K. (1981). On asymptotic accuracy of Efron's bootstrap. *Annals of Statistics,* 9, *1187-1195.*

[24] Tu, D. S. (1987). Contributions to the analysis of censored samples and the bootstrap methods. Ph.D. thesis, Institute of Systems Sciences, Academia Sinica, Beijing.

[25] Zheng, Z. G. (1987). Expanding the random weighting distribution for the error of mean in independently but not identically distributed situation. *Chinese J. Appl. Prob. and Statist., 3,* 159-168.

STATISTICAL DATA ANALYSIS AND INFERENCE
Y. Dodge (Editor)

DUAL POINCARÉ-TYPE INEQUALITIES VIA THE CRAMÉR-RAO AND THE CAUCHY-SCHWARZ INEQUALITIES AND RELATED CHARACTERIZATIONS*

by Theophilos CACOULLOS

Athens University and William Paterson College, N.J.
Panepistemiopolis, Athens 157 84, Greece

Upper and lower bounds for the variance of a function $g(\underset{\sim}{X})$ of a random vector $\underset{\sim}{X}=(X_1,\ldots,X_n)'$ are revisitied. Under independence of the X_i, an elementary derivation of the inequalities (1.1) and (1.4) is given, exhibiting also the nature of their duality, in the sense that replacing E and g_i^2 in the upper bound by E^2 and g_i, respectively, gives the lower bound. The same functions $w_i(x_i)$ appear in both bounds and they are shown to characterize the distribution of $\underset{\sim}{X}$. Equality is attained in both bounds iff g is linear.

1. INTRODUCTION AND SUMMARY

Let $\underset{\sim}{X}=(X_1,\ldots,X_n)'$ be a random (column) vector of independent, not necessarily identically distributed, components with means and variances

$$E(X_i)=\mu_i,\quad V(X_i)=\sigma_i^2,\quad i=1,\ldots,n$$

and $g: R^n \to R$ a Borel-measurable function with partial derivatives $g_i(\underset{\sim}{x})=\frac{\partial}{\partial x_i}g(x)$, such that $V[g(X)]<\infty$. Cacoullos [1] and Cacoullos and Papathanasiou [2], motivated by the upper-bound inequalities of Chernoff [3] and Chen [4], concerning the special case of normal $X_1,\ldots,X_n$, derived general upper-and lower-bound inequalities for the variance of $g(\underset{\sim}{X})$.

Specifically, if X_i is continuous with density $f_i(\cdot)$ $(i=1,\ldots,n)$, then ([1],[2] and Cacoullos and Papathanasiou [5])

$$\sum_{i=1}^{n}\sigma_i^2E^2[w_i(X_i)g_i(X)]\le V[g(X)]\le\sum_{i=1}^{n}\sigma_i^2E[w_i(X_i)g_i^2(X)], \tag{1.1}$$

where $w_i(x_i)$ is given by

$$\sigma_i^2w_i(x_i)f_i(x_i)=\int_{-\infty}^{x_i}(\mu_i-t)f_i(t)dt,\quad (i-1,\ldots,n). \tag{1.2}$$

and equality holds iff $g(\underset{\sim}{x})$ is linear.

It is noted that the inequality of Chernoff [3] and Chen [4] for a normal $\underset{\sim}{X}$ turned out to be a special case of inequalities for log concave density functions $f(\underset{\sim}{x})$, obtained by Brascamp and Lieb [6].

*Work supported in part by the Governor's Challenge Grant awarded to William Paterson College, while (Fall 1988) the author was visiting the Mathematics Department.

As regards a discrete integer-valued $\underset{\sim}{X}$ and g defined on the integer lattice points in R^n, (1.1) holds, provided the partial derivative g_i of g is replaced by the partial forward difference $\Delta_i g$, that is,

$$\Delta_i g(x) = g(x_1,\ldots,x_i+1,\ x_{i+1},\ldots,x_n) - g(\underset{\sim}{x}),$$

whereas $w_i(x_i)$ is defined by (cf. Cacoullos and Papathanasiou [5])

$$\sigma_i^2 w_i(x_i)p_i(x_i) = \sum_{i=x_i+1}^{\infty}(k-\mu_i)p_i(k), \tag{1.3}$$

p_i denoting the probability function (p.f.) of X_i, $i=1,\ldots,n$. Thus the discrete analogue of (1.1) takes the form

$$\sum_{i=1}^{n}\sigma_i^2 E^2\{w_i(X_i)\Delta_i g(\underset{\sim}{X})\} \le V[g(\underset{\sim}{X})] \le \sum_{i=1}^{n}\sigma_i^2 E(w_i(X_i)[\Delta_i g(\underset{\sim}{X})]^2\}. \tag{1.4}$$

Remarkably, the same $w_i(x_i)$ appear in the upper- and lower-bound inequalities in (1.1) and (1.4). Also either of these inequalities with given w_i has been shown [5] and Srivastava and Sreehari [7] to characterize the distribution of $\underset{\sim}{X}$.

The first characterization in this direction was given by Borovkov and Utev [8] and was motivated by the inequality for the normal $N(0,1)$, [3]; (in the present formulation $n=1$, $w(x)\equiv 1$). This was based on showing that $\sup_g U_X \ge 1$, where U_X is the functional $U_X = V[g(X)]/\sigma^2 E[g'(X)]^2$ and $\sup_g U_X = 1$ implies X is normal. From this and the inequality in [4], Prakasa Rao and Sreehari [9] characterized the multivariate normal. They,[10], also gave a characterization of the Poisson distribution using the upper-bound inequality obtained in [2] (cf.[1]). Other characterizations for discrete distributions were obtained by Srivastava and Sreehari [7], using the upper-bound inequalities given in [2]. Recently, [5], certain lower bounds obtained in [1] via the Cramer-Rao (C-R) inequality, were extended to non-regular cases (e.g., uniform X on $(0,\theta)$) and were cast into the form of (1.1) or (1.4).

It should be made clear, however, that the lower-bound inequalities obtained by applying a version of the C-R inequality (see, Lemmas 2.2 and 2.4 of [1]) depend on the underlying parameter θ, for which the function $g(\underset{\sim}{X})$ serves as an estimate. Thus different bounds may arise for the same g and different θ's (cf. (3.3) and (3.4) of [1]). It follows that these inequalities differ, in general, from the corresponding ones obtained from (1.1) or (1.4),where equality obtains if and only if $g(\underset{\sim}{x})$ is linear (see also Remark 5.2). Chen [11],used the (sharp) lower-bound inequalities in [1] to derive Poincaré-type inequalities, that is, of the form $\int g^2 d\mu \le cE\{|\nabla g(\underset{\sim}{X})|^2\}$ when μ is a normal probability measure on R^n and $\int g d\mu = 0$ (see also the remarks in the concluding section here).

Furthermore, the main feature of these inequalities, as exhibited by (1.1) and (1.4), is their differential form (since they involve the first two moments of first-order derivatives or differences of g) while the underlying distribution of X is represented by $w(x)$, in a characteristic manner, as shown in section 5.

In this, partly expository, paper, we present an alternative rather unified and, presumably, more illuminating and direct approach to inequalities of the type (1.1) and (1.4); at the same time we point out (Remark 3.1) the nature of their duality, in the sense that replacing E and g_i^2 on the right-hand side (RHS) of (1.1) by E^2 and g_i, respectively, yields the LHS inequality. In the univariate case (section 3) a simple derivation (integration by parts) of (1.1) is given, dispensing, inter alia, with the Lagrange identity, used in [2] for the derivation of the RHS of (1.1). Moreover, the derivation of the lower-bound inequalities is parameter-free, unlike via the Cramér-Rao inequality, whenever, of course, they are the same (see [1], especially p. 803). The general case $(n > 1)$ can be treated by exploiting the above duality property,once either side of (1.1) is established. Even this, however, in virtue of Lemma 2.1, can be reduced to the univariate case, using induction (on n) and an elementary expression for the variance of $g(X_1,X_2)$, taking also advantage of the independence of $X_1,X_2,\ldots,X_n$ (Proposition 4.1). Another direct proof of the LHS of (1.1) is based on a multivariate C-S inequality, yielding actually the stronger result of Theorem 4.1 for a vector function $\tilde{g}(\tilde{X})$. The elementary result of Lemma 2.1 led to inequalities (Theorem 3.2) for the unconditional $V[g(X)]$ when X belongs to a parametric family of distributions and the parameter itself is regarded as a random variable (Bayesian setup). Similarly, Theorem 3.3 coversthe general case of a mixed distribution, including the special case of a censored X.

As for characterizations, it is shown (Section 5) how the proofs (cf [5], [7], [9], [10], [12]) can be simplified in view of the expression (3.2) for the covariance of X and $g(X)$ and (1.1); in fact, the multivariate case, due to independence, is essentially reduced to the univariate. Finally, some concluding remarks are made about the relation of these inequalities to the C-R and dual Poincaré-type inequalities under non-normal measures.

2. SOME PRELIMINARIES

For our purposes we require certain elementary results, which can easily be established.

Lemma 2.1. Let X have a mixture distribution with density $f(x)$ given by

$$f(x) = \int f(x|\theta)dH(\theta), \tag{2.1}$$

where H is a (mixing) distribution function on Θ and $f(x|\theta)$ is the (conditional) density of X given $\theta \in \Theta$. Let $\mu(\theta) = E(X|\theta)$ and $\sigma^2(\theta) = V[X|\theta]$. Then

$$\begin{aligned} V(X) &= E_H\{V(X|\theta)\} + V_H\{E(X|\theta)\} \\ &= \int_\Theta \sigma^2(\theta)dH(\theta) + \int_\Theta \mu^2(\theta)dH(\theta) - [\int_\Theta \mu(\theta)dH(\theta)]^2 \end{aligned} \tag{2.2}$$

In the case of a k-point θ mixture,(2.2) reduces to the following.

Corollary 2.1. Let X have a mixture-distribution with density function

$$f(x) = \sum_{i=1}^{k} \alpha_i f_i(x), \qquad 0 \le \alpha_i \le 1, \quad \sum_{i=1}^{n} \alpha_i = 1,$$

where f_i is the density function of a rv X_i, with

$$E(X_i) = \mu_i, \quad V(X_i) = \sigma_i^2, \qquad i = 1,2,\ldots,k.$$

Then

$$V(X) = \sum_{i=1}^{k} \alpha_i \sigma_i^2 + \sum_{i<j} \alpha_i \alpha_j (\mu_i - \mu_j)^2. \tag{2.3}$$

For our purposes, we also need the following C-S inequality, which follows from the fact that the dispersion (variance-covariance)matrix $D(\underset{\sim}{Z})$ of a random vector $\underset{\sim}{Z}' = (\underset{\sim}{\xi}', \underset{\sim}{\eta}')$ is nonnegative definite.

Lemma 2.2. (A Multivariate C-S inequality.) Let $\underset{\sim}{\xi} = (\xi_1,\ldots,\xi_m)'$ and $\underset{\sim}{\eta} = (\eta_1,\ldots,\eta_n)'$ be two random vectors and suppose the dispersion matrix of $\underset{\sim}{\eta}$, $D(\underset{\sim}{\eta})$ is positive definite (denoted by $D(\underset{\sim}{\eta}) > 0$). Then

$$D(\underset{\sim}{\xi}) \ge Cov(\underset{\sim}{\xi},\underset{\sim}{\eta})[D(\underset{\sim}{\eta})]^{-1}[Cov(\underset{\sim}{\xi},\underset{\sim}{\eta})]' \tag{2.4}$$

where $Cov(\underset{\sim}{\xi},\underset{\sim}{\eta}) = (Cov(\underset{\sim}{\xi}_i, \underset{\sim}{\eta}_j))$ $i = 1,\ldots,m,$ $j = 1,\ldots,n.$

3. THE MAIN RESULTS: UNIVARIATE CASE.

We present a unified and elementary derivation of the general inequalities (1.1) and (1.4) for $n = 1$ exhibiting, at the same time, the nature of the duality between the lower and upper bounds. Moreover, from this and Lemma 2.1, inequalities are given for the (unconditional) variance of $g(X)$ when the distribution of X depends on a parameter θ which is itself regarded as an r.v. (Theorem 3.2). The special case of a mixed distribution is also considered.

Lemma 3.1. Let X be a continuous r.v. with density $f(x)$ and $E(X) = \mu$, $V(X) = \sigma^2$. Let g and g' be real-valued functions defined on R such that g in an indefinite integral of g' and $V[g(X)] < \infty$. Then

$$\frac{1}{\sigma^2}\left[\int_{-\infty}^{\infty} g'(x)\left\{\int_{-\infty}^{x}(\mu-t)f(t)dt\right\}dx\right]^2 \le V[g(X)] \le \int_{-\infty}^{\infty}[g'(x)]^2\left\{\int_{-\infty}^{x}(\mu-t)f(t)dt\right\}dx \quad (3.1)$$

Proof: The LHS follows from the C-S inequality in the form $[\mathrm{Cov}X,g(X)]^2 \le \sigma^2 V[g(X)]$ and the following expression for the covariance of X and $g(X)$ (cf. [6], Lemma 3.1)

$$\mathrm{Cov}(X,g(X)) = \int_{-\infty}^{\infty}(x-\mu)g(x)f(x)dx = \int_{-\infty}^{\infty} g(x)d\int_{-\infty}^{x}(t-\mu)f(t)dt$$

$$= \int_{-\infty}^{\infty} g'(x)\int_{-\infty}^{x}(\mu-t)f(t)dt\, dx; \quad (3.2)$$

here, in integrating by parts, we used the fact that

$$\lim_{x\to\pm\infty} g(x)\int_{-\infty}^{x}(t-\mu)f(t)dt = 0,$$

since $E\{|g'(X)\omega(X)|\} < \infty$(see (3.3) below).
As regards the RHS, using now the C-S inequality for integrals, we have

$$V[g(X)] = V\left[\int_{\mu}^{X} g'(t)dt\right] \le E\left[\int_{\mu}^{X} g'(t)dt\right]^2 \le E\left[\int_{\mu}^{X}dt\int_{\mu}^{X}[g'(t)]^2dt\right]$$

$$= E\left[(X-\mu)\int_{\mu}^{X}[g'(t)]^2dt\right] = \int_{\infty}^{\infty}(x-\mu)f(x)\int_{\mu}^{x}[g'(t)]^2dtdx$$

$$= \int_{-\infty}^{\infty}\int_{\mu}^{x}[g'(t)]^2dt\,d\left[\int_{-\infty}^{x}(t-\mu)f(t)dt\right],$$

and after integrating by parts and taking into account that

$$\lim_{x\to\pm\infty}\int_{-\infty}^{x}(t-\mu)f(t)dt\int_{\mu}^{x}[g'(t)]^2dt = 0$$

the assertion follows (cf. proof of Lemma 2.1 in [1], where 0 in $\int_0^{\infty}$ and $\int_{-\infty}^{0}$ must be replaced by $\mu = E(X)$.

Remark 3.1. The dual character of (3.1) basically boils down to the following interplay in applying the C-S inequality:

$$E^2\left(\int_{\mu}^{X}dx\int_{\mu}^{X} g'(t)dt\right) \le \sigma^2 V[g(X)] \le \sigma^2 E\left(\int_{\mu}^{X} g'(t)dt\right)^2 \le \sigma^2 E\left(\int_{\mu}^{X}dt\int_{\mu}^{X}[g'(t)]^2dt\right),$$

which after the preceding elaboration gives (3.1)and, in turn, setting

$$\sigma^2 w(x)f(x) = \int_{-\infty}^{x}(\mu-t)f(t)dt, \quad (3.3)$$

yields the basic result of

Theorem 3.1. Under the assumptions of Lemma 3.1, we have

$$\sigma^2 E^2[w(X)g'(X)] \le V[g(X)] \le \sigma^2 E\{w(X)[g'(X)]^2\}, \tag{3.4}$$

where equality holds iff $g(x)$ in linear.

Taking $g(x)=x$ in (3.4), or directly from (3.3), we have

$$E[w(X)] = 1 \tag{3.5}$$

As pointed out, the proof of the discrete analogue of (3.4) (i.e. (1.4) with $n=1$) is similar and is omitted.

Let us now return to the situation of Lemma 2.1, where the density of X depends on a parameter θ regarded as a random variable, with d.f. $H(\theta)$. Then (3.4), in virtue of (2.2), yields the following extension of Theorem 3.1.

Theorem 3.2. Let $\gamma(\theta)=E[g(X)|\theta]=E_\theta[g(X)]$ be differentiable and $E(\theta)=\eta$, $V(\theta)=\tau^2$, $V[\gamma(\theta)]<\infty$, $V[g(X)|\theta]<\infty$. Then

$$V[g(X)] \le E_h\{\sigma^2(\theta)E_\theta(w_\theta(X)[g'(X)]^2)\}+\tau^2 E_h\{\omega(\theta)[\gamma'(\theta)]^2\}$$

$$V[g(X)] \ge E_h\{\sigma^2(\theta)E_\theta^2(w_\theta(X)g'(X))\}+\tau^2 E_h[\omega(\theta)\gamma'(\theta)],$$

where $\omega_\theta(x)$ is given by

$$\sigma^2(\theta)w_\theta(x)f(x|\theta) = \int_{-\infty}^{x}(\mu(\theta)-t)f(t|\theta)d\theta$$

and

$$\tau^2\omega(\theta)h(\theta) = \int_{-\infty}^{\theta}(\eta-t)h(t)dt,$$

h denoting the density of θ; equality in both sides is attained iff both functions g and γ are linear. Similarly, Corollary (2.1) in conjunction with (3.4) gives the following inequalities:

$$\begin{aligned} V[g(X)] &\le \sum_{i=1}^{k} \alpha_i\sigma_i^2 E_i\{w_i(X)[g'(X)]^2\}+\sum_{i<j}\alpha_i\alpha_j(\gamma_i-\gamma_j)^2 \\ V[g(X)] &\ge \sum_{i=1}^{k}\alpha_i\sigma_i^2 E_i^2\{w_i(X)[g'(X)]\}+\sum_{i<j}\alpha_i\alpha_j(\gamma_i-\gamma_j)^2 \end{aligned} \tag{3.6}$$

where E_i denotes expectation under F_i and thus

$$\gamma_i = E_i[g(X)], \quad \sigma_i^2 w_i(x)f_i(x) = \int_{-\infty}^{x}(\mu_i-t)f_i(t)dt, \quad i=1,\dots,k.$$

This is obtained by applying (3.4) to the $V_i[g]$, that is,

$$\sigma_i^2 E_i^2\{w_i(X)g'(X)\} \leq V_i[g(X)] \leq \sigma_i^2 E_i\{w_i(X)[g'(X)]^2\}.$$

Of special interest is the case of a rv X with a mixed distribution, that is, with d.f.

$$F(x) = \alpha F_1(x)+(1-\alpha)F_2(x) \tag{3.7}$$

where F_1 is the d.f. of a continuous rv X_1, with density $f_1(x)$, $E(X_1) = \mu_1$, $V[X_1] = \sigma_1^2$, and F_2 the d.f. of a discrete rv X_2 with $E(X_2) = \mu_2$, $V[X_2] = \sigma_2^2$ and p.f. $\{p_k, k = 0,\pm1,\ldots\}$. Then, by (3.6) and its discrete analogue (cf. (1.1) and (1.4) for $n = 1$) we conclude

Theorem 3.3. If X has mixed distribution, as in (3.7), then

$$V[g(X)] \geq \alpha\sigma_1^2 E_1^2\{\omega_1(X)g'(X_1)\}+(1-\alpha)\sigma_2^2 E_2^2\{\omega_2(X_2)\Delta g(X_2)\}+\alpha(1-\alpha)(\mu_1-\mu_2)^2$$

$$V[g(X)] \leq \alpha\sigma_1^2 E_1\{\omega_1(X)[g'(X_1)]^2\}+(1-\alpha)\sigma_2^2 E_2\{\omega_2(X_2)[\Delta g(X_2)]^2\}+\alpha(1-\alpha)(\mu_1-\mu_2)^2 \tag{3.8}$$

where

$$\sigma_1^2\omega_1(x)f_1(x) = \int_{-\infty}^{x}(\mu_1-t)f_1(t)dt, \quad \sigma_2^2\omega_2(x)p(x) = \sum_{k=-\infty}^{x}(\mu_2-k)p(k), \tag{3.9}$$

and equality obtains iff $g(x)$ is linear

Remark 3.1. Obviously, for $0<\alpha<1$, the last term in (3.8), which represents the variance of the means, vanishes if the two means are the same. Also, the second (discrete variance) term may also vanish, e.g., in the interesting situation of a continuous X censored at a fixed point c on the right, i.e., in (3.7) $\alpha = F(c)$, $\mu_2 = c$, $\sigma_2^2 = 0$. Then (3.8) (cf. (3.6) as well) reduces to

$$\begin{aligned} V[g(X)] &\geq F(c)\sigma_1^2 E^2\{\omega_1(X)g'(X)\}+F(c)(1-F(c)\ (\mu_1-c)^2, \\ V[g(X)] &\leq F(c)\sigma_1^2 E_1\{\omega_1(X)[g'(X)]^2\}+F(c)(1-F(c)) \end{aligned} \tag{3.10}$$

where (in the notation of (3.7)) E_1 denotes expectation under the density f_1 of X after censoring, i.e.,

$$f_1(x) = f(x)/F(c), \quad x<c$$

f denoting the density of the (uncensored) observation X (in (3.7) $X_2 = c$ with probability one). Thus, for example, if X is exponential with $f(x) = e^{-x}$, then

$$\alpha = F(c) = 1-e^{-c},\ \mu_1 = 1-c/(e^c-1),\ \sigma_1^2 = 1-c^2e^{-c}/(1-e^{-c})^2$$

and the resulting w_1-function is no longer, as usual (cf.[1] and [2]), a linear or quadratic function of x; it is easily verified that $w(x)=x$ for the uncensored X changes now to $w_1(x)$:

$$\sigma_1^2 w_1(x) = x-c(e^x-1)/(e^c-1) \quad (0<x<c).$$

4. THE MULTIVARIATE CASE WITH INDEPENDENCE.

Here we make use of (2.2) and (3.4) to obtain a new simple proof of inequalities (1.1), (cf. [2] and [5]). It suffices to consider the case $n=2$, since the treatment of the general case $(n>2)$ is similar or, equivalently, it follows by induction. In effect, the proof of (1.1), hence also of (1.4), amounts to a repeated application of (2.2) and (3.4). Thus we prove

Proposition 4.1. Let $g(x_1,x_2)$ with partial derivatives g_1,g_2 and X_1,X_2 independent rv's such that $V[g(X_1,X_2)]<\infty$. Then (1.1) holds for $n=2$.

Proof. By (2.2), denoting by E_i and V_i the expectation and variance operators under the distribution of X_i, respectively, we have

$$V[g(X_1,X_2)] = E_2\{V[g(X_1,X_2)|X_2]\}+V_2\{E[g(X_1,X_2)|X_2]\}; \tag{4.1}$$

but the independence of X_1 and X_2 implies that

$$V[g(X_1,X_2)|X_2] = V_1[g(X_1,X_2)], \quad E[g(X_1,X_2)|X_2] = E_1[g(X_1,X_2)] \tag{4.2}$$

so that, by the RHS of (3.4), (4.1) gives

$$V[g(X_1,X_2)] \le E_2\{\sigma_1^2 E_1[w_1(X_1)g_1^2(X_1,X_2)]\}+\sigma_2^2 E_2\{E_1[w_2(X_2)g_2^2(X_1,X_2)]\}$$

$$= \sigma_1^2 E[w_1(X_1)g_1^2(X_1,X_2)]+\sigma_2^2 E[\omega_2(X_2)g_2^2(X_1,X_2)].$$

Thus the RHS inequality of (1.1) is established. Similarly, the LHS of (1.1) follows from the LHS of (3.4) and the proof of our assertion is complete.

Note. Exactly the same proof holds for (1.4) if we use the discrete analogue of (3.4), that is, (1.4) with $n=1$.

An alternative derivation of the LHS of (1.1) can be achieved via the multivariate C-S inequality (2.4), which actually yields a more general result, stated as

Theorem 4.1. Let $\underset{\sim}{g}(\underset{\sim}{x}) = (g^{(1)}(\underset{\sim}{x}),\ldots,g_{(m)}(\underset{\sim}{x}))' : R^n \to R^m$ be a partially differentiable (vector) function of $\underset{\sim}{x} = (x_1,\ldots,x_n)'$, with

$$E\{|g_k^{(r)}(X)|\}<\infty, \quad r=1,\ldots,m, \quad k=1,\ldots,n.$$

Then the dispersion matrix $D[g(\underset{\sim}{X})] = (\text{Cov}(g^{(i)}, g^{(j)}))$ of $g(\underset{\sim}{X})$ satisfies the inequality

$$D[g(\underset{\sim}{X})] \geq \Gamma \tag{4.3}$$

where $\Gamma = (\gamma_{ij})$ and

$$\gamma_{ij} = \sum_{k=1}^{n} \sigma_k^2 E\{w_k(X_k) g_k^{(i)}(\underset{\sim}{X})\} E\{w_k(X_k) g_k^{(j)}(\underset{\sim}{X})\} \tag{4.4}$$

$g_k^{(r)}$ denoting the partial derivative of $g^{(r)}(\underset{\sim}{x})$ w.r.t. x_k.

Proof. In (2.4) take $\xi_i = g^{(i)}(\underset{\sim}{X})$, $i = 1,\ldots,m$ and $\underset{\sim}{\eta} = \underset{\sim}{X}$. First, by the independence of $X_1,\ldots,X_n$, $[D(\underset{\sim}{X})]^{-1} = \text{Diag}(\sigma_i^{-2})$ where $\text{Diag}(\alpha_i)$ denotes a diagonal matrix with diagonal elements $\alpha_1,\ldots,\alpha_n$. Then the elements $\tau_{ij} = \text{Cov}(g^{(i)}, X_j)$ of the covariance matrix $T = \text{Cov}(\underset{\sim}{g}, \underset{\sim}{X})$ between $\underset{\sim}{g}$ and $\underset{\sim}{X}$ are given by

$$\tau_{ij} = \text{Cov}\left(X_j, \int_{\mu_j}^{X_j} g_j^{(i)}(X_1,\ldots,X_{j-1},t,X_{j+1},\ldots,X_n) + R_j^{(i)}\right)$$

where we set

$$R_j^{(i)} = g^{(i)}(X_1,\ldots,X_{j-1},\mu_j,X_{j+1},\ldots,X_n).$$

Note however that since X_j is independent of $X_1,\ldots,X_{j-1}$, X_{j+1}, X_n,

$$\text{Cov}(X_j, R_j^{(i)}) = 0,$$

and, by analogy to (3.2),

$$\text{Cov}(X_j, g^{(i)}) = \sigma_j^2 E\{w_j(X_j) g_j^{(i)}(\underset{\sim}{X})\} \quad i = 1,\ldots,m,\ j=1,\ldots,n \tag{4.5}$$

with $w_j(x_j)$ as defined in (1.2) or (1.3).

Now (4.3) follows by applying (2.4) and verifying that

$$T\,\text{Diag}(\sigma_i^{-2})T' = (\gamma_{ij}),$$

with γ_{ij} as given by (4.4).

Note that the LHS of (1.1) and (1.4) is the special case $m = 1$ of (4.3).

5. CHARACTERIZATIONS OF DISTRIBUTIONS BY VARIANCE INEQUALITIES.

As mentioned in Section 1, the RHS of (3.4) and its discrete analogue have been used by several authors to characterize univariate distributions. Similarly, the special case of (1.1) with $w_i(x_i) \equiv 1$ $(i = 1,\ldots,n)$ for characterizing multivariate normality. The possibility of using either inequality in (3.4) (or (1.1)) to characterize the distribution of X (or $\underset{\sim}{X}$) has been demonstrated in [5]. It is more explicitly stated here as

Theorem 5.1. Under the preceding conditions on $\underset{\sim}{X}$ and g, the functions $\omega_1,\omega_2,\ldots,\omega_n$ in either

$$\text{(a)} \qquad \inf_g \frac{V[g(\underset{\sim}{X})]}{\sum_{i=1}^{n}\sigma_i^2 E^2[w_i(X_i)g_i(\underset{\sim}{X})]} = 1 \tag{5.1}$$

or

$$\text{(b)} \qquad \sup_g \frac{V[g(\underset{\sim}{X})]}{\sum_{i=1}^{n}\sigma_i^2 E[w_i(X_i)g_i^2(\underset{\sim}{X})]} = 1 \tag{5.2}$$

characterize the distribution of $\underset{\sim}{X}$.

The proofs given so far for the univariate case rested on proving that (5.2) (similarly (5.1)) for $n=1$ or (3.4), applied to the function $g(x) = x+\theta h(x)$, implies that

$$b \equiv \text{Cov}(X,h(X))-\sigma^2 E[w(X)h'(X)] = 0 \tag{5.3}$$

since the resulting quadratic in θ is nonnegative, namely,

$$V[X+\theta h(X)] = a\theta^2+2b\theta \geq 0 \qquad a>0.$$

A further special choice of h, such as the real and imaginary part of $\exp[itx]$ (for continuous X) or s^X (for discrete X) leads to a differential equation satisfied by the characteristic (or probability generating) function of X (see [5], [7], [10]).

However, a direct characterization of the distribution of X can be obtained by using (5.3), which in conjunction with (3.2) gives the relation

$$\sigma^2 w(x)f(x) = \int_{-\infty}^{x} (\mu-t)f(t)dt; \tag{5.4}$$

since the RHS of (5.4) is differentiable, setting $w(x)f(x) = h(x)$ and assuming either f or w differentiable we obtain

$$\sigma^2 h'(x) = (\mu-x)f(x),$$

from which the density f of X is uniquely determined. This completes the proof of Theorem (5.1) for $n=1$.

The general case $(n>1)$ can be treated similarly. We now use

$$g(\underset{\sim}{x}) = x_i+\theta h(\underset{\sim}{x})$$

to obtain, from either (5.1) or (5.2),

$$\text{Cov}(X_i,h(\underset{\sim}{X})) = \sigma_i^2 E[w_i(X_i)h_i(\underset{\sim}{X})] \quad (i=1,\ldots,n) \tag{5.5}$$

But (cf. (3.2), (3.3) and (4.5))

$$\mathrm{Cov}(X_i, h(\underset{\sim}{X})) = \int_{-\infty}^{\infty} h_i(\underset{\sim}{x})\left\{\int_{-\infty}^{x_i}(\mu_i - t)f_i(t)dt\right\}dt$$

so that, for each $i = 1,\ldots,n$

$$\sigma_i^2 w_i(x_i)f_i(x_i) = \int_{-\infty}^{x_i}(\mu_i - t)f_i(t)dt. \tag{5.6}$$

Hence (5.6), like (5.4) determines f_i and the proof of Theorem 5.1 is complete.

The proof of the discrete analogue of Theorem 5.1, with derivatives replaced by differences, is parallel; first-order differential equations are replaced by first-order difference equations.

Remark 5.1. Remarkably, in spite of the appearance of the ith partial derivative $h_i(\underset{\sim}{x})$ of $h(\underset{\sim}{x})$ in (5.5), the differential equation resulting from (5.6), which governs f_i, is simple. It is to be emphasized that this simplification, like the preceding ones in Section 4, rests very heavily on the independence of $X_1,\ldots,X_n$. From the statistical point of view, in a way, this is not a serious limiting assumption if we regard $g(\underset{\sim}{X})$ as a statistic-estimator-of a population parameter θ, when $\{f(x|\theta), \theta \in \Theta\}$ is a parametric family of distributions and, in particular, the $X_1,\ldots,X_n$ are i.i.d. r v 's. The problem however, of analogous inequalities for the non-independence case remains open.

Remark 5.2. It is also noteworthy that the preceding parameter-free derivation of variance bounds, in particular, lower ones, on the one hand, dispenses with the Cramér-Rao regularity conditions (cf. lower bounds in [1]) and, on the other hand, it yields bounds for $V(g)$ in terms of g' or Δg rather than the Fisher information, so that, in effect their sharpness is tailored for (attained by) linear g. Finally, their relation to Poincaré-type (upper bound) analogues as well as their dual (lower-bound) inequalities, deserve further investigation, especially, under non-normal measures (cf. Chen [11], [12]).

ACKNOWLEDGEMENT: I thank Professor E. Phadia for a discussion that led to (3.10).

REFERENCES

[1] Cacoullos, Theophilos, On upper and lower bounds for the variance of a function of a random variable., Ann. Probability 10 (1982) 799-809.

[2] Cacoullos, T. and Papathanasiou, Vasilis, On upper bounds for the variance of functions of random variables. Statistics & Probability Letters 3 (1985) 175-184.

[3] Chernoff, H., A note on an inequality involving the normal distribution. Ann. Probability 9 (1981) 533-535.

[4] Chen, Louis H.Y., An inequality for the multivariate normal distribution. Journal of Multivariate Anal. 12 (1982) 306-315.

[5] Brascamp, H.J. and Lieb, E.H., On extensions of the Brunn-Minkowski and Prekopa-Leindler theorems including inequalities for log concave functions and with an application to the diffusion equation. J. Funct. Anal. 22 (1976) 366-389.

[6] Cacoullos, T. and Papathanasiou, V., Characterizations of distributions by variance bounds. Statistics & Probability Letters 7 No.5 (1989) (in press).

[7] Srivastava, D.R. and Sreehari, M., Characterization of a family of discrete distributions via a Chernoff type inequality. Statistics & Probability Letters 5 (1987) 293-294.

[8] Borokov, A.A. and Utev, S.A., On a inequality and a related characterization of the normal distribution. Theory Prob. Appl., t. 28 (1984) 219-228.

[9] Prakasa Rao, B.L.S. and Sreehari, M., Another characterization of multivariate normal distribution. Statistics & Probability Letters 4 (1986) 209-210.

[10] Prakasa Rao, B.L.S. and Sreehari, M., On a characterization on a Poisson distribution through an inequality of Chernoff type. Aus. J. Of Statist. 29 (1987) 38-41.

[11] Chen, Louis H.Y., Poicaré-type inequalities via stochastic integrals. Z. Wahrscheinlichkeitstheorie verw. Gebiete, 69 (1985) 255-277.

STATISTICAL DATA ANALYSIS AND INFERENCE
Y. Dodge (Editor)

SOME OPTIMALITY RESULTS ON STEIN'S TWO-STAGE SAMPLING

J.K. GHOSH and Rahul MUKERJEE

Division of Theoretical Statistics and Mathematics, Indian Statistical Institute, 203 Barrackpore Trunk Road, Calcutta 700 035, INDIA

It has been shown that under Stein's two-stage sampling procedure, the confidence interval for the population mean, as proposed by him cannot be improved upon. Allied results in the context of point estimation have also been indicated.

1. INTRODUCTION AND MAIN RESULTS

Consider a normal population with unknown mean μ and unknown variance σ^2. In order to obtain a confidence interval for μ with (a) confidence coefficient at least $1-\alpha$ and (b) length at most $2d$, where $\alpha(0 < \alpha < 1)$ and $d(> 0)$ are pre-assigned, Stein [5] introduced the following well-known two-stage procedure. A sample $X_1,\dots,X_m$ of size $m(\geq 2)$, fixed in advance, is drawn and $\bar{X}_m = m^{-1}\Sigma_{i=1}^m X_i$ and $s^2 = (m-1)^{-1}\Sigma_{i=1}^m (X_i - \bar{X}_m)^2$ are computed. Let N be the smallest integer satisfying

$$N \geq \max(m, t_o^2 s^2/d^2), \tag{1}$$

where t_o is the upper $(\alpha/2)$ - point of the t-distribution with $m-1$ degrees of freedom (d.f.). Then a further sample of $N-m$ observations $X_{m+1},\dots,X_N$ is drawn. Let $\bar{X} = N^{-1}\Sigma_{i=1}^N X_i$ be the mean of all the observations. Then, as shown by Stein [5], by (1), $(u_1(\underline{X}), u_2(\underline{X}))$ is a confidence interval for μ with the required properties (a) and (b) where $u_1(\underline{X}) = \bar{X} - t_o\, s\, N^{-\frac{1}{2}}$, $u_2(\underline{X}) = \bar{X} + t_o\, s\, N^{-\frac{1}{2}}$.

It has been observed by several authors that the information from the second sample on σ^2 is not utilized in the above procedure and that there is a consequent possibility of improvement. This point was noted by B.K. Ghosh in a private communication and also by Rao [4, pp. 487-488]. Theorem 1 below essentially shows that such an improvement is not possible. This finding is further substantiated by Theorems 2 and 3. The proofs are given in the next section.

Let S be the sample space under Stein's two-stage sampling and $L_j(.)$ be the Lebesgue measure over the class of Borel sets in R^j. Recall that the Stein's confidence interval $(u_1(\underline{X}), u_2(\underline{X}))$ has length $2t_o s\, N^{-\frac{1}{2}}$, which does not exceed $2d$, and exact confidence coefficient $1 - \alpha$.

Theorem 1. Under Stein's two-stage sampling as specified by (1), let $C(\underline{X}), \underline{X} \in S$, be any family of confidence sets satisfying

(i) $P_{\mu,\sigma^2}(\mu \in C(\underline{X})) \geq 1 - \alpha \quad \forall\ \mu, \sigma^2$,

(ii) $L_1(C(\underline{X})) \leq 2t_o\, s\, N^{-\frac{1}{2}}$, almost everywhere (relative to Lebesgue measure).

Then $P_{\mu,\sigma^2}[L_1(C(\underline{X})) = 2t_o\, s\, N^{-\frac{1}{2}}] = 1, \quad \forall\ \mu, \sigma^2$.

Theorem 2. Under Stein's two-stage sampling, let $C(\underline{X})$, $\underline{X} \in S$, be any family of confidence sets satisfying the condition (i) in Theorem 1 and

(ii) $L_1(C(\underline{X})) = 2t_o\, s\, N^{-\frac{1}{2}}$ almost everywhere (relative to Lebesgue measure),

(iii) for each fixed $\sigma^2(> 0)$, $P_{\mu,\sigma^2}(\mu \in C(\underline{X}))$ is continuous in μ.

Then $P_{\mu,\sigma^2}(\mu \in C(\underline{X})) = 1 - \alpha, \quad \forall\ \mu, \sigma^2$.

Theorem 3. Under Stein's two-stage sampling, let $C(\underline{X})$, $\underline{X} \in S$, be any family of confidence sets satisfying the condition (i) in Theorem 1 and

(ii) $L_1(C(\underline{X})) \leq 2d^*$ almost everywhere (relative to Lebesgue measure).

Then $d^* \geq d$

Remark 1. Notwithstanding the above, it is possible that one may be able to use the second stage variance to provide a random confidence coefficient which is better than $1 - \alpha$ in the sense of Kiefer [3] - vide Berger [1].

Remark 2. Consider all equivariant bounded length confidence sets under the group of location change. Then Stein's procedure is best among those which have the same sampling rule and whose length is less than or equal to that of Stein's interval almost everywhere. Using admissibility results for best equivariant procedures it should be possible to prove Theorem 2.

Remark 3. It would be interesting to prove a version of Theorem 2 without the condition (iii). This, however, seems to be difficult.

2. PROOFS

By (1), the sample space S, in Stein's two-stage sampling is given by $S = \cup_{j=m}^{\infty} S_j$, where

$$S_m = \{(x_1,\ldots,x_m) : (m-1)^{-1} \Sigma_{i=1}^m (x_i - \bar{x}_m)^2 \leq md^2/t_o^2\},$$

$$S_j = \{(x_1,\ldots,x_m) : (j-1)d^2/t_o^2 < (m-1)^{-1} \Sigma_{i=1}^m (x_i - \bar{x}_m)^2 \leq jd^2/t_o^2\},\ j \geq m+1.$$

Given $\underline{X} \in S_j$ (i.e., $N = j$), the observations may be denoted by $(X_1,\ldots,X_j) = \underline{X}^{(j)}$, and accordingly a family of confidence sets $C(X)$ may be represented by $C_j(\underline{X}^{(j)})$, $j \geq m$. Let $\phi(.)$ and $\Phi(.)$ represent the standard normal p.d.f. and c.d.f. respectively. The following lemma which is easy to prove (e.g., by the Neyman Pearson lemma) will be helpful.

Lemma 1. For every Borel set B in R^1,

$$\int_B \phi(x)dx \leq 2\Phi\{\tfrac{1}{2} L_1(B)\} - 1.$$

Proof of Theorem 1. If possible suppose there exists a family of confidence sets $C(\underline{X})$ satisfying the conditions (i) and (ii) in Theorem 1 for which the conclusion of Theorem 1 does not hold. Then there exists $\delta(>0), \mu_o, \sigma_o^2$ such that

$$P_{\mu_o,\sigma_o^2}(2t_o\, s\, N^{-\frac{1}{2}} - L_1(C(X)) > \delta) > 0 ,$$

so that defining the sets

$$T_j = \{\underline{x}^{(j)} : \underline{x}^{(j)} \in S_j,\ 2t_o\, s\, j^{-\frac{1}{2}} - L_1(C_j(\underline{x}^{(j)})) > \delta\} , \quad j \geq m \tag{2}$$

there exists a positive integer $k(\geq m)$ such that

$$L_k(T_k) > 0 . \tag{3}$$

A Bayesian approach will be used in proving the theorem. For $-\infty < \mu < \infty$, $v > 0$, let $g_v(\mu) = (2\pi v)^{-\frac{1}{2}} \exp\{-(2v)^{-1} \mu^2\}$. By condition (i),

$$1 - \alpha \leq \int_{-\infty}^{\infty} P_{\mu,\sigma^2}(\mu \in C(\underline{X}))g_v(\mu)d\mu = \sum_{j=m}^{\infty} H_j(v,\sigma^2), \quad \forall\ v,\ \sigma^2 > 0 , \tag{4}$$

where for $j \geq m$,

$$H_j(v,\sigma^2) = \int_{-\infty}^{\infty} \underset{\substack{\mu\in C_j(\underline{x}^{(j)}) \\ \underline{x}^{(j)} \in S_j}}{\int \cdots \int} (2\pi\sigma^2)^{-j/2} \exp\{-(2\sigma^2)^{-1} \sum_{i=1}^{j}(x_i-\mu)^2\}g_v(\mu)d\,\underline{x}^{(j)}d\mu .$$

It can be seen after some algebra that for each j,

$$H_j(v,\sigma^2) = \underset{\underline{x}^{(j)} \in S_j}{\int \cdots \int} \Big[\int_{\mu \in C_j(\underline{x}^{(j)})} \beta_j(\mu|\underline{x}^{(j)})d\mu\Big]\nu_j(\underline{x}^{(j)})d\underline{x}^{(j)} . \tag{5}$$

In the above $\nu_j(\underline{x}^{(j)})$ is the p.d.f. of $\underline{X}^{(j)}$ having the j-variate normal distribution $N_j(\underline{0}, \sigma^2 I_j + vJ_j)$ and $\beta_j(\mu|\underline{x}^{(j)})$ is the (conditional) p.d.f. of μ where μ is normal with mean $\Delta(\underline{x}^{(j)})$ and variance $v\sigma^2/(\sigma^2 + jv)$. Also, $\Delta(\underline{x}^{(j)}) = v\, \underline{1}_j'(\sigma^2 I_j + vJ_j)^{-1}\, \underline{x}^{(j)}$, $\underline{1}_j$ is the jx1 vector with all elements unity, I_j is the jxj identity matrix and $J_j = \underline{1}_j\, \underline{1}_j'$.

By Lemma 1 and condition (ii),

$$\int_{\mu\in C_j(\underline{x}^{(j)})} \beta_j(\mu|\underline{x}^{(j)})d\mu \leq 2\, \Phi[\tfrac{1}{2}\, L_1\{C_j(\underline{x}^{(j)})\}/\{v\sigma^2/(\sigma^2 + jv)\}^{\frac{1}{2}}] - 1 \tag{6}$$

$$\leq 2\, \Phi[t_o\, s\{(\sigma^2 + jv)/(jv\sigma^2)\}^{\frac{1}{2}}] - 1$$

$$= W(s,\ j;\ v,\ \sigma^2) , \tag{7}$$

say, for all $\underline{x}^{(j)} \in S_j^* (\subset S_j)$ where $L_j(S_j - S_j^*) = 0$. Therefore by (5), (7), for $j \geq m$,

$$H_j(v,\sigma^2) \leq \int_{\underline{x}^{(j)} \varepsilon S_j} \cdots \int W(s,j;v,\sigma^2)\, \nu_j(\underline{x}^{(j)})d\,\underline{x}^{(j)} \ , \quad \forall \ v, \ \sigma^2 > 0. \tag{8}$$

Also, noting that $\Phi(y_1) - \Phi(y_2) \geq \phi(y_1)(y_1-y_2)$ for $y_1 > y_2 \geq 0$, it follows from (1), (2) and (6) that for $\underline{x}^{(k)} \varepsilon T_k$,

$$\int_{\mu \varepsilon C_k(\underline{x}^{(k)})} \beta_k(\mu|\underline{x}^{(k)})d\mu \leq W(s,k;v,\sigma^2)$$

$$- \phi[t_o s\{(\sigma^2+kv)/(kv\sigma^2)\}^{\frac{1}{2}}]\delta\{(\sigma^2+kv)/(v\sigma^2)\}^{\frac{1}{2}}$$

$$\leq W(s,k;v,\sigma^2) - Q(v,\sigma^2), \tag{9}$$

where

$$Q(v,\sigma^2) = \phi[d\{(\sigma^2+kv)/(v\sigma^2)\}^{\frac{1}{2}}]\ \delta\{(\sigma^2+kv)/(v\sigma^2)\}^{\frac{1}{2}} . \tag{10}$$

By (4) - (9),

$$1 - \alpha \leq \sum_{j=m}^{\infty} \int_{\underline{x}^{(j)} \varepsilon S_j} \cdots \int W(s,j;v,\sigma^2)\nu_j(\underline{x}^{(j)})d\underline{x}^{(j)}$$

$$- Q(v,\sigma^2) \int_{\underline{x}^{(k)} \varepsilon T_k} \cdots \int \nu_k(\underline{x}^{(k)})d\underline{x}^{(k)} \qquad \forall \quad v, \ \sigma^2 > 0 . \tag{11}$$

Writing $\nu_k(.)$ explicitly, and using the fact that $(\sigma^2 I_k+J_k)^{-1}-(\sigma^2 I_k+vJ_k)^{-1}$ is non-negative definite for $v \geq 1$, it can be seen that

$$\int_{\underline{x}^{(k)} \varepsilon T_k} \cdots \int \nu_k(\underline{x}^{(k)})d\underline{x}^{(k)} \geq [(\sigma^2+k)/(\sigma^2+kv)]^{\frac{1}{2}}\rho(\sigma^2), \quad \forall\, v \geq 1, \quad \sigma^2 > 0, \tag{12}$$

where

$$\rho(\sigma^2) = \int_{\underline{x}^{(k)} \varepsilon T_k} \cdots \int [(2\pi)^k(\sigma^2+k)\sigma^{2(k-1)}]^{-\frac{1}{2}} \exp\{-\tfrac{1}{2}\,\underline{x}^{(k)'}(\sigma^2 I_k+J_k)^{-1}\underline{x}^{(k)}\}d\underline{x}^{(k)}$$

is free from v and is positive for every $\sigma^2(> 0)$ by (3). Also from (7), and the definitions of s^2 and S_j's,

$$\sum_{j=m}^{\infty} \int_{\underline{x}^{(j)} \varepsilon S_j} \cdots \int W(s,j;v,\sigma^2)\nu_j(\underline{x}^{(j)})d\,\underline{x}^{(j)}$$

$$= P[|\xi|(\lambda/(m-1))^{-\frac{1}{2}} \leq t_o\{(\sigma^2+Zv)/(Zv)\}^{\frac{1}{2}}] \ , \quad \forall \quad v, \ \sigma^2 > 0 \tag{13}$$

where ξ is a standard normal variate, λ has the chi-square distribution with $m-1$ d.f., ξ and λ are independent, and

$$\begin{aligned} Z &= m \quad \text{if} \quad t_o^2 \lambda\, \sigma^2/\{(m-1)d^2\} \leq m \\ &= j \quad \text{if} \quad j-1 < t_o^2\, \lambda\sigma^2/\{(m-1)d^2\} \leq j(j \geq m+1). \end{aligned}$$

Thus, $Z \geq m$ with probability 1 so that by (13) and the definition of t_0,

$$\sum_{j=m}^{\infty} \int_{\underline{x}^{(j)} \in S_j} \cdots \int W(s,j;v,\sigma^2)\nu_j(\underline{x}^{(j)})d\underline{x}^{(j)}$$

$$\leq P[|t_{m-1}| \leq t_0\{1+(mv)^{-1}\sigma^2\}^{\frac{1}{2}}] \leq 1-\alpha+2f(t_0)t_0\sigma^2(mv)^{-1} , \ \forall \ v,\sigma^2 > 0, \tag{14}$$

where t_{m-1} has the t-distribution with m-1 d.f. and f(.) is the p.d.f. of t_{m-1}. By (10), (11), (12), (14),

$$\delta\{(\sigma^2+k)/(v\sigma^2)\}^{\frac{1}{2}} \rho(\sigma^2)\phi[d\{(\sigma^2+kv)/(v\sigma^2)\}^{\frac{1}{2}}]$$

$$\leq 2f(t_0)t_0 \sigma^2(mv)^{-1} , \qquad \forall \ v \geq 1, \quad \sigma^2 > 0 ,$$

which is easily seen to lead to a contradiction by allowing $v \to \infty$, with σ^2 fixed. This completes the proof of Theorem 1.

Although Theorem 2 does not readily follow as a corollary to Theorem 1, the proof of this theorem is similar to that of Theorem 1 and hence omitted here. The proof of Theorem 3 involves use of the following lemma.

Lemma 2. For every $d^*(> 0)$, under Stein's two-stage sampling

$$\lim_{\sigma^2 \to \infty} P_{\mu,\sigma^2}(|\bar{X} - \mu| \leq d^*) = P(|t_{m-1}| \leq t_0 d^*/d).$$

Proof. Let $\{\sigma_\omega^2\}$ be any positive sequence such that $\lim \sigma_\omega^2 = \infty$. By the definition of N, $t_0^2 s^2/d^2 \leq N \leq t_0^2 s^2/d^2 + 1$. Hence as in Stein [5],

$$P(|t_{m-1}| \leq t_0 d^*/d) \leq P_{\mu,\sigma_\omega^2}(|\bar{X} -\mu| \leq d^*) \leq P(Y_\omega \leq 0) \tag{15}$$

for every ω, where the distribution of the random variable Y_ω is the same as that of $N^{\frac{1}{2}} s^{-1}|\bar{X} - \mu| - d^*(t_0^2 d^{-2} + s^{-2})^{\frac{1}{2}}$ under $\sigma^2 = \sigma_\omega^2$. Clearly $\{Y_\omega\}$ converges in distribution to $|t_{m-1}| - t_0 d^*/d$ and the result follows from (15).

Proof of Theorem 3. Under the conditions (i) and (ii) in Theorem 2, proceeding as in the derivation of (7), one obtains

$$1-\alpha \leq \sum_{j=m}^{\infty} A_j(v,\sigma^2) \int_{\underline{x}^{(j)} \in S_j} \cdots \int \nu_j(\underline{x}^{(j)})d\,\underline{x}^{(j)} , \quad \forall \ v, \ \sigma^2 > 0 \tag{16}$$

where

$$A_j(v,\sigma^2) = 2\,\Phi[d^*\{(\sigma^2+jv)/(v\sigma^2)\}^{\frac{1}{2}}] - 1 . \tag{17}$$

By the definition of S_j and $\nu_j(.)$, it can be seen that

$$\int_{\underline{x}^{(j)} \in S_j} \cdots \int \nu_j(\underline{x}^{(j)})d\,\underline{x}^{(j)} = P[\lambda \leq m(m-1)\{d/(t_0\sigma)\}^2] \quad \text{for } j = m ,$$

$$= P[(j-1)(m-1)\{d/(t_0\sigma)\}^2 < \lambda \tag{18}$$

$$\leq j(m-1)\{d/(t_0\sigma)\}^2] \quad \text{for } j \geq m+1,$$

where λ has the chi-square distribution with $m-1$ d.f. Hence allowing $v \to \infty$ in (16), it follows from (17), (18) that

$$1-\alpha \leq \sum_{j=m}^{\infty} \{2\Phi(d^* j^{\frac{1}{2}}/\sigma) - 1\} \int \dots \int_{\underline{x}^{(j)} \in S_j} \nu_j(\underline{x}^{(j)}) d\,\underline{x}^{(j)}$$

$$= P_{\mu,\sigma^2}(|\overline{X} - \mu| \leq d^*) , \quad \forall \quad \sigma^2 > 0 . \tag{19}$$

Note that the right-hand member of (19) is free from μ. If one allows $\sigma^2 \to \infty$ in (19) and makes use of Lemma 2, then the theorem follows.

It is also possible to derive results on the admissibility and minimaxity of $\overline{X}$ as a point estimator of μ, under squared error loss. The results follow essentially along the line of Ferguson [2, pp.141] and involve use of techniques similar to the ones employed in proving Theorems 1,3. The details may be obtained from the authors.

REFERENCES

[1] Berger, J., The frequentist viewpoint and conditioning, in : Proc. Berkeley Conference in Honor of J. Neyman and J. Kiefer (Wadsworth, Belmont, California, 1985) pp. 15-44.

[2] Ferguson, T.S., Mathematical Statistics, A Decision Theoretic Approach (Academic Press, New York, 1967).

[3] Kiefer, J., Conditional confidence statements and confidence estimators (with discussion), J. Amer. Statist. Assoc. 72 (1977) 789-827.

[4] Rao, C.R., Linear Statistical Inference and Its Applications, 2nd ed. (John Wiley, New York, 1973).

[5] Stein, C., A two-samples test for a linear hypothesis whose power is independent of variance, Ann. Math. Statist. 16 (1945) 243-258.

STATISTICAL DATA ANALYSIS AND INFERENCE
Y. Dodge (Editor)

THE CRAMÉR-FRÉCHET-RAO INEQUALITY FOR SEQUENTIAL ESTIMATION IN NON-REGULAR CASE

Z. GOVINDARAJULU* and I. VINCZE

Department of Statistics, University of Kentucky, Lexington, KY, USA and Mathematical Institute of the Hungarian Academy of Sciences, Budapest, H-1364, P.O.B. 127

Summary: A Cramér-Fréchet-Rao lower bound is investigated for sequential estimation which is general in the sense that in its derivation no regularity assumptions are made, the parameter space is completely arbitrary; moreover there is no restriction to independent and identically distributed random variables. Turning to the regular case a lower bound is obtained which is attainable and in general, is better than the classical result of Wolfowitz (1947). The non-regular translation parameter case is treated for certain situations giving the right order of magnitude, but not the best constants.

1. INTRODUCTION

In the last decades several authors investigated Cramér-Fréchet-Rao type lower bounds for fixed sample as well as sequential estimates in non-regular cases. Several peculiar phenomena became clear. For example, non-regularity can lead to some gain in sample size compared to related regular cases. For relevant literature we refer the reader to the work of Polfeldt (1970), Linnik and Romanovski (1972), Wijzman (1973), Ibrahimov and Khazminski (1974) and Simons (1980).

For considering a unified and general treatment of the problem of CFR type lower bound for the sequential estimation we begin with describing the sequential procedure, then we turn to the method leading to a lower bound which is general in the sense that its derivation does not need any regularity assumptions, the parameter space is completely arbitrary, and there is no restriction to independent or to identically distributed random variables. This lower bound is given for the variance of the mixture of two distributions, but it can be extended to a mixture of arbitrary (finite or infinite) number of distributions, for instance see Vincze (1979), Khatri (1980), Móri (1983). The origin of derivation goes back to the ingenious idea of M. Fréchet (1943), C. R. Rao (1945) and H. Cramér (1946) which was considered independently by the preceding authors. After the general framework the regular case will be considered, including the investigation of the attainability of the bound. In the non-regular case attainability can be expected to occur for the order of magnitude in the (mean) sample

*Presently, a visiting Professor at Columbia University for the year 1988-1989.

size only. For independent observations in the regular case we arrive at a better bound than the classical result of Wolfowitz (1947). The estimation of the translation parameter will then be considered as a nonregular situation in some cases. Finally a discussion — comparison with certain known results — is given.

2. A GENERAL CRAMÉR-FRÉCHET-RAO INEQUALITY FOR SEQUENTIAL ESTIMATION

2.1. Formulation of the sequential procedure

Let $X_1,X_2,\ldots$ be a sequence of random variables such that the vector $(X_1,X_2,\ldots,X_i)$ $(i=1,2,\ldots)$ have density function $f_\theta^{(i)}(x_1,x_2,\ldots,x_i)$ $(i==1,2,\ldots)$ with respect to a σ-finite measure μ_i on R^i. We shall assume that the dominating measures μ_i $(i=1,2,\ldots)$ are restrictions to R^i of a σ-finite measure μ on R^∞. Let θ be an element of the (arbitrary) parameter set Θ. We wish to estimate unbiasedly a real valued function $g(\theta)$ of $\theta\in\Theta$, by some functions $t_i(X_1,X_2,\ldots,X_i)$ $(i=1,2,\ldots)$ using a stopping rule which is closed in the following sense: When N denotes the random variable giving the sample size required for termination of the experiment, we assume that for every $\theta\in\Theta$ $P_\theta(N\leqq n)\to 1$ as $n\to\infty$. (We do not assume that this holds uniformly in θ.)

The sample space is $E_1\cup E_2\cup\ldots$ where $E_i\subset R^i$ and E_i consists of those points $(x_1,x_2,\ldots,x_i)$ which constitute stopping points. We assume that the relevant conditional probabilities exist.

Let

$$f_\theta(\underline{x}) = f_\theta(x_1,x_2,\ldots) = \sum_{i=1}^{\infty} f_\theta^{(i)}(x_1,x_2,\ldots,x_i)I_{E_i} \tag{2.i.1}$$

and

$$t(\underline{x}) = t(x_1,x_2,\ldots) = \sum_{i=1}^{\infty} t_i(x_1,x_2,\ldots,x_i)I_{E_i}\ , \tag{2.1.2}$$

where I_A denotes the indicator function of the event A. The system $\{E_i\ ,\ i=1,2,\ldots\}$ will be denoted by E. Concerning our assumptions see [1].

2.2. Non-regular (non-parametric) CFR relations

For fixed $\theta,\theta'\in\underline{\Theta}$ let

$$f_\alpha(\underline{x};\theta,\theta') = (1-\alpha)f_\theta(\underline{x}) + \alpha f_{\theta'}(\underline{x})\ ,\quad 0<\alpha<1\ .$$

Following [15] when considering for a moment α as an unknown parameter which is to be estimated, we can see that

$$\hat{\alpha} = \frac{t(\underline{x})-g(\theta)}{g(\theta')-g(\theta)} \tag{2.2.1}$$

is an unbiased estimator for α. Since the density $f_\alpha(\underline{x};\theta,\theta')$ is linear in α

one can obtain simply the regular CFR inequality.

$$\mathrm{Var}(\hat{\alpha}|\alpha) = \underline{D}_{\alpha}^{2}(\hat{\alpha}) \geqq \frac{\alpha(1-\alpha)}{I_{\alpha}[\theta,\theta']} . \tag{2.2.2}$$

Here the information quantity has the form $(0<\alpha<1)$:

$$I_{\alpha}(\theta,\theta') = \alpha(1-\alpha)\int \frac{(f_{\theta}(\underline{x})-f_{\theta'}(\underline{x}))^{2}}{(1-\alpha)f_{\theta}(\underline{x})+\alpha f_{\theta'}(\underline{x})}\, d\mu \tag{2.2.3}$$

(see [4]) and $0 \leqq I_{\alpha}(\theta,\theta') \leqq 1$. Substituting (2.2.1) and (2.2.3) into (2.2.2) and after a slight calculation we obtain

$$(1-\alpha)\underline{D}_{\theta}^{2}(t(\underline{x}))+\alpha\underline{D}_{\theta'}^{2}(t(\underline{x})) \geqq \alpha(1-\alpha)(g(\theta')-g(\theta))^{2}\left\{\frac{1}{I_{\alpha}(\theta,\theta')}-1\right\} , \tag{2.2.4}$$

which is a CFR inequality for the mixed variance.

We consider as a main result of our considerations *the amalgamation of the sequential procedure with a general inequality for a mixed variance getting a unified form, i.e. a Cramér-Fréchet-Rao inequality which is non-parametric, non-singular, valid for sequential and non-sequential estimation.*

Before turning to regular and non-regular examples some simple general statements will be given.

PROPOSITION 2.2.1. The CFR lower bound is a functional of the likelihood ratio $f_{\theta'}(\underline{x})/f_{\theta}(\underline{x})$, which will be denoted by $L(x;\theta,\theta') = [L(x;\theta',\theta)]^{-1}$ due to the following relation:

$$\frac{1}{I_{\alpha}(\theta,\theta')} - 1 = \frac{\int \frac{L(x;\theta,\theta')}{(1-\alpha)+\alpha L(x;\theta,\theta')}\, dP_{\theta}}{1-\int \frac{L(x;\theta,\theta')}{(1-\alpha)+\alpha L(x;\theta,\theta')}\, dP_{\theta}} = \frac{\int \frac{L(x;\theta',\theta)}{(1-\alpha)L(x;\theta',\theta)+\alpha}\, dP_{\theta'}}{1-\int \frac{L(x;\theta',\theta)}{(1-\alpha)L(x;\theta',\theta)+\alpha}\, dP_{\theta'}} \tag{2.2.5}$$

PROOF. An easy calculation results in

$$\begin{aligned} 1-I_{\alpha}(\theta,\theta') &= \int (1-\alpha)f_{\theta'}(\underline{x})+\alpha f_{\theta}(\underline{x})\, d\mu-\alpha(1-\alpha)\int \frac{(f_{\theta}(\underline{x})-f_{\theta'}(\underline{x}))^{2}}{(1-\alpha)f_{\theta}(\underline{x})+f_{\theta'}(\underline{x})}\, d\mu \\ &= \int \frac{f_{\theta}(\underline{x})f_{\theta'}(\underline{x})}{(1-\alpha)f_{\theta}(\underline{x})+\alpha f_{\theta'}(\underline{x})}\, d\mu = \int \frac{L(\underline{x};\theta,\theta')}{(1-\alpha)+\alpha L(\underline{x};\theta,\theta')}\, dP_{\theta} \end{aligned} \tag{2.2.6}$$

giving (2.2.5).

We introduce now the following notations:

$B=\{\underline{x}: f_{\theta}(\underline{x})f_{\theta'}(\underline{x})>0\}$, $C=\{\underline{x}: f_{\theta}(\underline{x})>0,\ f_{\theta'}(\underline{x})=0\}$, $D=\{\underline{x}: f_{\theta'}(\underline{x})>0,\ f_{\theta}(\underline{x})=0\}$,

further

$B^{(1)}=B\cap\{\underline{x}: L(\underline{x};\theta,\theta')>1\}$, $B^{(2)}=B\cap\{\underline{x}: L(\underline{x};\theta';\theta)>1\}$, $B^{(0)}=\{\underline{x}: L(\underline{x};\theta,\theta')=1\}$.

Whenever $\underline{x} \in R^{i}$ the notations B_i, C_i etc. will be used.

Our information quantity relates to the total variation and to the Hellinger distance in the following way:

PROPOSITION 2.2.2. The inequalities

$$I_{\frac{1}{2}}(\theta,\theta') \le \|P_\theta - P_{\theta'}\| = \frac{1}{2}\int |f_\theta(\underline{x}) - f_{\theta'}(x)|d\mu$$

$$= \frac{1}{2}P_\theta(C) + \frac{1}{2}P_{\theta'}(D) + \frac{1}{2}\int_{B^{(1)}}(f_{\theta'}(\underline{x}) - f_\theta(\underline{x}))d\mu + \frac{1}{2}\int_{B^{(2)}}(f_\theta(\underline{x}) - f_{\theta'}(\underline{x}))d\mu \qquad (2.2.7)$$

and

$$\frac{1}{2}H^2(\theta,\theta') \le I_{\frac{1}{2}}(\theta,\theta') \le h^2(\theta,\theta') = H^2(\theta,\theta') - \frac{1}{2}(P_\theta(C) + P_{\theta'}(D)) \qquad (2.2.8)$$

hold, where, as it is well known, for the Hellinger distance that

$$0 \le H^2(\theta,\theta') = \int (\sqrt{f_\theta(\underline{x})} - \sqrt{f_{\theta'}(\underline{x})})^2 \le 2 .$$

In connection with the proof of Proposition 2.2.2 we refer concerning (2.2.8) to [11], while (2.2.7) is also known (see e.g. [6]). Nevertheless we give the following hint for the justification of the latter: according to (2.2.6)

$$I_{\frac{1}{2}}(\theta,\theta') = 1 - 2\int \frac{L(\underline{x};\theta,\theta')}{1+L(\underline{x};\theta,\theta')}dP_\theta = 1 - 2\int_{B^{(1)}} \frac{L(\underline{x};\theta,\theta')}{1+L(\underline{x};\theta,\theta')}dP_\theta -$$

$$-2\int_{B^{(2)}} \frac{L(\underline{x};\theta',\theta)}{1+L(\underline{x};\theta',\theta)}dP_{\theta'} \le 1 - P_\theta(B^{(1)}) - P_{\theta'}(B^{(2)}) - \frac{1}{2}(P_\theta(B^{(0)}) + P_{\theta'}(B^{(0)})) .$$

Substituting 1 by

$$1 = \frac{1}{2}(P_\theta(B^{(0)}) + P_\theta(B^{(1)}) + P_\theta(B^{(2)}) + P_\theta(C)) + \frac{1}{2}(P_{\theta'}(B^{(0)}) + P_{\theta'}(B^{(1)}) + P_{\theta'}(B^{(2)}) + P_{\theta'}(D)) ,$$

we easily come to our statement (2.2.7).

As a consequence of (2.2.4) and of Proposition 2.2.2 we arrive at the following CFR relations

$$\frac{1}{2}(D_\theta^2(t(x)) + D_{\theta'}^2(t(x)) \ge \frac{1}{4}(g(\theta') - g(\theta))^2 \left\{\frac{1}{\|P_{\theta'} - P_\theta\|} - 1\right\} \qquad (2.2.9)$$

$$\frac{1}{2}(D_\theta^2(t(x)) + D_{\theta'}^2(t(x)) \ge \frac{1}{4}(g(\theta') - g(\theta))^2 \left\{\frac{1}{h^2(\theta,\theta')} - 1\right\} \qquad (2.2.10)$$

which are slightly weaker than (2.2.4) but we may obtain the right order of magnitude. This is clear in the case of (2.2.10) because according to (2.2.8) $I_{\frac{1}{2}}(\theta,\theta')$ and $H^2(\theta,\theta')$ are of the same order of magnitude. Moreover in the case of i.i.d. variables $h^2(\theta,\theta')$ can be computed in a simpler way.

3. THE REGULAR CASE

3.1. Let now θ be real taking on values in an open interval $(a,b)\subset R^1$. The following conditions are assumed:

$g(\theta)$ is differentiable in its domain (a,b),

$f_\theta^{(i)}(x_1,x_2,\ldots,x_i)$ $i=1,2,\ldots$ and $f_0(\underline{x})$ are differentiable with respect to θ for all $\underline{x}$,

the validity of the following relation will also be assumed; when $\theta'\to\theta$

$$\frac{1}{(\theta'-\theta)^2}\int\frac{(f_{\theta'}(\underline{x})-f_\theta(\underline{x}))^2}{(1-\alpha)f_\theta(\underline{x})+\alpha f_{\theta'}(\underline{x})}d\mu\to\int\frac{1}{f_\theta(\underline{x})}\left(\frac{\partial f_\theta(\underline{x})}{\partial\theta}\right)^2 d\mu\,, \tag{3.1.1}$$

the latter expression being the Fisher information for sequential procedure in the regular case.

Our relation (2.2.4) will now have the following form corresponding to the classical CFR inequality

$$\underline{D}_\theta^2(t(\underline{x}))\geqq\frac{g'^2(\theta)}{\int\frac{1}{f_\theta(\underline{x})}\left(\frac{\partial f_\theta(\underline{x})}{\partial\theta}\right)^2 d\mu}=\frac{g'^2(\theta)}{\int\frac{\sum_{i=1}^\infty I_{E_i}\left(\frac{\partial f_\theta^{(i)}(x_1,x_2,\ldots,x_i)}{\partial\theta}\right)^2}{\sum_{i=1}^\infty I_{E_i}f_\theta^{(i)}(x_1,x_2,\ldots,x_i)}d\mu} \tag{3.1.2}$$

where the relations $I_{E_i}^2=I_{E_i}$ and $I_{E_i}\cap I_{E_j}=\emptyset$ for $i\neq j$ were taken into consideration.

Turning to the i.i.d. case, the Fisher information has the form

$$J(\theta)=\int\frac{\sum_{i=1}^\infty I_{E_i}\left(\frac{\partial f_\theta^{(i)}(x_1,\ldots,x_i)}{\partial\theta}\right)^2}{\sum_{i=1}^\infty I_{E_i}f_\theta^{(i)}(x_1,\ldots,x_i)}d\mu=\int\frac{\sum_{i=1}^\infty I_{E_i}\left[\prod_{j=1}^i f_\theta(x_j)\sum_{k=1}^i\frac{\partial\log f_\theta(x_k)}{\partial\theta}\right]^2}{\sum_{i=1}^\infty I_{E_i}\prod_{j=1}^i f_\theta(x_j)}d\mu\,. \tag{3.1.3}$$

We now claim the following theorem:

THEOREM 3.1. *Using the above notations and assumptions the CFR inequality in the regular i.i.d. case has the form*

$$D^2(t(\underline{X}))\geqq\frac{g'^2(\theta)}{J_{\underline{X}}(\theta)} \tag{3.1.4}$$

moreover

$$J_{\underline{X}}(\theta)\leqq I_X(\theta)\underline{E}\,N\,, \tag{3.1.5}$$

where $I_X(\theta)$ *denotes the Fisher information belonging to a single observation.*

REMARK. (3.1.4) along with (3.1.5) implies that our CFR lower bound in the sequential case is better than the lower bound given by Wolfowitz [18].

PROOF OF THEOREM 3.1. Since (3.1.4) is a consequence of (3.1.2) and (3.1.3), it suffices to prove (3.1.5). For the numerator in the middle term in (3.1.3) we can now write

$$\sum_{i=1}^{\infty} I_{E_i}\left[\prod_{j=1}^{i} f_\theta(x_j)\sum_{k=1}^{i}\frac{\partial \ln f_\theta(x_k)}{\partial\theta}\right]^2$$

$$=\left[\sum_{i=1}^{\infty}\left\{I_{E_i}^{\frac{1}{2}}\sum_{k=1}^{i}\frac{\partial}{\partial\theta}\ln f_\theta(x_k)\sqrt{\prod_{j=1}^{i} f_\theta(x_j)}\right\}\sqrt{I_{E_i}\prod_{j=1}^{i} f\,(x_j)}\right]^2$$

$$\leq\left[\sum_{i=1}^{\infty} I_{E_i}\left(\sum_{k=1}^{i}\frac{\partial}{\partial\theta}\ln f_\theta(x_k)\right)^2\prod_{j=1}^{i} f_\theta(x_j)\right]\left[\sum_{i=1}^{\infty} I_{E_i}\prod_{j=1}^{i} f_\theta(x_j)\right] \tag{3.1.6}$$

where the Cauchy–Schwarz inequality was applied. Introducing the notation

$$S_N=\sum_{i=1}^{N}\frac{\partial}{\partial\theta}\ln f_\theta(x_i)=\sum_{i=1}^{\infty} I_{E_i}\sum_{j=1}^{i}\frac{\partial}{\partial\theta}\ln f_\theta(x_j)\ ,$$

we obtain from (3.1.3) and (3.1.6)

$$J_{\underline{X}}(\theta)\leq\sum_{i=1}^{\infty}\int_{E_i}\left(\sum_{k=1}^{i}\frac{\partial}{\partial\theta}\ln f_\theta(x_k)\right)^2\prod_{j=1}^{i} f_\theta(x_j)d\mu=\underline{E}\,S_N^2\ .$$

But from the above expression of S_N in the i.i.d. case we have

$$\underline{E}\,S_N^2=(\underline{E}\,N)\,\underline{E}\left[\frac{\partial}{\partial\theta}f_\theta(x)\right]^2=\underline{E}\,N\ I_X(\theta)\ .$$

Hence we establish the validity of Theorem 3.1.

3.2. Here we examine the *conditions for the equality in the regular case*. We begin with the remark that in the general inequality (2.2.4) we cannot have equality for certain (non-regular) distributions; this can be seen from the derivation of the original relation (see [15]). Now we shall prove the following

THEOREM 3.2. *In the case of the sequential CFR lower bound the equality is achieved only in the regular, exponential case.*

PROOF. Using the above notations (2.1.1), (2.1.2) we have

$$\underline{E}\,t(\underline{X})=g(\theta)=\int t(\underline{X})f_\theta(\underline{X})d\mu\ .$$

Under the regularity conditions we can obtain

$$g'^2(\theta) \leq D_\theta^2(t(\underline{x}))\int f_\theta(\underline{x})\left(\frac{\partial f_\theta(\underline{x})}{\partial\theta}\right)^2 d\mu .$$

Equality holds here whenever

$$t(\underline{x})-g(\theta) = \lambda(\theta)\frac{1}{f_\theta(\underline{x})}\frac{\partial f_\theta(x)}{\partial\theta}$$

or

$$[t(\underline{x})-g(\theta)]f_\theta(\underline{x}) = \lambda(\theta)\frac{\partial f_\theta(\underline{x})}{\partial\theta} . \qquad (3.2.1)$$

Due to the regularity conditions $f_\theta(\underline{x})$ is continuous in θ for all values of $\underline{x}$. Replacing now the expressions (2.1.1) and (2.1.2) into both sides of (3.2.1), we obtain

$$\left[\sum_{i=1}^{\infty} I_{E_i}(t_i(x_1,\ldots,x_i)-g(\theta)\right]\left[\sum_{i=1}^{\infty} I_{E_i} f^{(i)}(x_1,\ldots,x_i)\right]$$

$$= \lambda(\theta)\sum_{i=1}^{\infty} I_{E_i}\frac{\partial f_\theta^{(i)}(x_1,\ldots,x_i)}{\partial\theta}$$

or

$$\sum_{i=1}^{\infty} I_{E_i}(t_i(x_1,\ldots,x_i)-g(\theta))f^{(i)}(x_1,\ldots,x_i)$$

$$= \lambda(\theta)\sum_{i=1}^{\infty} I_{E_i}\frac{\partial f_\theta^{(i)}(x_1,\ldots,x_i)}{\partial\theta} ,$$

from which we can conclude for $i=1,2,\ldots$ the relation

$$t_i(x_1,\ldots,x_i)-g(\theta) = \lambda(\theta)\frac{\partial \ln f_\theta^{(i)}(x_1,\ldots,x_i)}{\partial\theta}$$

or

$$f_\theta^{(i)}(x_1,\ldots,x_i) = \exp\int\frac{t_i(x_1,\ldots,x_i)-g(\theta)}{\lambda(\theta)}d\theta + c^{(i)}(x_1,\ldots,x_i). \qquad (3.2.2)$$

The converse statement, i.e. that from (3.2.2) the equality in the corresponding sequential CFR inequality follows can be obtained in the usual way.

From our considerations it follows that *the CFR bound given in* (2.2.4) *can be sharp in the regular case only.*

4. THE CASE OF A TRANSLATION PARAMETER

4.1. Now let θ be a translation parameter which is to be estimated. As a consequence of (2.2.4), *for those unbiased estimators* $t(\underline{x})$ *the variance of which does not depend on* θ *the following relation holds:*

$$\underline{D}^2(t(\underline{x})) \geq \max_{\alpha} \sup_{\theta'} \alpha(1-\alpha)(\theta'-\theta)^2 \left\{\frac{1}{I_\alpha(\theta,\theta')} - 1\right\} . \tag{4.1.1}$$

Towards an application, let X be distributed according to the law $F(x-\theta)$, $-\infty < x < \infty$, where $F(x)$ is a distribution function for which $F(0)=0$, $F(1)=1$ having density function $f(x)$ which is positive for $x \in (0,1)$ and $f(x)=0$ for $x \notin (0,1)$. We shall assume that the observations $X_1, X_2, \ldots$ are independent random variables, i.e. $f_\theta^{(i)}(x_1,x_2,\ldots,x_i) = \prod_{j=1}^{i} f(x_j-\theta)$. We assume that the parameter θ is to be estimated; we claim the following:

THEOREM 4.1. *Using the above notations and assumptions the inequality*

$$D^2(t(\underline{X})) \geq \frac{1}{4} \max_{\delta} \delta^2 \frac{\underline{E}\left[2\varphi(\delta)^N - \frac{1}{2}(1-F(\delta))^N - \frac{1}{2}(F(1-\delta))^N\right]}{1 - \underline{E}\left[2\varphi(\delta)^N - \frac{1}{2}(1-F(\delta))^N - \frac{1}{2}(F(1-\delta))^N\right]} \tag{4.1.2}$$

holds, where

$$\varphi(\delta) = \int_\delta^1 \sqrt{f(x)f(x-\delta)}\,dx = 1 - \frac{1}{2} H_1^2(\delta) .$$

Our theorem implies the following

COROLLARY 4.1. *Whenever the function*

$$\psi(Z) = 2\varphi(\delta)^z - \frac{1}{2}(1-F(\delta))^z - \frac{1}{2}(F(1-\delta))^z$$

is convex function of $z>0$ *and* $0<\delta<\delta_0$ *for some* $\delta_0>0$, *then we can write*

$$D^2(t(x)) \geq \frac{1}{4} \max_{\delta} \delta^2 \frac{2\varphi(\delta)^{\overline{EN}} - \frac{1}{2}(1-F(\delta))^{\overline{EN}} - \frac{1}{2} F(1-\delta)^{\overline{EN}}}{1-2\varphi(\delta)^{\overline{EN}} + \frac{1}{2}(1-F(\delta))^{\overline{EN}} + \frac{1}{2}[F(1-\delta)]^{\overline{EN}}} \tag{4.1.3}$$

for $0<\delta<\delta_0$.

This statement is an immediate consequence of Jensen's inequality for convex functions.

PROOF OF THEOREM 4.1. The version (2.2.10) of our CFR inequality is applicable because the Hellinger expression is the same order of magnitude as our $I_{\frac{1}{2}}$. — Without loss of generality we can set $\theta=0$, $\theta'=\delta$, which implies that

$g(\theta')-g(\theta)=\delta$. This way we can have simplified notations:

$$h^2(\delta) = H^2(\delta) - \frac{1}{2} P_0(C) - \frac{1}{2} P_\delta(D) .$$

We have to determine these quantities for the sequential case when the observations are independent.

$$H^2(\delta) = \int (\sqrt{f_0(\underline{x})} - \sqrt{f_\delta(\underline{x})})^2 d\mu = 2(1 - \int \sqrt{f_0(\underline{x}) f_\delta(\underline{x})} d\mu)$$

$$= 2\left(1 - \sum_{i=1}^{\infty} \int_{I_{E_i}} \prod_{j=1}^{i} \sqrt{f(x_i) f(x_i - \delta)} d\mu\right) = 2\left[1 - \left(\int_\delta^1 \sqrt{f(x) f(x-\delta)} dx\right)^N\right]$$

$$= 2[1 - (1 - \frac{1}{2} H_1(\delta))^N] .$$

Further

$$P_0(C) = 1 - P_0(B) = 1 - \sum_{i=1}^{\infty} I_{E_i} P_0(B_i) = 1 - (1 - F(\delta))^N ,$$

similarly

$$P_\delta(D) = [F(1-\delta)]^N .$$

A slight calculation results in (4.1.2) and thus the proof of Theorem 4.1 and of Corollary 4.1 readily follow.

4.2. Here we should like to point out how useful our CFR bound is in the independent case. We shall consider (4.1.3), but our considerations are applicable for (4.1.2). We shall assume that $\delta_0 > \frac{1}{\underline{E}N^\alpha}$ when α is necessarily taking small values.

Namely beyond the values $F(\delta)$ and $F(1-\delta)$ we may obtain a bound which depends only on the measure of the linear sets:

$$B_1^{(0)} = \left\{x: f(x-\delta) = f(x)\right\} , \ B_1^{(1)} = \left\{x: \frac{f(x-\delta)}{f(x)} > 1\right\} \text{ and } B_1^{(2)} = \left\{x: \frac{f(x)}{f(x-\delta)} > 1\right\} .$$

Thus we obtain from the relations

$$\int_{B_1^{(1)}} \sqrt{\frac{f(x-\delta)}{f(x)}} dF(x) \geq P_0(B_1^{(1)}) \text{ and } \int_{B_1^{(2)}} \sqrt{\frac{f(x)}{f(x-\delta)}} dF(x-\delta) \geq P_\delta(B_1^{(2)})$$

the inequality

$$D^2(t(x)) \geq \frac{1}{4} \max \delta^2 \frac{2(P_0(B_1^{(1)})+P_\delta(B_1^{(2)})^{\underline{EN}} - \frac{1}{2}(1-F(\delta))^{\underline{EN}} - \frac{1}{2}[F(1-\delta)]^{\underline{EN}}}{1-2(P_0(B_1^{(1)})+P_\delta(B_1^{(2)})^{N} - \frac{1}{2}(1-F(\delta))^{\underline{EN}} - \frac{1}{2}[F(1-\delta)]^{\underline{EN}}} . \quad (4.1.4)$$

As an immediate consequence we have the following

COROLLARY 4.2. *Let* $f(x)$ *be monotonically non-decreasing (non-increasing) in* $[0,1]$, *then for any unbiased sequential estimator of* θ *the variance of which does not depend on* θ , *we have the inequality*

$$\begin{aligned} D^2(t(x)) &\geq c(\underline{E}N)\left[1-F^{-1}\left(1-\frac{1}{\underline{E}N}\right)\right]^2 \quad \text{if } f(x) \text{ is non-decreasing,} \\ D^2(t(x)) &\geq c(\underline{E}N)\left[F^{-1}\left(\frac{1}{\underline{E}N}\right)\right]^2 \quad \text{if } f(x) \text{ is non-increasing,} \end{aligned} \quad (4.1.5)$$

and for the quantity $c(\underline{E}N)$ *we have*

$$c(\underline{E}N) = \frac{1}{4} \frac{\left(1-\frac{1}{\underline{E}N}\right)^{\underline{E}N}}{1-\left(1-\frac{1}{\underline{E}N}\right)^{\underline{E}N}} < \frac{1}{4}\frac{1}{e-1} .$$

REMARK. Since $c(y)$ is an increasing function of y, we cannot replace in (4.1.5) $c(\underline{E}N)$ by $c(\infty)$; $\frac{1}{6,88} < c(\infty) < \frac{1}{6,87}$. On the other hand when we assume e.g. that $\underline{E}N \geq 20$ or $\underline{E}N \geq 50$, then we can apply the constants $c(20) = \frac{1}{7,158...} > \frac{1}{7,1581}$, or $c(50) = \frac{1}{6,9838...} > \frac{1}{6,9839}$, respectively. This means e.g. that

$$D^2(t(\underline{X})) \geq \frac{1}{7,1581} \frac{1}{(\underline{E}N)^2} , \text{ whenever } \underline{E}N \geq 20 .$$

We also have the following

COROLLARY 4.3. *Let* X_i , $i=1,2,...$ *be distributed uniformly in* $[\theta,\theta+1]$, $-\infty<\theta<\infty$. *In the case of independent observations we have that for any unbiased estimator* $t(\underline{X})$ *of* θ *the variance of which does not depend on* θ *the following inequality holds:*

$$D^2(t(\underline{X})) \geq c(\underline{E}N) \frac{1}{(\underline{E}N)^2} .$$

Here $c(\underline{E}N)$ agrees with the quantity occuring in Corollary 4.2. — Wald ([16] pp. 164-167; see also Govindarajulu [4] pp. 400-401) considered sequential estimation of the uniform translation parameter. Our result has some interest and relevance in this context.

Corollary 4.3 readily follows by setting $F(t)=t$, $0 \leq t \leq 1$ in Corollary 4.2. Concerning Corollary 4.2 this follows from (4.1.4) as in non-decreasing (increasing case) $B_1^{(2)}=\emptyset$ and $B_1^{(1)}=\{x: 0 \leq x \leq 1-\delta\}$ ($B_1^{(1)}=\emptyset$ and $B_1^{(2)} = \{x: \delta \leq x \leq 1\}$). In all these cases the choice was taken $\delta=1-F^{-1}(1-\frac{1}{\underline{E}N})$ and

$\delta = F^{-1}(\frac{1}{\underline{EN}})$, resp.

4.3. Relation (4.1.3) solves the problem generally and instead of considering special examples we pay attention to the order of magnitude in several cases. The quantities $F(\delta)$, $1-F(1-\delta)$ are crucial, i.e. the tail-characteristics, and $1-\frac{1}{2}H_1(\delta) = \int_\delta^1 \sqrt{f(x)f(x-\delta)}dx$ giving the behaviour of the middle part is also important. The largest of them will be competent for the order of magnitude in the mean-sample size. This way among $F(\delta)$ and $1-F(1-\delta)$ the more singular point has to be considered. Let this be $x=0$; Akahira [1] assumed that $f(x)$ is twice differentiable and in the neighbourhood of zero $f(x) \sim A \cdot x^{\alpha-1}$, $0<\alpha$ with $f'(x) \sim Bx^{\alpha-2}$; for $\alpha>2$ the boundedness of $f''(x)$ is also assumed. In this case he obtains for $0<\alpha<2$

$$I_{\frac{1}{2}}(\delta) \sim H_1^2(\delta) = O(\delta^\alpha) .$$

In his treatment the behaviour of $f(x)$ at $x=1$ is similar for $\beta>\alpha$ as at $x=1$. For $\alpha=\beta=1$, we have the order of magnitude for the CFR lower bound $(\underline{EN})^{-2}$.

Whenever $\alpha \geqq 2$ it can be seen easily that the Fisher information exists, i.e.

$$\lim_{\delta\to 0} \delta^{-2} I_{\frac{1}{2}}(\delta) = \lim_{\delta\to 0} \frac{1}{2\delta^2} \int_0^{1+\delta} \frac{(f(x)-f(x-\delta))^2}{f(x)+f(x-\delta)} dx = \frac{1}{4I_f} ,$$

I_f being the Fisher information. In this case $f(x-\delta)=f(x)-\delta f'(x)+o(\delta)$ which gives simply the above relation. For this case we can claim what was given in Section 3 on the regular case.

5. SOME COMMENTS

5.1. In fixed sample size case Vincze's (1979) result reduces to the classical CFR lower bound even in the i.i.d. case, while in the sequential case our bound is better than that given by Wolfowitz (1974).

5.2. Simons and Woodroofe (1983) obtain an inequality which holds for almost everywhere $\theta \subset (a,b)$; our general rule (2.2.4) is valid for all $\theta \in \Theta$. For the regular, i.i.d. case this is guaranteed by Assumption (3.1.1). (See also the condition $P_\theta(N<\infty)=1$, for all $\theta \subset \Theta$.) The first "counterexample" by Simons [3] fails to satisfy this requirement for $\theta \neq 0$.

5.3. The CFR relation for mixed variance given by Simons and Woodroofe [5] is a special case of the CFR relation given in [6] in the following sense: the authors restrict themselves to real parameter only, also to the value $\alpha=\frac{1}{2}$ instead of $0<\alpha<1$. Finally their bound is based on the Hellinger distance $H^2(\theta,\theta')$, hence it may be negative; at any case due to the relation

$$I_{\frac{1}{2}}(\theta,\theta') \leq H^2(\theta,\theta') - \frac{1}{2}(P_\theta(C)+P_{\theta'}(D)) \leq H^2(\theta,\theta') \leq 2$$

it is weaker than $I_{\frac{1}{2}}(\theta,\theta')$ or $h^2(\theta,\theta')$.

5.4. For our general results an arbitrary sequence of random variables $\{X_1,X_2,\ldots\}$ is considered, whereas authors (e.g. Linnik and Romanovski (1972), Ibrahimov and Khasminskii (1974), Simons and Woodroofe (1983)) usually assume i.i.d. sequences.

REFERENCES

[1] Akahira, M. (1975), Asymptotic theory for estimation of location in non-regular cases. I. Order of convergence of consistent estimators, Rep. Stat. Appl. JUSE, 22, 8-26.

[2] Cramér, H. (1946), *Mathematical Methods of Statistics*. Princeton University Press, Princeton.

[3] Fréchet, M. (1943), Sur l'extension de certaines evaluation statistiques au case de petits echantillons. *Rev. Inst. Internat. Statist.* 11. 182-205.

[4] Govindarajulu, Z. (1987), *The Sequential Statistical Analysis*. American Sciences Press Inc. Columbus, Ohio.

[5] Ibragimov, I. A. - R. Z. Khasminskii (1974), On sequential estimation of the location parameter for families of distributions with discontinuous densities. Theor. Probability Appl. 19 (1974) 669-682.

[6] Le Cam, L. (1986), *Asymptotic Methods in Statistical Decision Theory*, Springer-Verlag, Berlin, New York, Tokyo.

[7] Linnik, Yu. V. - I. V. Romanovsky (1972), Some new results in sequential estimation theory. Proceedings of the Sixth Berkeley Symposium on Math. Statist. and Probability. Vol. 1, 85-96.

[8] Khatri, C.G. (1980), Unified treatment of the Cramér-Rao bound for the non-regular density functions. *J. Statist. Plann. Inference*. 4, 75-79.

[9] Mőri, T. F. (1983), Note on the Cramér-Rao inequality in the non-regular case. The family of uniform distributions. *J. Statist. Plann. Inference*. 7, 353-358.

[10] Polfeldt, Th. (1970), Asymptotic results in non-regular estimation. *Skandinavisk Aktuarietidskrift* 1.-2. Supplement.

[11] Puri, M. L. - I. Vincze (1985), On the Cramér-Fréchet-Rao Inequality for Translation Parameter in the Case of Finite Support. Statistics. 16 (1985) 4, 495-506.

[12] Rao, C. R. (1945), Information and accuracy attainable in the estimation of statistical parameter. *Bull. Calcutta Math. Soc.* 37, 81-91.

[13] Simons, G. (1980), Sequential estimators and the Cramér-Rao lower bound. J. Statist. Planning and Inference. Vol. 4, 1980, pp. 67-74.

[14] Simons, G. - M. Woodroofe (1987), The Cramér-Rao inequality holds almost everywhere. - In *Recent Advances in Statistics* (Hermann Chernoff Volume), Academic Press, New York, 1987, pp. 69-93.

[15] Vincze I. (1979), On the Cramér-Fréchet-Rao inequality in the non-regular case. In *Contributions to Statistics*, Hajek Memorial Volume. Ed. Jana Jureckova. Academia, Prague, pp. 258-262.

[16] Wald, A. (1980), *Statisticsl Decision Functions*. Wiley, New York - London.

[17] Wijsman, R. A. (1973), On the attainement of the Cramér-Rao lower bound. Ann. Statist. 1. 538-542.

[18] Wolfowitz, J. (1947), The efficiency of sequential estimates and Wald's equation for sequential process. Ann. Math. Statist. 18 (1947), 215-230.

STATISTICAL DATA ANALYSIS AND INFERENCE
Y. Dodge (Editor)

WHITHER DELETE-K JACKKNIFING FOR SMOOTH STATISTICAL FUNCTIONALS?

Pranab Kumar SEN

Departments of Statistics and Biostatistics, University of North Carolina, Chapel Hill, NC 27599-3260, USA

The classical jackknifing based on a (without replacement) resampling scheme with deletion of one observation at a time serves the dual purpose of bias reduction and variance estimation. Delete-k jackknifing, for some $k > 1$, is a variant of this scheme. For not so smooth functionals, such delete-k jackknifing schemes may provide a consistent variance estimator while the classical jackknifing may fail to do so. On the other hand, in delete-k jackknifing, there may be a far greater computational complication. For these reasons, it is quite natural to inquire whether for a smooth statistical functionals, delete-k jackknifing is that desirable or not? In the light of second order asymptotic distributional representations, it is shown that for second order (compact) differentiable statistical functionals, for any k, small compared to n (the sample size), delete-k jackknifing behaves very similar to the classical one. Thus, for smooth functionals there may not be any incentive for delete-k jackknifing.

AMS *Subject Classifications: 60F05, 62E20, 62F12.*

1. INTRODUCTION

Consider a general *statistical functional* $\theta = T(F)$, where F is a distribution function (d.f.) defined on the real line R. Consider a sample $X_1,\ldots,X_n$ of n independent and identically distributed random variables (i.i.d.r.v.) with the d.f. F. Let F_n be the *sample (empirical) d.f.* based on $X_1,\ldots,X_n$. Since F_n is an efficient (and unbiased) estimator of F, a natural (nonparametric) estimator of θ is the functionals $T_n = T(F_n)$. Note that in general $T(\cdot)$ need not be a linear functional, and hence, the unbiasedness of F_n may not be transmitted on to that of T_n. In a majority of the cases, T_n is *biased*, and the expression for this bias (as well as the *mean square error*) may be quite involved; we also need suitable regularity conditions under which the mean square error and bias of T_n exist. To use T_n in drawing some statistical conclusions about θ, it may therefore be necessary to estimate the bias and mean square error from the given sample $\{X_1,\ldots,X_n\}$. The classical *jackknifing method* is a simple resampling scheme and it serves a dual purpose of reducing the possible bias of T_n and providing an estimator of its mean square error. It consists in identifying the n subsamples of size n-1 each (by eliminating one observation at a time from the base sample), and incorporating them in the formulation of the *pseudovariables* $T_{n,1},\ldots,T_{n,n}$ (to be precisely defined later on), which provide the *jackknifed estimator* $T^*_{n,1}(=(T_{n,1}+\ldots+T_{n,n})/n)$ and an estimator of its asymptotic mean square error; we denote the latter by $V^*_{n,1}(=(n-1)^{-1}\ \Sigma^n_{i=1}(T_{n,i} - T^*_{n,1})^2)$. When the bias of T_n has the leading term of the order n^{-1}, then under quite general regularity conditions, the leading term of bias of $T^*_{n,1}$ is of the order n^{-2}. Thus, the classical jackknifing reduces the bias of the original estimator T_n. Secondly, under similar regularity conditions, T_n and $T^*_{n,1}$ have the same asymptotic mean square (in the sense that $n^{\frac{1}{2}}(T_n-\theta)$ and $n^{\frac{1}{2}}(T^*_{n,1}-\theta)$ both have asymptotically the same normal distribution with 0 mean and variable σ_1^2), and $V^*_{n,1}$ consistently estimates σ_1^2. Thus, the classical jackknifing serves a dual purpose of bias reduction and variance estimation. However, the regularity conditions needed for achieving the dual objectives of jackknifing generally include some sort of *differentiability* of $T(\cdot)$ at F and some other related compactness conditions which may not hold universally. A simple example to illustrate this point is the following: Consider the population *p-quantile* $\theta=F^{-1}(p)$, for some $p\varepsilon(0,1)$, where the d.f. F may even be assumed to have a positive and finite density function $f(\theta)$. In this case, a natural estimator of θ is the sample p-quantile, $T_n=F_n^{-1}(p)$, and it is well known that $n^{\frac{1}{2}}(T_n-\theta)$ is asymptotically normal with 0 mean and variance $\sigma_1^2 = p(1-p)/f^2(\theta)$. Unfortunately, in this problem, $V^*_{n,1}$ does not converge (in probability or almost surely) to σ_1^2, but it has a nondegenerate limiting distribution with a centering constant σ_1^2. Hence, the second role (of variance estimation) of the classical jackknifing does not hold in this problem. A similar situation may arise in general with other functionals which may not be enough smooth (or differentiable). To broaden the scope of jackknifing so as to encompass some 'nonsmooth' functionals, a variant of the classical jackknifing method has been considered in the literature: For some $k\geq 1$, subsamples of size n-k are obtained by deleting k observations at a time from the base sample, and these subsample estimators are then incorporated in the formulation of the pseudovariables on which the jackknifed estimator of θ and its estimated mean square error rest. This is termed the *delete-k jackknifing method*. For $k>1$, there may be a difference in this setup depending on whether one would choose all possible $\binom{n}{k}$ subsamples of size n-k each from the base sample of [n/k] subsamples of size n-k each resulting from the deletion of distinct sets of k observations from the base sample. For the case of sample quantiles, either way, if k is not small but k/n is small, then a consistent estimator of σ_1^2 is provided by the delete-k jackknifing method. We may refer to Shao and Wu [15] and Shao [16] who have considered the usefulness of delete-k jackknifing for a class of nonsmooth functionals [in the context of variance estimation]. It may be noted that for $k = 1$, the number of subsamples of size $n-1$ each is equal to n, while for $k > 1$, this number is $O(n^k)$. Thus, from the computational aspect, delete-k jackknifing may be quite cumbrous, especially, when n and k are both large. For this reason, some workers (viz. Efron [2] and Ghosh et al. [4]) have advocated the use of *bootstrap methods* instead of delete-k jackknifing. For smooth functionals, none of these alternatives may not be that necessary. We plan to address this issue adequately in this study.

Under quite general regularity conditions (viz., Parr [8,9] and Sen [12]), for $T^*_{n,1}$, the delete-1 jackknifed version of T_n, as has been defined earlier,

$$(n-1)\{T_n - T^*_{n,1}\} \to T^*_2(F) \text{ almost surely (a.s.), as } n \to \infty, \tag{1.1}$$

where $T^*_2(F)$ is a suitable functional of F to be defined more precisely later on. Also, let $V^*_{n,1}$ be the usual (Tukey) jackknifed estimator of σ_1^2, the (asymptotic) mean square error of $n^{\frac{1}{2}}(T_n-\theta)$. Then, under the same regularity conditions,

$$V^*_{n,1} \to \sigma_1^2 \quad \text{a.s., as } n \to \infty. \tag{1.2}$$

Further, if $T_1(F;x)$ stands for the *influence function* corresponding to $T(F)$ and if we let $\bar{T}_{1n} = n^{-1} \Sigma_{i=1}^n T_1(F;X_i)$ [so that $n^{\frac{1}{2}}\bar{T}_{1n}$ is asymptotically normal with mean 0 and variance σ_1^2], then the following *second order asymptotic distributional representation (SOADR)* result holds (viz., Sen [12]):

$$n\{T^*_{n,1} - T(F) - \bar{T}_{1n}\} \overset{\mathcal{D}}{\to} \Sigma_{k\geq 1} \lambda_k\{Z_k^2 - 1\}, \tag{1.3}$$

where the Z_k are i.i.d.r.v.'s having the standard normal d.f., and the eigenvalues λ_k, $k \geq 1$, depend on the underlying d.f. F and the functional $T(\cdot)$. Note that

$$n^{\frac{1}{2}}\{T^*_{n,1} - \theta\} \underset{p}{\sim} n^{\frac{1}{2}}(T_n - \theta) \overset{\mathcal{D}}{\to} \mathcal{N}(0,\sigma_1^2), \tag{1.4}$$

even under weaker regularity conditions. Note further that (1.2) and (1.4) are first order properties, while (1.1) and (1.3) are second order ones.

The object of the present study is to focus on general delete-k jackknifed estimators $\{T^*_{n,k}, k \geq 1\}$ and the related versions of the Tukey variance estimators (of σ_1^2), and to examine how far these first and second order properties hold? In this context, we may note that for $k \geq 2$, one may have either a resampling scheme of $\lceil n/k \rceil$ deletion of distinct sets of k observations from the basic sample or the more natural case of $\binom{n}{k}$ possible subsamples of size n-k from the basic sample of size n. The first scheme has some arbitrariness in the partitioning, and we shall mainly consider the second case. Along with the preliminary notions, the proposed estimators are considered in

Section 2. The main results are then presented in Section 3 and their derivations are sketched in Section 4. The last section deals with some useful remarks and general conclusions.

2. PRELIMINARY NOTIONS

Note that the sample d.f. F_n is defined by

$$F_n(x) = n^{-1} \Sigma_{i=1}^{n} I(X_i \leq x), \; x \in R, \tag{2.1}$$

where I(A) stands for the indicator function of the set A. Thus, we have

$$\theta = T(F) \quad \text{and} \quad T_n = \hat{\theta}_n = T(F_n), \tag{2.2}$$

for a suitable functional T(·). Since we are dealing with statistical functionals, the delete-k jackknifed estimators may all be defined solely in terms of the sub-sample empirical d.f.s. Towards this, we let for every $\underset{\sim}{i} \in I$,

$$F_{n-k}^{(\underset{\sim}{i})}(x) = (n-k)^{-1} \Sigma_{j \in S_{n,k}^{(\underset{\sim}{i})}} I(X_j \leq x); \; S_{n,k}^{(\underset{\sim}{i})} = \underset{\sim}{n} \setminus \underset{\sim}{i}, \tag{2.3}$$

where

$$\underset{\sim}{i} = (i_1,\dots,i_k), \; \underset{\sim}{n} = \{1,\dots,n\} \text{ and } I = \text{set of all possible } \binom{n}{k} \; \underset{\sim}{i}. \tag{2.4}$$

Let then

$$T_{n-k}^{(\underset{\sim}{i})} = T(F_{n-k}^{(\underset{\sim}{i})}), \text{ for every } \underset{\sim}{i} \in I; \; \# \text{ of } I = \binom{n}{k}. \tag{2.5}$$

The pseusovariables generated by the delete-k jackknifing are defined as

$$T_{n,\underset{\sim}{i}}^{(k)} = k^{-1}\{n \, T_n - (n-k) \, T_{n-k}^{(\underset{\sim}{i})}\}, \quad \text{for } \underset{\sim}{i} \in I. \tag{2.6}$$

Then, the delete-k jackknifed estimator of θ is defined as

$$T_{n,k}^{*} = \binom{n}{k}^{-1} \Sigma_{\{\underset{\sim}{i} \in I\}} T_{n,\underset{\sim}{i}}^{(k)}\}, \text{ for } k > 1. \tag{2.7}$$

Side by side, we also introduce the variance functions

$$V_{n,k}^{*} = \binom{n}{k}^{-1} \Sigma_{\{\underset{\sim}{i} \in I\}} \{T_{n,i}^{(k)} - T_{n,k}^{*}\}^2, \quad \text{for } k \geq 1. \tag{2.8}$$

Next, we introduce some regularity conditions on T(·). Let A be a topological vector space (with respect to the topology generated by the sup-norm) and K be a class of compact subsets of A, such that every subset consisting of a single point belongs to K. We say that a functional τ(·) is *Hadamard-continuous* at $F \in A$, if

$$|\tau(G) - \tau(F)| \to 0 \quad \text{with } \|G-F\| \to 0 \text{ [uniformly on } G \in K \subseteq K], \tag{2.9}$$

where $\|G-F\|$ refers to the usual sup-norm [i.e., $\sup_x |G(x)-F(x)|$]. Next, as in Fernholz ([3], pp. 16-17), we say that τ(·) is *first order compact (Hadamard-) differentiable* at $F \in A$, if

$$\tau(G) = \tau(F+G-F) = \tau(F) + \int \tau_1(F;x)d[G(x)-F(x)] + R_1(F;G-F) \quad (2.10)$$

where

$$|R_1(F,G-F)| = o(\|G-F\|), \text{ uniformly in } G \in K, \quad (2.11)$$

and $\tau_1(F;x)$, the *first order compact derivative* of $\tau(\cdot)$ at F, may be so normalized that

$$\int \tau_1(F;x)dF(x) = 0. \quad (2.12)$$

Similarly, $\tau(\cdot)$ is *second order compact (Hadamard-) differentiable* at $F \in A$, if

$$\tau(G) = \tau(F) + \int \tau_1(F;x)d[G(x)-F(x)] + \quad (2.13)$$
$$\tfrac{1}{2}\int\int \tau_2(F;x,y)d[G(x) - F(x)]d[G(y)-F(y)] + R_2(F;G-F),$$

where

$$|R_2(F;G-F)| = o(\|G-F\|^2), \text{ uniformly in } G \in K, \quad (2.14)$$

and $\tau_2(F;\cdot)$, the *second order compact derivative* of $\tau(\cdot)$ at F, may be so normalized that

$$\int \tau_2(F;x,y)dF(y) = 0 = \int \tau_2(F;x,y)dF(x) \quad \text{a.e.} \quad (2.15)$$

Finally, we let

$$\tau_2^*(G) = \int \tau_2(G;x,x)dG(x), \; G \in K. \quad (2.16)$$

Other notations will be introduced as and when necessary.

3. THE MAIN RESULTS

For the delete-k jackknifing, we may either set k to be an arbitrary positive number or may even allow k $(=k_n)$ to increase with the sample size n, in such a way that as n increases,

$$k_n = O(n^{1-\eta}), \quad \text{for some } \eta > 0, \quad (3.1)$$

so that k_n is nondecreasing with $(n-1)/(n-k_n)$ converging to 1, as $n \to \infty$. First, we consider the following theorem relating to the first order properties.

THEOREM 3.1. *If* T(F) *is first order Hadamard-differentiable at* F *and* $T_1^{**}(G) = \int T_1^2(G,x)dG(x)$ $(<\infty)$ *is Hadamard-continuous at* F, *then for every* k *satisfying* (3.1), *defining the* $V_{n,k}^*$ *as in* (2.8), *we have* for $\sigma_1^2 = T_1^{**}(F)$,

$$k(n-1)(n-k)^{-1} V_{n,k}^* = V_{n,1}^* + o(1) \quad \text{a.s.} \quad (3.2)$$
$$= \sigma_1^2 + o(1) \quad \text{a.s., as } n \to \infty.$$

Thus, when adjusted by the scale factor $k(n-1)/(n-k)$, *the Tukey estimator of the variance in the delete-k jackknifing is strongly consistent for* σ_1^2.

THEOREM 3.2. *If* $T(F)$ *is second-order Hadamard-differentiable at* F *and* $T_2^*(\cdot)$ *defined by* (2.16) *is Hadamard-continuous at* F, *then*

$$(n-1)\{T_n - T_{n,k}^*\} \to (\tfrac{1}{2})T_2^*(F) \quad \text{a.s., as } n \to \infty. \tag{3.3}$$

Thus the second order bias property of the classical jackknifing is shared by delete-k jackknifing, for every $k \geq 1$. *Actually,* $(n-1)(T_{n,k}^* - T_{n,1}^*) = o(1)$ a.s., $n \to \infty$, *for every* k *satisfying* (3.1).

The next theorem relates to second order properties.

THEOREM 3.3. *Assume that*

$$E_F T_2^2(F;X_1,X_1) = \iint T_2^2(F;x,y)dF(x)dF(y) < \infty. \tag{3.4}$$

Then, under the hypothesis of Theorem 3.2, *for every* k *satisfying* (3.1),

$$n\{T_{n,k}^* - T(F) - \bar{T}_{1n}\} \xrightarrow{\mathcal{D}} \Sigma_{r\geq 1}\, \lambda_r\{Z_r^2 - 1\}, \tag{3.5}$$

where the Z_r *are i.i.d.r.v.'s with the standard normal d.f. and the eigenvalues* λ_r *correspond to the orthonormal functions* $\tau_r(\cdot)$, $r \geq 1$, defined by

$$\int T_2(F;x,y)\tau_r(x)dF(x) = \lambda_r\, \tau_r(y) \quad \text{a.e. } (F),\ r \geq 1; \tag{3.6}$$

$$\int \tau_r(y)\tau_s(y)dF(y) = 1 \ \text{ or } 0 \text{ accordings as } r = s \text{ or not.} \tag{3.7}$$

Thus, for any k *satisfying* (3.1), *delete-*k *jackknifing leads to the same SOADR as in the case of* $k = 1$.

4. PROOFS OF THE THEOREMS

Using (2.3), it readily follows that

$$\|F_{n-k}^{(\underset{\sim}{i})} - F_n\| \leq (k/n), \text{ with probability 1, for every } \underset{\sim}{i} \in I. \tag{4.1}$$

However, this result is not strong enough for our purpose (when we allow k to increase, subject to (3.1)). For this reason, we consider the following lemma.

Lemma 4.1. Whenever $k = k_n = o(n^{1-\eta})$, for some $\eta > 0$,

$$\binom{n}{k}^{-1} \Sigma_{\{\underset{\sim}{i}\in I\}} \|F_{n-k}^{(\underset{\sim}{i})} - F_n\|^2 = O(k(n-k)^{-2}) \text{ a.s., as } n \to \infty. \tag{4.2}$$

Proof. Parallel to (2.3), we define for every $\underset{\sim}{i} \in I$,

$$F_{\underset{\sim}{i},k}(x) = k^{-1}\, \Sigma_{j=1}^{k}\, I(X_{i_j} \leq x),\ x \in R. \tag{4.3}$$

Then, from (2.3) and (4.3), we obtain that

$$k^{-1}(n-k)\, \|F_{n-k}^{(\underset{\sim}{i})} - F_n\| = \|F_{\underset{\sim}{i},k} - F_n\|,\ \forall\, \underset{\sim}{i} \in I. \tag{4.4}$$

Thus, the left hand side of (4.2) can be written as

$$k(n-k)^{-2}\{\binom{n}{k}^{-1}\Sigma_{\{\underset{\sim}{i}\varepsilon I\}}k\|F_{\underset{\sim}{i},k}-F_n\|^2\} \tag{4.5}$$

$$\leq 2k(n-k)^{-2}\{\binom{n}{k}^{-1}\Sigma_{\{\underset{\sim}{i}\varepsilon I\}}k\|F_{\underset{\sim}{i},k}-F\|^2+(k/n)n\|F_n-F\|^2\}.$$

Now, by the classical results on the Kolmogorov-Smirnov goodness of fit statistic,

$$n\|F_n-F\|^2 = O((\log\log n)) \quad \text{a.s., as } n\to\infty, \tag{4.6}$$

while $k/n = o(n^{-\eta})$, so that $k\|F_n-F\|^2 = o(1)$ a.s., as $n\to\infty$. Further, we may identify $\binom{n}{k}^{-1}\Sigma_{\{\underset{\sim}{i}\varepsilon I\}}k\ \|F_{\underset{\sim}{i},k}-F\|^2$ as the U-statistic (Hoeffding [5]) corresponding to the kernel $\phi_k(X_1,\dots,X_k) = k\|F_k-F\|^2$. Using the basic results of Kiefer [7] [on the finite-sample behavior of the Kolmogorov-Smirnov type statistics], it follows that for every finite k (≥ 1),

$$E_F\{\phi_k(X_1,\dots,X_k)\} \leq D = \textstyle\int_0^\infty 4x\exp\{-2x^2\}dx = 1, \tag{4.7}$$

and for every finite integer r (≥ 1),

$$E_F\{[\phi_k(X_1,\dots,X_k)]^{2r}\} \leq c_r < \infty, \text{ uniformly in } k\geq 1. \tag{4.8}$$

As such, using the results in Sen and Ghosh [14] (viz., Sen [11], p. 56), it follows that for every (fixed) positive integer r, the 2r-th central moment of

$\binom{n}{k}^{-1}\Sigma_{\underset{\sim}{i}\varepsilon I}k\|F_{\underset{\sim}{i},k}-F\|^2$ is $O((k/n)^r) = o(n^{-r\eta})$, for every $k = o(n^{1-\eta})$, $\eta>0$.

Combining this with the Markov inequality and the Borel-Cantelli lemma, we immediately obtain by choosing $r: r\eta>1$, that for $k = o(n^{1-\eta})$, for some $\eta>0$,

$$\binom{n}{k}^{-1}\Sigma_{\{\underset{\sim}{i}\varepsilon I\}}k\|F_{\underset{\sim}{i},k}-F\|^2 = O(1) \text{ a.s., as } n\to\infty, \tag{4.9}$$

Hence, (4.2) follows from (4.5) and the above discussions. Q.E.D.

<u>Lemma 4.2.</u> For any $k: k = o(n^{1-\eta})$, for some $\eta>0$,

$$\binom{n}{k}^{-1}\Sigma_{\{\underset{\sim}{i}\varepsilon I\}}\|F_{n-k}^{(\underset{\sim}{i})}-F_n\| = O(k^{\frac{1}{2}}(n-k)^{-1}) \quad \text{a.s., as } n\to\infty. \tag{4.10}$$

The proof follows directly from (4.5) and the elementary moment inequality, and hence, is omitted.

Let us now start with the proofs of the theorems. First, by an appeal to (2.10)-(2.12), for $G = F_{n-k}^{(\underset{\sim}{i})}$ and $F = F_n$, we immediately obtain that for first order differentiable $T(\cdot)$,

$$T_{n-k}^{(\underset{\sim}{i})} = T(F_{n-k}^{(\underset{\sim}{i})}) = T(F_n + F_{n-k}^{(\underset{\sim}{i})} - F_n) \tag{4.11}$$

$$= T(F_n) + \int T_1(F_n;x)dF_{n-k}^{(\underset{\sim}{i})}(x) + R_1(F_n;\, F_{n-k}^{(\underset{\sim}{i})} - F_n),\ \forall \underset{\sim}{i} \in I,$$

where

$$|R_1(F_n;\, F_{n-k}^{(\underset{\sim}{i})} - F_n)| = o(\|F_{n-k}^{(\underset{\sim}{i})} - F_n\|),\ \forall \underset{\sim}{i} \in I. \tag{4.12}$$

In view of (2.12), the second term on the right hand side of (4.11) can be expressed as $-(n-k)^{-1}\,\Sigma_{j=1}^k\, T_1(F_n;X_{i_j})$. Therefore, by (2.6) and (4.11), we have

$$T_{n,\underset{\sim}{i}}^{(k)} = T_n + k^{-1}\,\Sigma_{j=1}^k\, T_1(F_n;X_{i_j}) + o(k^{-1}(n-k)\|F_{n-k}^{(i)} - F_n\|),\ \underset{\sim}{i} \in I. \tag{4.13}$$

From (2.7), (4.10) and (4.13), we obtain that

$$T^*_{n,k} = T_n + 0 + o(k^{-\frac{1}{2}}) \quad \text{a.s., as } n \to \infty; \tag{4.14}$$

in this context, it may be noted that by writing $\bar{T}_{1,\underset{\sim}{i}}^{(n)} = k^{-1}\,\Sigma_{j=1}^k T_1(F_n;X_{i_j}), \underset{\sim}{i}\in I$,

$$\binom{n}{k}^{-1}\,\Sigma_{\{\underset{\sim}{i}\in I\}}\,\bar{T}_{1,\underset{\sim}{i}}^{(n)} = n^{-1}\,\Sigma_{i=1}^n\, T_1(F_n;X_i) = \int T_1(F_n;x)dF_n(x) = 0, \tag{4.15}$$

[by (2.12)], which accounts for the second term on the right hand side of (4.14). It follows similarly that

$$\binom{n}{k}^{-1}\,\Sigma_{\{\underset{\sim}{i}\in I\}}\,(\bar{T}_{1,\underset{\sim}{i}}^{(n)})^2 = \binom{n}{k}^{-1}k^{-2}\,\Sigma_{\{\underset{\sim}{i}\in I\}}[\Sigma_{j=1}^k\, T_1(F_n;X_{i_j})]^2$$

$$= \binom{n}{k}^{-1}\,k^{-2}\,\{\Sigma_{\{\underset{\sim}{i}\in I\}}[\Sigma_{j=1}^k\, T_1^2(F_n;X_{i_j}) + \Sigma_{j\neq j'=1}^k\, T_1(F_n;X_{i_j})T_1(F_n;X_{i_{j'}})]\}$$

$$= (nk)^{-1}\,\Sigma_{i=1}^n\, T_1^2(F_n;X_i) + [(k-1)/kn(n-1)]\,\Sigma_{i\neq j=1}^n\, T_1(F_n;X_i)T_1(F_n;X_j)$$

$$= (kn)^{-1}\,\Sigma_{i=1}^n\, T_1^2(F_n;X_i) - [(k-1)/kn(n-1)]\,\Sigma_{i=1}^n\, T_1^2(F_n;X_i) \quad [\text{by } (4.15)]$$

$$= [(n-k)/k(n-1)]\; n^{-1}\,\Sigma_{i=1}^n\, T_1^2(F_n;X_i)$$

$$= [(n-k)/k(n-1)]\,\{\int T_1^2(F_n;x)dF_n(x)\}$$

$$= [(n-k)/k(n-1)]\; T_1^{**}(F_n)\,. \tag{4.16}$$

where $\|F_n - F\| \to 0$ a.s., as $n \to \infty$, and hence, by the assumed Hadamard-continuity of $T_1^{**}(\cdot)$, we obtain that $T_1^{**}(F_n) \to T_1^{**}(F) = \sigma_1^2$ a.s., as $n \to \infty$. Thus, from (4.16), it follows that

$$\binom{n}{k}^{-1}\,\Sigma_{\{\underset{\sim}{i}\in I\}}\,(\bar{T}_{n,\underset{\sim}{i}}^{(n)})^2 = [(n-k)/k(n-1)]\,\sigma_1^2 + o(k^{-1}) \text{ a.s.,} \quad \text{as } n \to \infty. \tag{4.17}$$

Next, from (4.13) and (4.14), we obtain that

$$\binom{n}{k}^{-1} \Sigma_{\{\underset{\sim}{i}\varepsilon I\}}\{T_{n,\underset{\sim}{i}}^{(k)} - T_{n,k}^{*} + \bar{T}_{1,\underset{\sim}{i}}^{(n)}\}^2 = o(k^{-1} + \frac{n-k}{k}\binom{n}{k}^{-1} \Sigma_{\underset{\sim}{i}\varepsilon I} \|F_{n-k}^{(i)} - F_n\|^2)$$

$$= o(k^{-1} + O((n-k)^{-1})) \quad \text{a.s. [by Lemma 4.1]} \tag{4.18}$$

$$= o(k^{-1}) \quad \text{a.s., as } n \to \infty, \text{ as } k = o(n^{1-\eta}), \text{ for some } \eta > 0.$$

At this stage, we may recall that (viz., Lemma 1 of Sen [10]) if $\{X_{Ni}\}$ and $\{Y_{Ni}\}$ are two sequences of r.v.'s (not necessarily independent), such that (i) $N^{-1} \Sigma_{i=1}^{N} X_{Ni}^2 \to A(<\infty)$ a.s. and (ii) $N^{-1} \Sigma_{i=1}^{N}(X_{Ni}-Y_{Ni})^2 \to 0$ a.s., as $N \to \infty$, then $N^{-1} \Sigma_{i=1}^{N} Y_{Ni}^2 \to A$ a.s., as $N \to \infty$. Letting $N = \binom{n}{k}$, we obtain from (2.8), (4.17) and (4.18) that under the hypothesis of Theorem 3.1, as $n \to \infty$,

$$k(n-1)(n-k)^{-1} V_{n,k}^{*} \to \sigma_1^2 \quad \text{a.s., for every } k\text{: } k = o(n^{1-\eta}), \text{ for some } \eta > 0. \tag{4.19}$$

This completes the proof of Theorem 3.1.

Next, we consider the proof of (3.3). By using (2.13)-(2.15), we obtain that

$$T_{n-k}^{(\underset{\sim}{i})} = T_n - (n-k)^{-1} \Sigma_{j=1}^{k} T_1(F_n;X_{i_j}) + [2(n-k)^2]^{-1} \Sigma_{j=1}^{k} \Sigma_{\ell=1}^{k} T_2(F_n;X_{i_j},X_{i_\ell})$$

$$+ o(\|F_{n-k}^{(\underset{\sim}{i})} - F_n\|^2), \ \forall \underset{\sim}{i} \in I. \tag{4.20}$$

Further, note that by virtue of (2.15), $\Sigma_{j=1}^{n} T_2(F_n;X_i,X_j) = 0$, for every $i(=1,\ldots,n)$, and hence, by (2.7), (4.20) and Lemma 4.1, we have

$$T_{n-k}^{*} = T_n - 0 - [2(n-k)]^{-1} \{n^{-1} \Sigma_{i=1}^{n} T_2(F_n;X_i,X_i)\} -$$

$$(k-1)[2(n-k)]^{-1} \{n(n-1)\}^{-1} \{\Sigma_{i\neq j=1}^{n} T_2(F_n;X_i,X_j)\} + o((n-k)^{-1}) \text{ a.s.}$$

$$= T_n - [2(n-k)]^{-1} \{1 - (k-1)/(n-1)\} \{n^{-1} \Sigma_{i=1}^{n} T_2(F_n;X_i,X_i)\} + o((n-k)^{-1})$$

$$= T_n - [2(n-1)]^{-1} \int T_2(F_n;x,x)dF_n(x) + o((n-k)^{-1}) \text{ a.s., as } n\to\infty, \tag{4.21}$$

so that

$$(n-1)\{T_n-T_{n-k}^{*}\} = \tfrac{1}{2} \int T_2(F_n;x,x)dF_n(x) + o(n/(n-k)) \quad \text{a.s.}$$

$$= \tfrac{1}{2} T_2^{*}(F_n) + o(1) \text{ a.s., as } n\to\infty, \ \forall k = o(n^{1-\eta}), \eta > 0. \tag{4.22}$$

Again $\|F_n-F\| \to 0$ a.s., as $n\to\infty$, while by the assumed Hadamard-continuity of $T_2^{*}(\cdot)$ at F, $T_2^{*}(F_n) \to T_2^{*}(F)$ a.s., as $n\to\infty$. This completes the proof of (3.3).

Finally, to prove (3.5), we note that by virtue of (3.3),

$$(n-1)\{T_{n,k}^{*} - T_{n,1}^{*}\} \to 0 \text{ a.s., as } n\to\infty, \forall k\text{: } k = o(n^{1-\eta}), \ \eta > 0. \tag{4.23}$$

As such, it suffices to show that (3.5) holds for the particular case of k=1, and this has already been proved in Theorem 2.2 of Sen (1988). Hence, we omit the details here.

5. CONCLUDING REMARKS.

From Theorems 3.1 and 3.2 it is quite clear that if instead of the classical jackknifing, one uses the delete-k jackknifing procedure for some $k \geq 1$, as regards the two estimators $T^*_{n,1}$ and $T^*_{n,k}$ are concerned, by virtue of (4.22), they are a.s. equivalent upto the order n^{-1}. Thus, upto the order n^{-1}, there is no change in the bias reduction picture due to higher order delete-k jackknifing. In practice, bias of the order n^{-2} (or higher) are of not much interest. As regards the Tukey form of the variance estimators are concerned, for delete-k jackknifing, one needs the adjustment factor $k(n-1)/(n-k)$, and with that (3.2) ensures their asymptotic (a.s.) equivalence upto the first order. On the other hand, from the computational aspect, delete-k jackknifing involves labor of the order n^k, and hence, with larger values of k, the computational complexities may increase rather abruptly. From all these considerations, it may be concluded that if jackknifing is to be considered for bias reduction and variance estimation for an estimator of a smooth functional T(F), then there is not much attraction for adaptation of delete-k jackknifing. The picture may, however, change considerably if we deal with so called non-smooth functionals where (2.11) or (2.14) may not hold.

In the absence of (2.11) or (2.14), the present analysis may not work out, and a different method of attack is therefore necessary to study the relative picture. For nonsmooth functionals, SOADR results are of different types (viz., Jurečkova and Sen [6]) and entail (generally) slower rates of convergence [viz., $n^{-3/4}$ instead of n^{-1}, for quantiles]. In the particular case of a sample quantile, the celebrated Bahadur [1] representation of sample quantiles provides an alternative way for the study of consistency of $V^*_{n,k}$ when k is large (but k/n is small). In this case, Ghosh et al. [4] have provided a rigorous proof of the consistency of (Efron's [2]) bootstrap estimator of the mean square error; their proof rests on an additional (although very mild) moment condition. For the delete-k jackknifed estimator, Shao and Wu [15] and Shao [16] have provided parallel results. Sen [13] has recently considered various resampling plans in the context of estimating a statistical functional. It follows from his findings that for smooth functionals, there may not be any particular advantage of Efron's [2] bootstrapping over the classical jack-knifing method. Even for smooth functionals, in the multiparameter case, for the Stein-rule or shrinkage estimation problem, neither jackknifing nor boot-strapping has any utility for large sample sizes! Incidentially, even for smooth functionals, their Stein-rule versions are generally not smooth, and this causes the main difficulty. From this perspective, for nonsmooth functionals, even when delete-k jackknifing and/or bootstrap methods may work out, there remains a genuine concern: Is there any optimal procedure (even in an asymptotic sense)? More work along this line is needed to draw a neat picture.

REFERENCES

[1] Bahadur, R.R. (1966). A note on quantiles for large samples. *Ann. Math. Statist. 37*, 577-580.

[2] Efron, B. (1982). *The Jackknife, the Bootstrap and Other Resampling Methods.* (SIAM Publication, Philadelphia).

[3] Fernholz, L.T. (1983). *Von Mises Calculus for Statistical Functionals.* (Lecture Notes in Statistics: 19; Springer-Verlag, New York.)

[4] Ghosh, M., Babu, G.J., Parr, W.C. and Singh, K. (1984). A note on bootstrapping the sample median. *Ann. Statist. 12,* 1130-1135.

[5] Hoeffding, W. (1948). A class of statistics with asymptotically normal distribution. *Ann. Math. Statist. 19,* 293-325.

[6] Jurečkova, J. and Sen, P.K. (1987). A second order asymptotic distributional representation of M-estimators with discontinuous score functions. *Ann. Probability 15,* 814-823.

[7] Kiefer, J. (1961). On large deviations of the empiric D.F. of vector chance variables and a law of iterated logarithm. *Pacific. J. Math., 11,* 649-660.

[8] Parr, W.C. (1983). A note on the jackknife, the bootstrap and delta method estimators of bias and variance. *Biometrika 70,* 719-722.

[9] Parr, W.C. (1985). Jackknifing differentiable statistical functions. *J. Roy. Statist. Soc., B 47,* 56-66.

[10] Sen, P.K. (1960). On some convergence properties of U-statistics. *Calcutta Statist. Assoc. Bull. 10,* 1-18.

[11] Sen, P.K. (1981). *Sequential Nonparametrics: Invariance Principles and Statistical Inference.* John Wiley, New York.

[12] Sen, P.K. (1988). Functional jackknifing: Rationality and general asymptotics. *Ann. Statist. 16,* 450-469.

[13] Sen, P.K. (1988). Functional approaches in resampling plans: A review of some recent developments. *Sankhyā, Ser. A 50,* in press.

[14] Sen, P.K. and Ghosh, M. (1981). Sequential point estimation of estimable parameters based on U-statistics. *Sankhya, Ser. A 43,* 331-344.

[15] Shao, J. and Wu, C.F.J. (1987). A general theory for jackknife variance estimation. Tech. Rep., Univ. Wisconsin, Dept. of Statistics.

[16] Shao, J. (1988). Consistency of the jackknife estimators of the variances of sample quantiles. *Commun. Statist. Theor. Math. 17,* 3017-3027.

Part Three
DISTRIBUTIONS, WEIGHTED DISTRIBUTIONS AND CHARACTERIZATION

While the Rose blows along the River Brink,
With old Khayyám the Ruby Vintage drink:
And when the Angel with his darker Draught
Draws up to Thee-take that, and do not shrink.

Khayyam Naishapuri-Rubáiyát

STATISTICAL DATA ANALYSIS AND INFERENCE
Y. Dodge (Editor)

THREE USEFUL EXPRESSIONS FOR EXPECTATIONS INVOLVING A WISHART MATRIX AND ITS INVERSE

George P.H. STYAN
Department of Mathematics and Statistics
McGill University, Montréal, Québec, Canada

Concise expressions in matrix notation are obtained for the expectations of $\mathbf{SAS}$, $\mathbf{SAS}^{-1}$ and $\mathbf{S}^{-1}\mathbf{AS}^{-1}$, where the random matrix $\mathbf{S}$ follows a central Wishart distribution and the nonrandom matrix $\mathbf{A}$ need not be symmetric. These results extend some special cases obtained by Das Gupta (1968) and have proved to be useful in finding the second-order moments of the maximum likelihood estimators in a multivariate normal population with missing data on several variates (Legault-Giguère 1974; Giguère and Styan, 1978). We also obtain the variance of the trace $\mathrm{tr}\mathbf{S}_1^{-1}\mathbf{S}_2$, where $\mathbf{S}_1$ and $\mathbf{S}_2$ are independently distributed central Wishart matrices with the same associated covariance matrix, and we show that each element of (or bilinear form in) $\mathbf{S}_1+\mathbf{S}_2$ is uncorrelated with each element of (or bilinear form in) $\mathbf{S}_1^{-1}\mathbf{S}_2$; that $\mathbf{S}_1+\mathbf{S}_2$ and $\mathbf{S}_1^{-1}\mathbf{S}_2$ are not independent, however, follows from a result of Olkin and Rubin (1964).

1. PRELIMINARIES

Our main purpose in this paper is to present concise expressions in matrix notation for the expectations of $\mathbf{SAS}$, $\mathbf{SAS}^{-1}$ and $\mathbf{S}^{-1}\mathbf{AS}^{-1}$, where the random matrix $\mathbf{S}$ follows a central Wishart distribution and the nonrandom matrix $\mathbf{A}$ need not be symmetric. We also obtain the variance of the trace $\mathrm{tr}\mathbf{S}_1^{-1}\mathbf{S}_2$, where $\mathbf{S}_1$ and $\mathbf{S}_2$ are independently distributed central Wishart matrices with the same associated covariance matrix, and we show that each element of (or bilinear form in) $\mathbf{S}_1+\mathbf{S}_2$ is uncorrelated with each element of (or bilinear form in) $\mathbf{S}_1^{-1}\mathbf{S}_2$; that $\mathbf{S}_1+\mathbf{S}_2$ and $\mathbf{S}_1^{-1}\mathbf{S}_2$ are not independent, however, follows from a result of Olkin and Rubin (1964). The main results in this paper were announced by Styan (1973) and included in Appendix B of the M.Sc. thesis by Legault-Giguère (1974). See also Giguère and Styan (1978).

The central Wishart distribution was introduced by Wishart (1928); following Anderson (1984, Section 7.2), we have the following:

Definition. *Let the $p \times 1$ random vectors $\mathbf{z}_1, \mathbf{z}_2, ..., \mathbf{z}_k$ be independently and identically distributed $\mathbf{N}_p(\mathbf{0}, \mathbf{V})$, where the nonnegative definite covariance matrix $\mathbf{V}$ is possibly singular. Then the random matrix $\mathbf{S}$ is said to follow the central Wishart distribution $\mathbf{W}_p(\mathbf{V}, k)$, with associated covariance matrix $\mathbf{V}$ and k degrees of freedom.*

If we assemble the random vectors $\mathbf{z}_1, \mathbf{z}_2, ..., \mathbf{z}_k$ into the $p\times k$ matrix $\mathbf{Z}' = (\mathbf{z}_1, ..., \mathbf{z}_k)$, then

$$\mathbf{S} = \mathbf{Z}'\mathbf{Z} = \sum_{\alpha=1}^{k} \mathbf{z}_\alpha \mathbf{z}_\alpha' \sim \mathbf{W}_p(\mathbf{V}, k). \tag{1}$$

We begin with the following lemma due to Ogasawara and Takahashi (1951).

Lemma 1. *Let $\mathbf{A}$ be a nonrandom $k\times k$ symmetric matrix. Then $\mathbf{S} = \mathbf{Z}'\mathbf{AZ} \sim \mathbf{W}_p(\mathbf{V}, r)$, with $\mathbf{V}$ possibly singular and $\mathbf{Z}$ defined as in* (1) *above, if and only if $\mathbf{A} = \mathbf{A}^2$, and then $r = \mathrm{rank}(\mathbf{A}) = \mathrm{tr}\mathbf{A}$, and $\mathcal{E}(\mathbf{S}) = r\mathbf{V}$.*

We will also use the following results:

Lemma 2. *If* $\mathbf{S} \sim \mathbf{W}_p(\mathbf{V}, k)$ *with* $\mathbf{V}$ *possibly singular, and if* $\mathbf{A}$ *is a* $q \times p$ *nonrandom matrix, then* $\mathbf{ASA'} \sim \mathbf{W}_q(\mathbf{AVA'}, k)$.

Lemma 3. *Let* $\mathbf{S} = \{s_{ij}\} \sim \mathbf{W}_p(\mathbf{V}, k)$ *with* $\mathbf{V}$ *nonsingular and* $k > p-1$, *and let* σ^{pp} *and* s^{pp} *denote the* (p, p) *elements of* $\mathbf{V}^{-1}$ *and* $\mathbf{S}^{-1}$, *respectively. Then* σ^{pp}/s^{pp} *is distributed as* χ^2_{k-p+1}, *independently of* s_{ij}; $i, j = 1, \ldots, p-1$.

For a proof of Lemma 3 see the proof of Lemma B4 in Legault-Giguère (1974). We now have:

Corollary 1. *Let* $\mathbf{S} \sim \mathbf{W}_p(\mathbf{V}, k)$ *with* $\mathbf{V}$ *nonsingular and* $k > p-1$. *Then for any nonrandom vector* $\mathbf{a} \neq \mathbf{0}$,

$$\mathbf{a'V^{-1}a/a'S^{-1}a} \sim \chi^2_{k-p+1}\,. \tag{2}$$

Proof. There exists a nonsingular matrix $\mathbf{A}$ with $\mathbf{A}^{-1}\mathbf{e}_p = \mathbf{a}$. Since $\mathbf{ASA'} \sim \mathbf{W}_p(\mathbf{AVA'}, k)$ the result (2) follows directly from Lemma 3. ❐

Corollary 2. *Let* $\mathbf{S} \sim \mathbf{W}_p(\mathbf{I}, k)$ *with* $k > p-1$. *Then* $1/s^{pp} \sim \chi^2_{k-p+1}$.

Corollary 3. *Let* $\mathbf{S} \sim \mathbf{W}_p(\mathbf{I}, k)$ *with* $k > p+3$. *Then* $\mathcal{E}[(s^{pp})^2] = (k-p-1)^{-1}(k-p-3)^{-1}$. (3)

The following result was proved by Das Gupta (1968).

Lemma 4. *Let* $\mathbf{S} \sim \mathbf{W}_p(\mathbf{I}, k)$, *with* $k > p+1$. *Then* $\mathcal{E}(\mathbf{S}^{-1}) = (k-p-1)^{-1}\mathbf{I}$.

Lemma 5. *Let* $\mathbf{S} \sim \mathbf{W}_p(\mathbf{V}, k)$, *with* $\mathbf{V}$ *nonsingular and* $k > p+1$. *Then* $\mathcal{E}(\mathbf{S}^{-1}) = (k-p-1)^{-1}\mathbf{V}^{-1}$.

Proof. Let $\mathbf{V} = \mathbf{TT'}$. Then by Lemma 2 $\mathbf{H} = \mathbf{T}^{-1}\mathbf{S}(\mathbf{T'})^{-1} \sim \mathbf{W}(\mathbf{I}, k)$, and so

$$\mathcal{E}(\mathbf{S}^{-1}) = \mathcal{E}[(\mathbf{T'})^{-1}\mathbf{H}^{-1}\mathbf{T}^{-1}] = (\mathbf{T'})^{-1}\mathcal{E}(\mathbf{H}^{-1})\mathbf{T}^{-1} = (\mathbf{T'})^{-1}(k-p-1)^{-1}\,\mathbf{T}^{-1} = (k-p-1)^{-1}\mathbf{V}^{-1},$$

using Lemma 4, which completes the proof. ❐

2. MAIN RESULTS

Our objective now is to derive closed form expressions for

$$\mathcal{E}(\mathbf{SAS}),\ \mathcal{E}(\mathbf{SAS}^{-1}),\ \text{and}\ \mathcal{E}(\mathbf{S}^{-1}\mathbf{AS}^{-1}),$$

where $\mathbf{S} \sim \mathbf{W}_p(\mathbf{V}, k)$ and $\mathbf{A}$ is a square nonrandom matrix, not necessarily symmetric. The special cases where $\mathbf{A} = \mathbf{I}$ or is symmetric rank one have been derived by Das Gupta (1968). Our approach uses a result in Olkin and Rubin (1962).

Lemma 6. *Let* $\mathbf{S} = \{s_{ij}\} \sim \mathbf{W}_p(\mathbf{I}, k)$, *with* $k > p+3$, and let $\mathbf{S}^{-1} = \{s^{\ell m}\}$. *Then for* $i, j, \ell, m = 1, \ldots, p$,

$$\text{(i)} \quad \mathcal{E}(s_{ij}s_{\ell m}) = b_1\delta_{ij}\delta_{\ell m} + b_2(\delta_{i\ell}\delta_{jm} + \delta_{im}\delta_{j\ell})$$

$$\text{(ii)} \quad \mathcal{E}(s_{ij}s^{\ell m}) = c_1\delta_{ij}\delta_{\ell m} + c_2(\delta_{i\ell}\delta_{jm} + \delta_{im}\delta_{j\ell})$$

$$\text{(iii)} \quad \mathcal{E}(s^{ij}s^{\ell m}) = d_1\delta_{ij}\delta_{\ell m} + d_2(\delta_{i\ell}\delta_{jm} + \delta_{im}\delta_{j\ell}),$$

where b_1, b_2, c_1, c_2, d_1 *and* d_2 *are scalar constants and* δ_{ij} *is the Kronecker delta.*

Proof. Since $\mathbf{S} \sim \mathbf{QSQ}'$ for any $p \times p$ orthogonal matrix $\mathbf{Q}$, Lemma 6 follows directly from (6) on page 1274 of Olkin and Rubin (1962). ❐

Corollary 4. *Let* $\mathbf{S} \sim \mathbf{W}_p(\mathbf{I}, k)$ *and suppose that* $\mathbf{A}$ *is a* $p \times p$ *nonrandom matrix, not necessarily symmetric, and that the scalar constants* b_1, b_2, c_1, c_2, d_1 *and* d_2 *are as defined in Lemma 6. Then*

$$\text{(i)} \quad \mathcal{E}(\mathbf{SAS}) = b_1\mathbf{A} + b_2\mathbf{A}' + b_2(\text{tr}\mathbf{A})\mathbf{I}$$

$$\text{(ii)} \quad \mathcal{E}(\mathbf{SAS}^{-1}) = c_1\mathbf{A} + c_2\mathbf{A}' + c_2(\text{tr}\mathbf{A})\mathbf{I}$$

$$\text{(iii)} \quad \mathcal{E}(\mathbf{S}^{-1}\mathbf{AS}^{-1}) = d_1\mathbf{A} + d_2\mathbf{A}' + d_2(\text{tr}\mathbf{A})\mathbf{I}.$$

Proof. We will only obtain (i) since the proofs of (ii) and (iii) are based on the same argument. The (i, m)th element of $\mathcal{E}(\mathbf{SAS})$ is:

$$\mathcal{E}(\mathbf{SAS})_{im} = \mathcal{E}[\textstyle\sum_{j\ell} s_{ij}\, a_{j\ell} s_{\ell m}] = \sum_{j\ell} a_{j\ell}\, \mathcal{E}(s_{ij}\, s_{\ell m}) = \sum_{j\ell} a_{j\ell}\, [b_1\delta_{ij}\delta_{\ell m} + b_2(\delta_{i\ell}\delta_{jm} + \delta_{im}\delta_{j\ell})],$$

from Lemma 6. Hence

$$\begin{aligned}\mathcal{E}(\mathbf{SAS})_{im} &= b_1 \textstyle\sum_{j\ell} \delta_{ij}\, a_{j\ell}\, \delta_{\ell m} + b_2 \sum_{j\ell} \delta_{mj}\, a_{j\ell}\, \delta_{\ell i} + b_2\, \delta_{im} \sum_{j\ell} \delta_{j\ell} a_{j\ell} \\ &= b_1\, a_{im} + b_2\, a_{mi} + b_2\, \delta_{im}\, \text{tr}\mathbf{A},\end{aligned}$$

and so (i) follows. ❐

To find explicit expressions for $\mathcal{E}(\mathbf{SAS})$, $\mathcal{E}(\mathbf{SAS}^{-1})$, and $\mathcal{E}(\mathbf{S}^{-1}\mathbf{AS}^{-1})$, it now remains only to find the constants $b_1, b_2, c_1, c_2, d_1, d_2$.

Theorem 1. *Let* $\mathbf{S} \sim \mathbf{W}_p(\mathbf{I}, k)$ *and suppose that the* $p \times p$ *nonrandom matrix* $\mathbf{A}$ *is not necessarily symmetric. Then*

$$\mathcal{E}(\mathbf{SAS}) = k[\mathbf{A}' + (\text{tr}\mathbf{A})\mathbf{I}] + k^2\mathbf{A}. \tag{4}$$

Proof. From Corollary 4(i),

$$\mathcal{E}(\mathbf{SAS}) = b_1\mathbf{A} + b_2\mathbf{A}' + b_2(\text{tr}\mathbf{A})\mathbf{I}. \tag{5}$$

Since $s_{ii} \sim \chi^2_k$, we have $\mathcal{E}[\mathbf{e}_i'\mathbf{S}\mathbf{e}_i\mathbf{e}_i'\mathbf{S}\mathbf{e}_i] = \mathcal{E}(s_{ii}^2) = 2k + k^2$. Substituting $\mathbf{A} = \mathbf{e}_i\mathbf{e}_i'$ in (5) yields

$$\mathcal{E}(s_{ii}^2) = \mathbf{e}_i'[b_1\mathbf{e}_i\mathbf{e}_i' + b_2\mathbf{e}_i\mathbf{e}_i' + b_2\mathbf{I}]\mathbf{e}_i = b_1 + 2b_2 = 2k + k^2.$$

Also $\mathcal{E}[\mathbf{e}_i'\mathbf{S}\mathbf{e}_i\mathbf{e}_j'\mathbf{S}\mathbf{e}_j] = \mathcal{E}(s_{ii})\mathcal{E}(s_{jj}) = k^2 = b_1$, since s_{ii} and s_{jj} are independent ($i \neq j$). Thus $b_2 = k$, which completes the proof. ❒

Corollary 5. *Let* $\mathbf{S} \sim \mathbf{W}_p(\mathbf{V}, k)$ *with* $\mathbf{V}$ *nonsingular, and suppose that the* $p{\times}p$ *matrix* $\mathbf{A}$ *is nonrandom and not necessarily symmetric. Then*

$$\mathcal{E}(\mathbf{SAS}) = k[\mathbf{VA'V} + (\mathrm{tr}\mathbf{AV})\mathbf{V}] + k^2\mathbf{VAV}. \tag{6}$$

Proof. Pre- and postmultiplying $\mathbf{S}$ by $\mathbf{V}^{-1/2}$ yields $\mathbf{V}^{-1/2}\mathbf{S}\mathbf{V}^{-1/2} \sim \mathbf{W}_p(\mathbf{I}, k)$ from Lemma 2. Using Theorem 1, we have $\mathcal{E}[(\mathbf{V}^{-1/2}\mathbf{S}\mathbf{V}^{-1/2})\,\mathbf{B}(\mathbf{V}^{-1/2}\mathbf{S}\mathbf{V}^{-1/2})] = k[\mathbf{B}' + (\mathrm{tr}\mathbf{B})\mathbf{I}] + k^2\mathbf{B}$, where $\mathbf{B}$ is a nonrandom $p{\times}p$ matrix, not necessarily symmetric. Thus

$$\mathcal{E}[\mathbf{S}\mathbf{V}^{-1/2}\mathbf{B}\mathbf{V}^{-1/2}\mathbf{S}] = k[\mathbf{V}^{1/2}\mathbf{B}'\mathbf{V}^{1/2} + (\mathrm{tr}\mathbf{B})\mathbf{V}] + k^2\mathbf{V}^{1/2}\mathbf{B}\mathbf{V}^{1/2}. \tag{7}$$

Let $\mathbf{A} = \mathbf{V}^{-1/2}\mathbf{B}\mathbf{V}^{-1/2}$. Then $\mathbf{B} = \mathbf{V}^{1/2}\mathbf{A}\mathbf{V}^{1/2}$ and (7) reduces to (6). ❒

The next two results were obtained by Das Gupta (1968) with $\mathbf{V} = \mathbf{I}$.

Corollary 6. *Let* $\mathbf{S} \sim \mathbf{W}_p(\mathbf{V}, k)$. *Then* $\mathcal{E}(\mathbf{S}^2) = k[(k+1)\mathbf{V} + (\mathrm{tr}\mathbf{V})\mathbf{I}]\mathbf{V}$.

Corollary 7. *Let* $\mathbf{S} \sim \mathbf{W}_p(\mathbf{V}, k)$, *and suppose that* $\mathbf{a}$ *is a nonrandom* $p{\times}1$ *vector. Then*

$$\mathcal{E}[\mathbf{Saa'S}] = k[(k+1)\mathbf{Vaa'} + (\mathbf{a'Va})\mathbf{I}]\mathbf{V}.$$

These results lead to:

Corollary 8. *Let* $\mathbf{S} \sim \mathbf{W}_p(\mathbf{V}, k)$, *and suppose that* $\mathbf{a}$ *is a nonrandom* $p{\times}1$ *vector. Then*

$$\mathrm{Var}(\mathbf{Sa}) = k[\mathbf{Vaa'V} + (\mathbf{a'Va})\mathbf{V}].$$

Proof. Clearly $\mathrm{Var}(\mathbf{Sa}) = \mathcal{E}[\mathbf{Saa'S}] - \mathcal{E}[\mathbf{Sa}]\mathcal{E}[\mathbf{a'S}]$; the first term is given by Corollary 7, while the second term, using Lemma 1, equals $k^2\mathbf{Vaa'V}$. Subtracting the two completes the proof. ❒

Corollary 9. *Let* $\mathbf{S} \sim \mathbf{W}_p(\mathbf{V}, k)$, *and suppose that* $\mathbf{a}$ *is a nonrandom* $p{\times}1$*vector. Then*

$$\mathrm{Var}(\mathbf{a'Sa}) = 2k(\mathbf{a'Va})^2.$$

Proof. This result follows directly from Lemma 2, since $\mathbf{a'Sa} \sim \mathbf{a'Va}\,\chi_k^2$. ❒

Corollary 10. *Suppose that* $\mathbf{S}_1$ *and* $\mathbf{S}_2$ *are distributed independently as* $\mathbf{W}_p(\mathbf{V}, k_1)$ *and* $\mathbf{W}_p(\mathbf{V}, k_2)$, *respectively, with* $\mathbf{V}$ *nonsingular and* $k_1 > p + 1$. *Then*

$$\mathcal{E}(\mathbf{S}_2\mathbf{S}_1^{-1}\mathbf{S}_2) = [k_2(k_2+p+1)/(k_1-p-1)]\mathbf{V}. \tag{8}$$

Proof. We have that

$$\mathcal{E}(\mathbf{S}_2\mathbf{S}_1^{-1}\mathbf{S}_2) = \mathcal{E}_{\mathbf{S}_1}[\mathcal{E}_{\mathbf{S}_2}(\mathbf{S}_2\mathbf{S}_1^{-1}\mathbf{S}_2 \mid \mathbf{S}_1)]$$

$$= \mathcal{E}_{\mathbf{S}_1}\{k_2[\mathbf{V}\mathbf{S}_1^{-1}\mathbf{V} + (\mathrm{tr}\mathbf{S}_1^{-1}\mathbf{V})\mathbf{V}] + k_2^2\mathbf{V}\mathbf{S}_1^{-1}\mathbf{V}\}$$

$$= k_2\mathbf{V}[\mathcal{E}(\mathbf{S}_1^{-1})]\mathbf{V} + k_2[\mathrm{tr}\mathcal{E}(\mathbf{S}_1^{-1}\mathbf{V})]\mathbf{V} + k_2^2\mathbf{V}[\mathcal{E}(\mathbf{S}_1)]\mathbf{V}, \tag{9}$$

using Corollary 5. From Lemma 5, (9) becomes

$$k_2\left[\frac{1}{k_1-p-1}\mathbf{V} + \frac{p}{k_1-p-1}\mathbf{V}\right] + \frac{k_2^2}{k_1-p-1}\mathbf{V} = [k_2(k_2+p+1)/(k_1-p-1)]\mathbf{V}.$$

Corollary 11. *Let* $\mathbf{S}_1$ *and* $\mathbf{S}_2$ *be distributed independently as* $\mathbf{W}_p(\mathbf{V}, k_1)$ *and* $\mathbf{W}_p(\mathbf{V}, k_2)$, *respectively, with* $\mathbf{V}$ *nonsingular and* $k_1 > p+1$. *Then*

$$\mathcal{E}(\mathrm{tr}\mathbf{S}_1^{-1}\mathbf{S}_2) = k_2p(k_1-p-1)^{-1}.$$

Proof. $\mathcal{E}(\mathrm{tr}\mathbf{S}_1^{-1}\mathbf{S}_2) = \mathrm{tr}[\mathcal{E}(\mathbf{S}_1^{-1}\mathbf{S}_2)] = \mathrm{tr}[\mathcal{E}(\mathbf{S}_1^{-1})\mathcal{E}\mathbf{S}_2] = \mathrm{tr}[(k_1-p-1)^{-1}\mathbf{V}^{-1}k_2\mathbf{V}] = k_2p(k_1-p-1)^{-1}$. ❐

Theorem 2. *Let* $\mathbf{S} \sim \mathbf{W}_p(\mathbf{I}, k)$ *with* $k > p+1$, *and suppose that the* $p\times p$ *nonrandom matrix* $\mathbf{A}$ *is not necessarily symmetric. Then*

$$\mathcal{E}(\mathbf{S}\mathbf{A}\mathbf{S}^{-1}) = (k-p-1)^{-1}[k\mathbf{A} - \mathbf{A}' - (\mathrm{tr}\mathbf{A})\mathbf{I}].$$

Proof. From Corollary 4(ii),

$$\mathcal{E}(\mathbf{S}\mathbf{A}\mathbf{S}^{-1}) = c_1\mathbf{A} + c_2\mathbf{A}' + c_2(\mathrm{tr}\mathbf{A})\mathbf{I}. \tag{10}$$

If we let $\mathbf{A} = \mathbf{I}$ in (10) then $\mathcal{E}(\mathbf{I}) = \mathbf{I} = c_1\mathbf{I} + c_2\mathbf{I} + c_2(\mathrm{tr}\mathbf{I})\mathbf{I}$; thus $1 = c_1 + c_2(p+1)$. Also

$$\mathcal{E}[\mathbf{e}_p'\mathbf{S}\mathbf{e}_p\, \mathbf{e}_1'\mathbf{S}^{-1}\mathbf{e}_1] = \mathcal{E}[s_{pp}s^{11}] = \mathcal{E}(s_{pp})\mathcal{E}(s^{11}) = k(k-p-1)^{-1}, \tag{11}$$

using (4) and Lemma 3. Substituting $\mathbf{A} = \mathbf{e}_p\mathbf{e}_1'$ in (10) leads to

$$\mathcal{E}[\mathbf{e}_p'\mathbf{S}\mathbf{e}_p\, \mathbf{e}_1'\mathbf{S}^{-1}\mathbf{e}_1] = \mathbf{e}_p'[c_1\mathbf{e}_p\mathbf{e}_1' + c_2\mathbf{e}_1\mathbf{e}_p']\mathbf{e}_1 = c_1.$$

Therefore $c_1 = k(k-p-1)^{-1}$, and by substituting for c_1 in (11), we obtain $c_2 = -(k-p-1)^{-1}$. ❐

Corollary 12. *Let* $\mathbf{S} \sim \mathbf{W}_p(\mathbf{V}, k)$ *with* $\mathbf{V}$ *nonsingular and* $k > p+1$, *and suppose that the* $p\times p$ *nonrandom matrix* $\mathbf{A}$ *is not necessarily symmetric. Then*

$$\mathcal{E}(\mathbf{S}\mathbf{A}\mathbf{S}^{-1}) = (k-p-1)^{-1}[k\mathbf{V}\mathbf{A}\mathbf{V}^{-1} - \mathbf{A}' - (\mathrm{tr}\mathbf{A})\mathbf{I}]. \tag{12}$$

Proof. From Theorem 2,

$$\mathcal{E}(\mathbf{V}^{-1/2}\mathbf{S}\mathbf{V}^{-1/2}\mathbf{B}\mathbf{V}^{1/2}\mathbf{S}^{-1}\mathbf{V}^{1/2}) = (k-p-1)^{-1}[k\mathbf{B} - \mathbf{B}' - (\mathrm{tr}\mathbf{B})\mathbf{I}],$$

where $\mathbf{B}$ is a nonrandom $p \times p$ matrix, not necessarily symmetric. Thus

$$\mathcal{E}[\mathbf{S}\mathbf{V}^{-1/2}\mathbf{B}\mathbf{V}^{-1/2}\mathbf{S}^{-1}] = (k-p-1)^{-1}[k\mathbf{V}^{1/2}\mathbf{B}\mathbf{V}^{-1/2} - \mathbf{V}^{1/2}\mathbf{B}'\mathbf{V}^{-1/2} - (\mathrm{tr}\mathbf{B})\mathbf{I}]. \tag{13}$$

Let $\mathbf{A} = \mathbf{V}^{-1/2}\mathbf{B}\mathbf{V}^{1/2}$. Then $\mathbf{B} = \mathbf{V}^{1/2}\mathbf{A}\mathbf{V}^{-1/2}$ and (13) reduces to (12). ❐

Corollary 13. *Let* $\mathbf{S} \sim \mathbf{W}_p(\mathbf{V}, k)$, *with* $\mathbf{V}$ *nonsingular and* $k > p+1$, *and suppose that the* $p \times p$ *nonrandom matrix* $\mathbf{A}$ *is not necessarily symmetric. Then*

$$\mathcal{E}(\mathbf{S}^{-1}\mathbf{A}\mathbf{S}) = (k-p-1)[k\mathbf{V}^{-1}\mathbf{A}\mathbf{V} - \mathbf{A}' - (\mathrm{tr}\mathbf{A})\mathbf{I}].$$

If $\mathbf{V} = \mathbf{I}_p$, *then*

$$\mathcal{E}(\mathbf{S}^{-1}\mathbf{A}\mathbf{S}) = \mathcal{E}(\mathbf{S}\mathbf{A}\mathbf{S}^{-1}) = (k-p-1)^{-1}[k\mathbf{A} - \mathbf{A}' - (\mathrm{tr}\mathbf{A})\mathbf{I}].$$

Proof. We have that $\mathcal{E}[(\mathbf{S}^{-1}\mathbf{A}\mathbf{S})'] = \mathcal{E}(\mathbf{S}\mathbf{A}'\mathbf{S}^{-1}) = (k-p-1)^{-1}[k\mathbf{V}\mathbf{A}'\mathbf{V}^{-1} - \mathbf{A} - (\mathrm{tr}\mathbf{A}')\mathbf{I}]$. Taking transposes gives the result. ❐

We will need the following lemma to obtain $\mathcal{E}(\mathbf{S}^{-1}\mathbf{A}\mathbf{S}^{-1})$ in Theorem 3 below.

Lemma 7. *Let* $\mathbf{S} \sim \mathbf{W}_p(\mathbf{I}, k)$ *with* $k > p+3$. *Then for all* $i \neq j$, $\mathcal{E}(s^{ii}s^{jj}) = (k-p-2)/[(k-p)(k-p-1)(k-p-3)]$.

Proof. Let $\mathbf{S} = \mathbf{Z}'\mathbf{Z}$, with $\mathbf{Z} = (\mathbf{z}_1, ..., \mathbf{z}_k)' = (\mathbf{Y}_1, \mathbf{y}_p)$, where $\mathbf{Y}_1$ is $k \times (p-1)$. Then partitioned matrix inversion using the Schur complement, cf. e.g., Styan (1985), yields

$$\mathcal{E}(s^{11}s^{pp}) = \mathcal{E}\{s^{pp}\mathbf{e}_1'(\mathbf{Y}_1'\mathbf{Y}_1)^{-1}\mathbf{e}_1 + [s^{pp}\mathbf{e}_1'(\mathbf{Y}_1'\mathbf{Y}_1)^{-1}\mathbf{Y}_1'\mathbf{y}_p]^2\}, \tag{14}$$

where

$$1/s^{pp} = \mathbf{y}_p'\mathbf{M}\mathbf{y}_p = \mathbf{y}_p'[\mathbf{I} - \mathbf{Y}_1(\mathbf{Y}_1'\mathbf{Y}_1)^{-1}\mathbf{Y}_1']\mathbf{y}_p.$$

From Lemma 3, we know that $\mathbf{y}_p'\mathbf{M}\mathbf{y}_p$ is distributed independently of $\mathbf{Y}_1$. Thus we can write the first term of the right-hand side of (14) as

$$\mathbf{e}_1'[\mathcal{E}(\mathbf{Y}_1'\mathbf{Y}_1)^{-1}]\mathbf{e}_1\mathcal{E}[1/\mathbf{y}_p'\mathbf{M}\mathbf{y}_p] = (k-p)^{-1}(k-p-1)^{-1}, \tag{15}$$

since $\mathbf{Y}_1'\mathbf{Y}_1 \sim \mathbf{W}_{p-1}(\mathbf{I}, k)$, $\mathcal{E}(\mathbf{Y}_1'\mathbf{Y}_1)^{-1} = (k-p)^{-1}\mathbf{I}$ from Lemma 4, and $\mathbf{y}_p'\mathbf{M}\mathbf{y}_p \sim \chi^2_{k-p+1}$. To find the second term on the right-hand side of (14) we first fix $\mathbf{Y}_1$. Then $\mathbf{y}_p'\mathbf{M}\mathbf{y}_p$ is distributed as χ^2_{k-p+1} independently of $\mathbf{Y}_1$, and of $\mathbf{e}_1'(\mathbf{Y}_1'\mathbf{Y}_1)^{-1}\mathbf{Y}_1'\mathbf{y}_p$, because $\mathbf{e}_1'(\mathbf{Y}_1'\mathbf{Y}_1)^{-1}\mathbf{Y}_1'\mathbf{M} = \mathbf{0}'$; moreover $\mathbf{e}_1'(\mathbf{Y}_1'\mathbf{Y}_1)^{-1}\mathbf{Y}_1'\mathbf{y}_p \sim \mathrm{N}(0, \mathbf{e}_1'(\mathbf{Y}_1'\mathbf{Y}_1)^{-1}\mathbf{e}_1)$. The second term becomes

$$\mathcal{E}_{\mathbf{Y}_1}\{\mathcal{E}_{\mathbf{y}_p}[\mathbf{e}_1'(\mathbf{Y}_1'\mathbf{Y}_1)^{-1}\mathbf{Y}_1'\mathbf{y}_p]^2 \mid \mathbf{Y}_1]\}\mathcal{E}_{\mathbf{Y}_p}[1/(\mathbf{y}_p'\mathbf{M}\mathbf{y}_p)^2] \tag{16}$$

$$= \mathcal{E}_{\mathbf{Y}_1}[\mathbf{e}_1'(\mathbf{Y}_1'\mathbf{Y}_1)^{-1}\mathbf{e}_1](k-p-1)^{-1}(k-p-3)^{-1}$$

$$= (k-p)^{-1}(k-p-1)^{-1}(k-p-3)^{-1},$$

using Corollary 3. Thus,

$$\mathcal{E}(s^{11}s^{pp}) = (k-p)^{-1}(k-p-1)^{-1} + (k-p)^{-1}(k-p-1)^{-1}(k-p-3)^{-1}$$

$$= (k-p-2)/[(k-p)(k-p-1)(k-p-3)], \tag{17}$$

which clearly equals $\mathcal{E}(s^{ii}s^{jj})$ for all $i \neq j$. ❐

Theorem 3. *Let* $\mathbf{S} \sim \mathbf{W}_p(\mathbf{I}, k)$ *with* $k > p+3$, *and suppose that the* $p \times p$ *nonrandom matrix* $\mathbf{A}$ *is not necessarily symmetric. Then*

$$\mathcal{E}(\mathbf{S}^{-1}\mathbf{A}\mathbf{S}^{-1}) = \frac{(k-p-2)\mathbf{A} + \mathbf{A}' + (\mathrm{tr}\mathbf{A})\mathbf{I}}{(k-p)(k-p-1)(k-p-3)} .$$

Proof. From Corollary 4(iii) $\mathcal{E}(\mathbf{S}^{-1}\mathbf{A}\mathbf{S}^{-1}) = d_1\mathbf{A} + d_2\mathbf{A}' + d_2(\mathrm{tr}\mathbf{A})\mathbf{I}$. Now $\mathcal{E}[\mathbf{e}_i'\mathbf{S}^{-1}\mathbf{e}_i\mathbf{e}_i'\mathbf{S}^{-1}\mathbf{e}_i] = \mathcal{E}[(s^{ii})^2] = (k-p-1)^{-1}(k-p-3)^{-1}$ from Corollary 3, while substituting $\mathbf{A} = \mathbf{e}_i\mathbf{e}_i'$ yields

$$\mathcal{E}[\mathbf{e}_i'\mathbf{S}^{-1}\mathbf{e}_i\mathbf{e}_i'\mathbf{S}^{-1}\mathbf{e}_i] = \mathbf{e}_i'[d_1\mathbf{e}_i\mathbf{e}_i' + d_2\mathbf{e}_i\mathbf{e}_i' + d_2\mathbf{I}]\mathbf{e}_i$$

$$= d_1 + 2d_2 = (k-p-1)^{-1}(k-p-3)^{-1} . \tag{18}$$

Substituting $\mathbf{A} = \mathbf{e}_i\mathbf{e}_j'$, with $i \neq j$, gives

$$\mathcal{E}[\mathbf{e}_i'\mathbf{S}^{-1}\mathbf{e}_i\mathbf{e}_j'\mathbf{S}^{-1}\mathbf{e}_j] = \mathbf{e}_i'[d_1\mathbf{e}_i\mathbf{e}_j' + d_2\mathbf{e}_j\mathbf{e}_j']\mathbf{e}_j$$

$$= d_1 = \mathcal{E}(s^{ii}s^{jj}) = (k-p-2)/[(k-p)(k-p-1)(k-p-3)] ,$$

from Lemma 7, and substituting for d_1 we obtain $d_2 = (k-p)^{-1}(k-p-1)^{-1}(k-p-3)^{-1}$.

Corollary 14. *Let* $\mathbf{S} \sim \mathbf{W}_p(\mathbf{V}, k)$, *with* $\mathbf{V}$ *nonsingular and* $k > p+3$, *and suppose that the* $p \times p$ *nonrandom matrix* $\mathbf{A}$ *is not necessarily symmetric. Then*

$$\mathcal{E}(\mathbf{S}^{-1}\mathbf{A}\mathbf{S}^{-1}) = \frac{(k-p-2)\mathbf{V}^{-1}\mathbf{A}\mathbf{V}^{-1} + \mathbf{V}^{-1}\mathbf{A}'\mathbf{V}^{-1} + (\mathrm{tr}\mathbf{A}\mathbf{V}^{-1})\mathbf{V}^{-1}}{(k-p)(k-p-1)(k-p-3)} . \tag{19}$$

Proof. From Theorem 3 we may write

$$\mathcal{E}(\mathbf{V}^{1/2}\mathbf{S}^{-1}\mathbf{V}^{1/2}\mathbf{B}\mathbf{V}^{1/2}\mathbf{S}^{-1}\mathbf{V}^{1/2}) = \frac{(k-p-2)\,\mathbf{B} + \mathbf{B}' + (\mathrm{tr}\mathbf{B})\mathbf{I}}{(k-p)(k-p-1)(k-p-3)}$$

where $\mathbf{B}$ is a nonrandom $p \times p$ matrix, not necessarily symmetric. Thus

$$\mathcal{E}(\mathbf{S}^{-1}\mathbf{V}^{1/2}\mathbf{B}\mathbf{V}^{1/2}\mathbf{S}^{-1}) = \frac{(k-p-2)\mathbf{V}^{-1/2}\mathbf{B}'\mathbf{V}^{-1/2} + \mathbf{V}^{-1/2}\mathbf{B}'\mathbf{V}^{-1/2} + (\mathrm{tr}\mathbf{B})\mathbf{V}^{-1}}{(k-p)(k-p-1)(k-p-3)}$$

Let $\mathbf{A} = \mathbf{V}^{1/2}\mathbf{B}\mathbf{V}^{1/2}$. Then $\mathbf{B} = \mathbf{V}^{-1/2}\mathbf{A}\mathbf{V}^{-1/2}$ and (19) follows. ❐

Corollary 15. *Let* $\mathbf{S} \sim \mathbf{W}_p(\mathbf{V}, k)$, *with* $\mathbf{V}$ *nonsingular and* $k > p+3$. *Then*

$$\mathcal{E}(\mathbf{S}^{-2}) = \frac{(k-p-1)\mathbf{V}^{-2} + (\mathrm{tr}\mathbf{V}^{-1})\mathbf{V}^{-1}}{(k-p)(k-p-1)(k-p-3)}$$

The special case of Corollary 15 when $\mathbf{V} = \mathbf{I}$ was obtained in a different way by Das Gupta (1968).

Corollary 16. *Let* $\mathbf{S} \sim \mathbf{W}_p(\mathbf{I}, k)$ *with* $k > p+3$. *Then*

$$\mathcal{E}(\mathrm{tr}\mathbf{S}^{-2}) = \frac{(k-1)p}{(k-p)(k-p-1)(k-p-3)}$$

and

$$\mathcal{E}[(\mathrm{tr}\mathbf{S}^{-1})^2] = \frac{p[(k-p-2)p+2]}{(k-p)(k-p-1)(k-p-3)}.$$

Proof. The first formula follows from Corollary 15. To establish the second formula, we use Theorem 3:

$$\begin{aligned}
\mathcal{E}[(\mathrm{tr}\mathbf{S}^{-1})^2] &= \mathcal{E}[\mathrm{tr}\mathbf{S}^{-1}\mathrm{tr}\mathbf{S}^{-1}] \\
&= \mathcal{E}[\sum_{i=1}^{p} \mathbf{e}_i' \mathbf{S}^{-1}\mathbf{e}_i \sum_{j=1}^{p} \mathbf{e}_j' \mathbf{S}^{-1}\mathbf{e}_j] \\
&= \sum_{i=1}^{p} \sum_{j=1}^{p} \mathbf{e}_i' \mathcal{E}[\mathbf{S}^{-1}\mathbf{e}_i\mathbf{e}_j'\mathbf{S}^{-1}]\mathbf{e}_j \\
&= \sum_{i=1}^{p} \sum_{j=1}^{p} \mathbf{e}_i' \left[\frac{(k-p-2)\,\mathbf{e}_i\mathbf{e}_j' + \mathbf{e}_j\mathbf{e}_i' + \delta_{ij}\mathbf{I}}{(k-p)(k-p-1)(k-p-3)} \right] \mathbf{e}_j
\end{aligned}$$

$$= \sum_{i=1}^{p} \sum_{j=1}^{p} \frac{(k-p-2) + 2\delta_{ij}}{(k-p)(k-p-1)(k-p-3)}$$

$$= \frac{(k-p-2)p^2 + 2p}{(k-p)(k-p-1)(k-p-3)}$$

$$= \frac{p[(k-p-2)p + 2]}{(k-p)(k-p-1)(k-p-3)} \, .$$

Corollary 17. *Let* $\mathbf{S} \sim \mathbf{W}_p(\mathbf{I}, k)$ *with* $k > p+3$. *Then*

$$\mathrm{Var}(\mathrm{tr}\mathbf{S}^{-1}) = \frac{2p(k-1)}{(k-p)(k-p-1)^2(k-p-3)} \, .$$

Corollary 18. *Let* $\mathbf{S} \sim \mathbf{W}_p(\mathbf{I}, k)$ *with* $k > p+3$, *and suppose that the* $p \times p$ *nonrandom matrix* $\mathbf{A}$ *is not necessarily symmetric. Then*

$$\mathcal{E}[(\mathrm{tr}\mathbf{A}\mathbf{S}^{-1})]\mathbf{S}^{-1}] = \frac{(k-p-2)(\mathrm{tr}\mathbf{A})\mathbf{I} + \mathbf{A}' + \mathbf{A}}{(k-p)(k-p-1)(k-p-3)} \, .$$

Proof. We write the (j, ℓ)th element of $\mathcal{E}[(\mathrm{tr}\mathbf{A}\mathbf{S}^{-1})]\mathbf{S}^{-1}]$ as $\mathcal{E}[(\mathrm{tr}\mathbf{A}\mathbf{S}^{-1})]s^{j\ell}]$. Then

$$\mathcal{E}[(\mathrm{tr}\mathbf{A}\mathbf{S}^{-1})]s^{j\ell}] = \sum_i \mathbf{e}_i'\mathbf{A}[\mathcal{E}(\mathbf{S}^{-1}\mathbf{e}_i\mathbf{e}_j'\mathbf{S}^{-1})]\mathbf{e}_\ell$$

$$= c\sum_i \mathbf{e}_i'\mathbf{A}[(k-p-2)\mathbf{e}_i\mathbf{e}_j' + \mathbf{e}_j\mathbf{e}_i' + \delta_{ij}\mathbf{I}]\mathbf{e}_\ell$$

$$= c[k-p-2)(\mathrm{tr}\mathbf{A})\delta_{j\ell} + \mathbf{e}_j'\mathbf{A}'\mathbf{e}_\ell + \mathbf{e}_j'\mathbf{A}\mathbf{e}_\ell],$$

using Theorem 3 with $c = 1/[(k-p)(k-p-1)(k-p-3)]$. The result follows. ❐

3. SOME APPLICATIONS

Suppose now that the $p \times p$ random matrices $\mathbf{S}_1$ and $\mathbf{S}_2$ are distributed independently as $\mathbf{W}_p(\mathbf{V}, k_1)$ and $\mathbf{W}_p(\mathbf{V}, k_2)$, respectively. When $p = 1$ it is well known that $\mathbf{S}_1+\mathbf{S}_2$ is independent of $\mathbf{S}_1^{-1}\mathbf{S}_2$. When $p > 1$, this property does not hold (Theorem 4 below). To prove this we use the following lemma which follows directly from a theorem on sufficient statistics due to Basu (1955); see also Lehmann (1986, Chapter 5, Theorem 2).

Lemma 8. *Let the* $p\times p$ *random matrices* $\mathbf{S}_1$ *and* $\mathbf{S}_2$ *be distributed independently as* $\mathbf{W}_p(\mathbf{V}, k_1)$ *and* $\mathbf{W}_p(\mathbf{V}, k_2)$, *respectively, with* $\mathbf{V}$ *nonsingular, and let* $\mathbf{G}$ *be a matrix function of* $\mathbf{S}_1$ *and* $\mathbf{S}_2$. *Then* $\mathbf{S}_1+\mathbf{S}_2$ *is independently distributed of* $\mathbf{G}$ *if and only if the distribution of* $\mathbf{G}$ *does not depend on* $\mathbf{V}$.

The following essentially equivalent result (Corollary 19), which was originally obtained by Olkin and Rubin (1964, Theorem 6.1), follows immediately from Lemma 8.

Corollary 19. *Let the* $p\times p$ *random matrices* $\mathbf{S}_1$ *and* $\mathbf{S}_2$ *be distributed independently as* $\mathbf{W}_p(\mathbf{V}, k_1)$ *and* $\mathbf{W}_p(\mathbf{V}, k_2)$, *respectively, with* $\mathbf{V}$ *nonsingular, and let* $\mathbf{G}$ *be a matrix function of* $\mathbf{S}_1$ *and* $\mathbf{S}_2$. *Then* $\mathbf{S}_1+\mathbf{S}_2$ *is independently distributed of* $\mathbf{G}$ *if the matrix function* $\mathbf{G}\{\mathbf{S}_1, \mathbf{S}_2\} \sim \mathbf{G}\{\mathbf{A}\mathbf{S}_1\mathbf{A}', \mathbf{A}\mathbf{S}_2\mathbf{A}'\}$ *for some nonsingular* $p\times p$ *matrix* $\mathbf{A}$. *Moreover, if* $\mathbf{S}_1+\mathbf{S}_2$ *is independent of* $\mathbf{G}$ *then for every nonsingular* $p\times p$ *matrix* $\mathbf{A}$, *the matrix* $\mathbf{G}\{\mathbf{S}_1, \mathbf{S}_2\} \sim \mathbf{G}\{\mathbf{A}\mathbf{S}_1\mathbf{A}', \mathbf{A}\mathbf{S}_2\mathbf{A}'\}$.

This leads to:

Theorem 4. *Let the* $p\times p$ *random matrices* $\mathbf{S}_1$ *and* $\mathbf{S}_2$ *be distributed independently as* $\mathbf{W}_p(\mathbf{V}, k_1)$ *and* $\mathbf{W}_p(\mathbf{V}, k_2)$, *respectively, with* $k_1-3 > p > 1$ *and* $\mathbf{V}$ *nonsingular. Then the distribution of* $\mathbf{S}_1^{-1}\mathbf{S}_2$ *depends on* $\mathbf{V}$, *and the* $p\times p$ *random matrices* $\mathbf{S}_1+\mathbf{S}_2$ *and* $\mathbf{S}_1^{-1}\mathbf{S}_2$ *are not independently distributed.*

Proof. It will suffice to show that $\mathcal{E}(\mathbf{S}_1^{-1}\mathbf{S}_2\mathbf{A}\mathbf{S}_1^{-1}\mathbf{S}_2)$ depends on $\mathbf{V}$ where $\mathbf{A}$ is a $p\times p$ nonrandom matrix, not necessarily symmetric. We have

$$\begin{aligned}\mathcal{E}(\mathbf{S}_1^{-1}\mathbf{S}_2\mathbf{A}\mathbf{S}_1^{-1}\mathbf{S}_2) &= \mathcal{E}(\mathbf{V}^{-1/2}\mathbf{H}_1^{-1}\mathbf{V}^{-1/2}\mathbf{V}^{1/2}\mathbf{H}_2\mathbf{V}^{1/2}\mathbf{A}\mathbf{V}^{-1/2}\mathbf{H}_1\mathbf{V}^{-1/2}\mathbf{V}^{1/2}\mathbf{H}_2\mathbf{V}^{1/2}) \\ &= \mathbf{V}^{-1/2}\mathcal{E}(\mathbf{H}_1^{-1}\mathbf{H}_2\mathbf{A}_0\mathbf{H}_1^{-1}\mathbf{H}_2)\,\mathbf{V}^{1/2},\end{aligned}$$

where $\mathbf{A}_0 = \mathbf{V}^{1/2}\mathbf{A}\mathbf{V}^{-1/2}$ and where $\mathbf{H}_1 \sim \mathbf{W}_p(\mathbf{I}, k_1)$ independently of $\mathbf{H}_2 \sim \mathbf{W}_p(\mathbf{I}, k_2)$, with $k_1 > p+3$. First, conditioning on $\mathbf{H}_1$, we get

$$\mathbf{V}^{-1/2}\{\mathcal{E}_{\mathbf{H}_1}[\mathcal{E}_{\mathbf{H}_2}(\mathbf{H}_1^{-1}\mathbf{H}_2\mathbf{A}_0\mathbf{H}_1^{-1}\mathbf{H}_2 \mid \mathbf{H}_1)]\}\mathbf{V}^{1/2} \tag{20}$$

$$= \mathbf{V}^{-1/2}\{\mathcal{E}_{\mathbf{H}_1}[\mathbf{H}_1^{-1}\{k_2\mathbf{H}_1^{-1}\mathbf{A}_0' + k_2(\mathrm{tr}\mathbf{A}_0\mathbf{H}_1^{-1})\mathbf{I} + k_2^2\mathbf{A}_0\mathbf{H}_1^{-1}\}]\}\mathbf{V}^{1/2}$$

$$= k_2\mathbf{V}^{-1/2}\mathcal{E}(\mathbf{H}_1^{-2})\mathbf{A}_0'\mathbf{V}^{1/2} + k_2\mathbf{V}^{-1/2}\mathcal{E}[(\mathrm{tr}\mathbf{A}_0\mathbf{H}_1^{-1})\mathbf{H}_1^{-1}]\mathbf{V}^{1/2} + k_2^2\mathbf{V}^{-1/2}\mathcal{E}(\mathbf{H}_1^{-1}\mathbf{A}_0\mathbf{H}_1^{-1})\mathbf{V}^{1/2},$$

using Theorem 1. From Corollary 15 we see that the first term in (20) is

$$\frac{k_2(k_1-1)\mathbf{V}^{-1/2}\mathbf{A}_0'\mathbf{V}^{1/2}}{(k_1-p)(k_1-p-1)(k_1-p-3)} = \frac{k_2(k_1-1)\mathbf{V}^{-1}\mathbf{A}'\mathbf{V}}{(k_1-p)(k_1-p-1)(k_1-p-3)}, \tag{21}$$

while the second term is

$$\frac{k_2\mathbf{V}^{-1/2}(k_1-p-2)[(\mathrm{tr}\mathbf{A}_0)\mathbf{I}+\mathbf{A}_0'+\mathbf{A}_0]\mathbf{V}^{1/2}}{(k_1-p)(k_1-p-1)(k_1-p-3)} = \frac{k_2(k_1-p-2)(\mathrm{tr}\mathbf{A})\mathbf{I}+\mathbf{V}^{-1}\mathbf{A}'\mathbf{V}+\mathbf{A}}{(k_1-p)(k_1-p-1)(k_1-p-3)}, \quad (22)$$

using Corollary 18. Finally, from Theorem 3, the last term of (20) is

$$\frac{k_2^2\,\mathbf{V}^{-1/2}[(k_1-p-2)\mathbf{A}_0+\mathbf{A}_0'+(\mathrm{tr}\mathbf{A}_0)\mathbf{I}]\mathbf{V}^{1/2}}{(k_1-p)(k_1-p-1)(k_1-p-3)};$$

substituting $\mathbf{A}_0 = \mathbf{V}^{1/2}\mathbf{A}\mathbf{V}^{-1/2}$ gives

$$\frac{k_2^2\,[(k_1-p-2)\mathbf{A}+\mathbf{V}^{-1}\mathbf{A}'\mathbf{V}+(\mathrm{tr}\mathbf{A})\mathbf{I}]}{(k_1-p)(k_1-p-1)(k_1-p-3)}. \quad (23)$$

Adding (21), (22), and (23) yields

$$\mathcal{E}(\mathbf{S}_1^{-1}\mathbf{S}_2\mathbf{A}\mathbf{S}_1^{-1}\mathbf{S}_2) = \frac{k_2\{(k_1+k_2)\mathbf{V}^{-1}\mathbf{A}\mathbf{V}+(k_1+k_2-p-2)(\mathrm{tr}\mathbf{A})\mathbf{I}+[k_2(k_1-p-2)+1]\mathbf{A}\}}{(k_1-p)(k_1-p-1)(k_1-p-3)},$$

which, in general, depends on $\mathbf{V}$ when $p > 1$ and so the proof is complete. ❐

Theorem 5. *Let the $p\times p$ random matrices $\mathbf{S}_1$ and $\mathbf{S}_2$ be distributed independently as $\mathbf{W}_p(\mathbf{V}, k_1)$ and $\mathbf{W}_p(\mathbf{V}, k_2)$, respectively, with $k_1 \geq p$ and $\mathbf{V}$ nonsingular. Then the distribution of the characteristic roots of $\mathbf{S}_1^{-1}\mathbf{S}_2$ does not depend on $\mathbf{V}$, and the $p\times p$ random matrix $\mathbf{S}_1+\mathbf{S}_2$ is distributed independently of the characteristic roots of $\mathbf{S}_1^{-1}\mathbf{S}_2$.*

Proof. The result follows immediately from Corollary 19, since for any nonsingular $p\times p$ matrix $\mathbf{A}$ we have that the characteristic roots

$$\mathrm{ch}[(\mathbf{A}\mathbf{S}_1\mathbf{A}')^{-1}\mathbf{A}\mathbf{S}_2\mathbf{A}'] = \mathrm{ch}[(\mathbf{A}')^{-1}\mathbf{S}_1^{-1}\mathbf{A}^{-1}\mathbf{A}\mathbf{S}_2\mathbf{A}'] = \mathrm{ch}[(\mathbf{A}')^{-1}\mathbf{S}_1^{-1}\mathbf{S}_2\mathbf{A}'] = \mathrm{ch}[(\mathbf{S}_1^{-1}\mathbf{S}_2)$$

and so the proof is complete. ❐

Corollary 20. *Suppose $\mathbf{S}_1$ and $\mathbf{S}_2$ are distributed independently as $\mathbf{W}_p(\mathbf{V}, k_1)$ and $\mathbf{W}_p(\mathbf{V}, k_2)$, respectively, with $k_1 > p+3$ and $\mathbf{V}$ nonsingular. Then the distributions of the trace of $\mathbf{S}_1^{-1}\mathbf{S}_2$ and of* $\det(\mathbf{S}_2)/\det(\mathbf{S}_1)$ *do not depend on $\mathbf{V}$. Moreover, $\mathbf{S}_1+\mathbf{S}_2$ is distributed independently of* $\mathrm{tr}(\mathbf{S}_1^{-1}\mathbf{S}_2)$, *and of* $\det(\mathbf{S}_2)/\det(\mathbf{S}_1)$.

To illustrate Corollary 20, we now prove:

Lemma 9. *Suppose $\mathbf{S}_1$ and $\mathbf{S}_2$ are distributed independently as $\mathbf{W}_p(\mathbf{V}, k_1)$ and $\mathbf{W}_p(\mathbf{V}, k_2)$, respectively, with $k_1 > p+3$ and $\mathbf{V}$ nonsingular. Then*

$$\mathrm{Var}[\mathrm{tr}(\mathbf{S}_1^{-1}\mathbf{S}_2)] = \frac{2pk_2(k_1-1)(k_1+k_2-p-1)}{(k_1-p)(k_1-p-1)^2(k_1-p-3)}.$$

Proof. It suffices to consider $\mathbf{S}_1$ and $\mathbf{S}_2$ with associated covariance matrix $\mathbf{I}$ because of Corollary 20. We write

$$\mathrm{Var}[\mathrm{tr}(\mathbf{S}_1^{-1}\mathbf{S}_2)] = \mathcal{E}_{\mathbf{S}_1}[\mathrm{Var}_{\mathbf{S}_2}(\mathrm{tr}\mathbf{S}_1^{-1}\mathbf{S}_2 \,|\mathbf{S}_1)] + \mathrm{Var}_{\mathbf{S}_1}[\mathcal{E}_{\mathbf{S}_2}(\mathrm{tr}\mathbf{S}^{-1}\mathbf{S}_2 \,|\mathbf{S}_1)] \,. \tag{24}$$

We obtain the first term on the right-hand side of (24) as

$$\begin{aligned}\mathcal{E}_{\mathbf{S}_1}[\mathrm{Var}_{\mathbf{S}_2}(\mathrm{tr}\mathbf{S}_1^{-1}\mathbf{S}_2\,|\mathbf{S}_1)] &= \mathcal{E}_{\mathbf{S}_1}[\mathrm{Var}_{\mathbf{S}_2}(\mathrm{tr}\mathbf{S}_1^{-1}\sum_{\alpha=1}^{k_2}\mathbf{z}_\alpha\mathbf{z}_\alpha'|\mathbf{S}_1)]\\ &= \mathcal{E}_{\mathbf{S}_1}[\mathrm{Var}_{\mathbf{z}_\alpha}(\sum_{\alpha=1}^{k_2}\mathbf{z}_\alpha'\mathbf{S}_1^{-1}\mathbf{z}_\alpha\,|\mathbf{S}_1)]\\ &= k_2\mathcal{E}_{\mathbf{S}_1}[\mathrm{Var}_{\mathbf{z}_\alpha}(\mathbf{z}_\alpha'\mathbf{S}_1^{-1}\mathbf{z}_\alpha\,|\mathbf{S}_1)]\\ &= 2k_2\mathcal{E}_{\mathbf{S}_1}(\mathrm{tr}\mathbf{S}_1^{-2}),\end{aligned} \tag{25}$$

where $\mathbf{z}_1, ..., \mathbf{z}_{k_2}$ are independently distributed $\mathbf{N}(\mathbf{0}, \mathbf{I}_p)$, cf. Searle (1971, p. 57). Using Corollary 16, we see that (25) reduces to

$$\frac{2k_2p(k_1-1)}{(k_1-p)(k_1-p-1)(k_1-p-3)}\,.$$

To find the last term of (24) we write

$$\mathrm{Var}_{\mathbf{S}_1}[\mathcal{E}_{\mathbf{S}_2}(\mathrm{tr}\mathbf{S}_1^{-1}\mathbf{S}_2\,|\mathbf{S}_1)] = \mathrm{Var}_{\mathbf{S}_1}\{\mathrm{tr}[\mathcal{E}_{\mathbf{S}_2}(\mathbf{S}_1^{-1}\mathbf{S}_2\,|\mathbf{S}_1)]\} = k_2^2\mathrm{Var}(\mathrm{tr}\mathbf{S}_1^{-1}),$$

using Lemma 1. Then

$$k_2^2\mathrm{Var}(\mathrm{tr}\mathbf{S}_1^{-1}) = \frac{2pk_2^2(k_1-1)}{(k_1-p)(k_1-p-1)^2(k_1-p-3)}\,,$$

from Corollary 17. Thus

$$\begin{aligned}\mathrm{Var}(\mathrm{tr}\mathbf{S}_1^{-1}\mathbf{S}_2) &= \frac{2k_2p(k_1-1)}{(k_1-p)(k_1-p-1)(k_1-p-3)} + \frac{2pk_2^2(k_1-1)}{(k_1-p)(k_1-p-1)^2(k_1-p-3)}\\ &= \frac{2k_2p(k_1-1)(k_1-p-1) + 2k_2^2p(k_1-1)}{(k_1-p)(k_1-p-1)^2(k_1-p-3)} = \frac{2k_2p(k_1-1)(k_1+k_2-p-1)}{(k_1-p)(k_1-p-1)^2(k_1-p-3)}\,,\end{aligned}$$

as claimed. ❒

While $\mathbf{S}_1+\mathbf{S}_2$ is not distributed independently of $\mathbf{S}_1^{-1}\mathbf{S}_2$ (Theorem 4), we find (Corollary 21 below) that any element in $\mathbf{S}_1+\mathbf{S}_2$ is uncorrelated with any element in $\mathbf{S}_1^{-1}\mathbf{S}_2$. We first prove a more general result:

Theorem 6. *Let the* $p\times p$ *random matrices* $\mathbf{S}_1$ *and* $\mathbf{S}_2$ *be distributed independently as* $\mathbf{W}_p(\mathbf{V}, k_1)$ *and* $\mathbf{W}_p(\mathbf{V}, k_2)$, *respectively, with* $k_1 > p+1$ *and* $\mathbf{V}$ *nonsingular. Then any bilinear form in* $\mathbf{S}_1+\mathbf{S}_2$ *is uncorrelated with any bilinear form in* $\mathbf{S}_1^{-1}\mathbf{S}_2$.

Proof. It suffices to show that $\mathrm{cov}[\mathbf{a}_0'\mathbf{S}_1^{-1}\mathbf{S}_2\mathbf{b}_0,\ \mathbf{c}_0'(\mathbf{S}_1+\mathbf{S}_2)\mathbf{d}_0] = 0$ for any $p\times 1$ vectors $\mathbf{a}_0$, $\mathbf{b}_0$, $\mathbf{c}_0$, and $\mathbf{d}_0$. Now $\mathbf{H}_1 = \mathbf{V}^{-1/2}\mathbf{S}_1\mathbf{V}^{-1/2} \sim \mathbf{W}_p(\mathbf{I}, k_1)$ independently of $\mathbf{H}_2 = \mathbf{V}^{-1/2}\mathbf{S}_2\mathbf{V}^{-1/2} \sim \mathbf{W}_p(\mathbf{I}, k_2)$. Thus

$$\mathrm{cov}[\mathbf{a}_0'\mathbf{S}_1^{-1}\mathbf{S}_2\mathbf{b}_0,\ \mathbf{c}_0'(\mathbf{S}_1+\mathbf{S}_2)\mathbf{d}_0] = \mathrm{cov}[\mathbf{a}_0'\mathbf{V}^{-1/2}\mathbf{H}_1^{-1}\mathbf{V}^{-1/2}\mathbf{V}^{1/2}\mathbf{H}_2\mathbf{V}^{1/2}\mathbf{b}_0,\ \mathbf{c}_0'\mathbf{V}^{1/2}(\mathbf{H}_1+\mathbf{H}_2)\mathbf{V}^{1/2}\mathbf{d}_0],$$

$$= \mathrm{cov}[\mathbf{a}'\mathbf{H}_1^{-1}\mathbf{H}_2\mathbf{b},\ \mathbf{c}'(\mathbf{H}_1+\mathbf{H}_2)\mathbf{d}], \tag{26}$$

where $\mathbf{a} = \mathbf{V}^{-1/2}\mathbf{a}_0$, $\mathbf{b} = \mathbf{V}^{1/2}\mathbf{b}_0$, $\mathbf{c} = \mathbf{V}^{1/2}\mathbf{c}_0$, and $\mathbf{d} = \mathbf{V}^{1/2}\mathbf{d}_0$. We may write (26) as

$$\mathcal{E}(\mathbf{a}'\mathbf{H}_1^{-1}\mathbf{H}_2\mathbf{b}\mathbf{c}'\mathbf{H}_1\mathbf{d}) + \mathcal{E}(\mathbf{a}'\mathbf{H}_1^{-1}\mathbf{H}_2\mathbf{b}\mathbf{c}'\mathbf{H}_2\mathbf{d}) - \mathcal{E}(\mathbf{a}'\mathbf{H}_1^{-1}\mathbf{H}_2\mathbf{b})\mathcal{E}[\mathbf{c}'(\mathbf{H}_1+\mathbf{H}_2)\,\mathbf{d})]. \tag{27}$$

The first term in (27) is

$$\mathcal{E}_{\mathbf{H}_2}[\mathcal{E}_{\mathbf{H}_1}(\mathbf{a}'\mathbf{H}_1^{-1}\mathbf{H}_2\mathbf{b}\mathbf{c}'\mathbf{H}_1\mathbf{d} \mid \mathbf{H}_2)] = \frac{\mathbf{a}'\mathcal{E}_{\mathbf{H}_2}[k_1\mathbf{H}_2\mathbf{b}\mathbf{c}' - \mathbf{c}\mathbf{b}'\mathbf{H}_2 - (\mathbf{c}'\mathbf{H}_2\mathbf{b})\mathbf{I}]\mathbf{d}}{k_1-p-1}$$

$$= \frac{k_1k_2\mathbf{a}'\mathbf{b}\mathbf{c}'\mathbf{d} - k_2\mathbf{a}'\mathbf{c}\mathbf{b}'\mathbf{d} - k_2\mathbf{a}'\mathbf{d}\mathbf{c}'\mathbf{b}}{k_1-p-1},$$

using Theorem 2 and Corollary 2. Since $\mathbf{H}_1$ is independent of $\mathbf{H}_2$, we can write the second term of (27) as

$$\mathcal{E}(\mathbf{a}'\mathbf{H}_1^{-1})\mathcal{E}(\mathbf{H}_2\mathbf{b}\mathbf{c}'\mathbf{H}_2\mathbf{d}) = \frac{\mathbf{a}'\mathcal{E}(\mathbf{H}_2\mathbf{b}\mathbf{c}'\mathbf{H}_2)\mathbf{d}}{k_1-p-1} = \frac{\mathbf{a}'\{k_2[\mathbf{c}\mathbf{b}'+(\mathbf{b}'\mathbf{c})\mathbf{I}] + k_2^2\mathbf{b}\mathbf{c}'\}\mathbf{d}}{k_1-p-1}$$

$$= \frac{k_2\mathbf{a}'\mathbf{c}\mathbf{b}'\mathbf{d} + k_2\mathbf{b}'\mathbf{c}\mathbf{a}'\mathbf{d} + k_2^2\mathbf{a}'\mathbf{b}\mathbf{c}'\mathbf{d}}{k_1-p-1}.$$

The first two terms in (27) add to

$$\frac{k_1k_2\mathbf{a}'\mathbf{b}\mathbf{c}'\mathbf{d} + k_2^2\mathbf{a}'\mathbf{b}\mathbf{c}'\mathbf{d}}{k_1-p-1} = \frac{k_2(k_1+k_2)\mathbf{a}'\mathbf{b}\mathbf{c}'\mathbf{d}}{k_1-p-1}, \tag{28}$$

while the last term in (27) is found, using Corollary 1 and Lemma 4, to equal

$$\frac{\mathbf{a}'\mathbf{b}k_2\mathbf{c}'\mathbf{d}(k_1+k_2)}{k_1-p-1}. \tag{29}$$

Subtracting (29) from (28) establishes the theorem. ❐

We end this paper with the following interesting result:

Corollary 21. *Let* $\mathbf{S}_1 \sim \mathbf{W}_p(\mathbf{V}, k_1)$ *independently of* $\mathbf{S}_2 \sim \mathbf{W}_p(\mathbf{V}, k_2)$, *with* $\mathbf{V}$ *nonsingular and* $k_1 > p+1$ *Then each element of* $\mathbf{S}_1^{-1}\mathbf{S}_2$ *is uncorrelated with each element of* $\mathbf{S}_1+\mathbf{S}_2$.

ACKNOWLEDGEMENTS

This research was begun in 1973 when Monique Andrée Legault-Giguère was completing her M.Sc. thesis under my guidance at McGill University; the main results in this paper were first announced (Styan 1973) at the IMS Annual Meeting (New York City, December 1973) and appeared in Appendix B of the thesis (Legault-Giguère 1974). I am most grateful to Monique Andrée Legault-Giguère for many stimulating and extremely helpful discussions, which also led to our two joint papers Giguère and Styan (1976, 1978).

This research was supported in part by the Natural Sciences and Engineering Research Council of Canada and by the Fonds pour la Formation de Chercheurs et l'Aide à la Recherche du Gouvernement du Québec.

REFERENCES

Anderson, T.W. (1984). *An Introduction to Multivariate Statistical Analysis.* Second Edition. Wiley, New York.

Basu, D. (1955). On statistics independent of a complete sufficient statistic. *Sankhyā,* 15, 377-380.

Das Gupta, Somesh (1968). Some aspects of discrimination function coefficients. *Sankhyā Ser. A,* 30, 387-400.

Giguère, Monique Andrée, and Styan, George P.H. (1976). Comparisons between maximum likelihood and naïve estimators in a multivariate normal population with data missing on one variate. *Bull. Inst. Internat. Statist.*, 46(3), 303-308.

Giguère, Monique Andrée, and Styan, George P.H. (1978). Multivariate normal estimation with missing data on several variates. *Trans. Seventh Prague Conference on Information Theory, Statistical Decision Functions, Random Processes and of the Eighth European Meeting of Statisticians (Tech. Univ., Prague, August 1974).* Academia, Prague, and D. Reidel, Dordrecht, vol.B, 129-139.

Legault-Giguère, Monique Andrée (1974). Multivariate Normal Estimation with Missing Data. M.Sc. thesis, Dept. Mathematics, McGill University, Montréal, vii + 123 pp.

Lehmann, E.L. (1986). *Testing Statistical Hypotheses.* Second Edition. Wiley, New York.

Ogasawara, Tōzirō, and Takahashi, Masayuki (1951). Independence of quadratic quantities in a normal system. *J. Sci. Hiroshima Univ. Ser. A*, 15, 1-9.

Olkin, Ingram, and Rubin, Herman (1962). A characterization of the Wishart distribution. *Ann. Math. Statist.*, 33, 1272-1280.

Olkin, Ingram, and Rubin, Herman (1964). Multivariate beta distributions and independence properties of the Wishart distribution. *Ann. Math. Statist.*, 35, 261-269.

Searle, S.R. (1971). *Linear Models.* Wiley, New York.

Styan, George P.H. (1973). Expectations of second degree expressions in a Wishart matrix and its inverse. (Abstract 141-9.) *The IMS Bulletin*, 2, page 216.

Styan, George P.H. (1985). Schur complements and linear statistical models. *Proc. First International Tampere Seminar on Linear Statistical Models and their Applications* (T. Pukkila and S. Puntanen, *eds.*), Dept. of Mathematical Sciences, University of Tampere, Finland, pp. 37-75.

Wishart, John (1928). The generalized product moment distribution in samples from a normal multivariate population. *Biometrika*, 20A, 32-52, 424.

STATISTICAL DATA ANALYSIS AND INFERENCE
Y. Dodge (Editor)

CHARACTERIZATIONS OF DISTRIBUTIONS VIA MOMENTS OF ORDER STATISTICS: A SURVEY AND COMPARISON OF METHODS

Gwo Dong LIN

Institute of Statistical Science, Academia Sinica, Taipei 11529, Taiwan, Republic of China

Recently, many interesting results about the characterizations of distributions using the moments of order statistics have been developed. We review these characteristic properties and compare the useful methods. The counterpart using the moments of record values is also investigated.

1. INTRODUCTION

Let $X_{k:n}$ be the k-th smallest order statistic of a random sample of size n from a distribution $F(x) = Pr(X \le x)$ with $E|X| < \infty$. Then Hoeffding [13] proved that the triangular array of all expected order statistics

$$\{ EX_{k:n} |\ k = 1,2,\ldots,n,\ n = 1,2,3,\ldots \} \qquad (1)$$

uniquely determines the distribution F in the whole class of distributions with finite mean. Since then many contributions have been made to improve this result, namely, to find subsets of (1) which still characterize F. In Section 2 we shall review these results. Section 3 will be devoted to the case of higher moments of order statistics. In Section 4 we shall review the characterization results using only two moments (of different order) of order statistics. Section 5 will be concerned with the characteristic properties of specific distributions using the relationships between product moments of order statistics. The counterpart using the moments of record values is also investigated in Section 6. Finally, Section 7 will concern the latest characterization result using the moments of population distribution.

2. CHARACTERIZATION OF DISTRIBUTION USING EXPECTED ORDER STATISTICS

Let $\mathcal{F}_m$ be the class of distributions F with $E|X|^m < \infty$.

Then as mentioned above, Hoeffding [13] proved that the set (1) characterizes F in $\mathcal{F}_1$. Actually, he obtained a somewhat more general result: the distribution F_n which assigns equal weights to the n values $EX_{k:n}$ converges weakly to F as $n \to \infty$.

Define the inverse function of a distribution F by

$$F^{-1}(t) = \inf \{ x \mid F(x) \geq t \},\ t \in (0, 1).$$

Then by the uniqueness of the Fourier transform of F^{-1} Chan [5] (see also Gupta [11]) proved that the set of all expected maximal order statistics

$$\{ EX_{n:n} \mid n = 1, 2, 3, \dots \} \qquad (2)$$

characterizes F in $\mathcal{F}_1$.

Using the transformation X' = -X, it is clear from (2) that the set of all expected minimal order statistics

$$\{ EX_{1:n} \mid n = 1, 2, 3, \dots \}$$

also characterizes F in $\mathcal{F}_1$. Konheim [22], who obtained the same result (2) independently by another approach, may be the first to mention Hoeffding's characterization result.

Applying the uniqueness of the moment problem Ali [1] (see also Wang [36]) extended Chan's result (2) to the following: for any fixed integer $r \geq 0$, the subset of (1)

$$\{ EX_{n-r:n} \mid n = r+1, r+2, \dots \} \qquad (3)$$

characterizes F in $\mathcal{F}_1$.

On the other hand, Pollak [33] proved that the subset of (1) (formed by choosing any element from each horizontal line of (1))

$$\{ EX_{k_n:n} \mid n = 1, 2, \dots \} \text{ with } k_n \in \{1,2,\dots,n\} \qquad (4)$$

is enough to characterize F in $\mathcal{F}_1$. To do this, Pollak first proved that the set (4) determines the set (1) and then used the following fact to prove independently that the set (1) characterizes F in $\mathcal{F}_1$.

LEMMA 1. Let $F \in \mathcal{F}_1$ and let $\lim_{n\to\infty} k_n/n = \alpha \in (0, 1)$ at which F^{-1} is continuous. Then

$$EX_{k_n:n} \longrightarrow F^{-1}(\alpha) \quad \text{as } n \to \infty.$$

Actually, Hoeffding [13] also gave this lemma in a different form. The first part of Pollak's proof can be derived by the triangular equality:

$$(n-k)EX_{k:n} + kEX_{k+1:n} = nEX_{k:n-1} \,. \tag{5}$$

The meaning of (5) is that the knowledge of any two of the three expected order statistics in (5) determines the third. Hence the set (4) determines the whole set (1) and then characterizes F by Hoeffding's result.

In the same spirit as (5), it is seen that for $1 \le k < k' \le n$, the set

$$\{ EX, EX_{2:2}, \ldots, EX_{n-2:n-2}, EX_{k:n}, EX_{k':n} \}$$

determines $EX_{n-1:n-1}$. Hence each set of the form

$$\{ EX \} \cup \{ EX_{k_j:2j+1} \}_{j=1}^{\infty} \cup \{ EX_{k_j':2j+1} \}_{j=1}^{\infty}, \tag{6}$$

where $1 \le k_j < k_j' \le 2j+1$, characterizes F in $\mathcal{F}_1$ (Lin [23]). In the same paper Lin pointed out that each set

$$\{ EX_{k_j:n_j} \}_{j=1}^{\infty} \text{ with } \{ k_j/n_j \}_{j=1}^{\infty} \text{ dense in } (0, 1) \tag{7}$$

also characterizes F in $\mathcal{F}_1$. This is in fact a by-product of Lemma 1 by noting that the inverse function F^{-1} is nondecreasing and left-continuous on (0, 1). Also, this is an extension of the following Hwang's [17] result: the set

$$\{ EX_{k:n_j} \mid k=1,2,\ldots,n_j \,,\ j=1,2,\ldots \} \text{ with } \lim_{j\to\infty} n_j = \infty \tag{8}$$

characterizes F in $\mathcal{F}_1$. Based on the triangular equality (5), Arnold and Meeden [2] derived a sufficient condition for a proper subset of (1) to determine F.

It should be mentioned that the m-th moment of order statistic $X_{k:n}$ can be explicitly written as

$$EX_{k:n}^m = k\binom{n}{k} \int_0^1 (F^{-1}(t))^m t^{k-1}(1-t)^{n-k}dt \tag{9}$$

provided either side exists. Hence $E|X_{k:n}^m| < \infty$ if $E|X^m| < \infty$. The existence of LHS of (9) has been studied by Sen [34] (see also Azlarov and Volodin [3], p.28).

Huang [14] introduced the Müntz-Szász theorem, a very powerful analytical tool, into the topic on characterization of distribution. It is an extension of the well-known Weierstrass approximation theorem (see, e.g., Boas [4], p.235).

MÜNTZ-SZASZ THEOREM. Let $\{ n_j \}_{j=1}^{\infty}$ be a sequence of real numbers satisfying

$$0 < n_1 < n_2 < \dots \quad \text{and} \quad \sum_{j=1}^{\infty} \frac{1}{n_j} = \infty. \tag{10}$$

Then the sequence of functions $\{ t^{n_j} \}_{j=1}^{\infty}$ is complete in $L(0, 1)$ (the space of all Lebesgue integrable functions on $(0, 1)$). Namely, for $f \in L(0, 1)$, the equalities

$$\int_0^1 f(t)\, t^{n_j} dt = 0, \; j = 1, 2, \dots, \tag{11}$$

imply

$$f(t) = 0 \quad \text{a.e. (almost everywhere) on } (0, 1). \tag{12}$$

Applying the formula (9) and the theorem above, Huang [14] extended Chan's result (2) to the following: each set

$$\{ EX_{n_j:n_j} \}_{j=1}^{\infty} \quad \text{with } \{ n_j \}_{j=1}^{\infty} \text{ satisfying (10)} \tag{13}$$

characterizes F in $\mathcal{F}_1$. Huang and Hwang [16] also improved Pollak's result (4) by excluding the first term EX. To do this, they extended the Müntz-Szász theorem by proving that the conclusion (12) still holds if the RHS of (11) is replaced by any fixed constant c. For other extensions of the Müntz-Szász theorem, see Hwang and Lin [19]. It remains open that whether any other proper subset of (4) (e.g., excluding the first two terms EX and $EX_{k_2:2}$ from the set (4)) still characterizes F in $\mathcal{F}_1$ (Huang [15]).

We now return to the triangular array (1), which characterizes F in $\mathcal{F}_1$ because the sequence of functions

$$\{ t^{k-1}(1-t)^{n-k} \mid 1 \le k \le n, \; n = 1, 2, \dots \} \tag{14}$$

is complete in $L(0, 1)$. For $m = 1$, rewrite (9) as

$$EX_{k:n} = k\binom{n}{k} \int_0^1 [F^{-1}(t)t^{k_0-1}(1-t)^{n_0-k_0}] \cdot$$
$$\cdot\; t^{k-k_0}(1-t)^{(n-n_0)-(k-k_0)}dt$$

where $1 \le k_0 \le n_0$ are fixed integers and

$$(k, n) \in \Lambda \equiv \{ (k, n) \mid k_0 \le k \le n-n_0+k_0, \; n \ge n_0 \}. \tag{15}$$

Since the sequence of functions

$$\{ t^{k-k_0}(1-t)^{(n-n_0)-(k-k_0)} \mid (k, n) \in \Lambda \}$$

is complete in $L(0, 1)$, the sub-triangular array of (1)

$$\{ EX_{k:n} \mid (k, n) \in \Lambda \} \tag{16}$$

characterizes F in $\mathcal{F}(k_0, n_0) \supset \mathcal{F}_1$, where

$$\mathcal{F}(k_0, n_0) \equiv \{ F \mid F \text{ satisfies } E|X_{k_0:n_0}| < \infty \}.$$

Similarly, all the results (2)-(4), (6)-(8) and (13) can be modified properly, as subsets of (16), to characterize F in $\mathcal{F}(k_0, n_0)$. This kind of improvement was first observed by Huang [14].

3. CHARACTERIZATION OF DISTRIBUTION USING HIGHER MOMENTS OF ORDER STATISTICS

Let $m \geq 3$ be an odd integer and F be the distribution of a random variable (r.v.) X. Then the distribution of X^m determines F and $(X^m)_{k:n}$ is distributed as $X^m_{k:n}$ $(\equiv (X_{k:n})^m)$. Therefore, all the characteristic properties of Section 2 can be extended to the case of higher odd moments of order statistics. For example, each set (a duplicate of (13))

$$\{ EX^m_{n_j:n_j} \}_{j=1}^{\infty} \text{ with } \{ n_j \}_{j=1}^{\infty} \text{ satisfying (10)}$$

characterizes F in $\mathcal{F}_m$.

However, the case of even moments cannot be done in the same way. Consider the following example: Let Pr (X = 1) = Pr (Y = -1) = 1, then for each even integer $m \geq 2$

$$EX^m_{k:n} = EY^m_{k:n} \quad \text{for all } 1 \leq k \leq n \text{ and } n \geq 1.$$

Hence the set of all even moments of all order statistics cannot characterize a distribution in the class $\bigcap_{m=1}^{\infty} \mathcal{F}_m$. But surprisingly, by adding one more odd moment of any order statistic, the results of Section 2 can be extended to the case of even moments. For example, the set (compared to (13))

$$\{EX\} \cup \{EX^2_{n_j:n_j}\}_{j=1}^{\infty} \text{ with } \{n_j\}_{j=1}^{\infty} \text{ satisfying (10)} \tag{17}$$

characterizes F in $\mathcal{F}_2$. To prove this, Hwang and Lin [18] used the following lemma.

LEMMA 2. Let F, G $\in \mathcal{F}_2$ have the same mean and let $(F^{-1})^2 = (G^{-1})^2$. Then F = G.

Roughly speaking, the second part of the set (17) characterizes $(F^{-1})^2$ by the Müntz-Szász theorem; then the first part and $(F^{-1})^2$ together characterize F by Lemma 2.

4. CHARACTERIZATION OF DISTRIBUTION USING TWO MOMENTS OF ORDER STATISTICS

The results reviewed so far involve infinitely many moments (of the same order) of order statistics. It is somowhat surprising that two suitable moments of order statistics together characterize the unit uniform distribution in $\mathcal{F}_2$ (Lin [26]). Specifically, the relations

$$EX^2 = \frac{1}{3} \quad \text{and} \quad EX_{2:2} = \frac{2}{3} \tag{18}$$

hold if and only if $F(x) = x$ on $(0, 1)$. This result is later extended to the following (see Lin [29] for the general case).

THEOREM 1. Assume that G is a nondecreasing and right-continuous function from $R \equiv (-\infty, \infty)$ to R. Then the relations

$$EG^2(X) = \frac{1}{3} \quad \text{and} \quad EG(X_{2:2}) = \frac{2}{3} \tag{19}$$

hold if and only if $F(x) = G(x)$ on $(\ F^{-1}(0^+), F^{-1}(1^-)\ )$ and F is continuous on R.

The keys to prove Theorem 1 are Lemma 3 below and the Cauchy-Schwarz inequality.

LEMMA 3. Let F and G be two distributions on R. Then the relation

$$G(F^{-1}(t)) = t \quad \text{a.e. on } (0, 1)$$

holds if and only if $F = G$ and F is continuous on R.

From (18) we see that the set $\{ EX^2, EX_{2:2} \}$ characterizes the unit uniform distribution on $(0, 1)$. It is natural to ask the question: for given $n \geq 3$, does the set $\{ EX^2, EX_{n:n} \}$ still characterize the unit uniform distribution in $\mathcal{F}_2$? The answer is negative due to Too and Lin [35]. But, for $n \geq 2$, the set $\{ EX^2, EX_{n:n} \}$ characterizes the distribution $F(x) = x^{1/(n-1)}$ on $(0, 1)$. In the same paper, requiring only one equality Too and Lin improve the results (18). Specifically, for $F \in \mathcal{F}_2$, the relation

$$EX_{2:2} - EX^2 = \frac{1}{3} \tag{20}$$

holds if and only if $F(x) = x$ on $(0, 1)$. The proof depends on solving directly the equation (20). Further, applying Lemma 3 the general continuous distributions can be characterized in the same way. Namely, let G be the same as in Theorem 1 and let the distribution of $G(X)$ be in $\mathcal{F}_2$, then the relation

$$EG(X_{2:2}) - EG^2(X) = \frac{1}{3} \tag{21}$$

holds if and only if $F(x) = G(x)$ on $(\ F^{-1}(0^+), F^{-1}(1^-)\ )$ and F is continuous on R.

For given integers $1 \leq k \leq n$ and $1 \leq \ell \leq m$, there are infinitely many distributions F satisfying

$$EX_{k:n} = k/(n+1) \quad \text{and} \quad EX_{\ell:m} = \ell/(m+1)$$

(Hwang and Lin [20]). This means that no pair of the expected order statistics (two moments of the same order) can characterize the unit uniform distribution in the class $\mathcal{F}_1$. However, restricting ourselves to a special class of distributions, we obtain some interesting results. For example, given $F_0 \in \mathcal{F}_1$, consider the class of distributions

$$\mathcal{F}_{F_0} = \{ F_{\mu,\sigma} | F_{\mu,\sigma}(x) = F_0((x-\mu)/\sigma) \text{ on } R, \mu \in R, \sigma > 0 \}.$$

In general, the set of two expected order statistics

$$\{ EX_{k:n}, EX_{\ell:m} \} \tag{22}$$

characterizes F in $\mathcal{F}_{F_0}$ if and only if $EX_{k:n}^{(0)} \neq EX_{\ell:m}^{(0)}$, where the superscript " 0 " means that the corresponding random sample is drawn from F_0. For example, setting $F_0(x) = x$ on (0, 1) the set (22) characterizes F in $\mathcal{F}_{F_0}$ if and only if $k/(n+1) \neq \ell/(m+1)$. And let $F_0(x) = 1 - \exp(-x)$ on $(0, \infty)$, then for any two distinct pairs (k, n) and (ℓ, m), the set (22) characterizes F in $\mathcal{F}_{F_0}$. For further results, see Hwang and Lin [20].

5. CHARACTERIZATION OF DISTRIBUTION USING PRODUCT MOMENTS OF ORDER STATISTICS

In the previous sections we review the characterization result using the moments of order statistics. It is interesting that using the relationships between the product moments of order statistics, some specific distributions can also be characterized. For example, Theorem 2 below extends Govindarajulu's [7] result about the normal distribution (see Lin [28] for other cases). Compared to (9), the product moment of order statistics can be explicitly written as

$$E(X_{r:n} X_{s:n}) = \frac{n!}{(r-1)!(s-r-1)!(n-s)!} \int_0^1 \int_u^1 F^{-1}(u)F^{-1}(v)\, u^{r-1} \cdot (v-u)^{s-r-1}(1-v)^{n-s} dv du, \quad \text{for } 1 \leq r < s \leq n,$$

provided either side of the equality exists. This is also a powerful formula when deriving the recurrence relations and identities for product moments of order statistics (Lin [28]).

THEOREM 2. Let X be a r.v. having distribution $F \in \mathcal{F}_2$ and $\{ n_j \}_{j=1}^{\infty}$ be a sequence of integers satisfying (10). Assume that EX = 0 and F^{-1} is differentiable on (0, 1). Then for any given integer $r \geq 1$, the relation

$$\sum_{s=1}^{n_j} E(X_{r:n_j} X_{s:n_j}) = 1 \quad \text{for all } n_j \geq r+1$$

holds if and only if F is the standard normal distribution Φ.

The key to prove Theorem 2 is the following.

LEMMA 4. Let F be a distribution with the inverse function F^{-1} differentiable on (0, 1). Then the relation

$$(F^{-1})'(u) \int_u^1 F^{-1}(v)dv = 1 \quad \text{a.e. on } (0, 1)$$

holds if and only if

$$F(x) = \{ \Phi(x)-\Phi(A) \}/\{ 1-\Phi(A) \} \quad \text{on } (A, \infty)$$

for some extended real number $A \geq -\infty$.

Extending some of Govindarajulu's ([8], [9]) results, Lin [27] provided the characteristic properties of the uniform, exponential, logistic and Pareto distributions using the relationships between the single moments of order statistics. And the characterizations of general distributions using the linear combinations of moments of order statistics (e.g., the expected spacings between order statistics) are investigated by Govindarajulu, Huang and Saleh [10] and Hwang and Lin [18]. (See also Galambos and Kotz [6], p.58.)

6. CHARACTERIZATION OF DISTRIBUTION USING MOMENTS OF RECORD VALUES

Let $\{ X_n \}_{n=1}^{\infty}$ be a sequence of independent and identically distributed r.v.'s having continuous distribution F. Define the sequence of record times $\{ L(n) \}_{n=0}^{\infty}$ by $L(0) = 1$ and $L(n) = \min\{ j | X_j > X_{L(n-1)} \}$ for $n \geq 1$. Then the sequence $\{X_{L(n)}\}_{n=0}^{\infty}$ is called the sequence of record values of $\{ X_n \}_{n=1}^{\infty}$. Note that for every sample point, $\{ X_{L(n)} \}_{n=0}^{\infty}$ forms a subsequence of $\{ X_{n:n} \}_{n=1}^{\infty}$. For convenience, denote by $\mathcal{F}_p^c$ the class of continuous distributions F with $E|X|^p < \infty$.

The existence of moments of record values was first studied by Nagaraja [32]. Kirmani and Beg [21] proved that for any fixed integer $n_0 \geq 0$, the set of expected record values $\{ EX_{L(n)} \}_{n=n_0}^{\infty}$ characterizes F in $\mathcal{F}_p^c$ with $p > 1$. This is the counterpart of Chan's result (2). However, a result that would stand in the same relation to Kirmani and Beg's result as does (13) to (2) need not hold (Lin and Huang [31]). Lin [25] then investigated (a) the characterization of continuous distribution using the higher moments of record values and (b) the relationships between the weak convergence of continuous distributions and the convergence of their moments of record values.

Due to the continuity of F which implies the strict monotonicity of F^{-1}, the set of even moments of record values, $\{ EX^m_{L(n)} \}^{\infty}_{n=0}$ say, characterizes F in $\mathcal{F}^c_p$ with p > m. The odd moment as in (17) is no longer needed when characterizing a continuous distribution. There are two keys to prove the characterization results above. The first is the explicit formula for the m-th moment of record value:

$$EX^m_{L(n)} = \frac{1}{n!} \int_0^1 (F^{-1}(t))^m (\log \frac{1}{1-t})^n dt, \text{ for } n \geq 0,$$

provided either side exists. The second is the complete sequence of functions $\{ (\log \frac{1}{1-t})^n \}^{\infty}_{n=0}$ in L(0, 1) (see Lin [24] for an extension of this fact).

Gupta [12] proved that for given integers $k \geq 1$ and $n_0 \geq 0$, the expected spacings between record values

$$\{ EX_{L(n+k)} - EX_{L(n)} \}^{\infty}_{n=n_0}$$

characterizes, up to a location parameter, F in $\mathcal{F}^c_p$ with p > 1. In fact, this result can be modified as follows. For given integers $k \geq 1$, $n_0 \geq 0$ and $m \geq 1$, the set

$$\{ (n+k)!EX^m_{L(n+k)} - n!EX^m_{L(n)} \}^{\infty}_{n=n_0}$$

characterizes F in $\mathcal{F}^c_p$ with p > m (Lin [27]).

Finally, we review the characterization results based on two record values. Theorem 3 below is the counterpart of Theorem 1. (See Lin [29] for the general case.)

THEOREM 3. Let X be a r.v. having continuous distribution F. Assume that G is nondecreasing and right-continuous function from R to $(-\infty, 1]$. Then F(x) = G(x) on $(F^{-1}(0^+), F^{-1}(1^-))$ if and only if

$$E\{ -\log(1-G(X_{L(1)})) \} = E\{ -\log(1-G(X)) \}^2 = 2. \quad (23)$$

Further, assuming the finiteness of the middle term of (23), the necessary and sufficient condition (23) can be replaced by the following:

$$2E\{ -\log(1-G(X_{L(1)})) \} - E\{ -\log(1-G(X)) \}^2 = 2. \quad (24)$$

We can prove this fact by solving directly the equation (24).

7. CHARACTERIZATION OF DISTRIBUTION USING MOMENTS

It is well known that to characterize the distribution F of a r.v. X using the set of moments $\{ EX^n \}^{\infty}_{n=1}$, some suitable conditions should be assumed (e.g., the Carleman's condition which requires the finiteness of moments of all orders). However, for characterizing distributions on (0, ∞), we have the following result in which $\mathcal{F}_\alpha^+$ denotes the class of distributions F on (0, ∞) with $EX^\alpha < \infty$.

THEOREM 4. Let $F \in \mathcal{F}_\alpha^+$ for some $\alpha > 0$. Let $\{ s_m \}_{m=1}^\infty$ be a sequence of positive and distinct numbers converging to some point $s_0 \in (0, \alpha)$. Then the set of moments $\{ EX^{s_m} \}_{m=1}^\infty$ characterizes F in $\mathcal{F}_\alpha^+$.

The proof of Theorem 4 is based on the uniqueness of analytic function on a domain (Lin [30]).

REFERENCES

[1] Ali, M.I., Canad. J. Statist. 4 (1976) 151-153.

[2] Arnold, B.C. and Meeden, G., Ann. Statist. 3 (1975) 754-758.

[3] Azlarov, T.A. and Volodin, N., Characterization Problems Associated with the Exponential Distribution (Translated by M. Stein and Edited by I. Olkin, Springer-Verlag, Berlin - New York, 1986).

[4] Boas, R.P., Entire Functions (Academic Press., New York, 1954).

[5] Chan, L.K., Amer. Math. Monthly 74 (1967) 950-951.

[6] Galambos, J. and Kotz, S., Characterizations of Probability Distributions (Lecture Notes Math., New York, 1978).

[7] Govindarajulu, Z., Ann. Math. Statist. 37 (1966) 1011-1015.

[8] Govindarajulu, Z., Skank. Aktuarietidskr. 49 (1966) 132-136.

[9] Govindarajulu, Z., Characterization of the Exponential Distribution Using Lower Moments of Order Statistics, in : Patil, G.P., Kotz, S. and Ord, J.K., (eds.) Statistical Distributions in Scientific Work (Dordrecht-Holland, D. Rediel Publishing Co. 1975) Vol. 3, pp. 117-129.

[10] Govindarajulu, Z., Huang, J.S. and Saleh, A.K. Md. E., Expected Values of the Spacings Between Order Statistics, in the same edited volume as in Reference [9], Vol. 3, pp. 143-147.

[11] Gupta, R.C., Commun. in Statist. 3 (1974) 287-289.

[12] Gupta, R.C., J. Appl. Prob. 21 (1984) 425-430.

[13] Hoeffding, W., Ann. Math. Statist. 24 (1953) 93-100.

[14] Huang, J.S., Ann. Inst. Statist. Math. 27 (1975) 87-93.

[15] Huang, J.S., Amer. Math. Monthly 85 (1978) 107-108.

[16] Huang, J.S. and Hwang, J.S., L_1-completeness of a Class of Beta Distributions, in the same edited volume as in Reference [9], Vol. 3, pp. 137-141.

[17] Hwang, J.S., Ann. Inst. Statist. Math. 30 (1978) 167-176.

[18] Hwang, J.S. and Lin, G.D., Bull. Inst. Math. Acad. Sinica 12 (1984) 179-202.

[19] Hwang, J.S. and Lin, G.D., Analysis 4 (1984) 143-160.

[20] Hwang, J.S. and Lin, G.D., Metrika 35 (1988) 329-338.

[21] Kirmani, S.N.U.A. and Beg, M.I., Sankhya 46A (1984) 463-465.

[22] Konheim, A.G., Amer. Math. Monthly 78 (1971) 524.
[23] Lin, G.D., Ann. Inst. Statist. Math. 36 (1984) 451-453.
[24] Lin, G.D., Tôhoku Math. J. 38 (1986) 595-598.
[25] Lin, G.D., Probab. Th. Rel. Fields 74 (1987) 479-483.
[26] Lin, G.D., Sankhya 50A (1988) 64-69.
[27] Lin, G.D., J. Statist. Plann. Inference 19 (1988) 73-80.
[28] Lin, G.D., J. Statist. Plann. Inference 21 (1989) 395-406.
[29] Lin, G.D., To appear in Sankhya 51A (1989).
[30] Lin, G.D., Characterizations of Distributions via Moments. Tech. Report B-87-3, Inst. of Statist. Science, Academia Sinica, R.O.C.
[31] Lin, G.D. and Huang, J.S., Sankhya 49A (1987) 272-273.
[32] Nagaraja, H.N., Austral. J. Statist. 20 (1978) 176-182.
[33] Pollak, M., Ann. Statist. 1 (1973) 180-182.
[34] Sen, P.K., Calcutta Statist. Assoc. Bull. 9 (1959) 1-19.
[35] Too, Y.-H. and Lin, G.D., To appear in Statist. & Probab. Letters (1989).
[36] Wang, Y.H., On a Characterization of the Distribution Functions by the Expected Values of the Order Statistics. Tech. Report 7 (1971) pp. 1-10, Division of Statistics, The Ohio State University.

STATISTICAL DATA ANALYSIS AND INFERENCE
Y. Dodge (Editor)

CONSTANCY OF REGRESSION OF A POLYNOMIAL OF SAMPLE AVERAGE OF POSITIVE RANDOM VARIABLES ON THEIR RATIOS CHARACTERIZES GAMMA DISTRIBUTION

Abram KAGAN

Department of Mathematics, University of Maryland
College Park, MD 20742, U.S.A.

Let $X_1, X_2, \ldots, X_n$ be i.i.d. positive random variables with a distribution function $F(x)$ and $P(\overline{X}) = A_k\overline{X}^k + \ldots + A_0$, $A_k \neq 0$ be a polynomial of degree $k \geq 2$ in $\overline{X}$, the sample average. It is proved that if $n \geq 2k$ and

$$\int_0^\infty x^{k+\varepsilon} dF < \infty, \qquad \int_0^\infty x^{-\varepsilon} dF < \infty$$

for an $\varepsilon > 0$, then

$$E\{P(\overline{X}) | X_1/\overline{X}, \ldots, X_n/\overline{X}\} = \text{constant}$$

if and only if F is gamma. The case of $k = 1$ was investigated by Khatri and Rao (1968) under the minimal necessary conditions $n \geq 3$ and $E(X_1) < \infty$.
If $F(x)$ contains a scale parameter $\sigma > 0$, $F(x) = F(x/\sigma)$, and $P(\overline{X})$ is used as an unbiased estimator of the parameter polynomial $\pi(\sigma) = E_\sigma P(\overline{X}) = A_k\sigma^k + \ldots + A_0$, then under the conditions

$$\int_0^\infty x^{2k} dF < \infty, \qquad \int_0^\infty x^{-\varepsilon} dF < \infty$$

for an $\varepsilon > 0$, $P(\overline{X})$ is the best unbiased estimator of $\pi(\sigma)$ with respect to quadratic loss if and only if F is gamma.

1. INTRODUCTION

Let $X_1, X_2, \ldots, X_n$ be i.i.d. positive random variables with F as its d.f. (distribution function), and

$$\overline{X} = n^{-1}(X_1 + X_2 + \ldots + X_n), \quad U = (X_1/\overline{X}, \ldots, X_n/\overline{X}).$$

Khatri and Rao (1968) showed that if $n \geq 3$ and $E(X_1) < \infty$ then $E(\overline{X}|U) =$ const if and only if F is gamma. In Kagan (1968) (see also Kagan, Linnik and Rao (1973), Lemma 7.13.1) it is shown that if $E(X_j^k) < \infty$ and a (non-constant) polynomial $P(\overline{X}) = A_k\overline{X}^k + \ldots + A_0$, $A_k \neq 0$ has constant regression on U for sample sizes $n = m, m+1, \ldots, m+k-1$ for some $m \geq 3$, i.e.,

$$E[P(\overline{X})|U] = c_n \text{ (constant)}, \quad n = m, m+1, \ldots, m+k-1, \tag{1.1}$$

then F is gamma.

The method used in a recent paper by Kagan and Rao (1988) for characterizing a Gaussian distribution by constancy of regression of a polynomial of the sample average on residuals makes it possible to replace the k conditions in (1.1) by the single condition

$$E[P(\bar{X})|U] = c_n \text{ (constant)} \tag{1.2}$$

for some $n \geq 2k$, provided that the minimal necessary condition $E(X_1^k) < \infty$ is replaced by

$$\int x^{k+\varepsilon} dF < \infty, \quad \int x^{-\varepsilon} dF < \infty \tag{1.3}$$

for a (small) $\varepsilon > 0$.

As a simple corollary of the result it is proved that if $X_1, X_2, \ldots, X_n$ are i.i.d. observations from a d.f. $F(x/\sigma)$, $F(+0) = 0$, depending on a scale parameter $\sigma > 0$, and a polynomial $P(\bar{X})$ is used as an unbiased estimator of its expectation $\pi(\sigma) = E_\sigma[P(\bar{X})] = A_k\sigma^k + \ldots + A_0$ and if further $n \geq 2k$ and

$$\int_0^\infty x^{2k} dF < \infty, \quad \int_0^\infty x^{-\varepsilon} dF < \infty \tag{1.4}$$

for some $\varepsilon > 0$, then $P(\bar{X})$ is the best unbiased estimator of $\pi(\sigma)$ with respect to quadratic loss if and only if F is gamma (cf. Theorem 7.13.1 in Kagan, Linnik and Rao (1973)).

2. THE MAIN THEOREMS

Theorem 1. Let $X_1, X_2, \ldots, X_n$ be i.i.d. positive random variables with a d.f. F and $P(\bar{X}) = A_k\bar{X}^k + \ldots + A_0$, $A_k \neq 0$, be a polynomial in $\bar{X}$ of degree $k \geq 2$. If $n \geq 2k$ and for some $\varepsilon > 0$ the conditions (1.3) are fulfilled, then the condition

$$E\{P(\bar{X})|X_1/\bar{X}, \ldots, X_n/\bar{X}\} = \text{constant} \tag{2.1}$$

holds if and only if F is gamma.

Note that any gamma d.f. F satisfies (1.3) for a suitable $\varepsilon > 0$. It is not known if Theorem 1 remains true when $\varepsilon = 0$ in (1.3), i.e., under the minimal necessary condition $E(X_1^k) < \infty$.

Theorem 2. Let $X_1, X_2, \ldots, X_n$ be i.i.d. positive observations from a d.f. $F(x/\sigma)$ depending on a scale parameter $\sigma > 0$. Suppose that $n \geq 2k$ and the conditions (1.4) are satisfied. A polynomial $P(\bar{X}) = A_k\bar{X}^k + \ldots + A_0$, $A_k \neq 0$, of degree $k \geq 2$ is the best unbiased estimator of the parametric function $\pi(\sigma) = E_\sigma[P(\bar{X})] = A_k\sigma^k + \ldots + A_0$ with respect to quadratic loss if and only if F is gamma.

Proof of Theorem 1. On setting

$$Y_j = \log X_j, \quad \bar{Y} = n^{-1}(Y_1 + \ldots + Y_n), \quad j = 1,2,\ldots,n$$

and noticing that the σ-algebra $\sigma(X_1/\bar{X}, \ldots, X_n/\bar{X})$ becomes $\sigma(Y_1-\bar{Y}, \ldots, Y_n-\bar{Y})$ one can rewrite (1.2) as

$$E[P(\sum_1^n e^{Y_j})|Y_1-\bar{Y},\ldots,Y_n-\bar{Y}] = \text{constant} = c \text{ (say)}. \tag{2.2}$$

The condition (1.3) takes the form

$$E[e^{(k+\varepsilon)Y_j}] < \infty, \quad E[e^{-\varepsilon Y_j}] < \infty. \tag{2.3}$$

Let

$$f(t) = E[\exp(tY_j)] = \int e^{ty}dG(y)$$

where G is the d.f. of Y_j. The function $f(t)$ is well defined for $-\varepsilon < t < k+\varepsilon$ and therefore $f(z)$ is analytic in the strip $-\varepsilon < \operatorname{Re} z < k+\varepsilon$.

Multiplying both sides of (2.2) by $\exp[\sum_1^n t_j(Y_j-\bar{Y})]$ for sufficiently small $|t_1|,\ldots,|t_n|$ and taking expectations we have

$$E\{[A_k(\sum_1^n \exp Y_j)^k + \ldots + A_0][\exp \sum_1^n t_j(Y_j-\bar{Y})]\} = c \prod_1^n f(\tau_j) \tag{2.4}$$

where

$$\tau_j = t_j-\bar{t}, \quad \bar{t} = n^{-1}(t_1 + \ldots + t_n), \quad j = 1,2,\ldots,n. \tag{2.5}$$

For an arbitrary $\psi(t)$ we set

$$\bar{\psi} = \bar{\psi}(t_1,t_2,\ldots,t_n) = \sum_1^n \psi(t_j)$$

and use a special notation for $\phi(t) = f(t+1)/f(t)$

$$\bar{\phi} = \sum_1^n \frac{f(t_j+1)}{f(t_j)}.$$

Lemma 1. For any integer m such that $1 \le m \le k$ and sufficiently small $|t_1|,|t_2|,\ldots,|t_n|$

$$E[(\sum_1^n e^{Y_j})^m \exp(\sum_1^n t_jY_j)] = \prod_1^n f(t_j)[\bar{\phi}^m + Q_m(\bar{\psi}_1,\ldots,\bar{\psi}_\ell)] \tag{2.6}$$

for some $\psi_1,\ldots,\psi_\ell$, $Q_m(u_1,u_2,\ldots,u_\ell)$ being a polynomial of degree $\le m-1$.

Proof of Lemma 1. We use induction. For $m = 1$,

$$E[(\sum_1^n e^{Y_j}) \exp(\sum_1^n t_jY_j)] = f(t_1+1) \prod_{j\neq 1} f(t_j) + \ldots + f(t_n+1) \prod_{j\neq n} f(t_j)$$

$$= \prod_1^n f(t_j)\,\bar{\phi}.$$

Introducing now the operator

$$\Delta = \Delta_1 + \ldots + \Delta_n$$

where

$$\Delta_j\Psi(t_1,\ldots,t_n) = \Psi(t_1,\ldots,t_{j-1},t_j+1,t_{j+1},\ldots,t_n)$$

the left hand side of (2.6) can be written as

$$E[(\sum_1^n e^{Y_j})^m \exp(\sum_1^n t_jY_j)] = \Delta^m[\prod_1^n f(t_j)]. \tag{2.7}$$

According to the induction postulate

$$\Delta^m[\prod_1^n f(t_j)] = \prod_1^n f(t_j)[\overline{\phi}^m + Q(\overline{\psi}_1,\ldots,\overline{\psi}_\ell)], \quad m \le k-1 \tag{2.8}$$

for some ℓ, $\psi_1,\ldots,\psi_\ell$ and a polynomial Q, all depending in general on m with the degree of $Q(u_1,\ldots,u_\ell) \le m-1$.

On applying Δ to both sides of (2.8) we get

$$\begin{aligned}\Delta^{m+1}\{\prod_1^n f(t_j)\} &= f(t_1+1)\prod_{j\neq 1} f(t_j)\{(\phi(t_1+1)+\sum_{j\neq 1}\phi(t_j))^m +\\ &\quad + Q(\psi_1(t_1+1)+\sum_{j\neq 1}\psi_1(t_j),\ldots,\psi_\ell(t_1+1)+\sum_{j\neq 1}\psi_\ell(t_j))\}+\ldots+\\ &\quad + f(t_n+1)\prod_{j\neq n} f(t_j)\{(\phi(t_n+1)+\sum_{j\neq n}\phi(t_j))^m +\\ &\quad + Q(\psi_1(t_n+1)+\sum_{j\neq 1}\psi_1(t_j),\ldots,\psi_\ell(t_n+1)+\sum_{j\neq n}\psi_\ell(t_j))\}\\ &= \prod_1^n f(t_j)\left\{\frac{f(t_1+1)}{f(t_1)}\left(\overline{\phi}+\phi(t_1+1)-\phi(t_1))^m +\right.\right.\\ &\quad \left. +Q(\overline{\psi}_1+\psi_i(t_1+1)-\psi_1(t_1),\ldots,\overline{\psi}_\ell+\psi_\ell(t_1+1)-\psi_\ell(t_1))\right) +\ldots+\\ &\quad + \frac{f(t_n+1)}{f(t_n)}\left((\overline{\phi}+\phi(t_n+1)-\phi(t_n))^m +\right.\\ &\quad \left.\left. + Q(\overline{\psi}_1+\psi_1(t_n+1)-\psi_1(t_n),\ldots,\overline{\psi}_\ell+\psi_\ell(t_n+1)-\psi_\ell(t_n))\right)\right\}.\end{aligned} \tag{2.9}$$

We have

$$\begin{aligned}&\sum_{j=1}^n \frac{f(t_j+1)}{f(t_j)}[\overline{\phi}+\phi(t_j+1)-\phi(t_j)]^m\\ &= \overline{\phi}^{m+1} + \sum_{k=1}^m \binom{m}{k}\overline{\phi}^{m-k}\sum_{j=1}^n \frac{f(t_j+1)}{f(t_j)}[\phi(t_j+1)-\phi(t_j)]^k\\ &= \overline{\phi}^{m+1} + q(\overline{\phi},\overline{\Psi}_1,\ldots,\overline{\Psi}_m)\end{aligned} \tag{2.10}$$

where

$$\Psi_k(t) = \phi(t)[\phi(t+1)-\phi(t)]^k, \quad k = 1,\ldots,m,$$

$$q(u_0,u_1,\ldots,u_m) = \sum_{k=1}^m \binom{m}{k}u_0^{m-k}u_k$$

so that degree of q is $\le m$.

Further

$$\sum_{j=1}^{n} \frac{f(t_j+1)}{f(t_j)} [\overline{\psi}_1+\psi_1(t_j+1)-\psi(t_j)]^{i_1} \ldots [\overline{\psi}_\ell+\psi_\ell(t_j+1)-\psi_\ell(t_j)]^{i_\ell}$$

$$= \sum_{\substack{k_1,\ldots,k_\ell \\ 0\le k_1\le i_1,\ldots,0\le k_\ell\le i_\ell}} \binom{i_1}{k_1} \cdots \binom{i_\ell}{k_\ell} \overline{\psi}_1^{k_1} \ldots \overline{\psi}_\ell^{k_\ell}$$

$$\cdot \sum_{j=1}^{n} \phi(t_j)[\psi_1(t_j+1)-\psi_1(t_j)]^{i_1-k_1} \ldots [\psi_\ell(t_j+1)-\psi_\ell(t_j)]^{i_\ell-k_\ell}. \tag{2.11}$$

On setting

$$\Psi_{m_1\ldots m_\ell}(t) = \phi(t)[\psi_1(t+1)-\psi_1(t)]^{m_1} \ldots [\psi_\ell(t+1)-\psi_\ell(t)]^{m_\ell}$$

we see from (2.7) and (2.11) that the left hand side of (2.8) has the required form (2.6).

Lemma 2. All the functions $\phi(t)$, $\psi_1(t),\ldots,\psi_\ell(t)$ appearing in (2.6) are infinitely differentiable for sufficiently small $|t|$.

Proof of Lemma 2. The result follows from the fact that $f(z)$ is analytic for $-\varepsilon < \mathrm{Re}\, z < k+\varepsilon$ and the functions $\psi_1,\ldots,\psi_\ell$ from (2.6) are polynomials in $\phi(t)$, $\phi(t+1)$, ..., $\phi(t+k)$.

Returning to the proof of Theorem 1 we shall use the method suggested in Kagan and Rao (1988).

By virtue of Lemma 1 we rewrite (2.4) as

$$\prod_1^n f(\tau_j)[A_k\overline{\phi}^k + Q(\overline{\psi}_1,\ldots,\overline{\psi}_m)] = c \prod_1^n f(\tau_j)$$

or, since $\prod_1^n f(\tau_j) \neq 0$, as

$$A_k\overline{\phi}^k + Q(\overline{\psi}_1,\ldots,\overline{\psi}_m) = c \tag{2.12}$$

with τ_j's as defined in (9) and restricted by the condition

$$\tau_1 + \ldots + \tau_n = 0,$$

and $Q(u_1,\ldots,u_m)$ being a polynomial of degree $\le k-1$.

We shall treat $\tau_1,\ldots,\tau_{n-1}$ as unrestricted variables substituting $-(\tau_1 + \ldots + \tau_{n-1})$ for τ_n in what follows. Also restrict τ_j's to the condition $|\tau_j| < \varepsilon$, $j = 1,\ldots,n$ for $\varepsilon > 0$ small enough.

Lemma 2 enables us to differentiate (2.12) with respect to any variable τ_j, $j = 1,\ldots,n-1$.

Let us differentiate (2.12) first with respect to τ_1, then with respect to τ_2 and subtract the first expression from the second. We shall have

$$0 = \{kA_k\bar{\phi}^{k-1}[\phi'(\tau_2)-\phi'(\tau_n)] + \sum_{j=1}^{m} [\psi_j(\tau_2)-\psi_j(\tau_n)](\frac{\partial Q}{\partial u_j})\Big|_{u_1=\bar{\psi}_1,\ldots,u_m=\bar{\psi}_m}\}$$

$$- \{kA_k\bar{\phi}^{k-1}[\phi'(\tau_1)-\phi'(\tau_n)] + \sum_{j=1}^{m} [\psi_j(\tau_1)-\psi_j(\tau_n)](\frac{\partial Q}{\partial u_j})\Big|_{u_1=\bar{\psi}_1,\ldots,u_m=\bar{\psi}_m}\}$$

so that

$$kA_k\bar{\phi}^{k-1}[\phi'(\tau_2)-\phi'(\tau_1)] + \sum_{j=1}^{m} [\psi_j(\tau_2)-\psi_j(\tau_1)](\frac{\partial Q}{\partial u_j})\Big|_{u_1=\bar{\psi}_1,\ldots,u_m=\bar{\psi}_m} = 0 \quad (2.13)$$

The polynomials $\partial Q(u_1,\ldots,u_m)/\partial u_j$, $j = 1,\ldots,m$ are of degrees $\leq$ k-2.

Now differentiate (2.13) first with respect to τ_3, then with respect to τ_4 and again subtract the first expression from the second. We repeat the procedure k-2 times involving the arguments $\tau_1,\tau_2,\ldots,\tau_{2k-5},\tau_{2k-4}$.

Since each differentiation reduces the degrees of the polynomials Q, $\partial Q/\partial u_j$, $\partial^2 Q/\partial u_j \partial u_\ell$, ..., after k-2 steps we shall have

$$k(k-1)\cdots 4\cdot 3A_k\bar{\phi}^2 \prod_{j=1}^{k-2} [\phi'(\tau_{2j})-\phi'(\tau_{2j-1})] +$$

$$+ \sum_{j=1}^{m} C_j(\tau_1,\tau_2;\ldots;\tau_{2k-5},\tau_{2k-4})\bar{\psi}_j = 0 \quad (2.14)$$

for some $C_1,\ldots,C_m$ depending on the indicated arguments. Differentiation of (2.14) with respect to τ_{2k-3} gives the final relation

$$k!A_k\bar{\phi} \prod_{j=1}^{k-2} [\phi'(\tau_{2j})-\phi'(\tau_{2j-1})][\phi'(\tau_{2k-3})-\phi'(\tau_n)] +$$

$$+ \sum_{j=1}^{m} C_j(\tau_1,\tau_2;\ldots;\tau_{2k-5},\tau_{2k-4})[\psi_j'(\tau_{2k-3})-\psi_j'(\tau_n)] = 0 \quad (2.15)$$

Suppose there exists a sequence

$$\{t_j;\ j = 1,2,\ldots\},\quad t_j \to 0 \ \text{as}\ j \to \infty \quad (2.16)$$

such that

$$\phi'(t_j) = \phi'(0),\quad j = 1,2,\ldots . \quad (2.17)$$

Due to analyticity of $\phi(z)$ for $-\varepsilon < \operatorname{Re} z < \varepsilon$ we conclude from (2.17) that $\phi(z)$ is linear. In terms of the d.f. G of Y_j it means that

$$\int e^{y+ity}dG = (A+Bt)\int e^{ity}dG$$

for all $t \in R^1$. Applying a lemma from Khatri and Rao (1968) (see also Lemma 6.1.1 in Kagan, Linnik and Rao (1973)) we conclude that $X_j = e^{Y_j}$ has gamma distribution.

If there is no sequence (2.16) satisfying (2.17), then there exists a (small) $\varepsilon_1 > 0$ such that

$$|\phi'(t) - \phi'(0)| \neq 0, \quad |t| < \varepsilon_1.$$

Fix $\tau^*_{2k-3} = 0$, $\tau^*_1, \tau^*_2, \ldots, \tau^*_{2k-4}$ in such a way that

$$\prod_{j=1}^{k-2} [\phi'(\tau^*_{2j}) - \phi'(\tau^*_{2j-1})] \neq 0.$$

Then in a neighborhood

$$|\tau_{2k-2}| < \varepsilon_2, \ldots, |\tau_n| < \varepsilon_2$$

with $\varepsilon_2 > 0$ small enough, the relation (2.15) becomes the Cauchy equation

$$\sum_{j=2k-2}^{n-1} \phi(\tau_j) = \Psi(\tau_{2k-2} + \ldots + \tau_{n-1}). \tag{2.18}$$

Since $n-1 - (2k-2) = n-2k+1 \geq 1$ due to $n \geq 2k$ there are at least two free variables in (2.18) so that the solution $\phi(t)$ of (2.18) is linear and again X_j's have gamma distribution.

If X_j's have gamma distribution then it is well known that $\overline{X}$ and the vector $(X_1/\overline{X}, \ldots, X_n/\overline{X})$ are independent. Thus, the relation (2.1) holds.

Proof of Theorem 2. Suppose $P(\overline{X})$ is the best unbiased estimator of $E_\sigma P = A_k\sigma^k + \ldots + A_0$, $\sigma > 0$. It is well known (see, for example, Lemma 7.2.1 in Kagan, Linnik and Rao (1973)) that in this case we must have

$$E_\sigma(Ph) \equiv 0, \quad \sigma > 0 \tag{2.19}$$

for any $h = h(X_1, \ldots, X_n)$ with

$$E_\sigma(h) \equiv 0, \quad \sigma > 0. \tag{2.20}$$

We shall use this condition only for

$$h = h(X_1/\overline{X}, \ldots, X_n/\overline{X}).$$

For such h, the left hand side of (2.19) becomes a polynomial in σ with the leading coefficient $A_k E_1(\overline{X}^k h)$. The condition

$$E_1(\overline{X}^k h) = 0$$

for all (bounded) h subject to (2.20) implies that

$$E_1(\overline{X}^k | X_1/\overline{X}, \ldots, X_n/\overline{X}) = \text{constant}.$$

Applying Theorem 1 we conclude that F is gamma.

Conversely, if $X_1, X_2, \ldots, X_n$ are i.i.d. observations from $F(x/\sigma)$, $\sigma > 0$ where F is gamma, then $\overline{X}$ is the complete sufficient statistic and thus $P(\overline{X})$ is the best unbiased estimator of its expectation.

REFERENCES

Kagan, A. M., Linnik, Yu. V. and Rao, C. R. (1973). *Characterization Problems in Mathematical Statistics*. Wiley.

Kagan, A. M. and Rao, C. R. (1988). Constancy of regression of a polynomial of sample average on residuals characterizes normal distribution. Technical Report No. 88-30, Center for Multivariate Analysis, University of Pittsburgh.
Kagan, F. M. (1968). Conditions of optimality of some estimators for families with a scale parameter (in Russian). *Doklady Akad. Nauk. Uzbek. SSR*, 6.
Khatri, C. G. and Rao, C. R. (1968). Some characterizations of the gamma disdribution. *Sankhyā*, A30, 1, 157-166.

STATISTICAL DATA ANALYSIS AND INFERENCE
Y. Dodge (Editor)

PROBING ENCOUNTERED DATA, META ANALYSIS AND WEIGHTED DISTRIBUTION METHODS

G. P. Patil and C. Taillie

Department of Statistics
Center for Statistical Ecology and Environmental Statistics
The Pennsylvania State University
University Park, PA 16802 USA

Fisher (1934) recognized that the method of ascertainment can influence the form for the distribution of recorded observations, a concept that was later formalized by Rao (1965) into what is termed "weighted distribution." This paper has two parts. The first reviews various structural properties of weighted distributions, both univariate and bivariate. The effect of weighted observations on Bayesian inference is also considered. The second part deals with three applications of weighted distributions: (1) in stochastic population dynamics, the exploited population size distribution is a weighted form of the natural (unexploited) size distribution; (2) Diaconis and Efron have introduced the double exponential family (DEF) for addressing overdispersion in data. The DEF is shown to be a weighted distribution with the unusual feature that the weight function involves the parameters (specifically, the mean) of the original distribution; (3) weight functions have been employed by Iyengar and Greenhouse (1988) to model the effects of publication bias in a meta-analysis study. We extend their model so as to account for heterogeneity as well as publication bias. For the particular data examined by Iyengar and Greenhouse, likelihood ratio tests indicate that heterogeneity, and not publication bias, is responsible for the observed pattern of effect sizes.

SETTING THE SCENE: THE ROLE OF C. R. RAO

Science, technology, and society are increasingly recognizing the expanding need for the quantification of random phenomena. The need to think clearly about random phenomena, whether in conceptualization and modeling, in decision making, in system simulation or in data analysis leads increasingly to the consideration of a host of problems arising in a wide variety of different statistical distributions, and to an appreciation of the pressing need for finding timely solutions to these problems. Professor Rao has displayed characteristic insight into this need and the matching creative vision.

In his fitting Foreword to the highly commendable three-volume set on the Dictionary and Classified Bibliography of Statistical Distributions in Scientific Work by Patil et al. (1984a,b,c) Rao provides his overall perspective and enthusiasm for the fascinating subject area of statistical distributions in model building, structural investigation, characterizations, and applications in various fields.

The innovative paper of Rao (1965) presented at the First International Symposium on Classical and Contagious Discrete Distributions in 1963 is a landmark paper. It has generated a stream of publications, and recently commanded a series of advanced research conferences on weighted distributions, observational studies, and representativeness issues. See, for example, Patil, Rao, and Zelen (1988a). See also Simon (1980), Stene (1981), Warren (1981) and Zelen (1971).

As Rao (1985) has himself put it, "Specification, or the choice of a model, is of great importance in data analysis. An appropriate specification for given data can be arrived at on the basis of past experience, information on the stochastic nature of events, a detailed knowledge of how observations are ascertained and recorded, and an exploratory analysis of current data itself using graphical displays, preliminary tests, and cross validation studies.

"Inaccuracies in specification can lead to wrong inference. It is therefore worthwhile to review the data under different possible specifications (models) to determine how variant the conclusions could be.

"Every data set has its own unique features which may be revealed in an initial scrutiny of data and/or during statistical analysis, which may have to be taken into account in interpreting data. Routine data analysis based on textbook methods or software packages can be misleading."

Looking ahead, one challenge stands out for everyone appreciative of statistical distributions in scientific work. That is the effective utilization of the modern unifying theories of statistical distributions and their applications in the classroom setting in the form of well identified courses. These would then become creative springboards for graduate research and teaching in statistics, probability, and quantitative studies. Now that a sufficient number of monographs and edited volumes are available, the preparation of such courses is not only feasible but of the utmost importance. This is still another inspiration that one can draw from Rao's pioneering work on weighted distributions and their applications.

1. INTRODUCTION AND BACKGROUND

1.1 Probing Encountered Data Using Weighted Distributions.

Traditional statistics theory and practice have been occupied largely with statistics involving randomization and replication. But in ecological and environmental work, observations most often fall in the non-experimental, non-replicated, and non-random categories. Additionally, the problems of model specification and data interpretation acquire special importance and great concern. In several situations of diverse nature in statistical ecology and environmental statistics, the theory of weighted distributions provides a perceptive and unifying approach for the problems of model specification and data interpretation. For the growing literature, see Patil (1984), Rao (1985), and Patil, Rao, and Zelen (1988a,b).

1.2 Meta Analysis - Combining Data for Stronger Conclusions.

Meta-analysis refers to the statistical integration of the results, and perhaps of the data, from separate studies. This quantitative approach to the synthesis of research efforts is a response to the growing need for broader and more rigorous summarization and integration of research literatures. See Hedges and Olkin (1985) for an overview. Also, Iyengar and Greenhouse (1988) and related discussion, Landau, Patil and Rawlings (1985), and Patil et al. (1986).

2. BIASED SAMPLING, WEIGHTED DISTRIBUTIONS APPROACH, AND UNBIASED INFERENCE

2.1 Introduction.

The concept of weighted distributions can be traced to the study of the effect of methods of ascertainment upon estimation of frequencies by Fisher (1934). In extending the basic ideas of Fisher, Rao (1965) saw the need for a unifying concept and identified various sampling situations that can be modeled by what he called weighted distributions. Within the biomedical context of cell kinetics and the early detection of disease, Zelen (1974) introduced weighted distributions to represent what he broadly perceived as length biased sampling. The initial idea of length biased sampling appears in Cox (1962).

Patil (1984) has discovered weighted distributions as stochastic models in the equilibrium study of populations subject to harvesting and predation. Mahfoud and Patil (1981) and Patil, Rao and Ratnaparkhi (1986) have identified a Bayesian analogue to the theory of weighted distributions through the relationship of the posterior distribution to the prior distribution via the likelihood. Several papers have appeared that deal with weighted distributions and their applications in a variety of fields. See the bibliography by Patil, Rao, and Zelen (1988a).

2.2 Univariate Weighted Distributions.

Suppose X is a non-negative observable random variable (rv) with its natural pdf $f(x;\theta)$, where the natural parameter $\theta \in \Omega$, the parameter space. Suppose a realization x of X under $f(x;\theta)$ enters the investigator's record with probability proportional to $w(x,\beta)$, so that

$$\mathrm{Prob}(\mathrm{Recording}|X=y)/\mathrm{Prob}(\mathrm{Recording}|X=x) = w(y,\beta)/w(x,\beta).$$

Here the recording (weight) function $w(x,\beta)$ is a non-negative function with parameter β representing the recording (sighting) mechanism.

Clearly, the recorded x is not an observation on X, but on the rv X^w, say, having pdf,

$$f^{w}(x;\theta,\beta) = \frac{w(x,\beta)f(x;\theta)}{\omega}, \quad (1)$$

where ω is the normalizing factor obtained to make the total probability equal to unity by choosing $\omega = E[w(X,\beta)]$. The rv X^{w} is called the weighted version of X, and its distribution in relation to that of X is called the weighted distribution with weight function w. Note that the weight function $w(x,\beta)$ need not lie between zero and one, and actually may exceed unity, as for example, when $w(x,\beta) = x$, in which case, $X^{*} = X^{w}$ is called the size-biased version of X. The distribution of X^{*} is called the size-biased distribution with pdf

$$f^{*}(x;\theta) = \frac{xf(x;\theta)}{\mu}, \quad (2)$$

where $\mu = E[X]$. The pdf f^{*} is called the length-biased or size-biased version of f, and the corresponding observational mechanism is called length- or size-biased sampling.

The concept of weighted distributions has been used during the last 25 years as a useful tool in the selection of appropriate models for observed data, especially when samples are drawn without a proper frame.

In many situations the model given in (1) is appropriate, and the statistical problems that arise are the determination of a suitable weight function $w(x,\beta)$ and drawing inference on θ.

The following examples from Patil and Rao (1977) may help illustrate a few situations generating weighted distributions.

(i) Truncation: The distribution of a random variable truncated to a set T is a weighted distribution with weight function $w(x) = 1$ for $x \in T$ and zero elsewhere.

(ii) Missing Data: If a proportion $1 - w(x)$, $0 \leq w(x) \leq 1$, out of the natural frequency of x is missing, that is, if the probability of x missing is $1 - w(x)$, the pdf to use for the analysis of the observed data is with the weight function $w(x)$.

(iii) Damaged Observations: Consider a damage model where an observation $X = x$ is reduced to y by a destructive process with pdf $d(y|x)$. See Rao (1965). Then the probability that the observation $X = x$ is undamaged is $d(x|x)$, and the distribution of the undamaged observation is the weighted distribution with $w(x) = d(x|x)$. For example, under binomial survival model, $d(x|x) = \theta^{x}$, $0 < \theta < 1$. An investigator recording only undamaged observations will need to work with a corresponding weighted distribution.

(iv) Analysis of Family Data: Various demographic and social studies involve family size and sex ratio as important factors which have some bearing on the main study. This example shows how a weighted distribution arises as a result of the size-biased sampling. The discussion is based on Rao (1965).

Consider the following data which relates to brothers and sisters in families of 104 boys who were admitted to a postgraduate course at the Indian Statistical Institute.

Family Size:	1	2	3	4	5	6	7	8	9	10	11	12	13	15	Total
No. of Families:	1	6	6	13	12	7	14	11	12	8	6	5	2	1	104
Brothers:	1	8	12	34	34	29	59	50	54	46	32	31	16	8	414
Sisters:	0	4	6	18	26	13	39	38	54	34	34	29	10	7	312

Let us assume that in families of given size n, the probability of a family with x boys coming into the record is proportional to x. Also, suppose that the number of boys follows a binomial distribution with probability parameter π. Then

$$f(x;\pi) = \binom{n}{x}\pi^x(1-\pi)^{n-x}$$

$$w(x) = x, \quad E[w(X)] = \omega = n\pi$$

$$f^w(x;\pi) = \binom{n-1}{x-1}\pi^{x-1}(1-\pi)^{n-x}$$

$$E[X^w/n] = \pi + (1-\pi)/n > \pi$$

$$E[(X^w-1)/(n-1)] = \pi .$$

If k boys representing families of size $n_1, n_2, \ldots, n_k$ report $x_1, x_2, \ldots, x_k$ boys, an unbiased estimate of π is

$$\tilde{\pi} = (\Sigma x_i - k)/(\Sigma n_i - k) = (414-104)/(726-104) = \frac{1}{2} .$$

Whereas if one wrongly treats x_i's as observations on k randomly drawn families with at least one boy, i.e. as arising from a truncated binomial, then the estimate of π will have an upward bias.

2.3 Weight Functions and Properties.

The following forms of weight functions $w(x)$ have appeared in the scientific and statistical literature:

(1) $w(x) = x^{\beta}$ for $\beta = 1,2,3,1/2,2/3,\ 0 < \beta < 1$.

(2) $w(x) = \binom{x}{2}^{\beta}$ for $\beta = 1,1/2,\ 0 < \beta < 1$.

(3) $w(x) = (x)^{(\beta)} = x(x-1)\ldots(x-\beta+1)$.

(4) $w(x) = e^{\beta x},\ \theta^{x}$.

(5) $w(x) = x+1,\ \alpha x+\beta$

(6) $w(x) = 1-(1-\beta)^{x}$ for $0 < \beta < 1$.

(7) $w(x) = (\alpha x+\beta)/(\delta x+\gamma)$.

(8) $w(x) = \text{Prob}(Y \le x)$ for some rv Y.

(9) $w(x) = \text{Prob}(Y > x)$ for some rv Y.

(10) $w(x) = r(x)$, where $r(x)$ is the probability of 'survival' of observation x.

We note that these weight functions are monotone increasing or decreasing functions of x. The following results provide useful comparisons of X^{w} with X.

Result 2.3.1 (Patil, Rao and Ratnaparkhi, 1986): The weighted version X^{w} is stochastically greater or smaller than the original rv X according as the weight function $w(x)$ is monotone increasing or decreasing in x. As a consequence, the expected value of the weighted version X^{w} is greater or smaller than the expected value of the original rv X according as the weight function $w(x)$ is monotone increasing or decreasing in x.

Result 2.3.2 (Zelen, 1974; Patil and Rao, 1978): The expected value of the size-biased version X^{*} is $E[X^*] = \mu[1+\sigma^2/\mu^2]$, where $E[X] = \mu$ and $V(X) = \sigma^2$. Further, the harmonic mean of X^* is equal to the mean of the original rv X, i.e. $E[1/X^*] = 1/\mu$. Another way of expressing these results is that

$$E[X^*]E[1/X^*] = 1 + \sigma^2/\mu^2.$$

Result 2.3.3 (Patil and Ord, 1975): Let the rv X have pdf $f(x;\theta)$ and have size-bias with weight function x^β. Then a necessary and sufficient condition for $f^w(x;\theta,\beta) = f(x;\eta)$, where $\eta = \eta(\theta,\beta)$, is that $f(x;\theta) = x^\theta a(x)/m(\theta)$. In this case, $f^w(x;\theta,\beta) = f(x;\theta+\beta)$. This result holds under certain mild regularity conditions.

Result 2.3.4 (Mahfoud and Patil, 1981): Consider X^w to be a rv X subject to a size-bias with the weight function x^β. Then X is lognormally distributed if and only if $V(\log X) \equiv V(\log X^w)$ for $\beta > 0$, where V stands for variance. Thus the invariance of the logarithmic variance under size-bias of order β characterizes the lognormal distribution.

Example: In sedimentology and various other fields, the diameter distribution of particles is usually analyzed by mass rather than frequency. Sieve analysis is a good example, which provides data consisting of sizes of sieves and corresponding masses of all particles retained by those sieves. It is interesting to note that the mass-size density is nothing but the weighted version of size-bias of order 3 of the pdf of the particle size (diameter). It can be verified that, if the diameter X is lognormal with parameters μ and σ^2, then X^w with $\beta = 3$ is lognormal with parameters $\mu + 3\sigma^2$ and σ^2. This property has been empirically noticed and utilized for inference in sedimentology literature. See, for example, Krumbein and Pettijohn (1938), and Herdan (1960).

2.4 *Bivariate Weighted Distributions.*

Let (X,Y) be a pair of non-negative rv's with a joint pdf $f(x,y)$ and let $w(x,y)$ be a non-negative weight function such that $E[w(X,Y)]$ exists. The weighted version of $f(x,y)$ is

$$f^w(x,y) = \frac{w(x,y)f(x,y)}{E[w(X,Y)]} .$$

The corresponding weighted version of (X,Y) is denoted by $(X,Y)^w$. The marginal and conditional distributions of $(X,Y)^w$ are:

$$f^w(x) = \frac{E[w(x,Y)|x]f(x)}{E[w(X,Y)]} ,$$

$$f^w(y|x) = \frac{w(x,y)f(y|x)}{E[w(x,Y)|x]} .$$

Clearly, both are weighted versions of the corresponding marginal and conditional distributions of (X,Y).

Examples of bivariate weight functions of practical interest are:

(1) $w(x,y) = x^\alpha$ (2) $w(x,y) = w(y)$

(3) $w(x,y) = x+y$ (4) $w(x,y) = x^\alpha y^\beta$

(5) $w(x,y) = \max(x,y)$ (6) $w(x,y) = \min(x,y)$.

The following results are of some interest; see Patil, Rao and Ratnaparkhi (1986) for more results.

Result 2.4.1 (Patil and Rao, 1978): Let (X,Y) be a pair of non-negative rv's with pdf $f(x,y)$. Let wf $w(x,y) = w(y)$, as is the case in sample surveys involving sampling with probability proportional to size. Then the random variables X and X^w are related by

$$f^w(x) = \frac{E[w(Y)|x]f(x)}{E[w(Y)]} .$$

Note that X^w is a weighted version of X, and the regression of $w(Y)$ on X serves as the weight function.

Result 2.4.2 (Mahfoud and Patil, 1982): Let (X,Y) be a pair of non-negative independent rv's with joint pdf $f(x,y) = f_X(x)f_Y(y)$, and let $w(x,y) = \max(x,y)$. Then the rv's $(X,Y)^w$ are dependent. Further, the regression of Y^w on X^w by $E[Y^w|X^w=x]$ is a decreasing function of x.

Example: Let (X,Y) be statistically independent lifetimes of two components in parallel forming a kit with lifetime $w = \max(x,y)$. Consider a renewal system of kits. If at time t, one records the lifetimes of the two components of the kit in action, their joint pdf is the weighted version of their natural pdf with weight function $w = \max(x,y)$. The result shows that while there is underlying independence in the true lifetimes of the two components/organs, the data so obtained will not reveal that; actually the data will show negative dependence.

2.5 Posterior and Weighted Distributions.

There is a Bayesian analogue to the theory of weighted distributions. (See Mahfoud and Patil, 1981; Patil, Rao and Ratnaparkhi, 1986).

Result 2.5.1 (Mahfoud and Patil, 1981): Consider the usual Bayesian inference in conjunction with (X,θ) having joint pdf $f(x,\theta) = f(x|\theta)f(\theta) = f(\theta|x)f(x)$. The posterior pdf $f(\theta|x) = f(x|\theta)f(\theta)/f(x) = \ell(\theta|x)f(\theta)/E[\ell(\Theta|x)]$ is a weighted version of the prior pdf $f(\theta)$. The weight function is the likelihood function of θ for the observed x.

Result 2.5.2 (Patil, Rao and Ratnaparkhi, 1986): Consider the usual Bayesian inference in conjunction with (X,θ) with pdf $f(x,\theta) = f(x|\theta)f(\theta) = f(\theta|x)f(x)$. Let $w(x,\theta) = w(x)$ be the weight function for the distribution of $X|\theta$, so that the pdf of $X^w|\theta$ is $w(x)f(x|\theta)/\omega(\theta)$, where $\omega(\theta) = E[w(X)|\theta]$. Then the original and the weighted posteriors are related by

$$f(\theta|x) = \frac{\omega(\theta)f^w(\theta|x)}{E[\omega(\theta)|X^w=x]} .$$

Further, the weighted posterior rv $\theta^w|X^w = x$ is stochastically greater or smaller than the original posterior rv $\theta|X = x$ according as $\omega(\theta)$ is monotonically decreasing or increasing function of θ.

Examples: The following tabular form may be a convenient format.

$X\|\Theta=\theta$	Θ	$\Theta\|X=x$	$w(x)$	$[\Theta\|X^w=x]$
Poisson(θ)	Gamma(k,λ)	Gamma($k+x,\frac{1}{\lambda+1}$)	x	Gamma($k+x-1,\frac{1}{\lambda+1}$)
Binomial(n,θ)	Beta(a,b)	Beta($x+a, n-x+b$)	x	Beta($x-1+a, n-x+b$)
Neg-Bin(k,θ)	Beta(a,b)	Beta($k+a, x+b$)	x	Beta($k+1+a, x-a+b$)
Exponential(θ)	Gamma(k,λ)	Gamma($k+1,\frac{\lambda}{\lambda x+1}$)	x	Gamma($k+2,\frac{\lambda}{\lambda x+1}$)

3. STOCHASTIC POPULATION DYNAMICS AND WEIGHTED DISTRIBUTIONS

Consider the stochastic differential equation,

$$\frac{1}{x}\frac{dx}{dt} = r(x,t) = g(x) + \gamma(t),$$

where, $x(t)$ = the population size at time t,

$r(x,t)$ = the per capita growth rate of population of size x at time t,

$g(x)$ = the biological part of the per capita growth rate dependent on population size x, but independent of time t, and

$\gamma(t)$ = the environmental part of the per capita growth rate dependent on time t, but independent of the population size x. Let $\gamma(t)$ be a white noise process with environmental unpredictability (i.e., variance) parameter σ^2.

The population size $x(t)$ is then a stochastic integral and its equilibrium pdf, when it exists, is given by

$$f(x) = \frac{\text{const}}{V(x)} \exp[2 \int \frac{M(x)}{V(x)} dx]$$

where $M(x) = xg(x)$ and $V(x) = \sigma^2 x^2$.

Further, if the population is subjected to exploitation (harvesting, predation, etc.) with per capita exploitation rate $h(x)$, the equilibrium pdf $f_h(x)$ of the exploited population size simplifies to

$$f_h(x) = \frac{w(x)f(x)}{E[W(X)]}$$

where $f(x)$ is the natural population equilibrium pdf and

$$w(x) = \exp[-2 \int \frac{xh(x)}{\sigma^2 x^2} dx].$$

For more discussion, see Patil (1984), Dennis and Patil (1984), and Patil et al. (1986).

4. MODELING CLUMPED SAMPLING, HETEROGENEITY, AND EXTRANEOUS VARIATION USING WEIGHTED DISTRIBUTION METHODS

4.1 Introduction.

During their examination of the problem of toxoplasmosis, Diaconis and Efron (Diaconis and Efron, 1985; Efron, 1986) looked at the data sets involved and found that there was more dispersion in the data sets than the existing models could accommodate and therefore they introduced a model called the double exponential family (DEF). This family, which is not to be confused with the Laplace distribution, enjoys the exponential family properties simultaneously for the mean and the dispersion parameters. It allows the data analyst to model overdispersion while carrying out the usual regression analyses for the mean as a function of the predictors. The overdispersion may be due to one or more possible causes, such as clumped sampling, heterogeneity, selection bias, etc. The issues pertaining to these kinds of causes arise in environmental and ecological research and management. Examples include soil sampling, water sampling, litter sampling, publication bias, etc.

4.2 Double Exponential Family as a Model for Clumped Data.

We write the linear exponential family with mean μ and variance $V(\mu)$ in the form

$$f_{\mu}(y) = \exp\{yA(\mu) - B(\mu) + D(y)\} \tag{1}$$

where $\mu A'(\mu) - B'(\mu) \equiv 0$ and $A'(\mu) = 1/V(\mu)$. The double exponential family (DEF) is then defined by

$$f_{\mu,\theta}(y) = c(\mu,\theta)\theta^{\frac{1}{2}}\{f_{\mu}(y)\}^{\theta}\{f_{y}(y)\}^{1-\theta} , \tag{2}$$

where the constant $c(\mu,\theta)$ is a normalizing constant and the motivation is to use the densities (2) as constituents of a regression analysis, in which the unknown parameters μ and θ are estimated from the data.

We (Patil, 1987; Patil and Taillie, 1987a) observe the DEF to be a weighted distribution of the following form:

$$f_{\mu,\theta}(y) = w_{\mu,\theta}(y)f_{\mu}(y)/E[w_{\mu,\theta}(Y)] \tag{3}$$

where the weight function is given by

$$w_{\mu,\theta}(y) = \left\{\frac{f_{y}(y)}{f_{\mu}(y)}\right\}^{1-\theta} \tag{4}$$

Furthermore, the weight function has the following interesting form

$$w_{\mu,\theta}(y) = \exp\{(1-\theta)I(y,\mu)\} , \tag{5}$$

where $I(y,\mu)$ is the Kullback-Leibler distance function between y and μ for the density function f. Notice that the Kullback-Leibler distance increases with the distance from the mean μ, thus allowing a more distant observation larger weight and accommodating extra dispersion in the data set when $1 - \theta > 0$.

Figure 1 provides a picturesque presentation of the makeup of the density function of the DEF at x. Suppose that we observe x as the first observation under the natural pdf $f_\mu(\bullet)$. Suppose that there is some clumping and that the additional x values occur under the pdf $f_x(\bullet)$ with mean x and not with mean μ. Then the pdf of the DEF at x turns out to be the θ-geometric mean of $f_\mu(x)$ and $f_x(x)$ where θ is a measure of clumping. Figures 2 and 3 plot Poisson and double Poisson probability functions with μ=10.5 and θ=.1,.5 . The increasing dispersion of the double Poisson with decreasing θ is evident in these pictures. For comparison, we have also plotted the negative binomial probability function having the same mean and variance as the double Poisson. Figures 4 and 5 plot the corresponding logarithmic weight functions. Also shown in these plots is the Pearson logarithmic weight function, $(1-\theta)(x-\mu)^2$, although the latter is too explosive in the right tail to be validly applied to the Poisson distribution. It is encouraging to observe that the DEF keeps the mean value about the same and yet accommodates overdispersion or underdispersion of the data set.

Figure 6 provides a brief tabular excerpt of the toxoplasmosis data which prompted Diaconis and Efron to introduce the double exponential family to deal with the issues arising from the overdispersion of the data set. Efron (1986) attributes the overdispersion to cluster sampling, but we have argued that heterogeneity (city-to-city variation) is a more likely explanation and that it can also be incorporated into the DEF framework (Patil and Taillie, 1987b).

PDF $f_{\mu,\vartheta}$ OF DOUBLE EXPONENTIAL FAMILY

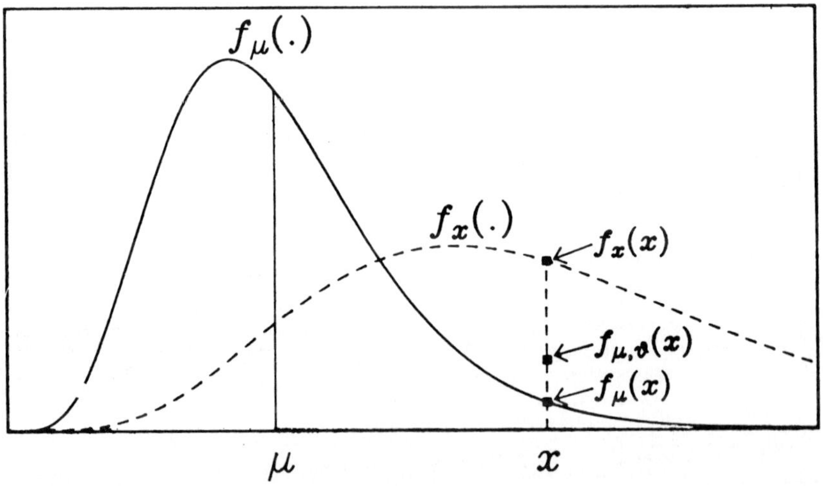

FIGURE 1

COMPARISON OF PROBABILITY FUNCTIONS

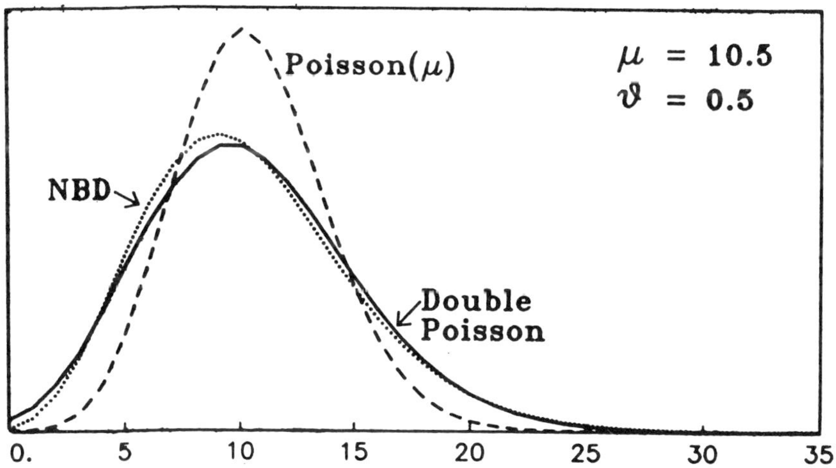

FIGURE 2

COMPARISON OF PROBABILITY FUNCTIONS

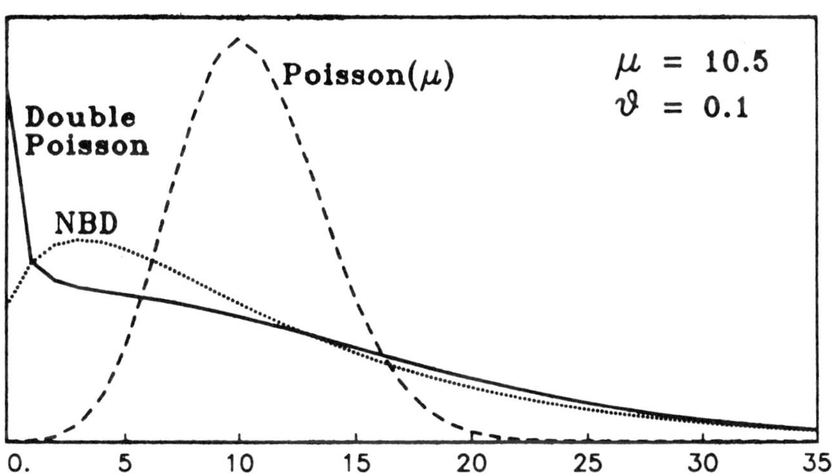

FIGURE 3

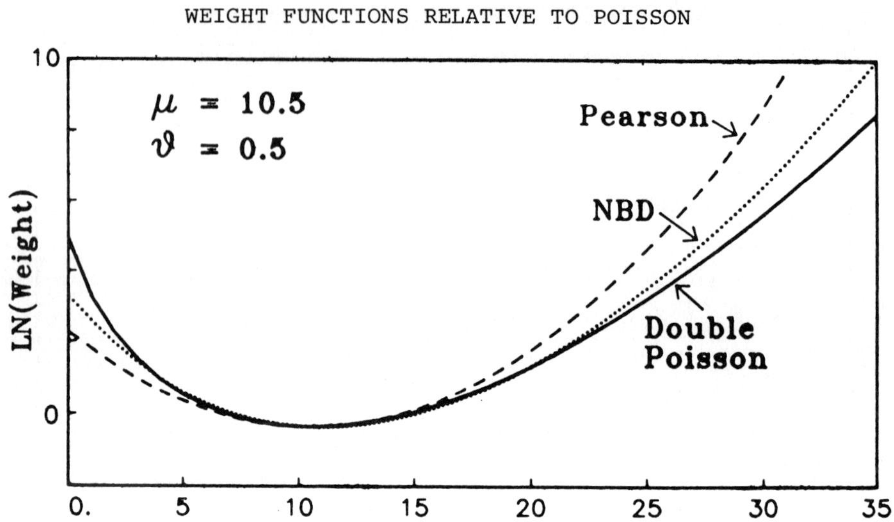

FIGURE 4

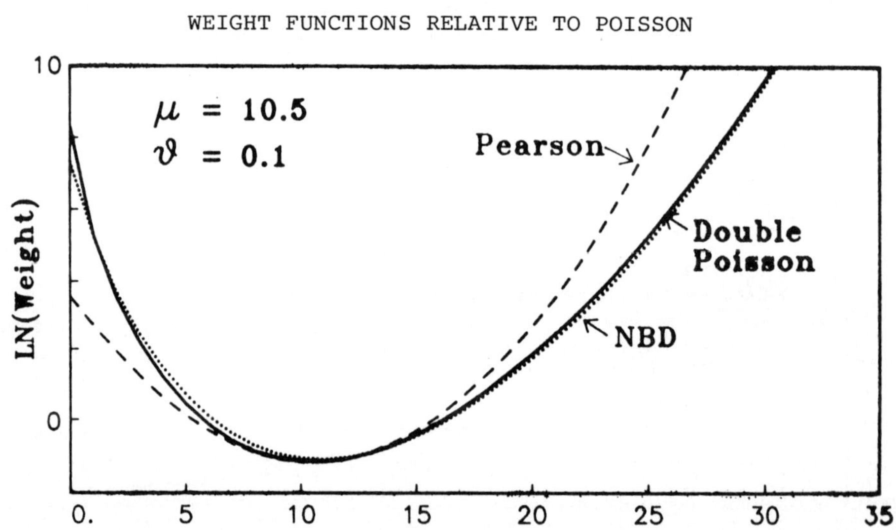

FIGURE 5

TOXOPLASMOSIS DATA
SHOWING THE PROPORTIONS OF SUBJECTS
TESTING POSITIVE (Y_j) and
NUMBERS TESTED (n_j)
IN 34 CITIES
(Efron, 1978, 1986)

City No.	Predictor (Rainfall)	Response Prop. Pos.	Original Sample Size	Effective Sample Size
j	x_j	Y_j	n_j	$n_j\theta_j$
1	1735	.500	4	2.4
•				
•				
4	1973	.300	10	6.9
•				
•				
20	1620	.278	18	13.8
•				
•				
24	1796	.532	77	15.3
•				
•				
34	2292	.622	37	29.1

- To assess the effect of rainfall on the proportions positive.
- Cities with large sample sizes n_j have large influence.
- The subjects may have been accrued in clumps, so the statistician should be using smaller values of n_j, not necessarily proportionately smaller in each city.
- Inadequacy of binomial model for this overdispersion.

FIGURE 6

5. META ANALYSIS INCORPORATING HETEROGENEITY AND PUBLICATION BIAS WITH WEIGHTED DISTRIBUTIONS

Meta-analysis consists of quantitative methods for combining evidence from different studies about a particular issue. As Iyengar and Greenhouse (1988) put it, "The application of statistical procedures to collections of results from individual studies for integrating, synthesizing and advancing a research domain is commonly known as meta-analysis. The objective of a meta-analysis is to summarize quantitatively a research literature with respect to a particular question and to examine systematically the manner in which a collection of studies contributes to knowledge about that question... Applications of meta-analysis as a tool for investigating scientific questions and for guiding public policy decisions are diverse and include, for example, the analysis of the efficacy of psychotherapy (Smith, Glass and Miller, 1980), the assessment of human lung cancer risks from various environmental emissions (DuMouchel and Harris, 1983), and the U. S. Department of Education study on school desegregation and black achievement." See also, for example, Fisher (1932), Hedges and Olkin (1985), Laird, Patil and Taillie (1988), Light and Pillemer (1984), Rosenthal (1984), and Wolf (1986).

5.1 An Example of Meta-Analysis due to Iyengar and Greenhouse.

Iyengar and Greenhouse (1988) consider the following example given in Hedges and Olkin (1985, p. 303). The data come from ten studies comparing the effects of experimental open classroom education with traditional education on student creativity and are reproduced in Table 1. Following Hedges and Olkin, Iyengar and Greenhouse assume that all studies are estimating the same effect size, denoted by θ. For the ith study, the second column gives the sample size, N_i, in each of the two samples. The third column gives the effect size estimate, $\hat{\theta}_i$, where $\hat{\theta}_i$ is the difference between the means for each education type divided by a pooled estimate of the standard deviation. The fourth and fifth columns give, respectively, the t-statistics, t_i, and the corresponding df, $q_i = 2N_i - 2$ for each study.

TABLE 1: Studies of effects of open vs. traditional education on creativity (Iyengar and Greenhouse, 1988).

i	N_i	$\hat{\theta}_i$	t_i	q_i
1	90	-0.583	-3.91	178
2	40	0.535	2.39	78
3	36	0.779	3.31	70
4	20	1.052	3.33	38
5	22	0.563	1.87	42
6	10	0.308	0.69	18
7	10	0.081	0.18	18
8	10	0.598	1.34	18
9	39	-0.178	-0.79	76
10	50	-0.234	-1.17	98

Denote the density of a noncentral t distribution with noncentrality η and with q df by $f(t;\eta,q)$. With no selection bias in the reporting of the studies in Table 1, the t_i have density $f(t;(N_i/2)^{1/2}\theta,q_i)$ under assumptions of independence and normality. Letting the weight function $w(t)$ model the selection bias in reporting a result for which the value, t, of the t-statistic is observed, the likelihood function for θ is given by

$$L(\theta,w) = \frac{\Pi_{i=1}^{10} f(t_i;(N_i/2)^{1/2}\theta,q_i)w(t_i)}{\Pi_{i=1}^{10} A((N_i/2)^{1/2}\theta,w,q_i)}, \tag{1}$$

where

$$A(\eta,w,q) = \int_{-\infty}^{\infty} f(t;\eta,q)w(t)dt\ . \tag{2}$$

Four of the ten studies in Table 1 yield results that are significant at the .05 level, so an appropriate weight function for this data should be non-zero everywhere. To examine different possible selection schemes and to examine the effect of the choice of the weight function upon the inferences about θ, Iyengar and Greenhouse consider the following two parametric families of weight functions:

$$w_1(x;\beta;q) = \begin{cases} \dfrac{|x|^\beta}{t(q,.05)^\beta}, & \text{if } |x| \le t(q,.05), \\ 1 & \text{otherwise}, \end{cases} \tag{3}$$

and

$$w_2(x;\gamma,q) = \begin{cases} e^{-\gamma}, & \text{if } |x| \le t(q,.05), \\ 1, & \text{otherwise}\ . \end{cases} \tag{4}$$

The parameters β and γ are nonnegative and $t(q,.05)$ is the .05 two-sided critical point of the central T-distribution with q degrees of freedom.

Both families of weight functions imply that all studies showing statistically significant results will be reported. In addition, these families have the following features: when β and γ are zero, the weight functions indicate no selection bias; when β and γ are infinite, the weight functions become the Hedges-Olkin scheme in which only statistically significant results are reported. The two weight function families differ only for nonsignificant results: w_1 says that the reporting probability increases as the outcome approaches statistical significance, whereas w_2 says that the reporting probability is constant for all non-significant results.

The maximum likelihood estimates, $(\hat{\theta},\hat{\beta})$ and $(\hat{\theta},\hat{\gamma})$, of the vector parameters (θ,β) and (θ,γ), respectively, along with their estimated standard deviations and their estimated covariance matrices (V) are presented in Table 2.

TABLE 2: MLEs of effect size and weight function parameters (Iyengar and Greenhouse, 1988).

$(\hat{\theta},\hat{\beta}) = (0.026, 1.33)$	$(\hat{\theta},\hat{\gamma}) = (0.022, 2.53)$
$(SD(\hat{\theta}), SD(\hat{\beta})) = (0.052, 0.59)$	$(SD(\hat{\theta}), SD(\hat{\gamma})) = (0.049, 0.65)$
$V(\hat{\theta},\hat{\beta}) = \begin{bmatrix} 0.003 & -0.003 \\ -0.003 & 0.348 \end{bmatrix}$	$V(\hat{\theta},\hat{\gamma}) = \begin{bmatrix} 0.002 & 0.000 \\ 0.000 & 0.417 \end{bmatrix}$

5.2 Publication Bias Versus Heterogeneity in the Iyengar and Greenhouse Example.

Patil and Taillie (1988) revisit the Iyengar and Greenhouse data set and its analysis in what follows. Additional comments appear in Laird, Patil and Taillie (1988). The reanalysis had three goals: (i) to examine additional weight functions and, in particular, to let the weight function depend on the p-value instead of the test statistic; (ii) to examine the feasibility of using a normal approximation to the noncentral T-distribution; and (iii) to incorporate heterogeneity of the effects parameter, as well as publication bias, into the model. Heterogeneity and publication bias are two competing explanations for the overdispersion in the t-scores of Table 1, and a joint model would permit likelihood ratio tests of their relative importance. For this particular data set, we find that the normal approximation is quite adequate, and that heterogeneity accounts for nearly all of the overdispersion with virtually no evidence of publication bias. Further investigation with additional data sets would be needed to judge the general suitability of the normal approximation and to determine if such decisive discrimination between heterogeneity and publication bias is always possible. Sensitivity to the choice of weight function is largely an open question since we found no evidence of publication bias for these data.

<u>Weight Functions Examined</u>

$$w_1 = (x/xcrit)^{\beta} \quad \text{if } |x| < xcrit$$
$$= 1 \quad \text{otherwise}$$

$$\text{Half-Normal} = \exp[\,-\beta * p(x)^2\,]$$

$$\text{Negative Exponential} = \exp[\,-\beta * p(x)\,]$$

Here p(x) is the p-value when the test statistic takes value x. The weight function w_1 is the same as that considered by Iyengar and Greenhouse (cf. equation 3). The corresponding weighted distributions, as estimated from the data, are plotted in Figure 7 against the p-values and also against the value of the test statistic (see below for description of the different models employed for parameter estimation). Asterisks along the horizontal axes indicate the ten observed p-values/T-scores.

Distribution Theory

The noncentral T-distribution with η = 2N-2 degrees of freedom used by Iyengar and Greenhouse is replaced by a normal distribution whose first two moments are given by

$$\text{mean} = \theta * A$$

$$\text{variance} = B + C*\theta^2 + D*\tau^2$$

where

$$A = \text{SQRT}(N/2)*\text{SQRT}(\eta/2)*\Gamma((\eta-1)/2)/\Gamma(\eta/2)$$

$$B = \eta/(\eta-2)$$

$$D = B*N/2$$

$$C = D - A^2 .$$

When $\tau = 0$, the preceding reduce to the mean and variance of the noncentral T with η degrees of freedom and with noncentrality parameter θ*SQRT(N/2). The heterogeneity parameter τ is introduced by randomizing the mean and variance of the noncentral T with respect to a distribution on the effects parameter which has mean θ and variance τ^2. Our parametric modelling is entirely in terms of the first two moments. We have not assumed a specific distributional form for the study-to-study variation in effect sizes; thus, the form of randomized T-distribution is indeterminate. Likelihoods are generated by using weighted normal distributions. Choice of the normal distribution is largely arbitrary, and offers no particular analytical convenience over any other 'closed form' pdf; this is because a numerical integration is required to determine the normalizer for the weighted pdf.

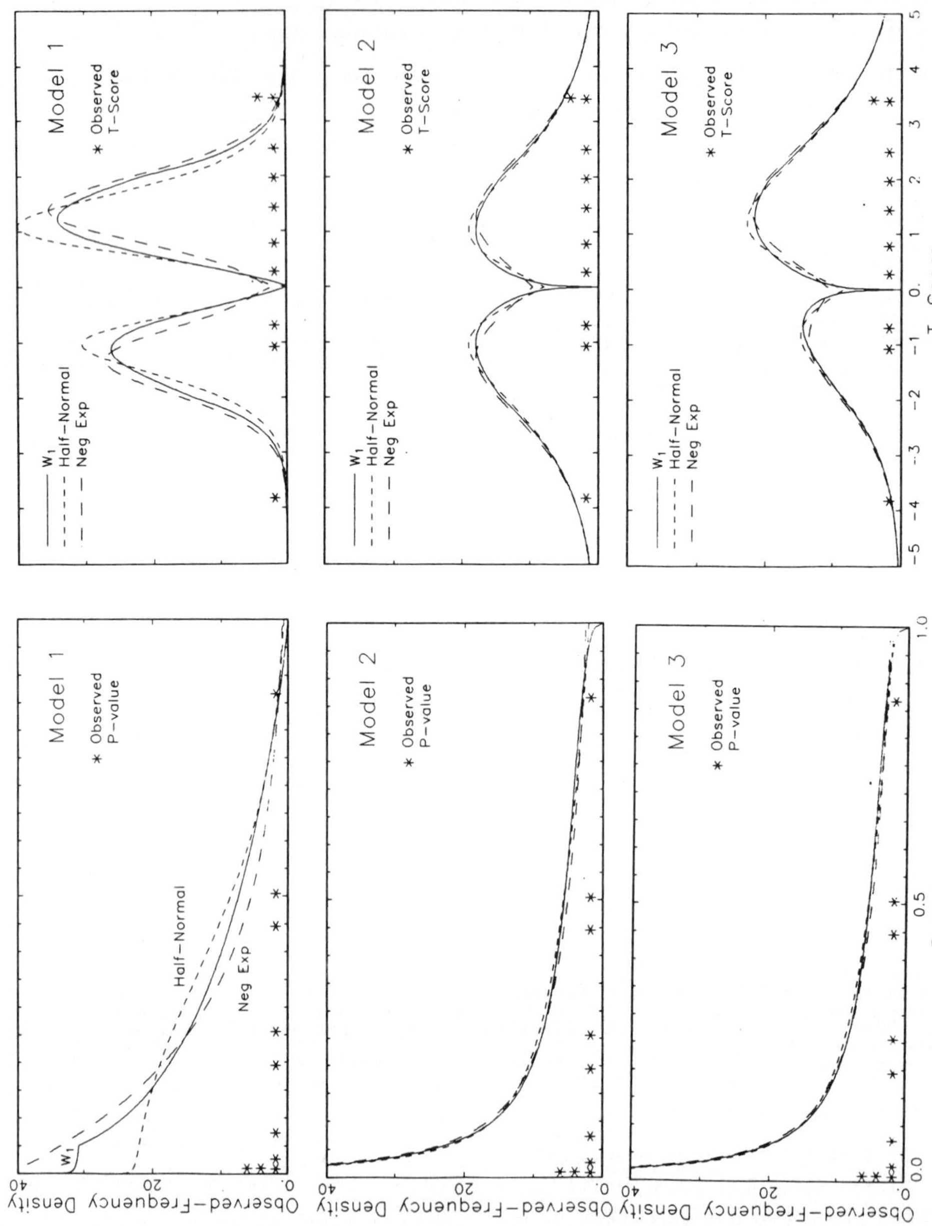
Model 1
Model 2
Model 3
W_1
Half-Normal
Neg Exp
Observed T-Score
Observed P-value
T-Score
P-Value
Observed-Frequency Density

FIGURE 7

Summary of Parameter Estimates

We consider three models depending on whether one or both of the population parameters (θ, τ) is allowed to be nonzero. Two additional models are examined in which there is no publication bias (β=0) so that all three weight functions are equal to unity. The parameter estimates are summarized in Table 3. For the first three models, the weight function was used to obtain an estimate for the number of unpublished studies. This number, when it is implausibly large, can be taken as evidence against the model and/or the choice of weight function; compare with the discussion of "fail-safe sample size" in Iyengar and Greenhouse.

Discussion

For the case of the w_1 weight function with $\tau = 0$ (Model 1), our estimates (θ=.029, β=1.31) using normal distributions compare favorably with the Iyengar and Greenhouse estimates (θ=.026, β=1.33) using noncentral T-distributions. But note that the noncentrality parameters are not very large and the degrees of freedom are not very small. We should also remember that the T-distributions (central and noncentral) are themselves only approximate since they assume normality and homoscedasticity within each study.

Reading across each row, the estimates for the population parameters (θ and τ) are quite consistent, indicating an insensitivity to the weight function. Similarly for the log-likelihoods. As might be expected, the estimates for the number of unpublished studies are sensitive to the choice of weight function.

The log-likelihoods (in the second decimal place for Model 3) favor slightly the negative exponential weight function. This is evidently due to the negative exponential's sharp peak at p=0, permitting a close match with the cluster of small observed p-values. A sharp peak at p=0 does not correspond to the usual meaning of 'publication bias' but may model other types of visibility bias including (i) preferential selection of experimental material on the part of the original investigator(s) and (ii) inaccessability to the meta-analyst of some moderately significant results (e.g., technical reports and government publications instead of journals).

Comparing Model 1 and Model 3, the difference in -2 LN(likelihood) is about 20 for each weight function. Upon comparison with the chi-square having one df, we see that the heterogeneity parameter is highly significant and of such a magnitude (τ=.4) as to account for most of the study-to-study variation in recorded effect sizes ($\hat{\theta}_i$ in Table 1). (Note that the formula for the variance permits negative values for τ^2 so that τ=0 is not on the boundary of parameter space.) Further, in view of the smallness for the estimated number of unpublished studies when heterogeneity is incorporated, it would appear that publication bias is relatively unimportant for this collection of studies. This is borne out by a likelihood ratio test of $\beta=0$ (Model 3 versus Model 4).

The estimated overall effect size increases markedly when heterogeneity is incorporated into the models. Nonetheless, a likelihood ratio test of Model 3 versus Model 2 indicates that the overall effect size is still not significant (p-value is about 0.17). Similarly for Model 4 versus Model 5.

Variance-Covariance Matrices

Table 4 gives the SD-Corr matrices for each of the three models and for each of the three weight functions (see Section 5.4 for derivations). A SD-Corr matrix has standard deviations on the diagonal and correlations on the offdiagonal. In the case of Model 1 and weight function w_1, our standard deviations for $\hat{\theta}$ and $\hat{\beta}$ using normal approximation (.055 and .60) correspond closely to the values (.052 and .59) obtained by Iyengar and Greenhouse using noncentral T-distribution (Table 2). Similarly, we obtain $\text{cov}(\hat{\theta},\hat{\beta}) = -.004$ as compared with the Iyengar and Greenhouse value of $-.003$.

5.3 Joint Weight Function for Heterogeneity and Publication Bias.

In case of normal distributions, heterogeneity can be represented by a Pearson type of weight function

$$w_H(x) = e^{\alpha(x-\theta)^2}, \quad \alpha > 0 , \tag{5}$$

where θ is the mean of the original normal distribution. As discussed in Section 4, Diaconis and Efron have incorporated overdispersion into more general (non-normal) exponential families by replacing $(x-\theta)^2$ in (5) by the Kullback-Liebler distance function. On the other hand, the various weight functions for publication bias can be written in a unified form as

$$w(x) = w(x,\beta) = \exp(-\beta K(x)) \tag{6}$$

where

$$K(x) = \{\log(x/x_{crit})\}^- \quad \text{for } w = w_1$$

$$K(x) = p(x) \quad \text{for neg. exp.}$$

$$K(x) = \{p(x)\}^2 \quad \text{for half-normal .}$$

TABLE 3: MLEs of effect size (θ), heterogeneity (τ), and weight function parameter (β) for five models and three weight functions.

	Weight Function		
Parameter	w_1	Half-Normal	Neg.Exponential
Model 1 (τ=0)			
θ	.029	.034	.026
β	1.31	3.66	3.97
Unpubl. Studies	20.1	11.6	30.1
-2 LN(likelihood)	60.0	61.5	58.2
Model 2 (θ = 0)			
τ^2 (τ)	.184(.43)	.192(.44)	.169(.41)
β	.339	1.16	1.15
Unpubl. Studies	2.5	2.0	3.7
-2 LN(likelihood)	42.4	42.3	42.2
Model 3 (full model)			
θ	.204	.206	.191
τ^2 (τ)	.168(.41)	.173(.42)	.159(.40)
β	.289	1.03	.981
Unpubl. Studies	2.1	1.7	3.0
-2 LN(likelihood)	40.5	40.4	40.4
Model 4 (β = 0)			
θ	.242		
τ^2 (τ)	.199(.45)		
Unpubl. Studies	-		
-2 LN(likelihood)	40.9		
Model 5 (θ = 0, β = 0)			
τ^2 (τ)	.235(.48)		
Unpubl. Studies	-		
-2 LN(likelihood)	42.9		

TABLE 4: Estimated SD–corr *matrices.*

Parameter	Weight Function								
	w_1			Half-Normal			Neg. Exponential		
Model 1 (τ=0)									
θ	.055		−.11	.059		−.10	.052		−.10
β			.60			1.89			1.52
Model 2 (θ = 0)									
τ^2		.13	−.48		.13	−.41		.12	−.56
β			.49			1.64			1.46
Model 3 (full model)									
θ	.16	.14	−.29	.16	.11	−.25	.15	.19	−.34
τ^2		.12	−.41		.12	−.35		.11	−.49
β			.47			1.62			1.43
Model 4 (β=0)									
θ	.17	−.01							
τ^2		.13							
Model 5 (θ = 0, β = 0)									
τ^2		.14							

Note that $K(x)$ may have an implicit dependence on the degrees of freedom—either through the critical point, x_{crit}, or through the p-value, $p(x)$.

Multiplying (5) by $\exp[-\beta K(x)]$, the combined effect of heterogeneity and publication bias is expressed by a joint weight function,

$$w(x) = e^{\alpha(x-\theta)^2-\beta K(x)} . \tag{7}$$

This form may account for our success in distinguishing heterogeneity from publication bias, at least when θ is small. For large x, $\beta K(x) \simeq 0$ so the quadratic is the dominant term in (7). Conversely, $\beta K(x)$ is the more important term for small x. Thus, publication bias can be described as a short-range effect while heterogeneity is long-range.

5.4 Properties of Fisher Information Matrix for the Publication Bias/Heterogeneity Model.

Consider the unified weight function $\exp[-\beta K(x)]$ defined by equation (6). The function $K(x)$ has several properties that will allow us to draw conclusions concerning the qualitative features of the correlations between estimated parameters. These properties are:

(i) $K(x) \geq 0$;

(ii) $K(x) = K(-x)$; and

(iii) $K(x)$ is a decreasing function of $|x|$.

Property (i) is necessary if $w(x)$ is to be interpreted as a *probability* of publication. (However, negative $K(x)$ may make mathematical sense. Similarly negative β lie in the valid parameter range provided $E[w]$ is finite. This can justify the use of likelihood ratio tests for the hypothesis that $\beta = 0$). The last two properties, of symmetry and monotonicity, are reasonable descriptors of publication bias, though one can envision circumstances in which either would fail. All three properties hold for the weight functions considered in Section 5.2.

The recorded t-scores, x, are modelled as observations from the weighted distribution

$$w(x,\beta)f(x,\theta,\tau^2)/w(\theta,\tau^2,\beta),$$

where $f(x,\theta,\tau^2)$ is a normal pdf with

$$\text{mean} = \mu = A\bullet\theta$$

$$\text{variance} = V = B + C\theta^2 + D\tau^2 \ .$$

Here A, B, C, D are positive constants that depend only on the degrees of freedom. For a given degree of freedom, the log-likelihood is

$$\ell = -\beta K(x) - \frac{(x-\mu)^2}{2V} + c_0 ,$$

where c_0 depends on the parameters but not on the datum x. We shall obtain the information matrix as the variance-covariance matrix of the score functions. Differentiating the log-likelihood yields the scores:

$$S_{\theta} = \partial\ell/\partial\theta = \frac{A}{V}(x-\mu) + \frac{C\theta}{V^2}(x-\mu)^2 + c_1 ,$$

$$S_{\tau^2} = \partial\ell/\partial\tau^2 = \frac{D}{2V^2}(x-\mu)^2 + c_2 ,$$

$$S_{\beta} = \partial\ell/\partial\beta = -K(x) + c_3 ,$$

where c_1, c_2, c_3 do not depend on x. Let

$$M_r = E[(x-\mu)^r], \quad r = 1,2,3,4,$$

$$\lambda_r = E[(x-\mu)^r K(x)], \quad r = 0,1,2,$$

$$\kappa = E[\{K(x)\}^2],$$

where the expectation is with respect to the weighted distribution. These constants (eight per degree of freedom) have to be calculated by numerical integration. The Fisher information matrix is now obtained as the covariance matrix of the scores:

$$\iota_{\theta\theta} = \frac{A^2}{V^2}(M_2-M_1^2) + \frac{2AC\theta}{V^3}(M_3-M_1M_2) + \frac{C^2\theta^2}{V^4}(M_4-M_2^2)$$

$$\iota_{\theta\tau^2} = \frac{AD}{2V^3}(M_3-M_1M_2) + \frac{CD\theta}{2V^4}(M_4-M_2^2)$$

$$\iota_{\theta\beta} = -\frac{A}{V}(\lambda_1-\lambda_0M_1) - \frac{C\theta}{V^2}(\lambda_2-\lambda_0M_2)$$

$$\iota_{\tau^2\tau^2} = \frac{D^2}{4V^4}(M_4-M_2^2)$$

$$\iota_{\tau^2\beta} = \frac{D}{2V^2}(\lambda_2-\lambda_0M_2)$$

$$\iota_{\beta\beta} = \kappa - \lambda_0^2 .$$

Note that these expressions have to be summed over all the degrees of freedom in the data set, counting multiplicity.

Thereom. Assume the effect size θ is close to zero and $K(x)$ has the three properties described above. Then

(i) $\text{corr}(\hat{\theta},\hat{\tau}^2)$ is close to zero,

(ii) $\text{corr}(\hat{\theta},\hat{\beta})$ is close to zero,

(iii) $\text{corr}(\hat{\tau}^2,\hat{\beta})$ is negative.

Remarks. Part (ii) of the theorem affirms a numerical finding of Iyengar and Greenhouse (1988); see our Table 2 and Table 4. When the true effect size is small, then the estimated effect size is not much influenced by the estimated weight function. This is generally not the case when the true effect size is large. Part (iii) of the theorem is to be expected, since publication bias and heterogeneity are competing explanations (in the absence of true effects) for any apparent overdispersion in the recorded t-scores.

Proof: We show that the information matrix has the form

$$J(\theta,\tau^2,\beta) = \begin{bmatrix} + & 0 & 0 \\ 0 & + & + \\ 0 & + & + \end{bmatrix} ,$$

the theorem then follows upon inversion of such a matrix. When $\theta = 0$, all but the first terms in $\iota_{\theta\theta}$, $\iota_{\theta\tau^2}$, $\iota_{\theta\beta}$ drop out. Also $\mu = 0$ and the weighted distribution is symmetric about zero (since w(x) is symmetric). Thus the odd moments M_1, M_3 and λ_1 are zero, implying that $\iota_{\theta\tau^2} = \iota_{\theta\beta} = 0$. Finally,

$$\begin{aligned} \iota_{\tau^2\beta} &= \frac{D}{2V^2}\,\text{cov}((x-\mu)^2,\ -K(x)) \\ &= \frac{D}{2V^2}\,\text{cov}(x^2,\ -K(x)) \ . \end{aligned}$$

But $-K(x)$ is an increasing function of x^2 so that $\iota_{\tau^2\beta} \geq 0$.

When $|\theta|$ is large, $J(\theta,\tau^2,\beta)$ is a full 3×3 matrix making it difficult to say much about the correlations. But if we restrict attention to situations and models in which $\tau^2 = 0$, then arguments similar to the preceding suggest that $\text{cor}(\hat{\theta},\hat{\beta})$ has the opposite sign as θ, which says that apparent significance can be mitigated by a strong publication bias. In such instances, the estimate of θ is potentially sensitive to the choice of weight function.

REFERENCES

Cox, D. R. (1962). *Renewal Theory.* Methuen's Monograph. Barnes and Noble, Inc., New York.

Dennis, B. and Patil, G. P. (1984). The gamma distribution and weighted multimodal gamma distributions as models of population abundance. *Math. Biosc.*, 68, 187-212.

Diaconis, P. and Efron, B. (1985). Testing the independence of a two-way table: New interpretations of the chi-square statistic (with discussion). *Ann. Statist.*, 13, 845-913.

DuMouchel, W. and Harris, J. (1983). Bayes methods for combining the results of cancer studies in humans and other species (with discussion). *J. Amer. Statist. Assoc.*, 78, 293-315.

Efron, B. (1978). Regression and ANOVA with zero-one data: Measures of residual variation. *J. Amer. Statist. Assoc.*, 73, 113-121.

Efron, B. (1986). Double exponential families and their use in generalized linear regression. *J. Amer. Statist. Assoc.*, 81, 709-721.

Fisher, R. A. (1932). *Statistical Methods for Research Workers*, 4th ed. Oliver and Boyd, London.

Fisher, R. A. (1934). The effects of methods of ascertainment upon the estimation of frequencies. *Ann. Eugenics*, 6, 13-25.

Hedges, L. V. and Olkin, I. (1985). *Statistical Methods for Meta-Analysis.* Academic Press.

Herdan, G. (1960). *Small Particle Statistics.* Elsevier, New York.

Iyengar, S. and Greenhouse, J. B. (1988). Selection models and the file drawer problem (with discussion). *Statist. Sci.*, 1, 109-135.

Krumbein, W. C. and Pettijohn, F. J. (1938). *Manual of Sedimentory Petrography.* D. Appleton-Century Co., Inc., New York

Laird, N., Patil, G. P. and Taillie, C. (1988). Comment on 'Selection models and the file drawer problem by S. Iyengar and J. B. Greenhouse.' *Statist. Sci.*, 3, 126-128.

Landau, E., Patil, G. P. and Rawlings, J. O. (1985). ASA Reviews of EPA-funded acid precipitation research. *Amer. Statist.*, 39, 243-273.

Light, R. and Pillemer, D. (1984). *Summing Up: The Science of Reviewing Research.* Harvard Univ. Press, Cambridge, MA.

Mahfoud, M. and Patil, G. P. (1981). In *Statistics in Theory and Practice: Essays in Honor of Bertil Matern.* Bo Ranneby, ed. Swedish Univ. Agri. Sci., Umea, Sweden, pp. 173-187.

Mahfoud, M. and Patil, G. P. (1982). On weighted distributions. In *Statistics and Probability Essays in Honor of C. R. Rao*, G. Kallianpur et al., eds. North Holland Publ. Co., pp. 479-492.

Patil, G. P. (1984). Studies in statistical ecology involving weighted distributions. In *Statistics Applications and New Directions: Proceedings of ISI Golden Jubilee International Conference*, December 1981, J. K. Ghosh and J. Roy, eds. Statistical Publ. Soc., Calcutta, India, pp. 478-503.

Patil, G. P. (1987). Encountered data, statistical ecology, environmental risk analysis, and weighted distributions. Special Invited General Methodology Lecture-I. Annual Meetings of the American Statistical Association, San Francisco.

Patil, G. P., Babu, G. J., Boswell, M. T., Chatterjee, K., Linder, E., and Taillie, C. (1986). Statistical issues in combining ecological and environmental studies with examples in marine fisheries research and management. In *Proceedings ASA/EPA Conference on Statistical Issues in Combining Environmental Studies*, EPA, Washington, DC.

Patil, G. P., Boswell, M. T., Joshi, S. W., and Ratnaparkhi, M. V. (1984). *Dictionary and Classified Bibliography of Statistical Distributions in Scientific Work, Vol. 1: Discrete Models.* International Co-operative Publishing House, Fairland, MD.

Patil, G. P., Boswell, M. T., and Ratnaparkhi, M. V. (1984). *Dictionary and Classified Bibliography of Statistical Distributions in Scientific Work, Vol. 2: Continuous Univariate Models.* International Co-operative Publishing House, Fairland, MD.

Patil, G. P., Boswell, M. T., Ratnaparkhi, M. V., and Roux, J. J. J. (1984). *Dictionary and Classified Bibliography of Statistical Distributions in Scientific Work, Vol. 3: Multivariate Models.* International Co-operative Publishing House, Fairland, MD.

Patil, G. P. and Ord, J. K. (1975). On size-biased sampling and related form-invariant weighted distributions. *Sankhya*, 38, 48-61.

Patil, G. P. and Rao, C. R. (1977). Weighted distributions and a survey of their applications. In *Applications of Statistics*, P. R. Krishnaiah, ed. North Holland Publishing Co., pp. 383-405.

Patil, G. P. and Rao, C. R. (1978). Weighted distributions and size-biased sampling with applications to wildlife populations and human families. *Biometrics*, 34, 179-184.

Patil, G. P., Rao, C. R. and Ratnaparkhi, M. V. (1986). On discrete weighted distributions and their use in model choice for observed data. *Comm. Statist.-Theor. Meth.*, 15(3), 907-918.

Patil, G. P., Rao, C. R. and Zelen, M. (1988a). A computerized bibliography of weighted distributions and related weighted methods for statistical analysis and interpretation of encountered data, observational studies, representativeness issues, and resulting inferences. Center for Statistical Ecology and Environmental Statistics, Pennsylvania State University.

Patil, G. P., Rao, C. R. and Zelen, M. (1988b). Weighted distributions. In *Encyclopedia of Statistical Sciences*, Vol. 9, S. Kotz and N. L. Johnson, eds. John Wiley, New York, pp. 565-571.

Patil, G. P. and Taillie, C. (1987a). The double exponential families and some of their properties and applications. MS.

Patil, G. P. and Taillie, C. (1987b). Incorporating heterogeneity into binomial regression. MS.

Patil, G. P. and Taillie, C. (1988). Distinguishing between heterogeneity and publication bias in meta-analysis using weighted distributions. MS.

Rao, C. R. (1965). On discrete distributions arising out of methods of ascertainment. In *Classical and Contagious Discrete Distributions*, G. P. Patil, ed. Pergamon Press and Statistical Publishing Society, Calcutta, pp. 320-332.

Rao, C. R. (1985). Weighted distributions arising out of methods of ascertainment: What population does a sample represent?. In *A Celebration of Statistics*, Chap. 24, A. C. Atkinson and S. E. Fienberg, eds. Springer-Verlag, New York, pp. 543-569.

Rosenthal, R. (1984). *Meta-Analytic Procedures for Social Research.* Sage Press, Beverly Hills, CA.

Simon, R. (1980). Length biased sampling in etiologic studies. *Amer. J. Epidemiology*, 111(4), 444-452.

Smith, M., Glass, G. and Miller, T. (1980). *The Benefits of Psychotherapy.* Johns Hopkins University Press, Baltimore, MD.

Stene, J. (1981). Probability distributions arising from the ascertainment and the analysis of data on human families and other groups. In *Statistical Distributions in Scientific Work, Vol. 6*, C. Taillie, G. P. Patil and B. Baldessari, eds. Reidel Publ. Co., Dordrecht and Boston, pp. 233-264.

Warren, W. G. (1975). Statistical distributions in forestry and forest products research. In *Statistical Distributions in Scientific Work*, G. P. Patil, S. Kotz and J. K. Ord, eds. Reidel Publ. Co., Dordrecht and Boston, pp. 369-384.

Wolf, F. (1986). *Meta-analysis: Quantitative Methods for Research Synthesis.* Sage Press, Beverly Hills, CA.

Zelen, M. (1971). Problems in the early detection of disease and the finding of faults. *Bull. Int. Statist. Inst.*, Proc. 38, Session I, pp. 649-661.

Zelen, M. (1974). Problems in cell kinetics and the early detection of disease. In *Reliability and Biometry*, F. Proschan and R. J. Serfling, eds. SIAM, Philadelphia, PA., pp. 701-706.

STATISTICAL DATA ANALYSIS AND INFERENCE
Y. Dodge (Editor)

FURTHER RESULTS ON IDENTIFICATION WHEN THE PARAMETERS ARE PARTIALLY UNKNOWN

Sana EL KHATTABI and Franz STREIT

Section de Mathématiques de l'Université de Genève, Case postale 240, 1211 Genève 24, Switzerland

Amoh and Kocherlakota [1] have studied identification rules which are useful to distinguish by discriminant analysis two different inverse Gaussian distributions with density functions

$$f(x : \mu_i, \lambda_i) = \left(\frac{\lambda_i}{2\pi x^3}\right)^{1/2} \exp\left[-\frac{\lambda_i (x-\mu_i)^2}{2\mu_i^2 x}\right] 1_{(0,\infty)}(x) \quad [i = 1,2]$$

They also analyzed the probabilities of errors (caused by misclassification) of these rules. These investigations were carried out for the case

$$\mu_1 \neq \mu_2 \text{ and } \lambda_1 = \lambda_2 = \lambda$$

In this paper the analogous problem with

$$\mu_1 = \mu_2 = \mu \text{ and } \lambda_1 \neq \lambda_2$$

is treated. Decision rules are derived according to standard methods and some of their exact and asymptotic properties are examined.

1. INTRODUCTION

The inverse Gaussian distributions are frequently applied in applications involving Brownian motion, sequential analysis, reliability theory, radio technique, electronics, and are recommended as alternatives for the use of lognormal distributions in statistical data analysis [3,pp.263-264 ; 4, pp. 150,152]. It seems therefore reasonable to establish identification rules yielding discrimination between two populations G_1 and G_2 distributed according to different inverse Gaussian probability laws with density functions.

$$f_{X|G_i}(x) = f_X(x : \mu_i, \lambda_i) = \left(\frac{\lambda_i}{2\pi x^3}\right)^{1/2} exp\left[-\frac{\lambda_i(x-\mu_i)^2}{2\mu_i^2 x}\right] 1_{(0,\infty)}(x)$$

$$[i = 1,2]$$
$$[\mu_i > 0, \lambda_i > 0]$$

with $1_{(0,\infty)}(x) = 1$ if $x > 0$ and $1_{(0,\infty)}(x) = 0$ otherwise. Amoh and Kocherlakota [1] obtained results for the case $\mu_1 \neq \mu_2$ and $\lambda_1 = \lambda_2 = \lambda$. We shall study the analogous problem with $\mu_1 = \mu_2 = \mu$ and $\lambda_1 \neq \lambda_2$ (In view of the fact that μ_i is a measure of location while λ_i is related to the shape of the distributions [i = 1,2] it seems at least as possible that we have to distinguish between λ_1 and λ_2 for known common μ than to have to distinguish between μ_1 and μ_2 for known common λ).

2. OPTIMAL IDENTIFICATION RULES FOR KNOWN μ, λ_1 AND λ_2

If p_i denotes the a-priori probability that a unit comes from $G_i[i = 1,2]$ and $C(2 \mid 1) > 0$ respectively $C(1 \mid 2) > 0$ the costs of assigning wrongly to G_2 respectively to G_1, then an optimal identification rule is given by attributing the unit with observation x to G_1 if

$$\frac{f_X(x : \mu, \lambda_1)}{f_X(x : \mu, \lambda_2)} \geq \frac{C(1 \mid 2)p_2}{C(2 \mid 1)p_1} : = d$$

and to G_2 otherwise [2, p. 201;5].

If we suppose (without loss of generality [due to the possibility to renumber the populations]) that $\lambda_2 > \lambda_1 > 0$ we have in our case

$$\frac{f_X(x : \mu, \lambda_1)}{f_X(x : \mu, \lambda_2)} \geq d \longleftrightarrow (\lambda_2 - \lambda_1)\frac{(x-\mu)^2}{2\mu^2 x} \geq \ell n \left(\left(\frac{\lambda_2}{\lambda_1}\right)^{1/2} d\right)$$

$$\longleftrightarrow x^2 - (2\mu + \kappa)x + \mu^2 \geq 0$$

where $\kappa : = \dfrac{2\mu^2}{\lambda_2 - \lambda_1}\ell n \left(\left(\dfrac{\lambda_2}{\lambda_1}\right)^{1/2} d\right)$ Putting $\triangle : = \kappa\,(4\mu + \kappa)$ and

$$x_1 : = 2^{-1}\left\{2\mu + \kappa - [\kappa(4\mu+\kappa)]^{1/2}\right\}$$

and

$$x_2 : = 2^{-1}\left\{2\mu + \kappa + [\kappa(4\mu+\kappa)]^{1/2}\right\}$$

for $\triangle \geq 0$, we find as region of attribution of the unit to G_1

$$R_1 = (0, x_1] \cup [x_2, \infty) \quad \text{if } \triangle \geq 0$$

and

$$R_1 = (0, \infty) \qquad \text{if } \triangle < 0$$

and as region of attribution to G_2

$$R_2 = (x_1, x_2) \qquad \text{if } \triangle \geq 0$$

and

$$R_2 = \emptyset \qquad \text{if } \triangle < 0.$$

If $C(1 \mid 2)$ and $C(2 \mid 1)$ are unknow one works usually with the assumption $C(1 \mid 2) = C(2 \mid 1)$, if p_1 and thus p_2 are unknown, it is customary to assume $p_1 = p_2$.

3. CALCULATION OF THE PROBABILITIES OF MISCLASSIFICATION FOR KNOWN λ_1, λ_2 AND μ.

Let $e_{2|1}(\mu, \lambda_1, \lambda_2)(e_{1|2}(\mu, \lambda_1, \lambda_2))$ be the probability of incorrectly assigning a unit from G_1 to G_2 (respectively from G_2 to G_1). We obtain

$$e_{2|1}(\mu, \lambda_1, \lambda_2) = F(x_2 : \mu, \lambda_1) - F(x_1 : \mu, \lambda_1) \qquad \text{if } \triangle \geq 0$$

and

$$e_{2|1}(\mu, \lambda_1, \lambda_2) = 0 \qquad \text{if } \triangle < 0$$

and

$$e_{1|2}(\mu, \lambda_1, \lambda_2) = 1 - [F(x_2 : \mu, \lambda_2) - F(x_1 : \mu, \lambda_2)] \quad \text{if } \triangle \geq 0$$

and

$$e_{1|2}(\mu, \lambda_1, \lambda_2) = 1 \qquad \text{if } \triangle < 0$$

For the concrete numerical evaluation the formula

$$F(z : \mu, \lambda) = \Phi\left(\left(\frac{\lambda}{z}\right)^{1/2}\left(\frac{z}{\mu} - 1\right)\right) + e^{\frac{2\lambda}{\mu}}\,\Phi\left(-\left(\frac{\lambda}{z}\right)^{1/2}\left(1 + \frac{z}{\mu}\right)\right)$$

is very useful [3 , pp 265, f.(7)], where Φ designates the distribution function of a standard normal variable.

4. CONDITIONAL PROBABILITES OF MISCLASSIFICATION APPLICABLE WHEN SOME OF THE PARAMETERS ARE UNKNOWN

If some of the parameters λ_1, λ_2 are actually unknown the optimal identification rule of paragraph 2 is replaced by the substitution rule based on (a) training sample(s) $X_{i1}, \ldots, X_{in_i}$ from a population with inverse Gaussian distribution with parameters μ and

λ_i. This substitution rule and its conditional probabilities of errors of misclassification is obtained by replacing λ_i by the realization of a consistent estimator $\widehat{\lambda}_i$ of λ_i in the formulae of the paragraphs 2 and 3. If μ is also unknown, μ could be replaced by the observed value of $(n_1+n_2)^{-1}(n_1\bar{X}_1+n_2\bar{X}_2)$.

5. ASYMPTOTIC DISTRIBUTION OF THE CONDITIONAL PROBABILITIES OF MISCLASSIFICATION WHEN μ IS KNOWN, λ_1 AND λ_2 ARE UNKNOWN AND $d=1$ (FOR INSTANCE $C(2\mid 1)=C(1\mid 2)$ AND $p_1=p_2$)

If $d=1$ we have $\triangle \geq 0$. We reparametrize the problem by using $\lambda_1^*=\lambda_1^{-1}$ and $\lambda_2^*=\lambda_2^{-1}$ instead of λ_1 and λ_2. Defining $e_{2|1}^*(\lambda_1^*,\lambda_2^*):=e_{2|1}\left(\frac{1}{\lambda_1^*},\frac{1}{\lambda_2^*}\right)=e_{2|1}(\lambda_1,\lambda_2)$, we propose to use the estimators

$$\widehat{\lambda_i^*}=n_i^{-1}\sum_{\ell=1}^{n}\frac{(X_{i\ell}-\mu)^2}{\mu^2 X_{i\ell}} \qquad [i=1,2]$$

According to [3. pp. 266-268] $\widehat{\lambda_i^*}\sim\frac{1}{\lambda_i n_i}\chi^2_{n_i}$. It follows that $E\left[\widehat{\lambda_i^*}:\mu_i,\lambda_i\right]=\frac{1}{\lambda_i}$ and $Var\left[\widehat{\lambda_i^*}:\mu_i,\lambda_i\right]=\frac{2}{\lambda_i^2 n_i}$. Thus $\widehat{\lambda_i^*}$ is an unbiased and consistent estimator of $\lambda_i^*=\frac{1}{\lambda_i}$.

Taylor development yields when neglecting higher order terms

$$e_{2|1}^*\left(\widehat{\lambda_1^*},\widehat{\lambda_2^*}\right)\approx e_{2|1}(\lambda_1,\lambda_2)+\left(\widehat{\lambda_1^*}-\frac{1}{\lambda_1}\right)\left.\frac{\partial e_{2|1}^*(\lambda_1^*,\lambda_2^*)}{\partial\lambda_1^*}\right|_{\substack{\lambda_1^*=\frac{1}{\lambda_1}\\ \lambda_2^*=\frac{1}{\lambda_2}}}$$

$$+\left(\widehat{\lambda_2^*}-\frac{1}{\lambda_2}\right)\left.\frac{\partial e_{2|1}^*(\lambda_1^*,\lambda_2^*)}{\partial\lambda_2^*}\right|_{\substack{\lambda_1^*=\frac{1}{\lambda_1}\\ \lambda_2^*=\frac{1}{\lambda_2}}}+\frac{1}{2}\left(\widehat{\lambda_1^*}-\frac{1}{\lambda_1}\right)^2\cdot$$

$$\left.\frac{\partial^2 e_{2|1}(\lambda_1^*,\lambda_2^*)}{\partial\lambda_1^{*2}}\right|_{\substack{\lambda_1^*=\frac{1}{\lambda_1}\\ \lambda_2^*=\frac{1}{\lambda_2}}}+\frac{1}{2}\left(\widehat{\lambda_2^*}-\frac{1}{\lambda_2}\right)^2\left.\frac{\partial^2 e_{2|1}^*(\lambda_1^*,\lambda_2^*)}{\partial\lambda_2^{*2}}\right|_{\substack{\lambda_1^*=\frac{1}{\lambda_1}\\ \lambda_2^*=\frac{1}{\lambda_2}}}$$

$$+\left(\widehat{\lambda_1^*}-\frac{1}{\lambda_1}\right)\left(\widehat{\lambda_2^*}-\frac{1}{\lambda_2}\right)\cdot\left.\frac{\partial^2 e_{2|1}^*(\lambda_1^*,\lambda_2^*)}{\partial\lambda_1^*\partial\lambda_2^*}\right|_{\substack{\lambda_1^*=\frac{1}{\lambda_1}\\ \lambda_2^*=\frac{1}{\lambda_2}}}$$

Denote by

$$\bar{k}_i := \left.\frac{\partial e_{2|1}^*(\lambda_1^*,\lambda_2^*)}{\partial\lambda_i^*}\right|_{\substack{\lambda_1^*=\frac{1}{\lambda_1}\\ \lambda_2^*=\frac{1}{\lambda_2}}}$$

and by

$$\bar{k}_{i,j} := \left.\frac{\partial^2 e_{2|1}^*(\lambda_1^*,\lambda_2^*)}{\partial\lambda_i^*\partial\lambda_j^*}\right|_{\substack{\lambda_1^*=\frac{1}{\lambda_1}\\ \lambda_2^*=\frac{1}{\lambda_2}}} \qquad [i,j\in\{1,2\}];$$

these quantities are straightforward to evaluate, but their formula are too long to reproduce in this text. We find

$$E\left[e_{2|1}^*(\lambda_1^*,\lambda_2^*)\right] = e_{2|1}(\lambda_1,\lambda_2)+\frac{\bar{k}_{1,1}}{2}\mathrm{Var}\left[\widehat{\lambda_1^*}:\frac{1}{\lambda_1}\right]$$

$$+\frac{\bar{k}_{2,2}}{2}\mathrm{Var}\left[\widehat{\lambda_2^*}:\frac{1}{\lambda_2}\right]+o(\min(n_1,n_2))$$

$$= e_{2|1}(\lambda_1,\lambda_2)+\frac{\bar{k}_{1,1}}{\lambda_1^2 n_1}+\frac{\bar{k}_{2,2}}{\lambda_2^2 n_2}+o(\min(n_1,n_2)).$$

and

$$\mathrm{Var}\left[e_{2|1}^*\left(\widehat{\lambda_1^*},\widehat{\lambda_2^*}\right)\ :\ \lambda_1,\lambda_2\right] = \frac{2\bar{k}_1^2}{\lambda_1^2 n_1}+\frac{2\bar{k}_2^2}{\lambda_2^2 n_2}+o\left(\min\left(n_1,n_2\right)\right)$$

In view of the fact that $\begin{pmatrix}\widehat{\lambda_1^*}\\ \widehat{\lambda_2^*}\end{pmatrix}$ follows an asymptotic bivariate normal distribution and that $e_{2|1}^*\left(\widehat{\lambda_1^*},\widehat{\lambda_2^*}\right)$ follows an asymptotic univariate normal distribution for $n_1,n_2\to\infty$ taking into account the variability in the training samples we get for $n_1,n_2\to\infty$

$$P\left(e_{2|1}\left(\left(\widehat{\lambda_1^*}\right)^{-1},\left(\widehat{\lambda_2^*}\right)^{-1}\right)\leq z:\mu,\lambda_1,\lambda_2\right)\approx$$

$$\Phi\left(\left[\frac{2\bar{k}_1^2}{\lambda_1^2 n_1}+\frac{2\bar{k}_2^2}{\lambda_2^2 n_2}\right]^{-\frac{1}{2}}\left[z-\frac{\bar{k}_{1,1}}{\lambda_1^2 n_1}-\frac{\bar{k}_{2,2}}{\lambda_2^2 n_2}\right]\right)$$

The asymptotic distribution of $e_{1|2}\left(\left(\widehat{\lambda_1^*}\right)^{-1},\left(\widehat{\lambda_2^*}\right)^{-1}\right)$ can be obtained in a similar way.

6. FINAL REMARK

Problems of identification when only parts of the parameters are known arise in many contexts. For a study of such a situation involving multivariate normal distributions see Streit, F.[6,7].

REFERENCES

[1] Amoh, R.K. and Kocherlakota, K. , Errors of Misclassification Associated with the Inverse Gaussian Distribution, Commun. Statist.-Theor.Meth. 15(2)(1986) 589-612.
[2] Anderson, T.W. An Introduction to Multivariate Statistical Analysis (J. Wiley, New York, 1984).
[3] Folks, J.R. and Chhikara, R.S. , The Inverse Gaussian Distribution and Its Statistical Application - A Review, J.R. Statist. Soc. B 40(3) (1978), 263-289.
[4] Johnson, N. and Kotz., S., Continuous Univariate Distributions. Vol. I. Chapter 15 (Houghton Mifflin, Boston, 1970).
[5] Rao, C.R., Linear Statistical Inference and Its Applications. Chapters 7d and 8e (J. Wiley, New York, 1973).
[6] Streit, F. Multivariate Linear Discrimination When the Covariance Matrix is Unknown, South African Statist. J. 13 (1979) 93-101 and 14 (1980) 76.
[7] Streit, F. , Etude des probabilités de classification erronnée d'une règle d'identification, C.R. Acad. Sc. Paris 303(I,8) (1986) 371-374.

Part Four
LINEAR MODELS AND MATRICES

"How Sweet is mortal Sovranty! "–think some:
Others–"How blest the Paradise to come! "
Ah, take the Cash in hand and waive the Rest,
Oh, the brave Music of a distant Drum!

Khayyam Naishapuri-Rubáiyát

STATISTICAL DATA ANALYSIS AND INFERENCE
Y. Dodge (Editor)

WEIGHTED-LEAST-SQUARES ESTIMATION IN THE GENERAL GAUSS-MARKOV MODEL

Jerzy K. BAKSALARY
Department of Mathematics, Pedagogical University, PL 65-069 Zielona Góra
and Department of Mathematical and Statistical Methods
Academy of Agriculture in Poznań, PL-60-637 Poznań, Poland

and

Simo PUNTANEN
Department of Mathematical Sciences, University of Tampere, P.O. Box 607
SF-33101 Tampere, Finland

Following a wider interpretation proposed by Rao (1971), weighted-least-squares estimators of $\mathbf{X}\boldsymbol{\beta}$ under the general Gauss-Markov model $\{\mathbf{Y}, \mathbf{X}\boldsymbol{\beta}, \sigma^2\mathbf{V}\}$ are considered in this paper as the family of all statistics of the form $\mathbf{X}\mathbf{b}_\mathbf{W} = \mathbf{X}(\mathbf{X}'\mathbf{W}\mathbf{X})^-\mathbf{X}'\mathbf{W}\mathbf{Y}$, where $\mathbf{W}$ may be any matrix satisfying the condition $\mathcal{R}(\mathbf{X}'\mathbf{W}\mathbf{V}) \subset \mathcal{R}(\mathbf{X}'\mathbf{W}\mathbf{X}) \neq \{\mathbf{0}\}$. Several properties of such estimators are discussed, beginning with their unbiasedness and invariance to the choice of a generalized inverse and with the requirement that the dispersion matrix of $\mathbf{X}\mathbf{b}_\mathbf{W}$ is below $\sigma^2\mathbf{V}$ with respect to the Löwner ordering. These two auxiliary properties are then applied to derive a number of criteria for $\mathbf{X}\mathbf{b}_\mathbf{W}$ to coincide with the best linear unbiased estimator of $\mathbf{X}\boldsymbol{\beta}$. Certain related questions, such as the problem of linear sufficiency and the concepts of unified-least-squares matrices, are also investigated. Finally, the estimators $\mathbf{X}\mathbf{b}_\mathbf{W}$ are examined from the point of view of their admissibility, thus providing a preliminary evaluation of their usefulness as biased estimators of $\mathbf{X}\boldsymbol{\beta}$.

1. INTRODUCTION

Consider the general Gauss-Markov model, denoted by

$$\mathrm{M} = \{\mathbf{Y}, \mathbf{X}\boldsymbol{\beta}, \sigma^2\mathbf{V}\}, \tag{1.1}$$

in which $\mathbf{Y}$ is an observable random vector with expectation $E(\mathbf{Y}) = \mathbf{X}\boldsymbol{\beta}$ and with dispersion matrix $D(\mathbf{Y}) = \sigma^2\mathbf{V}$, where $\mathbf{X}$ is an $n \times p$ known nonnull matrix, $\boldsymbol{\beta}$ is a $p \times 1$ vector of unknown parameters, $\mathbf{V}$ is an $n \times n$ known symmetric nonnegative definite matrix, and σ^2 is an unknown positive scalar. The matrices $\mathbf{X}$ and $\mathbf{V}$ are both allowed to be of arbitrary rank, but it is assumed, throughout the paper, that the model (1.1) is consistent [cf. Rao (1971, p. 378; 1973a, p. 297)], or, in other words, that the inference base is not self-contradictory [cf. Feuerverger and Fraser (1980, p. 44)], i.e.,

$$\mathbf{Y} \in \mathcal{R}(\mathbf{X} : \mathbf{V}), \tag{1.2}$$

where $\mathcal{R}(\mathbf{X} : \mathbf{V})$ denotes the range of the partitioned matrix $(\mathbf{X} : \mathbf{V})$. In the sequel we utilize Lemma 2.1 of Rao (1974), which asserts that $\mathcal{R}(\mathbf{X} : \mathbf{V})$ may be decomposed as

$$\mathcal{R}(\mathbf{X} : \mathbf{V}) = \mathcal{R}(\mathbf{X}) \oplus \mathcal{R}(\mathbf{V}\mathbf{X}^{\perp}), \tag{1.3}$$

where $\mathbf{X}^{\perp}$ is any matrix such that $\mathcal{R}(\mathbf{X}^{\perp}) = \mathcal{N}(\mathbf{X}')$, the null space of the transpose of $\mathbf{X}$. See also surveys of various decompositions of linear subspaces related to the model (1.1) in Rao (1984) and Nordström (1985).

Customarily, weighted-least-squares estimation is based on minimizing the (squared) seminorm of $\mathbf{Y} - \mathbf{X}\boldsymbol{\beta}$,

$$S_{\mathbf{W}} = (\mathbf{Y} - \mathbf{X}\boldsymbol{\beta})'\mathbf{W}(\mathbf{Y} - \mathbf{X}\boldsymbol{\beta}), \tag{1.4}$$

specified by a given symmetric nonnegative definite matrix $\mathbf{W}$. As is well known, this leads to the normal equations

$$\mathbf{X}'\mathbf{W}\mathbf{X}\mathbf{b}_{\mathbf{W}} = \mathbf{X}'\mathbf{W}\mathbf{Y}. \tag{1.5}$$

Notice that if $\mathbf{W}$ is as assumed above, then $\mathcal{R}(\mathbf{X}'\mathbf{W}\mathbf{X}) = \mathcal{R}(\mathbf{X}'\mathbf{W})$, and hence the normal equations are consistent for every $\mathbf{Y}$. A statistic $\mathbf{X}\mathbf{b}_{\mathbf{W}}$, with $\mathbf{b}_{\mathbf{W}}$ being any solution to the equations (1.5), is called a $\mathbf{W}$-weighted-least-squares estimator of $\mathbf{X}\boldsymbol{\beta}$. It has the form

$$\mathrm{WLSE}_{\mathbf{W}}(\mathbf{X}\boldsymbol{\beta}) = \mathbf{X}(\mathbf{X}'\mathbf{W}\mathbf{X})^{-}\mathbf{X}'\mathbf{W}\mathbf{Y}, \tag{1.6}$$

where $(\mathbf{X}'\mathbf{W}\mathbf{X})^{-}$ is any generalized inverse of $\mathbf{X}'\mathbf{W}\mathbf{X}$ in the sense of Rao (1962).

Two interesting particular choices of $\mathbf{W}$ are $\mathbf{W} = \mathbf{I}_n$, the identity matrix of order n, and $\mathbf{W} = \mathbf{V}^{+}$, the Moore-Penrose inverse of $\mathbf{V}$. The former leads to the *ordinary least-squares estimator* of $\mathbf{X}\boldsymbol{\beta}$,

$$\mathrm{OLSE}(\mathbf{X}\boldsymbol{\beta}) = \mathbf{P}_{\mathbf{X}}\mathbf{Y}, \tag{1.7}$$

where $\mathbf{P}_{\mathbf{X}} = \mathbf{X}(\mathbf{X}'\mathbf{X})^{-}\mathbf{X}' = \mathbf{X}\mathbf{X}^{+}$ is the orthogonal projector onto $\mathcal{R}(\mathbf{X})$, and the latter leads to a *generalized least-squares estimator* of $\mathbf{X}\boldsymbol{\beta}$,

$$\mathrm{GLSE}(\mathbf{X}\boldsymbol{\beta}) = \mathbf{X}(\mathbf{X}'\mathbf{V}^{+}\mathbf{X})^{-}\mathbf{X}'\mathbf{V}^{+}\mathbf{Y}. \tag{1.8}$$

Rao (1971, p. 372) pointed out that equations of the form (1.5) may also be considered in a wider sense, with $\mathbf{W}$ not necessarily being a symmetric nonnegative definite matrix. The right-hand side of (1.6) represents a solution in any case where the equations (1.5) are consistent. It is clear that the desired consistency holds for every $\mathbf{Y}$ satisfying (1.2) if and only if $\mathcal{R}(\mathbf{X}'\mathbf{W}\mathbf{V}) \subset \mathcal{R}(\mathbf{X}'\mathbf{W}\mathbf{X})$. On the other hand, to eliminate the trivial case where (1.5) holds for every $\mathbf{b}_{\mathbf{W}} \in \Re^{p}$, it is assumed that $\mathbf{X}'\mathbf{W}\mathbf{X} \neq \mathbf{0}$. The extended freedom in the choice of $\mathbf{W}$ was substantially exploited by Rao (1971, Section 4) to build the unified theory of least squares, including a characterization of all **W** for which a statistic of the form (1.6) represents the best linear unbiased estimator of $\mathbf{X}\boldsymbol{\beta}$ under the general Gauss-Markov model (1.1) as one of the main points.

The present paper follows the general interpretation of weighted-least-squares estimation proposed by Rao. However, since an estimator $\mathbf{X}(\mathbf{X}'\mathbf{W}\mathbf{X})^{-}\mathbf{X}'\mathbf{W}\mathbf{Y}$ which utilizes any $n \times n$ matrix **W** satisfying

$$\mathcal{R}(\mathbf{X}'\mathbf{W}\mathbf{V}) \subset \mathcal{R}(\mathbf{X}'\mathbf{W}\mathbf{X}) \neq \{\mathbf{0}\} \tag{1.9}$$

may no longer be interpretable through minimizing an expression of the form (1.4), we adopt the term *pseudo-***W***-weighted-least-squares estimator* of $\mathbf{X}\boldsymbol{\beta}$ and the notation P-WLSE$_{\mathbf{W}}(\mathbf{X}\boldsymbol{\beta})$. In Section 2, we show that a result of Searle and Pukelsheim (1986, Theorem 1), indicating a relationship between the unbiasedness of (1.6) and its invariance with respect to the choice of the generalized inverse involved, remains valid for the extended set of **W**-matrices. In the same section, we investigate an elementary, but necessary for regarding any linear estimator as reasonable, property of the dispersion matrix of P-WLSE$_{\mathbf{W}}(\mathbf{X}\boldsymbol{\beta})$. These two auxiliary results are then applied to derive a number of criteria for P-WLSE$_{\mathbf{W}}(\mathbf{X}\boldsymbol{\beta})$ to be identical with BLUE($\mathbf{X}\boldsymbol{\beta}$) under the model (1.1), including a modification of Theorem 4.1 of Rao (1971). Certain other aspects of best linear unbiased estimation, such as the problem of linear sufficiency and the concepts of unified-least-squares matrices, are also discussed. In the final section, we examine admissibility of pseudo-**W**-weighted-least-squares estimators, thus providing a preliminary evaluation of their usefulness as biased estimators of $\mathbf{X}\boldsymbol{\beta}$.

2. TWO AUXILIARY CHARACTERIZATIONS

In general, the product $\mathbf{X}(\mathbf{X}'\mathbf{W}\mathbf{X})^{-}\mathbf{X}'\mathbf{W}\mathbf{Y}$ is not invariant with respect to the choice of $(\mathbf{X}'\mathbf{W}\mathbf{X})^{-}$. In the case of a symmetric nonnegative definite **W**, Searle and Pukelsheim (1986, Theorem 1) noted an interesting fact that the invariance of (1.6) is equivalent to its unbiasedness, and showed that this is the case if and only if $\mathcal{R}(\mathbf{X}') = \mathcal{R}(\mathbf{X}'\mathbf{W})$. Lemma 1 below extends this result to any P-WLSE$_{\mathbf{W}}(\mathbf{X}\boldsymbol{\beta})$.

The class of *unbiased estimators* of $\mathbf{X}\boldsymbol{\beta}$ is specified here as

$$\mathcal{U}(\mathbf{X}\boldsymbol{\beta}) = \{\mathbf{A}\mathbf{Y}: \mathbf{A}\mathbf{X} = \mathbf{X}\}, \tag{2.1}$$

which actually corresponds to regarding $\boldsymbol{\beta}$ in the model (1.1) as free to vary over $\Re^p$. In other words, this is equivalent to neglecting restrictions on $\boldsymbol{\beta}$ that arise due to singularity of **V** when $\mathcal{R}(\mathbf{X}) \not\subset \mathcal{R}(\mathbf{V})$ but are not completely specified prior to an observation on **Y**; cf. comments in Seely and Zyskind (1971, p. 693) and Rao (1979, p. 1354). In this context, it should be emphasized, following Rao (1973a, pp. 297-298; 1976, p. 1033; 1985, p. 20), that there is no loss in generality in adopting the stronger unbiasedness condition as in (2.1), for if $\mathbf{A}\mathbf{Y}$ is an unbiased estimator of $\mathbf{X}\boldsymbol{\beta}$ not satisfying $\mathbf{A}\mathbf{X} = \mathbf{X}$, then there exists $\mathbf{B}\mathbf{Y}$ such that $\mathbf{B}\mathbf{X} = \mathbf{X}$ and $\mathbf{B}\mathbf{Y} = \mathbf{A}\mathbf{Y}$ with probability one.

LEMMA 1. *For* P-WLSE$_W$(Xβ) = $\mathbf{X(X'WX)^-X'WY}$, *with* $\mathbf{W}$ *satisfying* $\mathcal{R}(\mathbf{X'WV}) \subset \mathcal{R}(\mathbf{X'WX}) \neq \{\mathbf{0}\}$, *the following statements are equivalent:*

(a) P-WLSE$_W$(Xβ) *is invariant with respect to the choice of* $(\mathbf{X'WX})^-$ *for all* $\mathbf{Y} \in \mathcal{R}(\mathbf{X} : \mathbf{V})$,

(b) P-WLSE$_W$(Xβ) $\in \mathcal{U}(\mathbf{X}\beta)$, i.e., $\mathbf{X(X'WX)^-X'WX} = \mathbf{X}$,

(c) $r(\mathbf{X'WX}) = r(\mathbf{X})$.

Proof. In view of (1.2) and (1.9), the condition (a) is equivalent to the requirement of the invariance of the matrix $\mathbf{X(X'WX)^-X'WX}$. According to Rao and Mitra (1971, pp. 21 and 43) [see also Rao, Mitra, and Bhimasankaram (1972, Lemma 1)], the product $\mathbf{AB^-C}$, with $\mathbf{A} \neq \mathbf{0}$ and $\mathbf{C} \neq \mathbf{0}$, is invariant with respect to the choice of $\mathbf{B}^-$ if and only if $\mathcal{R}(\mathbf{A'}) \subset \mathcal{R}(\mathbf{B'})$ and $\mathcal{R}(\mathbf{C}) \subset \mathcal{R}(\mathbf{B})$. Consequently, (a) is equivalent to $\mathcal{R}(\mathbf{X'}) \subset \mathcal{R}(\mathbf{X'W'X})$, and hence to (c). Noting that the part (c) $\Rightarrow$ (b) follows directly from Lemma 2.2.6(c) in Rao and Mitra (1971) and that the part (b) $\Rightarrow$ (c) is obvious completes the proof. $\square$

It is clear that the dispersion matrix of P-WLSE$_W$(Xβ) is

$$D[\text{P-WLSE}_W(\mathbf{X}\beta)] = \sigma^2\mathbf{X(X'WX)^-X'WVW'X(X'WX)^{-\prime}X'}, \tag{2.2}$$

Lemma 2 below characterizes all those pseudo-W-weighted-least-squares estimators of $\mathbf{X}\beta$ whose dispersion matrices are below $\sigma^2\mathbf{V}$ with respect to the Löwner partial ordering, i.e., those for which $D(\mathbf{Y}) - D[\text{P-WLSE}_W(\mathbf{X}\beta)]$ is nonnegative definite. Notice that since $\mathbf{Y}$ is unbiased for $\mathbf{X}\beta$, the requirement formulated above must be fulfilled by any reasonable estimator.

LEMMA 2. *For* P-WLSE$_W$(Xβ) = $\mathbf{X(X'WX)^-X'WY}$, *with* $\mathbf{W}$ *satisfying* $\mathcal{R}(\mathbf{X'WV}) \subset \mathcal{R}(\mathbf{X'WX}) \neq \{\mathbf{0}\}$, *the following statements are equivalent:*

(a) $D[\text{P-WLSE}_W(\mathbf{X}\beta)] \leq D(\mathbf{Y})$ *under* $M = \{\mathbf{Y}, \mathbf{X}\beta, \sigma^2\mathbf{V}\}$,

(b) $\mathbf{X(X'WX)^-X'WV} = \mathbf{VW'X(X'WX)^{-\prime}X'}$,

(c) $\mathcal{R}(\mathbf{VW'X}) = \mathcal{R}[\mathbf{X(X'WX)^-X'WV}]$,

(d) $\mathcal{R}(\mathbf{VW'X}) \subset \mathcal{R}[\mathbf{X(X'WX)^-X'WX}]$.

Moreover, if P-WLSE$_W$(Xβ) $\in \mathcal{U}(\mathbf{X}\beta)$, *then these conditions are also equivalent to*

(e) $\mathcal{R}(\mathbf{VW'X}) \subset \mathcal{R}(\mathbf{X})$.

Proof. In view of (2.2), the condition (a) is equivalent to the nonnegative definiteness of

$$\mathbf{V} - \mathbf{X}(\mathbf{X}'\mathbf{W}\mathbf{X})^{-}\mathbf{X}'\mathbf{W}\mathbf{V}\mathbf{W}'\mathbf{X}(\mathbf{X}'\mathbf{W}\mathbf{X})^{-\prime}\mathbf{X}'.$$

Corollary 1 of Baksalary, Kala, and Kłaczyński (1983), which generalizes Proposition 2 of Taylor (1976), asserts that if an $n \times n$ matrix $\mathbf{B}$ and an $n \times n$ symmetric nonnegative definite matrix $\mathbf{C}$ are such that $\mathbf{CB}^2 = \mathbf{CB}$, then the Löwner ordering $\mathbf{B}'\mathbf{CB} \leq \mathbf{C}$ holds if and only if $\mathbf{B}'\mathbf{C} = \mathbf{CB}$, otherwise the difference $\mathbf{C} - \mathbf{B}'\mathbf{CB}$ being an indefinite matrix. Using the assumption (1.9), it can easily be verified that

$$\mathbf{V}[\mathbf{W}'\mathbf{X}(\mathbf{X}'\mathbf{W}\mathbf{X})^{-\prime}\mathbf{X}']^2 = \mathbf{V}\mathbf{W}'\mathbf{X}(\mathbf{X}'\mathbf{W}\mathbf{X})^{-\prime}\mathbf{X}',$$

and then the result quoted above shows that (a) $\Leftrightarrow$ (b). Postmultiplying in (b) by $\mathbf{W}'\mathbf{X}$ yields

$$\mathbf{X}(\mathbf{X}'\mathbf{W}\mathbf{X})^{-}\mathbf{X}'\mathbf{W}\mathbf{V}\mathbf{W}'\mathbf{X} = \mathbf{V}\mathbf{W}'\mathbf{X}, \tag{2.3}$$

thus establishing that (b) implies (c). The part (c) $\Rightarrow$ (d) is a straightforward consequence of (1.9). Further, it can easily be verified that (d) entails (2.3). Noting that postmultiplication in (2.3) by $(\mathbf{X}'\mathbf{W}\mathbf{X})^{-\prime}\mathbf{X}'$ leads to (b) concludes the proof that the statements (a) through (d) are all equivalent. The possibility of replacing (d) by (e) when P-WLSE$_\mathbf{W}(\mathbf{X\beta})$ is unbiased follows directly from Lemma 1. □

3. BEST LINEAR UNBIASED ESTIMATION

The auxiliary results of Section 2 are utilized here to derive a number of criteria for P-WLSE$_\mathbf{W}(\mathbf{X\beta})$ to be a representation of the *best linear unbiased estimator* of $\mathbf{X\beta}$ under the general Gauss-Markov model (1.1). Still regarding $\mathbf{\beta}$ as free to vary over the entire $\Re^p$,

$$\mathbf{AY} = \text{BLUE}(\mathbf{X\beta}) \text{ under } \text{M} = \{\mathbf{Y}, \mathbf{X\beta}, \sigma^2\mathbf{V}\} \;\Leftrightarrow\; \mathbf{A}(\mathbf{X} : \mathbf{VX}^{\perp}) = (\mathbf{X} : \mathbf{0}); \tag{3.1}$$

cf. Drygas (1970, p. 55), Rao (1973b, p. 282). See also Rao (1978, p. 1202) for a general explicit form of $\mathbf{A}$ satisfying the matrix equation on the right-hand side of (3.1).

THEOREM 1. *For* P-WLSE$_\mathbf{W}(\mathbf{X\beta}) = \mathbf{X}(\mathbf{X}'\mathbf{W}\mathbf{X})^{-}\mathbf{X}'\mathbf{W}\mathbf{Y}$, *with* $\mathbf{W}$ *satisfying* $\mathcal{R}(\mathbf{X}'\mathbf{W}\mathbf{V}) \subset \mathcal{R}(\mathbf{X}'\mathbf{W}\mathbf{X}) \neq \{\mathbf{0}\}$, *the following statements are equivalent:*

(a) P-WLSE$_\mathbf{W}(\mathbf{X\beta}) = \text{BLUE}(\mathbf{X\beta})$ *for all* $\mathbf{Y} \in \mathcal{R}(\mathbf{X} : \mathbf{V})$ *under* $\text{M} = \{\mathbf{Y}, \mathbf{X\beta}, \sigma^2\mathbf{V}\}$,

(b) $r(\mathbf{X}'\mathbf{W}\mathbf{X}) = r(\mathbf{X})$ *and* $\mathcal{R}(\mathbf{V}\mathbf{W}'\mathbf{X}) \subset \mathcal{R}(\mathbf{X})$,

(c) P-WLSE$_\mathbf{W}(\mathbf{X\beta}) \in \mathcal{U}(\mathbf{X\beta})$ *and* $D[\text{P-WLSE}_\mathbf{W}(\mathbf{X\beta})] \leq D(\mathbf{Y})$ *under* $\text{M} = \{\mathbf{Y}, \mathbf{X\beta}, \sigma^2\mathbf{V}\}$,

(d) $\mathbf{W}$ *is any matrix such that*

$$\mathbf{X}'\mathbf{W}(\mathbf{V} + \mathbf{X}\mathbf{U}\mathbf{X}') = \mathbf{X}' \tag{3.2}$$

for some $p \times p$ *matrix* $\mathbf{U}$,

(e) $\mathbf{W}$ *is any matrix of the form*

$$\mathbf{W} = (\mathbf{V} + \mathbf{X}\mathbf{U}\mathbf{X}')^- + \mathbf{K}, \tag{3.3}$$

where $\mathbf{U}$ *is some* $p \times p$ *matrix satisfying any of the following equivalent conditions:*

$$\mathcal{R}(\mathbf{X}) \subset \mathcal{R}(\mathbf{V} + \mathbf{X}\mathbf{U}\mathbf{X}'), \tag{3.4}$$

$$\mathcal{R}(\mathbf{X}) \subset \mathcal{R}(\mathbf{V} + \mathbf{X}\mathbf{U}'\mathbf{X}'), \tag{3.5}$$

$$\mathcal{R}(\mathbf{X} : \mathbf{V}) = \mathcal{R}(\mathbf{V} + \mathbf{X}\mathbf{U}\mathbf{X}'), \tag{3.6}$$

$$\mathcal{R}(\mathbf{X} : \mathbf{V}) = \mathcal{R}(\mathbf{V} + \mathbf{X}\mathbf{U}'\mathbf{X}'), \tag{3.7}$$

and $\mathbf{K}$ *is an arbitrary solution to the equation*

$$\mathbf{X}'\mathbf{K}(\mathbf{X} : \mathbf{V}) = (\mathbf{0} : \mathbf{0}). \tag{3.8}$$

Proof. Substituting $\mathbf{A} = \mathbf{X}(\mathbf{X}'\mathbf{W}\mathbf{X})^-\mathbf{X}'\mathbf{W}$ into (3.1) yields $\mathbf{X}(\mathbf{X}'\mathbf{W}\mathbf{X})^-\mathbf{X}'\mathbf{W}\mathbf{X} = \mathbf{X}$ and $\mathbf{X}(\mathbf{X}'\mathbf{W}\mathbf{X})^-\mathbf{X}'\mathbf{W}\mathbf{V}\mathbf{X}^\perp = \mathbf{0}$. The equivalence of the former equality to

$$\mathrm{r}(\mathbf{X}) = \mathrm{r}(\mathbf{X}'\mathbf{W}\mathbf{X}) \tag{3.9}$$

follows from Lemma 1; the equivalence of the latter equality to $\mathbf{X}'\mathbf{W}\mathbf{V}\mathbf{X}^\perp = \mathbf{0}$, and hence to

$$\mathcal{R}(\mathbf{V}\mathbf{W}'\mathbf{X}) \subset \mathcal{R}(\mathbf{X}), \tag{3.10}$$

is clear from the assumption (1.9); and the equivalence of (b) to (c) follows from Lemmas 1 and 2. According to Rao and Mitra (1971, Theorem 2.3.2), the equation $\mathbf{AZB} = \mathbf{C}$ is consistent with respect to $\mathbf{Z}$ if and only if $\mathbf{A}\mathbf{A}^-\mathbf{C}\mathbf{B}^-\mathbf{B} = \mathbf{C}$. Alternatively [cf., e.g., Baksalary and Kala (1980, Lemma 1)], this condition may be split into $\mathcal{R}(\mathbf{C}) \subset \mathcal{R}(\mathbf{A})$ and $\mathcal{R}(\mathbf{C}') \subset \mathcal{R}(\mathbf{B}')$. Hence it follows that a solution $\mathbf{U}$ to the equation (3.2) exists if and only if

$$\mathcal{R}(\mathbf{X}' - \mathbf{X}'\mathbf{W}\mathbf{V}) \subset \mathcal{R}(\mathbf{X}'\mathbf{W}\mathbf{X}) \ \text{ and } \ \mathcal{R}(\mathbf{X} - \mathbf{V}\mathbf{W}'\mathbf{X}) \subset \mathcal{R}(\mathbf{X}). \tag{3.11}$$

The latter relation in (3.11) is clearly equivalent to (3.10) and, in view of (1.9), the former relation in (3.11) reduces to $\mathcal{R}(\mathbf{X}') \subset \mathcal{R}(\mathbf{X}'\mathbf{W}\mathbf{X})$ which is clearly equivalent to (3.9). Reversing now the roles of $\mathbf{W}$ and $\mathbf{U}$ in (3.2), it follows that this equation is consistent with respect to $\mathbf{W}$ if and only if $\mathbf{U}$ satisfies (3.5). In view of (1.3) and the obvious inclusions $\mathcal{R}(\mathbf{V}\mathbf{X}^\perp) \subset \mathcal{R}(\mathbf{V} + \mathbf{X}\mathbf{U}\mathbf{X}')$

$\subset \mathcal{R}(\mathbf{X} : \mathbf{V})$, it is clear that (3.5) is equivalent to (3.7). Similarly, (3.4) is equivalent to (3.6). The latter entails $r(\mathbf{X} : \mathbf{V}) = r(\mathbf{V} + \mathbf{XUX}')$, and since $\mathcal{R}(\mathbf{V} + \mathbf{XUX}') \subset \mathcal{R}(\mathbf{X} : \mathbf{V})$, it is seen that (3.6) implies (3.7). The reverse implication follows by analogous arguments. If $\mathbf{U}$ fulfils any of the conditions (3.4) through (3.7), then a general representation of $\mathbf{W}$ satisfying (3.2) is

$$\mathbf{W} = \mathbf{X}'^{-}\mathbf{X}'(\mathbf{V} + \mathbf{XUX}')^{-} + \mathbf{K}, \tag{3.12}$$

where $\mathbf{K}$ is an arbitrary solution to the homogeneous matrix equation

$$\mathbf{X}'\mathbf{K}(\mathbf{V} + \mathbf{XUX}') = \mathbf{0}. \tag{3.13}$$

Noting that $(\mathbf{I}_n - \mathbf{X}'^{-}\mathbf{X}')(\mathbf{V} + \mathbf{XUX}')$ is a particular solution to (3.13), we may simplify (3.12) to the form (3.3). The equivalence of (3.13) to (3.8) is an obvious consequence of (3.6). □

The part (a) ⇔ (e) of Theorem 1 is a modification of the result originally established by Rao (1971, Theorem 4.1); see also the corresponding result for a symmetric $\mathbf{W}$ in Rao (1973c, Theorem 2.1). The modification consists in stressing the equivalences between the conditions (3.4) through (3.7) and in deleting the condition (3.9). The redundancy of (3.9) follows by noting that (3.5) entails

$$\mathbf{X}'(\mathbf{V} + \mathbf{XUX}')^{-}\mathbf{VX}^{\perp} = \mathbf{X}'(\mathbf{V} + \mathbf{XUX}')^{-}(\mathbf{V} + \mathbf{XUX}')\mathbf{X}^{\perp} = \mathbf{X}'\mathbf{X}^{\perp} = \mathbf{0},$$

and therefore, in view of (1.3) and (3.5), every $\mathbf{W}$ of the form (3.3) fulfils

$$\begin{aligned} r(\mathbf{X}'\mathbf{WX}) &= r[\mathbf{X}'(\mathbf{V} + \mathbf{XUX}')^{-}\mathbf{X}] = r[\mathbf{X}'(\mathbf{V} + \mathbf{XUX}')^{-}(\mathbf{X} : \mathbf{VX}^{\perp})] \\ &= r[\mathbf{X}'(\mathbf{V} + \mathbf{XUX}')^{-}(\mathbf{V} + \mathbf{XUX}')] = r(\mathbf{X}). \end{aligned}$$

For further discussion on the unified least-squares approach to linear estimation in the general Gauss-Markov model see Khatri (1968, 1982) and Mitra (1973). See also Zyskind and Martin (1969), who proved that the set of matrices $\mathbf{W}$ satisfying the conditions of Theorem 1 always contains certain generalized inverses of $\mathbf{V}$; cf. a recent discussion of this point in Searle and Pukelsheim (1989, Section 5).

The part (a) ⇔ (c) of Theorem 1 shows that, combined with the unbiasedness condition, the property of $D[\text{P-WLSE}_{\mathbf{W}}(\mathbf{X}\beta)]$ considered in Lemma 2 is surprisingly forceful. Actually, it appears that, irrespective whether $\text{P-WLSE}_{\mathbf{W}}(\mathbf{X}\beta) \in \mathcal{U}(\mathbf{X}\beta)$ or not, if

$$D[\text{P-WLSE}_{\mathbf{W}}(\mathbf{X}\beta)] \leq D(\mathbf{Y}) \text{ under } M = \{\mathbf{Y}, \mathbf{X}\beta, \sigma^2\mathbf{V}\}, \tag{3.14}$$

then necessarily

$$D[\text{P-WLSE}_{\mathbf{W}}(\mathbf{X}\beta)] \leq D(\text{BLUE}(\mathbf{X}\beta) \text{ under } M = \{\mathbf{Y}, \mathbf{X}\beta, \sigma^2\mathbf{V}\}. \tag{3.15}$$

In fact, on account of the part (a) ⇒ (b), (d) of Lemma 2, (3.14) entails

$$D[\text{P-WLSE}_{\mathbf{W}}(\mathbf{X\beta})] = \sigma^2\mathbf{X}(\mathbf{X'WX})^-\mathbf{X'WV} = \sigma^2\mathbf{VW'X}(\mathbf{X'WX})^{-\prime}\mathbf{X'} \tag{3.16}$$

and

$$\mathbf{Q_X VW'X} = \mathbf{0}, \tag{3.17}$$

where $\mathbf{Q_X} = \mathbf{I}_n - \mathbf{P_X}$ is the orthogonal projector onto $\mathcal{N}(\mathbf{X'})$. Moreover, it is known that

$$D[\text{BLUE}(\mathbf{X\beta})] = \sigma^2[\mathbf{V} - \mathbf{VQ_X}(\mathbf{Q_X VQ_X})^-\mathbf{Q_X V}];$$

cf., e.g., Puntanen (1986, p. 428) and Baksalary, Puntanen, and Styan (1989, Section 3). In view of this representation and (2.2), (3.16), and (3.17), it follows that

$$\begin{aligned} \mathbf{0} &\le \sigma^2[\mathbf{I}_n - \mathbf{X}(\mathbf{X'WX})^-\mathbf{X'W}][\mathbf{V} - \mathbf{VQ_X}(\mathbf{Q_X VQ_X})^-\mathbf{Q_X V}][\mathbf{I}_n - \mathbf{W'X}(\mathbf{X'WX})^{-\prime}\mathbf{X'}] \\ &= D[\text{BLUE}(\mathbf{X\beta})] - D[\text{P-WLSE}_{\mathbf{W}}(\mathbf{X\beta})], \end{aligned}$$

as claimed in (3.15).

In the particular case of $\mathbf{W} = \mathbf{I}_n$, i.e., when the P-WLSE$_{\mathbf{W}}$($\mathbf{X\beta}$) reduces to OLSE($\mathbf{X\beta}$) defined in (1.7), the part (a) ⇔ (b) ⇔ (c) of Theorem 1 leads to the following.

COROLLARY 1. *Under the general Gauss-Markov model* $\text{M} = \{\mathbf{Y}, \mathbf{X\beta}, \sigma^2\mathbf{V}\}$, *the following statements are equivalent:*

(a) OLSE($\mathbf{X\beta}$) = BLUE($\mathbf{X\beta}$) *for all* $\mathbf{Y} \in \mathcal{R}(\mathbf{X} : \mathbf{V})$,

(b) $\mathcal{R}(\mathbf{VX}) \subset \mathcal{R}(\mathbf{X})$,

(c) $D[\text{OLSE}(\mathbf{X\beta})] \le D(\mathbf{Y})$.

The condition (b) in Corollary 1 is a well-known criterion for the equality between OLSE($\mathbf{X\beta}$) and BLUE($\mathbf{X\beta}$). For the general Gauss-Markov model $\text{M} = \{\mathbf{Y}, \mathbf{X\beta}, \sigma^2\mathbf{V}\}$, with both $\mathbf{X}$ and $\mathbf{V}$ possibly deficient in rank, it was first established by Rao (1967, p. 364); see also Zyskind (1967, Theorem 2). The condition (c) was originally derived in Baksalary and Puntanen (1989, Corollary) using a different approach. See also Alalouf and Styan (1984), Puntanen (1987), Puntanen and Styan (1989) for exhaustive discussions of the problem of the equality between OLSE($\mathbf{X\beta}$) and BLUE($\mathbf{X\beta}$).

In the particular case of $\mathbf{W} = \mathbf{V}^+$, i.e., when P-WLSE$_{\mathbf{W}}$($\mathbf{X\beta}$) represents GLSE($\mathbf{X\beta}$) defined in (1.8), the statement (b) in the part (a) ⇔ (b) ⇔ (c) of Theorem 1 may be simplified to $\mathcal{R}(\mathbf{P_V X}) = \mathcal{R}(\mathbf{X})$, and hence to $\mathcal{R}(\mathbf{X}) \subset \mathcal{R}(\mathbf{V})$, thus leading to the following.

COROLLARY 2. *Under the general Gauss-Markov model* $M = \{\mathbf{Y}, \mathbf{X\beta}, \sigma^2\mathbf{V}\}$, *the following statements are equivalent:*

(a) $\mathrm{GLSE}(\mathbf{X\beta}) = \mathrm{BLUE}(\mathbf{X\beta})$ *for all* $\mathbf{Y} \in \mathcal{R}(\mathbf{X} : \mathbf{V})$,

(b) $\mathcal{R}(\mathbf{X}) \subset \mathcal{R}(\mathbf{V})$,

(c) $\mathrm{GLSE}(\mathbf{X\beta}) \in \mathcal{U}(\mathbf{X\beta})$ *and* $D[\mathrm{GLSE}(\mathbf{X\beta})] \leq D(\mathbf{Y})$.

The part (b) ⇒ (a) of Corollary 2 was noted by Goldman and Zelen (1964, p. 165). Mitra and Rao (1968, p. 286) pointed out that if (b) holds, then any generalized inverse of **V** may be used in (1.8) instead of the Moore-Penrose inverse $\mathbf{V}^+$. The converse implications were established by Zyskind and Martin (1969, Corollary 1.1).

Drygas (1983, p. 90; 1985, p. 178) introduced the concept of a *unified-least-squares (ULS) matrix*, defining a symmetric nonnegative definite matrix **T** to be a ULS matrix for the model $M = \{\mathbf{Y}, \mathbf{X\beta}, \sigma^2\mathbf{V}\}$ if

$$\mathcal{R}(\mathbf{X}) \subset \mathcal{R}(\mathbf{T}) \tag{3.18}$$

and

$$\mathcal{R}(\mathbf{VT^-X}) \subset \mathcal{R}(\mathbf{X}) \ \text{ for some } \mathbf{T}^- \in \mathbf{T}\{1\}, \tag{3.19}$$

where $\mathbf{T}\{1\}$ denotes the set of all generalized inverses of **T**. Motivated by some Drygas's remarks, Baksalary (1985) introduced the concept of a *strong ULS matrix*, replacing (3.19) by the requirement that

$$\mathcal{R}(\mathbf{VT^-X}) \subset \mathcal{R}(\mathbf{X}) \ \text{ for every } \mathbf{T}^- \in \mathbf{T}\{1\}. \tag{3.20}$$

The results of this section lead fairly straightforwardly to statistical interpretations of the conditions involved in these concepts. In view of Corollary 2, the condition (3.18) means that every estimator from the class

$$\{\mathbf{X}[\mathbf{X}'(\mathbf{T}^-)'\mathbf{X}]^-\mathbf{X}'(\mathbf{T}^-)'\mathbf{Y} : \ \mathbf{T}^- \in \mathbf{T}\{1\}\} \tag{3.21}$$

represents the best linear unbiased estimator of $\mathbf{X\beta}$ under the model $\{\mathbf{Y}, \mathbf{X\beta}, \sigma^2\mathbf{T}\}$, while in view of the part (a) ⇔ (b) of Theorem 1, the condition (3.19) then means that at least one estimator from the class (3.21) represents $\mathrm{BLUE}(\mathbf{X\beta})$ also under the original model $M = \{\mathbf{Y}, \mathbf{X\beta}, \sigma^2\mathbf{V}\}$, and the condition (3.20) extends this requirement to every estimator from the class (3.21). Hence it is seen that the two concepts of a ULS matrix correspond to two (from among three) approaches to the problem of the equality of best unbiased linear estimators of $\mathbf{X\beta}$ under two general Gauss-Markov models with different dispersion matrices; cf. Mitra and Moore (1973, Problems (2) and (3)). See also Rao (1968, Section 5; 1971, Theorems 5.2 and 5.5) and Kala (1981).

In the final part of the section dealing with best unbiased linear estimation suppose that P-WLSE$_{\mathbf{W}}$($\mathbf{X\beta}$) does not coincide with BLUE($\mathbf{X\beta}$) and consider the question if it still can reasonably be utilized in unbiased estimation of $\mathbf{X\beta}$ under the model (1.1). This leads to the problem of existence of a linear transformation of P-WLSE$_{\mathbf{W}}$($\mathbf{X\beta}$) such that the resulting statistic is identical with BLUE($\mathbf{X\beta}$). If the answer were positive, P-WLSE$_{\mathbf{W}}$($\mathbf{X\beta}$) would be called *linearly sufficient.* Notice that Gourieroux and Montfort (1980, Section 3F) and Baksalary and Kala (1981) investigated such a property not using explicitly the term "linear sufficiency". In the sense specified above, this term was adopted by Drygas (1983, Section 3), who also made comparisons, in his Section 6, with the definition of linear sufficiency given by Barnard (1963); cf. Joshi (1985). Müller, Rao, and Sinha (1984, Section 4) and Rao (1984, Section 6) indicated several alternative ways to define the concept of linear sufficiency, and Baksalary and Kala (1986) discussed this concept in the case where interest is restricted to a fixed vector of estimable parametric functions. See also Rogers (1987), who considers the same problem (without any reference to the earlier results, however).

To examine P-WLSE$_{\mathbf{W}}$($\mathbf{X\beta}$) from the above point of view we apply Corollary 2 of Baksalary and Kala (1986), stating that $\mathbf{AY}$ is linearly sufficient for $\mathbf{X\beta}$ if and only if

$$\mathcal{R}(\mathbf{X'A'}) = \mathcal{R}(\mathbf{X'}) \text{ and } \mathcal{R}(\mathbf{AX}) \cap \mathcal{R}(\mathbf{AVX}^{\perp}) = \{\mathbf{0}\}. \tag{3.22}$$

THEOREM 2. *A statistic* P-WLSE$_{\mathbf{W}}$($\mathbf{X\beta}$) $= \mathbf{X(X'WX)^{-}X'WY}$, *with* $\mathbf{W}$ *satisfying* $\mathcal{R}(\mathbf{X'WV}) \subset \mathcal{R}(\mathbf{X'WX}) \neq \{\mathbf{0}\}$, *cannot be linearly transformed into* BLUE($\mathbf{X\beta}$) *unless it is just identical with* BLUE($\mathbf{X\beta}$).

Proof. For $\mathbf{A} = \mathbf{X(X'WX)^{-}X'W}$, the former condition in (3.22) takes the form $\mathcal{R}(\mathbf{X'}) = \mathcal{R}(\mathbf{X'W'X})$. Lemma 1 asserts that if this equality holds, then $\mathbf{AX} = \mathbf{X(X'WX)^{-}X'WX} = \mathbf{X}$, and hence $\mathcal{R}(\mathbf{AVX}^{\perp}) = \mathcal{R}[\mathbf{X(X'WX)^{-}X'WVX}^{\perp}]$ is a subspace of $\mathcal{R}(\mathbf{AX})$. Consequently, from the latter condition in (3.22) it follows that $\mathbf{AVX}^{\perp} = \mathbf{0}$, and, on account of (3.1), combining this equality with $\mathbf{AX} = \mathbf{X}$ yields the result. □

4. ADMISSIBILITY

In view of the conclusion of Theorem 2, a possible interest in P-WLSE$_{\mathbf{W}}$($\mathbf{X\beta}$) which differs from BLUE($\mathbf{X\beta}$) is limited to biased estimation. For any $\mathbf{FY}$ regarded as an estimator of $\mathbf{X\beta}$ under the model $\mathrm{M} = \{\mathbf{Y}, \mathbf{X\beta}, \sigma^2\mathbf{V}\}$, the mean square error is defined as

$$\begin{aligned} MSE(\mathbf{FY}; \mathbf{X\beta}) &= E[(\mathbf{FY} - \mathbf{X\beta})'(\mathbf{FY} - \mathbf{X\beta})] \\ &= \sigma^2 \mathrm{trace}(\mathbf{FVF'}) + \boldsymbol{\beta}'\mathbf{X}'(\mathbf{F} - \mathbf{I}_n)'(\mathbf{F} - \mathbf{I}_n)\mathbf{X\beta}, \end{aligned}$$

and $\mathbf{AY}$ is said to be *admissible* among the class of all linear estimators of $\mathbf{X\beta}$ if there does not exist $\mathbf{FY}$ such that the inequality

$$MSE(\mathbf{FY}; \mathbf{X\beta}) \leq MSE(\mathbf{AY}; \mathbf{X\beta})$$

holds for all $p \times 1$ vectors $\mathbf{\beta}$ and $\sigma^2 > 0$, being strict for some $\mathbf{\beta}$ and σ^2. The fundamental work on admissibility of linear estimators was done by Rao (1976), and papers dealing with this problem in the context of the general Gauss-Markov model are Mathew, Rao, and Sinha (1984), Baksalary and Markiewicz (1988), Baksalary and Mathew (1988), and Klonecki and Zontek (1988).

THEOREM 3. *A statistic* P-WLSE$_\mathbf{W}(\mathbf{X\beta}) = \mathbf{X}(\mathbf{X'WX})^-\mathbf{X'WY}$, *with* $\mathbf{W}$ *satisfying* $\mathcal{R}(\mathbf{X'WV}) \subset \mathcal{R}(\mathbf{X'WX}) \neq \{\mathbf{0}\}$, *is admissible among the class of all linear estimators of* $\mathbf{X\beta}$ *under the general Gauss-Markov model* $\mathrm{M} = \{\mathbf{Y}, \mathbf{X\beta}, \sigma^2\mathbf{V}\}$ *if and only if*

$$\mathbf{X}(\mathbf{X'WX})^-\mathbf{X'WV} = \mathbf{VW'X}(\mathbf{X'WX})^{-\prime}\mathbf{X'} \tag{4.1}$$

and

$$\mathcal{R}[\mathbf{X}(\mathbf{X'WX})^-\mathbf{X'WX} - \mathbf{X}] = \mathcal{R}[\mathbf{X}(\mathbf{X'WX})^-\mathbf{X'WS} - \mathbf{S}]\ , \tag{4.2}$$

where $\mathbf{S}$ *is any matrix such that* $\mathcal{R}(\mathbf{S}) = \mathcal{R}(\mathbf{X}) \cap \mathcal{R}(\mathbf{V})$.

Proof. According to Corollary 3 in Baksalary and Markiewicz (1988), P-WLSE$_\mathbf{W}(\mathbf{X\beta})$ is admissible for $\mathbf{X\beta}$ if and only if the conditions (4.1) and (4.2) hold along with

$$\mathcal{R}(\mathbf{VW'X}) \subset \mathcal{R}(\mathbf{X}) \tag{4.3}$$

and

$$\mathbf{X}(\mathbf{X'WX})^-\mathbf{X'WVW'X}(\mathbf{X'WX})^{-\prime}\mathbf{X'} \leq \mathbf{X}(\mathbf{X'WX})^-\mathbf{X'WV}. \tag{4.4}$$

Since both (4.3) and (4.4) are implied by (4.1), the result follows. □

According to Lemma 2, the condition (4.1) is equivalent to the Löwner ordering D[P-WLSE$_\mathbf{W}(\mathbf{X\beta})] \leq D(\mathbf{Y})$ which, as was already pointed out, must be fulfilled by any reasonable estimator. Moreover, it is clear that if the model $\mathrm{M} = \{\mathbf{Y}, \mathbf{X\beta}, \sigma^2\mathbf{V}\}$ is nonsingular or weakly singular, i.e., if $\mathcal{R}(\mathbf{X}) \subset \mathcal{R}(\mathbf{V})$, then the condition (4.2) holds trivially. Consequently, it follows that biased pseudo-$\mathbf{W}$-weighted-least-squares estimators satisfy the admissibility criterion in many cases. It should be recalled, however, that according to Lemma 1, a biased P-WLSE$_\mathbf{W}(\mathbf{X\beta}) = \mathbf{X}(\mathbf{X'WX})^-\mathbf{X'WY}$ depends not only on the choice of $\mathbf{W}$ but also on the choice of $(\mathbf{X'WX})^-$.

ACKNOWLEDGEMENTS

This research was begun during the Second International Tampere Conference in Statistics in June 1987, continued while the authors were visiting McGill University in Montréal in May 1988, and completed while the first author was visiting the University of Dortmund in March 1989. Hospitality of Professors George P.H. Styan and Götz Trenkler is gratefully acknowledged.

This research was supported in part by the Natural Sciences and Engineering Council of Canada and by the Fonds pour la Formation de Chercheurs et l'Aide à la Recherche du Gouvernement du Québec. The research of the first author was partially supported also by the Government of Poland Grant No. CPBP 01.02. IV.

REFERENCES

Alalouf, I.S. and Styan, G.P.H. (1984). Characterizations of the conditions for the ordinary least squares estimator to be best linear unbiased. *Topics in Applied Statistics* (Y.P. Chaubey and T.D. Dwivedi, *eds.*). Dept. of Mathematics, Concordia University, Montreal, 331-344.

Baksalary, J.K. (1985). Strong unified-least-squares matrices for a general linear model. *Linear Algebra Appl.* **70**, 61-65.

Baksalary, J.K. and Kala, R. (1980). The matrix equation $AXB + CYD = E$. *Linear Algebra Appl.* **30**, 141-147.

Baksalary, J.K. and Kala, R. (1981). Linear transformations preserving best linear unbiased estimators in a general Gauss-Markoff model. *Ann. Statist.* **9**, 913-916.

Baksalary, J.K. and Kala, R. (1986). Linear sufficiency with respect to a given vector of parametric functions. *J. Statist. Plann. Inference* **14**, 331-338.

Baksalary, J.K., Kala, R., and Kłaczyński, K. (1983). The matrix inequality $M \geq B^*MB$. *Linear Algebra Appl.* **54**, 77-86.

Baksalary, J.K. and Markiewicz, A. (1988). Admissible linear estimators in the general Gauss-Markov model. *J. Statist. Plann. Inference* **19**, 349-359.

Baksalary, J.K. and Mathew, T. (1988). Admissible linear estimation in a general Gauss-Markov model with an incorrectly specified dispersion matrix. *J. Multivariate Anal.* **27**, 53-67.

Baksalary, J.K. and Puntanen, S. (1989). Characterizations of the best linear unbiased estimator in the general Gauss-Markov model with the use of matrix partial orderings. *Linear Algebra Appl.*; to appear.

Baksalary, J.K., Puntanen, S., and Styan, G.P.H. (1989). A property of the dispersion matrix of the best linear unbiased estimator in the general Gauss-Markov model. *Sankhyā Ser. A*; to appear.

Barnard, G.A. (1963). The logic of least squares. *J. Roy. Statist. Soc. Ser. B* **25**, 124-127.

Drygas, H. (1970). *The Coordinate-Free Approach to Gauss-Markov Estimation.* Springer, Berlin.

Drygas, H. (1983). Sufficiency and completeness in the general Gauss-Markov model. *Sankhyā Ser. A* **45**, 88-98.

Drygas, H. (1985). On the unified theory of least squares. *Probab. Math. Statist.* **5**, 177-186.

Feuerverger, A. and Fraser, D.A.S. (1980). Categorical information and the singular linear model. *Canad. J. Statist.* **8**, 41-45.

Goldman, A.J. and Zelen, M. (1964). Weak generalized inverses and minimum variance unbiased estimation. *J. Res. Nat. Bur. Standards Sect. B* **68**, 151-172.

Gourieroux, C. and Montfort, A. (1980). Sufficient linear structures: econometric applications. *Econometrica* **48**, 1083-1097.

Joshi, V.M. (1985). Linear sufficiency. *Encyclopedia of Statistical Sciences*, Vol. 5 (S. Kotz, N.L. Johnson and C.B. Read, *eds.*). Wiley, New York, 64-65.

Kala, R. (1981). Projectors and linear estimation in general linear models. *Comm. Statist. Theory Methods* **10**, 849-873.

Khatri, C.G. (1968). Some results for the singular normal multivariate regression models. *Sankhyā Ser. A* **30**, 267-280.

Khatri, C.G. (1982). A representation of a matrix and its use in the Gauss-Markoff model. *J. Indian Statist. Assoc.* **20**, 89-98.

Klonecki, W. and Zontek, S. (1988). On the structure of admissible linear estimators. *J. Multivariate Anal.* **24**, 11-30.

Mathew, T., Rao, C.R., and Sinha, B.K. (1984). Admissible linear estimation in singular linear models. *Comm. Statist. Theory Methods* **13**, 3033-3045.

Mitra, S. K. (1973). Unified least squares approach to linear estimation in a general Gauss-Markov model. *SIAM J. Appl. Math.* **25**, 671-680.

Mitra, S. K. and Moore, B.J. (1973). Gauss-Markov estimation with an incorrect dispersion matrix. *Sankhyā Ser. A* **35**, 139-152.

Mitra, S.K. and Rao, C.R. (1968). Some results in estimation and tests of linear hypotheses under the Gauss-Markoff model. *Sankhyā Ser. A* **30**, 281-290

Müller, J., Rao, C.R., and Sinha, B.K. (1984). Inference on parameters in a linear model: a review of recent results. *Experimental Design, Statistical Models, and Genetic Statistics* (K. Hinkelmann, *ed.*). Dekker, New York, 277-295.

Nordström, K. (1985). On a decomposition of the singular Gauss-Markov model. *Linear Statistical Inference* (T. Caliński and W. Klonecki, *eds.*). Springer, Berlin, 231-245.

Puntanen, S. (1986). Properties of the covariance matrix of the BLUE in the general linear model. *Pacific Statistical Congress* (I.S. Francis, B.F.J. Manly and F.C. Lam, *eds.*). North-Holland, Amsterdam, 425-430.

Puntanen, S. (1987). On the relative goodness of ordinary least squares estimation in the general linear model. Ph.D. dissertation in *Acta Univ. Tamper. Ser. A*, Vol. 216.

Puntanen, S. and Styan, G.P.H. (1989). On the equality of the ordinary least squares estimator and the best linear unbiased estimator. *Amer. Statist.* **43**; in press.

Rao, C.R. (1962). A note on a generalized inverse of a matrix with applications to problems in statistics. *J. Roy. Statist. Soc. Ser. B.* **24**, 152-158.

Rao, C.R. (1967). Least squares theory using an estimated dispersion matrix and its application to measurement of signals. *Proceedings of the Fifth Berkeley Symposium on Mathematical Statistics and Probability*, Vol. 1 (L. Le Cam and J. Neyman, *eds.*). University of California Press, Berkeley, CA, 355-372.

Rao, C.R. (1968). A note on a previous lemma in the theory of least squares and some further results. *Sankhyā Ser. A* **30**, 245-252.

Rao, C.R. (1971). Unified theory of linear estimation. *Sankhyā Ser. A* **33**, 371-394. [Corrigenda: **34** (1972), pp. 194 and 477.]

Rao, C.R. (1973a). *Linear Statistical Inference and Its Applications*. Second Edition. Wiley, New York. [First Edition: 1965.]

Rao, C.R. (1973b). Representations of best linear unbiased estimators in the Gauss-Markoff model with a singular dispersion matrix. *J. Multivariate Anal.* **3**, 276-292.

Rao, C.R. (1973c). Unified theory of least squares. *Comm. Statist.* **1**, 1-8.

Rao, C.R. (1974). Projectors, generalized inverses and the BLUE's. *J. Roy. Statist. Soc. Ser. B.* **36**, 442-448.

Rao, C.R. (1976). Estimation of parameters in a linear model. (The 1975 Wald Memorial Lectures.) *Ann. Statist.* **4**, 1023-1037. [Corrigendum: **7** (1979), p. 696.]

Rao, C.R. (1978). Choice of best linear estimators in the Gauss-Markoff model with a singular dispersion matrix. *Comm. Statist. Theory Methods* **7**, 1199-1208.

Rao, C.R. (1979). Estimation of parameters in the singular Gauss-Markoff model. *Comm. Statist. Theory Methods* **8**, 1353-1358.

Rao, C.R. (1984). Inference from linear models with fixed effects: recent results and some problems. *Statistics: An Appraisal.* (H.A. David and H.T. David, *eds.*). The Iowa State University Press, Ames, IA, 345-369.

Rao, C.R. (1985). A unified approach to inference from linear models. *Proceedings of the First International Tampere Seminar on Linear Statistical Models and their Applications* (T. Pukkila and S. Puntanen, *eds.*). Dept. of Mathematical Sciences, University of Tampere, 9-36.

Rao, C.R. and Mitra, S.K. (1971). *Generalized Inverse of Matrices and its Applications.* Wiley, New York.

Rao, C.R., Mitra, S.K., and Bhimasankaram, P. (1972). Determination of a matrix by its subclasses of generalized inverses. *Sankhyā Ser. A* **34**, 5-8.

Rogers, A.J. (1987). On the derivation of efficiency results in linear estimation. *Austral. J. Statist.* **29**, 60-65.

Searle, S.R. and Pukelsheim, F. (1986). Estimation of the mean vector in linear models. Institut für Mathematik, Universität Augsburg, Preprint Nr. 106.

Searle, S.R. and Pukelsheim, F. (1989). On least squares and best linear unbiased estimation. Biometrics Unit, Cornell University, Paper No. BU-997-M.

Seely, J. and Zyskind, G. (1971). Linear spaces and minimum variance unbiased estimation. *Ann. Math. Statist.* **42**, 691-703.

Taylor, W.E. (1976). Prior information on the coefficients when the disturbance covariance matrix is unknown. *Econometrica* **44**, 725-739.

Zyskind, G. (1967). On canonical forms, non-negative covariance matrices and best and simple least squares linear estimators in linear models. *Ann. Math. Statist.* **38**,1092-1109.

Zyskind, G. and Martin, F.B. (1969). On best linear estimation and a general Gauss-Markov theorem in linear models with arbitrary nonnegative structure. *SIAM J. Appl. Math.* **17**, 1190-1202.

STATISTICAL DATA ANALYSIS AND INFERENCE
Y. Dodge (Editor)

EVALUATING PRE–TEST PREDICTORS OF SUCCESS IN LINEAR REGRESSION MODELS.

Arthur Cohen
Rutgers University

Harold B. Sackrowitz

Department of Statistics, Rutgers University, New Brunswick, NJ USA 08903

Consider a linear regression problem where it is desired to predict (at a setting of the fixed variables whether a future response (the dependent variable) will exceed a given threshold. We study loss functions that are 1 for incorrect predictions and 0 otherwise. We first show that such a problem is equivalent, in a sense, to a hypothesis testing problem. The sense is that the admissible rules for the two problems are the same. For the prediction problem we consider the procedures that make a terminal decision after a preliminary test. That is, for example, our final prediction is based on a statistic involving p or (p–1) independent variables depending on the outcome of a preliminary test. The preliminary test method is a popular variable selection procedure. We give necessary and sufficient conditions for the admissibility of such procedures.

1. INTRODUCTION

One of statistics most useful functions is to predict. Oftentimes a regression model is assumed where a response variable y is linearly related to a set of independent variables $(x_1,x_2,\ldots,x_k)' = \mathbf{x}$. At a particular value of $\mathbf{x}$, say $\mathbf{x}^*$, it may be desired to predict whether a future value of y, say y^*, will exceed (be less than) a preassigned threshold value, say T. For example, $\mathbf{x}$ may be fixed levels of treatments and y may be blood pressure and one is interested in whether the predicted blood pressure at $\mathbf{x} = \mathbf{x}^*$ will be less than a dangerous level. Many real life problems can be formulated so that y exceeding a particular value, represents success, whereas y, not exceeding the value represents failure. We consider the problem of predicting success for a random loss function which is 1, when we predict failure but success should have been predicted, and the loss is one when we erroneously predict success. That is, if T represents the threshold value the loss function is

$$(1.1)\qquad L(a,y^*) = 1, \qquad \text{if } a = 0,\ y^* > T$$
$$= 1, \qquad \text{if } a = 1,\ y^* \le T,$$

where a = 0 represents the action of predicting failure and a = 1 represents the action of predicting success. The loss is zero otherwise.

Ultimately our goal is to study variable selection techniques for the above loss function. Our first result is to recognize that under the usual regression model assumptions the above stated problem is equivalent in a sense to a hypothesis testing problem with a 0–1 loss function. The sense in which the problems are equivalent is that the class of admissible rules for each problem is the same. This result is in Section 2.

We then study the popular prediction procedure based on a preliminary test procedure. Preliminary testing is a typical if not the most popular variable selection method. We find necessary and sufficient conditions for the admissibility of the prediction procedure based on preliminary testing. The conditions depend on the sample size n, x^*, the elements of the design matrix, and critically upon the sizes of the pre and post tests. The conditions are explicit so that one can easily know what sizes of the pre and post tests to choose so that an admissible procedure will be assured. The method of proof is similar to that used by Cohen [1] but here it is somewhat more complicated. We conclude with some remarks. The admissibility results and remarks are in Section 3.

Variable selection methods with prediction as the ultimate goal have been studied extensively. We mention only a few references since most others can be found in these. Seber [2] has a full chapter on the subject as does Draper and Smith [3]. An extensive treatment also appears in Leamer [4] and in a recent handbook by Siotani, Hayakawa, and Fujikoshi [5].

2. EQUIVALENCE OF PREDICTION AND TESTING PROBLEMS.

Consider the linear regression model

$$y_i = \boldsymbol{\beta}' \mathbf{x}_i + \varepsilon_i, \quad i = 1,2,\dots,n,$$

where $\boldsymbol{\beta}$ is a k×l vector of parameters $\mathbf{x}_i$ is a k×l vector of fixed values, n > k, $X = (\mathbf{x}_1,\dots,\mathbf{x}_k)$ is full rank, and $\boldsymbol{\varepsilon} = (\varepsilon_1,\dots,\varepsilon_n)'$, is multivariate normal with mean vector $\mathbf{0}$ and covariance matrix $\sigma^2 I$, and σ^2 is unknown. Assume without loss of generality for this paper that $(X'X)$ is diagonal. Consider first a prediction problem, called Problem I, where it is desired to predict whether a future observation y^* taken at $\mathbf{x} = \mathbf{x}^*$ will exceed a threshold value T, say. It will be easy to recognize that without loss of generality we can set T = 0. (Simply take the observations to be $y_i - T$.) The random loss function is assumed to be 1 if one predicts $y^* > 0$ when $y^* < 0$, 1 when one predicts $y^* < 0$ when $y^* > 0$, and 0 otherwise.

Next consider Problem II which is simply a test of the hypothesis H: $\boldsymbol{\beta}' \mathbf{x}^* \leq 0$ vs K: $\boldsymbol{\beta}' \mathbf{x}^* > 0$, with the usual 0–1 loss function. Let $\mathbf{y} = (y_1,\dots,y_n)'$ and denote a decision procedure for problem II by $\delta(\mathbf{y})$, the probability of rejection. The risk function for Problem II for $\delta(\mathbf{y})$ is

$$R(\delta,(\boldsymbol{\beta},\sigma)) = E\,\delta(\mathbf{y}) \qquad \text{for } \boldsymbol{\beta}'\mathbf{x}^* \leq 0 \tag{2.1}$$
$$= E\,(1-\delta(\mathbf{y})) \quad \text{for } \boldsymbol{\beta}'\mathbf{x}^* > 0$$

From (2.1) the risk is the size function for $\boldsymbol{\beta}'\mathbf{x}^* \leq 0$ and it is 1 minus the power function for $\boldsymbol{\beta}'\mathbf{x}^* > 0$.

Let $\delta(\mathbf{y})$ also represent a decision procedure for Problem I, only this time it represents the probability of predicting $y^* > 0$. (Both Problems I and II are two action problems so we may use $\delta(\mathbf{y})$ in either case.) The risk function for Problem I is

$$\text{(2.2)} \qquad R(\delta,(\boldsymbol{\beta},\sigma)) = E_{\mathbf{y}}E_{y^*}\{\delta(\mathbf{y})I_{(-\infty,0)}(y^*) + (1-\delta(\mathbf{y}))I_{(0,\infty)}(y^*)\}$$

$$= E_{\mathbf{y}}\{\delta(\mathbf{y})\Phi(-\boldsymbol{\beta}'\mathbf{x}^*/\sigma) + (1-\delta(\mathbf{y}))(1-\Phi(-\boldsymbol{\beta}'\mathbf{x}^*/\sigma))\}$$

$$= E_{\mathbf{y}}\{\delta(\mathbf{y})[2\Phi(-\boldsymbol{\beta}'\mathbf{x}^*/\sigma)-1]\} + \Phi(\boldsymbol{\beta}'\mathbf{x}^*/\sigma).$$

Now let $\delta^*(\mathbf{y})$ be any other procedure. Then the difference in risks between $\delta(\mathbf{y})$ and $\delta^*(\mathbf{y})$ from (2.2) is

$$\text{(2.3)} \qquad R(\delta,(\boldsymbol{\beta},\sigma)) - R(\delta^*,(\boldsymbol{\beta},\sigma)) = [2\Phi(-\boldsymbol{\beta}'\mathbf{x}^*/\sigma) - 1][E\,\delta(\mathbf{y}) - E\,\delta^*(\mathbf{y})].$$

Note that (2.3) ≥ 0 for all $\boldsymbol{\beta}'\mathbf{x}^*/\sigma \leq 0$ if and only if $E\delta(\mathbf{y}) \geq E\,\delta^*(\mathbf{y})$. Also (2.3) ≥ 0 for all $\boldsymbol{\beta}'\mathbf{x}^* > 0$ if and only if $E\delta^*(\mathbf{y}) \geq E\,\delta(\mathbf{y})$. In other words $\delta(\mathbf{y})$ is inadmissible for Problem I if and only if $\delta(\mathbf{y})$ is inadmissible for Problem II.

3. NECESSARY AND SUFFICIENT CONDITIONS FOR ADMISSIBILITY.

In this section we study the variable selection method based on preliminary testing. We find necessary and sufficient conditions for admissibility of tests based on preliminary tests. The results of Section 2 imply that these conditions are appropriate also for the prediction problem described in Section 2.

The regression model we study is as follows: Let $\mathbf{Y}^{n\times 1} \sim N(\alpha\mathbf{X}_1^{n\times 1} + \beta\,\mathbf{X}_2^{n\times 1}, \sigma^2 I)$, where we treat $\mathbf{X}_1$ and $\mathbf{X}_2$ as fixed values of independent variables. Without loss of generality we assume $\mathbf{X}_1'\mathbf{X}_2 = 0$. Let $r_1 = \|\mathbf{X}_1\|$ and $r_2 = \|\mathbf{X}_2\|$. Let $\mathbf{x}^{*\prime} = (x_1^*, x_2^*)$ be fixed. We wish to test the null hypothesis H: $\alpha x_1^* + \beta x_2^* \leq 0$ vs K: $\alpha x_1^* + \beta x_2^* > 0$. We may without loss of generality let $x_1^* = 1$ and set $x_2^* = x^*$ so that the testing problem reduces to H: $\alpha + \beta x^* \leq 0$ vs K: $\alpha + \beta x^* > 0$. Let $\mathbf{U} = \mathbf{A}\mathbf{Y}$, where $\mathbf{A}$ is a matrix with first row $\mathbf{X}_1'/r_1$, second row $\mathbf{X}_2'/r_2$, and last (n–2) rows denoted by $A^{(2)}$ where $A^{(2)}$ consists of orthogonal rows of unit length that are orthogonal to the first two rows as well. $A^{(2)}$ has rank (n–2). Hence $\mathbf{U} = (r_1\hat{\alpha}, r_2\hat{\beta}, \mathbf{u}_2')' \sim N((r_1\alpha, r_2\beta, \mathbf{0}')', \sigma^2 I)$ where $\hat{\alpha},\hat{\beta}$ are least squares estimators. Also let $\mathbf{Z} = \mathbf{B}\mathbf{U}$ where

$$B = \begin{pmatrix} [r_2/(r_2^2 + r_1^2 x^{*2})^{1/2}] & [r_1 x^*/(r_2^2 + r_1^2 x^{*2})^{1/2}] & 0' \\ [r_1 x^*/(r_2^2 + r_1^2 x^{*2})^{1/2}] & [-r_2/(r_2^2 + r_1^2 x^{*})^{1/2}] & 0' \\ 0 & 0 & I \end{pmatrix}$$

so that

(3.1) $$Z_1 = r_1 r_2(\hat{\alpha} + \hat{\beta} x^*)/(r_2^2 + r_1^2 x^{*2})^{1/2},$$

$$Z_2 = (r_1^2 x^* \hat{\alpha} - r_2^2 \hat{\beta})/(r_2^2 + r_1^2 x^{*2})^{1/2}, \quad \text{and}$$

$$Z \sim N((r_1 r_2(\alpha + \beta x^*)/(r_2^2 + r_1^2 x^{*2})^{1/2}, (r_1^2 x^* \alpha - r_2^2 \beta)/(r_2^2 + r_1^2 x^{*2})^{1/2}, 0')', \sigma^2 I)$$

Clearly, if we let $\gamma = (r_1 x^*/r_2)$ then

(3.2) $$r_2\hat{\beta} = U_2 = r_1 x^*(Z_1 - Z_2/\gamma)/(r_2^2 + r_1^2 x^{*2})^{1/2},$$

and

(3.3) $$r_1\hat{\alpha} = U_1 = r_2(Z_1 + \gamma Z_2)/(r_2^2 + r_1^2 x^{*2})^{1/2}.$$

At this point let s^2 denote the residual sum of squares for the regression model under consideration, so that $s^2 = Z'Z - Z_1^2 - Z_2^2$. It is well known that $(s^2/\sigma^2) \sim \chi^2_{n-2}$ and s^2 is independent of Z_1, Z_2. We let $R = Z'Z$ and now consider the following class of preliminary test tests (PTT):

(3.4) $$\begin{cases} \text{(a) If } (r_2\hat{\beta}) \le C\, s^2, \text{ then reject if} \\ \quad (r_1\hat{\alpha}) > \sqrt{K_1 s^2}\,; \\ \text{(b) If } (r_2\hat{\beta})^2 > C\, s^2, \text{ then reject if} \\ \quad [(r_1^2 r_2^2)/(r_2^2 + r_1^2 x^{*2})]^{1/2} (\hat{\alpha} + \hat{\beta} x^*) > \sqrt{K_2 s^2} \end{cases}$$

Note C, K_1, K_2 are always assumed positive. In terms of the variables in Z, (3.4) becomes

(a) If

(3.5) $$(Z_1 - Z_2/\gamma)^2 \leq C[(1+\gamma^2)/\gamma^2](R - Z_1^2 - Z_2^2),$$
then reject if

(3.6) $$(Z_1 + \gamma Z_2)^2 > (1+\gamma^2)K_1(R - Z_1^2 - Z_2^2)$$
and $(Z_1 + \gamma Z_2) > 0$;

(b) If

(3.7) $$(Z_1 - Z_2/\gamma)^2 > C\,[(1+\gamma^2)/\gamma^2](R - Z_1^2 - Z_2^2),$$
then reject if

(3.8) $$Z_1^2 > K_2(R - Z_1^2 - Z_2^2)$$
and $Z_1 > 0$.

Using the identity

(3.9) $$(1+\gamma^2)(Z_1^2 + Z_2^2) = \gamma^2(Z_1 - Z_2/\gamma)^2 + (Z_1 + \gamma Z_2)^2,$$
we rewrite (3.5) and (3.7) as

(3.10) $$\frac{(Z_1 - Z_2/\gamma)^2}{\left[\frac{(1+\gamma^2)CR}{\gamma^2(1+C)}\right]} + \frac{(Z_1+\gamma Z_2)^2}{(1+\gamma^2)R} \lesseqgtr 1.$$

Also rewrite (3.6) and (3.8) respectively as

(3.11) $$\frac{(Z_1 - Z_2/\gamma)^2}{\left[\frac{(1+\gamma^2)R}{\gamma^2}\right]} + \frac{(Z_1+\gamma Z_2)^2}{\left[\frac{(1+\gamma^2)K_1 R}{K_1 + 1}\right]} > 1,$$

(3.12) $$\frac{Z_1^2}{\left[\frac{K_2 R}{K_2+1}\right]} + \frac{Z_2^2}{R} > 1\ .$$

Before giving the necessary and sufficient conditions for the above tests to be admissible we note first:

THEOREM 3.1: *A necessary and sufficient condition for a test of* H vs K *to be admissible is that for each fixed* ($Z_2 = z_2$, $Z'Z = z'z$) *the acceptance section must be an interval of the form* (L, $z_1(z_2, z'z)$), *where* L= *infimum of the support of the conditional distribution of* Z_1 *given* Z_2, $Z'Z$.

PROOF: Write the joint density of **Z** as its form of a multivariate exponential family. That is, denote E **Z** = $\boldsymbol{\nu} = (\nu_1,\nu_2,0,\dots,0)'$, and then

$$f_Z(z;\nu,\sigma^2) = C(\nu,\sigma^2)e^{z_1\nu_1/\sigma^2 + z_2\nu_2/\sigma^2 - z'z/2\sigma^2}\,d\mu(z). \tag{3.13}$$

Recall that the hypothesis testing problem is H: $\nu_1 \le 0$ vs K: $\nu_1 > 0$ which is equivalent to H: $\nu_1/\sigma^2 \le 0$ vs. K: $\nu_1/\sigma^2 > 0$. Now apply the result of Matthes and Truax [6], especially Section 4, to complete the proof of the theorem. ||

Next note that when the left hand sides of (3.10), (3.11), and (3.12) are equal to 1, these relations describe 3 ellipses which we label as ellipse I,II,III respectively. The ellipses satisfy the following:

LEMMA 3.2: *The four intersection points for ellipse* I *and* II *are*

(i) $\sqrt{R/(K_1+C+1)(1+\gamma^2)}\,[\gamma\sqrt{C} + \sqrt{K_1}, -\sqrt{C} + \gamma\sqrt{K_1}\,] \equiv (z_1^{(1)},z_2^{(1)})$

(ii) $\sqrt{R/(K_1+C+1)(1+\gamma^2)}\,[\gamma\sqrt{C} - \sqrt{K_1}, -\sqrt{C} - \gamma\sqrt{K_1}\,] \equiv (z_1^{(2)},z_2^{(2)})$

(iii) $\sqrt{R/(K_1+C+1)(1+\gamma^2)}\,[-\gamma\sqrt{C} + \sqrt{K_1}, \sqrt{C} + \gamma\sqrt{K_1}\,] \equiv (z_1^{(3)},z_2^{(3)})$

(iv) $\sqrt{R/(K_1+C+1)(1+\gamma^2)}\,[-\gamma\sqrt{C} - \sqrt{K_1}, \sqrt{C} - \gamma\sqrt{K_1}\,] \equiv (z_1^{(4)},z_2^{(4)})$

When $Z_1 = 0$, *the positive value of* Z_2 *on ellipse* III *exceeds the positive value of* Z_2 *of ellipse* I, *whereas the negative value of* Z_2 *on ellipse* III *is less than the negative value of* Z_2 *of ellipse* I. *Furthermore the maximum value of the second coordinate of ellipse I is*

$\sqrt{R/(1+\gamma^2)}\,\sqrt{(\gamma^2+\gamma^2C+C)/(1+C)} \equiv z_2^{(0)}$ *and it occurs when*

$Z_1 = \gamma\sqrt{R}/\sqrt{(1+C)(1+\gamma^2)(\gamma^2+\gamma^2C+C)} \equiv z_1^{(0)}$.

PROOF: To find the maximum of ellipse I write the ellipse as a quadratic form $z'Gz = 1$ and recognize that the maximum occurs when a line of support has 0 slope or when this line is orthogonal to the Z_2 axis. This then implies that the maximum is the element in the second row, second column of G^{-1}. (See for example, Scheffe [7] p.410.) The other parts of the lemma are straightforward.

The main result of this section is

THEOREM 3.3: Suppose

(3.14) $C/K_1 \geq 1/\gamma^2$. *Then the* PTT *of* (3.4) *is admissible if and only if*

(3.15) $(\sqrt{K_1} + \gamma\sqrt{C})^2 \geq K_2(1+\gamma^2)$

and

(3.16) $CK_2 \leq \gamma^2/(1+\gamma^2)$.

Suppose

(3.17) $C/K_1 < 1/\gamma^2$,

then the PTT *of* (3.4) *is admissible if and only if* (3.15) *holds and*

(3.18) $[(\sqrt{K_1} - \gamma\sqrt{C})^2 - K_2(1+\gamma^2)][CK_2 - \gamma^2/(1+\gamma^2)] \geq 0$.

PROOF: Thc proof relies on Theorem 3.1. Therefore for fixed $(Z_2, Z'Z)$, admissibility is true if and only if acceptance sections in Z_1 are intervals of the form (L, z_1). From (3.4), (3.10) (3.11) and (3.12) these sections are unions of two sets. The first set is determined by (3.4(a)) which involves ellipse I and II. The second set is determined by (3.4(b)) which involves ellipse I and III. At this point we recognize that without loss of generality that we may take $\gamma > 0$. This is true since from (3.10), (3.11) and (3.12) an acceptance section for fixed (z_2,γ) with $\gamma > 0$ is the same as an acceptance section with $(-z_2,-\gamma)$. Hence assume $\gamma > 0$. We will argue the case (3.17) in some detail noting that the other cases can be handled in a similar manner. When (3.17) holds the intersection point $(z_1^{(3)},z_2^{(3)})$ lies in the first quadrant. The other relevant intersection point $(z_1^{(1)},z_2^{(1)})$ lies in either the first or fourth quadrant but this does not effect the proof. Also $(z_1^{(0)},z_2^{(0)})$ lies in the first quadrant and can lie either to the right or left of $(z_1^{(3)},z_2^{(3)})$. (This does not effect the proof either except that the roles of $(z_1^{(0)},z_2^{(0)})$, $(z_1^{(3)},z_2^{(3)})$ would be interchanged). In Figure 3.1 we depict ellipse I and ellipse II in the situation where $(z_1^{(0)},z_2^{(0)})$, $(z_1^{(1)},z_2^{(1)})$, and $(z_1^{(3)},z_2^{(3)})$ all lie in the first quadrant and $z_1^{(0)} > z_1^{(3)}$.

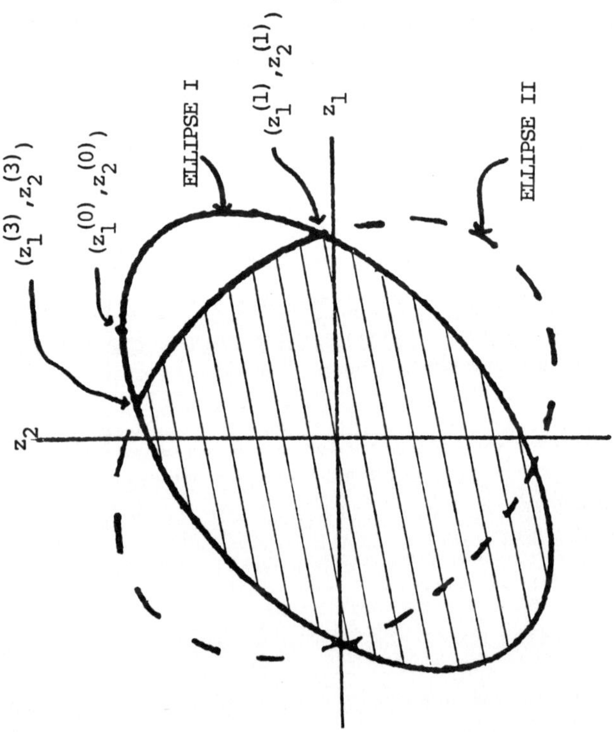

Figure 3.1

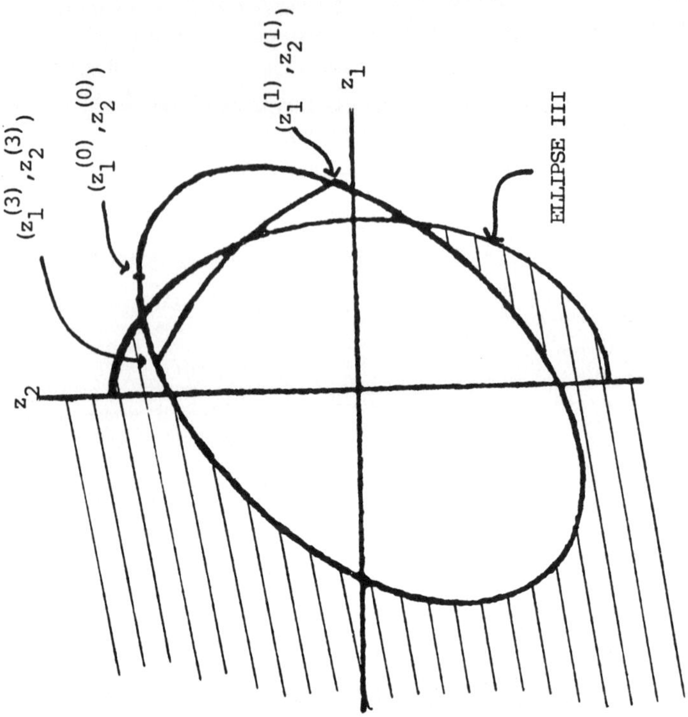

Figure 3.2

For this situation the first interval of the acceptance section is determined by the intersection of ellipses I and II unioned with the lower part of ellipse I. The second interval to be added to the first interval to form the acceptance section is determined by intersecting the outside of ellipse I with the set to the left of the right boundary part of ellipse III. Recall from Lemma 3.2 that when $Z_1 = 0$ the positive value of Z_2 on ellipse III exceeds the positive value of Z_2 on ellipse I. See Figure 3.2. Note that the acceptance section will be an interval of the form $[L,z_1)$ if and only if $(z_1^{(3)},z_2^{(3)})$ lies inside ellipse III while $(z_1^{(0)},z_2^{(0)})$ and $(z_1^{(1)},z_2^{(1)})$ lie outside of ellipse III. To determine whether points $(z_1^{(1)},z_2^{(1)})$ and $(z_1^{(3)},z_2^{(3)})$ lie in or out of ellipse III we note first that

$z_1^{(1)^2} + z_2^{(1)^2} = z_1^{(3)^2} + z_2^{(3)^2} = R(C+K_1)/(C+K_1+1)$. Also the left side of (3.12) can be written as $Z'QZ = Z_1^2(1 + 1/K_2)(1/R) + Z_2^2/R = (C+K_1)/(C+K_1+1) + Z_1^2/K_2R$. Hence the point $(z_1^{(1)},z_2^{(1)})$ lies outside the ellipse III when $z_1^{(1)} > \sqrt{(K_2R)/(C_1+K_1+1)}$, which is equivalent to (3.15). The point $(z_1^{(3)},z_2^{(3)})$ lies inside ellipse III if and only if

$(K_1-2\gamma\sqrt{CK_1}+\gamma^2C) < K_2(1+\gamma^2)$, while $(z_1^{(0)},z_2^{(0)})$ lies outside if and only if $CK_2 < \gamma^2/(1+\gamma^2)$. These latter facts are equivalent to (3.18). (Note that if $z_1^{(0)} < z_1^{(3)}$ (3.18) would still be the correct condition.) The above argument is unchanged even if $(z_1^{(1)},z_2^{(1)})$ is not in the first quadrant. This completes the proof for the case $C/K_1 < 1/\gamma^2$. The other case can be argued in a similar fashion. ||

The necessary and sufficient conditions for admissibility simplify when $K_1 = K_2 = K$. The following facts are easily established.

(3.19) $C \le K$ implies $(z_1^{(3)},z_2^{(3)})$ is inside ellipse III

(3.20) $C \le K$ implies $(z_1^{(1)},z_2^{(1)})$ is outside ellipse III
if and only if $(\gamma^2/(1+\gamma^2)) \le 4CK/(K+C)^2$

(3.21) $(z_1^{(0)},z_2^{(0)})$ is outside ellipse III if and only if $CK \le \gamma^2/(1+\gamma^2)$

(3.22) $z_1^{(0)} \ge z_1^{(3)}$ if and only if $CK \le \gamma^2(1+C)^2$.

From (3.21) and (3.22) it follows that

(3.23) $z_1^{(0)} \geq z_1^{(3)}$ and $(z_1^{(0)},z_2^{(0)})$ is outside ellipse III if and only if $CK \leq \gamma^2/(1+\gamma^2)$.

(3.24) $C \geq K$ implies if $(z_1^{(3)},z_2^{(3)})$ is in the first quadrant (i.e. $K \geq \gamma^2 C$) then $(z_1^{(3)},z_2^{(3)})$ is inside ellipse III

(3.25) $C \geq K$ implies $(z_1^{(1)},z_2^{(1)})$ is outside ellipse III.

COROLLARY 3.4: *If* $K_1 = K_2 = K$ *and* $C \leq K$ *then the* PTT *of* (3.4) *is admissible if and only if*

(3.26) $CK \leq \gamma^2/(1+\gamma^2) \leq 4CK/(K+C)^2$.

PROOF: Use (3.19), (3.20), and (3.23)

COROLLARY 3.5: *If* $K_1 = K_2 = K$ *and* $C \geq K$ *then the* PTT *of* (3.4) *is admissible if and only if*

(3.27) $CK \leq \gamma^2/(1+\gamma^2)$.

PROOF: Use (3.23), (3.24), and (3.25).

Remark: Corollaries 3.4 and 3.5 imply that if $CK \geq 1$ the PTT is inadmissible.

Remark: As $n \to \infty$, $(C/K) \to [\Phi^{-1}(1-\alpha_1/2)/\Phi^{-1}(1-\alpha_2)]^2$ where α_1 is the size of the pretest and α_2 is the common size of the post tests. Φ is the standard normal c.d.f. This can be used as a guide for choosing sizes that yield admissible tests.

REMARK 3.6: The development of Section 3 is extendible to the case of a two sided alternative for the post test. The situation is slightly more complicated. Another extension is to the case where the model is $Ey = \beta' \mathbf{x}$ and β is a k×l vector. We assume again that the design matrix $\mathbf{X}$ is of full rank and $\mathbf{X}'\mathbf{X}$ is diagonal. The pre–test is concerned with whether $\beta_k = 0$ or not. If we let $r_i = \|x_i\|$, x_i the ith column of the design matrix, let $V = 1/\sqrt{\Sigma x_i^{*2}/r_i^2}$, $\gamma = x_{k+1}^* V/r_{k+1}$, we may proceed as in the earlier part of Section 3. Still another extension would be to the case where in the latter model the pre–test is concerned with $\beta^{(2)} = 0$, where $\beta^{(2)}$ is a subvector of the vector β.

REMARK 3.7: Computer simulation indicated that the size of the PTT is approximately the size of the post tests when in fact the post tests had the same size. In other words the size of the post test was not effected very much by the size of the pre–test.

REFERENCES

[1] Cohen, A. (1974). To pool or not to pool in hypothesis testing. *Journal of the American Statistical Association*, *69*, 721–725.

[2] Seber, G.A.F. (1977). *Linear Regression Analysis*, Wiley, New York.

[3] Draper, N. and Smith, H. (1981). *Applied Regression Analysis*, Second Edition, Wiley, New York.

[4] Leamer, E.E. (1978). *Specification Searches: Ad Hoc Inference with Nonexperimental Data.* Wiley, New York.

[5] Siotani, M., Hayakawa, T., and Fujikoshi, Y. (1985). Modern Multivariate Statistical Analysis: A Graduate Course and Handbook. American Science Press, Columbus, Ohio.

[6] Matthes, T.K. and Truax, D.R. (1967). Tests of composite hypotheses for the multivariate exponential family. *Ann. Math. Statist.* *38*,681–97.

[7] Scheffé, H. (1959). *The Analysis of Variance*, Wiley, New York

STATISTICAL DATA ANALYSIS AND INFERENCE
Y. Dodge (Editor)

OPTIMUM INVARIANT TESTS IN MIXED LINEAR MODELS WITH TWO VARIANCE COMPONENTS

Thomas Mathew*

Department of Mathematics and Statistics
University of Maryland Baltimore County
Baltimore, MD 21228, U.S.A.

In a mixed linear model with two variance components, optimum invariant tests are derived for the significance of a variance component for one-sided and two-sided alternatives. When the alternative is one-sided, a locally best invariant (LBI) test always exists, which is uniformly most powerful invariant (UMPI) under a condition. Under the same condition, a uniformly most powerful invariant unbiased test exists for the two sided alternative and without this condition, a locally best invariant unbiased test can be obtained. The LBI and UMPI character of Wald's variance component test is also investigated. The results are applied to some Anova models, a model from genetics and the nearest neighbor correlation model.

1. INTRODUCTION AND PRELIMINARIES

Consider the $n \times 1$ normally distributed random vector $\mathbf{y}$ with

$$E(\mathbf{y}) = X\beta \quad \text{and} \quad Cov(\mathbf{y}) = \sigma^2(I + \theta V) \tag{1.1}$$

where X and V are known matrices of orders $n \times m$ and $n \times n$ respectively, β is an $m \times 1$ vector of unknown parameters and $\sigma^2 > 0$ and θ are unknown scalar parameters. It is assumed that $I + \theta V$ is positive definite. However, θ is *not* assumed to be non-negative, nor is it assumed that the matrix V is non-negative definite (nnd). The problem is to derive an optimum test, via invariance, for testing the hypothesis $H_0 : \theta = 0$ against one-sided or two-sided alternatives.

Clearly, any Anova model with exactly one random effect is a special case of (1.1). If (1.1) corresponeds to an Anova model, then θ is non-negative and V is nnd, and testing the significance of the random effect is equivalent to testing $H_0 : \theta = 0$ vs $H_1 : \theta > 0$. An example of a situation where V may not be nnd and θ could be negative is the nearest neighbor correlation model, considered by Kiefer and Wynn (1981). If (1.1) corresponds to the nearest neighbor correlation model for block designs (with fixed effects), then θ is the correlation between adjacent plots in a block. Here it is of interest to test the significance of the correlation between neighboring plots, since if the correlation is negligible, then ordinary least squares estimators (OLSE) can be used in the analysis. Indeed, Kiefer and Wynn (1981) use the OLSE in the optimal design problem under the nearest neighbor model.

The paper is organized as follows: In the next section, a group which leaves the testing problem invariant is exhibited. We then derive the ratio of the non-null to null densities of

*Research sponsored by Grant AFOSR 89-0237.

a maximal invariant. For testing $H_0 : \theta = 0$ against the one-sided alternative, a condition is obtained under which a uniformly most powerful invariant (UMPI) test will exist. The UMPI test, when it exists, turns out to be an exact F-test. For the same problem, a locally best invariant (LBI) test always exists. Similarly, for the two-sided alternative, one can derive a uniformly most powerful invariant unbiased (UMPIU) or locally best invariant unbiased (LBIU) test. For balanced Anova models, the UMPI test (for testing $H_0 : \theta = 0$ vs $H_1 : \theta > 0$) is the Anova F-test (see also Mathew and Sinha (1988a) and Seifert (1979)).

In the context of Anova models (balanced or unbalanced), an exact F-test (due to Wald (1949)) always exists for testing $H_0 : \theta = 0$ vs $H_1 : \theta > 0$. In Section 3 it is shown that when a UMPI test exists for this problem, it coincides with the Wald test. However, if a UMPI test does not exist, then the LBI test turns out to be different from the Wald test.

In the last section we have applied the results to several Anova models, a model from genetics and the nearest neighbor correlation model. The problem has been solved by Das and Sinha (1987) for the one-way random effects model and by Mathew and Sinha (1988b) for the two-way mixed effects model.

In the literature, models have been considered where the inverse of the covariance matrix is of the form $\sigma^2(I + \theta V)$, and the problem is to derive an optimum test for testing $H_0 : \theta = 0$. Such models arise in several applications, and for the relevant results and applications we refer to Kariya (1977, 1980) and King (1980).

2. DERIVATION OF THE OPTIMUM TEST

In (1.1), we assume without loss of generality that the $n \times m$ matrix X has rank m. The testing problem $H_0 : \theta = 0$ (against one-sided or two-sided alternatives) is invariant under the group G of transformations $\mathbf{y} \longrightarrow c(\mathbf{y} + X\alpha)$, $c > 0$ and α, an arbitrary m-component vector. Let $P = X(X'X^{-1})X'$ be the orthogonal projection matrix onto the column space of X. Write $I - P = ZZ'$, where Z is $n \times n - m$ satisfying $Z'Z = I_{n-m}$. It is readily verified that a maximal invariant (induced by the group G) is $\mathbf{T} = Z'\mathbf{y} / \parallel Z'\mathbf{y} \parallel$. It can also be shown that the ratio, say R, of the non-null to null density of $\mathbf{T}$ is

$$R = |I + \theta V|^{-\frac{1}{2}} |X'X|^{\frac{1}{2}} |X'(I + \theta V)X|^{-\frac{1}{2}} \left\{ \frac{\mathbf{y}'Z(I + \theta Z'VZ)^{-1}Z'\mathbf{y}}{\mathbf{y}'ZZ'\mathbf{y}} \right\}^{-\frac{n-m}{2}} \tag{2.1}$$

The above expression of R follows from the proof of Theorem 1 in King (1980). (2.1) can also be derived by using Wijsman's (1967) representation theorem, noting that a left invariant measure on G is $d\alpha\ dc/c$.

Let $r = \text{rank}(Z'VZ) = \text{rank}((I - P)V(I - P))$ (recall that $ZZ' = I - P$). Suppose, $0 < r < n - m = \text{rank}(I - P)$ and suppose $(I - P)V(I - P)$ has equal non-zero eigenvalues, equal to say λ. Then the nonzero eigenvalues of $Z'VZ$ are all equal to λ and let $C = (C_1, C_2)$ be an orthogonal matrix such that $Z'VZ = \lambda C_1C_1'$. Then $Z(I + \theta Z'VZ)^{-1}Z' = (1+\lambda\theta)^{-1}ZC_1C_1'Z' + ZC_2C_2'Z' = (1+\lambda\theta)^{-1}Q_1 + Q_0$, where $Q_1 = ZC_1C_1'Z'$ and $Q_0 = ZC_2C_2'Z'$.

It is readily verified that Q_1 is the orthogonal projection matrix associated with the eigenvalue λ of $(I-P)V(I-P)$ and $Q_0=(I-P)-Q_1$. Note that $Q_0\neq 0$, since $r<n-m$. Under these assumptions, we get

$$\mathbf{y}'Z(I+\theta Z'VZ)^{-1}Z'\mathbf{y}=(1+\lambda\theta)^{-1}\mathbf{y}'Q_1\mathbf{y}+\mathbf{y}'Q_0\mathbf{y}=\mathbf{y}'(I-P)\mathbf{y}-\frac{\lambda\theta}{1+\lambda\theta}\mathbf{y}'Q_1\mathbf{y}.$$

Hence (2.1) simplifies to

$$R=h(\theta)\left\{\frac{\mathbf{y}'(I-P)\mathbf{y}-\frac{\lambda\theta}{1+\lambda\theta}\mathbf{y}'Q_1\mathbf{y}}{\mathbf{y}'(I-P)\mathbf{y}}\right\}^{-\frac{n-m}{2}}$$

$$=h(\theta)\left\{1-\frac{\lambda\theta}{1+\lambda\theta}(1+\frac{\mathbf{y}'Q_0\mathbf{y}}{\mathbf{y}'Q_1\mathbf{y}})^{-1}\right\}^{-\frac{n-m}{2}} \tag{2.2}$$

where,

$$h(\theta)=|I+\theta V|^{-\frac{1}{2}}|X'X|^{\frac{1}{2}}|X'(I+\theta V)X|^{-\frac{1}{2}} \tag{2.3}$$

Consider the testing problem $H_0:\theta=0$ vs $H_1:\theta>0$. We note that the problem is non-trivial only when $\lambda>0$. If the density ratio R admits the representation (2.2) and if $\lambda>0$, then clearly, R is increasing in $\frac{\mathbf{y}'Q_1\mathbf{y}}{\mathbf{y}'Q_0\mathbf{y}}$. Consequently, in this case, the UMPI test for $H_0:\theta=0$ vs $H_1:\theta>0$ rejects H_0 for large values of

$$F_0=\frac{n-m-r}{r}\;\frac{\mathbf{y}'Q_1\mathbf{y}}{\mathbf{y}'Q_0\mathbf{y}} \tag{2.4}$$

It is easy to verify that F_0 in (2.4) is distributed as central F, under H_0.

If the non-zero eigenvalues of $(I-P)V(I-P)$ are not all equal, then an LBI test can be derived by expanding R in (2.1), locally around $\theta=0$. This gives,

$$R=1+\frac{n-m}{2}\;\theta\;\frac{\mathbf{y}'(I-P)V(I-P)\mathbf{y}}{\mathbf{y}'(I-P)\mathbf{y}}-\frac{\theta}{2}\;[trV+tr(X'VX)(X'X)^{-1}]\;+o(\theta) \tag{2.5}$$

uniformly in $\mathbf{y}$. Hence, for testing $H_0:\theta=0$ vs $H_1:\theta>0$, the LBI test rejects H_0 for large values of

$$\frac{\mathbf{y}'(I-P)V(I-P)\mathbf{y}}{\mathbf{y}'(I-P)\mathbf{y}}$$

We summarize these results in the following theorem.

THEOREM 2.1. *In the model (1.1), let* $P=X(X'X)^{-1}X'$ *and* $r=rank((I-P)V(I-P))$ *and assume* $r>0$.

[a] *Suppose* $r<n-m$ *and the non-zero eigenvalues of* $(I-P)V(I-P)$ *are all equal to* λ, *where* $\lambda>0$, *with associated projection matrix* Q_1. *Then for testing* $H_0:\theta=0$ *vs* $H_1:\theta>0$, *the UMPI test rejects* H_0 *for large values of*

$$F_0=\frac{n-m-r}{r}\;\frac{\mathbf{y}'Q_1\mathbf{y}}{\mathbf{y}'Q_0\mathbf{y}},$$

where $Q_0=(I-P)-Q_1$. *Further, under* H_0, F_0 *is distributed as central* F *with degrees of freedom* $(r,n-m-r)$.

[b] *If the non-zero eigenvalues of* $(I-P)V(I-P)$ *are not all equal, then for testing* $H_0 : \theta = 0$ *vs* $H_1 : \theta > 0$, *the LBI test rejects* H_0 *for large values of*

$$\frac{\mathbf{y}'(I-P)V(I-P)\mathbf{y}}{\mathbf{y}'(I-P)\mathbf{y}}.$$

REMARK 2.1. The UMPI test in Theorem 2.1[a] can also be derived from the results in Spjøtvoll (1967), see also ElBassiouni and Seely (1988, p.159).

REMARK 2.2. In the notations of Theorem 2.1, if $0 < r < n - m$ and if the non-zero eigenvalues of $(I-P)V(I-P)$ are all equal, then for testing $H_0 : \theta = 0$ vs $H_1 : \theta \neq 0$, the UMPIU test rejects H_0 if F_0 is large or small. When a UMPIU test does not exist, an LBIU test (which is not so explicit) can be characterized as in Ferguson (1967, Section 5.5).

REMARK 2.3. Under H_0, the maximal invariant $\mathbf{T} = Z'\mathbf{y}/ \parallel Z'\mathbf{y} \parallel$ has the uniform distribution on $\{\mathbf{w} : \mathbf{w} \in R^{n-m},\ \ \mathbf{w}'\mathbf{w} = 1\}$. Hence, the optimum tests obtained above are obviously null robust if normality is replaced by any spherically symmetric distribution. The tests are also non-null and optimality robust, from the results in Section 2 of Kariya and Sinha (1985).

3. WALD'S VARIANCE COMPONENT TEST

In (1.1), suppose θ is non-negative and V is a non-negative definite matrix of rank s. Write $V = BB'$, where B is an $n \times s$ matrix of rank s. Then for testing $H_0 : \theta = 0$ vs $H_1 : \theta > 0$, Wald's test rejects H_0 for large values of

$$\frac{\mathbf{y}'(I-P)B(B'(I-P)B)^{-}B'(I-P)\mathbf{y}}{\mathbf{y}'(I-P)\mathbf{y}}$$

or equivalently, for large values of

$$\frac{\mathbf{y}'(I-P)B(B'(I-P)B)^{-}B'(I-P)\mathbf{y}}{\mathbf{y}'(I-P)\mathbf{y} - \mathbf{y}'(I-P)B(B'(I-P)B)^{-}B'(I-P)\mathbf{y}}$$

Here '–' denotes generalized inverse. The latter statistic, suitably normalized, is distributed as central F under H_0. (For a derivation of the test and its properties, we refer to Seely and ElBassiouni (1983), Seely and Rady (1988), and ElBassiouni and Seely (1988)). Comparing with Theorem 2.1[b], it follows that Wald's test (which is invariant) is equivalent to the LBI test if and only if $(I-P)V(I-P)$ is a positive scalar multiple of $(I-P)B(B'(I-P)B)^{-}B'(I-P)$. Using $I - P = ZZ'$, the condition is equivalent to

$$Z'VZ = \delta Z'B(B'ZZ'B)^{-}\ B'Z$$

for some $\delta > 0$, which holds if and only if the non-zero eigenvalues of $Z'VZ$ are all equal and equal to δ, in which case a UMPI test exists (from Theorem 2.1[a]) and Wald's test coincides with the UMPI test. However, if a UMPI test does not exist, then Wald's test never coincides with the LBI test.

4. EXAMPLES

(*i*) *The One-Way Random Effects Model*:
Let k be the number of treatments, n_i be the number of observations for the ith treatment and $n = \sum_{i=1}^{k} n_i$. The treatment effects are assumed to be random, resulting in the one-way random effects model. Let J_m denote the $m \times m$ matrix of ones. Then, in the notations of Sections 1 and 2, $V = diag\ (J_{n_1}, \ldots J_{n_k})$ and $P = \frac{1}{n} J_n$. It can be verified that $\mathbf{y}'(I-P)V(I-P)\mathbf{y} = \sum_{i=1}^{k} n_i^2\ (\bar{y}_{i.} - \bar{y}_{..})^2$ and $\mathbf{y}'(I-P)\mathbf{y} = \sum_{i=1}^{k} \sum_{j=1}^{n_i} (y_{ij} - \bar{y}_{i.})^2 + \sum_{i=1}^{k} n_i\ (\bar{y}_{i.} - \bar{y}_{..})^2$, where y_{ij} is the jth observation corresponding to the ith treatment, $\bar{y}_{i.} = \frac{1}{n_i} \sum_{j=1}^{n_i} y_{ij}$ and $\bar{y}_{..} = \frac{1}{n} \sum_{i=1}^{k} \sum_{j=1}^{n_i} y_{ij}$.

Let σ_τ^2 denotes the variance of the treatment effects and $\theta = \sigma_\tau^2/\sigma^2$. Then for testing $H_0 : \theta = 0$ vs $H_1 : \theta > 0$, applying Theorem 2.1[b], the LBI test rejects H_0 for large values of

$$\frac{\sum_{i=1}^{k} n_i^2 (\bar{y}_{i.} - \bar{y}_{..})^2}{\left\{\sum_{i=1}^{k} \sum_{j=1}^{n_i} (y_{ij} - \bar{y}_{i.})^2 + \sum_{i=1}^{k} n_i (\bar{y}_{i.} - \bar{y}_{..})^2\right\}}$$

Furthermore, the non-zero eigenvalues of $(I-P)V(I-P)$ are equal if and only if the n_i's are all equal. In other words, a UMPI test exists only in the balanced case. The LBI test was first derived by Das and Sinha (1987). It should be noted that Wald's test is an exact F-test for this problem even in the unbalanced case.

(*ii*) *A Two-Way Mixed Model*:
Consider a block design with fixed block effects and random treatment effects (and no interaction). Let σ_τ^2 and θ be as defined in *(i)* above and let v be the number of treatments, b, the number of blocks and k_i, the ith block size. Then P in Theorem 2.1 is the orthogonal projection onto the column space of $diag\ (\mathbf{1}_{k_1}, \ldots \mathbf{1}_{k_b})$. Also, $V = FF'$, where F is the design matrix for the treatment effects. Let y_{ijl}, $\bar{B}_j$, $\mathbf{Q}$ and C respectively denote the lth observation for the ith treatment in the jth block, average of the observations in the jth block, the vector of adjusted treatment totals and the C-matrix of the design ($i = 1, 2, \ldots v;\ j = 1, 2, \ldots b; l = 1, 2, \ldots n_{ij}$, where n_{ij} is the number of observations in the jth block for the ith treatment). Then the sum of squares within blocks, say SS_W, is given by $SS_W = \sum_{j=1}^{b} \sum_{i=1}^{v} \sum_{l=1}^{n_{ij}} (y_{ijl} - \bar{B}_j)^2$. It can now be verified that $\mathbf{y}'(I-P)V(I-P)\mathbf{y} = \mathbf{Q}'\mathbf{Q}$ and $\mathbf{y}'(I-P)\mathbf{y} = SS_W + \mathbf{Q}'C^-\mathbf{Q}$ where C^-denotes a generalized inverse of C. Thus the LBI test for testing $H_0 : \theta = 0$ vs $H_1 : \theta > 0$ rejects H_0 for large values of

$$\frac{\mathbf{Q}'\mathbf{Q}}{\{SS_W + \mathbf{Q}'C^-\mathbf{Q}\}}$$

It also follows that the non-zero eigenvalues of $(I-P)V(I-P)$ are equal if and only if the non-zero eigenvalues of the C-matrix are equal, in which case $\mathbf{Q}'\mathbf{Q}$ is a multiple of $\mathbf{Q}'C^-\mathbf{Q}$ and the UMPI test rejects H_0 for large values of $\frac{\mathbf{Q}'\mathbf{Q}}{SS_W}$ which can be normalized to get an F-test. Thus, there are some unbalanced situations where a UMPI test exists. For example, a BIBD admits a UMPI test for testing $H_0 : \theta = 0$ vs $H_1 : \theta > 0$. Also, in the case of only two treatments (ie, v=2), a UMPI test always exists, regardless of the unbalancedness in the design.

The above arguments can be easily extended to a three-way classification model or higher order classification models. One can show for instance, that the Generalized Youden Design defined in Kiefer (1975) (with fixed row and column effects, and random treatment effects) admits a UMPI test for testing the significance of treatment effects.

(iii) A Model from Genetics:

Let $X_i = diag\ (\mathbf{1}_{n_{i1}}, \ldots \mathbf{1}_{n_{ib}})$ for $i = 1, 2, \ldots v$ (where n_{ij}'s are positive integers), $X' = (X_1', X_2', \ldots X_v')$, $F = I_v \otimes \mathbf{1}_n$ ($n = \sum_{j=1}^{b} n_{ij}$ for each i). Consider the model

$$\mathbf{y} = X\beta + F\tau + \mathbf{e}, \tag{4.1}$$

where $\mathbf{y}$ is a $vn \times 1$ vector of observations, β is a $b \times 1$ vector of unknown parameters, τ and $\mathbf{e}$ are respectively $v \times 1$ and $vn \times 1$ random vectors distributed as $N(\mathbf{0}, \sigma_\tau^2 I_v)$ and $N(\mathbf{0}, \sigma^2 I_{vn})$. Such a model arises in genetic experiments and for background information we refer to Thompson (1976, 1977a, 1977b) and Gnot (1979). The problem is to test $H_0 : \theta = 0$ vs $H_1 : \theta > 0$, where, $\theta = \frac{\sigma_\tau^2}{\sigma^2}$. With $V = FF'$ and P denoting the orthogonal projector onto the column space of X, the non-zero eigenvalues of $(I - P)V(I - P)$ are clearly same as those of $F'(I - P)F$. If $D = X'X$ (which is a diagonal matrix) and $N = ((n_{ij}))$, then $F'(I - P)F = n\ (I_v - \frac{1}{n} ND^{-1}N')$. Hence, a UMPI test exists if the eigenvalues of $\frac{1}{n} ND^{-1}N'$, different from 1, are all equal. Otherwise an LBI test exists. If the n_{ij} 's are all equal then a UMPI test obviously exists. One can also construct examples where the n_{ij} 's are not all equal and still a UMPI test exists.

(iv) The Nearest Neighbor Correlation Model:

Consider a complete block design for comparing v treatments in b blocks of v plots each. The nearest neighbor correlation model is the usual block- treatment additive model for the observations with the additional assumption that observations from neighboring plots in a block are correlated and otherwise they are uncorrelated (see Kiefer and Wynn(1981)). As in Kiefer and Wynn, we assume that the correlation has a common value, say ρ, and the hypothesis to be tested is $H_0 : \rho = 0$ against one sided or two sided alternatives. The hypothesis is of considerable interest, since if H_0 is accepted, then the use of the ordinary least squares procedure is justified in analyzing the block design. Let y_{ij} denote the observation from the jth plot in the ith block, j=1, 2,....v; i=1, 2, ...b and $\mathbf{y} = (y_{11}, y_{12}, \ldots y_{1v}, y_{21}, y_{22}, \ldots y_{2v}, \ldots\ldots y_{b1}, y_{b2}, \ldots y_{bv})'$. If β is the $b \times 1$ vector of block effects, τ is the $v \times 1$ vector of treatment effects, F is the $bv \times v$ design matrix for treatment effects and $\mathbf{e}$ is the $bv \times 1$ vector of experimental error terms, then the nearest neighbor correlation model is

$$\mathbf{y} = \mu\mathbf{1}_{bv} + (I_b \otimes \mathbf{1}_v)\beta + F\tau + \mathbf{e}, \tag{4.2}$$

with $\mathrm{cov}(\mathbf{y}) = \sigma^2[I_{bv} + \rho(I_b \otimes A)]$, where

$$A = \begin{pmatrix} 0 & 1 & 0 & 0 & \ldots & 0 & 0 & 0 \\ 1 & 0 & 1 & 0 & \ldots & 0 & 0 & 0 \\ 0 & 1 & 0 & 1 & \ldots & 0 & 0 & 0 \\ .. & .. & .. & .. & .. & .. & .. & .. \\ 0 & 0 & 0 & 0 & \ldots & 1 & 0 & 1 \\ 0 & 0 & 0 & 0 & \ldots & 0 & 1 & 0 \end{pmatrix}$$

We note that $F' = (F_1'\ F_2' \ldots F_b')$, where each F_i is a $v \times v$ permutation matrix. Hence, in the notations of Section 1, $X = [\mathbf{1}_{bv}, I_b \otimes \mathbf{1}_v, F]$ and it is easily verified that

$$I - P = [I_b \otimes (I_v - \frac{1}{v}J_v)][I_{bv} - \frac{1}{b}FF'].$$

With A as defined above, the nonzero eigenvalues of $(I - P)V(I - P)$ are same as those of $(I_{bv} - \frac{1}{b}FF')(I_b \otimes (I_v - \frac{1}{v}J_v)A)$. Preliminary investigations show that the design (ie, the matrix F) cannot be chosen to satisfy the above eigenvalue condition. Hence a uniformly most powerful test does not exist for testing $H_0 : \rho = 0$ agaist $H_1 : \rho > 0$ (or $H_1 : \rho \neq 0$) in the class of invariant tests (respectively invariant unbiased tests). An LBI (respectively LBIU) test can obviously be obtained as described in Section 2.

ACKNOWLEDGEMENT

The author expresses his gratitude to Professor Bimal Kumar Sinha whose comments resulted in the elimination of an error and improved presentation.

REFERENCES

[1] Das, R. and Sinha, B.K., Robust Optimum Invariant Unbiased Tests for Variance Components, in: Pukkila, T. and Puntanen, S., (eds.), Proceedings of the Second International Tampere Conference in Statistics (University of Tampere, Tampere, Finland, 1987) pp. 317-342.

[2] ElBassiouni, M.Y. and Seely, J.F., On the Power of Wald's Variance Component Test in the Unbalanced Random One-Way Model, in: Dodge, Y., Federov, V. V. and Wynn, H. P., (eds.) Optimal Design and Analysis of Experiments (North-Holland, Amsterdam, 1988) pp. 157-165.

[3] Ferguson, T.S., Mathematical Statistics, A Decision Theoretic Approach (Academic Press, New York, 1967).

[4] Gnot, S., Estimation of Parameters in Certain Hierarchical Genetic Models, Mathematische Operationsforschung und Statistik, Series Statistics 10 (1979) pp. 529–539.

[5] Kariya, T., A Robustness Property of the Tests for Serial Correlations, The Annals of Statistics 5 (1977) pp. 1212–1220.

[6] Kariya, T., Locally Robust Tests for Serial Correlation in Least Squares Regression, The Annals of Statistics 8 (1980) pp. 1065–1070.

[7] Kariya, T. and Sinha, B. K., Nonnull and Optimality Robustness of Some Tests, The Annals of Statistics 13 (1985) pp. 1182–1197.

[8] King. M.L., Robust Tests for Spherical Symmetry and Their Applications to Least Squares Regression, The Annals of Statistics 8 (1980) pp. 1265–1271.

[9] Kiefer, J., Construction and Optimality of Generalized Youden Designs, in: Srivastava, J., (ed.), A Survey of Statistical Designs and Linear Models (North-Holland, Amsterdam, 1975) pp. 333-353.

[10] Kiefer, J. and Wynn, H. P., Optimum Balanced Block and Latin Square Designs for Correlated Observations, The Annals of Statistics 9 (1981) pp. 737–757.

[11] Mathew, T. and Sinha, B.K., Optimum Tests for Fixed Effects and Variance Components in Balanced Models, Journal of the American Statistical Association 83 (1988a) pp. 133–135.

[12] Mathew, T. and Sinha, B.K., Optimum Tests in Unbalanced Two-Way Models Without Interaction, The Annals of Statistics 16 (1988b) pp. 1727–1740.

[13] Seely, J.F. and ElBassiouni, Y., Applying Wald's Variance Component Test, The Annals of Statistics 11 (1983) pp. 197–201.

[14] Seely, J.F and Rady, E.H., When can Random Effects be Treated as Fixed Effects for Computing a Test Statistic for a Linear Hypothesis? Communications in Statistics-Theory and Methods 17 (1988) pp. 1089–1109.

[15] Seifert, B., Optimal Testing for Fixed Effects in General Balanced Mixed Classification Models, Mathematische Operationsforschung und Statistik, Series Statistics 10 (1979) pp. 237-256.

[16] Spjøtvoll, E., Optimum Invariant Tests in Unbalanced Variance Components Models, The Annals of Mathematical Statistics 38 (1967) pp. 422–429.
[17] Thompson, R., Design of Experiments to Estimate Heritability when Observations are Available on Parents and Offsprings, Biometrics 32 (1976) pp. 283–304.
[18] Thompson, R., The Estimation of Heritability with Unbalanced Data I. Observations Available on Parents and Offsprings, Biometrics 33 (1977a) pp. 485–496.
[19] Thompson, R., The Estimation of Heritability with Unbalanced Data II. Data Available on More Than Two Generations, Biometrics 33 (1977b) pp. 497–507.
[20] Wald, A., A Note on Regression Analysis, The Annals of Mathematical Statistics 18 (1947) pp. 586–589.
[21] Wijsman, R.A., Cross-Sections of Orbits and Their Application to Densities of Maximal Invariants, Proceedings of the Fifth Berkeley Symposium on Mathematical Statistics and Probability 1 (1967) pp. 389–400.

STATISTICAL DATA ANALYSIS AND INFERENCE
Y. Dodge (Editor)

CONFIDENCE BOUNDS FOR PROPORTION OF σ_E^2 IN UNBALANCED TWO-WAY CROSSED MODELS

Mohammad K. Kazempour
Department of Statistics
University of Central Florida
Orlando, FL 32816

Franklin A. Graybill
Department of Statistics
Colorado State University
Ft. Collins, CO 80523

ABSTRACT

Consider the two-way crossed components of variance model given by
$Y_{ijk} = \mu + A_i + B_j + E_{ijk}(i=1,2,...a;\ j=1,2,...b;\ k=1,2,...n_{ij}\geq 0)$
where A_i, B_j and E_{ijk} are unobservable independent normal random variables with means zeros and variances σ_A^2, σ_B^2 and σ_E^2 respectively. In this paper procedures for finding confidence bounds on $\rho_E = \sigma_E^2/(\sigma_A^2+\sigma_B^2+\sigma_E^2)$ are derived. The performance of the bounds are evaluated by conducting a simulation study.

1. INTRODUCTION

The problem discussed here is that of obtaining confidence bounds for the proportion of total variances for unbalanced two-way crossed classification components-of-variance models where some of the cells may be empty. The model described in several text books i.e. Sheffé [7], Searle [8] and Graybill [1], is

$$Y_{ijk} = \mu + A_i + B_j + E_{ijk}(i = 1..a;\ j = 1,..b;\ k = 1,...n_{ij} \geq 0) \quad (1.1)$$

where $A_1, A_2, ...A_a; B_1, B_2, ...B_b; E_{111}, E_{112}, ... E_{11n_{11}}, E_{121}, ...E_{abn_{ab}}$ are unobservable random variables, jointly normally distributed, pairwise uncorrelated with zero means. The variances of A_i, B_j and E_{ijk} are σ_A^2, σ_B^2 and σ_E^2 respectively.

The model in (1.1) can be written in matrix form as

$$\underline{Y} = \mu X_0 + X\begin{bmatrix}\underline{\alpha}\\ \underline{\beta}\end{bmatrix} + \underline{E}$$

where $X = [X_1, X_2]$; $\underline{Y}$ is a vector of Nx1 and the elements are Y_{ijk}; μ is a constant; X_0 is a matrix of Nx1 of 1's; X_1 is a known Nxa; X_2 is a Nxb;

$\underline{E}$ is a Nx1 with elements E_{ijk}; $N = \sum_{i=1}^{a} \sum_{j=1}^{b} n_{ij}$; $\underline{\alpha} \sim MVN(\underline{0},\sigma_A^2 I_a)$;
$\underline{\beta} \sim MVN(\underline{0},\sigma_B^2 I_b)$, $\underline{E} \sim MVN(\underline{0},\sigma_E^2 I_N)$, $\underline{\alpha}$, $\underline{\beta}$ and $\underline{E}$ are independent. $E(\underline{Y}) = \mu X_0$, $var(\underline{Y}) = \Sigma = X_1 X_1' \sigma_A^2 + X_2 X_2' \sigma_B^2 + I_N \sigma_E^2$.

The problem is that of obtaining approximate confidence bounds on $\rho_E = \sigma_E^2/(\sigma_A^2+\sigma_B^2+\sigma_E^2)$. Several approximate methods have been developed to construct confidence bounds on ρ_E in the balanced complete models when $n_{ij} = m$ for all i and j, i.e. Satterthwaite [6], Howe and Myers [5]. Lu, Burdic and Graybill [4] developed approximate confidence bounds on the ratios of expected mean squares $(\Theta_1+d\Theta_2)/\Theta_3$ where $d \geq 0$ and $n_i S^2{}_i/\Theta_i$ are independent chi-square with n_i, i = 1,2,3 degrees of freedom. In this paper we study their method and then extend it for model (1.1) where the assumption of independence of $n_i S_i^2/\Theta_i$, i = 1,2,3; may no longer hold, also $n_i s_i^2/\Theta_i$ may not be distributed as a chi-square except for i = 3.

2. DERIVATION OF THE CONFIDENCE BOUNDS

For the case that $n_{ij} = m$ for all i and j Lu, Graybill and Burdic (LGB) derived good approximate upper, lower and two sided confidence bounds for ρ_E. (LGB) employed the Modified Large Sample (MLS) approximation derived by Graybill and Wang [2]. The MLS method is based on an unbiased estimator of δ, say $\hat{\delta}$, and an unbiased estimator of the variance of $\hat{\delta}$, say $\hat{var}(\hat{\delta})$. It can be proved, Lu (1986), that

$$\hat{\delta} = [(S_1^2+dS_2^2)/S_3^2](1-2/n_3)$$

is an unbiased estimator of $\delta = (\theta_1+d\theta_2)/\theta_3$, and

$$\hat{var}(\hat{\delta}) = c_1(S_1^2/S_3^2)^2 + c_2(dS_2^2/S_3^2)^2 + c_3(dS_1^2S_2^2/S_3^2)$$

is an unbiased estimator of $var(\hat{\delta})$, where c_1, c_2 and c_3 are functions of n_1, n_2 and n_3.

An approximate $1-\alpha$ upper confidence bound, U_M, is obtained as

$$U_M = \hat{\delta} + (a_U S_1^4+b_U S_2^4+c_U dS_1^2S_2^2)^{\frac{1}{2}}/S_3^2 , \tag{2.1}$$

and an approximate $1-\alpha$ lower confidence bound, L_M, is

$$L_M = \hat{\delta} - (a_L S_1^4+b_L S_2^4+c_L dS_1^2S_2^2)^{\frac{1}{2}}/S_3^2 , \tag{2.2}$$

where a_U, b_U, c_U, a_L, b_L and c_L are obtained by imposing limiting conditions on U_M and L_M, their values are

$$a_U = (1-2/n_3-F^{-1}_{1-\alpha:n_1,n_3})^2 , \qquad a_L = (1-2/n_3-F^{-1}_{\alpha:n_1,n_3})^2 ,$$

$$b_U = (1-2/n_3-F^{-1}_{1-\alpha:n_2,n_3})^2 , \qquad b_L = (1-2/n_3-F_{\alpha:n_2,n_3})^2 ,$$

$$c_U = [1-2/n_3-F^{-1}_{1-\alpha:n_1+n_2,n_3}]^2(n_1+n_2)^2/n_1n_2 - a_U n_1/n_2 - b_U n_2/n_1 ,$$

$$c_L = [1-2/n_3-F^{-1}_{\alpha:n_1+n_2,n_3}]^2(n_1+n_2)^2/n_1n_2 - a_Ln_1/n_2 - b_Ln_2/n_1 .$$

To extend the above method for the case that $n_{ij} \geq 0$ (model (1.1)) S^2_A, S^2_B and S^2_C are replaced for S^2_1, S^2_2 and S^2_3 respectively, where

$$S^2_A = \underline{Y}' \frac{XX^- - X_2X_2^-}{a-1} \underline{Y} , \qquad S^2_B = \underline{Y}' \frac{XX^- - X_1X_1^-}{b-1} \underline{Y} , \qquad S^2_E = \underline{Y}' \frac{I-XX^-}{N-a-b+1} \underline{Y} , \quad (2.3)$$

(Z^- is the Moore-Penrose inverse of Z). Also let $T_1 = X_1'(XX^- - X_2X_2^-)X_1$, $T_2 = X_2'(XX^- - X_1X_1^-)X_2$, $A_1 = tr(T_1)/(a-1)$, $A_2 = tr(T_2)/(b-1)$.
Then we have

$$\Theta_1 = E(S^2_A) = \sigma^2_E + A_1\sigma^2_A ,$$

$$\Theta_2 = E(S^2_B) = \sigma^2_E + A_2\sigma^2_B ,$$

$$\Theta_3 = E(S^2_E) = \sigma^2_E , \quad (2.4)$$

$$\rho_E = \sigma^2_E/(\sigma^2_A+\sigma^2_B+\sigma^2_E) \text{ and}$$

$$\rho_E^{-1} = \frac{A_2}{A_1A_2} \cdot \frac{\Theta_1+A_1A_2^{-1}\Theta_2}{\Theta_3} + \frac{A_1A_2-A_1-A_2}{A_1A_2} .$$

Notice that S^2_A and S^2_B are not scaled chi-squares for general unbalanced models. Let the first two moments of $\frac{KS^2}{E(S^2)}$ be equal to the first two moments of a chi-square with K degrees of freedom, this is what we shall call 'approximate' chi-square (Satterthwaite's technique).

Theorem 2.1. If $n_aS^2_A/E(S^2_A)$ is approximately a chi-square and the design is connected then

$$(a-1)/(\frac{(a-1)tr(T_1^2)}{(tr(T_1))^2}) \leq n_a \leq a-1$$

Proof: For any symmetric matrix G the following inequality holds

$$[tr(G)]^2 \leq rank(G) \cdot tr(G^2)$$

rank $(T_1) \leq (a-1)$, T_1 is symmetric then $(tr(T_1))^2 \leq (a-1) \cdot tr(T_1^2)$ and $\frac{(a-1)tr(T_1^2)}{(tr(T_1))^2} \geq 1$; since $n_aS^2_A/E(S^2_A)$ is 'approximately' distributed as a chi-square with n_a degrees of freedom we have

$$n_a = \frac{2[E(S_A^2)]^2}{\mathrm{var}(S_A^2)} = \frac{(a-1)}{1+[\frac{(a-1)tr(T_1^2)}{(tr(T_1))^2}-1][1/(1+(a-1)\lambda/tr(T_1))]}, \qquad (2.5)$$

where $\lambda = \frac{\sigma_E^2}{\sigma_A^2}$ and therefore $0 \le \lambda < \infty$.

n_a is a monotone increasing function in λ, so for $\lambda = 0$ n_a reaches its minimum, $\frac{a-1}{(a-1)tr(T_1^2)/(tr(T_1))^2}$, and if $\lambda \to \infty$, n_a reaches its maximum, a-1, this proves the theorem.

Note: If appropriate estimator for σ_E^2 and σ_A^2 are used in (2.5) then an estimator of n_a is obtained. S_E^2 and $(S_A^2-S_E^2)/(tr(T_1)/(a-1))$ are unbiased estimators for σ_E^2 and σ_A^2 respectively, replacing these in (2.5) Then $\tilde{n}_a$ is resulted which can be used to estimate n_a.

$$\tilde{n}_a = (a-1)/\left[1 + (\frac{(a-1)tr(T_1^2)}{(tr(T_1))^2} - 1)(\frac{S_A^2}{S_E^2} - 1)\right]$$

<u>Theorem 2.2.</u> If $\sigma_E^2 = 0$ then S_A^2 and S_B^2 are independent (Kazempour [3]).

If we let $d = A_1A_2^{-1}$ then ρ_E^{-1} will be a one-to-one function of δ. In (LGB) notation an unbiased estimator of $\frac{\Theta_1+A_1A_2^{-1}\Theta_2}{\Theta_3}$ is $(1-2/n_3)\frac{S_A^2+A_1A_2^{-1}S_B^2}{S_E^2}$ and the proposed approximate $1-\alpha$ upper confidence bound for ρ_E in unbalanced model is

$$U = \left[1-A_1^{-1}-A_2^{-1}+A_1((1-2/n_3)\frac{S_A^2+A_1A_2^{-1}S_B^2}{S_E^2} + \frac{(a_U'S_A^4+b_U'S_B^4+c_U'A_1A_2^{-1}S_A^2S_B^2)^{\frac{1}{2}}}{S_E^2})\right]^{-1} \qquad (2.6)$$

$1-\alpha$ lower confidence bound for ρ_E is

$$L = \left[1-A_1^{-1}-A_2^{-1}+A_1((1-2/n_3)\frac{S_A^2+A_1A_2^{-1}S_B^2}{S_E^2} - \frac{(a_L'S_A^4+b_L'S_B^4+c_L'A_1A_2^{-1}S_A^2S_B^2)^{\frac{1}{2}}}{S_E^2})\right]^{-1} \qquad (2.7)$$

$1-2\alpha$ two sided confidence bounds is (L, U). Where

$$a_U' = (1-2/n_3-F^{-1}_{1-\alpha:\hat{n}_1,n_3})^2, \qquad a_L' = (1-2/n_3-F^{-1}_{\alpha:\hat{n}_1,n_3}),$$

$$b_U' = (1-2/n_3 - F^{-1}_{1-\alpha:\hat{n}_2,n_3})^2 , \qquad b_L' = (1-2/n_3 - F_{\alpha:\hat{n}_2,n_3})^2 ,$$

$$c_U' = [1-2/n_3 - F^{-1}_{1-\alpha:\hat{n}_1+\hat{n}_2,n_3}]^2(\hat{n}_1+\hat{n}_2)^2/\hat{n}_1\hat{n}_2 - a_U'\hat{n}_1/\hat{n}_2 - b_U'\hat{n}_2/\hat{n}_1 ,$$

$$c_L' = [1-2/n_3 - F^{-1}_{\alpha:\hat{n}_1+\hat{n}_2,n_3}]^2(\hat{n}_1+\hat{n}_2)^2/\hat{n}_1\hat{n}_2 - a_L'\hat{n}_1/\hat{n}_2 - b_L'\hat{n}_2/\hat{n}_1 .$$

$\hat{n}_1$ and $\hat{n}_2$ are the degrees of freedom. They can be the minimum or the maximum values resulted from Theorem 2.1 or they can be calculated as $\tilde{n}_a$ by replacing the appropriate estimators for σ_A^2, σ_B^2 and σ_E^2 in (2.5). If $S_A^2 < S_E^2$ then the minimum value of (2.5) is used for $\tilde{n}_a$.

3. EVALUATION

A simulation study was conducted to determine how close the confidence coefficients of the proposed approximate confidence bounds for ρ_E are to the nominal values, 1-α.

Tables 1 to 4 present the results of this simulation study. The results are 'very good' in the sense that they cover the nominal 1-α value and also they are not too conservative or too liberal, they are around the ranges of the simulation error ($\alpha \pm 1.96\sqrt{\frac{\alpha(1-\alpha)}{1000}}$).

These tables are obtained by using 1000 runs for each design configuration and each 1-α level. As it can be seen from confidence bounds in (2.6) and (2.7) the distributions of the constructed bounds and the confidence coefficients for ρ_E depend on the unknown parameters σ_A^2, σ_B^2 and σ_E^2. To examine the effects of this dependency we let $\rho_A = \frac{\sigma_A^2}{\sigma^2}$, $\rho_B = \frac{\sigma_B^2}{\sigma^2}$, where $\sigma^2 = \sigma_A^2 + \sigma_B^2 + \sigma_E^2$ assume a subset of their allowable values; ρ_A and ρ_B were assigned .05, .2, .4, .8, .9, .99. For each chosen pair of ρ_A and ρ_B and each chosen design configuration the bounds were calculated 1000 times and the number of times which a hit occurred (a calculated bound covered the true parameter) were counted, the counts were divided by 1000. This was done for all 21 different combinations of values for ρ_A and ρ_B then the smallest and the largest values of these 21 values were recorded in the tables. As one can see from the tables, even though the distribution of the bounds and the confidence coefficients depend on the unknown parameters, changes in the unknown parameters over the allowable values do not affect the coverages.

TABLE 1

Range of Coefficients for 95% One-Sided Confidence Bounds on ρ_E for Two-Way Balanced Complete Models

Number of Blocks	Number of Treatments	Lower Bound	Upper Bound
3	3	.953 - .962	.951 - .983
3	7	.938 - .958	.895 - .954
3	11	.956 - .953	.874 - .954
5	3	.952 - .965	.909 - .957
5	11	.946 - .965	.932 - .966
7	3	.939 - .957	.899 - .942
7	11	.950 - .962	.937 - .963
9	3	.946 - .961	.876 - .946
9	11	.951 - .972	.929 - .963

TABLE 2

Range of Coefficients for 95% One-Sided Confidence Bounds on ρ_E for Two-Way Unbalanced Model

Design* Number	Number of Blocks	Number of Treatments	Number of Observation	Lower Bound	Upper Bound
D1	3	3	40	.939 - .963	.944 - .981
D2	3	8	55	.939 - .960	.875 - .956
D3	5	5	80	.949 - .970	.929 - .979
D4	5	3	44	.938 - .968	.863 - .955
D5	8	3	49	.950 - .966	.880 - .946
D6	3	5	35	.940 - .957	.873 - .946
D7	4	5	18	.952 - .958	.938 - .977
D8	3	8	79	.932 - .961	.819 - .949

*Incidence matrices of these designs are presented in the end of this article.

TABLE 3

Range of Coefficients for 90% Two-Sided Confidence Bounds on ρ_E for Two-Way Unbalanced Model

Design Number	Number of Blocks	Number of Treatments	Number of Observation	Range of Coefficient
D1	3	3	40	.885 - .944
D2	3	8	55	.818 - .898
D3	5	5	80	.883 - .935
D4	5	3	44	.802 - .913
D5	8	3	49	.832 - .901
D6	3	5	35	.825 - .889
D7	4	5	18	.893 - .926
D8	3	8	79	.823 - .887

TABLE 4

Range of Coefficients for 90% One-Sided Confidence Bounds on ρ_E for Two-Way Balanced Complete Models

Number of Blocks	Number of Treatments	Two-Sided
3	3	.878 - .940
3	7	.837 - .893
3	11	.822 - .886
5	3	.864 - .909
5	11	.888 - .931
7	3	.852 - .893
7	11	.889 - .918
9	3	.827 - .899
9	11	.881 - .920

Incidence matrices for different designs which were used in previous sections are presented here. Rows are blocks and columns are treatments. The entry is the number of observations in the cell.

Design #1 $$D_1 = \begin{bmatrix} 6 & 2 & 2 \\ 5 & 4 & 8 \\ 3 & 8 & 2 \end{bmatrix}$$

Design #2 $$D_2 = \begin{bmatrix} 5 & 1 & 3 & 4 & 7 & 4 & 2 & 9 \\ 6 & 1 & 3 & 7 & 1 & 10 & 1 & 7 \\ 1 & 5 & 1 & 4 & 1 & 1 & 3 & 2 \end{bmatrix}$$

Design #3 $$D_3 = \begin{bmatrix} 4 & 3 & 1 & 5 & 2 \\ 6 & 3 & 6 & 4 & 3 \\ 5 & 2 & 1 & 3 & 4 \\ 5 & 2 & 2 & 1 & 3 \\ 4 & 1 & 3 & 2 & 5 \end{bmatrix}$$

Design #4 $$D_4 = \begin{bmatrix} 1 & 3 & 4 \\ 2 & 2 & 5 \\ 1 & 1 & 2 \\ 2 & 5 & 4 \\ 3 & 6 & 3 \end{bmatrix}$$

Design #5 $$D_5 = \begin{bmatrix} 1 & 1 & 2 \\ 2 & 3 & 1 \\ 1 & 4 & 1 \\ 2 & 2 & 3 \\ 3 & 4 & 3 \\ 2 & 1 & 1 \\ 1 & 1 & 4 \\ 3 & 1 & 2 \end{bmatrix}$$

Design #6 $$D_6 = \begin{bmatrix} 1 & 1 & 2 & 3 & 1 \\ 2 & 4 & 2 & 1 & 3 \\ 5 & 2 & 4 & 3 & 1 \end{bmatrix}$$

Design #7 $$D_7 = \begin{bmatrix} 3 & 1 & 1 & 0 & 0 \\ 0 & 1 & 2 & 1 & 0 \\ 1 & 0 & 1 & 0 & 0 \\ 1 & 3 & 0 & 1 & 2 \end{bmatrix}$$

Design #8 $$D_8 = \begin{bmatrix} 0 & 0 & 2 & 3 & 0 & 2 & 40 & 3 \\ 3 & 2 & 0 & 0 & 2 & 4 & 5 & 2 \\ 4 & 3 & 3 & 1 & 0 & 0 & 0 & 0 \end{bmatrix}$$

REFERENCES

[1] Graybill, F. A. (1976). Theory and Application of the Linear Model. Duxbury Press, North Scituate, Massachusetts.

[2] Graybill, F. A. and Wang, C. (1979). Confidence intervals for proportions of variability in two-factor nested variance component models. Journal of the American Statistical Association 74, 368-374.

[3] Kazempour, M. K. (1987). Confidence bounds on functions of variance components in unbalanced models. Ph.D. thesis, Department of Statistics, Colorado State University.

[4] Lu, T. C., Graybill, F. A. and Burdic, R. K. (1987). Confidence intervals on the ratio of expected mean squares $(\Theta_1+d\Theta_2)/\Theta_3$.

[5] Myers, R. H. and Howe, R. B. (1971). On alternative approximate F test for hypothesis involving variance components. Biometrika, 58, 393-396.

[6] Satterthwaite, F. E. (1946). An approximate distribution of estimates of variance components. Biometric Bulletin 2, 110-114.

[7] Scheffé, H. (1970). The analysis of variance. New York: John Wiley.

[8] Searle, S. R. (1970). Linear models. New York: John Wiley.

STATISTICAL DATA ANALYSIS AND INFERENCE
Y. Dodge (Editor)

A LEMMA ON OPTIMIZATION OF A MATRIX FUNCTION AND A REVIEW OF THE UNIFIED THEORY OF LINEAR ESTIMATION

C. Radhakrishna Rao

Center for Multivariate Analysis, Penn State University
University Park, PA 16802

Abstract: A method of computing the minimum of the matrix valued function $A'VA - F'A - A'F$ of a matrix argument A subject to a constrainst of the form $X'A = P$ is proposed. The solultion is used to develop a unified theory of estimation in the mixed effects linear model $\underset{\sim}{Y} = X\underset{\sim}{\beta}+U\underset{\sim}{\xi}+\epsilon$ without making any assumptions on the ranks of the matrices X, U or of the dispersion matrices of $\underset{\sim}{\xi}$ and $\underset{\sim}{\epsilon}$. The estimates of $X\underset{\sim}{\beta}$, $U\underset{\sim}{\xi}$ and $\underset{\sim}{\epsilon}$ depend on the elements of a g-inverse of a partitioned matrix as in the case of the fixed effects model (see Rao (1971, 1972, 1973, 1985)).

A New concept of conditioned equations (similar to that of normal equations) is introduced for simultaneous estimation of fixed effects $(X\underset{\sim}{\beta})$, random effects $(U\underset{\sim}{\xi})$, and the error $(\underset{\sim}{\epsilon})$. The paper is self contained and simplifies the approach presented in earlier papers of the author on a unified theory of linear estimation based on minimizing a quadratic form.

AMS Classification Index: 62J05

1. INTRODUCTION

We provide a unified theory of estimation in linear models with fixed or mixed effects by first establishing a lemma on minimizing a matrix function. We need the following notations and results in matrix algebra, most of which can be found in the first chapter of the author's book (Rao (1973)).

1.1 *Notations*

A matrix A of order $n\times m$ is denoted by $A: n\times m$, its rank by $\rho(A)$ and its range, i.e., the space spanned by its column vectors, by $R(A)$. A generalized inverse of $A: n\times m$, denoted by $A^-: m\times n$, which is not necessarily unique, satisfies the equation $AA^-A = A$. Note that $AA^-\underset{\sim}{a} = \underset{\sim}{a}$ if $\underset{\sim}{a} \in R(A)$.

Let $V: n\times n$ be n.n.d., $X: n\times m$ be any given matrix and choose $K: m\times m$ symmetric such that

$$R(T) = R(V:X), \text{ with } T = V+XKX', \tag{1.1.1}$$

where $(V:X)$ denotes a partitioned matrix. [$K = I$ is a choice which satisfies (1.1.1) whatever V and X may be. K can be zero when $R(X) \subset R(V)$.] Denote by T^- a g-inverse of T. Since T is symmetric, $(T^-)'$ and also $(T^- + (T^-)')/2$ are g-inverses of T, of which the latter is symmetric. In this paper, we shall consider T^- wherever it occurs to be symmetric without loss of generality. Consider $X'T^-X$ which is also symmetric with $\rho(X) = \rho(X') = \rho(X'T^-X)$. To see this, observe that $X = TA$ for some A since $R(X) \subset R(T)$. Then

$$X'T^-X = A'TT^-TA = A'TA. \tag{1.1.2}$$

If there exists a vector $\underset{\sim}{a}$ such that

$$a'X'T^-X = \underset{\sim}{a}'A'TA = 0 \Rightarrow a'A'X = 0,\ a'A'VA = 0 \Rightarrow a'A'X = 0,\ a'A'V = 0$$
$$\Rightarrow a'A'T = 0 \Rightarrow a'X' = 0$$

so that $R(X') \subset R(X'T^-X)$ or $\rho(X') \leq \rho(X'T^-X)$. But $\rho(X'T^-X) \leq \rho(X')$, which proves the required result. Let $(X'T^-X)^-$ be a symmetric g-inverse of $X'T^-X$ without loss of generality. Then

$$(X'T^-X)(X'T^-X)^-X = X,\ X' = X'(X'T^-X)^-(X'T^-X) \tag{1.1.3}$$

since $R(X) = R(X'T^-X)$. We also need in the sequel the result (see Rao (1973, p. 34))

$$\rho(A) = \text{trace } A = \text{tr } A, \text{ if } A^2 = A \tag{1.1.4}$$

i.e., when A is idempotent.

1.2 A matrix equation

The computational problem that arises in our discussion is the solution of the matrix equation

$$\begin{aligned} VA + XB &= F \\ X'A &= P \end{aligned} \tag{1.2.1}$$

where $V: n\times n$ is n.n.d., $X: n\times m$, $F: n\times k$ and $P: m\times k$ are such that $R(F) \subset R(V:X)$ and $R(P) \subset R(X')$.

Under the conditions on F and P, the equation (1.2.1) admits a solution for $A: n\times k$ and $B: m\times k$. To show this, let there exist a row vector $(\underset{\sim}{a}', \underset{\sim}{b}')$ such that

$$(\underset{\sim}{a}', \underset{\sim}{b}')\begin{pmatrix} V & X \\ X' & 0 \end{pmatrix} = 0 \tag{1.2.2}$$

$$\Rightarrow \underset{\sim}{a}'V + \underset{\sim}{b}'X' = 0, \quad \underset{\sim}{a}'X = 0 \tag{1.2.3}$$

$$\Rightarrow \underset{\sim}{a}'V\underset{\sim}{a} + \underset{\sim}{b}'X'\underset{\sim}{a} = 0, \text{ multiplying (1.2.3) by } \underset{\sim}{a}$$

$$\Rightarrow \underset{\sim}{a}'V\underset{\sim}{a} = 0 \text{ or } \underset{\sim}{a}'V = 0, \text{ hence } \underset{\sim}{b}'X' = 0.$$

Then

$$(\underset{\sim}{a}' \ \underset{\sim}{b}')\binom{F}{P} = \underset{\sim}{a}'F + \underset{\sim}{b}'P = 0 \tag{1.2.4}$$

since $R(F) \subset R(V:X)$ and $R(P) \subset R(X')$. Thus (1.2.2) $\Rightarrow$ (1.2.4), which shows that the equation (1.2.1) is consistent.

Let

$$\begin{pmatrix} V & X \\ X' & 0 \end{pmatrix}^{-} = \begin{pmatrix} C_1 & C_2 \\ C_3 & -C_4 \end{pmatrix} \tag{1.2.5}$$

be any choice of the g-inverse. Because of the symmetry of the left-hand side matrix in (1.2.5)

$$\begin{pmatrix} C_1 & C_2 \\ C_3 & -C_4 \end{pmatrix}'$$

is also a g-inverse and so also

$$\frac{1}{2}\begin{pmatrix} C_1 & C_2 \\ C_3 & -C_4 \end{pmatrix} + \frac{1}{2}\begin{pmatrix} C_1 & C_2 \\ C_3 & -C_4 \end{pmatrix}'$$

which is symmetric so that, we can assume the choice of g-inverse in (1.2.5) to be symmetric and write it as

$$\begin{pmatrix} C_1 & C_2 \\ C_2' & -C_4 \end{pmatrix} \tag{1.2.6}$$

with C_1 and C_4 as symmetric matrices.

Using (1.2.6), we can write a solution to (1.2.1) as

$$A_* = C_1F+C_2P, \quad B_* = C_2'F-C_4P. \tag{1.2.7}$$

Considering the special case $F = 0$, and $P = X'$ and substituting A_* and B_* in the equation (1.2.1), we have

$$VC_2X' = XC_4X', \; X'C_2X' = X' \tag{1.2.8}$$

so that C_2' is a g-inverse of X. Considering the special case $F = V$ and $P = 0$, and substituting A_* and B_* in the equation (1.2.1), we have

$$VC_1V+XC_2'V = V, \; X'C_1V = 0 \tag{1.2.9}$$

$$\Rightarrow VC_1VC_1V = VC_1V. \tag{1.2.10}$$

1.3 *Projection operator*

Consider the subspaces $R(X)$ and $R(VZ)$, where V: $n \times n$, n.n.d. and X: $n \times m$ are given matrices and $Z = X^{\perp}$, i.e., a matrix of maximum rank such that $X'X^{\perp} = 0$. Then the intersection of $R(X)$ and $R(VZ)$ consists of the null vector only, for otherwise let

$$X\underset{\sim}{a} = VZ\underset{\sim}{b} \neq 0 \tag{1.3.1}$$

for some vectors $\underset{\sim}{a}$ and $\underset{\sim}{b}$. Then

$$0 = Z'X\underset{\sim}{a} = Z'VZ\underset{\sim}{b} \Rightarrow b'Z'VZ\underset{\sim}{b} = 0 \Rightarrow VZ\underset{\sim}{b} = 0$$

which is a contradiction. In such a case any vector $\underset{\sim}{y} \in R(X: VZ)$ can be uniquely decomposed as

$$\underset{\sim}{y} = \underset{\sim}{y}_1+\underset{\sim}{y}_2, \ \underset{\sim}{y}_1 \in R(X), \ \underset{\sim}{y}_2 \in R(VZ). \tag{1.3.2}$$

The mapping P_X: $\underset{\sim}{y} \rightarrow \underset{\sim}{y}_1$ is called a projection of $\underset{\sim}{y}$ on $R(X)$ along $R(VZ)$ (see Rao (1974)). Note that

$$Xa+VZb \in R(X: VZ) \quad \text{for arbitrary } \underset{\sim}{a} \text{ and } \underset{\sim}{b}. \tag{1.3.3}$$

If P_X is such that $P_X(Xa+VZb) = Xa$, then

$$P_X X\underset{\sim}{a} = X\underset{\sim}{a} \quad \forall\ \underset{\sim}{a} \Rightarrow P_X X = X$$

$$P_X VZ\underset{\sim}{b} = 0 \quad \forall\ \underset{\sim}{b} \Rightarrow P_X VZ = 0 \quad \text{or} \quad VP_X' \in R(X).$$

We then have two equations satisfied by P_X'

$$\begin{aligned} VP_X'+XB &= 0 \\ X'P_X' &= X' \end{aligned} \tag{1.3.3}$$

which is a special case of (1.2.1) with $F = 0$ and $P = X'$. Applying the formula (1.2.7), we have an explicit expression for P_X' as

$$P_X' = C_2X' \quad \text{or} \quad P_X = XC_2'. \tag{1.3.4}$$

Note that P_X as chosen is idempotent since

$$P_X P_X = XC_2'XC_2' = XC_2' \tag{1.3.5}$$

using (1.2.8). [There can be other choices of P_X which may not be idempotent (see Rao (1974)).]

1.4 *Direct solution of the matrix equation*

We obtain a direct solution of the matrix equation (1.2.1) without computing a g-inverse first, which will also provide expressions for C_1, C_2, C_3 and C_4 in the g-inverse (1.2.6) in terms of V and X.

Consider the equation

$$\begin{aligned} VA+XB &= F \\ X'A &= P \end{aligned} \tag{1.4.1}$$

(where $R(F) \subset R(V:X)$ and $R(P) \subset R(X')$. Let $T = V+XKX'$ where K is a symmetric matrix chosen such that $R(T) = R(V:X)$. The equation (1.4.1) is equivalent to

$$\begin{aligned} (V+XKX')A+XB &= F+XKP \\ X'A &= P. \end{aligned} \tag{1.4.2}$$

From the first equation in (1.4.2), one choice of A is

$$A = T^-(F+XKP-XB). \tag{1.4.3}$$

Substituting in the second equation

$$(X'T^-X)B = X'T^-(F+XKP)-P \tag{1.4.4}$$

giving a solution for B, A as

$$B = (X'T^-X)^-X'T^-F+(X'T^-X)^-(X'T^-XK-I)P \tag{1.4.5}$$

$$A = (T^- - T^-X(X'TX)^-X'T^-)F+T^-X(X'T^-X)^-P. \tag{1.4.6}$$

It may be verified that A and B as above satisfy (1.4.1). Comparing the solultion (1.4.5, 1.4.6) withn (1.3.7) we find expressions for C_1, C_2, C_3 and C_4 in terms of V and X as

$$C_1 = T^- - T^- X(X'TX)^- X'T^- = T^-(I-P_X)$$
$$C_2' = (X'T^-X)^- X'T^-, \quad -C_4 = (X'T^-X)^-(X'T^-XK-I). \tag{1.4.7}$$

[Note that T can be chosen as V (with K = 0) if V is nonsingular or when $R(X) \subset R(V)$.] An explicit expression for the projection operator on R(X) along R(VZ) given in (1.3.4) is

$$P_X = XC_2' = X(X'T^-X)^- X'T^-. \tag{1.4.8}$$

Note that, we have chosen T^- and $(X'T^-X)^-$ as symmetric g-inverses for convenience. As a matter of fact, it may be verified that in all the expressions where these inverses occur, any g-inverse could be used without altering the value.

1.5 *Simultaneous estimation of several parameters*

Let $\underset{\sim}{\theta}' = (\theta_1,\ldots,\theta_p)$ be a p-vector parameter. An estimator $\hat{\underset{\sim}{\theta}}' = (\hat{\theta}_1,\ldots,\hat{\theta}_p)$ is said to be a minimum dispersion unbiased estimator (MUDE) of $\underset{\sim}{\theta}$ if $E(\hat{\theta}) = \theta$ and

$$D(\hat{\underset{\sim}{\theta}}) = E(\hat{\underset{\sim}{\theta}}-\underset{\sim}{\theta})(\hat{\underset{\sim}{\theta}}-\underset{\sim}{\theta})' \le E(\tilde{\underset{\sim}{\theta}}-\underset{\sim}{\theta})(\tilde{\underset{\sim}{\theta}}-\underset{\sim}{\theta})' \tag{1.5.1}$$

for every unbiased estimator $\tilde{\underset{\sim}{\theta}}$ of $\underset{\sim}{\theta}$.

It is well known that if T is a minimum variance unbiased estimator (MVUE) of a scalar function $f(\underset{\sim}{\theta})$, then it is necessary and sufficient that $E(T) = f(\underset{\sim}{\theta})$ and

$$\operatorname{cov}(T,t) = 0 \quad \forall\ t \ni E(t) = 0 \text{ and } V(t) < \infty \tag{1.5.2}$$

(see Rao (1973), p. 317). From this result, it follows that $\hat{\underset{\sim}{\theta}}$ is the MDUE of $\underset{\sim}{\theta}$ if and only if

$$\operatorname{cov}(\hat{\theta}_i, t) = 0 \quad \forall\, t \ni E(t) = 0 \text{ and } V(t) < \infty$$
$$i = 1,\ldots,p \tag{1.5.3}$$

i.e., if and only if each $\hat{\theta}_i$ is MVUE of θ_i, $i = 1,\ldots,p$. The 'if part' follows since for any vector $\underset{\sim}{b}$

$$\operatorname{cov}(\underset{\sim}{b}'\hat{\underset{\sim}{\theta}}, t) = 0 \Rightarrow \underset{\sim}{b}'D(\hat{\underset{\sim}{\theta}})\underset{\sim}{b} \le \underset{\sim}{b}'D(\tilde{\underset{\sim}{\theta}})b \;\forall\; \underset{\sim}{b} \Rightarrow D(\hat{\underset{\sim}{\theta}}) \le D(\tilde{\underset{\sim}{\theta}})$$

for any alternative unbiased estimator $\tilde{\underset{\sim}{\theta}}$ of $\underset{\sim}{\theta}$. The 'only if' part is obvious.

Further if $\hat{\underset{\sim}{\theta}}$ is the MDUE of $\underset{\sim}{\theta}$, then

$$P'\hat{\underset{\sim}{\theta}} \text{ is the MDUE of } P'\underset{\sim}{\theta} \tag{1.5.4}$$

for any P': k×p.

2. A BASIC LEMMA

We state and prove a basic lemma on optimization of a matrix function. We say that G_*: n×n is a minimum in a given set of matrics {G: n×n} if G_* belongs to the set, and for any element G in the set, $G - G_*$ is n.n.d. (which may also be denoted by $G_* \le G$).

<u>Lemma</u>. Let V: n×n be n.n.d., X: n×m, F: n×k and P: m×k be given matrices such that $R(F) \subset R(V:X)$ and $R(P) \subset R(X')$, and consider the matrix function of A: n×k

$$f(A) = A'VA - A'F - F'A. \tag{2.1}$$

Then

$$\min_{X'A=P} f(A) = F(A_*)$$

where (A_*, B_*) is a solultion of the equation

$$\begin{aligned} VA+XB &= F \\ X'A &= P. \end{aligned} \tag{2.2}$$

Furthermore,

$$f(A_*) = \min_{X'A=P} f(A) = -A_*'F-B_*'P = -F'A_*-P'B_* \tag{2.3}$$

and is unique for any solution of (2.2).

Proof. Let (A_*,B_*) be a solulltion of (2.2), whose existence is shown in Section 1.2. Any A such that $X'A = P$ can be written as A_*+ZC where $Z = X^{\perp}$ and C is arbitrary. Then

$$\begin{aligned} f(A) &= A'VA-A'F-F'A = (A_*+ZC)'V(A_*+ZC)-(A_*+ZC)'F-F'(A_*+ZC) \\ &= (A_*'VA_*-A_*'F-F'A_*)+C'Z'VZC + (A_*'V-F')ZC+C'Z'(VA_*-F) && (2.4) \\ &= (A_*'VA_*-A_*'F-F'A_*)+C'Z'VZC = f(A_*)+C'Z'VZC && (2.5) \end{aligned}$$

since, using the equation (2.2)

$$VA_*+XB_* = F \;\Rightarrow\; \begin{cases} C'Z'(VA_*-F) = 0 \\ (A_*'V-F')ZC = 0 \end{cases}$$

so that the last two terms in (2.4) are zero. The difference $f(A)-f(A_*)$ $= C'Z'VZC$ which is n.n.d.; this proves the optimization part. Now

$$\begin{aligned} f(A_*) &= A_*'VA_*-A_*'F-F'A_* \\ &= A_*'(VA_*-F)-F'A_* = -A_*'XB_*-F'A_* \\ &= -P'B_*-F'A_* = -A_*'F-B_*'P \end{aligned} \tag{2.6}$$

which proves (2.3).

Using the solution (1.2.7) of the equation (2.2)

$$A_* = C_1F+C_2P, \quad B_* = C_2'F-C_4P, \tag{2.7}$$

$$f(A_*) = -P'(C_2'F-C_4P)-F'(C_1F+C_2P)$$

$$= P'C_4P-P'C_2'F-F'C_2P-F'C_1F \tag{2.8}$$

$$= P'C_4P \quad \text{if} \quad F = 0 \tag{2.9}$$

$$= -F'C_1F \quad \text{if} \quad P = 0. \tag{2.10}$$

3. LINEAR MODEL WITH FIXED EFFECTS

Consider the Gauss-Markoff linear model

$$\underset{n\times 1}{\underset{\sim}{Y}} = \underset{n\times m}{X}\ \underset{m\times 1}{\underset{\sim}{\beta}} + \underset{n\times 1}{\underset{\sim}{\epsilon}} \tag{3.1}$$

where $E(\underset{\sim}{\epsilon}) = 0$ and $D(\underset{\sim}{\epsilon}) = \sigma^2V$. We discuss the estimation of the vector parameter $\underset{\sim}{\theta} = X\underset{\sim}{\beta}$ and the estimation (prediction) of $\underset{\sim}{\epsilon}$, the "signal" and "noise" parts respectively of the model (3.1). It may be noted that in the model (3.1), $\underset{\sim}{Y} \in R(V:X)$ and $\underset{\sim}{\epsilon} \in R(V)$ with prob. 1.

3.1 *Estimation of* $X\underset{\sim}{\beta}$ *(signal)*

Let $A'\underset{\sim}{Y}$ be an unbiased estimator of $X\underset{\sim}{\beta}$, which implies

$$E(A'\underset{\sim}{Y}) = A'X\underset{\sim}{\beta} = X\underset{\sim}{\beta} \ \forall\ \underset{\sim}{\beta} \ \Rightarrow\ X'A = X'. \tag{3.1.1}$$

The dispersion matrix of $A'\underset{\sim}{Y}$ is

$$E[(A'\underset{\sim}{Y}-X\underset{\sim}{\beta})(A'\underset{\sim}{Y}-X\underset{\sim}{\beta})'] = E(A'\underset{\sim}{\epsilon}\underset{\sim}{\epsilon}'A) = \sigma^2A'VA. \tag{3.1.2}$$

To find the MDUE of $X\underset{\sim}{\beta}$, we solve the problem

$$\min_{X'A=X'} A'VA \tag{3.1.3}$$

which is a special case of (1.2.1) with $F = 0$ and $P = X'$. The optimum A_* is a solution of

$$\begin{aligned} VA+XB &= 0 \\ X'A \quad &= X'. \end{aligned} \tag{3.1.4}$$

Using the g-inverse (1.2.6)

$$A^* = C_2X', \quad B_* = -C_4X' \tag{3.1.5}$$

so that the MDE of $\underset{\sim}{\theta} = X\underset{\sim}{\beta}$ is

$$\begin{aligned} \hat{\underset{\sim}{\theta}} &= A_*'\underset{\sim}{Y} = XC_2'\underset{\sim}{Y} = X(X'T^-X)^-X'T^-\underset{\sim}{Y} \\ &= P_X\underset{\sim}{Y} = X\hat{\underset{\sim}{\beta}} \quad (\text{defining } \hat{\underset{\sim}{\beta}} = (X'T^-X)^-X'T^-\underset{\sim}{Y}) \end{aligned} \tag{3.1.6}$$

using the expression (1.4.7) for C_2' and the expression (1.4.8) for the projection operator P_X. The dispersion matrix of $\hat{\underset{\sim}{\theta}}$ is, using the expression (2.5) for $A_*'VA_*$,

$$\begin{aligned} D(\hat{\underset{\sim}{\theta}}) &= -\sigma^2XB_* = \sigma^2XC_4X' \\ &= \sigma^2X(X'T^-X)^-(I-X'T^-XK) = \sigma^2[X(X'T^-X)^-X'-XKX'] \\ &= \sigma^2X[X'T^-X)^--K]X' \end{aligned} \tag{3.1.7}$$

using the expression (1.4.7) for C_4.

Suppose that we want to estimate k parametric functions $P'\underset{\sim}{\beta}$, where P: m×k is a given matrix. For $P'\underset{\sim}{\beta}$ to be estimable unbiasedly, we must have $R(P) \subset R(X')$, or $P = X'M$ for some M: n×k. Then $P'\underset{\sim}{\beta} = M'X\underset{\sim}{\beta} = M'\underset{\sim}{\theta}$. By the formula (1.5.4) the MDE of $M'\underset{\sim}{\theta}$ $(= P'\underset{\sim}{\beta})$ is

$$M'\hat{\underset{\sim}{\theta}} = M'XC_2'\underset{\sim}{Y} = P'C_2\underset{\sim}{Y} = P'\hat{\underset{\sim}{\beta}} \tag{3.1.8}$$

with the dispersion matrix

$$D(M'\hat{\theta}) = M'D(\hat{\theta})M = \sigma^2 M'XC_4X'M$$
$$= \sigma^2 P'C_4P = \sigma^2 P'((X'T^-X)^- - K)P. \qquad (3.1.9)$$

This completes the estimation of estimable linear functions of β.

Note: $\hat{\beta}$ as defined in (3.1.6) is not an estimate of β unless $\rho(X) = m$ (full). If $\rho(X) < m$, then β as such is not estimable unbiasedly. However, if $P\beta$ is estimable unbiasedly, i.e., $R(P) \subset R(X')$, then $P\hat{\beta}$ with $\hat{\beta}$ as in (3.1.6) provides the MDUE of $P\beta$. In this sense $\hat{\beta}$ is a pseudo estimate of β. The parameter that explicitely occurs in the model (3.1) is $\theta = X\beta$ which only implies that $\theta \in R(X)$. Indeed, the model can be written as $Y = \theta+\epsilon$ where the admissible region of the parameter θ is the space spanned by the columns of the X matrix. β is introduced only as a structural parameter, which may not be identifiable unless X is of full rank. The model when written as $Y = \theta+\epsilon$ with $\theta \in R(X)$ is said to have a coordinate free formulation.

3.2 *Estimation of* ϵ *(the error)*

Let $A'Y$ be an unbiased estimator (or predictor) of ϵ. Then

$$E(A'Y-\epsilon) = A'X\beta = 0 \Rightarrow X'A = 0$$
$$D(A'Y-\epsilon) = D[(A'-I)\epsilon] = \sigma^2(A'-I)V(A-I)$$
$$= \sigma^2(A'VA-A'V-VA+V). \qquad (3.2.1)$$

The problem is to find

$$\min_{X'A=0} (A'VA-A'V-VA+V). \qquad (3.2.2)$$

This is a special case of (1.2.1) with $F = V$ and $P = 0$. The optimum A is then a solultion of

$$\begin{aligned} VA+XB &= V \\ X'A &= 0. \end{aligned} \tag{3.2.3}$$

Using the g-inverse (1.2.6), a solution of (3.2.3) is

$$A_* = C_1 V, \quad B_* = C_2' V$$

giving the MDE of $\underset{\sim}{\varepsilon}$ as

$$\begin{aligned} \hat{\underset{\sim}{\varepsilon}} &= A_*' \underset{\sim}{Y} = VC_1 \underset{\sim}{Y} = VT^-(I-P_X)\underset{\sim}{Y} \\ &= TT^-(I-P_X)\underset{\sim}{Y} = (I-P_X)\underset{\sim}{Y} \end{aligned} \tag{3.2.4}$$

using the expression (1.4.7) for C_1. Further

$$\begin{aligned} D(\hat{\underset{\sim}{\varepsilon}}-\underset{\sim}{\varepsilon}) &= E[(\hat{\underset{\sim}{\varepsilon}}-\underset{\sim}{\varepsilon})(\hat{\underset{\sim}{\varepsilon}}-\underset{\sim}{\varepsilon})'] \\ &= \sigma^2(-VA_*+V) = \sigma^2(V-VC_1V), \text{ using } (2.5). \end{aligned} \tag{3.2.5}$$

The estimator of $Q'\underset{\sim}{\varepsilon}$, any specified set of linear functions of $\underset{\sim}{\varepsilon}$, is

$$Q'\hat{\underset{\sim}{\varepsilon}} = Q'VC_1Y, \tag{3.2.6}$$

with

$$E[(Q'\hat{\underset{\sim}{\varepsilon}}-Q'\underset{\sim}{\varepsilon})(Q'\hat{\underset{\sim}{\varepsilon}}-Q'\underset{\sim}{\varepsilon})'] = \sigma^2 Q'(V-VC_1V)Q. \tag{3.2.7}$$

This completes the discussion of the problems associated with the prediction (estimation) in the linear model (3.1).

Note 1. It is seen that the estimates of the signal and noise are

$$X\hat{\underset{\sim}{\beta}} = P_X\underset{\sim}{Y}, \quad \hat{\underset{\sim}{\varepsilon}} = (I-P_X)\underset{\sim}{Y} \tag{3.2.8}$$

providing a decomposition of the observed vector

$$\underset{\sim}{Y} = P_X\underset{\sim}{Y}+(I-P_X)\underset{\sim}{Y} \tag{3.2.9}$$

where $P_X\underset{\sim}{Y}$ is the projection of $\underset{\sim}{Y}$ on $R(X)$ parallel to $R(VZ)$ as discussed in Section 1.3.

<u>Note 2</u>. The estimates of signal and noise are uncorrelated since

$$\begin{aligned} \operatorname{cov}(X\hat{\underset{\sim}{\beta}},\hat{\underset{\sim}{\epsilon}}) &= \operatorname{cov}(P_X\underset{\sim}{Y},(I-P_X)\underset{\sim}{Y}) \\ &= \sigma^2 P_X V(I-P_X') = \sigma^2 P_X T(I-P_X') \\ &= \sigma^2 X(X'T^-X)^-X'(I-P_X') = 0. \end{aligned} \tag{3.2.10}$$

<u>Note 3</u>. The MDE of $P'X\underset{\sim}{\beta}+Q'\underset{\sim}{\epsilon}$ a linear function of signal and noise, is

$$P'X\hat{\underset{\sim}{\beta}} + Q'\hat{\underset{\sim}{\epsilon}} \tag{3.2.11}$$

using the result proved in Section 1.5. The dispersion matrix of error in the estimate (3.2.11) is

$$P'D(X\hat{\beta})P+Q'D(\hat{\epsilon}-\epsilon)Q. \tag{3.2.12}$$

The expressions for $D(X\hat{\underset{\sim}{\beta}})$ and $D(\hat{\underset{\sim}{\epsilon}}-\underset{\sim}{\epsilon})$ are given in (3.1.7) and (3.2.5) respectively.

3.3 *Estimation of* σ^2

The MDE of $\underset{\sim}{\epsilon}$ is

$$\hat{\underset{\sim}{\epsilon}} = (I-P_X)\underset{\sim}{Y} = VC_1\underset{\sim}{Y} \tag{3.3.1}$$

with the dispersion matrix

$$D(\hat{\underset{\sim}{\varepsilon}}) = D(VC_1\underset{\sim}{Y}) = \sigma^2 VC_1VC_1V = \sigma^2 VC_1V \tag{3.3.2}$$

using (1.2.10). Consider the quadratic form $\hat{\underset{\sim}{\varepsilon}}'(VC_1V)^-\hat{\underset{\sim}{\varepsilon}}$ whose expected value is

$$\begin{aligned} &E(\underset{\sim}{Y}'C_1V(VC_1V)^-VC_1\underset{\sim}{Y} \\ &= E(\underset{\sim}{\varepsilon}'C_1V(VC_1V)^-VC_1\underset{\sim}{\varepsilon}), \text{ since } VC_1X = 0 \\ &= \operatorname{tr} C_1V(VC_1V)^-VC_1E(\underset{\sim}{\varepsilon}\underset{\sim}{\varepsilon}') \\ &= \sigma^2 \operatorname{tr} C_1V(VC_1V)^-VC_1V = \sigma^2 \operatorname{tr} C_1V. \end{aligned} \tag{3.3.3}$$

Now

$$\begin{aligned} \operatorname{tr} C_1V &= \operatorname{tr} C_1T = \operatorname{tr}(T^-T-T^-X(X'T^-X)^-X'T^-T) \\ &= \rho(T^-T)-\rho(T^-X(X'T^-X)^-X') \\ &= \rho(T)-\rho(X) = \rho(V:X)-\rho(X) \end{aligned} \tag{3.3.4}$$

observing that the trace and rank of an idempotent matrix are the same. Then an unbiased estimate of σ^2 is $\underset{\sim}{Y}'C_1\underset{\sim}{Y}/[\rho(V:X)-\rho(X)]$.

3.4 *The conditioned equation*

It is seen that the MDE of $X\underset{\sim}{\beta}$ is

$$X\hat{\underset{\sim}{\beta}} \text{ with } \hat{\underset{\sim}{\beta}} = C_2\underset{\sim}{Y} \tag{3.4.1}$$

and that of $\underset{\sim}{\varepsilon}$ is

$$\hat{\underset{\sim}{\varepsilon}} = V\hat{\underset{\sim}{\alpha}} \text{ with } \hat{\underset{\sim}{\alpha}} = C_1\underset{\sim}{Y}. \tag{3.4.2}$$

The form of $(\hat{\underset{\sim}{\alpha}},\hat{\underset{\sim}{\beta}})$ suggests that it is a solution of the equation

$$\begin{aligned} V\underset{\sim}{\alpha}+X\underset{\sim}{\beta} &= \underset{\sim}{Y} \\ X'\underset{\sim}{\alpha} &= \underset{\sim}{0}. \end{aligned} \tag{3.4.3}$$

Thus, in estimating the unknown signal $X\beta$ and noise ϵ in th model (3.1), we may start with the equation (3.4.3) in (α,β), which can be set up in terms of given V, X and Y without making any prior computations. [We call this the conditioned equation. Note that the first equation in (3.4.3) is just the model equation since ϵ can be written as $V\alpha$. The second equation in (3.4.3) is a condition on α.] The next step is to obtain a solution $(\hat{\alpha},\hat{\beta})$ of the conditioned equation (3.4.3) and estimate $X\beta$ by $X\hat{\beta}$, ϵ by $V\hat{\alpha}$ and σ^2 by $\hat{\alpha}'Y/[\rho(V:X)-\rho(X)]$.

We can go a step further and write down the equations in terms of ϵ and β, the two unknowns appearing in the model (3.1). To do this, we write the equation (3.4.3) in the form

$$\begin{aligned} T\alpha+X\beta &= Y \\ X'T^-T\alpha &= 0 \end{aligned} \tag{3.4.4}$$

where $T = V+XKX'$ as defined earlier. Now, replacing $T\alpha = V\alpha$ by ϵ, we have

$$\begin{aligned} \epsilon+X\beta &= Y \\ X'T^-\epsilon &= 0 \end{aligned} \tag{3.4.5}$$

for any choice of T^-. The estimates $\hat{\beta}$ and $\hat{\epsilon}$ given in (3.4.1) and (3.4.2) respectively are direct solutions of (3.4.5). The disadvantage of (3.4.5) over (3.4.3) is that it needs the prior computation of T^-.

4. MIXED EFFECTS MODEL

The mixed effects linear model is of the form

$$\underset{n\times 1}{Y} = \underset{n\times m}{X}\ \underset{m\times 1}{\beta} + \underset{n\times s}{U}\ \underset{s\times 1}{\xi} + \underset{n\times 1}{\epsilon} \tag{4.1}$$

with

$$E(\underset{\sim}{\epsilon}) = 0, \quad D(\underset{\sim}{\epsilon}) = V, \quad E(\underset{\sim}{\xi}) = 0, \quad D(\underset{\sim}{\xi}) = \Gamma, \quad \mathrm{cov}(\underset{\sim}{\epsilon},\underset{\sim}{\xi}) = 0. \tag{4.2}$$

The problem in this case is one of estimating the unknown elements $X\underset{\sim}{\beta}$, $U\underset{\sim}{\xi}$ and $\underset{\sim}{\epsilon}$ in the model, which are the fixed effects, random effects and random error respectively.

4.1 *Estimation of* $X\underset{\sim}{\beta}$ *(the fixed effect)*

Let $A'Y$ be an unbiased estimate of $X\underset{\sim}{\beta}$. Then $E(A'Y) = X\underset{\sim}{\beta}\ \forall\ \beta \Rightarrow$ $A'X = X'$, and

$$\begin{aligned} E[(A'Y-X\underset{\sim}{\beta})(A'Y-X\underset{\sim}{\beta})'] &= E[A'(U\underset{\sim}{\xi}+\underset{\sim}{\epsilon})(U\underset{\sim}{\xi}+\underset{\sim}{\epsilon})'A] \\ &= A'(U\Gamma U'+V)A = A'V_*A \end{aligned} \tag{4.1.1}$$

where $V_* = U\Gamma U'+V$. Then the problem is that of finding

$$\min_{X'A=X'} A'V_*A.$$

The optimum A is a solution of the equation

$$\begin{aligned} V_*A+XB &= 0 \\ X'A \quad &= X' \end{aligned}$$

which is of the same form as in the fixed effects case except that V_* takes the place of V. If

$$\begin{pmatrix} V_* & X \\ X' & 0 \end{pmatrix}^- = \begin{pmatrix} C_1 & C_2 \\ C_3 & -C_4 \end{pmatrix} \tag{4.1.2}$$

then

$$A_* = C_2X', \quad B_* = -C_4X' \tag{4.1.3}$$

giving the MDE of $X\beta$ as

$$A_*'Y = XC_2'Y \tag{4.1.4}$$

with the dispersion matrix

$$XC_4X' \tag{4.1.5}$$

4.2 *Prediction of* $U\xi$ *(the random effect)*

Let $A'Y$ be a predictor of $U\xi$ such that

$$E(A'Y-U\xi) = A'X\beta = 0 \Rightarrow A'X = 0.$$

Then

$$\begin{aligned} E(A'Y-U\xi)(A'Y-U\xi)' &= E[(A'-I)U\xi\xi'U'(A-I)] + E[A'\epsilon\epsilon'A] \\ &= (A'-I)U\Gamma U'(A-I)+A'VA \\ &= A'V_*A-A'U\Gamma U'-U\Gamma U'A+U\Gamma U'. \end{aligned} \tag{4.2.1}$$

The problem is to find

$$\min_{X'A=0} (A'V_*A-A'W-WA+W) \tag{4.2.2}$$

where $W = U\Gamma U'$. Applying the basic lemma of Section 2, the minimum is attained at a solution A_* of the equation

$$\begin{aligned} V_*A+XB &= W \\ X'A &= 0. \end{aligned} \tag{4.2.3}$$

In terms of the elements of the g-inverse (4.1.2), a solution is

$$A_* = C_1W, \quad B_* = C_2'W \tag{4.2.4}$$

giving the MDP of $U\underset{\sim}{\xi}$ as

$$A_*'\underset{\sim}{Y} = WC_1\underset{\sim}{Y} \tag{4.2.5}$$

with the dispersion of prediction error

$$-A_*'W+W = W-WC_1W. \tag{4.2.6}$$

4.3 *Estimation of* $\underset{\sim}{\epsilon}$

Let $A'\underset{\sim}{Y}$ be an estimate of ϵ such that

$$E(A'\underset{\sim}{Y}-\underset{\sim}{\epsilon}) = A'X\underset{\sim}{\beta} = \underset{\sim}{0} \Rightarrow A'X = \underset{\sim}{0}.$$

Then

$$\begin{aligned} E[(A'\underset{\sim}{Y}-\underset{\sim}{\epsilon})(A'Y-\underset{\sim}{\epsilon})'] &= E[A'U\underset{\sim}{\xi}\underset{\sim}{\xi}'U'A]+E[(A'-I)\underset{\sim}{\epsilon}\underset{\sim}{\epsilon}'(A-I)] \\ &= A'V_*A-A'V-VA+V. \end{aligned} \tag{4.3.1}$$

Proceeding as in Section 4.2, the optimum

$$A_* = C_1V \tag{4.3.2}$$

giving the MDP of $\underset{\sim}{\epsilon}$ as

$$A_*'\underset{\sim}{Y} = VC_1\underset{\sim}{Y} \tag{4.3.3}$$

with the dispersion of prediction error

$$-A_*'V+V = V-VC_1V. \tag{4.3.4}$$

The expressions (4.1.4), (4.2.4) and (4.3.3) for the estimators (predictors) of $X\underset{\sim}{\beta}$, $U\underset{\sim}{\xi}$ and $\underset{\sim}{\epsilon}$ suggest an alternative procedure of computing them through a conditioned equation as in the case of the fixed effects model. Consider the equation

$$
\begin{aligned}
V_* \underset{\sim}{\alpha} + X \underset{\sim}{\beta} &= \underset{\sim}{Y} \\
X' \underset{\sim}{\alpha} &= \underset{\sim}{0}
\end{aligned}
\tag{4.3.5}
$$

and solve for $(\underset{\sim}{\alpha}, \underset{\sim}{\beta})$. If $(\hat{\underset{\sim}{\alpha}}, \hat{\underset{\sim}{\beta}})$ is a solution, then the estimate of $X\underset{\sim}{\beta}$ is $X\hat{\underset{\sim}{\beta}}$, of $U\underset{\sim}{\xi}$ is $W\hat{\underset{\sim}{\alpha}}$ and that of $\underset{\sim}{\epsilon}$ is $V\hat{\underset{\sim}{\alpha}}$.

The estimators obtained in this section involve the matrices V and Γ which may not be known in practice. They cannot in general be estimated unless they have a simple structure involving a smaller number of parameters as in the case of variance components. In such a case, we may first estimate V and Γ by some method such as those described in the case of variance components by Rao and Kleffe (1988) and use them in the second stage for the estimation of $X\underset{\sim}{\beta}$, $U\underset{\sim}{\xi}$ and $\underset{\sim}{\epsilon}$.

For literature on the estimation of fixed and random effects on classical lines under some assumptions on the ranks of V and X, the reader is referred to Harville (1976) and Henderson (1984).

We discussed separately the estimation of $X\underset{\sim}{\beta}$, $U\underset{\sim}{\xi}$ and $\underset{\sim}{\epsilon}$. If we need an estimate of a joint function of the fixed and random effects and the random error such as

$$
P' X\underset{\sim}{\beta} + Q' U\underset{\sim}{\xi} + R' \underset{\sim}{\epsilon}
\tag{4.3.6}
$$

we need only substitute the separate estimates for the unknowns and obtain the estimate

$$
P' X\hat{\underset{\sim}{\beta}} + Q' U\hat{\underset{\sim}{\xi}} + R' \hat{\underset{\sim}{\epsilon}}
\tag{4.3.7}
$$

which is the MDE for (4.3.6) by virtue of what is proved in Section 1.5.

5. REFERENCES

1. Harville, D. A. (1976). Extension of Gauss-Markoff theorem to include the estimation of random effects. *Ann. of Statistics* 4, 348-395.
2. Henderson, C. R. (1984). *Application of Linear Models in Animal Breeding*. University of Guelph.
3. Rao, C. Radhakrishna (1971). Unified theory of linear estimation. *Sankhya* A 33, 370-396 and *Sankhya* A 34, 477.
4. Rao, C. Radhakrishna (1972). A note on the IPM method in the unified theory of linear estimation. *Sankhya* A 34, 285-288.

5. Rao, C. Radhakrishna (1973). *Linear Statistical Inference and Its Applications*, (Second Edition). Wiley, New York.
6. Rao, C. Radhakrishna (1974). Projectors, generalized inverses and BLUE's. *J. Roy. Statist. Soc.* 36, 442-448.
7. Rao, C. Radhakrishna (1985). A unified approach to inference from linear models, *Proc. First International Tampere Seminar on Linear Models*, (eds. T. Pukkila and S. Puntanen), 361 383.
8. Rao, C. Radhakrishna (1988). Estimation in linear models with mixed effects: a unified theory. *Proc. Second International Tampere Conference in Statistics*, (eds. T. Pukkila and S. Puntanen), 73-98.
9. Rao, C. Radhakrishna and Kleffe Jürgen (1988). *Estimation of Variance Components and Applications*. North Holland.
10. Rao, C. Radhakrishna (1987). A unified approach to estimation in linear models with fixed and mixed effects. Tech. Rept. No. 88-44, Center for Multivariate Analysis, Univ. of Pittsburgh.

STATISTICAL DATA ANALYSIS AND INFERENCE
Y. Dodge (Editor)
Elsevier Science Publishers B.V. (North-Holland), 1989

ROBUST LOWER RANK APPROXIMATION OF MATRICES

Gérard ANTILLE

Département d'Econométrie, Université de Genève
CH - 1211 Genève 4

Mathematical solution of the problem of reduced rank approximation with singular row and column-weights matrices is given.
An iteratively reweighted algorithm based on the previous result is proposed to construct low rank approximation of matrices resistant to spurious data. Examples are discussed.

1. INTRODUCTION

The problem of reduced rank approximation of matrices has arisen out of factorial studies in psychology. Solutions were given in Eckart and Young [1936], Householder and Young [1938] and methods of computation were developed in Hotelling [1933], [1936]. Earlier, Fisher and Mackenzie [1923], in a theory of fitting a product formula gave a first approach to the problem of matrix approximation.

Mirsky [1960] proposed a solution for any unitarily invariant norm; Maitre [1968] used the Gastinel-norms; and Bargmann and Baker [1977] the L_∞-norm. Professor Rao [1979] applied separation theorems for singular values of a matrix to problems of approximating a given random variable by a r.v. in a specified class. In Rao [1980] the emphasis is on approximating a given matrix by a member of a given class.

De Leeuw [1984] defined the approximation by using orthogonally invaviant norms in combination with row and column-weights which can be singular. In section 2 of that paper, we present a closed form to de Leeuw's result, and we give general forms of the best rank k approximation matrix.

The approximation of data matrices by matrices of lower rank is used in a large variety of methods in multivariate analysis as for instance in principal components analysis, in factor analysis, in canonical correlation analysis, in Procuste analysis, in the h-plot display, in fitting ANOVA models for non additive data. But all those techniques are usually constructed with unweighted least squares. Hence, if some elements of the data matrix lie far out from the general pattern they greatly influence the fit. So alternative methods resistant to outliers are desirable.

Weighted least squares for lower rank approximation has been studied in Gabriel and Zamir [1979]. But weighted least squares solutions, as classical ones, are not resistant to bad data unless fortuitous choice of weights reduces their influence. Iteratively reweighted least squares (IRLS) select weights as to reduce the influence of outliers, but problems of convergence can occur if the algorithms get off with a bad start. Gabriel and Odoroff [1984], in an approach analogous to M-estimation of multiple regression, proposed several methods based on IRLS algorithms using innately resistant methods as initialization processes.

In section 3 of that article, we propose an iteratively reweighted procedure for any orthogonally invariant norm such that the best rank approximation with row and column-weights is computed at each stage of the procedure, which is not the case for many IRLS algorithms. Two examples are discussed. In the first one, we choose a matrix given in Gabriel and Odoroff [1984], for which many IRLS procedures, for a rank one approximation, fit a matrix of zeroes. In the second example, we choose the Bradu55 data set (Hawkins & al [1984], Galpin [1986]) which contains 25% of outliers. The study of weights matrices allows us to detect aberrant values in rows or in columns.

The procedure proposed in this paper performs well, independently of the type of outliers, i.e. few very large values or several outliers in one row or column of the data matrix. Obviously, several fits should be computed with different tuning constants or different initialization matrices, hence entries of the data matrix with low weights in all those fits can be considered as outliers.

2. MATRIX APPROXIMATIONS

Let X be a $n \times m$ real matrix of rank $r(X) = r$. The additive form of the singular value decomposition (SVD) of X is defined as

$$X = \sum_{i=1}^{r} s_i \, \mathbf{g}_i \, \mathbf{d}_i'$$

where $s_1 \geq s_2 \geq \ldots s_r > 0$ are the singular values of X, $\{\mathbf{g}_1, \ldots \mathbf{g}_r\}$ and $\{\mathbf{d}_1, \ldots \mathbf{d}_r\}$ are orthonormal bases of respectively the column space of X and the row space of X.

As it is well known, the $\mathbf{g}_i$'s are eigenvectors of XX' with eigenvalues $s_1^2, \ldots, s_r^2$ and the $\mathbf{d}_i$'s are eigenvectors of $X'X$ with the same eigenvalues. If we denote by G and D the matrices whose columns are respectively the $\mathbf{g}_i$'s and the $\mathbf{d}_i$'s the multiplicative form of the SVD of X is given by $X = GSD'$. An extended form of the SVD is $X = \overline{G}\,\overline{S}\,\overline{D}'$ where $\overline{G}$ and $\overline{D}$ are square orthonormal matrices, $\overline{G} = \begin{bmatrix} G & G_o \end{bmatrix}$; $\overline{D} = \begin{bmatrix} D & D_o \end{bmatrix}$ and $\overline{S} = \begin{bmatrix} S & O \\ O & O \end{bmatrix}$.

Let us recall that a symetric gauge function is a real valued function ϕ, defined on $\mathbb{R}^n$, which satisfies the following conditions for $\mathbf{x}$ and $\mathbf{y} \in \mathbb{R}^n$

i) $\phi(\mathbf{x}) > 0$, $\mathbf{x} \neq 0$

ii) $\phi(c\mathbf{x}) = |c| \, \phi(\mathbf{x})$, $c \in \mathbb{R}$

iii) $\phi(\mathbf{x}+\mathbf{y}) \leq \phi(\mathbf{x}) + \phi(\mathbf{y})$

iv) $\phi(\mathbf{x}_\pi) = \phi(\mathbf{x})$

v) $\phi(J\mathbf{x}) = \phi(\mathbf{x})$

where J is any sign diagonal matrix and π is any permutation of the coordinates.

It is well known (Von Neumann [1937]) that any orthogonally invariant norm can be defined by $\|X\| = \phi(\mathbf{s})$, where $\mathbf{s}$ is the vector whose components are the ordered singular values of X. For instance $\phi(\mathbf{s}) = s_{max}$ is the euclidean norm, $\phi(\mathbf{s}) = (\sum s_i^2)^{1/2}$ is the Frobenius norm of X.

In the following part of that paper we consider orthogonally invariant norms denoted by $\|\cdot\|$. In that context, theorem 1 due to Mirsky [1960] gives a solution of the lower rank approximation problem :

Theorem 1 : Let $X = \sum^{r} s_i \, \mathbf{g}_i \, \mathbf{d}_i'$ be the SVD of X.

$$\inf_{\{Y | r(Y) \leq k\}} \|X - Y\| = \sum_{i=k+1}^{r} s_i^2$$

The minimum is attained if and only if $Y = \sum^{k} s_i \, \mathbf{g}_i \, \mathbf{d}_i'$

Theorem 2 due to Rao [1980] gives a solution with positive definite weights matrices :

Theorem 2 : Let M and N be positive definite matrices of order m and n.

Let $\sum^{\ell} t_i h_i e_i'$ the SVD of $M^{1/2}X^{N1/2}$

$$\inf_{\{Y|r(Y)\le k\}} \|M^{1/2}(X-Y)N^{1/2}\| = \|\sum_{i=k+1}^{\ell} t_i h_i e_i'\|$$

A generalization to singular row and column-weights matrices is given in de Leeuw [1984]. In theorem 3 and 4, we propose closed forms of de Leeuw's results.

Theorem 3 : Let $A = FTK'$ and $B = LVP'$ be the SVD of a pxn and a qxm matrix with $\min(r(A);r(B)) \ge k$

A solution of $\inf_{\{Y|r(Y) \le k} \|A(X - Y)B'\|$ is given by $\hat{Y} = KT^{-1}\hat{Z}V^{-1}P'$ where $\hat{Z}$ is Mirsky's best rank k approximation of T K'XPV .

Proof. By the orthogonality of F and L we have

$$\|A(X - Y)B'\| = \|TK'XPV - TK'YPV\|$$

$$=: \|U - Z\|$$

$$\inf_Y \|A(X - Y)\ B'\| = \inf_Z \|U - Z\|$$

Let $\hat{Z}$ be Mirsky's best rank approximation of U , hence $\hat{Y} = K\ T^{-1}\hat{Z}V^{-1}P'$ minimizes $\|A(X - Y)B'\|$.

Theorem 4 : Let $A = \bar{F}\ \bar{T}\ \bar{K}'$ and $B = \bar{L}\ \bar{V}\ \bar{P}'$ be the extended SVD of the matrices given in theorem 3, and let $\hat{Y}$ be the approximation defined in theorem 2. Forms of general solutions of $\inf\|A(X-Y)B'\|$, for any nxn matrix C preserving the rank k of the solution, are given by :

1) $\hat{Y} + K(C - I_k C I_k)\ \bar{P}'$;
   $I_k = (d_i)$ being the diagonal matrix whose elements are defined by $d_i = 1$ for $1 \le i \le k$, $d_i = 0$ for $k < i \le n$.

2) $\hat{Y} + (I - A^- A)\ C$;
   $\hat{Y} + C\ (I - B'B'^-)$;
   $\hat{Y} + (I - A^- A)C(I - B'B'^-)$;

   A^- being any generalized inverse of A such that $A = AA^- A$, idem for B'^- .

3) $\hat{Y} + Q_A C$; $\hat{Y} = CQ_B$; $\hat{Y} + Q_A C Q_B$;
   the columns of Q_A , respectively Q_B , forming a basis of the kernel of A respectively B'.

Proof. Let $\tilde{Y} = \hat{Y} + M$ be one of the general solutions :

$$\|A(X - \hat{Y})B'\| \le \|A(X - \tilde{Y})B'\| \le \|A(X - \hat{Y})B'\| + \|AMB'\|$$

As it is easily shown below, $\|AMB'\| = 0$, hence $\|A(X - \hat{Y})B'\| = \|A(X-\tilde{Y})B'\|$, so $\tilde{Y}$ is a best rank k approximation of X .

1) $M = \overline{K}(C - I_n\ C\ I_n)\ \overline{P}'$ so $K'\overline{K}(C - I_n\ C\ I_n)\ \overline{P}'P = 0$ then $\|AMB'\| = 0$;

2) $M = (I - A^{-}A)C$ so $A(I - A^{-}A)C = 0$, then $\|AMB'\| = 0$; similarly for the two other choices of M .

3) if $M = K_A C$ so $AK_A C = 0$, then $\|AMB'\| = 0$; similarly for the other choices of M .

3. ROBUST APPROXIMATIONS

Huber [1964] proposed to generalize maximum likelihood estimators in defining M-estimators as the solution of the minimization, over the parameter space Θ, of the objective function $\sum \rho(x_i ; \theta)$, where ρ is some function on $\chi \times \Theta$, χ being the sample space. If $\psi(x_i , \theta) = \frac{\partial}{\partial\theta}\rho(x_i ; \theta)$ exists, the estimate $\hat{\theta}$, of θ , satisfies the implicit equation $\sum\psi(x_i;0) = 0$.

A variant of M-estimators is given by fully iterated W-estimators, those being usually defined by the choice of $\psi(x;\theta) = x \cdot w(x_i;\theta)$, $w(x ; \theta)$ being a weight function. Hence starting the iterative procedure by $\hat{\theta}_o$, one has to compute $\hat{\theta}_{j+1} = \sum_i x_i w(x_i ; \hat{\theta}_j)$ until the sequence $\hat{\theta}_j$ converges.

An approach analogous to W-estimation for matrix approximations is given in Gabriel and Odoroff [1983]. Gabriel [1978] pointed out the equivalence between fitting a matrix of rank k or less and fitting a matrix product AB' of ranks k or less. Gabriel and Zamir [1979] used the objective function $\sum w_{ij}(x_{ij} - \mathbf{a}_i\ \mathbf{b}_j')^2$, where $\mathbf{a}_i$ and $\mathbf{b}_j$ are rows of A and B, for lower rank approximations by least squares with any choice of weights.

Let us now restrict our attention to objective functions based on orthogonally invariant norms. By theorems 2, 3 and 4, we have exact mathematical forms of lower rank matrix approximations with row and column-weights. For robustification purposes we have to define the weights from a robust fit to initialize an iteratively reweighted procedure for a gauge function ϕ ; such a process will be called an IRϕ-algorithm. Then at the jth iteration, the results of theorem 2, 3, or 4 can be applied with weight matrices based on the (j-1)th iteration.

The IRϕ-algorithm has to be initialized by an approximation that is insensitive to aberrant observations otherwise outliers could have rather small residuals and would not be downweighted in later iterations. Methods based on medians, fractiles or trimmed means are known to be resistant to outlying values. Tukey's median polish [1977] is used to initialize the algorithm and provides R°, the residual matrix used to compute weight matrices to be implemented in the IRϕ-algorithm.

Weights are given by Tukey's bisquare function :

$$w^{\ell}_{ij} = \begin{cases} [1 - (r^{\ell}_{ij}/T)^2]^2 & \text{if } |r_{ij}| \le T \\ 0 & \text{if } |r_{ij}| > T \end{cases}$$

where r^{ℓ}_{ij} is the scaled residual at the ℓ-th iteration. According to the row or column-weights matrix under construction, the scale parameter is the mean of the absolute row or column residuals at the $(\ell-1)$th iteration. IRϕ should be performed several times with different values of T, the tuning constant; this allows us to detect different types of outliers, i.e. elements far away from the other entries of the data matrix, or rows and columns with several aberrant observations.

The matrix A given in table 1, whose unweighted rank one approximation is given in table 2, has been studied in Gabriel and Odoroff [1983]

Table 1

5	-1	0	-2	-2
-7	0	8	0	-1
4	1	0	-5	0
8	-3	2	-7	0
4	1	0	-5	0
9	-4	-3	0	-2
10	-5	0	-4	-1

Table 3

4	-1	0	-2	0
0	0	0	0	0
4	-1	0	-2	0
8	-2	0	-4	0
4	-1	0	-2	0
0	0	0	0	0
8	-2	0	-4	0

Table 2

4.8016	-1.4549	-0.7766	-2.1583	-0.4119
-6.1837	1.8736	1.0001	2.7796	0.5304
4.4784	-1.3569	-0.7243	-2.0130	-0.3841
8.8385	-2.6781	-1.4295	-3.9729	-0.7581
4.4784	-1.3569	-0.7243	-2.0130	-0.3841
8.1882	-2.4810	-1.3243	-3.6806	-0.7024
10.0942	-3.0585	-1.6325	-4.5373	-0.8658

For the rank one approximation of that matrix some IRLS procedures fit a matrix of zeroes, some other do not converge. A procedure proposed by Siegel in a personal communication to Gabriel and Odoroff provides the rank one approximation given in table 3. The rank one approximation computed with the IRϕ-algorithm with $t = 4$ is in table 4.

Table 4

6.2091	-1.8812	0.7039	-4.0595	-0.9723
-3.4583	1.0478	-0.3920	2.2610	0.5415
6.5218	-1.9759	0.7393	-4.2640	-1.0212
6.4822	-1.9639	0.7348	-4.2380	-1.0150
6.5218	-1.9759	0.7393	-4.2640	-0.0212
5.7796	-1.7511	0.6552	-3.7787	-0.9050
6.1485	-1.8628	0.6970	-4.0199	-0.9628

The study of the corresponding weights matrices has pointed out the (2 ; 3) entry of the matrix A as an aberrant value, whatever the tuning value in the interval [2 , 8] . For the rank 2 approximation with $T = 2$ entries (2 ; 1), (2 ; 3), (4 ; 4), 6 ; 4) are column-outliers and (2 ; 3), (7 ; 1), (7 , 2) are row-outliers.

The second matrix studied in this article is the Bradu55 data set (Hawkins & al [1984], Galpin [1986]). We choose these data to examine the ability of the IRϕ-algorithm to deal with a data set near the breakdown point for regression procedure. By construction the first 10 entries of column 1 and the first 14 entries of colums 2, 3, 4 are outliers. For the rank 1 approximation with $T = 2$, the column-aberrant values pointed out by the weights matrix of the last stage of the IRϕ values are entries 1 to 10 of the first column and entries 2, 3, 5, 6, 7, 9,

11, 12, 13, 14 of the 2nd, 3rd and 4th column. Entries 1, 3, 8 and 10 of the 2nd and 4th column get weights between .4 and .8; weights for other entries of the matrix being larger than .9. So we consider that in those cases, the IRϕ-algorithm performs correctly. Moreover, the IRϕ-rank one and rank two approximations do not breakdown.

4. CONCLUSION

Reduced rank appproximation has been widely studied and main results are due to Mirsky [1960], Rao [1980] and de Leeuw [1984]. Our contribution given in theorem 3 and 4 provides an extension of Professor Rao's result for approximations with singular row or column-weights matrices.

In section 3, we propose an algorithm to compute robust low rank approximations. The main difference or improvement, with respect to IRLS procedures, is based on the computation at each stage of the procedure of the best rank approximation with row and column-weights matrices.

NOTE:
All the computations have been performed with a MATLAB version for personal computer.

REFERENCES

Bargmann, R.E. and F.D. Baker [1977], **A Minimax Approach to Component Analysis**, in P.R. Krisnaiah (ed.), Applications of Statistics, Amsterdam, North-Holland Publishing Company.

Eckart, C. and G. Young [1936], **The Approximation of one Matrix by another of Lower Rank**, *Psychometrica*, 1, pp. 211-218.

Fisher, R.A. and W.A. MacKenzie [1923], **Studies in Crop Variation**, *Journal of Agricultural Sciences*, 13, pp. 311-320.

Gabriel, K.R. [1978], **Least Squares Approximation of Matrices by Additive and Multiplicative Models**, *Journal of the Royal Statistical Society B*, 40, pp. 186-196.

Gabriel, K.R. and S. Zamir [1979], **Lower Rank Approximation of Matrices by Least Squares with any Choice of Weight**, *Technometrics*, 21, pp. 489-498.

Gabriel, K.R. and C.L. Odoroff [1984], **Resistant Lower Rank Approximation of Matrices**, Department of Statistics and Division of Biostatistics, The University of Rochester, Technical Report 83/02, Rochester, New York.

Galpin, J.S. [1986], **Robust and Bounded Influence Regression**, National Research Institute for Mathematical Sciences, Technical Report, Pretoria, pp- 135-257.

Hawkins, D.M., D. Bradu and G.V. Kass [1984], **Location of Several Outliers in Multiple Regression Data using Elemental Sets**, *Technometrics*, 26, pp. 197-208.

Hotelling, H. [1933], **Analysis of a Complex of Statistical Variables into Principal Components**, *Journal of Educational Psychology*, 24, pp. 417-441, 498-236.

Hotelling, H. [1936], **Simplified Calculation for Principal Components**, *Psychometrica*, 1, pp. 27-35.

Householder, A.S. and G. Young [1938], **Matrix Approximation and Latent Roots**, *American Mathematical Monthly*, 45, pp. 165-171.

Huber, P.J. [1964], **Robust Estimation of a Location Parameter**, *Annals of Mathematical Statistics*, 35, pp. 73-101.

Leeuw (de), J. [1984], **Fixed Rank Matrix Approximation with Singular Weights Matrices**, *Computational Statistics Quarterly*, 1, pp. 3-12.

Maitre, J.F. [1968], **Approximation de Rang donné dans un espaxe de matrices**, in Programmation en Mathématiques Numériques, Colloque International du CNRS, 165, Paris, Editions CNRS.

Mirsky, L. [1960], **Symmetric Gauge Functions and Unitary Invariant Norms**, *Quarterly Journal of Mathematics*, 11, pp. 50-59.

Neumann (von), J. [1937], **Some Matrix Inequalities and Metrization of Metric Spaces**, *Univ. Rev 1*, Tomsk, pp. 286-299.

Rao, C.R. [1979], **Separation Theorems for Singular Values of Matrices and their Applications in multivariate analysis**, *Journal of Multivariate Analysis*, 9, pp. 362-377.

Rao, C.R. [1980], **Matrix Approximations and Reduction of Dimensionality in Multivariate Statistical Analysis**, in P.R. Krisnaiah (ed.) Multivariate Analysis V, Amsterdam, North-Holland Publishing Company.

Tukey, J.W. [1977], **Exploratory Data Analysis**, Reading, Mass., Addison-Wesley.

STATISTICAL DATA ANALYSIS AND INFERENCE
Y. Dodge (Editor)

BLOCK INDEPENDENCE IN GENERALIZED INVERSE: A COORDINATE FREE LOOK

SUJIT KUMAR MITRA

Indian Statistical Institute
New Delhi-110016, INDIA

Block independence in generalized inverses was noticed by Hall. The concept was defined in a natural way. It was studied subsequently by Hall and Meyer and by Rao and Yanai with respect to the entire class of generalized inverses of the matrix. Apart from providing a coordinate free description of this phenomenon the study is now extended to other subclasses of generalized inverses. Certain other interesting decompositions of a similar kind of the class of generalized inverses of the matrix or its subclasses are also recorded.

1. INTRODUCTION

It was shown by Rao [20] that the conformable partitioned form

$$\begin{pmatrix} C_{11} & C_{12} \\ C_{21} & -C_{22} \end{pmatrix} \tag{1}$$

of the generalized inverse of a bordered matrix

$$F = \begin{pmatrix} V & X \\ X' & 0 \end{pmatrix} \tag{2}$$

contains all the necessary ingredients for the computations required in the conventional statistical analysis of a general Gauss-Markov model $(Y, X\beta, \sigma^2 V)$ with a possibly singular covariance matrix $\sigma^2 V$. Khatri [9] and Rao [20] have derived interesting properties of the various blocks C_{ij} of a g-inverse of F. Hall and Meyer [6] noted that the different blocks C_{ij} can be independently determined. The concept of block independence in the present context will shortly be made more precise in the sequel. The author [12] showed that the properties mentioned by Khatri [9] and Rao [20] are indeed true for a more general bordered matrix of the type

$$F = \begin{pmatrix} A & X \\ Y & 0 \end{pmatrix} \tag{3}$$

obeying a rank additivity condition

$$\text{Rank } F = \text{Rank } (A \quad X) + \text{Rank } (Y \quad 0) = \text{Rank} \begin{pmatrix} A \\ Y \end{pmatrix} + \text{Rank} \begin{pmatrix} X \\ 0 \end{pmatrix} \tag{4}$$

Rao and Yanai [22] have shown that the block independence property noted by Hall [5], Hall and Meyer [6] is true for a more general matrix of the type

$$V = \begin{pmatrix} V_{11} & V_{12} \\ V_{21} & V_{22} \end{pmatrix} \tag{5}$$

if and only if V obeys a condition similar to (4) namely

$$\text{Rank } V = \text{Rank}(V_{11} \quad V_{12}) + \text{Rank}(V_{21} \quad V_{22}) = \text{Rank}\begin{pmatrix} V_{11} \\ V_{21} \end{pmatrix} + \text{Rank}\begin{pmatrix} V_{12} \\ V_{22} \end{pmatrix} \tag{6}$$

The object of the present paper is to describe the block independence property in a more general setting and to indicate the role played by generalized inverse invariance in this context [see Lemma 1.1 below and also Carlson [2]). We first introduce some notations and terminology.

Matrices are denoted by capital lecters, column vectors by lower case letters. For a matrix A, $M(A)$, $N(A)$ and A' denote the column span, null space and transpose of A. F^n denotes the vector space of n-tuples and $F^{m\times n}$ the vector space of matrices defined on a field F. C represents the field of complex numbers. For a complex matrix A, A* denotes its complex conjugate transpose. $B = A \oplus (B-A)$ means

$$\text{Rank } B = \text{Rank } A + \text{Rank } (B-A) \tag{7}$$

and is read as A and (B-A) are disjoint. A^- denotes a generalized inverse (g-inverse) of A, that is a solution G of the matrix equation AGA = A. The reflexive g-inverse A_r^- is a solution G of the pair of equations AGA=A, GAG=G. For a square matrix A, A_R^- denotes a g-inverse of A satisfying the additional condition $M(G') \subset M(A')$. A_C^- is similarly defined through the conditions AGA=A, $M(G) \subset M(A)$. A g-inverse of A which is both A_R^- and A_C^- is denoted by A_{RC}^-. A_R^-, A_C^- and A_{RC}^- exist if and only if the matrix A has index 1 that is Rank A^2=Rank A [11]. The notations A_R^-, A_C^- and A_{RC}^- were changed respectively to A_ρ^-, A_χ^- and $A_{\rho\chi}^-$ in [21]. $A_{RC}^-(A_{\rho\chi}^-)$ is also known as the group inverse of A and denoted by $A^\#$ [4]. A g-inverse of A which commutes with A is denoted by A_{com}^-. A_{com}^- exists, if and only if the matrix A is of index 1. For a complex matrix A, a minimum norm g-inverse A_m^- is a matrix G that satisfies the pair of equations AGA=A, (GA)*=GA. A least-squares g-inverse A_ℓ^- is similarly defined through the equations AGA=A, (AG)*=AG. The Moore-Penrose inverse A^+ is the unique solution G of the simultaneous matrix equations

$$AGA=A, \quad GAG=G, \quad (AG)^*=AG, \quad (GA)^*=GA.$$

$\{A^-\}$ represents the class of all g-inverses of A. $\{A_m^-\}$, $\{A_\ell^-\}$ etc. are similarly interpreted. A_{mr}^-, $A_{\ell r}^-$, $A_{\ell m}^-$ are defined as follows:

$$\{A_{mr}^-\} = \{A_m^-\} \cap \{A_r^-\}, \{A_{\ell r}^-\} = \{A_\ell^-\} \cap \{A_r^-\}, \{A_{\ell m}^-\} = \{A_\ell^-\} \cap \{A_m^-\}.$$

The sum of two subsets S and T of a vector space is defined in the usual way. Thus if $S \subset F^{n\times m}$, $T \subset F^{n\times m}$

$$S + T = \{X+Y : X \in S,\ Y \in T\}$$

For a fixed matrix $A \in F^{m\times n}$, the product SAT is interpreted as follows

$$SAT = \{XAY : X \in S,\ Y \in T\}$$

Similar interpretation hold for the products PS and SQ when P and Q are fixed matrices. It is to be noted that it is this definition of the sum of S and T which captures the essence of 'independence' in the coordinate free treatment of 'block independence' given in the next section. Blocking is taken care of through projections. The following lemma is virtually contained in [21, p. 43].

A proof which is valid for any field appears in [18].

Lemma 1.1. *For nonnull matrices* A *and* B, AC^-B *is invariant under the choice of the* g-*inverse if and only if*

$$M(B) \subset M(C), \quad M(A') \subset M(C') \tag{8}$$

Lemma 1.2 is the well known duality theorem ([21], Theorem 3.2.4).

Lemma 1.2. *For a complex matrix* A.

$$\{A^-_m\} = \{[(A^*)^-_\ell]^*\} \tag{9}$$

Lemma 1.3 appears as Theorem 5.3 in [11]. The proof given in [11] is seen to be valid for any field.

Lemma 1.3. *Let* $A \in F^{m\times m}$ *and* A *be of index 1, then* $A^\#$ *is a polynomial in* A *with scalar coefficients.*

Definition [21]: A pair of matrices A and B are said to be *parallel summable* (p.s.) if $A(A+B)^-B$ is invariant under the choice of the g-inverse $(A+B)^-$.

When A and B are p.s., $A(A+B)^-B$ is called the parallel sum of A and B and is denoted by the symbol P(A,B).

The space preordering $A \overset{s}{<} B$ is defined as follows (cf. Marshall and Olkin [10], p. 13). We write

$$A \overset{s}{<} B \text{ if } M(A) \subset M(B),\ M(A') \subset M(B') \tag{10}$$

In a star semigroup with a proper involution (denoted by *), Drazin [3] introduced the concept of a star order which in the context of complex matrices of order m×n could be stated thus. We write

$$A \overset{*}{<} B \text{ if } BA^* = AA^*,\ A^*B = A^*A \tag{11}$$

It was shown that (11) is equivalent to the following definition

$$A \overset{*}{<} B \text{ if } BA^+ = AA^+,\ A^+B = A^+A \tag{12}$$

Since then starting from the minus order of Hartwig [7] and Nambooripad [19] a host of other partial orders have been introduced through various classes of g-inverse of A [14, 16]. A broad definition seems to be as follows. For a set S, let the collection of all subsets of S be denoted by $P(S)$.

Let G be a map from $F^{m\times n}$ to $P(F^{n\times m})$

$$A \xrightarrow[G]{} G_A \tag{13}$$

which maps every matrix A in $F^{m\times n}$ to a specified nonempty subset G_A of g-inverses of A. We write

$$A \overset{G}{<} B \text{ if } BA^-=AA^-,\ A^-B=A^-A \text{ for some } A^- \in G_A \tag{14}$$

The minus order ($\bar{<}$) corresponds to the choice $G_A = \{A^-\}$ and the star order ($\overset{*}{<}$) to the choice G_A = the single member set $\{A^+\}$. Since the minus order uses the largest class, we have

$$A \overset{G}{<} B \Longrightarrow A \bar{<} B \tag{15}$$

In fact if the maps G_1 and G_2 are such that $G_{2A} \subset G_{1A}$ $\forall$ A, then

$$A \overset{G_2}{<} B \Longrightarrow A \overset{G_1}{<} B \tag{16}$$

Lemma 1.4. *The following statements are equivalent*

(a) $A \bar{<} B$

(b) $B = A \oplus (B-A)$

(c) A and B-A are p.s. and $P(A,B-A) = 0$

(d) $\{B^-\} \subset \{A^-\}$

(e) $A = BP = QB$, where P and Q are projectors

(f) $A^-B = A^-A$ for some A^- and $M(A) \subset M(B)$

(g) $BA^- = AA^-$ for some A^- and $M(A') \subset M(B')$.

Equivalence of (a) and (b) is due to Hartwig [7,8], that of (a) with (c) or (d) is due to the author [13]. Sambamurty [23] established the equivalence of (d) and (e) and Baksalary and Mitra [1] that of (a) with (f) or (g).

Lemma 1.5 appears as Theorem 5.1 in [15].

Lemma 1.5. *Let* $A \bar{<} B$. *Then* $\{B^-AB^-,\ B^- \textit{ arbitrary}\} = \{A_r^-:\ BA_r^-=AA_r^-,\ A_r^-B = A_r^-A\}$.

For a subset G_A of $\{A^-\}$ define the subset $\tilde{G}_A$ as follows

$$\tilde{G}_A = \{G_o : AG_o = AG,\ G_oA = GA \text{ for some } G \in G_A\} \tag{17}$$

The set G_A is said to be *complete* if

$$G_A = \tilde{G}_A \tag{18}$$

Otherwise $\tilde{G}_A$ is called the *completion* of G_A. G_A is *semi-complete* (complete with respect to the reflexive g-inverses) if

$$G_A \cap \{A_r^-\} = \tilde{G}_A \cap \{A_r^-\} \tag{19}$$

Definition (*complete* and *semi-complete maps* [17]). A map G as in (13) is said to be *complete* (*semi-complete*) according as, for every matrix A, the image G_A is complete (semi-complete).

Put

$$G_{A|B} = \{GAG : G \in G_B\} \tag{20}$$

We define the relation $\underset{G}{<}$ as follows

We write $A \underset{G}{<} B$ if

$$A \overset{-}{<} B \text{ and } G_{A|B} \subset G_A \tag{21}$$

The relation $A \overset{G}{<} B$ as defined in (14), imitating the definition of the star order by Drazin [3], need not always correspond to a partial order unlike the relation $A \underset{G}{<} B$ which (if the comparability condition (21) is nonvacuous) is always so (see [17], Theorem 2.2). The following theorem which combines Theorems 2.1 and 2.3 in [17] gives a sufficient condition under which (14) defines a partial order

Theorem 1.1. *Let* $A \overset{G}{<} B \Longrightarrow G_{A|B} \subset G_A$. *Then*

(a) (14) *defines a partial order on* $F^{m\times n}$

(b) $A \overset{G}{<} B \Longleftrightarrow A \underset{G}{<} B$

We reproduce the following theorem from [17]

Theorem 1.2. *When the map* G *is semicomplete the following statements are equivalent*

(a) $A \underset{G}{<} B$

(b) $\tilde{G}_B \subset \tilde{G}_A$, $A \overset{s}{<} B$ (22)

Table 1 also reproduced from [17] lists the seven known partial orders and shows where each fits in the unified theory. Though the original definitions were based on condition (14) each could be equivalently defined through condition (21) on account of Theorem 1.1 and in fact also by (22) since the maps are semi-complete.

Table 1 : Seven known partial orders($\overset{G}{<}$)

Sl. No.	Name of the partial order	Notation	G_A	$\tilde{G}_A$
1	minus	$\overset{-}{<}$	$\{A^-\}$	$\{A^-\}$
2	minus ρ	$\overset{-}{<}_\rho$	$\{A^-_\rho\}$	$\{A^-_\rho + (I-A^-_\rho A)\ U\ (I-AA^-_\rho), U \text{ arb.}\}$
3	minus χ	$\overset{-}{<}_\chi$	$\{A^-_\chi\}$	$\{A^-_\chi + (I-A^-_\chi A)\ U\ (I-AA^-_\chi), U \text{ arb.}\}$
4	Sharp	$\overset{\#}{<}$	$\{A^\#\}$	$\{A^-_{com}\}$
5	minus m	$\overset{-}{<}_m$	$\{A^-_m\}$	$\{A^-_m\}$
6	minus ℓ	$\overset{-}{<}_\ell$	$\{A^-_\ell\}$	$\{A^-_\ell\}$
7	Star	$\overset{*}{<}$	$\{A^+\}$	$\{A^-_{\ell m}\}$

Note:

1. The first four partial orders are for any field, the last three require a complex field.
2. Serial No. 2,3,4 are for square matrices and for comparability, both the matrices have to be of index 1.
3. In Serial No. 1,4,5,6 and 7 the condition $A \overset{s}{<} B$ follows from the condition $\tilde{G}_B \subset \tilde{G}_A$. Hence each of these orders could be defined by the single equivalent condition $\tilde{G}_B \subset \tilde{G}_A$ (see [13] for ser.No.1, [14] for 4, [16] for 5 and 6 and [1] for 7).
4. In serial No. 2 and 3 the condition $A \overset{s}{<} B$ is necessary in addition to $\tilde{G}_B \subset \tilde{G}_A$. For the pair of matrices $A = \begin{pmatrix} 1 & 0 \\ 0 & 0 \end{pmatrix}$, $B = \begin{pmatrix} 1 & 0 \\ 1 & 0 \end{pmatrix}$ for example

$$\{A^-_\rho\} = \{B^-_\rho\} = \{\begin{pmatrix} 1 & 0 \\ x & 0 \end{pmatrix}, x \text{ arb.}\}$$

$$\widetilde{\{A^-_\rho\}} = \widetilde{\{B^-_\rho\}} = \{\begin{pmatrix} 1 & 0 \\ u & v \end{pmatrix}, u,v \text{ arb.}\}.$$ (See [16] for Ser.No.2 and 3)

2. BLOCK INDEPENDENCE IN g-INVERSE : THE COORDINATE FREE WAY

Let $B_i \in F^{m_i \times n}$, i = 1,2, and B be the partitioned matrix

$$B = \begin{pmatrix} B_1 \\ B_2 \end{pmatrix} \tag{23}$$

$B \in F^{m\times n}$, where $m = m_1+m_2$. Let $G \in \{B^-\}$ and G be conformably partitioned as

$$G = (G_1 : G_2) \tag{24}$$

where $G_i \in F^{n\times m_i}$, $i = 1,2$.

Definition [5]. The g-inverses of B are said to have the *block independence property* if for distinct choice $G^{(1)}$, $G^{(2)}$ of G

$$(G_1^{(1)} \vdots G_2^{(2)}) \in \{B^-\} \tag{25}$$

There is an obvious extension of this definition for a matrix B partitioned columnwise. We shall take this definition for granted even though we may not have stated the same explicitly. Such simple partitions were not considered by Hall [5] or by any other author. Nevertheless we ascribe the definition to him, keeping in view the spirit in which the definition was made

Theorem 2.1. *The following statements are equivalent*

(a) *Rank* B = *Rank* B_1 + *Rank* B_2

(b) $\begin{pmatrix} B_1 \\ 0 \end{pmatrix} \bar{<} B$

(c) *The* g-*inverses of* B *have the block independence property*

We shall state and prove a more general version of Theorem 2.1.

Theorem 2.2. *Let* Q *be a projector in* F^m. *Then the following statements are equivalent*

(a) $QB \bar{<} B$ (26)

(b) $\{B^-\} = \{B^-Q\} + \{B^-(I-Q)\}$, (27)

$\{B^-Q\} \subset \{(QB)^-\}$, $\{B^-(I-Q)\} \subset \{[(I-Q)B]^-\}$ (28)

Proof: To avoid trivially assume that $QB \neq 0$, $(I-Q)B \neq 0$. Let us write A for QB, so that $B-A = (I-Q)B$

(a) $\Longrightarrow$ (b). Note that by lemma 1.4

$$A \bar{<} B \Longrightarrow (B-A) \bar{<} B$$

Hence by the same lemma once again

$$BB^-A = A, \quad BB^-(B-A) = B-A$$

If $(B^-)_i \in \{B^-\}$, $i = 1,2$ we have therefore

$$B[(B^-)_1Q+(B^-)_2(I-Q)]B = B(B^-)_1A+B(B^-)_2(B-A)$$
$$= A + (B-A) = B$$

Further on account of Lemma 1.4

$$AB^-QA=AB^-A=A, \quad (B-A)B^-(I-Q)(B-A)=(B-A)B^-(B-A)=B-A$$

(b) $\Longrightarrow$ (a). Let $(B^-)_i \in \{B^-\}$, $i = 1,2$, (b) $\Longrightarrow$

$$B[(B^-)_1Q+(B^-)_2(I-Q)]B = B[(B^-)_1A+(B^-)_2(B-A)] = B$$

Noting that $B(B^-)_2A + B(B^-)_2(B-A) = B(B^-)_2B = B$, we have

$$B(B^-)_1A = B(B^-)_2A$$

which implies that BB^-A is invariant under the choice of B^-, which in turn in view of Lemma 1.1 leads to the conclusion

$$\mathcal{M}(A) = \mathcal{M}(QB) \subset \mathcal{M}(B) \Longrightarrow QB \overset{-}{<} B$$

by Lemma 1.4. □

Remark 1: In Theorem 2.2, Q and B are linked only through the condition (26) (27) and (28) but could be otherwise arbitrary. For any such arbitrary pair Q,B a decomposition of $\{B_r^-\}$ of the type implied by (27) and (28) may not hold. One could state this as a formal theorem and prove it but the reward is hardly worth the efforts. We shall be content with a counter example. Consider the following complex matrices

$$B = \begin{pmatrix} 1 & 0 \\ 0 & 0 \\ 0 & 1 \end{pmatrix} \quad , \; Q = \text{diag}\,(1,\,1,\,0)$$

For arbitrary complex numbers x and y

$$\begin{pmatrix} 1 & x & 0 \\ 0 & y & 1 \end{pmatrix} \quad \varepsilon\ \{B_r^-\} = \{B^-\}$$

However,

$$\begin{pmatrix} 1 & x & 0 \\ 0 & y & 0 \end{pmatrix} = \begin{pmatrix} 1 & x & 0 \\ 0 & y & 1 \end{pmatrix} Q \notin \{(QB)_r^-\} \text{ unless } y=0$$

since Rank QB = 1.

With suitable restrictions on the projector, the decomposition is valid for other classes of g-inverses, for example for square matrices B of index 1 the completions $\widetilde{\{B_\rho^-\}}$, $\widetilde{\{B_\chi^-\}}$ and $\{B_{com}^-\}$ and for complex matrices B the classes $\{B_m^-\}$, $\{B_\ell^-\}$ and $\{B_{\ell m}^-\}$ all complete classes.

Theorem 2.3. *Let* Q *be a projector and the matrix* B, *of index 1, commute with* Q. *Then*

$$\text{(a)}\ \widetilde{\{B_\rho^-\}} = \widetilde{\{B_\rho^-\}}Q + \widetilde{\{B_\rho^-\}}\,(I-Q) \tag{29}$$

$$\widetilde{\{B_\rho^-\}}Q \subset \widetilde{\{(QB)_\rho^-\}}\ ,\ \widetilde{\{B_\rho^-\}}(I-Q) \subset \widetilde{\{[(I-Q)B]_\rho^-\}} \tag{30}$$

$$\text{(b)}\ \widetilde{\{B_\chi^-\}} = Q\widetilde{\{B_\chi^-\}} + (I-Q)\widetilde{\{B_\chi^-\}} \tag{31}$$

$$Q\widetilde{\{B_\chi^-\}} \subset \widetilde{\{(BQ)_\chi^-\}},\ (I-Q)\widetilde{\{B_\chi^-\}} \subset \widetilde{\{[B(I-Q)]_\chi^-\}} \tag{32}$$

$$\text{(c)}\ \{B_{com}^-\} = \{B_{com}^-Q\} + \{B_{com}^-(I-Q)\} \tag{33}$$

$$\{B_{com}^-Q\} \subset \{(QB)_{com}^-\}, \{B_{com}^-(I-Q)\} \subset \{[(I-Q)B]_{com}^-\} \tag{34}$$

Proof: We shall prove the (a) part. The proof of other parts are similar. First we show that QB = BQ is of index 1.

$$(QB)^2 = QBQB = Q^2B^2 = QB^2$$

$\text{Rank } B^2 = \text{Rank } B \Longrightarrow M(B^2) = M(B) \Longrightarrow M[(QB)^2] = M(QB^2) = M(QB) \Longrightarrow$

$$\text{Rank } (QB)^2 = \text{Rank } QB.$$

Check that by Lemma 1.4, $QB = BQ \overset{-}{<} B$. Further

$$QBQ(QB^{\#}Q)QBQ = QBQB^{\#}QBQ = Q^2BB^{\#}BQ^2 = QBQ$$

Trivially $M(QB^{\#}Q) \subset M(QB) = M(QBQ)$

$$M[(QB^{\#}Q)'] \subset M(Q'B') = M[(QBQ)']$$

These show that $QB^{\#}Q = (QBQ)^{\#} = (QB)^{\#} = (BQ)^{\#}$

Since $\{\widetilde{B_\rho^-}\}Q \subset \{B^-Q\}, \{\widetilde{B_\rho^-}\}(I-Q) \subset \{B^-(I-Q)\}$, by Theorem 2.2

$$\{\widetilde{B_\rho^-}\}Q + \{\widetilde{B_\rho^-}\}(I-Q) \subset \{B^-Q\} + \{B^-(I-Q)\} = \{B^-\}.$$

Also Theorem 2.2 $\Longrightarrow$

$$\{\widetilde{B_\rho^-}\}Q \subset \{(QB)^-\}$$

which in view of

$$QB\{\widetilde{B_\rho^-}\}Q = \{QBB^{\#}Q\} = \{QBQB^{\#}Q\} = \{QB(QB)^{\#}\}$$

implies that

$$\{\widetilde{B_\rho^-}\}Q \subset \{\widetilde{(QB)_\rho^-}\}$$

Similarly $\qquad \{\widetilde{B_\rho^-}\}(I-Q) \subset \{\widetilde{[(I-Q)B]_\rho^-}\}$

Let $G_i \in \{\widetilde{B_\rho^-}\}$ $i = 1,2$, then $B[G_1Q+G_2(I-Q)] = BB^{\#}(Q+I-Q) = BB^{\#}$

This shows

$$\{\widetilde{B_\rho^-}\}Q + \{\widetilde{B_\rho^-}\}(I-Q) = \{\widetilde{B_\rho^-}\}$$

since $\quad G_1Q + G_1(I-Q) = G_1.$ □

Remark 2. The (b) part of Theorem 2.3 follows from the (a) part noting that the premises of the theorem imply that Q' is a projector, B' also of index 1 commutes with Q' and $Q'B' \overset{-}{<} B'$ and that $[(B')_\rho^-]' \in \{B_\chi^-\}$. For the (c) part one may have to use Lemma 1.3. This will imply that Q commutes with $B^{\#}$ as well.

Theorem 2.4. *Let* Q *be an orthogonal projector in* C^m *and* $B \in C^{m\times n}$. *Then*
$QB \overset{-}{<} B \Longrightarrow \{B_\ell^-\} = \{B_\ell^-Q\} + \{B_\ell^-(I-Q)\},\ \{B_\ell^-Q\} \subset \{(QB)_\ell^-\}, \{B_\ell^-(I-Q)\} \subset \{[(I-Q)B]_\ell^-\}$

Proof: As in the proof of Theorem 2.3 we first establish using Theorem 2.2 that

$$\{B_\ell^-Q\} + \{B_\ell^-(I-Q)\} \subset \{B^-\}, \{B_\ell^-Q\} \subset \{(QB)^-\}, \{B_\ell^-(I-Q)\} \subset \{[(I-Q)B]^-\}.$$

Further QBB_ℓ^-Q and $(I-Q)BB_\ell^-(I-Q)$ are both hermitian. Hence

$$\{B_\ell^-Q\} \subset \{(QB)_\ell^-\},\ \{B_\ell^-(I-Q)\} \subset \{[(I-Q)B]_\ell^-\}$$

Rest of the proof follows as in that of Theorem 2.3 noting that $BB_\ell^- = BB^+$.

□

The following theorem can be proved on similar lines or can be derived from Theorem 2.4 using Lemma 1.2.

Theorem 2.5. *Let* P *be an orthogonal projector in* C^n *and* $B \in C^{m\times n}$. *Then*
$BP \overset{-}{<} B \Longrightarrow \{B_m^-\} = \{PB_m^-\} + \{(I-P)B_m^-\}, \{PB_m^-\} \subset \{(BP)_m^-\}$

$$\{(I-P)B_m^-\} \subset \{[B(I-P)]_m^-\}$$

Premises for Theorem 2.4 and 2.5 could be restated without requiring the projectors Q and P to be orthogonal but requiring instead that $QB \bar{<}_\ell B$ and $BP \bar{<}_m B$ respectively. These follow from the following equivalences

$A \bar{<}_\ell B \Longleftrightarrow A = QB$ where Q is an orthogonal projector onto a subspace of $M(B)$

$A \bar{<}_m B \Longleftrightarrow A = BP$ where P is an orthogonal projector onto a subspace of $M(B^*)$.

[16, Theorems 4.2 and 4.3].

Theorem 2.6. *The following statements are equivalent*

(a) $A \stackrel{\#}{<} B$

(b) $A = BQ = QB$ *for some projector* Q *onto a subspace of* $M(B)$ *and* B *is of index 1.*

Proof: (a) $\Longrightarrow$ (b). $A \stackrel{\#}{<} B \Longrightarrow AA^\# = BA^\#$, $A^\#A = A^\#B$. Put $A^\#A = AA^\# = Q$ and check that $A = BQ = QB$.

(b) $\Longrightarrow$ (a). By Lemma 1.4, (b) $\Longrightarrow A \bar{<} B \Longrightarrow \{B^-\} \subset \{A^-\}$. Choose and fix an arbitrary matrix $(B_{com}^-)_o \in \{B_{com}^-\}$

$$A(B_{com}^-)_o = QB(B_{com}^-)_o = QBB^\# = QB^\#B$$

$(B_{com}^-)_o A = (B_{com}^-)_o BQ = B^\#BQ = QB^\#B = A(B_{com}^-)_o$, since Q commutes with both B and $B^\#$ (see Remark 2). This shows $(B_{com}^-)_o$ commutes with A. Noting that $(B_{com}^-)_o \in \{B^-\} \subset \{A^-\}$ we have thus $(B_{com}^-)_o \in \{A_{com}^-\} \Longrightarrow \{B_{com}^-\} \subset \{A_{com}^-\} \Longrightarrow A \stackrel{\#}{<} B$, the other condition $A \stackrel{s}{\prec} B$ being trivially true (see Table 1). □

Remark 3. In view of Theorem 2.6 one wonders if results similar to Theorem 2.4 and 2.5, but for the ρ- and χ-inverses are true under a weaker condition involving the minus ρ or minus χ order instead of the sharp order as in Theorem 2.3.

The following extension of Theorem 2.2, in so far as the equivalence of statements (a) and (b) is concerned, reminds us of Theorem 2.2 in Rao and Yanai [22].

Theorem 2.7. *Let* P *and* Q *be projectors in* F^n *and* F^m *respectively and* $B \in F^{m\times n}$. *Then the following statements are equivalent*

(a) $QB \bar{<} B$, $BP \bar{<} B$ (35)

(b) $\{PB^-Q\} + \{PB^-(I-Q)\} + \{(I-P)B^-Q\} + \{(I-P)B^-(I-Q)\} = \{B^-\}$ (36)

Further (a) $\Longrightarrow \{PB^-Q\} \cap \{(QBP)^-\} \neq \phi$, *if and only if*

$QBP \bar{<} B$ (37)

when the conditions (35) *and* (37) *hold*

$\{PB^-Q\} \subset \{(QBP)^-\}$ (38)

Proof: (a) $\Longrightarrow$ (b). (a) $\Longrightarrow$ $M(QB) \subset M(B)$, $M[(I-Q)B] \subset M(B)$, $M[(BP)'] \subset M(B')$, $M[(I-P)'B'] \subset M(B')$. Hence by Lemma 1.1, BPB^-QB, $BPB^-(I-Q)B$ etc. are invariant under the choice of the g-inverse B^-. If G_i, i = 1,2,3,4 are choices of B^- possibly all distinct, then

$$BPG_1QB + BPG_2(I-Q)B + B(I-Q)G_3QB + B(I-P)G_4(I-Q)B$$

$$= BPG_1QB + BPG_1(I-Q)B + B(I-P)G_1QB + B(I-P)G_1(I-Q)B$$

$$= [BP+B(I-P)]G_1[QB+(I-Q)B] = BG_1B = B \Longrightarrow \text{(b)}$$

(b) $\Longrightarrow$ (a). Choose $G_3 = G_1$, $G_4 = G_2$. Then (b) $\Longrightarrow$

$$\{D^-Q\} + \{B^-(I-Q)\} = \{B^-\}$$

$\Longrightarrow$ $QB \bar{<} B$, by Theorem 2.2. Similarly (b) $\Longrightarrow$

$$\{PB^-\} + \{(I-P)B^-\} = \{B^-\}$$

$\Longrightarrow$ $BP \bar{<} B$.

Let $\{PB^-Q\} \cap \{(QBP)^-\} \neq \phi$ and let in fact $(B^-)_o$ be a choice of B^- such that

$$P(B^-)_oQ \in \{(QBP)^-\}$$

Then $\qquad QBP(B^-)_oQBP = QBP$

However we have seen that when (a) holds BPB^-QB is invariant under the choice of B^-. Hence

$$QBPB^-QBP = QBP \ \forall \ B^-$$

$\Longrightarrow$ $\{B^-\} \subset \{(QBP)^-\} \Longleftrightarrow QBP \bar{<} B$.

The last part of Theorem 2.7 is trivial. □

Remark 4. The following example shows that $QB \bar{<} B$, $BP \bar{<} B \not\Longrightarrow QBP \bar{<} B$. Consider the complex matrix

$$B = \begin{pmatrix} 1 & 1 \\ 1 & 0 \end{pmatrix}$$

Let P = Q = diag (1, 0). Since B is nonsingular it is easily seen that (35) is true here. On the other hand

$$QBP = \begin{pmatrix} 1 & 0 \\ 0 & 0 \end{pmatrix}, \quad B-QBP = \begin{pmatrix} 0 & 1 \\ 1 & 0 \end{pmatrix}$$

Rank QBP = 1, Rank (B-QBP) = 2. Hence (37) is not true.

This example is interesting from another point of view. Here

$$(I-Q)BP \bar{<} B,\ QB(I-P) \bar{<} B \text{ and } (I-Q)B(I-P) \bar{<} B.$$

Remark 5. If (I-Q)B(I-P) = 0, then (35) $\Longrightarrow$

$$QB(I-P) = B(I-P) \bar{<} B \Longrightarrow \{(I-P)B^-Q\} \subset \{[QB(I-P)]^-\}$$

$$(I-P)BP = (I-Q)B \bar{<} B \Longrightarrow \{PB^-(I-Q)\} \subset \{[(I-Q)BP]^-\}$$

This condition (I-P)B(I-P) = 0 cannot be replaced by a weaker condition $(I-Q)B(I-P) \bar{<} B$. Choose B as a Remark 4, P = diag(1, 0), Q = diag(0, 1) for a counter example.

Remark 6. Condition (35) implies that the matrix equation

$$BG^{(1)}B = QBP$$

is consistent and so are individually the equations $BG^{(2)}B = (I-Q)BP$, $BG^{(3)}B = QB(I-P)$ and $BG^{(4)}B = (I-Q)B(I-P)$. Let the respective classes of solutions be denoted by $\{G^{(i)}\}$, $i = 1,2,3,4$. We have the decomposition

$$\{G^{(1)}\} + \{G^{(2)}\} + \{G^{(3)}\} + \{G^{(4)}\} = \{B^-\} \tag{39}$$

which is different from (36). One has only to compute both the decompositions (36) and (39) in respect of the matrix B given in Remark 4, to convince himself that they are indeed different. Both the decompositions involve the blocking operations in some form or the other. In (36) the involvement is direct. In (39) it enters through the right hand side of the equations one has to solve to determine the classes $\{G^{(i)}\}$, $i = 1,2,3,4$.

3. OTHER DECOMPOSITIONS OF $\{B^-\}$

In this section we consider certain other decompositions of $\{B^-\}$ which depend on the matrix A that B dominates under the minus order but not explicitly on projectors P and Q. Here for example it is possible to have a natural decomposition of $\{B_r^-\}$ unlike what we have seen in Remark 1. We first prove Theorem 3.1 which is closely related to lemma 1.5.

Theorem 3.1. *Let* $A \overset{-}{<} B$. *Then*

$$\{B^- - B^-(B-A)B^-,\ B^- \text{ arb.}\} = \{A^- : BA^- = AA^-,\ A^-B = A^-A\} \tag{40}$$

Proof: Put

$$\{A^-\}_B = \{A^- : BA^- = AA^-,\ A^-B = A^-A\} \tag{41}$$

Since $A \overset{-}{<} B$, it is easily seen using Lemma 1.4 that for arbitrary choice of B^-

$$B^- - B^-(B-A)B^- \ \varepsilon\ \{A^-\}_B$$

Conversely choose arbitrarily and fix $(A^-)_o \ \varepsilon\ \{A^-\}_B$ and $(B^-)_o \ \varepsilon\ \{B^-\}$. Observe that $A \overset{-}{<} B \Longrightarrow (B-A) \overset{-}{<} B \Longrightarrow$

$$B[(A^-)_o + (B^-)_o(B-A)(B^-)_o]B = A + (B-A) = B$$

$\Longrightarrow (B^-)_1 = (A^-)_o + (B^-)_o(B-A)(B^-)_o \ \varepsilon\ \{B^-\}$

Further $(B^-)_1(B-A) = (B^-)_o(B-A)(B^-)_o(B-A) = (B^-)_o(B-A)$

$$(B-A)(B^-)_1 = (B-A)(B^-)_o(B-A)(B^-)_o = (B-A)(B^-)_o$$

$\Longrightarrow \quad (B^-)_1(B-A)(B^-)_1 = (B^-)_o(B-A)(B^-)_o$

$\Longrightarrow \quad (B^-)_1 = (A^-)_o + (B^-)_1(B-A)(B^-)_1$

$\Longrightarrow \quad (A^-)_o \ \varepsilon\ \{B^- - B^-(B-A)B^-,\ B^- \text{ arb.}\}$ □

Theorem 3.2. *Let* $A \overset{-}{<} B$. *Then*

$$\{A^-\}_B + \{(B-A)^-\}_B = \{B^-\} \tag{42}$$

Proof: The '$\subset$' part is verified through straightforward computations as in the proof of Theorem 3.1. For the '$\supset$' part choose and fix an arbitrary matrix $(B^-)_o \ \varepsilon \ \{B^-\}$. Then

$$(B^-)_o-(B^-)_oA(B^-)_o \ \varepsilon \ \{(B-A)^-\}_B$$

By Lemma 1.5, $(B^-)_oA(B^-)_o \ \varepsilon \ \{A_r^- : BA_r^- = AA_r^-, \ A_r^-B = A_r^-A\} \subset \{A^-\}_B$.

Hence $\quad (B^-)_o = (B^-)_oA(B^-)_o+[(B^-)_o-(B^-)_oA(B^-)_o] \ \varepsilon \ \{A^-\}_B + \{(B-A)^-\}_B$.

We have virtually proved the following corollary. For a pair $A,B,A \bar{<} B$, put

$$\{A_r^-\}_B = \{A_r^-:BA_r^- = AA_r^-, \ A_r^-B=A_r^-A\} = \{B^-AB^-, B^- \text{ arb.}\} \tag{43}$$

Corollary 3.1. *Let* $A \bar{<} B$. *Then*

$$\{A_r^-\}_B + \{(B-A)^-\}_B = \{B^-\} \tag{44}$$

Further an arbitrary g-*inverse* B^- *can be decomposed as in* (44) *in one and only one way.*

Proof: We need only prove the uniqueness part. Let specific B^- be decomposed in two ways as follows:

$$B^- = (B^-)_1A(B^-)_1 + [(B^-)_2-(B^-)_2A(B^-)_2]$$

$$= (B^-)_3A(B^-)_3 + [(B^-)_4 - (B^-)_4A(B^-)_4]$$

Then $A \bar{<} B \Longrightarrow$

$$B^-A = (B^-)_1A = (B^-)_3A$$

$$AB^- = A(B^-)_1 = A(B^-)_3$$

$$\Longrightarrow B^-AB^- = (B^-)_1A(B^-)_1 = (B^-)_3A(B^-)_3$$

$$B^--B^-AB^- = (B^-)_2-(B^-)_2A(B^-)_2$$

$$= (B^-)_4 - (B^-)_4A(B^-)_4. \qquad \square$$

The proof of the following theorem is fairly straightforward. It involves arguments with which we are already familiar. We omit the proof.

Theorem 3.3. *Let* $A \bar{<} B$. *Then*

$$\{B_r^-\} = \{A_r^-\}_B + \{(B-A)_r^-\}_B \tag{45}$$

Further any arbitrary reflexive g-*inverse* B_r^- *can be decomposed as in* (45) *in one and only one way.*

We note that $A \bar{<} B$ implies that the equations

$$BG_1B = A, \quad BG_2B = (B-A)$$

are consistent. Let the class of solutions be denoted by $\{G_1\}$ and $\{G_2\}$ respectively. Observe that

$$B\{A^-\}_DB = A, \ B\{(B-A)^-\}_DB = (B-A)$$

Hence $\quad \{A^-\}_B \subset \{G_1\}, \ \{(B-A)^-\}_B \subset \{G_2\}$

Let $(B^-)_o$ be a fixed chosen g-inverse of B. Then it is easily seen that

$$\{G_1\} = \{B^- - (B^-)_o(B-A)(B^-)_o, \ B^- \text{ arb.}\}$$

$$\{G_2\} = \{B^- - (B^-)_oA(B^-)_o, B^- \text{ arb.}\}$$

Hence $\{G_1\} \not\subset \{A^-\}_B, \ \{G_2\} \not\subset \{(B-A)^-\}_B$

We nevertheless have the following theorem which reminds us of the decomposition in remark 6.

Theorem 3.4. *Let* $A \bar{<} B$. *Then*

$$\{B^-\} = \{G_1\} + \{G_2\} \tag{46}$$

Further $\{G_1\} \subset \{A^-\}, \ \{G_2\} \subset \{(B-A)^-\}$

Proof: The first part is trivial. For the second part noting that $A \bar{<} B$, choose an arbitrary $(A^-)_o \in \{A^-\}_B$. Then

$$BG_1B = A \Longrightarrow A(A^-)_oBG_1B(A^-)_oA = A(A^-)_oA(A^-)_oA$$
$$\Longrightarrow AG_1A = A.$$

Similarly $BG_2B = (B-A) \Longrightarrow (B-A)G_2(B-A) = (B-A)$. □

REFERENCES

[1] J.Baksalary and S.K.Mitra, Left-Star and Right-Star partial orderings (preprint).

[2] D.Carlson, Generalized inverse invariance, partial orders and rank minimisation problems for matrices, Current Trends in Matrix Theory (F.Uhlig and R.Grone Eds.) North Holland, New York, 81-89 (1987).

[3] M.P.Drazin, Natural structures on semigroups with involution, Bull.Amer. Math. Soc. 84 : 139-141 (1978).

[4] I.Erdelyi, On the matrix equation Ax = λBx. J.Math. Anal. Appl. 17 : 117-132, (1967).

[5] F.J.Hall,Generalized inverses of a bordered matrix of operators, SIAM J. Appl. Math. 29 : 152-163 (1975).

[6] F.J.Hall and C.D.Meyer, Generalized inverses of the fundamental bordered matrix used in linear estimation, Sankhya Ser.A 37: 428-438 (1975).

[7] R.E.Hartwig, How to order regular elements, Math.Japon 25: 1-13 (1980).

[8] R.E.Hartwig and G.P.H.Styan, On some characterization of the "star" partial orderings and rank substractivity, Linear Algebra Appl. 82 : 145-161 (1986).

[9] C.G.Khatri, Some results for the singular multivariate regression models, Sankhya Ser.A 30 : 268-280 (1968).

[10] A.W.Marshall and I.Olkin, Inequalities : Theory of Majorization and its Applications, Academic Press, Orlando (1979).

[11] S.K.Mitra, A new class of g-inverse of square matrices, Sankhya Ser.A 30 : 323-330 (1968).

[12] S.K.Mitra, Properties of the fundamental bordered matrix used in linear estimation in Statistics and Probability (Essays in honor of C.R.Rao), North Holland, Amsterdam, 1982, pp. 505-510.

[13] S.K.Mitra, The minus partial order and the shorted matrix, Linear Algebra Appl. 83 : 1-27 (1986).

[14] S.K.Mitra, On group inverses and the sharp order, Linear Algebra Appl. 92 : 17-37 (1987).

[15] S.K.Mitra, Infimum of a pair of matrices, Linear Algebra Appl. 105 : 163-182 (1988).

[16] S.K.Mitra, Shorted matrices in Star and related orderings, To appear in Circuits, Systems and Signal Processing (Special Issue).

[17] S.K.Mitra, Matrix partial orders through generalized inverses: Unified Theory. submitted to Linear Algebra Appl.

[18] S.K.Mitra and P.L.Odell, On parallel summability of matrices, Linear Algebra Appl. 74 : 239-255 (1986).

[19] K.S.S.Nambooripad, The natural partial order on a regular semigroup, Proc. Edinburgh Math. Soc. 23 : 249-260 (1980).

[20] C.R.Rao, Unified theory of linear estimation, Sankhya Ser.A 33 : 371-394 (1971).

[21] C.R.Rao and S.K.Mitra, Generalized inverse of matrices and its Applications, Wiley, New York (1971).

[22] C.R.Rao and H.Yanai, Generalized inverses of partitioned matrices useful in statistical applications, Linear Algebra Appl. 70 : 105-113 (1985).

[23] P.Sambamurty, Characterization of a matrix by its subclass of g-inverses, Sankhya Ser. A 48 : 412-414 (1987).

STATISTICAL DATA ANALYSIS AND INFERENCE
Y. Dodge (Editor)

GENERALIZED INVERSES AND GENERALIZED CONVEXITY.

Jean-Pierre Schellhorn

Department of Econometrics
University of Geneva
CH-1211 Genève-4 Switzerland

Statistics and Data analysis make an increasing use of the the concepts of mathematical programming and of convex analysis. This contribution shows how the linear and order-theoretic aspects of a derivation operator and of its right inverse, the anti-derivation (or integration) operator, enables one to subsume some generalized concepts of convexity under a comprehensive but simple theory.

1. INTRODUCTION

Convex analysis is that part of analysis which studies what can be done if we use convex instead of differentiable functions, namely in problems of mathematical programming and nonlinear evolution equations.

This contribution shows how to induce an inverse move, starting from an operational calculus with abstract derivation operator and then defining the convenient convexity concepts and subgradients. We introduce first the fundamental definitions of the derivation or differential operator D and of its right inverse, the integral or anti-derivation or summation operator S; the context is any vector space with some partial ordering. As examples of more advanced concepts, we present the derivative of a projector and the exponential of an endomorphism.

Next, for any abstract derivation operator D, we develop original concepts of D-convex functions as well as D-subgradients. We show, with complete proofs, that some important basic properties in mathematical programming are still valid.

2. DERIVATIONS AND ANTIDERIVATIONS

In operator theory, the term <u>operational</u> <u>calculus</u> generally indicates a way of defining "functions" f(T) of an operator T , so that a kind of algebraic homomorphism (a "transform") is established between a set of complex-valued functions f and the corresponding set of operators f(T), i.e.functions whose values are operators.

Example: Characteristic functions f of random variables as Fourier transforms of distribution functions (cf.Rao [6],p.99).

Among the many possible presentations let us choose the operational calculus studied by Bittner [1].

2.1

Let $V_0 \supseteq V_1 \supseteq V_2 \supseteq \ldots \supseteq V_n$ be an inclusion sequence of vector spaces with a common convex cone Q defining a compatible partial ordering.
Let the <u>derivation</u> <u>D</u> be a linear operator from V_1 to V_0, and the <u>anti-derivation</u> or <u>summation</u> operator S a linear operator from V_0 to V_1; besides linearity, the only condition is that each operator is a <u>one-sided</u> <u>inverse</u> of the other:

$$D\,S\,f = f \quad \text{for all } f \in V\ ;$$

of course, S D is then idempotent.

2.2

<u>Constants</u> are elements c of V_1 such that $D\,c = 0$ (kernel of D).
Two important sources of constants are the following ones:

i) The general solution of $D\,x = f$ is $x = S\,f + c$; then c is constant.

ii) $LOC = I - S\,D$ is the linear operator often called "limit condition"; we prefer to call it "<u>local</u> <u>condition</u>", because we usually work without explicit topological conditions. Of course, LOC(f), written LOC f is a constant.

Hypotheses: (H1) Kernel(D) contains other elements than 0.
(H2) Kernel(S) = {0} : if Sf = 0 , then f = 0.

V_n is the domain of the iterated operator $D^n = D\,D^{n-1}$, $D^0 = I$;
($f \in V_n$ means that $D^n f$ exists). $S^n = S\,S^{n-1}$, $S^0 = I$.

3. THE ABSTRACT TAYLOR SERIES EXPANSION

3.1

The following identity is elementary and valid for every $f \in V_n$:

$$f = (f - S\,Df) + S\,(Df - S\,D^2f) + S^2(D^2 - S\,D^3f) + \ldots +$$
$$+ S^{n-1}(D^{n-1}f - S\,D^nf) + S^nD^nf\ .$$

Write $\phi_0(f) = LOC\ f = (\ I - S\ D\ )\ f$ for the local condition of f,

and $\phi_i(f) = LOC(\ D^if\ ) = (I-SD)\ D^if = (\ D^i\ f - S\ D^{i+1}\ f\ )$.

Then the identity becomes a <u>Taylor</u> <u>(series)</u> <u>expansion</u> :

$$f = \phi_{(0)}(f) + S\,\phi_{(1)}(f) + S^2\phi_{(2)}(f) + \ldots + S^{n-1}\phi_{(n-1)}(f) + R_{(n)}(f)$$

$$f = LOC(f) + S\,LOC(Df) + S^2LOC(D^2f) + \ldots + S^{n-1}LOC(D^{n-1}f) + R_{(n)}(f)$$

$R_{(n)}(f) = S^n\,D^n\,f$ is the remainder.

This expansion does not use the fact that S is an right-inverse of D; actually it is valid for any pair (D,S) of linear operators. D and S are linear, so that their action on this Taylor expansion is "decoupled" as in the classical case:

$$Sf = S\,\phi_0(f) + S^2\,\phi_1(f) + \ldots + S^{n+1}D^nf$$
$$Df = D\,\phi_0(f) + D\,S\,\phi_1(f) + \ldots + D\,R_{(n)}f$$
$$= \phi_1(f) + \ldots + S^{n-1}D^nf \qquad \text{[if DSg=g and D LC=0]}$$
$$= LOC(Df) + S\,LOC(D^2f) + \ldots + S^{n-1}D^nf.$$

Most of the following properties of the pair (D,S) will depend on the property that S is a right-inverse of D.

3.2 Example 1: V_n is the vector space of n-times derivable real functions f(x) of a real variable.
$Df = [df/dx]_{x=a}$, the usual derivative in x=a, and
$Sf = \int_a^x f(t)\,dt$, the usual definite integral from a to x. Then

$S^n f = \int_a^x f(t)\,[\,(x-t)^{(n-1)}/((n-1)!)\,]\,dt$; this is a classical

fact,if we choose as primitive Sf that F(t) which is zero in a, as well as all first (n-1) derivatives of F(t).
Let us look at the first truncated Taylor expansions with the remainder in integral representation (Lagrange):

(0)

$$f(x) = f(a) + \int_a^x f'(t)\,dt = f(a) - S\,D\,f(x),$$

or $f(a) = (I-SD)f = \phi_0(f)$.

The <u>local condition is LOC(f) ≡ f(a)</u>.

(1)

$$f(x) = f(a) + f'(a)\,(x-a)/1! + \int_a^x \int_a^x f''(t)\,dt$$

$$\text{or } f'(a)\,(x-a)/1! = f(x) - f(a) - S^2D^2f$$

$$= [(I-SD)f+S(I-SD)Df+S^2D^2f] - (I-SD)f - S\,D^2f^2 = S\,\phi_1(f).$$

The <u>local condition is LOC(Df) ≡ f'(a) (x-a)</u> = "df(a)".

(2)

$$f(x) = f(a) + f'(a)\,(x-a)/1! + f''(a)\,(x-a)^2/2! + $$

$$+ \int_a^x \int_a^x \int_a^x f'''(t)\,dt$$

$$\text{or } f''(a)\,(x-a)^2/2! = f(x) - \phi_0(f) - S\,\phi_1(f) - S^3D^3f$$

$$= [(I-SD)f+S(I-SD)Df+S^2(I-SD)D^2f+S^3D^3f]$$

$$- [(I-SD)f] - S[(I-SD)Df] - S^3D^3f = S^2\,\phi_2(f)$$

The <u>local condition is LOC(D^2f) ≡</u> 0.5<u>$(x-a)^2$f''(a) (x-a)</u> = "$d^2f(a)$".

3.3 Example 2:
V_n is the vector space of n-times derivable real functions f(x) defined on an open set $\mathfrak{D}$ in R^n.
f <=> f(y) = f(x+w);
Df <=> transpose of the gradient vector of f at x.

$$\begin{aligned} f(x+w) &= f(x) + w'\text{grad } f(x) + 1/2\, w'[\text{hessian } f(x)]\, w + R_{(3)}(f) \\ &= \phi_0(f) + S\,\phi_1(f) + S^2\,\phi_2(f) + R_{(3)}(f) \\ &= LOC\,(f) + LOC\,(Df) + LOC\,(D^2f) + R_{(3)}(f) \end{aligned}$$

In this context,
LOC(f) is the value of f for the vector x,
LOC(Df) is the directional derivative of f at x in the direction w,
LOC(D^2f) is the value in w of the jacobian-quadratic-form associated to the gradient of f (at x).
Actually, in order to get the value of the components of the Taylor expansion in two different points (like x and x+w), it is wise to introduce a FOCALISATION operator next to the LOCALISATION operator.
LOC_y f : means that we already know this information locally in y;
FOC_z f : means that we are extracting this information locally in z.

$$f(x+w) = f(x) + w'\text{grad } f(x) + 1/2\, w'[\text{hessian } f(x)]\, w + R_{(3)}(f)$$

will be rewritten as

$$FOC_{x+w}(f) = LOC_x(f) + LOC_x(Df) + LOC_x(D^2f) + \ldots$$

In this example, Taylor expansion and direction w are in the foreground, and the pair (D,S) of operators is in the background.

3.4 Example 3:

An abstract notion of "ladder" of polynomials is often used in approximation theory and in combinatorics.
Let $\mathbb{R}[x]$ be the algebra of all polynomials in x with real coefficients $P_n(x)$; in $\mathbb{R}[x]$, any sequence $\{\Pi_0(x), \Pi_1(x), \Pi_2(x), \ldots\}$ beginning with $\Pi_0(x)=1$, $\Pi_q(0)=0$ and degree$(\Pi_q)=q$ (for $q>0$) is called a <u>normal family of polynomials</u> N.
A derivation associated to N is any linear operator D such that $D[\Pi_q(x)] = q\,\Pi_{q-1}(x)$ (for $q>0$) and $D[\Pi_0(x)] = 0$. In fact, such a D is unique for a given normal family:
$D[P_n(x)] = D[\,a_n \Pi_n + \ldots + a_1 \Pi_1 + a_0 \Pi_0\,] = n\,a_n \Pi_{n-1} + \ldots + a_1 \Pi_0$.
The anti-derivation S and the Taylor expansion are evident.
Orthogonal polynomials form useful normal families, because they obey a second order differential relation (equivalent to the orthogonality property): $g(x)\,D^2\Pi_n(x) + h(x)\,D\,\Pi_n(x) + J_n\,\Pi_n(x) = 0$.

4. SOME MORE EXAMPLES

4.1 Example 4:

$\mathcal{V}_n$ is the vector space of n-times derivable real matricial functions M(X) defined on an open set $\mathfrak{D}$ in $\mathbb{R}^{n\times m}$, i.e. rectangular matrices which are functions of rectangular matrices.
The explicit form of Taylor expansions is given in the paper of J.Vetter [10]. In this case, the derivative D and the Taylor ex-expansion are in the foreground; and the anti-derivative S is in the background.

It is likely that some important special cases of the following two examples can be subsumed under the above abstract theory.

4.2 Example 5:

The <u>fractional derivative</u> $D(\alpha;z)f$ of a function $f(z)$ is a generalization of the familiar derivative $D(n;z)f = d^n f(z)/(dz)^n$; the order n may be replaced by arbitrary (integral, rational, irrational or complex) α. Many definitions are available (a convenient reference is the book by Oldham and Spanier [3].
For many definitions, $D(-\alpha)$ is the <u>fractional integration</u> or antiderivation associated to $D(\alpha)$.
Taylor's series expansions for a complex-valued function f(z) have been studied, e.g. by Olsder [4]; he finds that they are nothing more than the Fourier series of a new function constructed from f(z).
Fractional derivatives have been employed to simplify the solution of ordinary, partial and integral equations. They are also useful in the study of special functions which appear namely as density functions of statistics; for instance:

i) the incomplete gamma function $\gamma(a,z)$ is
$\gamma(a,z) = \Gamma(a)\exp(-z)\;D(-a;z)[\exp(z)]$;

ii) the confluent hypergeometric function ${}_1F_1(a;c;z)$ is
${}_1F_1(a;c;z) = (\Gamma(c)/\Gamma(a))\,z^{1-c}\;D(a-c;z)[\exp(z)\,z^{a-1}]$;

iii) Gauss' hypergeometric function ${}_2F_1(a,b;c;z)$ is

$${}_2F_1(a,b;c;z) = (\Gamma(c)/\Gamma(a))\,z^{1-c}\,D(b-c;z)[\,z^{b-1}(1-z)^{-a}\,].$$

There is also a connexion with mixtures of distributions, Mandelbrot's fractals and the concept of non-integral dimension (fractals and fractal dimension have recently acquired value in Data compression, and may bring some more bonus in Data analysis).

4.3 Example 6:

In a formal approach to <u>stochastic</u> <u>integration</u>, Krener [2] defines integrals of the form $\int_b^a f(t, W(t))\, dW(t)$,

with W(t) a Wiener-Lévy process and f(t,w) a real-valued function, which is continuously differentiable in t and w.
For this, he uses the fundamental theorem of calculus:

$$\int_a^b dH(t,W) = H(t,W)\Big|_a^b \qquad \{ \text{ or } SD\, f = f - LC(f) \}$$

and a family of <u>chain</u> <u>rules</u> <u>for</u> <u>differentials</u>,
i.e. Taylor's expansions with specific multiplication rules like
$dt \times dt = 0 \qquad dW \times dW = dt \qquad dt \times dW = 0$.

5. DERIVATIVES AND PROJECTORS

5.1

S is but a very special generalized inverse of D: it is a right inverse. If, in some application, D is represented by a matrix A of order mxn, then S exists iff rank(A)=m; a general solution of A X = I is given in Rao and Mitra ([7],p.20).
As noted above, the generalized Taylor expansion do not depend on S being the right inverse of D; but most of the uses of the expansion do.
In a space of square matrices, with D=A and $S=A^-$ and for any compatible matrix f=M, the successive Taylor expansions are
$M = (I-A^-A)M + A^-A\, M$ (anti-projection of M + Taylor rest)
$M = (I-A^-A)M + A^-(I-A^-A)AM + A^-(A^-A)AM$
and so on.

5.2 Example 7:

Suppose that we are interested in all integral solutions of the vector equation A X B = C , with A, B and C given integral matrices (i.e.components are rational integers).
As the Moore-Penrose inverse of a real or complex matrix M follows easily from the singular (value) decomposition of M,
as a reflexive g-inverse $A^\#$ with integral projectors $AA^\#$and $A^\#A$ follows easily from the Smith (normal) decomposition of A.
But A X B = C will have an integral solution iff $A^\#\, C\, B^\#$ is an integral solution.
In order to use the general scheme of this paper, we have to solve problems like the following one.
Problem: Replace right-inverse anti-derivation S by reflexive-g-inverse S. and study the modified consequences.

5.3 Example 4 (continues).

More interpretations are needed for the matricial derivatives of matrices, as the following one. Let us consider:
W = a gramian (semi-positive-definite); $Z^- = W A'$; and
$P = W A' (AWA')^{-1} A = A^- A = Z^- Z$ = a <u>W-orthogonal</u> <u>projector</u>.
$D_{A+eB}(P) = \lim_{e\Rightarrow 0+} (1/e)[P(A+eB)-P(A)]$ = the <u>directional</u> <u>derivative</u> of P at A in the direction of B.
Then it comes out that this derivative is the (W^{-1}-orthogonal) sum $H \perp J_-$ of two matrices (direct effects + indirect effects) :
$H = A^- B (I-P)$, $J = (I-P) W B' Z$ and $H' W^{-1} J = 0$.

6. D-EXPONENTIALS OF ENDOMORPHISMS

Abstract exponentials of endomorphisms furnish solutions of initial value problems of linear differential equations for operator-valued functions; included are not only ordinary linear recurrence equations and linear differential equations, but also linear partial differential equations.

6.1

f is a D-analytic function if $S^m D^m f$ converges to 0 if $m \Rightarrow \infty$; some topology of course is implied here.
Let $\mathcal{A}$ be a space of D-analytic functions. Endomorphisms in $\mathcal{A}$ are linear and continuous operators. In $\mathcal{A}$ there is a natural shift (translation) operator T defined by

$$T[f] = T[\phi_{(0)}(f) + S\,\phi_{(1)}(f) + S^2\phi_{(2)}(f) + \dots]$$

$$= S\,\phi_{(0)}(f) + 2\,S^2\phi_{(1)}(f) + 3\,S^3\phi_{(2)}(f) + \dots$$

Let L be an endomorphism which commutes with D and S:

$$L\,D\,f = D\,L\,f \quad \text{and} \quad L\,S\,f = S\,L\,f\,.$$

Then the exponential operator exp(L T) transforms

$$f = \phi_{(0)}(f) + S\,\phi_{(1)}(f) + S^2\phi_{(2)}(f) + \dots$$

into $F = \Psi_{(0)}(f) + S\,\Psi_{(1)}(f) + S^2\Psi_{(2)}(f) + \dots$

with $\Psi_k = \sum_{k=0}^{n} \binom{n}{k} L^k \phi_{(n-k)}$.

Properties: $\exp(L_1T)\exp(L_2T) = \exp((L_1+L_2)T)$ if $L_1L_2=L_2L_1$;
$\exp(\emptyset\, T) = I$; $\exp(LT)\exp(-LT) = I$;
$D\exp(LT)f = \exp(LT)\,Df + L\exp(LT)\,f$;
$LC(\exp(LT)f) = LC(f)$.

6.2 Example 8:

If V_0 is the space of complex-valued functions x(n) of an integer $n \geq 0$, if $D[x(n)] = x(n+1)-x(n)$ and $T[x(n)] = n\,x(n-1)$, then $S[x(n)] = x(0)+x(1)+..+x(n)$, $LC[x(n)] = x(0)$ and

$$\exp(LT)[x(n) = \sum_{k=0}^{n} \binom{n}{k} L^k x(n-k)\,.$$

All functions in V_0 are D-analytic.

Problem: Try to generalize to these exponentials of endomorphisms due to Bittner [1] such useful properties of the usual exponential of a real square matrix $\exp(tA)=B(t)$, say:

i) A is M is Metzler ($a_{ij} \geq 0, i \neq j$) iff all (components of all) B(t) are ≥ 0 .

ii) A is Metzler and every row of A is a contrast iff all B(t) are row stochastic.

7. GENERALIZED CONCEPTS OF MONOTONICITY AND CONVEXITY

Generalized convex or subharmonic functions, based on differential operator-based inequalities, have been studied in functional analysis since many years (it should have been one of the topics in the never published second volume of "Inequalities" by Edwin Beckenbach and Richard Bellman). But , at the writer's knowledge, authors were not interested simultaneously in mathematical programming and operational calculus.

For instance, pseudo-convexity has emerged as an important sufficient condition to make global a local minimum: a function f having a gradient grad f(x) at the point y is called pseudo-convex in y iff (for all w) if $w'\text{grad } f(y) \geq 0$ then $f(y) \leq f(y+w)$.
The (necessary) condition grad $f(y) = 0$
is then sufficient to make f(y) not superior to ANY f(x+w), i.e. f(y) is a (non-unique) global minimum.
Unicity of the global minimum requires a notion of strict D-pseudo-convexity, which will not be discussed here.
Convexity properties of functions correspond to monotonicity properties of their gradient; because pseudo-convexity is important, pseudo-monotonicity has to be studied.

7.1 Definitions:

1) A S-mean-value theorem is valid for f in y if
for some $(q,\hat{x}=y+qw)$, $FOC_{y+w}\,[f - LOC_y(f)] = FOC_{\hat{x}}[S\ LOC_y(Df)]$
i.e. $FOC_{y+w}\,[\ f - \phi_0(f)\ ] = FOC_{\hat{x}}[\ \phi_1(f)\ ]$.
2) f is D-convex in y if, for any z=y+w,
$FOC_{y+w}\,[f-LOC_y(f)] \geq FOC_{y+w}\,[S\ LOC_y(Df)]$
i.e. $FOC_{y+\lambda w}\,[f - \phi_0(f)] \geq FOC_{y+\lambda w}\,[\phi_1(f)]$ for any $\lambda>0$.
3) f is D-pseudo-convex in y if, for any z=y+w,
$FOC_y\,[\ S\ LOC_{y+w}\,(Df)\ ] \geq 0$ implies $FOC_y(f) \leq FOC_{y+w}(f)$.
4) Df is S-monotone(increasing) if, for all pairs (y,z),
$FOC_y\,[\ S\ LOC_z\,(Df)\ ] + FOC_z\,[\ S\ LOC_y\,(Df)\ ] \geq 0$.
5) Df is S-pseudo-monotone(increasing) in y if, for any z=y+w,
$FOC_z\,[\ S\ LOC_y\,(Df)\ ] \geq 0$ implies $FOC_z\,[\ S\ LOC_z(Df)\ ] \geq 0$.

Note: If the derivative $D\psi$ of a function $\psi(\lambda)$ is defined so that
$LOC_g\,[\ D\psi\ ] = FOC_{y+gw}\,[\ S\ LOC_y(Df)\ ] = LOC_{y+gw}\,[\ S\ \phi_1(f)\ ]$,
then $\psi(\lambda)=f(y+\lambda w)$, $\psi(0)=f(y)$, $\psi(1)=f(y+w)$ and
if ψ is D-convex then f is D-convex.

7.2 Example 2 (continues):

2) f is convex in y iff for all w
$f(y+w) - f(y) \geq w'\text{grad } f(y)$....for scalar-valued f

Note: The convexity of f corresponds to the convexity of the epigraph of f; every (semi-)norm is a convex function.

3) f is pseudo-convex in y iff for all w
$w'\text{grad } f(y) \geq 0$ implies $f(y) \leq f(y+w)$.

Example: ln(1+x) is pseudo-convex but not convex for $x \geq 0$.

Note: If f(y) is a real number and not a vector, then the pseudo-convexity of f implies the quasi-convexity of f:
$f(y+w) \leq f(y)$ implies $w'\text{grad } f(y) \leq 0$;
The quasi-convexity of f corresponds to the convexity of the level sets of f; thus the set of global minimum points of f is convex. Every monotonic function is quasi-convex.

Note: Functions f which are pseudo-convex AND pseudo-concave are called pseudo-linear; if such a f is defined on a convex set K, then it assumes its minimum (maximum) on an extreme subset, which may be empty, of K.

4) F is monotone (increasing) iff for all y and w,
$w'\ F(y) \leq w'\ F(y+w)$.
5) F is pseudo-monotone in y iff for all w (and λw, $\lambda>0$)
if $w'F(y) \geq 0$ then $w'F(y+\lambda w) \geq 0$.

Example: 1/(1+x) is pseudo-monotone but not monotone for $x \geq 0$.

7.3

The four properties which are needed are the following ones:

i) if f is D-convex, then f is D-pseudo-convex;

ii) if Df is S-monotone, then Df is D-pseudo-monotone;

iii) f is D-convex iff Df is D-monotone;

iv) if a S-mean value theorem is valid (definition 1), then f is D-pseudo-convex iff Df is S-pseudo-monotone.

Proofs:

i) Suppose that $0 \le FOC_y\,[S\ LOC_{y+w}\,(Df)]$; because f is D-convex, $FOC_y\,[\ S\ LOC_{y+w}\,(Df)\ ] \le FOC_y\,[\ f - LOC_{y+w}\,(f)\ ]$; thus $FOC_y\,[f] \le FOC_{y+w}\,[f]$, i.e. $f(y) \le f(y+w)$.

ii) Put y=z into the definition of S-monotonicity.

iii) ==>

For arbitrary y and z, D-convexity implies that

$FOC_y\,[\ S\ LOC_z(Df)\ ] \ \ge\ FOC_y\,[\ f - LOC_z(f)\ ]$ and

$FOC_z\,[\ S\ LOC_y(Df)\ ] \ \ge\ FOC_z\,[\ f - LOC_y(f)\ ]$;

adding both inequalities gives the D-monotonicity of Df.

<==

After the definition of D-convexity, a function $\psi(\lambda)$ has been associated to f. The (easy) steps of the proof are now:

f is D-convex, because ψ is $\underline{D}$-convex, because $\underline{D}\psi$ is $\underline{S}$-monotone, because Df is S-monotone.

iv) ==>

If we have non pseudo-monoticity, i.e.

$FOC_y\,[\ S\ LOC_z\,(Df)\ ] + FOC_z\,[\ S\ LOC_y\,(Df)\ ] < 0$, and if

$FOC_y\,[\ S\ LOC_z\,(Df)\ ] \ge 0$. then $FOC_z\,[\ S\ LOC_y\,(Df)\ ] < 0$;

but $FOC_y\,[S\ LOC_z\,(Df)] \ge 0$ implies $FOC_y(f) \le FOC_z(f)$ (D-p.c.), which is contradictory with $FOC_z\,[\ S\ LOC_y\,(Df)\ ] < 0$.

<==

$FOC_y\,[S\ LOC_{y+w}(Df)] \ge 0$ and $f(y) > f(y+w)$ are contradictory if Df is S-pseudo-monotone; because (mean value theorem) there exists y+qw such that $f(y+w)-f(y) = FOC_{y+qw}\,[S\ LOC_y(Df)]$, thus with $FOC_{y+qw}\,[S\ LOC_y(Df)] < 0$; this and S-pseudo-monotonicity make $FOC_y\,[S\ LOC_{y+w}(Df)] < 0$.

7.4 Example 9:

The neighbourhood FOC_z of a given point LOC_y can consist only in transforms g(y)=z, g belonging to a group of transforms. It is then tempting to generalize the (differential aspects of) Schur-convex functions, or "Schur-increasing"functions for a better name.

Note: S-monotonicity and D-convexity with Loewner ordering for matrix-valued functions are also needed in Data analysis.

8 NECESSITY AND SUFFICIENCY OF GENERALIZED KUHN-TUCKER CONDITIONS

This section shows how some of the concepts studied in the previous section, allow us to generalize the Kuhn-Tucker conditions of mathematical programming, both the sufficient and the necessary ones.

Let us consider the <u>constrained maximum problem</u> max f(z) for $z \in K$.

8.1

Let K_o be any set in the domain of f;

$g: \mathbb{R}_m \succ \mathbb{R}_q$ be a function.;

$K = K_o \cup \{z: g(z) \ge 0\}$ be the <u>constraint set</u> K;

$z^\wedge \in K$ be a point compatible with the constraint.

"Convenient regularity conditions are valid" will mean that for every index i such that $g_i(z^\wedge)=0$, the property $[\ S\ \phi_1(g_i) \ge 0\ ]$ implies the property $[\ -S\ \phi_1(f) \ge 0\ ]$.

The proposed generalized conditions are as follows:

There exists m real numbers $\lambda_i \geq 0$ such that
(KT1) $\lambda_1 S\phi_1(g_1) +\ldots+ \lambda_m S\phi_1(g_m) = - S\phi_1(f)$ (cone of gradients)
(KT2) $\lambda_1 \phi_o(g_1) +\ldots+ \lambda_m \phi_o(g_m) = 0$ (complementarity);
or, in terms of "local conditions":
(KT1) $\lambda_1 LOC(Dg_1)+\ldots+ \lambda_m LOC(Dg_m) = - LOC(f)$
(KT2) $\lambda_1 LOC(g_1) +\ldots+ \lambda_m LOC(g_m) = 0$.

8.2

SUFFICIENCY: If f and all g's are D-concave [resp. D-convex] and if the generalized Kuhn-Tucker conditions are valid, then $f(x\^) \geq f(x)$, a maximum on K [resp. $f(x\^) \leq f(x)$, a mimimum on K].

Proof (in the case of D-concavity):

$$\begin{aligned} f - \phi_o(f) \leq S\phi_1(f) &= - \lambda_1 S\phi_1(g_1) -\ldots- \lambda_m S\phi_1(g_m) \\ &= \lambda_1 [-S\phi_1(g_1)] +\ldots+ \lambda_m [-S\phi_1(g_m)] \\ &\leq \lambda_1 [\phi_o(g_1)-g_1]+\ldots+ \lambda_m [\phi_o(g_m)-g_m] \\ &= [\lambda_1 \phi_o(g_1)+\ldots+ \lambda_m \phi_o(g_m)] - [\lambda_1 g_1+\ldots+ \lambda_m g_m] \\ &= [0] - [\geq 0] \leq 0, \text{ i.e. } f(x) - f(x\^) \leq 0. \end{aligned}$$

Note: More refined sufficient conditions are given below.

8.3

NECESSITY: If convenient regularity conditions are valid, and if $f(x\^) \geq f(x)$, a maximum on K [resp. $f(x\^) \leq f(x)$, a mimimum on K] , then the generalized Kuhn-Tucker conditions are valid.

Proof: Farkas theorem says that , for real matrices P and R, if $Rz\geq 0$ implies $Pz\geq 0$, then $P = Q R$ for some $Q \geq 0$.

If all $g_i(z\^)=0$, then
[$S\phi_1(g) \geq 0$] implies [$-S\phi_1(f) \geq 0$] (regularity condition),
thus $-S\phi_1(f) = \lambda_1 S\phi_1(g_1) +\ldots+ \lambda_m S\phi_1(g_m)$ with all $\lambda_i \geq 0$.

If some $g_e(z\^) >0$, then the same conclusion is valid with $\lambda_e=0$.

Thus $-S\phi_1(f) = \lambda_1 S\phi_1(g_1) +\ldots+ \lambda_m S\phi_1(g_m)$
with $\lambda_i \geq 0$ if $\phi_o(g_i) =0$
and $\lambda_i =0$ if $\phi_o(g_i) >0$;

And $\lambda_1 \phi_o(g_1)+\ldots+ \lambda_m \phi_o(g_m)$, as a sum of $(\geq 0)*(0)$ terms and of $(0)*(>0)$ terms, is zero (complementarity).

8.4 Exemple 2 (continuing):

The generalized conditions becomes the classical Kuhn-Tucker conditions. It is possible to formulate abstract (D-S) conditions analogous to the classical necessary, resp. sufficient conditions for vectorial extrema (= non-dominated , efficient, or Pareto-optimal solutions).

9. SUBGRADIENTS AND PSEUDO-SUBGRADIENTS

9.1

Suppose that the vector space $\mathcal{V}$ is a space of functions $f:\mathbb{R}_m \succ \mathbb{R}_n$; a linear function in $\mathcal{V}$, which is compatible with a function f whose range is $\mathbb{R}_p$ $(p \leq n)$, can be identified with a (nxp)-matrix.

Let K_o belong to the common domain of all functions f;
let h: $\mathbb{R}_m \succ \mathbb{R}_q$ be a (constraint-)function..
The <u>constraint set</u> K will be the set $K = K_o \cup \{z: h(z) \geq 0\}$.
We use LOC_z and $FOC_{z\^}$, and write $\Delta z = z - z\^ -$,
$\Delta f = f - f(z\^) = f - \phi_o(f)$, $\Delta h = h - h(z\^) = h - \phi_o(h)$.

9.2

The linear operator F, compatible with a function f, is a <u>D-subgradient</u> of f relative to z^ and K iff $F\,\Delta z \le \Delta f$.

NOTE: If F is D-convex, then $\phi_1(f)$ acts as a $F\,\Delta z$.
The linear operator H, compatible with a function h, is a <u>D-pseudo-subgradient</u> of h relative to z^ and K iff

$$\Delta h \ge 0 \quad \text{implies} \quad H\,\Delta z \ge 0 .$$

NOTE: If F is D-pseudo-convex, then $\phi_1(f)$ acts as a $H\,\Delta z$.
Every D-subgradient is a corresponding D-pseudo-subgradient.

9.3

A set of sufficient conditions, generalizing the above conditions by Kuhn and Tucker, but also the 1977 set by Seierstad and Sydsaeter, for the <u>constrained</u> <u>minimum</u> <u>problem</u>

$$\min f(z) \text{ for } z\in K , \; f \text{ scalar-valued,}$$

is the following one:

IF (1) $z^\wedge \in K$, i.e. if $z\in K_o$ and $h(z^\wedge) \ge 0$,
(2) f has a D-subgradient F relative to z^ and K,
(3) h has a D-pseudo-subgradient H relative to z^ and K,
(4) a nonnegative vector v ($v_i \ge 0$) exists such that $v'\phi_o(h)=0$,
(5) $w' = F - v'H$ verifies the variational inequality

$$w'\,\Delta z \ge 0 \quad \text{for all } z\in K \quad (\Delta z=z-z^\wedge);$$

THEN $f(z^\wedge) \le f(z)$ for all $z\in K$, i.e. $z^\wedge = z_{min}$.

Note: (2') f is D-convex and (3') h is D-subconvex
are sufficient for (2) and (3).

Proof:

i) Because (4), $v'\phi_o(h)=0$ with $v\ge 0$; $h(z)\ge 0$ for all $z\in K$; thus $v'\phi_o(h) \le v'h$.

ii) $v'\phi_o(h) = 0$ implies that $v'H\,\Delta z \ge 0$:
effectively, $h_j(z^\wedge)$ being the j_{th} component of $\phi_o(h)$,
either α) all $h_j(z^\wedge) = 0$, making all $v_j \ge 0$ and all $h_j(z)-h_j(z^\wedge) \ge 0$; this means that Δh is ≥ 0 and implies (H being a D-pseudo-subgradient) that $H\,\Delta z \ge 0$ and $v'H\,\Delta z \ge 0$;
or β) some $h_j(z^\wedge) > 0$, making $v_j=0$ and $v_j'H_j\,\Delta z=0$.

iii) $0 \le w'\,\Delta z = (F-v'H)\,\Delta z$ (variational inequality)
means $F\,\Delta z \ge v'H\,\Delta z \ge 0$ (because ii).

iv) $\Delta f \ge F\,\Delta z$ (F being a D-subgradient)
≥ 0 (because iii),
means that $f(z) \ge f(z^\wedge)$.

9.4

Problem: Interpretations should be studied for the D-subgradients and D-pseudo-subgradients in the case of orthogonal functions (example 3), fractional derivatives (example 5), and matricials derivatives of matrices (example 4),

Problem: Introduce, through a Galois correspondance between partially ordered sets, the notions of conjugacy of D-convex functions and of D-Legendre transforms [in the sense that

$$\text{Legendre}(f) = g \quad \text{iff} \quad \text{grad } f(x) = p \quad \text{and} \quad \text{grad } g(p) = x].$$

10. CONCLUSION

Apart Fourier transforms (characteristic functions), statisticians could live and prosper without any operational calculus.
This contribution advocates a more systematic use of such a calculus in regard of its interesting properties.

11. REFERENCES

[1] Bittner, R. (1961), Operational Calculus in Linear Spaces, Stud. Math. 20, 1-18.

[2] Krener, A.J. (1979), A formal approach to stochastic integration and differential equations, Stochastics 3, 105-25.

[3] Oldham, K.B. and Spanier, J.K. (1974), The Fractional Calculus (Acad. Pr., New York).

[4] Osler, T.J. (1971), Taylor's series generalized for fractional derivatives and aplications, SIAM J. Math. Anal. 2, 37-48.

[5] Osler, T.J. (1972), An integral analogue of Taylor's series and its use in computing Fourier transforms, Math. Comp. 26, 449-60.

[6] Rao, C.R. (1973), Linear Statistical Inference and its Applications (second edition, Wiley, New York).

[7] Rao, C.R. and Mitra, S.K. (1971), Generalized Inverses of Matrices and its Applications (Wiley, New York).

[8] Roberts, A.W. and Varberg, D.E. (1973), Convex functions (Acad. Pr., New York).

[9] Seierstad, A. and Sydsaeter, K. (1977), Sufficient conditions in optimal control theory, Internat. Econom. Rev. 18, 367-91.

[10] Vetter, W.J. (1973), Matrix calculus operations and Taylor expansions, SIAM Rev. 15, 352-69.

Part Five
MULTIVARIATE ANALYSIS

And look—a thousand Blossoms with the Day
Woke—and a thousand scatter'd into Clay,
And this first Summer Month that brings the Rose
Shall take jamshýd and Kaikobád away.

Khayyam Naishapuri-Rubáiyát

STATISTICAL DATA ANALYSIS AND INFERENCE
Y. Dodge (Editor)
© Elsevier Science Publishers B.V. (North-Holland), 1989

DISTANCE ANALYSIS IN DISCRIMINATION AND CLASSIFICATION USING BOTH CONTINUOUS AND CATEGORICAL VARIABLES

Carles M. CUADRAS

Departament d'Estadística
Universitat de Barcelona, Diagonal 645
08028 Barcelona, Spain

This paper is concerned with the application of distance functions, together with principal coordinate analysis, to some multivariate problems, namely multiple regression, manova and discriminant analysis. The main problem tackled is that of giving a distance-based rule to assign an individual to one of two distinct groups. The discriminant analysis with both continuous and discrete variables, as well as that with binary, ordinal and qualitative data, is also studied. Closed form expressions of distances between individuals are proposed and discussed.

1. INTRODUCTION

The treatment of statistical problems by the decision rule based on the notion of distance is a well-known methodology. Starting from the pioneering works of Pearson, Fisher and Rao, the first application of this approach seems to be due to Matusita [1]. Since then, many statisticians (Gupta, Dillon, Goldstein, Le Cam, Krzanowski, Amari, Burbea, Oller,...) have attempted to construct distance-based decision rules to discrimination, classification and other multivariate problems. The purpose of this paper is to provide an approach to multiple regression, manova and discriminant analysis from the same point of view.

We show how the application of distance functions, together with principal coordinate analysis, allow us to define a multiple regression model with both continuous and categorical variables. Moreover, using intradistance and interdistance formulas, we will be able to study and interpret some aspects of multivariate analysis of variance. However, the main problem treated here is to show how we can set up a discriminant rule by using an individual distance function. Both the sample case and the parametric case are considered. A solution to the discrimination with mixed variables is also given.

2. DISTANCE ANALYSIS IN MULTIPLE REGRESSION AND MANOVA

Let $\mathbf{Y}(n\text{x}p)$ be an observed matrix of p quantitative variables on each of n independent experimental units or individuals, $\mathbf{X}(n\text{x}m)$ a known design matrix, $\mathbf{B}(m\text{x}p)$ a parametric regression matrix and $\mathbf{U}$ a matrix of random errors whose rows are uncorrelated, each with mean $\mathbf{0}$ and common covariance matrix Σ. Consider the model

$$\mathbf{Y} = \mathbf{X}\,\mathbf{B} + \mathbf{U} \tag{1}$$

and a system of q estimable parametric functions

$$\Psi_i = P_i'\,\mathbf{B}, \qquad i = 1,\ldots,q. \tag{2}$$

Let us define the square distance

$$D^2(\Psi_i,\Psi_j) = (\Psi_i - \Psi_j)'\Sigma^{-1}(\Psi_i - \Psi_j). \tag{3}$$

Substituting Ψ and Σ by the usual estimates in the manova model (1) we obtain a suitable estimation of (3). Clearly, this estimated distance is a Euclidean distance (see [2, 3] for other desirable properties). The relationships among the estimable parametric functions can be explored geometrically by using principal coordinate analysis [4], which leads to a coordinate representation of the system (2) in a low-dimensional Euclidean space. In this way, we generalize the canonical variate analysis [5] to represent the levels of two factors in a two-way analysis [6] or all the levels in a k-way analysis [7, 8]. Some applications to Medicine, Genetics and Agriculture can be found in [9, 10, 11] respectively.

Now consider s independent manova models

$$\mathbf{Y}_i = \mathbf{X}\,\mathbf{B}_i + \mathbf{U}_i, \qquad i = 1,\ldots,s,$$

and suppose that the matrices have the same order for any i. Let us define the square distance

$$L^2(i,j) = \mathrm{tr}\,[\Sigma^{-1}(\mathbf{B}_i - \mathbf{B}_j)\mathbf{X}'\mathbf{X}(\mathbf{B}_i - \mathbf{B}_j)]. \tag{4}$$

Extensions of distance (4) to different order matrices as well as some inference properties have been studied by Cuadras and Rios [12, 13]. This distance could be used to test $H_0\!: \mathbf{X}\,\mathbf{B}_i = \mathbf{X}\,\mathbf{B}_j$. On the other hand, we may use principal coordinate analysis and cluster analysis for classification purposes. For example, we may wish to compare the polynomial growth trends for s populations and perform a classification of the populations. See [14, 15] for some applications. See also Burbea and Oller [16].

Principal coordinate analysis on an ultrametric matrix may also be used to elucidate the Euclidean dimensionality of hierarchical clustering scheme [17, 18].

Suppose now that we wish to relate a continuous variable y to a variable-vector W. The usual regression model would not be suitable if W was a mixture of continuous, binary and categorical variables. In order to overcome this problem, we consider n individuals {1,2,...,n} (say). Taking measurements on y and W we obtain the vector

$$\mathbf{y} = (y_1,...,y_n)',$$

and the distance matrix

$$\mathbf{D} = (d_{ij}), \qquad i,j = 1,...,n.$$

D can be computed in diferent ways. For example, Gower [19] has proposed an all-purpose measure of similarity c_{ij} taking into account mixed variables, such that $d_{ij}=\sqrt{c_{ii} + c_{jj} - 2c_{ij}}$ is a Euclidean distance.

Henceforth we suppose that $\mathbf{D}=(d_{ij})$ is a Euclidean distance matrix. Let $\mathbf{A}=(a_{ij})$, where $a_{ij}=-(d^2_{ij})/2$, and set

$$\mathbf{B} = \mathbf{HAH}, \tag{5}$$

where $\mathbf{H} = \mathbf{I}-n^{-1}\mathbf{11}'$ and $\mathbf{1} = (1,...,1)'$. Then $\mathbf{B} \geq \mathbf{0}$, rank $\mathbf{B} = m$, and

$$\mathbf{B} = \mathbf{X}\,\mathbf{X}',$$

$\mathbf{X}(n\times m)$ being a matrix of rank m. If, in addition,

$$\mathbf{X}'\mathbf{X} = \Lambda,$$

where $\Lambda = \mathrm{diag}(\lambda_1,...,\lambda_n)$ is the diagonal matrix of eigenvalues of **B**, then the rows of **X** are the principal coordinates of {1,2,...,n} with respect to the distance matrix **D** [4,5]. In other words, there exists a Euclidean configuration of points, namely, $\mathbf{x}_1,...,\mathbf{x}_n \in \mathbb{R}^m$, such that

$$d^2_{ij} = (\mathbf{x}_i - \mathbf{x}_j)'(\mathbf{x}_i - \mathbf{x}_j).$$

In order to relate y to W, or equivalently, y to **D**, we suggest the linear regression model

$$\mathbf{y} = \mu\mathbf{1} + \mathbf{X}\beta + \mathbf{u}. \tag{6}$$

Note that **X** is a centered matrix and both **1** and the columns of **X** are eigenvectors of **B**. Thereby it is easy to prove that the LS estimates of μ and β are

$$\hat{\mu} = \bar{y}, \qquad \hat{\beta} = \Lambda^{-1}\mathbf{X}'\mathbf{y}. \tag{7}$$

It can also be proved [20] that

$$R^2 = \mathbf{y}'\mathbf{X}\,\Lambda^{-1}\mathbf{X}'\mathbf{y}\ /\Sigma(y_i - \bar{y})^2 \tag{8}$$

is the square multiple correlation of y with **X**. Therefore R^2 provides a measure of association of the continuous variable y with the mixed vector W.

Next, suppose that a further individual n+1 (say) becomes available and we know the distance d_{in+1} (i=1,...,n). We wish to predict the measurement of variable y on n+1. Setting

$$\mathbf{d} = (d^2_{1n+1},\ldots,d^2_{nn+1})',$$

$$\mathbf{b} = (b_{11},\ldots,b_{nn})',$$

the prediction is given by

$$y_{n+1} = \bar{y} + \tfrac{1}{2}(\mathbf{b} - \mathbf{d})'\mathbf{B}^{-}\mathbf{y}, \tag{9}$$

where $\mathbf{B}^-$ is a g-inverse of **B** (see [20]).

We have checked these results for mixed qualitative and quantitative real data [21]. We find that model (6) improves the classical regression model by scoring qualitative variables and proceeding as if they were all continuous. It also improves the so-called MORALS algorithm [22]. However, the problem with this distance method is to define a suitable distance function.

3. A DISTANCE-BASED CLASSIFICATION RULE

Suppose we have a set C_i of n_i individuals belonging to Π_i, i=1,2. We propose a discriminant rule between Π_1 and Π_2 on the basis of a distance function d(.,.).

Suppose that d(.,.) can be calculated on the basis of a vector W of discrete and continuous variables and let $\mathbf{D}_1 = (d_{ij})$ be the $n_1 x n_1$ intradistance matrix on C_1. Let ω be an individual to be allocated and suppose that, using d(.,.) again, we can calculate the distances d_i, i=1,...,n_1, from each of the set C_1.

Let us define the function

$$f_1(\omega) = \frac{1}{n_1}\sum_i d_i^2 - \frac{1}{2n_1^2}\sum_{ij} d_{ij}^2 \ . \tag{10}$$

In an analogous way we define $f_2(\omega)$. In order to decide whether ω belongs to Π_1 or Π_2, we suggest the classification rule:

$$\text{allocate } \omega \text{ to } \Pi_i \quad \text{if} \quad f_i(\omega) = \min\{f_1(\omega), f_2(\omega)\}. \tag{11}$$

The function (10) and the classification rule (11) satisfy some properties.

3.1. Euclidean distance

Suppose that d(.,.) is a Euclidean distance. Then we can find a

Euclidean configuration $\mathbf{x}_1,\ldots,\mathbf{x}_{n_1},\mathbf{x} \in \mathbb{R}^p$ such that

$$d_i^2 = (\mathbf{x}_i - \mathbf{x})'(\mathbf{x}_i - \mathbf{x}),$$

$$d_{ij}^2 = (\mathbf{x}_i - \mathbf{x}_j)'(\mathbf{x}_i - \mathbf{x}_j).$$

Set

$$\bar{\mathbf{x}} = \frac{1}{n_1}\sum_i \mathbf{x}_i.$$

After some algebra we find that

$$\sum_{ij} d_{ij}^2 = 2n_1 \sum_i (\mathbf{x}_i - \bar{\mathbf{x}})'(\mathbf{x}_i - \bar{\mathbf{x}}),$$

$$\sum_i d_i^2 = n_1 d^2(\mathbf{x},\bar{\mathbf{x}}) + \sum_i (\mathbf{x}_i - \bar{\mathbf{x}})'(\mathbf{x}_i - \bar{\mathbf{x}}),$$

hence

$$f_1(\omega) = d^2(\mathbf{x},\bar{\mathbf{x}}).$$

Analogously, C_2 and ω are represented by $\mathbf{y}_1,\ldots,\mathbf{y}_{n_2},\mathbf{y} \in \mathbb{R}^q$ and we obtain

$$f_2(\omega) = d^2(\mathbf{y},\bar{\mathbf{y}}).$$

Note that $\mathbf{x} \neq \mathbf{y}$ is possible. Therefore rule (11) becomes

$$\text{allocate } \omega \text{ to } \Pi_1 \quad \text{if} \quad d^2(\mathbf{x},\bar{\mathbf{x}}) < d^2(\mathbf{y},\bar{\mathbf{y}}), \tag{12}$$

and to Π_2 otherwise.

3.2. Mahalanobis distance

Suppose that Π_k is the $\mathbf{N}_m(\mu_k,\Sigma_k)$ population, $k = 1,2$. Let $\mathbf{x}_1^{(k)},\ldots,\mathbf{x}_{n_k}^{(k)}$ be the observed sample known to come from Π_k, and let $\mathbf{x}$ be the observation to be allocated. If the distance function is the Mahalanobis distance

$$(\mathbf{x}_i^{(k)} - \mathbf{x}_j^{(k)})'\Sigma_k^{-1}(\mathbf{x}_i^{(k)} - \mathbf{x}_j^{(k)}), \tag{13}$$

the classification rule (11) is based on the quadratic discriminant function (QDF). Replacing the parameters by the unbiased estimation, the sample QDF is

$$Q(\mathbf{x}) = \frac{1}{2}\log\frac{|\mathbf{S}_2|}{|\mathbf{S}_1|} - \frac{1}{2}\log(\bar{\mathbf{x}}^{(1)'}\mathbf{S}_1^{-1}\bar{\mathbf{x}}^{(1)} - \bar{\mathbf{x}}^{(2)'}\mathbf{S}_2^{-1}\bar{\mathbf{x}}^{(2)})$$
$$+ \mathbf{x}'(\mathbf{S}_1^{-1}\bar{\mathbf{x}}^{(1)} - \mathbf{S}_2^{-1}\bar{\mathbf{x}}^{(2)}) - \frac{1}{2}\mathbf{x}'(\mathbf{S}_1^{-1} - \mathbf{S}_2^{-1})\mathbf{x}.$$

The QDF reduces to the linear discriminant function (LDF)

$$L(\mathbf{x}) = [\mathbf{x} - \tfrac{1}{2}(\bar{\mathbf{x}}^{(1)} + \bar{\mathbf{x}}^{(2)})]'\mathbf{S}^{-1}(\bar{\mathbf{x}}^{(1)} - \bar{\mathbf{x}}^{(2)}), \tag{14}$$

when it is assumed that $\Sigma_1 = \Sigma_2$, and $\mathbf{S}_1$ and $\mathbf{S}_2$ are replaced by the pooled sample covariance matrix $\mathbf{S}$.

3.3. Non-Euclidean distance

Suppose that the distance matrix $\mathbf{D}=(d_{ij})$ on C_1 is non-Euclidean. Then the matrix $\mathbf{B}$ defined in (5) has r positive and s negative eigenvalues, where $r+s = n_1-1$. Let

$$\mathbf{X} = \begin{pmatrix} x_{11} & \dots & x_{1r} \\ & \dots\dots & \\ x_{n_1 1} & \dots & x_{n_1 r} \end{pmatrix} \qquad \mathbf{Z} = \begin{pmatrix} z_{11} & \dots & z_{1s} \\ & \dots & \\ z_{n_1 1} & \dots & z_{n_1 s} \end{pmatrix}$$

be the real matrices such that the columns of $\mathbf{X}$ and $\mathbf{Z}$ are the eigenvectors with positive and negative eigenvalues, respectively. Then we can identify each individual of C_1 as a point with the coordinates

$$\begin{aligned}&(x_{i1},\dots,x_{ir},z_{i1}\sqrt{-1},\dots,z_{is}\sqrt{-1},\ 0)\\ &= (\mathbf{x}_i',\ \mathbf{z}_i'\sqrt{-1},\ 0), \qquad i=1,\dots,n_1,\end{aligned} \tag{15}$$

and it is verified that

$$d_{ij}^2 = (\mathbf{x}_i - \mathbf{x}_j)'(\mathbf{x}_i - \mathbf{x}_j) - (\mathbf{z}_i - \mathbf{z}_j)'(\mathbf{z}_i - \mathbf{z}_j), \quad i,j=1,\dots,n_1.$$

Note that the mean of the n_1 points (15) is $\mathbf{0}$.

Let Λ_x, Λ_z be the diagonal matrices with the positive and negative eigenvalues of $\mathbf{B}$, respectively. Extending the formulas given by Gower [23], the coordinates of the individual ω to be allocated, with respect to C_1, are

$$\mathbf{w}_1 = (x_1,\dots,x_r,z_1\sqrt{-1},\dots,z_s\sqrt{-1},y) = (\mathbf{x}',\mathbf{z}'\sqrt{-1},y),$$

where

$$\mathbf{x} = \tfrac{1}{2}\Lambda_x^{-1}\mathbf{X}'\mathbf{v}, \qquad \mathbf{z} = \tfrac{1}{2}\Lambda_z^{-1}\mathbf{Z}'\mathbf{v},$$

$$y^2 = \frac{1}{n_1}\left(\sum_i d_i^2 - \text{tra}\,\mathbf{B}\right) - (\mathbf{x}'\mathbf{x} - \mathbf{z}'\mathbf{z}), \tag{16}$$

$$\mathbf{v} = (b_{11} - d_1^2,\dots,b_{n_1 n_1} - d_{n_1}^2)'.$$

We obtain the coordinates $\mathbf{w}_2$ of ω with respect to C_2 analogously.

As $\mathbf{0}$ is the mean, a reasonable classification rule is:

$$\text{allocate } \omega \text{ to } \Pi_i \text{ if } d^2(\mathbf{w}_i,\mathbf{0}) = \min\{d^2(\mathbf{w}_1,\mathbf{0}),d^2(\mathbf{w}_2,\mathbf{0})\}. \tag{17}$$

But, on relating (10) to (16) we obtain

$$d^2(\mathbf{w}_i,\mathbf{0}) = f_i(\omega).$$

Thus, (17) becomes (11), and the suggested classification rule holds, even for a non-Euclidean distance.

4. A PROBABILISTIC FORMULATION

Suppose that the random variables are related to a random vector with a probability density function $p_i(\mathbf{x})$, with respect to a suitable measure λ, if $\mathbf{x}$ comes from Π_i, i=1,2. Let $\mathbf{x}_0$ be the observation to be allocated and let d(.,.) be a distance function. For fixed $\mathbf{x}_0$, $d^2(\mathbf{x}_0,\mathbf{x})$ is a random variable defined on Π_i and $d^2(\mathbf{x},\mathbf{y})$ is a random variable defined on $\Pi_i \times \Pi_i$.

Let us define

$$f_i(\mathbf{x}_0) = E_{\Pi_i}\{d^2(\mathbf{x}_0,\mathbf{x})\} - \tfrac{1}{2}E_{\Pi_i^2}\{d^2(\mathbf{x},\mathbf{y})\}, \tag{18}$$

where E means mathematical expectation, i.e.,

$$E_{\Pi_i}\{d^2(\mathbf{x}_0,\mathbf{x})\} = \int d^2(\mathbf{x}_0,\mathbf{x})p_i(\mathbf{x})d\lambda(\mathbf{x}),$$

$$E_{\Pi_i^2}\{d^2(\mathbf{x},\mathbf{y})\} = \int d^2(\mathbf{x},\mathbf{y})p_i(\mathbf{x})p_j(\mathbf{y})d\lambda(\mathbf{x})d\lambda(\mathbf{y}).$$

Consider now the classification rule:

$$\text{allocate } \mathbf{x}_0 \text{ to } \Pi_i \quad \text{if} \quad f_i(\mathbf{x}_0) = \min\{f_1(\mathbf{x}_0),f_2(\mathbf{x}_0)\}. \tag{19}$$

Remarks:

4.1. We define the diversity coefficient of Π_i as

$$H_i = E_{\Pi_i^2}\{d^2(\mathbf{x},\mathbf{y})\}.$$

Let $\Pi_0 = \{\mathbf{x}_0\}$ be. Then $H_0 = 0$ and the average difference between Π_0 and Π_i is

$$H_{i0} = E_{\Pi_i}\{d^2(\mathbf{x}_0,\mathbf{x})\}.$$

Hence

$$f_i(\mathbf{x}_0) = H_{i0} - \tfrac{1}{2}(H_i + H_0) \tag{20}$$

is the Jensen difference between Π_i and Π_0 .See Rao [24].

4.2. If Π_i is the $\mathbf{N}_m(\mu_i,\Sigma_i)$ population and we choose the Mahalanobis distance, then it is verified that

$$H_{i0} = (\mathbf{x}_0 - \mu_i)'\Sigma_i^{-1}(\mathbf{x}_0 - \mu_i) + m,$$

$$H_i = 2m,$$

hence

$$f_i(\mathbf{x}_0) = (\mathbf{x}_0 - \mu_i)'\Sigma_i^{-1}(\mathbf{x}_0 - \mu_i). \tag{21}$$

4.3. Suppose that $\mathbf{x} = (x_1,\ldots,x_m)$ has the multinomial distribution

$$p_1^{x_1}\ldots p_m^{x_m}, \qquad x_i\in\{0,1\},$$

if $\mathbf{x}$ comes from Π_1, and the multinomial distribution

$$q_1^{x_1}\ldots q_m^{x_m}, \qquad x_i\in\{0,1\},$$

if $\mathbf{x}$ comes from Π_2.

Suppose that $\mathbf{x} = (x_1,...,x_m)$, $\mathbf{y} = (y_1,...,y_m)$ come from Π_1 and $x_i=1$, $y_j=1$, i.e., $\mathbf{x}$ falls in the cell i and $\mathbf{y}$ falls in the cell j. In the next section, we justify the square distance

$$d^2(\mathbf{x},\mathbf{y}) = (1 - \delta_{ij})(p_i^{-1} + p_j^{-1}), \tag{22}$$

where δ_{ij} is the Kronecker delta. If $\mathbf{x}_0= (x_1^0,...,x_m^0)$, where $x_k^0=1$, we then obtain

$$H_{10} = \sum_{j\neq k} p_j(p_j^{-1} + p_k^{-1}) = (m - 2) + p_k^{-1},$$

$$H_1 = \sum_{i\neq j}\sum p_i p_j (p_i^{-1}+ p_j^{-1}) = 2(m - 1),$$

$$f_1(\mathbf{x}_0) = (1 - p_k)/p_k.$$

Analogously, we obtain

$$f_2(\mathbf{x}_0) = (1 - q_k)/q_k.$$

The classification rule (19) is therefore:

allocate $\mathbf{x}_0$ to Π_1 if $p_k > q_k$,

and to Π_2 otherwise.

4.4. Suppose that we have a random vector W = (X,Y), where X and Y have the multivariate normal and the multinomial distribution respectively. If it is known that X and Y are independent of each other, then we can make use of items 4.2 and 4.3 in order to obtain an allocation rule based on the sum of the square distances (13) and (22).

Generally, we may obtain a classification rule for mixed data if we are able to define a suitable distance function. Once again this is the problem.

5. DISTANCES BETWEEN INDIVIDUALS

Suppose that we have sample data arising from measuring q characteristics on each of N individuals. Suppose that the variables may be quantitative or qualitative (nominal, ordinal or dichotomous). If no probabilistic structure in the variables is defined, we can use the all-purpose measure of similarity between individuals proposed by Gower [19]. This similarity can be written as [5]

$$s_{ij} = \sum_{k=1}^{n_1} (1-|x_{ik}- x_{jk}|/R_k)+a+\alpha)/[n_1+(n_2-d)+n_3], \tag{23}$$

where n_1 is the number of continuous variables, a and d are the number of positive and negative matches, respectively, for the n_2

dichotomous variables, and α is the number of matches for the n_3 multistate variables. R_k is the range of variable k. If $q=n_2$ then s_{ij} reduces to Jaccard's coefficient, whereas if all the variables are binary, then s_{ij} reduces to the matching coefficient. In addition, it is necessary to introduce some corrections in (23) when there are some data missing. The square distance between individuals i,j is

$$d_{ij}^2 = 1 - s_{ij}. \tag{24}$$

The matrix $\mathbf{D} = (d_{ij})$ is a Euclidean distance matrix on the set of N individuals, except (possibly) when there are missing data.

Now suppose a probabilistic context. Then the populations Π_1 and Π_2 can be characterized by probability density functions $p_1(\mathbf{x}), p_2(\mathbf{x})$. We can use the Matusita distance [1] between $p_1(\mathbf{x})$ and $p_2(\mathbf{x})$ in order to define a distance between Π_1 and Π_2. An observation $\mathbf{x}_0$ to be allocated can be represented by a degenerate population with probability density function

$$p(\mathbf{x}) = 1 \qquad \text{if} \quad \mathbf{x} = \mathbf{x}_0,$$
$$p(\mathbf{x}) = 0 \qquad \text{otherwise.}$$

Adopting the Matusita distance between $p_i(\mathbf{x})$ and $p(\mathbf{x})$, we can obtain an allocation rule by the minimum distance. This allocation rule is studied by Krzanowski [25] on the basis of the location model for mixed data. However, it is not feasible to give a definition of distance between individuals in a similar way.

Now assume a parametrization, i.e., suppose we have a probability density function $p(\mathbf{x}|\theta)$, where $\theta = (\theta_1,\ldots,\theta_m)$ is a parametric vector belonging to a parameter space Θ. Suppose that each population Π_i is characterized by a particular point θ of Θ. We may interpret Θ as a manifold and consider the Fisher information matrix $\mathbf{G} = (g_{ij}(\theta))$ as a covariant symmetric tensor of secon order for all $\theta\in\Theta$, where

$$g_{ij}(\theta) = E[\frac{\partial}{\partial\theta_i}\log p(X|\theta)\frac{\partial}{\partial\theta_j}\log p(X|\theta)], \qquad i,j=1,\ldots,m.$$

Thereby, Θ is a Riemannian manifold and we can define a geodesic distance between two points θ_A and θ_B of Θ. The distance obtained is called the "information geodesic distance" or "the Rao distance" and is invariant under admissible transformations of the parameters as well as the random variables.

This distance was proposed by Rao [26] and studied later by Atkinson and Mitchell [27], Burbea and Rao [28], Oller and Cuadras [29], Burbea [30] and Mitchell [31].

The Rao distance method enables us to define a distance between two individuals $\mathbf{x}_i,\mathbf{x}_j$ for each $\theta\in\Theta$. Consider the random variables

$$Z_{\theta_i} = \frac{\partial}{\partial\theta_i}\log p(X|\theta), \qquad i=1,\dots,m. \tag{25}$$

Assuming certain regularity conditions it is verified that

$$E\,(Z_{\theta_i}) = 0, \qquad i=1,\dots,m.$$

Hence, the Fisher information matrix $\mathbf{G} = (g_{ij}(\theta))$ is given by

$$g_{ij}(\theta) = E\,(Z_{\theta_i} Z_{\theta_j}), \qquad i,j=1,\dots,m.$$

i.e., $\mathbf{G}$ is the covariance matrix of $Z_{\boldsymbol{\theta}} = (Z_{\theta_1},\dots,Z_{\theta_m})$ at θ. On the other hand [32], there exists a natural isomorphism between the tangent space $T_{\boldsymbol{\theta}}$ and the vectorial space $\langle Z_{\boldsymbol{\theta}}\rangle$ spanned by the variables (25). Thereafter, characterizing a population by a point θ and identifying each individual as an element of the dual space $\langle Z_{\boldsymbol{\theta}}\rangle^*$, we find a natural way to define a distance between individuals.

Let $\mathbf{x}_i,\mathbf{x}_j$ be two individuals and consider the observed values

$$\mathbf{z}_i = (z_{i1},\dots,z_{im})', \qquad \mathbf{z}_j = (z_{j1},\dots,z_{jm})',$$

of $Z_{\boldsymbol{\theta}}$. We define the square distance

$$d^2(\mathbf{x}_i,\mathbf{x}_j) = (\mathbf{z}_i - \mathbf{z}_j)'\mathbf{G}^{-1}(\mathbf{z}_i - \mathbf{z}_j). \tag{26}$$

Note that (26) is the local Euclidean distance on $\langle Z_{\boldsymbol{\theta}}\rangle^*$ induced by the metric $\mathbf{G}$ at θ. Some theoretical results on the above mentioned distance may be found in [33,34]. See also [35].

We now consider some specific examples and a generalization.

5.1. Normal distribution (Σ fixed)

If we have the $\mathbf{N}_m(\mu,\Sigma)$ population and Σ is considered as a fixed matrix, then

$$Z_\mu = \Sigma^{-1}(X - \mu),$$
$$\mathbf{G} = \Sigma^{-1}.$$

Therefore, the distance between $\mathbf{z}_i=\Sigma^{-1}(\mathbf{x}_i-\mu)$ and $\mathbf{z}_j=\Sigma^{-1}(\mathbf{x}_j-\mu)$ is the familiar Mahalanobis distance.

5.2. Multinomial distribution

Suppose that $(x_1,\dots,x_m,x_{m+1})$ has the multinomial distribution

$$p_1^{x_1}\dots p_m^{x_m}\, p_{m+1}^{x_{m+1}}, \qquad x_i\in\{0,1\},$$

with $0<p_i<1$, $i=1,\dots,m$, and $p_{m+1}=1-\Sigma p_i$, $x_{m+1}=1-\Sigma x_i$. The metric tensor on the manifold $\{(p_1,\dots,p_m) \mid \Sigma p_i<1\}$ is given by

$$g_{ij} = p_i\delta_{ij} + 1/p_{m+1}.$$

On using the transformation (25) and substituting **G** in (26), the distance already introduced in (22) is found.

Suppose that the variable is ordinal, that is, an ordered multistate variable, with each state scaled according to

$$0 < \alpha_1 < \dots < \alpha_{m+1}.$$

In this case, a generalization of (22) could be

$$\left|\frac{\alpha_i - \alpha_j}{\alpha_{m+1} - \alpha_1}\right| (1 - \delta_{ij})(p_i^{-1} + p_j^{-1}). \tag{27}$$

5.3. Negative binomial

Suppose that the probability density function is

$$p(x|\theta, r) = \frac{\Gamma(x + r)}{\Gamma(r).r!} \theta^r (1 - \theta)^x, \qquad x=0,1,2,\dots$$

with $0 < \theta < 1$, $r > 0$.

For a fixed r we obtain

$$Z = \frac{r}{\theta} - \frac{X}{1-\theta},$$

$$g(\theta) = \frac{r}{(1-\theta)\theta^2},$$

and the distance is

$$d(x_i, x_j) = \frac{\theta|x_i - x_j|}{\sqrt{r(1-\theta)}}.$$

If both r and θ are not fixed parameters, then (25) gives

$$Z_1 = \frac{r}{\theta} - \frac{X}{1-\theta},$$

$$Z_2 = \frac{1}{r} + \frac{1}{r+1} + \dots + \frac{1}{r+X-1} + \frac{r}{\theta}.$$

In this case, it is not easy to calculate the Fisher information matrix, but a suitable estimation may be obtained by using the results shown in the following section.

5.4. The general case

Suppose that we have a random vector W with a pdf $p(\mathbf{w}|\theta)$, where θ is a parametric vector. Assuming certain regularity conditions, consider the column random vector

$$Z = \frac{\partial}{\partial\theta} \log p(W|\theta), \tag{28}$$

the Fisher information matrix

$$\mathbf{G} = E(Z.Z'), \tag{29}$$

and the quadratic form

$$Z'\mathbf{G}^{-1}Z. \tag{30}$$

We now consider some specific situations in order to compute and estimate (29) and (30).

5.4.1. Suppose that θ can be partitioned into $\theta = (\theta^{(1)}, \theta^{(2)})$. Setting

$$Z_{(i)} = \frac{\partial}{\partial \theta^{(i)}} \log p(W|\theta^{(i)}), \qquad i=1,2, \tag{31}$$

then

$$\mathbf{G} = \begin{pmatrix} \mathbf{G}_{(11)} & \mathbf{G}_{(12)} \\ \mathbf{G}_{(21)} & \mathbf{G}_{(22)} \end{pmatrix}, \tag{32}$$

where

$$\mathbf{G}_{(ij)} = E(Z_{(i)} Z'_{(j)}).$$

5.4.2. Suppose that $W = (X_1, X_2)$, where X_1 and X_2 are random vectors with pdfs $p_1(\mathbf{x}_1|\theta_1)$ and $p_2(\mathbf{x}_2|\theta_2)$, respectively, and $\theta_i = (\theta_{i1}, \ldots, \theta_{im_i})$, i=1,2, are parametric vectors. For example, X_1 could be a vector of discrete variables and X_2 a vector of continuous variables. Setting

$$Z_i = \frac{\partial}{\partial \theta_i} \log p(X_i|\theta_i), \qquad i=1,2, \tag{33}$$

$$\mathbf{G}_{ij} = E(Z_i Z'_j),$$

$$\mathbf{G} = \begin{pmatrix} \mathbf{G}_{11} & \mathbf{G}_{12} \\ \mathbf{G}_{21} & \mathbf{G}_{22} \end{pmatrix},$$

then

$$(Z'_1, Z'_2)\ \mathbf{G}^{-1} (Z'_1, Z'_2)' \tag{34}$$

is a quadratic form for a pdf $p(\mathbf{w})$ with marginals $p_1(\mathbf{x}_1|\theta_1)$ and $p_2(\mathbf{x}_2|\theta_2)$.

Note the difference between (31) and (33) and that here $\mathbf{G}$ is not a Fisher information matrix. Despite this fact, (34) can be used to define a distance between individuals.

5.4.3. We now wish to estimate $\mathbf{G}$. Let $\mathbf{w}_i$, i=1,...,N, be a random sample from a population with pdf $p(\mathbf{w})$. Replacing the parameters by the ML estimates, and using the transformation (31) or (33), we obtain the data matrices $\mathbf{Z}_1, \mathbf{Z}_2$.

Let $\bar{\mathbf{z}}_i = \frac{1}{N}\mathbf{Z}'_i\mathbf{1}$, i=1,2, be the sample mean vectors. Then

$$\bar{\mathbf{z}}_1 = \mathbf{0}, \qquad \bar{\mathbf{z}}_2 = \mathbf{0},$$

and $N^{-1}\mathbf{Z}'_i\mathbf{Z}_j$ converges, for $N \to \infty$, in probability to $\mathbf{G}_{ij}$.

Thus, if we have two observations $\mathbf{w}_i$, $\mathbf{w}_j$ and we previously estimate the parameters, we obtain $\hat{\mathbf{z}}_i$, $\hat{\mathbf{z}}_j$ by using either (31) or (33). Then, the estimation of the square distance (26) is given by

$$(\hat{\mathbf{z}}_i - \hat{\mathbf{z}}_j)'\hat{\mathbf{G}}^{-1}(\hat{\mathbf{z}}_i - \hat{\mathbf{z}}_j), \tag{35}$$

with

$$\hat{\mathbf{G}} = \frac{1}{N}\begin{pmatrix} \mathbf{Z}_1'\mathbf{Z}_1 & \mathbf{Z}_1'\mathbf{Z}_2 \\ \mathbf{Z}_2'\mathbf{Z}_1 & \mathbf{Z}_2'\mathbf{Z}_2 \end{pmatrix}. \tag{36}$$

6. DISCRIMINANT ANALYSIS FOR MIXED DATA

Once again, we consider the classification problem. Suppose that we have a set C_k of N_k individuals belonging to Π_k, k=1,2. Using the previous results, we obtain the data matrices $\mathbf{Z}_1, \mathbf{Z}_2$ and from (36) we obtain the estimates $\hat{\mathbf{G}}_k$, k=1,2. Let $\mathbf{w}$ be an individual to be allocated. As the sample mean vectors are zero, we may define the square distance between $\mathbf{w}$ and C_k by

$$f_k(\mathbf{w}) = d^2(\mathbf{w},\mathbf{0}) = \hat{\mathbf{z}}_k'\hat{\mathbf{G}}_k^{-1}\hat{\mathbf{z}}_k, \tag{37}$$

where $\mathbf{w}$ is transformed to $\hat{\mathbf{z}}_k$.

We now suggest the classification rule:

$$\text{allocate } \mathbf{w} \text{ to } \Pi_i \text{ if } f_i(\mathbf{w}) = \min\{f_1(\mathbf{w}), f_2(\mathbf{w})\}. \tag{38}$$

Remarks:

6.1. The classification rule (38) is closely related to (11), (12) and (17).

6.2. Extensions to higher dimensions, namely $\theta=(\theta^{(1)},\ldots,\theta^{(m)})$, $W=(X_1,\ldots,X_m)$, m>2, do not present further difficulties.

6.3. Krzanowski [25,36] uses a distance-based classification rule for the mixed data case, based on the location model. The location model [37] is applicable when the data contain both binary and continuous variables. This model computes s LDFs, one for each s distinct pattern of the binary variables. We can also apply the classification rule (11) for mixed data using (24), or apply the more general rule (38) using the distance (26).

6.4. The rule (38) is essentially equivalent to (11) by using the universally accepted Mahalanobis distance. Nevertheless, it is worth pointing out that the classification rule (11) is valid even for a non-Euclidean distance (see section 3.3).

6.5. We may obtain a real extension, with respect to the classical results, by considering a family of pdfs $p(\mathbf{w}) = p(\mathbf{x}_1,\mathbf{x}_2)$ with given marginals $p_1(\mathbf{x}_1)$, $p_2(\mathbf{x}_2)$ [38,39,40].

6.6. Another extension with respect to the classification rule based on the Mahalanobis distance, would be obtained by considering some parameters as non-fixed parameters. For example, by considering that the covariance matrix Σ is non fixed in the normal distribution or that r is non fixed in the negative binomial distribution.

REFERENCES

[1] Matusita, K., Ann. Math. Stat. 26 (1955) 631-640.
[2] Cuadras, C.M., Trab. Est. I. Oper. 25 (1974) 3-31.
[3] Arenas, C., On a distance between estimable functions, Qüestiio (1989) in print.
[4] Gower, J.C., Biometrika 53 (1966) 315-328.
[5] Seber, G.A.F., Multivariate Observations (Wiley, N. York, 1984).
[6] Takeuchi, K., Yanai, H. and Mukherjee, B.N., The Foundations of Multivariate Analysis (Wiley El., New Delhi, 1982).
[7] Cuadras, C.M., Metodos de Analisis Multivariante (Eunibar, Barcelona, 1981).
[8] Cuadras, C.M., Analisis y representacion multidimensional de la variabilidad, in: Martinell, J. (ed.), Inter. Symp. Concept. Method Paleo., (Barcelona, 1981) pp. 287-297.
[9] Ballus, C., Cuadras, C.M., Malga, A., Sanchez-Turet, M. and Vallve, C., Rev. Dep. Psiquiatria, Univ. de Barcelona, 7 (1980) 107-122.
[10] Coll, M.D., Cuadras, C.M. and Egozcue, J., Genet. Res. 36 (1980) 219-234.
[11] Oller, J.M. and Cuadras, C.M., Qüestiio 6 (1982) 221-229.
[12] Cuadras, C.M. and Rios, M., A distance between multivariate linear models and its properties, in: Bonet, E. (ed.), II Catalan Inter. Symp. on Stat., Vol. 2, Cont. Papers (Barcelona, 1986) pp.81-84.
[13] Rios, C.M. and Cuadras, C.M., Qüestiio 10 (1986) 83-92.
[14] Cuadras, C.M., Oller, J.M., Arcas, A., and Rios, M., Qüestiio 9 (1985) 219-250.
[15] Rios, M. and Oller, J.M. Rao distance between multivariate normal models and their application to the classification of response curves, submitted to Biometrics.
[16] Burbea, J. and Oller, J.M., Statistics & Decisions 6 (1988) 209-221.
[17] Ohsumi, N. and Nakamura, T., Proc Inst. Statist. Mathem. 28 (1981) 117-133.
[18] Cuadras, C.M. and Oller, J.M., Qüestiio 11(3) (1987) 37-57.
[19] Gower, J.C., Biometrics 27 (1971) 857-871.
[20] Cuadras, C.M., Munibe, sup. no. 6 (1988) 25-33.
[21] Cuadras, C.M. and Arenas, C., Multiple linear regression based on distances, 47th ISI session (Paris, 1989).
[22] Young ,F.W., Psychometrika 46 (1981) 357-388.
[23] Gower, J.C., Biometrika 55 (1968) 582-585.
[24] Rao, C.R., Theor. Pop. Biology 21 (1982) 24-43.
[25] Krzanowski, W.J., Comp. and Maths. with Appls. 12A (1986) 179-185.
[26] Rao, C.R., Bull.Calcutta Math. Soc. 37 (1945) 81-91.
[27] Atkinson, C. and Mitchell, A.F.S., Sankhya 43 (1981) 345-465.
[28] Burbea, J. and Rao, C.R., J. Multivariate Anal. 12 (1982) 575-596.
[29] Oller, C.M. and Cuadras, C.M., Sankhya 47 (1985) 75-83.
[30] Burbea, J., Expo. Math. 4 (1986) 347-378.
[31] Mitchell, A.F.S., Internat. Stat. Review 56 (1988) 1-16.
[32] Amari, S., Differential geometrical methods in statistics, Lecture Notes in Statistics 28 (Springer Verlag, Berlin, 1985).

[33] Oller, J.M., Some geometrical aspects on data analysis and statistics, this volume.
[34] Miñarro, A. and Oller, J.M., A distance between statistical individuals based on the information metric, Manuscript, (Barcelona, 1989).
[35] Cuadras, C.M., Distancias Estadisticas (with discussion), Estadistica Española, vol.30 (119) (1988) in print.
[36] Krzanowski, W.J., J. of Classification 4 (1987) 73-84.
[37] Krzanowski, W.J., JASA 70 (1975) 782-790.
[38] Cuadras, C.M. and Auge, J., Comm. in Stat., Theor. Meth. A10 (1981) 339-353.
[39] Ruiz-Rivas, C. and Cuadras, C.M., J. Multivariate Anal. 27 (1988) 447-456.
[40] Marshall, A.W. and Olkin, I., JASA 83 (1988) 834-841.

STATISTICAL DATA ANALYSIS AND INFERENCE
Y. Dodge (Editor)

STEIN TYPE ESTIMATORS FOR REGRESSION COEFFICIENTS IN A MULTIVARIATE LINEAR MODEL UNDER ELLIPTICAL DISTRIBUTIONS+

C. G. Khatri*

Gujarat Universtiy, Ahmedabad-380 009, India

Abstract: In a subclass of elliptical distributions, the results of Srivastava and Bilodeau (1989) for a univariate linear regression model in estimating the regression coefficients by shrinkage (or Stein type) estimator are extended to a multivariate linear regression model. We propose two types of Stein type or Shrinkage estimators for multivariate regressions and they are considered over the usual unbiased least squares estimator using mean square error. unbiased estimates of bias and risk are also given for the proposed estimates.

AMS Subject Classifications: 62H12, 62J05

1. INTRODUCTION

The multivariate linear regression model is

$$Y = XCU + \epsilon \qquad (1.1)$$

where Y is a $p\times n$ matrix of observations, X and U are $p\times m$ and $k\times n$ known matrices of ranks m and k respectively, C is an $m\times k$ matrix of unknown regression coefficients and ϵ has an elliptical distribution given by

$$p(\epsilon) = \int_0^\infty \phi(o, t^{-1}\Sigma, I_n; \epsilon)\ W(dt). \qquad (1.2)$$

Here, given t, the column vectors of ϵ are independent normals with zero means and the covariance matrix $t^{-1}\Sigma$. Denote $W^+ - W^-$ be the Jordan decomposition of W in positive and negative parts. We assume

*Center for Multivariate Analysis, 123 Pond Laboratory, Penn State University.

+Air Force Office for Scientific Research under Grant AFOSR-89-0279.

$$\int_0^\infty W(dt) = 1, \int_0^\infty t^{-1}W^+(dt) < \infty \text{ and } \int_0^\infty t^{-1}W^-(dt) < \infty. \tag{1.3}$$

Khatri and Rao (1988) considered $W(t) = 1$ for $t \geq 1$ and $W(t) = 0$ for $t < 1$ and $\Sigma = X\Gamma X' + \sigma^2 I_p$. For the normal distribution, they showed that the mean square dispersion matrix D_e for an empirical Bayes estimate $B_{(e)}$ of $B = CU$ is less than or equal to that D_ℓ for the least squares estimate $B_\ell = (X'X)^{-1}X'Y$ of B if and only if $0 \leq b \leq 2(n-k-m-1)/(np-nm+2)$ $= 2/c$(say). Here, $B_{(e)}$ is defined by

$$B_{(e)} = B_\ell - bs_e^2(X'X)^{-1}S_b^{-1}B_\ell(I-U'(UU')^{-1}U) \tag{1.4}$$

and the mean square dispersion matrix is

$$D_e = E(B_{(e)}-B)A(B_{(e)}-B)' \tag{1.5}$$

where A is a non-negative definite matrix, and

$$s_e^2 = tr(I-X(X'X)^{-1}X')YY' \quad \text{and} \quad S_b = B_\ell(I-U'(UU')^{-1}U)B_\ell'. \tag{1.6}$$

It is obvious from the assumptions of (1.2) and (1.3) that the above result of Khatri and Rao (1988) is valid for the elliptical subclass of distributions. Further, an unbiased estimate of D_e is

$$\begin{aligned}\hat{D}_e &= (s_e^2/n(p-m))(tr\ A)(X'X)^{-1} \\ &- \frac{(2b-cb^2)(n-k-m-1)}{(n-k)(np-nm+2)} s_e^4\{tr\ A(I-U'(UU')^{-1}U)\}(X'XS_bX'X)^{-1}.\end{aligned} \tag{1.7}$$

Recently, Srivastava and Bilodeau (1989) considered the Stein type estimators for C when (i) $p = 1 = m$ and (ii) $X = I$ and $k = 1$. Here, we consider the following two Stein type (or shrinkage) estimators for C:

$$C_{s1} = C_\ell - bS_b(C_\ell UU'Q^{-1}UU'C_\ell')^{-1}C_\ell UU'Q^{-1} \quad \text{for} \quad k > m+1 \tag{1.8}$$

and

$$C_{s2} = C_{\ell} - b\{tr(S_b^{-1}C_{\ell}UU'Q^{-1}UU'C_{\ell}')\}^{-1}C_{\ell}UU'Q^{-1} \quad \text{for} \quad km > 2 \tag{1.9}$$

where

$$C_{\ell} = B_{\ell}U'(UU')^{-1} = (X'X)^{-1}X'YU'(UU')^{-1},$$

and the mean square errors are taken as

$$R_{si} = E\{vec(C_{si}-C)\}'(Q\otimes\Gamma_1^{-1})\{vec(C_{si}-C)\} = E\ tr\ \Gamma_1^{-1}(C_{si}-C)Q(C_{si}-C)' \tag{1.10}$$

for $i = 1,2$ where Q is any $k\times k$ positive definite matrix, $\Gamma_1 = \Gamma+\sigma^2(X'X)^{-1}$ and if $Y = (\underset{\sim}{y}_1,\underset{\sim}{y}_2,\ldots,\underset{\sim}{y}_n)$, then $(vec\ Y)' = (\underset{\sim}{y}_1',\underset{\sim}{y}_1',\ldots,\underset{\sim}{y}_n')$. Further, $P\otimes Q$ is defined as a Kronecker product of P with Q and is equal to $(p_{ij}Q)$ if $P = (p_{ij})$. We show that

$$R_{s1} \leq R_{\ell} \quad \text{if and only if} \quad 0 \leq b \leq 2(k-m-1)/(n-k+m+1), \tag{1.11}$$

while

$$R_{s2} \leq R_{\ell} \quad \text{if} \quad 0 \leq b \leq 2(km-2)/(n-k-m+q+2) \quad \text{with} \quad q = \min(m,k). \tag{1.12}$$

In (1.12), we require that $W(t)$ is a probability measure, but hope that it will be true for a general measure. We obtain the estimates of Biases and risks for C_{si}, $i = 1,2$. These results are extensions of Srivastava and Bilodeau (1989) to the multivariate regression model and we do not extend the general univariate regression of Stein type estimates considered by Berger (1975, 1976) and Brandwein (1979), because for $m \geq 2$, these extensions appear to be complicated for unknown Γ_1. Section 2 gives some necessary lemmas and Sections 3 and 4 give the results for C_{s1} and C_{s2} respectively.

2. SOME USEFUL LEMMAS

If $Z = (\underset{\sim}{z}_1,\underset{\sim}{z}_2,\ldots,\underset{\sim}{z}_n)$, then we have defined $vec(Z)$ as

$$(\text{vec } Z)' = (\underset{\sim}{z}_1', \underset{\sim}{z}_2', \ldots, \underset{\sim}{z}_n').$$

With the above definition, the following two results are easy to establish:

$$\text{tr } PZQZ' = (\text{vec } Z)'(Q \otimes P)(\text{vec } Z) \quad \text{for any symmetric matrices } P \text{ and } Q \tag{2.1}$$

and

$$\text{vec}(T_1 Z T_2) = (T_2' \otimes T_1) \text{ vec } Z \tag{2.2}$$

for any T_1 and T_2 matrices for which the product $T_1 Z T_2$ is defined.

Lemma 1. Let Z be an $m \times k$ matrix such that its elements are independent $N(\theta_{ij}, 1)$ where $(\theta_{ij}) = \theta = EZ$. Then

(i) $E(ZAZ')^{-1}Z(Z-\theta)' = (k-m-1)E(ZAZ')^{-1}$ for $k > m+1$ and A being a diagonal matrtix with positive diagonal elements, and for any positive definite matrix F of order $km \times km$,

(ii) $E[(\text{vec } Z)'F(\text{vec } Z)]^{-1}(\text{vec } Z)(\text{vec}(Z-\theta))'$

$$= E[\{(\text{vec } Z)'F(\text{vec } Z)\}^{-1} - 2\{(\text{vec } Z)'F(\text{vec } Z)\}^{-2}(\text{vec } Z)(\text{vec } Z)'F]$$

and in particular,

(iia) $E\{(\text{vec } Z)'F(\text{vec } Z)\}^{-1}(\text{vec}(Z-\theta))'(\text{vec } Z)$

$$= (km-2)E\{(\text{vec } Z)'F(\text{vec } Z)\}^{-1}.$$

Proof. Results (ii) and (iia) are established by Srivastava and Bilodeau (1989). Let us denote $Z = (\underset{\sim}{z}_1, \underset{\sim}{z}_2, \ldots, \underset{\sim}{z}_k)$ and $A = \text{diag}(a_1, a_2, \ldots, a_k)$. Let us consider

$$B_i = \int (ZAZ')^{-1} \underset{\sim}{z}_i (\underset{\sim}{z}_i - \underset{\sim}{\theta}_i)' \phi(Z)\, dZ \tag{2.3}$$

where

$$\phi(Z) = (2\pi)^{-mk/2} \exp(-\tfrac{1}{2} \sum_{j=1}^{k} (\underset{\sim}{z}_i - \underset{\sim}{\theta}_i)'(\underset{\sim}{z}_i - \underset{\sim}{\theta}_i)).$$

Note that $\frac{\partial}{\partial z_{\ell i}} \phi(Z) = -(z_{\ell i} - \theta_{\ell i})\phi(Z)$ for $\ell = 1,2,\ldots,m$ and $i = 1,2,\ldots,k$ with $Z = (z_{\ell i})$ and $\theta = (\theta_{\ell i})$. Hence for the ℓ-th column of B_i, we have

$$(B_i)_\ell = -\int (ZAZ')^{-1} \underset{\sim}{z}_i \left(\frac{\partial}{\partial z_{\ell i}} \phi(Z)\right) dZ$$

and integrating by parts for $z_{\ell i}$ element only, we get

$$(B_i)_\ell = E(ZAZ')^{-1} \underset{\sim}{e}_\ell - E(ZAZ')^{-1} a_i (\underset{\sim}{e}_\ell \underset{\sim}{z}_i' + \underset{\sim}{z}_i \underset{\sim}{e}_\ell')(ZAZ')^{-1} \underset{\sim}{z}_i$$

where $\underset{\sim}{e}_\ell$ denotes the ℓ-th column vector of I_m. From this, we get

$$B_i = E(ZAZ')^{-1} - E(ZAZ')^{-1}(a_i \underset{\sim}{z}_i \underset{\sim}{z}_i')(ZAZ')^{-1} - E(ZAZ')^{-1}\{tr(a_i \underset{\sim}{z}_i \underset{\sim}{z}_i'(ZAZ')^{-1})\}.$$

Now,

$$B = \sum_{i=1}^{k} B_i = E(ZAZ')^{-1} Z(Z-\theta)' = (k-m-1)E(ZAZ')^{-1}$$

because $\sum_{i=1}^{k} a_i \underset{\sim}{z}_i \underset{\sim}{z}_i'(ZAZ')^{-1} = I_m$. This proves (i) and Lemma 1.

<u>Corollary 1</u>. Let the column vectors of Z, an $m\times k$ matrix, be independent $N(\underset{\sim}{\theta}_i, \Sigma)$ with $\theta = (\underset{\sim}{\theta}_1, \ldots, \underset{\sim}{\theta}_k)$. Then

(i) $E(ZAZ')^{-1} Z(Z-\theta)' = (k-m-1)\{E(ZAZ')^{-1}\}\Sigma$ for any positive definite A,

and

(ii) $$E\{(\text{vec } Z)' F (\text{vec } Z)\}^{-1} (\text{vec } Z)(\text{vec}(Z-\theta))'$$
$$= E\{(\text{vec } Z)' F (\text{vec } Z)\}^{-1}\Sigma - 2E\{(\text{vec } Z)' F (\text{vec } Z)\}^{-2} (\text{vec } Z)(\text{vec } Z)' F\Sigma,$$

for any positive definite matrix F.

Proof for (i) follows by using $Z_1 = \Sigma^{-1/2}ZU$ with $A = UD_aU'$ and D_a being diagonal and for (ii) follows by using $Z_2 = \Sigma^{-1/2}Z$.

Lemma 2. Let S be distributed as Wishart denoted by $W_m(f,\Sigma)$ where f denotes the degrees of freedom and $ES = f\Sigma$. Then

(i) $E(SMS) = f(f+1)\Sigma M\Sigma + f\Sigma(\mathrm{tr}\,\Sigma M)$ for any symmetric matrix M,

(ii) $E(S\,\mathrm{tr}\,MS) = 2f\Sigma M\Sigma + f^2\Sigma(\mathrm{tr}\,\Sigma M)$ for any symmetric matrix M,

(iii) $ES(\underset{\sim}{x}'S^{-1}\underset{\sim}{x})^{-1} = f(f-m+1)\Sigma(\underset{\sim}{x}'\Sigma^{-1}\underset{\sim}{x})^{-1}+2(f-m+1)\underset{\sim}{x}\underset{\sim}{x}'(\underset{\sim}{x}\Sigma^{-1}\underset{\sim}{x})^{-2}$ for any nonzero vector $\underset{\sim}{x}$ and

(iv) $(\underset{\sim}{x}'\Sigma^{-1}\underset{\sim}{x})/(\underset{\sim}{x}'S^{-1}\underset{\sim}{x}) \sim \chi^2_{f-m+1}$ for any nonzero vector $\underset{\sim}{x}$.

Proof. Results (i) and (iv) are given in the text books, see for example, Anderson (1984), Srivastava and Khatri (1979) and Rao (1973). For (ii), we can write

$$\Sigma = AA' \quad \text{and} \quad M = (A')^{-1}D_\alpha A^{-1}, \quad \sum_{i=1}^{m} \alpha_i = \mathrm{tr}(M\Sigma), \tag{2.4}$$

where A is nonsingular and $D_\alpha = \mathrm{diag}(\alpha_1,\alpha_2,\ldots,\alpha_m)$. Then $A^{-1}SA'^{-1} = V \sim W_m(f,I_m)$ and $S(\mathrm{tr}\,MS) = A(V\sum_{i=1}^{m}\alpha_i v_{ii})A'$. Hence

$$ES(\mathrm{tr}\,MS) = \sum_{i=1}^{m} A(EVv_{ii})\alpha_i A' = (\sum_{i=1}^{m}\alpha_i AE(Vv_{ii})A'). \tag{2.5}$$

Since v_{ii} $(i = 1,2,\ldots,m)$ and $(r_{ij} = v_{ij}/\sqrt{v_{ii}v_{jj}})$ are independently distributed, $v_{ii} \sim \chi^2_f$ and $Er_{ij} = 0$ for $i \neq j$. Hence,

$$E(Vv_{ii}) = f^2 I_m + 2f\underset{\sim}{e}_i\underset{\sim}{e}_i'$$

where $\underset{\sim}{e}_i$ denotes the i-th column vector of I_m. Using this and (2.4) in (2.5), we get (ii).

For (iii) and (iv), let X_1 be an $m\times(m-1)$ matrix such that $\underset{\sim}{x}'X_1 = 0$ and $X_1'X_1 = I_{m-1}$. Let $X_0 = (\underset{\sim}{x}, X_1)$ and

$$X_0'SX_0 = W = \begin{pmatrix} \omega_{11} & \underset{\sim}{\omega}' \\ \underset{\sim}{\omega} & W_{11} \end{pmatrix} \sim W_m(f, \Sigma_0) \quad \text{with} \quad \Sigma_0 = X_0'\Sigma X_0 = \begin{pmatrix} \sigma_{11} & \underset{\sim}{\sigma}' \\ \underset{\sim}{\sigma} & \Sigma_{11} \end{pmatrix}$$

Then $\omega_{11\bullet 2} = \omega_{11} - \underset{\sim}{\omega}'W_{11}^{-1}\underset{\sim}{\omega} = \underset{\sim}{x}'[S-SX_1(X_1'SX_1)^{-1}X_1'S]\underset{\sim}{x} = (\underset{\sim}{x}'\underset{\sim}{x})^2/(\underset{\sim}{x}'S^{-1}\underset{\sim}{x})$, and $(\underset{\sim}{\omega}, W_{11})$ are independently distributed. We observe that

$$\omega_{11\bullet 2}/\sigma_{11\bullet 2} = \underset{\sim}{x}'\Sigma^{-1}\underset{\sim}{x}/\underset{\sim}{x}S^{-1}\underset{\sim}{x} \sim \chi^2_{f-m+1},$$

given W_{11}, $W_{11}^{-1}\underset{\sim}{\omega} \sim N(\Sigma_{11}^{-1}\underset{\sim}{\sigma}, \sigma_{11\bullet 2}W_{11}^{-1})$ and $W_{11} \sim W_{m-1}(f, \Sigma_{11})$. Using these, we get

$$\begin{aligned} ES(\underset{\sim}{x}'S^{-1}\underset{\sim}{x})^{-1} &= (X_0')^{-1}\{E\begin{pmatrix} \omega_{11\bullet 2} & 0 \\ 0 & 0 \end{pmatrix} \frac{\omega_{11\bullet 2}}{(\underset{\sim}{x}'\underset{\sim}{x})^2} + E\begin{pmatrix} \underset{\sim}{\omega}'W_{11}^{-1}\underset{\sim}{\omega} & \underset{\sim}{\omega}' \\ \underset{\sim}{\omega} & W_{11} \end{pmatrix} \frac{\omega_{11\bullet 2}}{(\underset{\sim}{x}'\underset{\sim}{x})^2}\} X_0^{-1} \\ &= (X_0')^{-1}\left\{\underset{\sim}{e}_1\underset{\sim}{e}_1' \frac{(f-m+1)(f-m+3)(\underset{\sim}{x}'\underset{\sim}{x})^2}{(\underset{\sim}{x}'\Sigma^{-1}\underset{\sim}{x})^2}\right. \\ &\quad \left. + \frac{(f-m+1)}{\underset{\sim}{x}'\Sigma^{-1}\underset{\sim}{x}} \begin{pmatrix} f\underset{\sim}{\sigma}'\Sigma_{11}^{-1}\underset{\sim}{\sigma} + (m-1)\sigma_{11\bullet 2} & f\underset{\sim}{\sigma}' \\ f\underset{\sim}{\sigma} & f\Sigma_{11} \end{pmatrix}\right\} X_0^{-1}. \end{aligned}$$

Note that $(X_0')^{-1} = (\underset{\sim}{x}/\underset{\sim}{x}'\underset{\sim}{x}, X_1)$ and hence, we get

$$ES(\underset{\sim}{x}'S^{-1}\underset{\sim}{x})^{-1} = 2(f-m+1)\frac{\underset{\sim}{x}\underset{\sim}{x}'}{(\underset{\sim}{x}'\Sigma^{-1}\underset{\sim}{x})^2} + \frac{f(f-m+1)}{\underset{\sim}{x}'\Sigma^{-1}\underset{\sim}{x}}\Sigma, \text{ which proves (iii).}$$

Lemma 3. Let S be distributed as Wishart, $W_m(f, \Sigma)$ and let M be an $m\times m$ non-negative definite matrix of rank q ($\le m$). Then

$$(\mathrm{tr}\ M\Sigma^{-1})E(\mathrm{tr}\ MS^{-1})^{-2} \leq (f-m+q+2)E(\mathrm{tr}\ MS^{-1})^{-1}$$

and the equality holds if and only if $q = 1$.

Proof. We can find a nonsingular matrix A such that

$$AA' = \Sigma \quad \text{and} \quad AD_\alpha A' = M$$

where $D_\alpha = \mathrm{diag}(D,0)$ and $D = \mathrm{diag}(\alpha_1,\alpha_2,\ldots,\alpha_q)$ with $\alpha_1 \geq \alpha_2 \geq \ldots \geq \alpha_q > 0$. Let

$$A^{-1}SA'^{-1} = V = \begin{pmatrix} V_{11} & V_{12} \\ V_{21} & V_{22} \end{pmatrix} \begin{matrix} q \\ m-q \end{matrix} \sim W_m(f,I_m).$$
$$\qquad\qquad q \quad m-q$$

Then

$$\mathrm{tr}\ MS^{-1} = \mathrm{tr}\ D_\alpha V^{-1} = \mathrm{tr}\ DV_{11\bullet 2}^{-1} \quad \text{with} \quad V_{11\bullet 2} = V_{11}-V_{12}V_{22}^{-1}V_{21} \sim W_q(f-m+q,I_q). \tag{2.6}$$

Let us denote $V_{11\bullet 2} = (v_{ij})$, $r_{ij} = v_{ij}/\sqrt{v_{ii}v_{jj}}$ for $i \neq j$ and $R^{-1} = (r_{ij})^{-1} = (r^{ij})$. Then (2.6) gives

$$\mathrm{tr}\ MS^{-1} = \sum_{i=1}^{q} (\alpha_i r^{ii}/v_{ii}) = (\sum_{i=1}^{q} \alpha_i r^{ii}) \sum_{i=1}^{q} (p_i/v_{ii}) \tag{2.7}$$

where

$$0 < p_i = \alpha_i r^{ii} / \sum_{j=1}^{q} \alpha_j r^{jj} < 1 \quad \text{and} \quad \sum_{i=1}^{q} p_i = 1$$

Further, $v_{11},v_{22},\ldots,v_{qq}$ and R are independently distributed and $v_{ii} \sim \chi^2_{f-m+q}$.

Now, we shall use Bhattacharya's result (1984), namely

$$E\left(\sum_{i=1}^{q} p_i/v_{ii}\right)^{-2} \le \max_i \left(\frac{Ev_{ii}^2}{Ev_{ii}}\right) E\left(\sum_{i=1}^{q} p_i/v_{ii}\right)^{-1} \quad \text{for fixed } p_1,\dots,p_q, \tag{2.8}$$

and hence for fixed value of R, we get

$$E\left(\sum_{i=1}^{q} p_i/v_{ii}\right)^{-2} \le (f-m+q+2)\, E\left(\sum_{i=1}^{q} p_i/v_{ii}\right)^{-1}$$

or for given R

$$E(\operatorname{tr} MS^{-1})^{-2} \le \left(\sum_{i=1}^{q} \alpha_i r^{ii}\right)^{-1}(f-m+q+2)\, E(\operatorname{tr} MS^{-1})^{-1}. \tag{2.9}$$

We observe that for $q = 1$, $r^{ii} = 1$ and the equality holds on account of Lemma 2(iv). For $q > 1$, $r^{ii} > 1$ for all $i = 1,2,\dots,q$ and hence

$$\left(\sum_{i=1}^{q} \alpha_i r^{ii}\right)^{-1} < \left(\sum_{i=1}^{q} \alpha_i\right)^{-1} = (\operatorname{tr} M\Sigma^{-1})^{-1}. \tag{2.10}$$

Use of (2.10) in (2.9) proves Lemma 3.

<u>Lemma 4</u>. Let S be distributed as noncentral Wishart denoted by $W_m(f,\Sigma,\Omega)$ where f denotes the degrees of freedom and $ES = f\Sigma + \Omega$. Then

$$0 < ES^{-1} \le (f-m-1)^{-1}\Sigma^{-1}.$$

<u>Proof</u>. We can find a nonsingular matrix A such that

$$\Sigma = AA' \quad \text{and} \quad \Omega = AD_\alpha A'$$

where $D_\alpha = \operatorname{diag}(\alpha_1,\alpha_2,\dots,\alpha_m)$ with $\alpha_i \ge 0$ for all i. Let $V = A^{-1}SA'^{-1}$. Then $V \sim W_m(f,I_m,D_\alpha)$ and

$$A'(ES^{-1})A = EV^{-1} = E(v^{ij}). \tag{2.11}$$

Let $V = ZZ'$ where the elements of Z are independent normals with unit variance and $EZ = \theta$ with $\theta\theta' = D_\alpha$. Thus, the distribution of V is invariant under the transformation $V \rightarrow \Delta V \Delta'$ for any matrix $\Delta = \text{diag}(\epsilon_1,\ldots,\epsilon_m)$ with $\epsilon_i = +1$ or -1, because $\Delta\theta\theta'\Delta' = D_\alpha$, $|\Delta V \Delta| = |V|$ and $\text{tr}(\Delta V \Delta) = \text{tr}\ V$.

Let $Z' = (\underset{\sim}{z}_1, Z_2)$ and $Z' = (\underset{\sim}{z}_1, \underset{\sim}{z}_2, Z_3)$. Then,

$$(v^{11})^{-1} = \underset{\sim}{z}_1'[I - Z_2(Z_2'Z_2)^{-1}Z_2']\underset{\sim}{z}_1$$

and

$$v^{12} = -\ \underset{\sim}{z}_1'[I - Z_3(Z_3'Z_3)^{-1}Z_3']\underset{\sim}{z}_2 |Z_3Z_3'|/|ZZ'|.$$

Notice that changing $\underset{\sim}{z}_1$ to $-\underset{\sim}{z}_1$, we see that v^{12} changes its sign. Since the distribution of V is invariant under this transformation, we must have

$$Ev^{12} = -Ev^{12} \quad \text{or} \quad Ev^{12} = 0. \tag{2.13}$$

Similarly, we can show that $Ev^{ij} = 0$ for all $i \neq j$. Now, given Z_2, the distribution of $(v^{11})^{-1}$ is noncentral chi-square with $(f-m+1)$ degrees of freedom and the noncentrality parameter will depend on Z_2. Given Z_2, it is easy to show that

$$E(v^{11}|Z_2) \leq (f-m-1)^{-1} \quad \text{and so} \quad Ev^{11} \leq (f-m-1)^{-1}. \tag{2.14}$$

Similarly, we can show that $Ev^{ii} \leq (f-m-1)^{-1}$ for $i = 1,2,\ldots,m$. Thus, using (2.13) and (2.14) in (2.11), we get

$$A'(ES^{-1})A = EV^{-1} \leq (f-m-1)^{-1}I_m \tag{2.15}$$

where the inequality is in the Lowner's sense; that is, if A and B be two symmetric matrices, then $A \leq B$ if and only if $B - A$ is non-negative definite. Thus, (2.15) establishes Lemma 4.

3. STEIN TYPE ESTIMATOR (1.8)

We observe that the estimator C_{s1} given by (1.8) depends on C_ℓ and S_b. Using the distribution ϵ given by (1.2), it can be shown that given t, C_ℓ and S_b are independently distributed, the density of C_ℓ given t is

$$\phi(C,t^{-1}\Gamma_1,(UU')^{-1};C_\ell)$$
$$= (2\pi)^{-km/2}|t^{-1}\Gamma_1|^{-k/2}|UU'|^{m/2}\exp(-\tfrac{t}{2}\mathrm{tr}\ \Gamma_1^{-1}(C_\ell-C)UU'(C_\ell-C)') \tag{3.1}$$

and that of S_b given t is

$$W_m(n-k,t^{-1}\Gamma_1;S_b)$$
$$= \{2^{m(n-k)/2}\Gamma_m(\tfrac{n-k}{2})\}^{-1}|t^{-1}\Gamma_1|^{-(n-k)/2}\ |S_b|^{(n-k-m-1)/2}\exp(-\tfrac{t}{2}\mathrm{tr}\ \Gamma_1^{-1}S_b), \tag{3.2}$$

where

$$\Gamma_m(f/2) = \pi^{m(m-1)/4}\prod_{i=1}^{m}\Gamma(\tfrac{f-i+1}{2}). \tag{3.3}$$

Then the joint density of C_ℓ and S_b is

$$p(C_\ell,S_b) = \int_0^\infty \phi(C,t^{-1}\Gamma_1,(UU')^{-1};C_\ell)W_m(n-k,t^{-1}\Gamma_1;S_b)W(dt). \tag{3.4}$$

We shall first obtain

$$R(C_{s1}) = E(C_{s1}-C)Q(C_{s1}-C)' = R(C_\ell)+b^2ES_b(C_\ell DC_\ell')^{-1}S_b$$
$$- bES_b(C_\ell DC_\ell')^{-1}C_\ell UU'(C_\ell-C)'-bE(C_\ell-C)UU'C_\ell'(C_\ell DC_\ell')^{-1}S_b \tag{3.5}$$

where $D = UU'Q^{-1}UU'$ and $R(C_\ell) = E(C_\ell-C)Q(C_\ell-C)'$. Using Corollary 1(i) and Lemma 2 for fixed t, we get

$$E_t(C_\ell DC_\ell')^{-1}C_\ell UU'(C_\ell - C)' = (k-m-1)E_t(C_\ell DC_\ell')^{-1}\Gamma_1/t,\ E_tS_b = (n-k)\Gamma_1/t$$

and

$$E_tS_b(C_\ell DC_\ell')^{-1}S_b$$
$$= (n-k)(n-k+1)E_t\Gamma_1(C_\ell DC_\ell')^{-1}\Gamma_1/t^2 + (n-k)\Gamma_1E_t(\mathrm{tr}\ \Gamma_1(C_\ell DC_\ell')^{-1})/t^2.$$

Using these in (3.5), we can write

$$R(C_{s1})-R(C_\ell) = (n-k)\{b^2(n-k+1)-2b(k-m-1)\}\Gamma_1^{1/2}\int_0^\infty\{E_t(ZAZ')^{-1}\}t^{-1}W(dt)\Gamma_1^{1/2}$$
$$+ b^2(n-k)\Gamma_1\int_0^\infty\{E_t\mathrm{tr}(ZAZ')^{-1}\}t^{-1}W(dt). \tag{3.6}$$

where given t, $E_tZ = \theta = \sqrt{t}\ \Gamma_1^{-1/2}C(UU')^{1/2} = (\theta_{ij})$, $A = (UU')^{1/2}Q^{-1}(UU')^{1/2}$, and given t, the elements of Z are independent $N(\theta_{ij},1)$. Further, using Lemma 4, we observe that

$$E_t(ZAZ')^{-1} \le (k-m-1)^{-1}I_m \quad\text{and}\quad E_t\mathrm{tr}(ZAZ')^{-1} \le m/(k-m-1).$$

and hence by assumption (1.3),

$$\left|\int_0^\infty\{E_t(ZAZ')^{-1}\}t^{-1}W(dt)\right| < \infty \quad\text{and}\quad \left|\int_0^\infty E_t\{\mathrm{tr}(ZAZ')^{-1}\}t^{-1}W(dt)\right| < \infty. \tag{3.7}$$

We have denoted $E_tg(C_\ell,S_b,t)$ as the expectation over C_ℓ and S_b when t is fixed and we shall denote

$$Eg(C_\ell,S_b,t) = \int_0^\infty E_tg(C_\ell,S_b,t)W(dt). \tag{3.8}$$

Note that

$0 < E\ \mathrm{tr}[S_b\Gamma_1^{-1}S_b(C_\ell DC_\ell')^{-1}]$ for any joint distribution of S_b and C_ℓ, and using conditional expectation by Lemma 2(i), we get

$$0 < (n-k)(n-k+m+1)\ E\{\mathrm{tr}\ \Gamma_1(C_\ell DC_\ell')^{-1}/t^2\} \tag{3.9}$$

Now, using (3.6) and the definition (1.10) for R_{s1}, we get

$$R_{s1}-R_\ell = (n-k)[b^2(n-k+m+1)-2b(k-m-1)]E\{\mathrm{tr}\ \Gamma_1(C_\ell DC_\ell')^{-1}/t^2\}. \tag{3.10}$$

Noting (3.9) and (3.7), we see that

$$R_{s1} \le R_\ell \quad \text{if and only if} \quad 0 \le b \le 2b_* \quad \text{and} \quad b_* = (k-m-1)/(n-k+m+1), \tag{3.11}$$

and the maximum reduction will be obtained when $b = b_*$. It is obvious that an unbiased estimate of $\mathrm{Bias}(C_{s1})$ is

$$\widehat{\mathrm{Bias}}(C_{s1}) = -bS_b(C_\ell DC_\ell')^{-1}C_\ell UU'Q^{-1}. \tag{3.12}$$

We shall obtain an unbiased estimate of $R(C_{s1})$. Use of Lemma 2(i) and (ii) gives unbiased estimates of $R(C_\ell)$, $E(\Gamma_1(\mathrm{tr}\ \Gamma_1T/t^2))$ and $E(\Gamma_1T\Gamma_1/t^2)$ respectively as

$$(n-k)^{-1}S_b(\mathrm{tr}\ Q(UU')^{-1}),\ \{(n-k+1)S_b(\mathrm{tr}\ S_bT)-2S_bTS_b\}/(n-k)(n-k+2)(n-k-1) \tag{3.13a}$$

and

$$\{S_bTS_b(n-k)-S_b(\mathrm{tr}\ S_bT)\}/(n-k)(n-k+2)(n-k-1). \tag{3.13b}$$

Observe that the unbiased estimate (given above) for $E(\Gamma_1T\Gamma_1/t^2)$ may not be positive while that for $E(\Gamma_1(\mathrm{tr}\ \Gamma_1T)/t^2)$ is always positive. Using (3.13) in (3.6), we get an unbiased estimate of $R(C_{s1})$ as

$$\hat{R}(C_{s1}) = (n-k)^{-1}S_b(\text{tr } Q(UU')^{-1})$$
$$+ 2b(k-m-1)S_b(\text{tr } S_b(C_\ell DC_\ell')^{-1})/(n-k+2)(n-k-1)$$
$$+ \{b^2(n-k)(n-k+2)-2b^2-2b(n-k)(k-m-1)\}S_b(C_\ell DC_\ell')^{-1}S_b/(n-k-1)(n-k+2). \tag{3.14}$$

When $m = 1$, we get the result of Srivastava and Bilodeau (1989).

4. STEIN TYPE ESTIMATOR (1.9)

As noted in Section 3, we shall work with the conditional expectations. Here

$$R_{s2} = R_\ell + b^2 E(\text{tr } S_b^{-1}C_\ell DC_\ell')^{-2}(\text{tr } \Gamma_1^{-1}C_\ell DC_\ell')$$
$$-2bE(\text{tr } S_b^{-1}C_\ell DC_\ell')^{-1}(\text{tr } \Gamma_1^{-1}C_\ell UU'(C_\ell - C)'). \tag{4.1}$$

Using Corollary 1 for conditional expectation,

$$E_t(\text{tr } S_b^{-1}C_\ell DC_\ell')^{-1}\{\text{tr } \Gamma_1^{-1}C_\ell UU'(C_\ell - C)'\}$$
$$= E_t\{(\text{vec } Z_1)'F(\text{vec } Z_1)\}^{-1}(\text{vec } Z_1)'(\text{vec}(Z_1-\theta_1))\}$$
$$= (km-2)t^{-1}E_t(\text{tr } S_b^{-1}C_\ell DC_\ell')^{-1} \tag{4.2}$$

where $\Gamma_1^{-1/2}C_\ell(UU')^{1/2} = Z_1$, $\theta_1 = \Gamma_1^{-1/2}C(UU')^{1/2}$ and $F = A \otimes (\Gamma_1^{1/2}S_b^{-1}\Gamma_1^{1/2})$. We observe that

$$0 < E(\text{tr } S_b^{-1}C_\ell DC_\ell')^{-2}(\text{tr } \Gamma_1^{-1}C_\ell DC_\ell') = E\{E_t(\text{tr } S_b^{-1}C_\ell DC_\ell')^{-2}(\text{tr } \Gamma_1^{-1}C_\ell DC_\ell')\}$$

and using conditional expectation and Lemma 3, we have

$$E_t\{(\text{tr } S_b^{-1}C_\ell DC_\ell')^{-2}(\text{tr } \Gamma_1^{-1}C_\ell DC_\ell')\} \le (n-k-m+q+2)E_t(\text{tr } S_b^{-1}C_\ell DC_\ell')^{-1}/t \tag{4.3}$$

but the question is the validity of

$$0 < E(\text{tr } S_b^{-1}C_\ell DC_\ell')^{-2}(\text{tr } \Gamma_1^{-1}C_\ell DC_\ell') \le (n-k-m+q+2)E[(\text{tr } S_b^{-1}C_\ell DC_\ell')^{-1}/t] \tag{4.3a}$$

where $q = \min(m,k)$. Further, taking $V = t\Gamma_1^{-1/2}S_b\Gamma_1^{-1/2}$ and $Z = (t\Sigma^{-1})^{1/2}C_\ell(UU')^{1/2}$,

$$E_t(\text{tr } S_b^{-1}C_\ell DC_\ell')^{-1} = E_t(\text{tr } V^{-1}ZAZ')^{-1} \le \lambda_1 E_t\{(\max_i v_{ii})(\text{tr } ZZ')^{-1}\} < \infty.$$

where λ_1 = maximum eigen value of A^{-1}. We observe at this stage that (4.3a) will be valid if $W^-(t) = 0$ for all t. Therefore, using (4.2) and (4.3a) in (4.1), we get

$$R_{s1} \le R_\ell + \{b^2(n-k-m+q+2)-2b(km-2)\}\int_0^\infty t^{-1}\{E_t(\text{tr } S_b^{-1}C_\ell DC_\ell')^{-1}\}W(dt) \tag{4.4}$$

and if $0 \le b \le 2(km-2)/(n-k-m+q+2)$ with $q = \min(k,m)$, $R_{s2} \le R_\ell$, provided $W(t)$ is the probability measure.

It is obvious that an estimate of $\text{Bias}(C_{s2})$ is

$$\widehat{\text{Bias}}(C_{s2}) = -b(\text{tr } S_b^{-1}C_\ell DC_\ell')^{-1}C_\ell UU'Q^{-1}. \tag{4.5}$$

Let $\Psi = E\{\text{vec}(C_{s2}-C)\}\{\text{vec}(C_{s2}-C)\}'$. We shall try to obtain an approximate estimate of Ψ for $k > 1$ and an unbiased estimate of Ψ for $k = 1$. Observe that

$$\begin{aligned}\Psi = {} & E((UU')^{-1} \otimes t^{-1}\Gamma_1) + b^2E(\text{tr } S_b^{-1}C_\ell DC_\ell')^{-2}G(\text{vec } C_\ell)(\text{vec } C_\ell)'G' \\ & - bE(\text{tr } S_b^{-1}C_\ell DC_\ell')^{-1}\{G(\text{vec } C_\ell)(\text{vec}(C_\ell-C))'+(\text{vec}(C_\ell-C))(\text{vec } C_\ell)'G'\},\end{aligned} \tag{4.6}$$

where $G = (Q^{-1}UU' \otimes I_m)$. Unbiased estimate of $E(t^{-1}\Gamma_1)$ is $S_b/(n-k)$. By Corollary 1(ii), for fixed t,

$$E_t[\{(\text{vec } C_\ell)'(D \otimes S_b^{-1})(\text{vec } C_\ell)\}^{-1}(\text{vec } C_\ell)(\text{vec}(C_\ell - C))']$$
$$= E_t(\text{tr } S_b^{-1}C_\ell DC_\ell')^{-1}((UU')^{-1} \otimes t^{-1}\Gamma_1)$$
$$- 2E_t\{(\text{tr } S_b^{-1}C_\ell DC_\ell')^{-2}(\text{vec } C_\ell)(\text{vec } C_\ell)'(UU'Q^{-1} \otimes S_b^{-1}t^{-1}\Gamma_1)\}. \quad (4.7)$$

It is very difficult to get an unbiased estimate of the above quantity if $\min(m,k) > 1$, while for $\min(m,k) = 1$, unbiased estimate of the above quantity can be obtained. Srivastava and Bilodeau (1989) gave an unbiased estimate for $m = 1$ only.

For $\min(m,k) > 1$, an approximate estimate of

$$E\{(\text{tr } S_b^{-1}C_\ell DC_\ell')^{-1}G(\text{vec } C_\ell)(\text{vec}(C_\ell - C))'\} \quad \text{is given by}$$

$$(n-k)^{-1}(\text{tr } S_b^{-1}C_\ell DC_\ell')^{-1}(Q^{-1} \otimes S_b)$$
$$- 2(n-k)^{-1}(\text{tr } S_b^{-1}C_\ell DC_\ell')^{-2}G(\text{vec } C_\ell)(\text{vec } C_\ell)'G'. \quad (4.8)$$

and hence for $\min(m,k) > 1$, an approximate estimate of Ψ is

$$\hat{\Psi}_A = (UU')^{-1} \otimes S_b/(n-k) - 2b(n-k)^{-1}(\text{tr } S_b^{-1}C_\ell DC_\ell')^{-1}(Q^{-1} \otimes S_b)$$
$$+ (b^2+4b(n-k)^{-1})(\text{tr } S_b^{-1}C_\ell DC_\ell')^{-2}G(\text{vec } C_\ell)(\text{vec } C_\ell)'G'. \quad (4.9)$$

For $m = 1$, we shall get Srivastava and Bilodeau's (1989) unbiased estimate of Ψ if we replace $b(n-k)^{-1}$ by $b(n-k+2)^{-1}$. Now, we shall obtain unbiased estimate of Ψ when $k = 1$. For this, we use Lemma 2(iii) and (iv).

When $k = 1$, we shall take $Q = 1$, and taking $\underset{\sim}{x} = C_\ell(UU')^{1/2}$ and $\theta = C(UU')^{1/2}$, we have

$$E_t(\text{tr } S_b^{-1}C_\ell DC_\ell')^{-1}(\text{vec } C_\ell)(\text{vec}(C_\ell - C))' = (UU')^{-2}E_t(\underset{\sim}{x}'S_b^{-1}\underset{\sim}{x})^{-1}\underset{\sim}{x}(\underset{\sim}{x}-\underset{\sim}{\theta})'$$
$$= (UU')^{-2}(n-m) \quad E_t(\underset{\sim}{x}'\Gamma_1^{-1}\underset{\sim}{x})^{-1}\underset{\sim}{x}(\underset{\sim}{x}-\underset{\sim}{\theta})'/t \quad \text{by Lemma 2(iv)}$$
$$= (tUU')^{-2}(n-m) \quad E_t\{\Gamma_1(\underset{\sim}{x}'\Gamma_1^{-1}\underset{\sim}{x})^{-1} - 2(\underset{\sim}{x}'\Gamma_1^{-1}\underset{\sim}{x})^{-2}\underset{\sim}{x}\underset{\sim}{x}'\}.$$

Using Lemma 2(iii) and (iv), unbiased estimates of $Et^{-2}\Gamma_1(\underset{\sim}{x}'\Gamma_1^{-1}\underset{\sim}{x})^{-1}$ and $Et^{-2}(\underset{\sim}{x}'\Gamma_1^{-1}\underset{\sim}{x})^{-2}$ are respectively given by

$$\frac{\{S_b(\underset{\sim}{x}'S_b^{-1}\underset{\sim}{x})^{-1}-2(n-m+2)^{-1}(\underset{\sim}{x}'S_b^{-1}\underset{\sim}{x})^{-2}\underset{\sim}{x}\underset{\sim}{x}'\}}{(n-1)(n-m)} \quad \text{and} \quad \{(n-m)(n-m+2)(\underset{\sim}{x}'S_b^{-1}\underset{\sim}{x})^2\}^{-1}. \tag{4.10}$$

Hence, using (4.10) we get an unbiased estimate of Ψ (for $k = 1$ and $Q = 1$) as

$$\hat{\Psi} = \frac{S_b}{(n-1)UU'} - \frac{2bS_b/(n-1)}{(UU')C_\ell'S_b^{-1}C_\ell} + \left(b^2 + \frac{4bn}{(n-1)(n-m+2)}\right)\frac{C_\ell C_\ell'}{(UU')^2(C_\ell'S_b^{-1}C_\ell)^2}. \tag{4.11}$$

Compare (4.11) with (4.9). If in (4.11), $n/(n-m+2) \simeq 1$, then we get (4.9). Thus, we expect a small bias in the estimation of Ψ by (4.9).

REFERENCES

Anderson, T. W. (1984). *Introduction to Multivariate Statistical Analysis* (Second Edition), John Wiley, New York.

Bhattacharya, C. G. (1984). Two inequalities with an application. *Ann. Inst. Statist. Math.*, 36, 129-134.

Berger, J. (1975). Minimax estimation of location vectors for a wide class of densities. *Ann. Statist.* 3, 1318-1328.

Berger, J. (1976). Minimax estimation of a multivariate normal mean under arbitrary quadratic loss. *J. Multivariate Analysis*, 6, 256-264.

Brandwein, A. C. (1979). Minimax estimation for the mean of spherically symmetric distributions under general quadratic loss. *J. Multivariate Analysis*, 9, 579-588.

Khatri, C. G. and Rao, C. R. (1988). Multivariate linear model with latent variables: Problems of estimation. *Technical Report No. 88-48*. Center for Multivariate Analysis, Pond Laboratory, Penn State University, University Park, PA 16802.

Rao, C. R. (1973). *Linear Statistical Inference and Its Applications* (Second Edition), Wiley, New York.

Srivastava, M. S. and Bilodeau, M. (1989). Stein estimation under elliptical distributions. *J. Multivariate Analysis*, 28, 247-259.

Srivastava, M. S. and Khatri, C. G. (1979). *Introduction to Multivariate Statistics*. North Holland, New York.

STATISTICAL DATA ANALYSIS AND INFERENCE
Y. Dodge (Editor)

CUTS, DEEP CUTS AND MINIMAL ELLIPSOIDS

Peter E. Nüesch

Ecole Polytechnique Fédérale de Lausanne
Département de Mathématiques
MA-Ecublens
CH - 1015 Lausanne

Properties of multivariate concentration ellipsoids are used to derive the minimum volume circumscribed ellipsoid of a given halfellipsoid, which, geometrically, describes the step from k to k+1 in what is called the ellipsoid algorithm. The proof hinges on the fact that all observations lie on the surface of the concentration ellipsoid if the minimum number of observations is taken to render the covariance structure regular.

1. INTRODUCTION

The concentration ellipsoid in multivariate statistical analysis is defined as

$$(\underset{\sim}{x} - \bar{\underset{\sim}{x}})^T S^{-1} (\underset{\sim}{x} - \bar{\underset{\sim}{x}}) = c \tag{1}$$

It is an ellipsoid centered at the sample mean $\bar{\underset{\sim}{x}}$, its form is given by the inverse of the sample covariance matrix or the second-order moment matrix S, defined as in Rao [6, (8a.5.2.)], its size by c. (Different authors differ about the constant c. However this fact will be of no importance here). In case of a barely regular covariance structure, which means that exactly n+1 observations of a n-variate continuous distribution have been taken, the ellipsoid (1) exhibits some remarkable properties if c is chosen appropriately :

i) It passes through the n+1 points in $\mathbb{R}^n$.

i i) It is the minimal volume ellipsoid to do so.

While minimal ellipsoids have been the object of several statistical investigations (see e.g. Helbling [3]) they are also of importance in nonlinear programming, where the so called ellipsoid algorithm has proved itself very efficient (Dziuban et al [2]).

We show in this paper that the simple properties of the concentration ellipsoid lead directly to the equation of the minimum volume ellipsoid that circumscribes a given half-ellipsoid. Construction ot this new ellipsoid is the step from k to k+1 in the above mentioned algorithm.

This algorithm is described in §2. In the following two paragraphs the optimal properties i) and ii) of the concentration ellipsoid are proven. (Some of the results have previously been published in Nüesch [5]). §5 presents the main idea, that is, how to replace the hyperellipsoid of the cut by a maximum volume inscribed hypersimplex so as to make the properties of the concentration ellipsoid applicable. In §6 the new ellipsoid is constructed.

2. THE ELLIPSOID ALGORITHM

To solve the nonlinear programming problems of the form

$$\text{minimise } f_0(\underset{\sim}{x})$$

$$\text{subject to } \underset{\sim}{x} \in S = \{\underset{\sim}{x} \in \mathbb{R}^n \mid f_i(\underset{\sim}{x}) \le 0, \quad i = 1,2,\ldots,m\} \tag{2}$$

f_i being convex functions.

Dziuban et al [2] describe the following ellipsoid algorithm :

Step 0. Select $\underset{\sim}{x}^0$ and a symmetric positive definite matrix Q_0 so that the ellipsoid

$$E_0 = \{\underset{\sim}{x} \in \mathbb{R}^n \mid (\underset{\sim}{x} - \underset{\sim}{x}^0)^T Q_0^{-1} (\underset{\sim}{x} - \underset{\sim}{x}^0) \le 1\} \tag{3}$$

contains a point x^* optimal for (2). Set $k = 0$.

Step 1. Let f_v be a constraint function for which $f_v(\underset{\sim}{x}^k) > 0$ or the objective function if $f_i(\underset{\sim}{x}^k) \le 0$, $i = 1,2,\ldots,m$.

Calculate the gradient $\underset{\sim}{g}_k$ of f_v at $\underset{\sim}{x}^k$.

If $\underset{\sim}{g}_k = \underset{\sim}{0}$, STOP because $\underset{\sim}{x}^k$ is a minimizing point.

Calculate $\underset{\sim}{d} = \dfrac{-Q_k \underset{\sim}{g}_k}{\sqrt{\underset{\sim}{g}_k^T Q_k \underset{\sim}{g}_k}}$ provided $\underset{\sim}{g}_k^T Q_k \underset{\sim}{g}_k > 0$, otherwise STOP.

Step 2. Let

$$\underset{\sim}{x}^{k+1} = \underset{\sim}{x}^k + \frac{1}{n+1}\underset{\sim}{d} \tag{4}$$

and

$$Q_{k+1} = \frac{n^2}{n^2-1}\left(Q_k - \frac{2}{n+1}\underset{\sim}{d}\underset{\sim}{d}^T\right). \tag{5}$$

Return to Step 1 with k+1 replacing k.

Starting with E_0, the algorithm generates a sequence of successively smaller ellipsoids so that if E_k denotes the current ellipsoid with matrix Q_k and center $\underset{\sim}{x}^k$, then the next ellipsoid E_{k+1} has its center at $\underset{\sim}{x}^{k+1}$ and a matrix Q_{k+1} as given in Step 2.

Geometrically, the ellipsoid E_{k+1} is generated as follows. Using the gradient $\underset{\sim}{g}_k$ of Step 1, construct the hyperplane

$$\underset{\sim}{g}_k^T (\underset{\sim}{x} - \underset{\sim}{x}^k) = 0 \tag{6}$$

passing through the center $\underset{\sim}{x}^k$ of E_k. When each f_i is convex, then $\underset{\sim}{x}^*$ belongs to the half-space

$$H_k = \{\underset{\sim}{x} \mid \underset{\sim}{g}_k{}^T (\underset{\sim}{x}-\underset{\sim}{x}^k) \le 0\}$$

and so $\underset{\sim}{x}^* \in E_k \cap H_k$. The ellipsoid E_{k+1} defined by the update formulae in Step 2 is the unique ellipsoid of minimum volume containing $E_k \cap H_k$ and so $\underset{\sim}{x}^* \in E_{k+1}$ (see Appendix B in Bland, Goldfarb and Todd [1]).

In the following a simpler proof than that in [1] is given using minimal circumscribed ellipsoids of n-simplexes.

3. THE HAT MATRIX H

A nondegenerate n-simplex $X = \{\underset{\sim}{x}_1, \underset{\sim}{x}_2, \ldots, \underset{\sim}{x}_{n+1}\}$ gives rise to a full rank nx(n+1)-matrix X. The matrix $H = X^T S^{-1} X$ with $S = XX^T$ is called hat matrix (see Hoaglin and Welsch [4]), since in linear regression the least squares fit $\hat{\underset{\sim}{Y}}$ (Y hat) is the projection of Y through H. H has the following properties : it is

- i) symmetric
- ii) idempotent
- iii) of order n+1
- iv) of rank n.

The first three properties are selfevident. Using those one gets the fourth, since

$$\begin{aligned} \text{rank }(H) &= \text{trace }(H) \\ &= \text{trace }(X^T S^{-1} X) \\ &= \text{trace }(XX^T S^{-1}) \\ &= \text{trace }(I_n) \\ &= n. \end{aligned}$$

The difference between order and rank of an idempotent matrix equals the number of zero eigenvalues, the others being equal to one. Thus n of the n+1 eigenvalues λ_i of H are 1, and $(n+1) - n = 1$ eigenvalue vanishes. Set $\lambda_i = 1 \quad i = 1,\ldots,n$ and $\lambda_{n+1} = 0$. If one shifts the origin into the center of the simplex one gets $X\underset{\sim}{1} = \underset{\sim}{0}$ or $H\underset{\sim}{1} = \underset{\sim}{0}$. In this case $\underset{\sim}{1}$ is the eigenvector of the eigenvalue 0. Call it (normed) $\underset{\sim}{c}_{n+1} = \frac{1}{\sqrt{n+1}}\underset{\sim}{1}$.

The spectral decomposition of H gives

$$\begin{aligned} H &= \sum_{i=1}^{n+1} \lambda_i \underset{\sim}{c}_i \underset{\sim}{c}_i{}^T = \sum^{n} \underset{\sim}{c}_i \underset{\sim}{c}_i{}^T = \sum^{n+1} \underset{\sim}{c}_i \underset{\sim}{c}_i{}^T - \underset{\sim}{c}_{n+1} \underset{\sim}{c}_{n+1}^T \\ &= I - \frac{1}{n+1} \underset{\sim}{1}\underset{\sim}{1}^T = I - \frac{1}{n+1} E \end{aligned}$$

$$h_{ii} = \frac{n}{n+1}, \quad h_{ij} = -\frac{1}{n+1}$$

One concludes that every matrix X with row sum $\underset{\sim}{0}$ yields the same matrix H. $H = X^T(XX^T)^{-1}X$ is thus independent of X.

4. THE MINIMAL CIRCUMSCRIBED ELLIPSOID

Since $h_{ij} = (X^TS^{-1}X)_{ij} = \underset{\sim}{x}_i{}^TS^{-1}\underset{\sim}{x}_j$ the ellipsoid

$$\underset{\sim}{x}^TS^{-1}\underset{\sim}{x} = \frac{n}{n+1} \tag{7}$$

circumscribes the simplex. (Observe that XX^T is the second-order moment matrix and differs from the unbiased sample covariance matrix by a factor n).

We now show that this ellipsoid is the minimal circumscribed ellipsoid. To see this, transform the simplex X into Y by $\underset{\sim}{x} = U\underset{\sim}{y}$ with a square root U of S i.e. $S = UU^T$. We get

$$\begin{aligned} |\underset{\sim}{y}_i - \underset{\sim}{y}_j|^2 &= (\underset{\sim}{x}_i - \underset{\sim}{x}_j)^TS^{-1}(\underset{\sim}{x}_i - \underset{\sim}{x}_j) \\ &= \underset{\sim}{x}_i{}^TS^{-1}\underset{\sim}{x}_i + \underset{\sim}{x}_j{}^TS^{-1}\underset{\sim}{x}_j - 2\underset{\sim}{x}_i{}^TS^{-1}\underset{\sim}{x}_j \\ &= 2(h_{ii} - h_{ij}) \\ &= 2. \end{aligned}$$

Note that the transformed simplex not only is a regular polyhedron but also has side lengths $\sqrt{2}$, independent of n. The sphere $\underset{\sim}{y}^T\underset{\sim}{y} = h_{ii}$ is the transformed circumscribed ellipsoid which is clearly minimal. Since the transformation used was linear, the minimal property holds also in $\mathfrak{X}$-space, that is for the original simplex.

5. THE MAXIMAL INSCRIBED SIMPLEX

While a given simplex has exactly one minimal circumscribed ellipsoid, this ellipsoid has infinitely many inscribed simplexes of maximal volume.

If one of the n+1 edges $\underset{\sim}{d}$ is given, the remaining n edges $\underset{\sim}{x}_i$ form a maximal volume inscribed hypersimplex in the hyperellipsoid cut out from the original ellipsoid by the hyperplane which i) passes through $\bar{\underset{\sim}{x}} = -\frac{1}{n}\underset{\sim}{d}$ and ii) is conjugate to $\underset{\sim}{d}$.

To see this, keep the origin in the center of the ellipsoid. Thus

$$\text{i)}\quad \underset{\sim}{0} = \sum^{n+1} \underset{\sim}{x}_i = \sum^{n} \underset{\sim}{x}_i + \underset{\sim}{d} = n\bar{\underset{\sim}{x}} + \underset{\sim}{d},$$

where $\bar{\underset{\sim}{x}}$ is the center of the (n−1)-hypersimplex.

$$\begin{aligned}\text{ii)}\quad \underset{\sim}{d}^T S^{-1}(\underset{\sim}{x}_i - \bar{\underset{\sim}{x}}) &= \underset{\sim}{d}^T S^{-1}(\underset{\sim}{x}_i + \frac{1}{n}\underset{\sim}{d}) \\ &= \underset{\sim}{d}^T S^{-1}\underset{\sim}{x}_i + \frac{1}{n}\underset{\sim}{d}^T S^{-1}\underset{\sim}{d} \\ &= h_{ij} + \frac{1}{n} h_{ii} \\ &= 0.\end{aligned}$$

6. CONSTRUCTION OF E_{k+1}

The definition of $\underset{\sim}{d}$ in step 1 of §2 shows that $\underset{\sim}{d}$ is a point of the surface of E_k since $\underset{\sim}{d}^T Q_k^{-1}\underset{\sim}{d} = 1$, and the hyperplane (6) $\underset{\sim}{g}_k{}^T(\underset{\sim}{x}-\underset{\sim}{x}^k) = 0$ through the center of E_k is conjugate to $\underset{\sim}{d}$ since

$$\underset{\sim}{d}^T Q_k^{-1}(\underset{\sim}{x}-\underset{\sim}{x}^k) = \frac{-1}{\sqrt{\underset{\sim}{g}_k{}^T Q_k^{-1}\underset{\sim}{g}_k}}\,\underset{\sim}{g}_k{}^T(\underset{\sim}{x}-\underset{\sim}{x}^k) = 0.$$

Geometrically the hypersimplex now is i) shifted through the center of E_k and ii) blown up to fill the cut. The center $\underset{\sim}{x}^k$ and the ellipsoid E_k become $\underset{\sim}{x}^{k+1}$ and E_{k+1}. These operations clearly preserve the minimal property.

Algebraically we have

$$\text{i)}\quad \underset{\sim}{y}_i = \underset{\sim}{x}_i + \frac{1}{n}\underset{\sim}{d} = \underset{\sim}{x}_i - \bar{\underset{\sim}{x}} \qquad i = 1,\ldots,n \qquad \text{(Fig. 1)}$$

$$\begin{aligned}(n+1)\underset{\sim}{x}^{k+1} &= \sum^{n} \underset{\sim}{y}_i + \underset{\sim}{d} = [\sum \underset{\sim}{x}_i + \underset{\sim}{d}] + \underset{\sim}{d} \\ &= (n+1)\underset{\sim}{x}^k + \underset{\sim}{d}\end{aligned}$$

$$\boxed{\underset{\sim}{x}^{k+1} = \underset{\sim}{x}^k + \frac{1}{n+1}\underset{\sim}{d}}$$

which is (4)

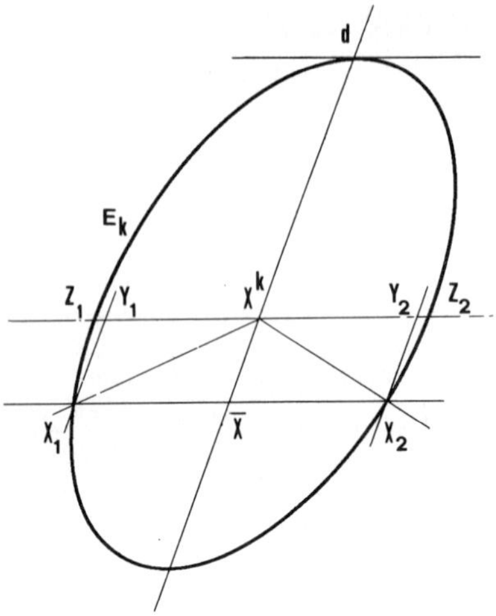

FIGURE 1

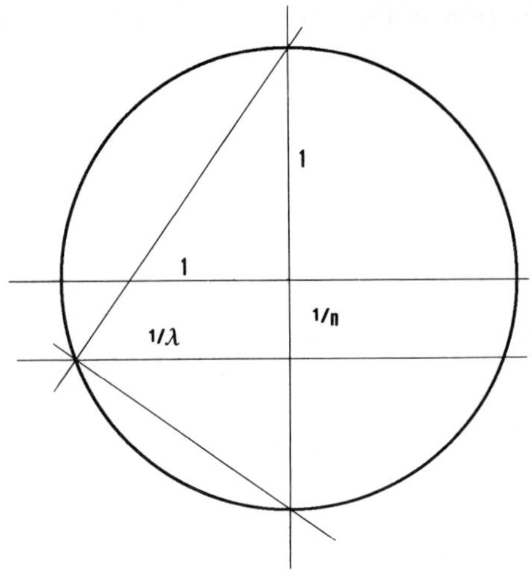

FIGURE 2

ii) $S_k = \Sigma \underset{\sim}{x}_i \underset{\sim}{x}_i^T + \underset{\sim}{d}\underset{\sim}{d}^T$

$$S_{k+1} = \Sigma \underset{\sim}{z}_i \underset{\sim}{z}_i^T + \underset{\sim}{d}\underset{\sim}{d}^T - (n+1)\, \underset{\sim}{x}^{k+1} \underset{\sim}{x}^{k+1\,T}$$

where $\underset{\sim}{z}_i = \lambda \underset{\sim}{y}_i$

the correction factor is necessary since E_{k+1} is not centered at $\underset{\sim}{0}$

$$\begin{aligned} \Sigma \underset{\sim}{z}_i \underset{\sim}{z}_i^T &= \lambda^2[\Sigma \underset{\sim}{x}_i \underset{\sim}{x}_i^T - n\bar{\underset{\sim}{x}}\bar{\underset{\sim}{x}}^T] \\ &= \lambda^2[\Sigma \underset{\sim}{x}_i \underset{\sim}{x}_i^T - \frac{1}{n} \underset{\sim}{d}\underset{\sim}{d}^T] \\ &= \lambda^2[S_k - \frac{n+1}{n} \underset{\sim}{d}\underset{\sim}{d}^T] \end{aligned}$$

The blow-up factor λ can easily be seen to be $\sqrt{n^2/(n^2-1)}$ (Fig. 2). Thus

$$\begin{aligned} S_{k+1} &= \Sigma \underset{\sim}{z}_i \underset{\sim}{z}_i^T + \underset{\sim}{d}\underset{\sim}{d}^T - (n+1)\, \underset{\sim}{x}^{k+1} \underset{\sim}{x}^{k+1\,T} \\ &= \Sigma \underset{\sim}{z}_i \underset{\sim}{z}_i^T + \frac{n}{n+1} \underset{\sim}{d}\underset{\sim}{d}^T \\ &= \lambda^2[S_k - \frac{n+1}{n} \underset{\sim}{d}\underset{\sim}{d}^T] + \frac{n}{n+1}\,\frac{n}{n-1}\,\frac{n-1}{n} \underset{\sim}{d}\underset{\sim}{d}^T \\ &= \lambda^2[S_k - (\frac{n+1}{n} - \frac{n-1}{n})\ \underset{\sim}{d}\underset{\sim}{d}^T] \end{aligned}$$

$$S_{k+1} = \frac{n^2}{n^2-1} [S_k - \frac{2}{n} \underset{\sim}{d}\underset{\sim}{d}^T].$$

Using (7) $\underset{\sim}{x}^T S^{-1} \underset{\sim}{x} = \frac{n}{n+1}$ and defining $Q \doteq \frac{n}{n+1} S$, one gets $\underset{\sim}{x}^T Q^{-1} \underset{\sim}{x} = 1$ as in (3). This leads to the relation

$$\boxed{Q_{k+1} = \frac{n^2}{n^2-1} [Q_k - \frac{2}{n+1} \underset{\sim}{d}\underset{\sim}{d}^T]}$$

which coincides with (5).

Remark

the construction of E_{k+1} for a deep cut situation is analogous, with a different shift and a different blow-up factor. The principle however remain the same.

REFERENCES

[1] Bland, R.G., Goldfarb, D., Todd, M.J., The ellipsoid method : A survey, Operations Research 29 (1981) 1039-1091.
[2] Dziuban, S.T., Ecker, J.G., Kupferschmid, M., Using deep cuts in an ellipsoid algorithm for nonlinear programming, Mathematical Programming Study 25 (1985) 93-107.
[3] Helbling, J.M., Ellipsoides minimaux de couverture en statistique multivariée, Ph.D. thesis, Ecole Polytechnique Fédérale de Lausanne, Switzerland (1983).
[4] Hoaglin, D.C., Welsch, R.E., The hat matrix in regression and ANOVA, American Statistician 32 (1978) 17-22.
[5] Nüesch, P.E., Steinerellipsoide, Elemente der Mathematik 38 (1983) 137-142.
[6] Rao, C.R., Linear Statistical Inference and its Applications, 2nd edition (John Wiley, New York, 1973).

STATISTICAL DATA ANALYSIS AND INFERENCE
Y. Dodge (Editor)

ON THE PRINCIPAL COMPONENT ANALYSIS OF THREE-WAY DATA MATRICES

Alfredo RIZZI

Dipartimento di Statistica, Probabilita' e Statistiche Applicate. Universita' "La Sapienza", Roma

This paper is concerned with the Principal Component Analysis of three-way data matrices. Principal matrices are a linear combination of K matrices observed on K different occasions or sources of data; they have maximum euclidean norm and they are incorrelated. After some remarks on centering three-way data matrices, the author introduces, also, some new definitions for the standardization of multiple matrices. Applications are found at the end of the paper.

1. INTRODUCTION

Consider N statistical units on which p characteristics are observed on K different occasions or sources of data. If the quantitatives characteristics are not homogeneous, in the sense that they are expressed in different units of measure, we must standardize variables to eliminate units of measure and to permit algebraic operations.
A statistical analysis will be completely defined,(Escouffier [5] é),when the statistician introduces a quadratic form, D, to compute the distances between the statistical units and select a set of masses, m_i, to be attributed to the statistical units.
The matrix :

$$M \equiv \text{diag} (m_1,m_2,...,m_n)$$

allows the definition of an inner product between the variables. So the triplet :

$$(X; D; M)$$

is the basic tool for describing a two- way data matrix.
For three way data matrices we can introduce a matrix :

$$N \equiv \text{diag} (n_1,n_2,...,n_K)$$

that allows the definition of an inner product between sources of data. So that :

$$({}_kX; {}_kM; {}_kD; N) \qquad (k=1,2,...,K)$$

is the basic tool for describing a three-way data matrix. In many situations ${}_kM \equiv I$, ${}_kD = 1/n\ I$ and $N \equiv I$. This is the case that we will consider in this paper.

2. TRANSFORMATIONS OF MULTI-WAY MATRICES

2.1 Once K data matrices are observed, we are faced with the problem of their eventual transformation. This can regard various aspects.

2.2 Quantification of the modalities of the statistical units, whether quantitatives or qualitatives.
In the case of qualitative variables, quantification allows one to apply the usual techniques of data analysis for quantitative characters. For qualitative ordinal characters, the approach of Herzel [7], based on optimization, is a tool of great flexibility for the researcher; it can be used in many different situations. From this point of view, a single qualitative character can be found to have two or more different quantifications in the same or in different research. As V. Castellano [1] stated, "there exist no characters which are presented to the researcher already neat and complete; all are the result of an act of will which is the choice between various possibilities, with the goal of a greater organization of knowledge, which is the objective of science."

Recent literature has been greatly concerned with the problem of the quantification of quantitative and qualitative characters; the sole bibliography of Nishisato [12] reports over 1000 references. We cannot say that an adequate solution has yet been found, at least not from the point of view of logic or of computational and interpretative algorithms.

2.3 Centering is done according to the definition of Kruskal [9]: "subtracting a constant term from every element so that resulting data values have a mean 0".

Centering can be done with regard to various aspects:

a) Total number of units analysed;

b) Different number of units in the K matrices ;

c) Different marginal totals in the K matrices;

d) Different mean values in the cells of the K matrices;

e) Different matrix margins.

a)Consider K two-way matrices:

$${}_1X_{n,p}\ ;\ {}_2X_{n,p}\ ;\dots;\ {}_KX_{n,p}$$

in which ${}_kx_{i,j}$ ($i=1,2,\dots,n$; $j=1,2,\dots,p$; $k=1,2,\dots,K$) is the generic term of the matrix ${}_kX_{n,p}$.

Suppose that all of the ${}_kx_{i,j}$ are expressed in the same unit of measure; these could be, for example, ammounts in dollars, quantities in tons or inhabitants per region in various countries. The ${}_kx_{i,j}$ instead, represent either quantities with a unit of measure or absolute frequencies, with the exclusion, therefore, of dichotomic (zero or one) type coding of qualitative variables, matrices of rank, matrices "mixtes", etc.

A three way matrix, ${}_kX_{n,p}$ is said centered if :

$$(1) \qquad \Sigma_i\Sigma_j\Sigma_k\ {}_kx_{i,j} = 0.$$

The relationship governing the modalities of the table centered on those of the original matrix is:

$${}_kx^*_{i,j} = {}_kx_{i,j} - {}_.x_{.,.} / (k\ n\ p)$$

where:

$${}_.x_{..} = \Sigma_i\Sigma_j\Sigma_k\ {}_kx_{i,j}$$

is the sum of all the modalities considered.
This type of transformation makes invariant the analyses of all data matrices ${}_kX_{n,p}$ in which :

$${}_kx^*_{i,j} = {}_kx_{i,j}+a$$

with a real number.

b)If one wishes to eliminate the influence one of the three-ways of matrix ${}_kX_{n,p}$ upon the other two, for example, of the different populations of K matrices, there must be:

$$(2) \qquad \Sigma_i\Sigma_j\ {}_kx_{i,j} = 0 \qquad (k=1,2,\ldots,K)$$

This effect is obtained by considering:

$$(3) \qquad {}_kx^*_{i,j} = {}_kx_{i,j} - {}_kx_{.,.} / (n\ p)$$

with:

$${}_kx_{.,.} = \Sigma_i\Sigma_j\ {}_kx_{i,j}$$

Obviously if the sum of elements in each of the K matrices is zero, the sum of the total number of elements will be zero. Therefore this type of centering includes the preceeding.

c) To eliminate the effect of the rows and columns (double centering), we apply:

$$(4) \qquad \Sigma_i\ {}_kx_{i,j} = \Sigma_j\ {}_kx_{i,j} = 0 \qquad (k=1,2,\ldots,K)$$

to every matrix.
This is done considering:

$${}_kx^*_{i,j} = {}_kx_{i,j} - {}_kx_{.,j}\ {}_kx_{i,.} / {}_kx_{.,.}$$

where:

$$_k x_{.,j} = \Sigma_i \; _k x_{i,j} \quad \text{and} \quad _k x_{i,.} = \Sigma_j \; _k x_{i,j}$$

In this centering, with the implication of 1 and 2, the quadratic Euclidean norm of the matrix , varying k, it is invariant.

$$\Sigma_i \Sigma_j (_k x^*_{i,j})^2 = \Sigma_i \Sigma_j \; (_k x_{i,j} - {_k x_{.j}} \; {_k x_{i,.}} / {_k x_{.,.}})^2 =$$

$$= B^2$$

where B^2 is Belson's index which varies between 0 and $n^2/4$ (D'Ambra and Marchetti [4]); this maximum is reached in the case of a perfect bilateral dependence in a 2x2 table. Marcotorchino's , [11] ,coefficient remains invariant beaucause one can write:

$$_k x^*_{i,j} = {_k x_{i,j}} - {_k x_{.,j}}/p - {_k x_{i,.}}/n + {_k x_{.,.}}/(n \; p)$$

$$\Sigma_i \Sigma_j ({_k x^*_{i,j}})^2 = I^2$$

d)To eliminate the effect due to differing mean values :

$$\Sigma_k \; {_k x^*_{i,j}} = 0 \text{ for every } (i,j)$$

one can consider one of two transformations:

$$_k x^*_{i,j} = {_k x_{i,j}} - {_. x_{i,j}}/k$$

or:

$$_k x^*_{i,j} = {_k x_{i,j}} - {_. x_{i,j}} \; {_k x_{.,.}} \; / \; {_. x_{..}}$$

for each combination of (i,j). This centering implies 1.

e)Finally, to eliminate the effect due to the three marginals of the three sets, one can consider :

$$\Sigma_i \; {_k x_{i,j}} = \Sigma_j \; {_k x_{i,j}} = \Sigma_k \; {_k x_{i,j}} = 0$$

wich is obtained considering :

$$_k x^*_{i,j} = {_k x_{i,j}} - {_k x_{.,j}}/ \; n - {_k x_{i,.}}/ \; p - {_. x_{i,j}}/ \; k +$$

$$+ {_k x_{.,.}}/(n \; p) + {_. x_{.,j}}/(k \; n) + {_. x_{i,.}}/(k \; p) - {_. x_{i,.}}/(k \; p \; n)$$

This centering implies all of the preceeding.

3. STANDARDIZATION

3.1 We will call simple standardization the process which transforms a succession of K data matrices into a succession of matrices with unitary variance and mean ugual zero, for each variable . This procedure is sometimes called normalization in literature.

One passes from a matrix $X_{n,p}$ to a matrix centered for column (matrix of deviation) via the transformation:

$$S_{n,p} = X_{n,p} - U_{n,p} X_{k,k}$$

where $U_{n,p}$ is the matrix of all ones and $X_{k,k}$ is the diagonal matrix of order K, with the average of the K variables on the principal diagonal.
The standardized matrix is given by:

$$Z_{n,p} = S_{n,p} D_{p,p}$$

where $D_{p,p}$ is the quadratic matrix of order p with the inverse of the standard deviation of the variables on it's principal diagonal.
If one applies the standardization procedure after the centering operation, as indicated, the centering of the matrices will naturally be lost.
Various iterative procedures have been introduced (Kruskal [9] ,D'Alessio [3]) for standardization with respect to more than one of the ways for three-way matrices. The final solution generally depends on the order in which the $_kX_{i,j}$ are placed in the standardization procedure. It is our opinion that the various attempts made are clearly far from a satisfactory solution.

3.2 Standardization for a two-way matrix.

Considering a two-way matrix of n statistical units and p variables, " full standardization " is the procedure which allows one to pass from matrices $X_{n,p}$ to matrices $Z_{n,p}$ with X_i column vectors of n elements with the following properties:

- the mean of each of the p variables is null;
- the variance is unitary .
- $r_{i,j}=1$ if $i=j$; $r_{i,j} = 0$ if $i \neq j$ $(i,j = 1,2...,p)$

Concerning a full standardized matrix, the matrix of correlation is the matrix I with all elements of the principal diagonal being unitary and the others equal to zero.
To obtain the " full standardized " two-way matrix $Y_{n,p}$ one can consider the matrix:

$$Y_{n,p} = Z_{n,p} Q_{n,p} \Omega^{-1/2}$$

where $Q_{p,p}$ is the matrix of the normalized eigenvectors of $X_{n,p}' X_{n,p}$ and $\Omega^{-1/2}$ is the diagonal matrix with the inverse of the square roots of eigenvalues on the principal diagonal.
In fact:

$$E (Y_{n,p}' Y_{n,p}) = E (\Omega^{-1/2} Q' Z' Z Q \Omega^{-1/2}) = I$$

insofar as if $A_{p,p}$ is a non-singular symetric matrix and Q is is orthogonal:

$$Q' A Q = \Omega.$$

Z' Z is not singular if $X_{n,p}$ is of full rank. Naturally there are other matrices which allow the passage from a matrix $X_{n,p}$ to a full standardized matrix.

3.3 Full standardization for three-way matrices.

Consider a three-way matrix:

$$_kX_{n,p} \qquad (k=1,2,\ldots,K)$$

We can say it is full standardized if:

$$_kX_{n,p} \qquad (k=1,2,\ldots,K)$$

is full standardised.

3.4 The symmetrical bilinear form < A,B > = tr (A B) defines an inner product on the set of matrices A such that M A = A M' of M- symmetric matrices . S (M) contains all operators of the form W M; an arbitrary element A of S (M) cannot be considered as an operator unless A M is positive definite (Escoufier [6]) .

Proposition: If a three-way matrix $_kX_{n,p}$ (k=1,2,...,K) is fully standardized, for the coefficient of Escouffier one has:

$$R_v ({}_rX_{n,p} \; {}_sX_{n,p}) = 1 \qquad (r,s =1,2,\ldots,K)$$

Demonstration: We have, for every (i j), (i,j=1,2,...p):

$$A = 1/n \;\; X' \; X = I$$
$$B = 1/n \;\; X' \; X = I$$

$$R_v = tr (A B)/ \sqrt{tr (A^2)} \sqrt{tr (B^2)} = 1$$

4. Principal Matrices

4.1 Principal matrices (D'Alessio [3] , Rizzi [14]), are given great importance in the ambit of the methods of synthesis of multi-way matrices. We can consider these to be extensions of the method of principal components to the case of three-way matrices. To synthesize the information contained in the matrix it is useful to consider a linear combination of the matrices $_kX_{n,p}$ (k=1,2,...,K) , in a number less than K which offers as much information as possible from the original matrices themselves.

Given the matrices $_kX_{n,p}$ (k=1,2,...K) relative to the same units and variables observed on K different occasions, let us consider the first principal matrix, where we will have:

$$_1E_{n,p} = \Sigma_k \; {}_1a \; {}_kX_{n,p}$$

whith:

$$\| \, {}_1E_{n,p} \, \| = \max$$

$${}_1a' \; {}_1a = 1$$

where ${}_1a'$ is the row vector of componentc

$$(\, {}_1a_1 \, ; {}_1a_2 \, ; \ldots ; {}_1a_k \,)$$

Analogously, the second principal matrix will show:

$${}_2E_{n,p} = \Sigma_k \; {}_2a_k \; {}_kX_{n,p}$$

$$\| \, {}_2E_{n,p} \, \| = \max$$

$${}_2a' \; {}_2a = 1$$

$$< {}_1E \; ; \; {}_2E > = 0$$

For the generic principal matrix we will have:

$${}_rE_{n,p} = \Sigma_k \; {}_ra_k \; {}_kX_{n,p}$$

$$\| \, {}_rE_{n,p} \, \| = \max$$

$${}_ra' \; {}_ra = 1$$

$$< {}_rE \; ; \; {}_sE > = 0 \qquad (r \neq s)$$

4.2 To calculate the coefficients we can proceed in the following way (D'Alessio [3]). We construct matrix A of the generic elements:

$$< {}_rE \; ; \; {}_sE > = tr \, (\, {}_rX' \; {}_sX \,)$$

Therefore, the trace is assumed to be an inner product. It is known that matrix A can be obtained as:

$$A = V'.V$$

where V is a matrix that juxtaposes the matrices ${}_kX_{n,p}$ which are opportunely vectorized.

$$V_{n,K} = (\, vec \; {}_1X; \;\; vec \; {}_2X; \ldots ; \; vec \; {}_kX \,)$$

where the operator vec X is the linear operator of vectorization which transforms matrix X into a vector comprised of the columns of matrix X located one above the other.

$$tr \, (\, {}_rX' \; {}_sX \,) = (\, vec \; {}_r \, X \,)' \; vec \; {}_sX$$

In essence, the analysis of the principal matrices ${}_hE$ is similar to the analysis of the principal components of matrix A. Matrix A is quadratic and will have r eigenvalues with which r eigenevectors are associated. A normalized eigenvector in which

the components are the coefficients of the linear combinations of the first principal matrix will correspond to maximum eigenvalue. Analogously, the other r-1 eigenvalues will be associated with r-1 normalized eigenvectors which furnish the coefficients allowing the construction of the other r-1 principal matrices.
To determine the number of principal matrices to conserve it is enough to evaluate, as in the method of principal components and with the same type of problems, the part of the inertia of matrices ${}_hX$ in which matrices ${}_hE$ can be taken account of.
Matrices ${}_hE$, inferior in number to K, synthesize the K matrices ${}_kX$; it is possible to effectuate the analyses of principal components for each of the matrices ${}_hE$.

4.3 For the principal matrices, because of the way in wich they are constructed, the trace of the products is null; this, naturally,does not imply that the product of the two matrices is the null matrix.

The examination of the eigenvalues of matrix A is not sufficient to decide on the number of principal components necessary for the analysis. The idea that if the eigenvalue is "small" the relative principal matrix can be ignored is rather vague. To evaluate the content of the information in the initial matrices we propose the analysis of Escoufier's coefficients between the principal matrix under construction and all the others. Essentialy, we are dealing with an analysis of the symmetrical matrix which contains Escoufier's coefficient between the original matrices and the principal matrices. The principal components obtained with a matrix, constitute a very powerful tool for data analysis; this analysis is sometimes limited when it is considered opportune to use a graphic representation which leads to an immediate loss of information because it is always limited to just two components.

In the analysis of principal matrices introduced here a graphic representation is not the essential moment. In general, it is not effective in applications where the number of statistical units is greater than a few dozen. A graphic representation, although useful as an instrument to facilitate communication with non-specialists, often hides information, leading to deceptions about the reality of the phenomenon; in particular when, as often occurs now, one is dealing with the output of a computer.

5. COMPARISON WITH THE STATIS METHOD

5.1 With K matrices observed in K different sources of data at our disposal, we can examine three problems (Escoufier [6]), which are different from those of the well-known STATIS method, introduced by the French school in the 70's.

First of all we need to make an overall comparison of the K different sources of data in order to decide which studies are nearest to reality (those that give similar representations of the statistical units). Next, we would be interested in a synthesis of the K matrices ${}_kX_{n,p}$. Lastly we return to the statistical units and the variables to make a detailed

exploration of the differences between the matrices observed on the different occasions.
The principal matrices method presents many similarities and differences when compared with the Statis method. Statis was introduced by considering a semidefinite quadratic matrix associated with each element of a set to be a characteristic element; in it's extension to rectanqular matrices it is reffered to contingency matrices and only one eigenvalue is considered to individuate the compromise matrix while in the principal matrices method the number of eigenvalues is a function of the percentage of the variabilities to be explained. In Statis a graphic representation is an essential element while in the principal matrices method it is not considered to be of practical assistance.

Each ${}_{k}X_{n,p}$ can be represented as a point of S_m, where m<K eigenvalues express a fixed percentage of the inertia of matrix A. The projection of "matrix points" on the s-th factorial axis is done with the relation:

$$1/\alpha_s \; V' \; V \; \mu_s$$

where α_s is the s-th eigenvalue and μ_s is the associated normalized eigenvector. We note that the principal matrices are not fully standardized.

6. APPLICATIONS

6.1 Consider the distribution of employees, in thousands, per region of Italy and branch of economic activity as published by Istat (the official institute of Italian statistics) for the years 1974, 76, 78, 80, 82 and 84 (Yearbook of General Accounts, 1986 edition, volumne 2).

The following economic activities were considered for each Italian region:

Agriculture, forestry and fishing;

Industry;

Services;

Services not for sale.

The data are represented by a three-way matrix; N=20 regions, p=4 variables and K=6 sources of data. To eliminate the influence of the different populations of the sources of data, we proceed to double centering in such a way that the frequency per row and column is equal to zero.

The quadratic norms, in thousands, of the matrices considered are:

year	quadratic norm
1974	1.91
76	1.89
78	1.84
80	1.87
82	1.80
84	1.75

Analysis of these values indicates a basic homogeneity in the number of workers per branch of economic activity with a slight tendency to diminuish in time. Principal components analysis of the data matrices yields the following eigenvalues:

α_1=10,9	98.770%
α_2= 0.13	1.175%
α_3= 0.003	0.002%
α_4= 0.002	
α_5= 0.001	
α_6= 0.0003	

Just the first two principal axes expose substantially all the variability.

The extracted coordinates of the matrix points on the factorial axes are the following:

-1373.07	169.97	-35.41	15.99	-0.90	-5.12
-1370.25	122.32	1.57	-13.02	10.58	13.85
-1355.30	63.36	16.55	-25.51	0.46	-11.49
-1367.88	6.15	23.09	11.93	-24.19	3.67
-1336.85	-116.52	20.33	21.42	18.82	-2.62
-1296.13	-261.93	-26.75	-11.13	-4.48	1.62

The sign of the coordinates of the matrices on the factorial axes is completly arbitrary, in the sense that all the "negatives" could be positive and the "positives" negative.

6.2 At the end of this paper are the essential elements for determining principal components matrices relating to a simulated situation of 9 double centered matrices of 7 units and 4 variables ${}_9X_{7,4}$. The procedure for the computer individualises two principal matrices,

7. CONCLUDING REMARKS

Principal component analysis of a matrix of quantitative data is a useful instrument to synthesize many data matrices. It depends on the preceding transformations of the observed matrices. As in principal component analysis of two way matrices, there are many problems in the interpretation of the meaning of the components. It can also be useful for clustering of three- way matrices and for many other multidimensional analyses in which we must eliminate the errors that are always found in observed data.

References

[1] Castellano V., Nota introduttiva alla discussione di alcuni problemi del metodo statistico interessanti in particolare in Sociologia, Istituto di Statistica e Ricerca Sociale "C.Gini", Lezioni del Corso per esperti sociali,Roma,(1965).

[2] Coppi, R., Analysis of Three-Way Data Matrices Based on Pairwaise Relation Measures, proceedings of Compstat 1986, Phisica-Verlag, Heidelberg Wien ,1986.

[3] D'Alessio, G., L'Analisi di successioni di matrici di dati qualitativi, Tesi di dottorato, Dipartimento di Statistica, Probabilita' e Statistiche Applicate, Universita' " La Sapienza " Roma, 1986

[4] D'Ambra ,L.,Marchetti, G.M., Un metodo per l'analisi interstrutturale di piu' matrici basato su misure di relazione tra le unita' statistiche ,Societa' Italiana di Statistica,XXXIII riunione, Bari 1986

[5] Escoufier,Y,Pages,J.P. Cazes,P. Operateurs et analyse des tableaux a plus de deux dimensions,Cahiers N. 25,du B.U.R.O Paris, 1976.

[6] Escoufier,Y. Principal components analysis whith respect to instruments variables, Methods for Multidimensional Data Analysis, Dipartimento di Matematica e di Statistica della Universita' di Napoli,1987.

[7] Herzel,A. Un criterio di quantificazione. Aspetti statistici. Metron, Vol. XXXII. 1974.

[8] Kroonemberg,P.M. Three mode-principal component analysis,DSWO Press, Leiden,1983.

[9] Kruskal,J.B. Multilinear models for data analysis.Behavior metrika,1981.

[10] Lauro,N., D'Ambra,L. L'analyse non symetrique des correspondances, in Data Analysis and Informatics III,North Holland, 1984.

[11] Marcotorchino, F. Utilisation des comparaison par paires en statistique des contingences,Etude du centre scientifique IBM France,1984.

[12] Nishisato, S. Quantification of categorital data : a bibliography 1975-86 ,Microstats Toronto 1986.

[13] Rizzi,A. Analisi dei dati. Nuova Italia Scientifica, Roma, 1985.

[14] Rizzi,A. On the syntesis of three-way data matrices. Proceedings of Multivariate Data Analysis, North Holland,1989.

[15] Saporta,G . About maximal association criteria in linear analysis and in cluster analysis, Classification and related analysis of data analysis,North Holland,1988.

Simulated situation of 9 data matrices of 7 units

DATA MATRIX

Source of data 1

40.000	35.000	27.000	38.000
12.000	14.000	10.000	15.000
41.000	36.000	28.000	39.000
22.000	24.000	22.000	26.000
8.000	7.000	9.000	6.000
32.000	29.000	25.000	31.000
20.000	15.000	12.000	11.000

Source of data 2

40.090	35.030	27.090	38.050
11.990	13.970	9.960	15.030
41.020	35.990	27.950	38.980
22.010	24.020	22.040	26.010
7.990	7.010	9.020	5.990
31.980	28.970	24.950	31.030
20.040	14.990	11.980	11.020

Source of data 3

40.050	35.020	26.990	37.990
11.030	14.010	10.010	14.990
40.990	35.990	27.030	37.990
21.990	24.010	21.990	26.020
8.030	7.010	9.010	6.020
31.980	28.990	24.970	30.990
20.040	14.990	12.030	11.010

Source of data 4

40.000	35.040	27.030	37.990
11.950	13.990	9.990	14.980
40.980	35.980	27.990	39.020
22.040	24.010	22.010	26.010
8.010	7.020	9.040	6.020
31.990	29.010	24.980	30.980
19.990	14.950	12.010	10.990

Source of data 5

39.980	34.970	26.990	38.010
12.000	13.990	10.010	15.020
41.000	36.970	27.980	39.010
22.010	24.010	21.990	25.980
7.990	7.020	9.010	6.010
31.980	29.010	24.020	30.990
20.010	15.030	11.990	10.980

Source of data 6

40.040	35.010	27.020	37.090
11.990	13.980	10.010	14.980
41.010	35.980	27.980	39.010
21.040	24.030	21.990	26.010
8.010	6.990	8.980	6.010
32.040	28.980	24.970	31.010
20.010	14.950	12.010	11.030

Source cf data 7

100.000	98.000	80.000	98.000
63.000	61.000	58.000	65.000
101.000	120.000	98.000	99.000
80.000	87.000	75.000	82.000
50.000	42.000	48.000	53.000
78.000	81.000	75.000	83.000
73.000	68.000	71.000	64.000

Source of data 8

100.100	98.040	80.030	97.990
62.990	60.980	57.980	64.930
101.010	119.990	97.990	98.970
79.900	86.900	75.010	81.980
50.080	41.970	48.030	52.980
78.020	81.030	75.020	83.040
74.930	68.010	70.990	63.980

Source of data 9

99.900	98.010	80.010	97.980
63.010	61.000	58.000	65.030
101.010	120.040	97.990	99.010
79.990	87.010	74.980	81.900
49.990	42.040	47.990	52.900
78.040	81.030	74.980	83.010
72.970	67.970	71.930	64.040

* DOUBLE CENTERED MATRIX *

Source of data 1

2.643	-0.214	-4.357	1.929
-3.107	1.036	0.893	1.179
2.643	-0.214	-4.357	1.929
-3.857	0.286	2.143	1.429
-1.857	-0.714	5.143	-2.571
0.393	-0.464	-0.607	0.679
3.143	0.286	1.143	-4.571

Source of data 2

2.658	-0.239	-4.324	1.905
-3.115	1.028	0.874	1.212
2.668	-0.199	-4.384	1.915
-3.877	0.296	2.171	1.410
-1.880	-0.697	5.169	-2.593
0.380	-0.467	-0.631	0.718
3.165	0.278	1.124	-4.567

Source of data 3

2.706	-0.311	-4.342	1.946
-3.811	1.182	1.180	1.449
3.159	0.172	-4.790	1.459
-3.844	0.189	2.168	1.486
-1.819	-0.826	5.173	-2.529
0.416	-0.561	-0.582	0.726
3.191	0.154	1.193	-4.539

Source of data 4

2.634	-0.189	-4.349	1.905
-3.129	1.048	0.898	1.182
2.636	-0.227	-4.367	1.957
-3.829	0.278	2.128	1.422
-1.864	-0.717	5.153	-2.573
0.399	-0.444	-0.624	0.670
3.154	0.251	1.161	-4.565

Source of data 5

2.638	-0.376	-4.212	1.950
-3.109	0.876	1.041	1.192
2.406	0.371	-4.474	1.697
-3.842	0.154	2.278	1.410
-1.872	-0.846	5.288	-2.570
0.626	-0.349	-1.194	0.917
3.153	0.169	1.273	-4.595

Source of data 6

2.950	-0.049	-4.187	1.286
-3.050	0.971	0.853	1.226
2.715	-0.284	-4.432	2.001
-4.527	0.494	2.305	1.728
-1.787	-0.776	5.065	-2.502
0.490	-0.539	-0.697	0.746
3.210	0.181	1.093	-4.484

Source of data 7

4.964	1.250	-9.331	3.107
0.214	-3.500	0.929	2.357
-4.536	12.750	-1.821	-6.393
-2.036	3.250	-1.321	0.107
0.714	-9.000	4.429	3.857
-2.286	-1.000	0.429	2.857
2.964	-3.750	6.679	-5.893

Source of data 8

4.801	1.328	-9.272	3.142
0.011	-3.412	0.998	2.402
-4.739	12.828	-1.762	-6.328
-2.306	3.281	-1.199	0.225
0.556	-8.967	4.503	3.907
-2.516	-0.919	0.481	2.955
4.194	-4.139	6.251	-6.305

Source of data 9

4.929	1.298	-9.385	3.158
0.254	-3.497	0.820	2.423
-4.498	12.790	-1.943	-6.350
-1.976	3.303	-1.410	0.083
0.764	-8.927	4.340	3.823
-2.221	-0.972	0.295	2.898
2.747	-3.995	7.283	-6.035

NORM² OF VECTORS Vt

SOURCE OF DATA	I^2	HISTOGRAM
1	164.3572	
2	165.2541	
3	176.9992	
4	164.3533	
5	166.6488	
6	169.4346	
7	618.0000	
8	632.7425	
9	628.6385	

Tab.1

MATRIX A = V'V

$$\begin{bmatrix} 164.36 & 164.80 & 169.89 & 164.35 & 164.93 & 166.26 & 107.79 & 114.40 & 108.37 \\ 164.80 & 165.25 & 170.38 & 164.79 & 165.42 & 166.74 & 107.71 & 114.38 & 108.29 \\ 169.89 & 170.38 & 177.00 & 169.89 & 170.82 & 171.78 & 116.44 & 123.16 & 117.07 \\ 164.35 & 164.79 & 169.89 & 164.35 & 164.93 & 166.25 & 107.47 & 114.10 & 108.08 \\ 164.93 & 165.42 & 170.82 & 164.93 & 166.65 & 166.82 & 119.68 & 126.34 & 120.43 \\ 166.26 & 166.74 & 171.78 & 166.25 & 166.82 & 169.43 & 106.85 & 113.69 & 107.35 \\ 107.79 & 107.71 & 116.44 & 107.47 & 119.68 & 106.85 & 618.00 & 624.18 & 623.02 \\ 114.40 & 114.38 & 123.16 & 114.10 & 126.34 & 113.69 & 624.18 & 632.74 & 628.76 \\ 108.37 & 108.29 & 117.07 & 108.08 & 120.43 & 107.35 & 623.02 & 628.76 & 628.64 \end{bmatrix}$$

EIGENVECTOR MATRIX

$$\begin{bmatrix} -0.162 & 0.372 & -0.002 & 0.044 & 0.130 & -0.413 & -0.049 & 0.091 & -0.797 \\ -0.162 & 0.373 & -0.015 & 0.034 & 0.127 & -0.338 & -0.051 & -0.774 & 0.319 \\ -0.172 & 0.381 & 0.187 & -0.650 & -0.579 & 0.172 & 0.049 & 0.011 & -0.009 \\ -0.162 & 0.372 & -0.001 & 0.038 & 0.138 & -0.407 & 0.062 & 0.621 & 0.509 \\ -0.172 & 0.363 & 0.083 & -0.130 & 0.646 & 0.628 & -0.042 & 0.032 & -0.017 \\ -0.162 & 0.378 & -0.225 & 0.682 & -0.434 & 0.355 & 0.022 & 0.021 & -0.006 \\ -0.524 & -0.237 & 0.202 & 0.057 & -0.067 & -0.005 & -0.785 & 0.055 & 0.045 \\ -0.532 & -0.224 & -0.754 & -0.231 & 0.045 & 0.004 & 0.207 & -0.007 & -0.016 \\ -0.528 & -0.240 & 0.546 & 0.185 & 0.012 & -0.012 & 0.572 & -0.049 & -0.028 \end{bmatrix}$$

EIGENVALUES

74.66 28.25 0.07 0.05 0.03 0.02 0.01 0.00 0.00

PRINCIPAL COMPONENTS

1°pc	2 °pc	3 °pc	4 °pc	5 °pc	6 °pc	7 °pc	8 °pc	9 °pc
-10.44	2.62	0.09	0.30	-0.16	0.04	-0.09	-0.00	0.01
2.95	-7.33	-0.06	0.50	0.28	-0.08	-0.03	-0.02	-0.00
4.59	9.28	0.33	-0.21	-0.37	0.01	0.02	-0.02	-0.00
7.26	-7.40	0.29	-0.48	0.20	-0.19	0.01	0.01	0.01
0.76	-4.61	0.08	0.05	-0.13	0.03	0.01	0.01	-0.01
3.26	2.65	0.24	0.07	0.12	0.21	-0.01	0.02	-0.01
-8.38	4.79	-0.98	-0.23	0.06	-0.02	0.09	-0.00	0.00
-1.82	-1.42	-0.11	0.20	-0.14	-0.08	0.05	0.03	0.01
4.48	4.73	0.01	-0.16	-0.09	-0.06	0.04	0.01	-0.00
-20.20	-9.12	0.02	-0.25	0.05	0.24	-0.03	-0.01	-0.00
-5.47	-1.66	-0.08	0.26	-0.14	-0.07	0.02	-0.02	0.00
14.95	4.58	0.03	-0.05	0.08	-0.01	0.10	-0.02	-0.01
1.99	-0.38	-0.04	-0.03	0.17	0.04	0.03	0.01	0.01
6.06	3.27	0.18	0.04	0.07	-0.07	-0.21	-0.01	-0.01
19.03	-3.08	-0.17	-0.11	-0.01	0.28	0.06	-0.02	0.00
-2.40	1.50	-0.02	-0.25	-0.04	0.12	-0.05	0.03	-0.01
7.35	-8.71	-0.30	0.12	0.10	-0.12	-0.03	0.00	-0.00
-0.11	5.84	-0.09	0.05	0.08	0.15	-0.04	-0.01	-0.00
-12.13	8.46	0.05	0.01	0.21	-0.02	-0.11	0.00	0.02
0.08	-1.89	-0.16	-0.04	-0.38	-0.38	-0.04	-0.01	-0.00
-11.82	-2.13	0.68	0.22	0.03	-0.03	0.21	0.00	-0.00
-6.77	1.87	0.19	-0.39	0.30	-0.24	-0.01	-0.01	-0.02
-5.02	1.10	0.06	-0.10	-0.15	0.02	0.04	-0.02	0.01
8.26	8.55	-0.06	0.34	0.22	-0.13	0.05	0.02	0.01
-1.69	3.22	-0.12	0.16	-0.14	0.11	0.01	0.02	-0.01
-3.58	-8.43	-0.16	0.00	-0.17	-0.00	-0.00	0.01	-0.00
-5.34	-0.37	-0.04	0.01	0.10	0.13	0.01	-0.03	0.00
14.14	-5.94	0.12	-0.03	-0.16	0.11	-0.09	0.00	0.01

Coordinates of point matrices on factor axes

-39.23	55.32	-0.01	0.27	0.63	-1.65	-0.10	0.04	-0.16
-39.27	55.52	-0.11	0.21	0.62	-1.34	-0.11	-0.35	0.07
-41.50	56.64	1.40	-3.94	-2.83	0.68	0.10	0.01	-0.00
-39.17	55.35	-0.01	0.23	0.67	-1.62	0.13	0.28	0.10
-41.52	54.08	0.62	-0.79	3.16	2.50	-0.09	0.02	-0.00
-39.31	56.33	-1.68	4.13	-2.12	1.41	0.05	0.01	-0.00
-126.70	-35.30	1.51	0.35	-0.33	-0.02	-1.65	0.03	0.01
-128.72	-33.38	-5.63	-1.40	0.22	0.02	0.43	-0.00	-0.00
-127.70	-35.71	4.08	1.12	0.06	-0.05	1.20	-0.02	-0.01

PERCENTAGE OF VARIANCE EXPLAINED BY FACTORIAL AXES

N.	% VARIANCE	% CUM VARIANCE	HISTOGRAM
1	72.428	72.428	
2	27.403	99.830	
3	0.069	99.900	
4	0.046	99.945	
5	0.030	99.975	
6	0.020	99.994	
7	0.005	100.000	
8	0.000	100.000	
9	0.000	100.000	

Tab.2

I Principal matrix

-10.438	-1.816	19.028	-6.773
2.946	4.478	-2.402	-5.023
4.587	-20.196	7.351	8.259
7.265	-5.469	-0.108	-1.688
0.758	14.953	-12.131	-3.581
3.260	1.992	0.084	-5.336
-8.377	6.057	-11.823	14.143

II Principal matrix

2.622	-1.418	-3.076	1.873
-7.330	4.729	1.501	1.100
9.280	-9.116	-8.712	8.548
-7.403	-1.664	5.844	3.222
-4.611	4.580	8.463	-8.432
2.647	-0.379	-1.893	-0.375
4.795	3.268	-2.127	-5.936

III Principal matrix

0.092	-0.114	-0.165	0.188
-0.059	0.013	-0.017	0.062
0.334	0.020	-0.297	-0.058
0.295	-0.081	-0.094	-0.120
0.081	0.028	0.050	-0.158
0.236	-0.045	-0.155	-0.036
-0.979	0.178	0.678	0.123

IV Principal matrix

0.297	0.198	-0.106	-0.388
0.502	-0.158	-0.246	-0.098
-0.207	-0.252	0.117	0.341
-0.479	0.264	0.053	0.161
0.046	-0.055	0.008	0.000
0.067	-0.034	-0.045	0.011
-0.227	0.037	0.218	-0.027

V Principal matrix

-0.159	-0.135	-0.005	0.299
0.281	-0.090	-0.039	-0.153
-0.370	0.051	0.103	0.216
0.198	-0.143	0.081	-0.136
-0.131	0.084	0.213	-0.166
0.119	0.167	-0.381	0.095
0.062	0.066	0.028	-0.156

VI Principal matrix

0.038	-0.077	0.283	-0.244
-0.083	-0.057	0.118	0.022
0.009	0.238	-0.118	-0.129
-0.190	-0.071	0.154	0.107
0.029	-0.006	-0.022	-0.001
0.214	0.038	-0.383	0.131
-0.018	-0.065	-0.031	0.115

VII Principal matrix

-0.094	0.047	0.057	-0.011
-0.028	0.043	-0.052	0.037
0.016	-0.034	-0.027	0.046
0.010	0.021	-0.040	0.008
0.013	0.100	-0.110	-0.003
-0.008	0.032	-0.039	0.014
0.091	-0.209	0.210	-0.092

VIII Principal matrix

-0.004	0.028	-0.015	-0.009
-0.021	0.012	0.031	-0.021
-0.017	-0.006	0.002	0.022
0.011	-0.017	-0.015	0.020
0.009	-0.023	0.002	0.012
0.023	0.013	-0.008	-0.028
-0.001	-0.007	0.002	0.005

IX Principal matrix

0.006	0.006	0.003	-0.015
-0.004	-0.000	-0.005	0.010
-0.001	-0.001	-0.005	0.007
0.014	0.003	-0.004	-0.013
-0.008	-0.005	0.016	-0.002
-0.007	0.006	-0.002	0.004
0.001	-0.008	-0.002	0.010

Part Six
SPECIAL TOPICS IN DATA ANALYSIS

Alas, that Spring should vanish with the Rose!
That Youth's sweet-scented Manuscript should close!
The Nightingale that in the Branches sang,
Ah, whence, and whither flown again, who knows!

Khayyam Naishapuri-Rubáiyát

STATISTICAL DATA ANALYSIS AND INFERENCE
Y. Dodge (Editor)

JOINT DETERMINATION OF AUTOREGRESSIVE ORDER AND DEGREE OF DIFFERENCING

Tarmo M. Pukkila

Department of Mathematical Sciences, University of Tampere
P.O. Box 607, SF-33101 Tampere, Finland

Box and Jenkins(1970) consider a method of building ARIMA(p,d,q) models to describe the behaviour of an observed time series. The basic idea is to describe the underlying stochastic structure parsimoniously but adequately. Model adequacy, among other things, means that the model transforms the data into white noise. Powerful white noise tests are of importance in practical time series analysis. In this paper we present an objective method to build ARI(p,d) models. The approach follows the model building principles of Box and Jenkins; find a parsimoniously parametrized model which transforms the data into white noise. The white noise hypothesis is tested by applying an autoregressive order determination criterion for the residuals. This approach leads to a new order determination criterion. Its statistical properties are illustrated with simulations. The method is also compared with traditional order determination methods.

1. INTRODUCTION

Let us assume that the observed time series $X_1,X_2,...,X_n$ is generated by an autoregressive integrated time series model of order (p,q), i.e. an ARI(p,d) model which can be expressed compactly in the form

$$\phi(\mathrm{B})(\nabla^d X_t - \mu) = a_t. \tag{1.1}$$

In this formula B is the backshift operator such that $B^k X_t = X_{t-k}$ for all k and $\nabla = 1-B$ is the differencing operator such that $\nabla X_t = X_t - X_{t-1}$. We will assume that the autoregressive operator

$$\phi(B) = 1- \sum_{k=1}^{p}\phi_k B^k \tag{1.2}$$

has its zeros outside the unit circle. The time series $\{a_t\}$ is a white noise process, i.e. a time series of uncorrelated random variables with mean zero and variance σ^2. Under the above assumptions $\{\nabla^d X_t\}$ is then a stationary AR(p) process, i.e. an autoregressive time series with $\mu = E\{\nabla^d X_t\}$ as its mean.

The process (1.1) can be considered as a special case of an autoregressive integrated moving average model of order (p,d,q), i.e. an ARIMA(p,d,q) model which can be written in the form

$$\phi(B)(\nabla^d X_t - \mu) = \theta(B)a_t. \tag{1.3}$$

Usually it is assumed that the moving average operator

$$\theta(B) = 1 - \sum_{k=1}^{q}\theta_k B^k \tag{1.4}$$

also has its zeros outside the unit circle. Under this assumption (1.4) defines then an invertible model. The process $\{\nabla^d X_t\}$ is said to obey ARMA(p,q) model which, under the above assumptions, is both stationary and invertible. The process $\{X_t\}$ itself is not stationary but it is, however, invertible.

Box and Jenkins(1970) introduced a method of building ARIMA(p,d,q) models. This method is comprised of three stages, i.e. identification, estimation and diagnostic checkings. In the identification stage selection of the orders d, p and q is made in a subjective manner by studying the graphs of the time series $\{\nabla^d X_t\}$ for various d as well as the corresponding estimated autocorrelation and partial autocorrelation functions. Then the actual parameters associated with p and q are estimated using the maximum likelihood method. To determine model adequacy several diagnostic tests are performed on the parameters of the model as well as the residuals of the fitted structure.

Aside from the estimated autocorrelations and partial autocorrelations a number of other tools have been suggested for the identification of ARIMA models. Chatfield(1979), Cleveland(1972), Bhansali(1980,1983), Hipel, McLeod and Lennox(1977), Abraham and Ledolter(1984) and others have studied the use of inverse autocorrelations for determining p and q. Beguin, Gourieroux and Monfort(1980) studied the use of the so-called corner method for the identification purposes. Woodward and Gray(1978) defined generalized autocorrelations and the corresponding generalized partial autocorrelations for the identification of ARMA models. Tiao and Tsay(1983a,b) and Tsay and Tiao(1984) considered the use of extended sample autocorrelations in the selection of the orders p and q. In all of these methods a pattern, depending on p and q, is sought in the behaviour of the function under consideration. Theoretically these patterns uniquely determine p and q. In practice, however, the sampling variation in the parameter estimates makes it usually very difficult to interpret the estimated functions. Therefore the estimation of the parameters p and q any of the above methods is difficult.

Pukkila and Kallinen(1987) proposed a new method to specify the orders d, p and q. This method is based on the white noise test proposed by Pukkila and Krishnaiah(1988a) for univariate time series. Pukkila and Krishnaiah(1988b) generalize the univariate white noise test to cover also multivariate time series. The method considered by Pukkila and Kallinen(1987) makes the identification of ARIMA(p,d,q) models completely automatic and objective.

Univariate ARIMA(p,d,q) models describe the behaviour of a time series only on its own past. This is one reason why the specification of an ARIMA(p,d,q) model might be done without a human intervention.

In this article we will consider the method proposed by Pukkila and Kallinen in the case of ARI(p,d) models. Our main focus is in the estimation of p and d. Using simulations we will compare the performance of the method with the results reported in Yajima(1985). In the paper we will will also consider the use of the method as a test of stationarity; an ARIMA model is nonstationary if $d>0$. Thus, the estimation of d also leads to a test of stationarity. Again using simulations we will compare the power of new test of stationarity(nonstationarity) with the tests considered by Dickey, Bell and Miller(1986). The method considered by Pukkila and Kallinen can be used as a test of stationarity for any value of p, and even more generally, for any value of p and q in ARMA(p,q) models.

The paper is organized in the following way: In Section 2 we will discuss the identification approach of time series models generally. We will also describe an autoregressive order selection method which is based on the white noise test of Pukkila and Krishnaiah(1988a). In Section 3 the method is is generalized to cover also the joint determination of p and d in ARI(p,d) models. In Section 4 we will give numerial results and Section 5 offers some concluding remarks.

2. SPECIFICATION OF AR(p) MODELS

Any invertible ARIMA(p,d,q) model can be expressed as an infinite autoregression of the form

$$X_t = \pi_0 + \sum_{k=1}^{\infty} \pi_k X_{t-k} + a_t. \tag{2.1}$$

The AR(∞) model (2.1), on the other hand, can be approximated as closely as necessary by a finite autoregression

$$X_t = \delta_0 + \sum_{k=1}^{m} \delta_k X_{t-k} + \varepsilon_t. \tag{2.2}$$

By selecting $\delta_1 = \delta_2 = ... = \delta_m = 0$ in (2.2) we can obtain the white noise process as a special case of an AR(m) model (2.2). Equivalently,we can consider an AR(0) process as a white noise process.

The order p of an autoregressive process (1.1) with d=0 can be estimated by applying an autoregressive order determination criterion which are generally of the form

$$\delta(p) = n\log\hat{\sigma}_p^2 + pg(n),\ p = 1,2,...,p^*, \tag{2.3}$$

where $\hat{\sigma}_p^2$ is the residual variance from the estimated AR(p) model and $g(n)$ is a non-negative penalty function. If $g(n)$ = 2, the formula (2.3) becomes the well-known AIC criterion. If $g(n)$ = logn, the BIC criterion is obtained as

$$\text{BIC}(p) = n\log\hat{\sigma}_p^2 + p\log n,\ p = 1,2,\ldots,p^*. \tag{2.4}$$

Using the BIC criterion the order p of an autoregressive model can be estimated by minimizing BIC(p) for $p = 1,2,\ldots,p^*$ where p^* is a predetermined integer.

For purely autoregressive models the selection of p^* is straightforward; simply select p^* to be large enough. On the other hand for mixed ARMA models the selection of the maximum order (p^*,q^*) is a difficult task. If the selected maximum orders p^* and q^* are too large, then there will be estimation problems due to the fact that many of the ARMA(p,q) models to be tested will be unidentified. The BIC criterion is consistent while AIC is not. In this paper we will use only the BIC criterion.

The basic idea in the Box and Jenkins model building procedure is to find the parsimonious model which transforms the observed time series $X_1,X_2,\ldots,X_n$ into white noise. For this reason in this modeling strategy diagnostic checkings are applied in the estimated residuals to see whether these are white enough. Equivalently, the residual series for an adequate time series model should be generated by an AR(0) model. There is nothing wrong in this modeling approach. This does not, however, mean that the strategy could not be developed further.

We have extended the modeling approach of Box and Jenkins by incorporating the use of order determination criterion BIC to aid in the selection of the order of the process. As a biproduct the new approach also produces a new order determination criterion for AR or more generally for ARMA models(see Pukkila and Kallinen(1987) and Pukkila and Krishnaiah(1988a). Following Pukkila and Krishnaiah(1988a) the notation TBIC is used in this article for the new order determination criterion. It selects the order for stationary AR models as follows.

Step 1. As the first step it is tested whether the observed time series $X_1,X_2,\ldots,X_n$ is generated by a white noise process. This test is performed by describing the behaviour of the observed time series by an autoregressive model and by selecting the order for the AR model by using the traditional BIC criterion. If the estimated order $\hat{p} = 0$ the hypothesis of white noise is not rejected; the observed time series is interpreted to be a series of white noise; move to Step 5. If, on the other hand $\hat{p} > 0$, the observed time series is not generated by a white noise; to describe the behavior of the time series at least an autoregressive model of order 1 will be needed.

Step 2. Fit an AR(1) model in the data and calculate the corresponding residual series $\hat{a}_t$. The behaviour of $\hat{a}_t$ is described by an AR model whose order is selected by using BIC. If BIC gives the order zero for the residual series, TBIC gives $\hat{p} = 1$ for the order of the autoregressive model describing the original time series $X_1,X_2,\ldots,X_n$; move to Step 5. If, on the other hand, BIC gives the order greater than zero descring the statistical behaviour of the residual series, the AR(1) model for the observed time series $X_1,X_2,\ldots,X_n$ is not adequate: move to Step 3.

Step 3. Fit an AR(2) model in the data and calculate the corresponding residual series $\hat{a}_t$. The behaviour of $\hat{a}_t$ is described by an AR model whose order is selected by using BIC. If BIC gives the order zero for the residual series, TBIC gives $\hat{p} = 2$ for the order of the autoregressive model describing the original time series $X_1,X_2,\ldots,X_n$; move to Step 5. If, on the other hand, BIC gives the order greater than zero descring the statistical behaviour of the residual series, the AR(2) model for the observed time series $X_1,X_2,\ldots,X_n$ is not adequate: move to Step 4.

Step 4. Increase the order $\hat{p}$ of the autoregression for the original series $X_1,X_2,\ldots,X_n$ to $\hat{p}+1$. Fit an AR($\hat{p}$+1) model in the data and calculate the corresponding residual series $\hat{a}_t$. The behaviour of $\hat{a}_t$ is described by an AR model whose order is selected by using BIC. If BIC gives the order zero for the residual series, TBIC gives $\hat{p}+1$ for the order of the autoregressive model describing the original time series $X_1,X_2,...,X_n$; move to Step 5. If, on the other hand, BIC gives the order greater than zero descring the statistical behaviour of the residual series, the autoregressive model of order $\hat{p}+1$ model for the observed time series $X_1,X_2,...,X_n$ is not adequate: move to Step 4 again.

Step 5. The order of the autoregressive model describing the behaviour of the observed time series $X_1,X_2,...,X_n$ has been selected by using the TBIC criterion.

Several remarks concerning the above procedure are worth of mentioning. First, to apply the ordinary BIC criterion, one has to specify $p*$ the maximum order of the autoregression for the time series $X_1,X_2,...,X_n$. On the other hand, to apply TBIC it is not necessary to define the maximum order of the autoregression for the observed time series $X_1,X_2,...,X_n$. Of course in TBIC we use the BIC criterion to test whether the residuals from the estimated AR models are white noise or not. At this stage the maximum order has to be selected. Pukkila and Krishnaiah(1988a) showed that this maximum order does change the statistical properties of the residual test if the maximum order is large enough.

We have estimated the residual variance σ^2 in (2.3) by using the Yule-Walker residual variance estimator(see e.g. Box and Jenkins(1970, pp. 274-284)

$$\hat{\sigma}_p^2 = c(0)(1 - \sum_{k=1}^{p} \hat{\phi}_k r(k)), \tag{2.5}$$

where $r(k) = c(k)/c(0)$ is the estimated autocorrelation at lag k, $k = 1,2,...,p$, and $c(k)$ is the estimated autocovariance calculated from

$$c(k) = \frac{1}{n}\sum_{t=1}^{n-k}(X_t - \bar{x})(X_{t+k} - \bar{x}),\ k= 0,\ 1,2,...p, \tag{2.6}$$

and $\bar{x}$ is mean of the observed series, i.e.

$$\bar{x} = \frac{1}{n}\sum_{t=1}^{n} X_t. \tag{2.7}$$

The estimates $\hat{\phi}_1, \hat{\phi}_2,...,\hat{\phi}_p$ are obtained by solving the Yule-Walker equations

$$\begin{pmatrix} 1 & r(1) & \dots & r(p\text{-}1) \\ r(1) & 1 & \dots & r(p\text{-}2) \\ . & . & \dots & . \\ r(p\text{-}1) & r(p\text{-}2) & \dots & 1 \end{pmatrix} \begin{pmatrix} \hat{\phi}_1 \\ \hat{\phi}_2 \\ . \\ \hat{\phi}_p \end{pmatrix} = \begin{pmatrix} r(1) \\ r(2) \\ . \\ r(p) \end{pmatrix}. \tag{2.8}$$

Thus we have followed Hannan and Quinn(1979) in the estimation of the innovation variance σ^2; the Yule-Walker estimator (2.5) is used to estimate σ^2. An alternative way to estimate σ^2 is to calculate the residuals $\hat{a}_t$ where

$$\hat{a}_t = (X_t - \bar{x}) - \sum_{k=1}^{p} \hat{\phi}_k (X_{t-k} - \bar{x}),\ t=p+1,p+2,\ldots,n. \tag{2.9}$$

The autoregressive estimates $\hat{\phi}_k$ in (2.9) shoud be obtained by applying e.g. the Burg estimates(see Burg(1967)) The residuals $\hat{a}_t$ lead to the residual variance estimator

$$\hat{\sigma}^2_{p,res} = \frac{1}{n-p} \sum_{t=p+1}^{n} \hat{a}^2_p. \tag{2.10}$$

Of course the difference between the two residual variance estimators $\hat{\sigma}^2_p$ and $\hat{\sigma}^2_{p,res}$ approaches to zero in probability as n, the number of observations increases. This means that for large samples the order determination criterion BIC should give similar estimates for p regardless the selection of $\hat{\sigma}^2_p$ or $\hat{\sigma}^2_{p,res}$ in the equation (2.5). In practice, however, depending on the generating process the results for BIC can depend heavily on the residual variance estimator as we will see later in this paper. We will also demonstrate by simulations that the order estimates produced by BIC can also depend on the estimation method of the autoregressive parameters. We will demonstrate that the Yule-Walker and Burg estimates for autoregressive parameters can lead to very different autoregressive order estimates.

It is evident that the new order selection method selects the AR model with the lowest order such that the residuals from the model are white noise. This means that the method leads both to a parsimonious and adequate model. This is also the leading principle of the Box and Jenkins modeling strategy. In fact this principle is followed in all statistical modeling; an explanation should be simple but at the same time adequate.

The new autoregressive order selection method leads to a consistent estimate $\hat{p}$ of the true order p, of course only in the case the true order exists. Compared with the BIC criterion for TBIC it is not necessary to specify the maximum order of autoregressive model for the generating autoregressive process. For autoregressive models this is not a big problem but for more general ARMA models it is a real problem for several reasons which are not considered more closely in this paper. A reader is referred e.g. to Pukkila and Krishnaiah(1988a).

3. SPECIFICATION OF ARI(p,d) MODELS

In the specification of ARI(p,d) models we will follow the same principle as in the specification of stationary AR models. This principle means that we will select the simplest model which transforms the observed time series into white noise, i.e. using TBIC we will select the parsimonious model which is adequate at the same time. The simplicity principle, or parsimony, means that the sum $p+d$ is as small as possible. For ARI(p,d) model the method works in the same way as for stationary models; we will increase the sum $p+d$ until the residuals from the corresponding

ARI(p,d) model are white noise. This principle automatically leads to the simplest possible explanation which is also adequate.

Suppose that $p+d = M$. There are several different ARI(p,d) models with M parameters. If we do not restrict d, the degree of differencing, there are $M+1$ different ARI(p,d) structures. If the residuals for some of these models appear to be white noise, in our approach we will estimate also all other ARI(p,d) models with $M = p+d$. It is possible that several different estimated ARI(p,d) models have white noise residuals according to the white noise test, i.e. several different ARI(p,d) models with $M = p+d$ can be adequate. In our approach we will select the adequate ARI(p,d) model for which the residual variance $\hat{\sigma}^2_{p,d}$ attains the minimum.

This method to select p and d also produces a stationarity or non-stationarity test as a by-product; if $\hat{d}>0$ the observed time series $X_1,X_2,...,X_n$ is interpreted to have been generated by a non-stationary process. On the other hand, if BIC gives $\hat{d} = 0$ we regard the observed time series as a realization from a stationary autoregressive process. The background of this reasoning is that we estimate autoregressive models by applying the Burg algorithm which gives autoregressive parameter estimates $\hat{\phi}_1, \hat{\phi}_2,...,\hat{\phi}_p$ always leading to a stationary autoregressive model.

The following setup clarifies the the order selection in the case of ARI(p,d) modeling. In this example $p+d$=3; there are 4 different ARI(p,d) models, marked by x, with 3 parameters. Of these parameters only p will be estimated by using the Burg algorithm. The rest of the 3-p parameters are not estimated;they will determined by the differencing operator of order 3-p. The number of the parameters, i.e. $p+d$ is increased from 0 in steps of 1 until the model is adequate.

$d \backslash p$	0	1	2	3	4	5	6	7	...
0	.	.	.	x	.	.	.	.	.
1	.	.	x	.	.	.	.	.	.
2	.	x	.	.	.	.	.	.	.
3	x	.	.	.	.	.	.	.	.
.	.	.	.	.	.	.	.	.	.
.	.	.	.	.	.	.	.	.	.
.	.	.	.	.	.	.	.	.	.
.	.	.	.	.	.	.	.	.	.
.	.	.	.	.	.	.	.	.	.

In practical applications d, the degree of differencing is usually not very large. Therefore in practice to save computing time it is possible to restrict the maximum degree of differencing d^* e.g. such that $d^*<5$. This saves computing time in practice but for other reasons it is not necessary.

An alternative method to estimate the order (p,d) of an ARI(p,d) model is to apply the BIC criterion. In this method we would then minimize the criterion

$$\text{BIC}(p,d) = n\log(\hat{\sigma}^2_{p,res}) + p\log n,\ p = 0,1,2,...,p^*;\ d = 0,1,2,...,d^*. \tag{3.1}$$

This approach implies that $(p^*+1)(d^*+1)$ models will be estimated to minimize BIC(p,d).

4. NUMERICAL RESULTS

Let us first consider the stationary AR(4) model

$$X_t = 2.7607X_{t-1} - 3.8106X_{t-2} + 2.6535X_{t-3} - 0.9238X_{t-4} + a_t. \tag{4.1}$$

This model has been considered by Beamish and Priestley(1981) and later by several other authors. The characteristic equation of the AR(4) process (4.1) has two pairs of complex roots. For this reason the spectrum of the process contains two sharp peaks which are close to each other(see Priestley(1981), p.609). Very large sample sizes would be needed in order that the Yule-Walker autoregressive spectrum estimate would show these two peaks. If the sample size is 100 the Yule-Walker autoregressive spectrum estimate reveals in most cases only one peak which is located between the true peaks. This can be seen also in Priestley(1981, p. 609).

Pukkila and Krishnaiah(1988a) used the model (4.1) to describe the performance of the order selection criterion TBIC. Pukkila(1988) used the same model to describe the statistical behaviour of the Yule-Walker estimates. It was observed that in the case of the model (4.1) the variance of the Yule-Walker estimates of the autoregressive parameters increase when the sample size is increased from 50 to 200. This was a very surprising observation. Naturally this phenomenon is unexpected because the Yule-Walker estimates are consistent. In the case of the model (4.1) and for the sample sizes from 50 to 200 the Yule-Walker estimates are extremely biased. For larger sample sizes, however, individual estimates tend to be closer to the true parameter values occasionally increasing the variance. It is clear that, due to the consistency of the Yule-Walker estimates, beyond a certain sample size, the variance begins to decrease when sample size is increased. For the above AR(4) model this limit is, however, larger than 200.

One of the goals of this study was to obtain more information on the order determination criteria BIC and TBIC as well as their performance relative to each other. We also wanted to study the behaviour of BIC in the case an AR model is estimated by applying a biased estimation method such as the Yule-Walker method. Furthermore, it is interesting to test whether an AR model whose order is selected by applying BIC, transforms the observed time series into white noise residuals. We summarize these results in Table 1.

Table 1 gives the distributions of the order identified by the BIC and TBIC criteria for sample sizes 50,100 and 200 in the case of the model (4.1). For BIC the distributions are given both for the Yule-Walker and Burg estimation methods. For BIC and for the Burg estimates of the autoregressive parameters the residual variance for BIC has been estimated by using (2.10). In the case of TBIC the autoregressive models for the observed time series have been estimated by using the Burg algorithm. Table 1 also gives the number of those replications(Sign.) for which the white noise test of Pukkila and Krishnaiah(1988a) rejected the white noise hypothesis for the residuals after the selection of the AR model order using the BIC criterion.

Table 1. The observed order distributions for the BIC and TBIC order determination criteria in the case of the fourth order autoregression (4.1) for the sample sizes 50,100 and 200. For BIC the distributions correspond to the Yule-Walker(YW) and Burg(Burg) estimation methods. For BIC the residual variance is calculated from the residuals.

	n=50			n=100			n=200		
	BIC		TBIC	BIC		TBIC	BIC		TBIC
Order	YW	Burg		YW	Burg		YW	Burg	
0	0	0	0	0	0	0	0	0	0
1	0	0	0	0	0	0	0	0	0
2	40	0	0	12	0	0	0	0	0
3	11	0	0	0	0	0	0	0	0
4	0	90	96	3	94	100	3	98	100
5	2	7	2	5	3	0	17	1	0
6	14	3	0	26	1	0	37	1	0
7	18	0	2	34	1	0	40	0	0
8	11	0	0	14	0	0	2	0	0
9	1	0	0	4	1	0	1	0	0
10	3	0	0	2	0	0	0	0	0
Sign.	100	3		99	0		100	0	

On the basis of Table 1 we see that the performance of BIC depends heavily on the estimation method. By comparing Table 1 with TABLE VIII of Pukkila and Krishnaiah(1988a) we can see that the distribution of the order estimate $\hat{p}$ also depends on whether the residual variance is calculated from the residuals or using the Yule-Walker type estimate (2.5). Table 1 reveals also that TBIC is the best of the three order determination criteria considered here.

The bottom row of Table 1 gives the number of those realizations for which the Pukkila and Krishnaiah test rejected the white noise hypothesis for the residuals after the degree of autoregression was estimated by applying the BIC criterion. These numbers also shed some light on the problems connected with the Yule-Walker estimation method; the Yule-Walker estimates are the poorer the further the zeros of the characteristic equation are located from the origin, or equivalently the more the process differs from white noise. Pukkila(1979) and Pukkila and Nyquist(1985) have also considered a related problem in the frequency domain.

We also compare the performance of the order selection criterion TBIC and the methods considered by Yajima(1985) in the selection of the degree of differencing. In order to do this comparison we carried out a simulation study using the same ARMA models and the same parameter combinations as in Yajima(1985). The results are given in Table 2.

In Table 2 we give the observed distributions of the degree of differencing in 100 realizations for five ARIMA(1,2,1) processes using the sample sizes 100, 200 and 500. Similar Monte Carlo simulations were also carried out by Yajima(1985) for the sample sizes 200 and 500. In practical applications we often have small sample sizes. Therefore we included also the sample size 100 in our simulations reported in Table 2.

Table 2. The distribution of the degree of differencing in 100 realizations for five ARIMA(1,2,1) processes using the sample sizes 100, 200 and 500 using TBIC.

ϕ	θ	$n\backslash d$	0	1	2	3	4	5
0.90	0.00	100	0	0	67	33	0	0
		200	0	1	99	0	0	0
		500	0	0	100	0	0	0
0.00	0.90	100	3	97	0	0	0	0
		200	0	100	0	0	0	0
		500	0	100	0	0	0	0
0.90	0.40	100	2	1	94	3	0	0
		200	3	0	97	0	0	0
		500	1	0	99	0	0	0
0.40	0.90	100	0	100	0	0	0	0
		200	0	100	0	0	0	0
		500	0	99	1	0	0	0
0.90	-0.90	100	1	2	52	45	0	0
		200	0	1	79	20	0	0
		500	1	0	96	3	0	0

Yajima(1985) also considers methods of joint estimation of p and d, the order of autoregression and the degree of differencing. Using simulations we compared these methods with TBIC. A summary of these results is given in Table 3 which gives the joint distributions of the degree of differencing $\hat{d}$ and the order of autoregression $\hat{p}$ three different ARI(p,d) models.

Table 3. The joint distributions of the degree of differencing and and order of autoregression for three different ARI(1,2) models using the sample sizes 100, 200 and 500.

		n=100			n=200			n=500		
ϕ	$\hat{d}\backslash\hat{p}$	0	1	2	0	1	2	0	1	2
0.90	0	0	0	0	0	0	0	0	0	0
	1	0	0	0	0	0	1	0	0	0
	2	0	67	0	0	99	0	0	100	0
	3	33	0	0	0	0	0	0	0	0
0.80	0	0	0	0	0	0	0	0	0	0
	1	0	0	1	0	0	1	0	0	0
	2	0	99	0	0	99	0	0	100	0
	3	0	0	0	0	0	0	0	0	0
0.30	0	0	0	0	3	0	0	0	0	0
	1	0	1	2	0	0	2	0	0	0
	2	19	75	0	0	98	0	0	100	0
	3	0	0	0	0	0	0	0	0	0

As can be seen TBIC very accurately determines the orders d and p. Yajima(1985) used the same models as examples to describe performance the method which was considered in the paper. The sample sizes considered by Yajima(1985) were 200 and 500. By comparing Table 3 of this paper with Table 2 of Yajima(1985) on can see TBIC to outperform the method considered by Yajima(1985). It can be seen that TBIC produces such estimates for p and d which an accurate order estimation method should produce.

The above considerations show that our method can be used to determine both the degree of autoregression and the degree of differencing. We also compare the performance of our method with the tests considered by Dickey, Bell and Miller(1986) in the case of AR(1) model. In these considerations the idea is to test whether an ARIMA model is nonstationary, i.e. whether $d>0$. If $d>0$, differencing the observed time series d times makes it stationary. Such tests are of great importance in practical time series analysis. The numerical results are given in Table 4.

Table 4. Estimated power of the test H_0: $d=1$ versus H_1: $d=0$ in the case of the model $X_t = \rho X_{t-1} + a_t$ for the sample sizes 50 and 100. The power is estimated on the basis of 2000 replications.

ρ	n=50	n=100
1.000	0.062	0.023
0.999	0.075	0.021
0.990	0.115	0.054
0.980	0.098	0.075
0.950	0.234	0.174
0.900	0.390	0.684
0.850	0.607	0.973
0.800	0.786	1.000
0.700	0.980	1.000
0.500	1.000	1.000

It is interesting to compare the numbers of Table 4 with Table 2 of Dickey, Bell and Miller(1986). The results of these tables are comparable especially in the case n=50. This is because the significance levels in the case n=50 in the two tables seem to be approximately the same. We can see that for n=50 our test is at least as powerful as the most powerful test considered by Dickey, Bell and Miller(1986). For the sample size 100 direct power comparisons are not so straightforward because in our test the significance level is, instead of 0.05, only about 0.025

5. CONCLUDING REMARKS

In this paper we have considered an objective method to identify ARI(p,d) models, i.e. the case where the behaviour of an observed time series can be described by an autoregressive model. The basic idea in this method is the same as in the time series modeling approach of Box and Jenkins(1970); find a parsimoniously parametrized model which transforms the observed time series into white noise. In this approach powerful white noise tests have an essential role.

A white noise process can be considered as an autoregressive process of order zero. For this reason a natural white noise test is obtained by applying an autoregressive order selection method;

if the estimated autoregressive order is zero, the time series is considered to have been generated by a white noise process. If, on the other hand, the estimated autoregressive order is greater than zero, the white noise hypothesis is rejected. When an order selection criterion is applied in the residual series from an estimated model, a new type of order selection criterion is obtained. In this paper we have shown that this approach leads results which are at least as good as traditional order determination criteria, in many cases even much better than the well-known order determination criteria.

ACKNOWLEDGEMENTS

The author is grateful to Dr. Sergio Koreisha for his helpful comments and criticism during the manuscript stage of this article.

REFERENCES

1. Abraham, B. and Ledolter, J.(1984). A note on inverse autocorrelations. *Biometrika* 71, pp. 609-614.
2. Beamish, N. and Priestley, M.B.(1981). A study of AR and window spectral estimation. *Applied Statistics*, 30, pp. 41 - 58.
3.Beguin, J.M., Gourieroux, C. and Monfort, A.(1980). Identification of a mixed autoregressive-moving average process: The corner method. In *Time Series*, Ed. O.D. Anderson, pp.423-436. Amsterdam: North-Holland.
4. Bhansali, R.J.(1980). The autoregressive and the window estimate of the inverse correlation function. *Biometrika* 67, pp. 511-566.
5. Bhansali, R.J.(1983). The inverse partial correlation function of a time series and its application. J. *Mult. Anal.* 13, pp. 310-327.
6. Box, G.E.P. and Jenkins, G.M.(1970). *Time Series Analysis, Forecasting and Control.* Holden-Day, San Francisco.
7. Burg, J.P.(1967). Maximum entropy spectral analysis. Paper presented at *the 37th Annual International Meeting. Soc. of Explor. Geophysics.* Oklahoma City, October 31, 1967.
8. Chatfield, C.(1979). Inverse autocorrelations. *J. R. Statist. Soc. A* 142, pp. 363-377.
9. Cleveland, W.S.(1972). The inverse autocorrelations of a time series and their applications. *Technometrics* 14, pp. 277-298.
10 Dickey, D.A., Bell, W.R. and Miller R.B.(1986). Unit roots in time series models: tests and implications. *The American Statistician* 40, pp. 12-31.
11. Hannan, E.J. and Quinn, B.G.(1979). The determination of the order of an autoregression. *J.R.Statist Soc. B* 41, pp. 190-195.
12. Hipel, K.W., McLeod, A.I. and Lennox, W.C.(1977). Advances in Box-Jenkins modelling 1. Model construction. *Water Resources Res.* 13, pp. 567-575.
13. Priestley, M.B.(1981). *Spectral Analysis and Time Series. Volume 1: Univariate Series.* Academic Press, London.
14. Pukkila, T.M.(1979). The bias in periodogram ordinates and the estimation of ARMA models in the frequency domain. *Australian Journal of Statistics* 21, pp. 121-128.

15. Pukkila, T.M.and Nyquist, H.(1985). On the frequency domain estimation of the innovation variance of a stationary univariate time series. *Biometrika* 72, pp. 317-323.
16. Pukkila ,T.M. and Kallinen, A.(1987) , On the order determination of time series models. *Proc. Second International Conference in Statistics*(Tampere, Finland, 1-4 June 1987, Tarmo Pukkila and Simo Puntanen, Editors. Dept. of Mathematical Sciences, Univ. of Tampere, pp. 275-288.
17. Pukkila ,T.M. and Krishnaiah, P.R.(1988) , On the use of autoregressive order determination criteria in univariate white noise tests, *IEEE Trans. Acoust., Speech, Signal Processing* 36, pp. 764-774.
18. Pukkila ,T.M. and Krishnaiah, P.R.(1988) , On the use of autoregressive order determination criteria in multivariate white noise tests, *IEEE Trans. Acoust., Speech, Signal Processing* 36, pp. 1396-1403.
19. Pukkila, T.M.(1988). An improved estimation method for univariate autoregressive models. *Journal of Multivariate Analysis* 27, pp. 422-433.
20. Tiao, G.C. and Tsay, R.S.(1983a). Consistency properties of least squares estimates of autoregressive parameters in ARMA models. *Ann. Statist.* 11, pp. 856-871.
21. Tiao, G.C. and Tsay, R.S.(1983b). Multiple time series modelling and extended sample cross-correlations. *J. Business Econ. Statist.* 1, pp. 43-56.
22. Tsay, R.S. and Tiao, G.C.(1984). Consistent estimates of autoregressive parameters and extended sample autocorrelation function for stationary and nonstationary ARMA models. *J. Am. Statist. Assoc.* 79, pp. 84-96.
23. Woodward, W.A. and Gray, H.L.(1978). New ARMA models for Wölfer's sunspot data. *Comm. Statist. B* 7, pp. 97-115.

STATISTICAL DATA ANALYSIS AND INFERENCE
Y. Dodge (Editor)

ANALYSIS OF DISCRETE LONGITUDINAL DATA: INFINITE-LAG MARKOV MODELS

Farhad MEHRAN

Bureau of Statistics
International Labour Office
CH-1211 Geneva 22

The purpose of this paper is to introduce a methodology based on Markov chains for fitting longitudinal data with categorical values. The methodology, referred to as infinite-lag Markov models, is a generalisation of the autoregressive Markov models developed in the context of stochastic reservoir theory by Pegram [1] and Raftery [2]. Infinite-lag Markov models have infinite memory and, therefore, can usefully serve to describe processes for which in principle the complete past should be taken into account and not just the recent past. Infinte-lag Markov models are not constrained by the length of the longitudinal sequence obtained from the survey. They can be used to fit longitudinal sequences of any length, possibly with missing values and non-consecutive data points.

1. INTRODUCTION

Repeated observations with a constant sample give rise to longitudinal data sets. Often the data consist of an outcome variable (say, employment status) and a set of covariates (e.g., sex, age, occupational group, branch of economic activity, etc.) measured repeatedly over time for a set of sample of individuals. When the outcome variable is categorical, like employment status with categories "employed", "unemployed" and "not-in-labour-force", the analysis may consist of describing the probability of the outcome as a function of the covariates and the past values of the outcome variable as well as the past values of those covariates which change over time. In the employment status example given here, the problem could be the analysis of the probability of employment of a currently unemployed worker given his (or her) socio-demographic characteristics and his (or her) past employment and unemployment experience.

One approach in analysing such data is to view the longitudinal data on the outcome variable as duration data (e.g., the duration of employment or unemployment) and to investigate the effects of the covariates on the time it takes before certain event occurs (e.g., the duration of unemployment before obtaining an employment). The standard methodology is the hazard rate model commonly used in biostatistics for analysing survival times of patients in clinical trials, and now increasingly applied in other areas of research including quality control, economics and psychology. Basic references are

Cox and Oakes [3] and Kalbfleisch and Prentice [4]. Applications to labour market data may be found e.g. in Heckman and Singer [5] and Hamerle [6].

Other approaches for the analysis of longitudinal data with categorical or binary outcomes are logistic regression models, Cox [7], probit models, Ochi and Prentice [8], random-effect models, Stiratelli et al. [9], log-linear models, Koch et al. [10], and generalised linear models, Liang and Zeger [11].

The present paper considers a different approach based on Markov chains. Pegram [1] and Raftery [2] developed parsimonious models for multi-lag Markov chains, called autoregressive Markov models, which preserve the limiting distribution and the autocorrelation structure of the underlying multi-lag Markov chain. The purpose of this paper is to generalise this methodology to situations where the number of lags may be arbitrarily large. Infinite-lag Markov models have infinite memory and, therefore, can usefully serve to describe processes for which in principle the complete past should be taken into account and not just the recent past. Another feature of infinite-lag Markov models is their flexibility for fitting sequences of any length, possibly with missing values and non-consecutive data points. With autoregressive Markov models, the treatment of missing values and non-consecutive sequences could be complicated and the number of lags that may be considered is constrained by the length of the longitudinal sequence obtained from the survey.

2. AUTOREGRESSIVE MARKOV MODELS

We begin with a simple example to motivate the problem and set the notation. Let Y_t be the employment status of a worker measured at the reference period t of the current round of a longitudinal survey; $Y_t=E$, if the worker is employed at time t, and $Y_t=U$, if unemployed. Measured also at time t are several socio-demographic characteristics X_t of the worker, such as sex, age, race, level of education, occupational group, branch of economic activity and income. These current observations jointly with the measurements made at earlier rounds of the survey generate a set of longitudinal data, $Y_t, Y_{t-1}, Y_{t-2},\ldots$ and $X_t, X_{t-1}, X_{t-2},\ldots$, describing the work experience and the socio-demographic characteristics of the worker as observed through the survey. Certain covariates X_t such as sex and race are time-invariant, but others such as income and occupation may change over time.

The probability of being employed or unemployed at time t is denoted by $P(Y_t=E)=p_t$ and $P(Y_t=U)=q_t$, respectively, with $p_t+q_t=1$. The conditional probability of the employment status of the worker given his (or her) past work experience and socio-demographic characteristics is expressed as

$$(1) \qquad P(Y_t|Y_{t-1},Y_{t-2},\ldots;X),$$

where X denotes the entire set of the covariates, both those that are time-invariant and those that are time-varying. Modelling the conditional probability (1) is the main concern of this paper.

Although the problem is expressed here in terms of employment and unemployment, it can be easily formulated in general terms to apply to longitudinal data with discrete valued outcomes, for which the outcome probability is to be studied as a function of past values and of the covariates. The outcome variable Y_t need not be dichotomous ("employed" or "unemployed"), but may take on an arbitrary number of values (say, $E_1,E_2,\ldots,E_n$).

The conditional probability (1) is a transition probability and if the transition depends only on the immediate past, it reduces to a first-order Markov chain characterised by the transition probability,

$$P(Y_t|Y_{t-1},Y_{t-2},\ldots;X) = P(Y_t|Y_{t-1};X). \tag{2}$$

In general, however, the process may have long memory and the conditional probability may depend not only on the immediate past but also on the outcomes in the more distant past. For this purpose, higher-order Markov chains should be used, but, in practice, the number of parameters becomes prohibitively large as the number of transition probabilities increases geometrically with the order of the multi-lag Markov chain to be used.

A model which preserves the essential properties of multi-lag Markov chains but involve a limited number of parameters is the autoregressive Markov model, where the probability distribution of the current outcome Y_t depends on the preceding outcomes in a particular additive form,

$$P(Y_t|Y_{t-1},Y_{t-2},\ldots;X) = a_1Q_X(Y_t|Y_{t-1}) + \ldots + a_kQ_X(Y_t|Y_{t-k}), \tag{3}$$

where $a_1,a_2,\ldots,a_k$ are lag parameters which may depend on X, with $a_1+a_2+\ldots+a_k=1$; and the Q_X elements are state parameters describing the state in which the process was at a preceding reference date t-h, h=1,...,k, but independently of its timing. The state parameters may be expressed as a time-independent transition matrix $Q_X=[Q_{ij}]$, where

$$Q_{ij}=Q_X(Y_t=E_j|Y_{t'}=E_i), \tag{4}$$

for any $t'<t$ and $i,j=1,\ldots,n$. The Q_{ij} elements may be interpreted as transition probabilities, namely, the probability of being in state $Y_t=E_j$ at time t having been in state $Y_t'=E_i$ throughout the past $t'<t$.

In general, an autoregressive Markov model has (k-1) lag parameters and n(n-1) state parameters, where n is the number of possible values of the outcome variable. Each lag thus adds just one parameter in the model. To these, of course, should be added the parameters associated to the covariates.

Although the autoregressive Markov model uses only a few number of parameters, it preserves two essentially properties of the underlying multi-lag Markov chain, namely, the marginal distribution and the autocorrelation structure. These results have been proved by Pegram [1] and Raftery [2] when the outcome variable takes on discrete ordinal values, $E_i=i$, $i=1,2,\ldots,n$. Adke and Deshmukh [12] have shown that the marginal distribution property holds for an arbitrary state space, not necessarily countable.

3. INFINITE-LAG MARKOV MODELS

3.1. The Model

The autoregressive Markov model may be generalised in a straightforward manner to an infinite-lag model of the form,

$$(5) \qquad P(Y_t|Y_{t-1},Y_{t-2},\ldots;X) = a_1Q_X(Y_t|Y_{t-1}) + a_2Q_X(Y_t|Y_{t-2}) + \cdots,$$

with $a_1+a_2+\ldots=1$. One advantage of the infinite-lag model is that it can account for the entire past of the subject under study, and not just the recent past. Furthermore, the choice of the model and the number of lags to be considered need not be restricted to the length of the longitudinal data at hand. In fact, the issue of determining the appropriate number of lags for fitting the model does not arise.

If it can be assumed that the effects of past outcomes are smaller the more distant they are from the current period, the lag parameters a_k may be considered to be a decreasing function of k. Two simple parametric functions,without excluding other possibilities, which reflect this feature are:

$$(6a) \qquad a_k=\rho^{k-1}(1-\rho), \qquad 0<\rho<1, \text{ and } k=1,2,\ldots$$

$$(6b) \qquad a_k=k^{-\alpha}-(k+1)^{-\alpha}, \qquad \alpha>0, \text{ and } k=1,2,\ldots$$

In both cases the lag parameters are decreasing functions of k and add up to unity. The use of a functional form for the lag parameters sharply reduces the number of parameters of the model.

3.2 Limiting Distribution

The limiting distribution of the infinite-lag Markov model is given by

$$(7) \qquad \lim_{k\to\infty} P(Y_{t+k}|Y_{t-1},Y_{t-2},\ldots) = p,$$

where p is the probability distribution defined by $\lim Q^k = pp'$. This result may be derived by treating the data points $Y_{t+k-1}, Y_{t+k-2}, \ldots, Y_t$ as missing or as in Adke and Desshmukh [12].

3.3. Missing Data

In general, since in practice longitudinal data are of finite length possibly with missing values and non-consecutive data points, it is necessary to derive the corresponding forms of the infinite-lag model for fitting the data. Missing data of concern here are those which creates a gap in the longitudinal sequence. For example, in the sequence $Y_t, Y_{t-1}, Y_{t-2}, \ldots,$ the measurement at time t-k may be missing. In this case the form of the model remains the same except that the unknown term $Q_X(Y_t|Y_{t-k})$ is replaced by the convolution,

$$Q(Y_t|Y_{t-k}) = a_1 Q^{(2)}(Y_t|Y_{t-k-1}) + a_2 Q^{(2)}(Y_t|Y_{t-k-2}) + \ldots, \tag{8}$$

where the terms $Q^{(2)}(.|.)$ denote the 2-step transition probabilities associated with the transition matrix Q^2. For the sake of convenience the index X is dropped here and in the remainder of the paper. The above result is obtained from the following argument,

$$\begin{aligned} P(Y_t|Y_{t-1},\ldots,Y_{t-k+1},Y_{t-k-1},\ldots) &= \sum_{Y_{t-k}} P(Y_t|Y_{t-1},\ldots,Y_{t-k},\ldots)P(Y_{t-k}|Y_{t-k-1},\ldots) \\ &= \sum_{Y_{t-k}} [\sum_{i=1} a_i Q(Y_t|Y_{t-i})][\sum_{j=1} a_j Q(Y_{t-k}|Y_{t-k-j})] \\ &= \sum_{i\neq k} a_i Q(Y_t|Y_{t-i}) + a_k[\sum_{j=1} a_j Q^{(2)}(Y_t|Y_{t-k-j})], \end{aligned}$$

where we have used the fact that $\sum_{Y_{t-k}} Q(Y_{t-k}|Y_{t-k-j}) = 1$ and $\sum_j a_j = 1$ and that

$$\sum_{Y_{t-k}} Q(Y_t|Y_{t-k})Q(Y_{t-k}|Y_{t-k-j}) = Q^{(2)}(Y_t|Y_{t-k}).$$

When there are more than one missing data, a similar argument applies and the unknown terms of the infinite-lag Markov model are to be replaced by corresponding convolutions in terms of the elements of the 2-step, 3-step transition matrices Q^2, $Q^3, \ldots$. Thus, when there are two missing data points, Y_{t-h} and Y_{t-k}, with h<k the terms $Q(Y_t|Y_{t-h})$ and $Q(Y_t|Y_{t-k})$ of the infinite-lag Markov model should be replaced as in (7), expect that the term

$Q^{(2)}(Y_t|Y_{t-k})$ corresponding to $j=k-h$ in the expression for $Q(Y_t|Y_{t-h})$ is itself unknown and should be replaced by

$$Q^{(2)}(Y_t|Y_{t-k}) = a_1Q^{(3)}(Y_t|Y_{t-k-1}) + a_2Q^{(3)}(Y_t|Y_{t-k-2}) + \ldots \tag{9}$$

The case for three or more missing data can similarly be deduced. In general,

$$Q^{(r)}(Y_t|Y_{t-k}) = a_1Q^{(r+1)}(Y_t|Y_{t-k-1}) + a_2Q^{(r+1)}(Y_t|Y_{t-k-2}) + \ldots, \tag{10}$$

where the terms $Q^{(r+1)}(.|.)$ correspond to the elements of the r-step transition matrix Q^{r+1}, $r=0,1,2,\ldots$ At the limit as r increases, $\lim Q^r = pp'$, where p is the eigenvector of Q.

3.4 Finite Sequences

A finite sequence may be treated as an infinite sequence in which the observations after a certain point are missing. Thus it can be shown that the infinite-lag Markov model for finite sequences XXX...X of length k may be expressed as

$$P(Y_t|Y_{t-1},\ldots,Y_{t-k}) = a_1Q(Y_t|Y_{t-1})+\ldots+a_kQ(Y_t|Y_{t-k}) + \theta_kP(Y_t), \tag{11}$$

where $\theta_k=1-a_1-a_2\ldots-a_k$ and $P(Y_t)$ is the stationary distribution p of the transition matrix Q, i.e., Qp=p. This is obtained by applying (9) repeatedly for the missing terms $Q(Y_t|Y_{t-k-i})$, $i=1,2,\ldots$ of the infinite-lag Markov model and noticing that as r tends to infinity $Q^{(r)}(Y_t|Y_{t-k-i})$, $i=1,2,\ldots$ tends to the elements of the limiting matrix $\lim Q^r=pp'$.

3.5 Non-consecutive Sequences

Gaps are created in longitudinal data not only with missing values but also when by design measurement is skipped at certain points in time, for example, when the longitudinal data are obtained from matched samples of certain rotation sampling schemes. Longitudinal data that can be generated by matching the overlapping samples of rotation sampling schemes are often of two types: consecutive such as XXXXXX or non-consecutive such as XXXX--------XXXX. The particular non-consecutive sequence illustrated here corresponds to the 4-8-4 rotation sampling scheme of the US Current Population Survey in which each unit in the sample is surveyed for 4 consecutive months, dropped from the sample for the next 8 months, and surveyed again for 4 more months before permanently leaving the sample.

The infinite-lag Markov model for fitting non-consecutive sequences of the general form k-m-k may be expressed by

$$\begin{aligned} P(Y_t|Y_{t-1},\ldots,Y_{t-k},Y_{t-k-m-1},\ldots,Y_{t-2k-m}) &= (1-a_1-\ldots-a_{2k+m})P(Y_t) \\ &\quad + a_1Q(Y_t|Y_{t-1})+ \ldots + a_{2k+m}Q(Y_t|Y_{t-2k-m}) \end{aligned} \tag{12a}$$

where the middle terms $Q(Y_t|Y_{t-k-i})$, $i=1,\ldots,m$, corresponding to the missing points in the non-consecutive sequence, k-m-k, are derived iteratively from

$$Q^{(r)}(Y_t|Y_{t-k-m+j}) = \sum_{i=1}^{k+j} a_i Q^{(r+1)}(Y_t|Y_{t-k-m+j-i}) + \theta_{k+j}P(Y_t), \tag{12b}$$

where $\theta_{k+j}=1-a_1-a_2-\ldots-a_{k+j}$, $j=0,1,\ldots,m-1$ and $r=1,\ldots,m$.

4. A NUMERICAL ILLUSTRATION

To illustrate the infinite-lag model described in the preceding section, consider the problem of determining the probability of employment of an unemployed worker who, according to the longitudinal data available from the survey, has been unemployed during the past four months with a spell of inactivity in the third month. Using (11) this probability may be expressed as,

$$\begin{aligned} P(E|UUNU) &= P(Y_t=E|Y_{t-1}=U, Y_{t-2}=U, Y_{t-3}=N, Y_{t-4}=U), \\ &= a_1Q(E|U)+a_2Q(E|U)+a_3Q(E|N)+a_4Q(E|U)+\theta_4P(E) \end{aligned} \tag{13}$$

where E represents "employment", U "unemployment" and N "inactivity" or "not-in-labour-force"; the Q terms are elements of the transition matrix

$$Q = \begin{vmatrix} Q(E|E) & Q(U|E) & Q(N|E) \\ Q(E|U) & Q(U|U) & Q(N|U) \\ Q(E|N) & Q(U|N) & Q(N|N) \end{vmatrix}$$

with eigenvector $p=(P(E),P(U),P(N))'$; and $\theta_4=(1-a_1-a_2-a_3-a_4)$. If θ_4 is set to zero, or, $a_1+a_2+a_3+a_4=1$ than (13) reduces to the autoregressive model presented in Raftery [2]. If the transition matrix is set as $Q=\pi I+(1-\pi)1.p'$, where I is the identity matrix, 1 the unit vector and p the vector of limiting probabilities, (13) reduces to the model introduced by Pegram [1].

Using the numerical values obtained on the basis of a set of four-month matched data from the the US Current Population Survey corresponding to December 1981, January, February and March 1982, described in Mehran [13], we have

$$Q = \begin{vmatrix} 0.974 & 0.014 & 0.012 \\ 0.116 & 0.791 & 0.093 \\ 0.016 & 0.012 & 0.972 \end{vmatrix}$$

for which the stationary distribution is $p=(0.521,\ 0.058,\ 0.421)'$. Setting $a_k=\delta^{k-1}-\delta^k$, with $= 0.709$, the probability (13) may be calculated as

$$\begin{aligned} P(E|UUNU) &= a_1 0.116 + a_2 0.116 + a_3 0.016 + a_4 0.116 + \theta_4 0.521 \\ &= 0.461. \end{aligned}$$

For a non-consecutive sequence, e.g., UU--NU, the probability is obtained on the basis of (12) as follows,

$$P(E|UU\text{--}NU) = a_1Q(E|U)+a_2Q(E|U)+a_3Q_3(E|-)+a4Q_4(E|-) +a_5Q(E|N)+a_6Q(E|U)+\theta_4P(E),$$

with

$$Q4(E|-) = a_1Q^{(2)}(E|N)+a_2Q^{(2)}(E|U)+\theta_2P(E),$$

$$Q_3(E|-) = a_1Q_4{}^{(2)}(E|-)+a_2Q^{(2)}(E|N)+a_3Q^{(2)}(E|U)+\theta_3P(E),$$

and

$$Q_4{}^{(2)}(E|-) = a_1Q^{(3)}(E|N)+a_2Q^{(3)}(E|U)+\theta_2P(E),$$

where the terms $Q^{(2)}$ and $Q^{(3)}$ are elements of the power matrices Q^2 and Q^3, respectively. For the numerical values given above, this gives

$$P(E|UU\text{--}NU) = 0.597,$$

which should be compared with the probability $P(E|UUNU)=0.461$ obtained earlier.

REFERENCES

[1] Pegram, G.G.S., "An Autoregressive Model for Multilag Markov Chains," J. Appl. Prob. 17 (1980) 350-362.

[2] Raftery, A.E., "A Model for Higher-order Markov Chains," J.R. Statist. Soc. B 47 (1985) 528-539.

[3] Cox, D.R. and Oakes, D., Analysis of Survival Data (Chapman and Hall, London, 1984).

[4] Kalbfleisch, J.D. and Prentice, R.L., The Statistical Analysis of Failure Time Data (Wiley, New York, 1980).

[5] Heckman, J.J. and Singer, B., "Socoal Science Duration Analysis," in: Heckman, J.J. & Singer, B., (eds.), Longitudinal Analysis of Labor Market Data (cambridge University Press, New York, 1985).

[6] Hamerle, A., "Multiple-spell Regression Models for Duration Data," Appl. Statist. 38 (1989) 127-138.

[7] Cox, D.R., Analysis of Binary Data (Chapman and Hall, London, 1970).

[8] Ochi, Y. and Prentice, R.L., "Likelihood Inference in Correlated Probit Regression," Biometrika 71 (1984) 531-543.

[9] Stiratelli, R., Laird, N. and Ware, J., "Random Effects Models for Serial Observations with Binary Responses," Biometrics 40 (1984) 961-971.

[10] Koch, G.C., Landis, J.R., Freeman, J.L., Freeman, D.H. and Lehman, R.G., "A General Methodology for the Analysis of Repeated Measurements of Categorical Data," Biometrics 33 (1977) 133-158.

[11] Ling, K.-Y. and Zeger, S.L., "Longitudinal Data Analysis Using Generalised Linear Models," Biometrika 73 (1986) 13-22.

[12] Adke, S.R. and Deshmukh, S.R., "Limit Distribution of a Higher Order Markov Chain," J.R. Statist. Soc. B 50 (1988) 105-108.

[13] Mehran, F., "Longitudinal Analysis of Employment and Unemployment Data Based on Matched Rotation Samples," Labour: Review of Labour Economics and Industrial Relations (1989) forthcoming.

STATISTICAL DATA ANALYSIS AND INFERENCE
Y. Dodge (Editor)
Elsevier Science Publishers B.V. (North-Holland), 1989

STATISTICS FOR STOCHASTIC PROCESSES : applications to engineering and finance

A.F. Gualtierotti

HEC-IDHEAP
UNIL CH-1015 Lausanne

Problems in engineering and finance often confront the statistician with large data sets which cannot be replicated. Usual statistical methods are then difficult, if not impossible, to apply. This paper is an attempt to explain and illustrate the challenge issued by these problems.

1 INTRODUCTION

Many problems in engineering and finance present two characteristics:

* the data sets are very large
* the data sets cannot be replicated

Here are some examples. The first we mention are listed in the preface of the book [12]. They involve modeling and signal processing of random pressure fields in underwater acoustics (sonar), and in wheat silos, for early detection of weevils, as well as the study of biological signals. It is mentioned there that a signal which is sampled at the rate of $\Delta = 1,25\ 10^{-4}$ seconds to describe frequences in the range from 0 to 3000 Hertz produces, within an hour, 10^7 observations! An example from finance can be found in [29]: *The source of the data is the database we have collected of interbank over-the -counter prices published by Reuters in a multiple contributors page (the FXFX page). Among the traders these prices are called ticks. They cover the market worldwide and 24 hours a day with approximately 200 contributors being major banks or brokers. Each exchange is automatically recorded with all its information by a computer ... It represents a considerable amout of data of the order of a million ticks for the most traded currencies ...*

As a consequence of these characteristics usual statistical practice results in a number of difficulties which we now illustrate. The problem considered to that end is that of detecting the presence of a signal perturbed and thus hidden by (additive) noise. This is really a hypothesis test with the following ingredients. The observations form the set

$$\{x(t), t \in I\},$$

I a set of time points or an interval. The noise values are

$$\{n(t), t \in I\}$$

and the signal values are

$$\{s(t), t \in I\}.$$

Thus the testing problem can be written

$$H_0 : x(t) = n(t), \; t \in I$$

$$H_1 : x(t) = s(t) + n(t), \; t \in I.$$

Let us suppose for a moment that the signal is not random and that the noise is Gaussian. One then automatically tries to use for that testing problem the likelihood ratio which is optimal in many respects [34]. The test functional is then

$$< \underline{x}, \Sigma_N^{-1} \underline{s} >_{\Re^n}$$

where $\underline{x}$ and $\underline{s}$ are the vectors with respective components $x(t)$ and $s(t)$, $t \in I$, Σ_N is the covariance matrix of the noise, and the formula represents the inner product in Euclidean space with dimension n, the cardinality of I. Let Φ denote the distribution of a standard normal random variable. Since

$$< \underline{s}, \Sigma_N^{-1} \underline{s} >_{\Re^n} > 0,$$

the **Type I** error is then

$$\Phi(c / \parallel \Sigma_N^{-1/2} \underline{s} \parallel_{\Re^n}).$$

If $(s(t), t \in I)$ is a sample from a function s lying in the reproducing kernel Hilbert space (RKHS) [20,30] of the (Gaussian) noise, then

$$\parallel \Sigma_N^{-1/2} \underline{s} \parallel_{\Re^n}^2$$

is the square of the norm of the projection of the signal onto the subspace of the RKHS generated by the functions $(R_N(t, \cdot), t \in I)$[30], where R_N is the covariance function of the noise. If s does not belong to the RKHS of the noise, then, as the sample size increases,

$$\parallel \Sigma_N^{-1/2} \underline{s} \parallel_{\Re^n}^2$$

tends to infinity, and consequently the error of **TypeI** *increases as sample size increases!* The simplest example is obtained by choosing for the noise a standard Wiener process and for the signal the function $s(t) = \sqrt{t}$, for it is well known that the RKHS of the Wiener process is made of absolutely continuous functions with square integrable derivative [20]. If one now tries to use a substitute statistic, like the deflection [3], which is used whenever the likelihood is not available, one must compute expressions of the form [2]

$$\Lambda(\underline{x}) = \underline{x}^t R_N^{-1} (R_{S+N} - R_N) R_N^{-1} \underline{x}$$

where R denotes the covariance and the subscript the process of which R is the covariance, that is invert and multiply huge matrices, with problems, beyond the numerical and computing ones, which are similar to those already mentioned in connection with the likelihood. The explanation for the phenomenon is simple: the likelihood ratio is reflecting the singularity of the measures generated by the noise and the signal-plus-noise processes, rather than the differences due to the presence of the signal, which would be the case in a problem for which the measures are equivalent. One may think that these pathologies are due to the countable additivity of probability measures which is a technical necessity rather than a modeling requirement [24], or to the use of the Wiener process, but the example we have given is a finite dimensional one, and it is the size of the sample which brings into the picture these pathologies.

These problems are those which may occur when it is possible to explicitely compute the required statistics. It is often the case that a problem which is intractable when treated as a finite discrete time problem may become easier when seen as a continuous time problem. It can also be argued that it makes more sense that it be handled that way [9]: *Where the stochastic signal case is considered, it is reduced to [a] time discrete version ... for the reason that the continuous time is too difficult to handle, and anyhow, in digital computer processing (as is the rule), it is so discretized in the analog-to-digital (A-D) conversion process. This is indeed true; but the authors invariably proceed to make the assumption that the noise samples $\{N_n\}$ are mutually independent. But this requires that the sampling rate (in the periodic sampling of the data) be not more than twice the postulated* bandwidth *of the noise, itself actually unknown. Indeed in most practical cases the sampling rate is far higher than twice the bandwidth. To meet this objection, one may then allow the $\{N_n\}$ to be correlated. But then the correlation function must be known, and anyone with experience in handling real data can easily appreciate that it is unrealistic to require that much knowledge of the noise process, even if the complication in the theory can be borne. We maintain, in any event, that it is much better to work with the time continuous model ... , allowing as high a sampling rate in the processing as the A-D converter is designed for.* Detection problems are typically in the latter category and are easily illustrated with applications in underwater acoustics and more generally in geophysical problems. Here is a description of the detection problem for reverberation-limited active sonar [6]. *In the noise-only situation, the principal noise component will be due to the superposition of reflections from a large number of small inhomogeneities (marine organisms, air bubbles, surface irregularities, etc.). Each of these inhomogeneities scatters a small amount of energy incident from the transmitted waveform back to the receiver, and each can be modeled as a random scatterer. These scatterers are typically assumed to have the same statistical properties, and to have a Poisson spatial distribution. The central limit theorem can then be applied to obtain a Gaussian reverberation process at the receiver ... When a target such* *as a submarine is present, it will have few scatterers compared to those producing the reverberation. Moreover, in many situations (depending on aspect angle, etc.) a small subset of the target scatterers will predominate. Each scatterer can again be modeled as producing a random scattered waveform from the incident transmitted waveform. However, due to the paucity of scatterers, the central limit theorem cannot be applied, and so the target return (received signal) can be a sample function from a highly non-Gaussian stochastic process. Since the waveform incident on the target is first scattered by the inhomogeneities producing the reverberation, as in the composite target reflection, the (received) signal and the noise will be statistically dependent.* Finally the processes under study are not stationary. If one insits in retaining a detection problem in which the index set of time instants I is discrete and finite, it is impossible to say anything about the law of $\underline{s} + \underline{n}$, and thus to carry out a statistical analysis. However if one is willing to take into account the continuous nature of pressure waves and assume that the discrete observations are samples at discrete times of adequately smooth processes, then one can obtain general qualitative forms for the likelihood and there is then some hope that a statistical analysis will prove fruitful. There is a price to pay for these assumptions: one is then working with measures on infinite dimensional spaces where probability measures are more likely to be singular (testing with no error) than equivalent (testing using the likelihood), and it raises the question of the robustness of the models used, rather, or before, that of the statistics used [21].

It thus seems that for very large data sets one must resort to the assumption that the data are generated by a stochastic process and consequently that one must develop statistical methods adequate for such situations. As already mentioned there are prerequisites to a statistical theory of stochastic processes: one must ascertain the equivalence of the probability measures involved, in order to obtain a likelihood; the latter must be derived and one must be able to use it. We shall summarize what is known and usable in this area in the following section.

2 EQUIVALENCE, LIKELIHOODS AND MUTUAL INFORMATION

Statistical methodology for stochastic processes, if one dare say so given the available litterature, is at the present time fairly limited (it is often restricted to general mathematical problems: central limit theorems, derivation of likelihoods, etc.) and there is not a great deal of experience in applying it [10,20,27,28] (for a recent effort, see [5,8]). One must generally take advantage of whatever pecularities the processes considered exhibit and the techniques used vary very greatly from one type of process to another. There are of course attempts at bridging the gaps as in the book *Linear Stochastic Systems with Constant Coefficients* [1] in which there is a blend of time series and stochastic differential equations techniques (linear and Gaussian case). A number of general tools exist however such as likelihood based methods which serve as well in testing as in estimation, the method of sieves [20] and quasi-likelihood methods [19] which are geared towards estimation problems. Mutual information [33] has the potential of becoming a very general tool, but so far has mainly interested engineers (the exception is [25]) , though it plays a significant role in large deviations results [11].We shall concentrate in what follows on likelihoods for signal-in-noise problems.

The oldest case for which fairly complete results are known is the Gaussian case (both signal and noise are Gaussian) and the most complete treatment can be found in [35] where conditions for the likelihood to be a quadratic form are derived (a very non trivial fact!). The tool is the theory of operators on Hilbert space, for signals have finite energy and thus square integrable paths, and covariances are kernels of integral operators with finite trace. For non Gaussian processes the most useful techniques come from stochastic calculus [28], for, through Ito's formula, it provides a mechanical tool for computing with stochastic processes. Stochastic calculus operates on martingales and continuous martingales look very much like Brownian motion [14] so that they are not always good models. It is possible however to sometimes retain the power of martingale methods while working with more realistic models and there can be some useful byproducts: thus many of the considerations revolving around the robustness of the filtering problem [15] seem to vanish with other models as shown by the form of the likelihood in [7]. The same methods allow one to go beyond the Gaussian noise assumption [8].

Mutual information is also known as Kullback-Leibler information [26] and is well known as a methodological tool in the theory of inference[25,32]. This is not surprising since there are very close connections between equivalence and mutual information [4]. Furthermore mutual information is a basic tool in information theory since its supremum determines the rate at which information can be reliably transmitted [33].

3 AN EXAMPLE FROM FINANCE

We shall illustrate some of the points raised above with a deceptively simple example drawn from the financial litterature. *Geometric Brownian motion is often used to model prices of assets, say shares of stock, that are traded in a perfect market*[22]. Empirical research [29,36] shows that certain features exhibited by financial data cannot be produced by geometric Brownian motion. So more general models have been used and we are going to consider what appears as a simple extension of it [13]: if S denotes the value of the asset, then the asset evolves in time according to the equation

$$dS_t = (A + BS_t)dt + CS_t^D dW_t$$

where W is a standard Brownian motion. This equation generalizes geometric Brownian motion in that the latter is obtained by setting $A = 0$ and $D = 1$. There are then two problems. The first is to estimate A, B, C and D, the second to find in the model the qualitative behaviour exhibited by the observed data.

Though in the case of geometric Brownian motion the equation for S is unambiguous when considered as a diffusion or when considered as the solution to a stochastic differential equation, this is no longer the case for the more general equation, essentially because the diffusion coefficient is, in the latter case, defined only on $(0,\infty)$, that is, the state space of the process is made of the positive reals, whereas ordinarily stochastic differential equations have the whole real line as state space. For a diffusion one must then worry about the behaviour of the process at the boundaries of the state space. This behaviour is basically determined by the integral

$$\int_{x_0}^{x} \mu/\sigma^2$$

and some integrals related to it, where μ is the drift and σ the diffusion coefficient [23]. The boundaries are then classified accordingly and in the present case one gets the classification given in the *APPENDIX* (NAT means a natural boundary, ENT an entrance boundary, REG a regular boundary, and EXT means an exit, trap, or absorbtion boundary: for an explanation see [23]; when $< C^2/2$ appears in a table, the corresponding values of A or B are strictly positive also!).

Thus a certain number of cases are naturally excluded and in particular those for which the process tends to ∞ in finite time. Which cases should be retained and which should not does not seem to be discussed in the litterature which is referenced in this paper. If now one would like to find the process in the family which fits best a given data set, there are two possible attitudes. One is to adjust within the whole family of parameters, or only within the family of parameters which make (economic) sense. It seems that the former is usually done. Thus in [13] a likelihood based estimation procedure is used, which is motivated by the intuitive idea that a diffusion is locally Gaussian [16,20]. One says that $(A=0)$

$$(dS/S) - Bdt \sim N(0, C^2 S^{2(D-1)}dt),$$

and then one writes the Gaussian likelihood for independent values (one for each time interval). Since one observes a sample path, that the process has not been shown to have independent increments, such a procedure can only have the value of convenience, at least as long as it has not been somehow confirmed. In fact it is argued in [20] that the argument leads to the standard form of the likelihood [28]. The odd thing about this derivation is that it does not make use of any of the properties the diffusion may have. The difficulty lies perhaps in that likelihoods for diffusions are only available in very specific cases, all requiring that the diffusion coefficient be one [31]. We shall suggest here another estimation method based on the properties of stochastic differential equations. Indeed the way the diffusion is written, i.e.

$$dS_t = (A + BS_t)dt + CS_t^D dW_t,$$

suggests that S is the solution of a stochastic differential equation. The problem lies in the fact that the usual formulation implies that σ in

$$dX(\omega,t) = \mu(X(\omega,t))dt + \sigma(X(\omega,t))W(\omega,dt)$$

has integrability and positivity properties which do not obtain here. In fact there can be no solution [17] except the trivial one, if σ can be 0. We are thus going to consider an "approximate" solution obtained by replacing σ by σ_n defined as follows. Let

$$\lambda_n(x) = \begin{cases} (1/n) & \text{if } x \leq (1/n) \\ x & \text{if } (1/n) < x < n \\ n & \text{if } x \geq n \end{cases}$$

σ_n is the defined to be $\sigma \circ \lambda_n$. The stochastic differential equation

$$dX(\omega,t) = \mu(X(\omega,t))dt + \sigma(X(\omega,t))W(\omega,dt)$$

where

$$\mu(x) = A + Bx$$

and

$$\sigma(x) = \sigma_n(x)$$

has a unique solution since the conditions

a) of growth ($x \in \Re$)

$$\mu^2(x) + \sigma_n^2(x) \leq K(1+x^2)$$

b) and Lipschitz (x *and* $y \in \Re$)

$$\mid \mu(x) - \mu(y) \mid + \mid \sigma_n(x) - \sigma_n(y) \mid \leq L \mid x - y \mid$$

are obviously satisfied. A likelihood is now obtained as follows. Let Y be the solution of the stochastic differential equation

$$dY(\omega,t) = \sigma_n(Y(\omega,t))W(\omega,dt)$$

Then, if P_X is the measure induced by X on the space of continuous functions and P_Y that induced by Y, one has

$$\frac{dP_X}{dP_Y}(f) = exp\left\{\int_0^T \frac{\mu(f(t))}{\sigma_n^2(f(t))} f(dt) - \frac{1}{2}\int_0^T \frac{\mu^2(f(t))}{\sigma_n^2(f(t))} dt\right\}$$

Conditions for equivalence [28] are satisfied since σ_n is bounded above and below and the processes have continuous sample paths.

The problem with this form of the likelihood is that the reference measure P_Y depends on the unknown parameters C and D and thus that it does not make sense to optimize with respect to it! Standard treatments of this problem assume that C and D are known [10,20], which is obviously not the case here. But one may use the quadratic variation of the process X. Indeed the solution of a stochastic differential equation is a semimartingale and its quadratic variation is that of the martingale part, that is

$$\int_0^t \sigma_n^2(X(\omega,t))dt.$$

Denote this function $V(\omega,t)$. Since X is continuous, V can be differentiated with respect to t, and since σ_n is by construction positive, one can take the logarithm of the derivative. One thus gets:

$$ln(\frac{dV(\omega,t)}{dt}) = 2lnC + 2Dln(\lambda_n(X(\omega,t))).$$

So C and D are obtainable by regression metods, though it is not yet clear what the distribution of the error is. Furthermore $V(\omega,t)$ can be approximated by finite sums as follows [14]:

$$V(\omega,t) = (p - lim)\sum_{i=1}^{n}(X(\omega,t_{n,(i+1)}) - X(\omega,t_{n,i}))^2.$$

Let $v(t_i) = ln(\frac{dV(\omega,t_i)}{dt})$ and $u(t_i) = ln(\lambda_n(X(\omega,t_i)))$, the t_i's being the observation times. The least squares equations for C and D are:

$$\frac{1}{2}\sum v(t_i) = nlnC + D\sum u(t_i)$$

and

$$\frac{1}{2}\sum u(t_i)v(t_i) = lnC\sum u(t_i) + D\sum u(t_i)^2.$$

Now

$$\frac{1}{n}\sum v(t_i) \sim \frac{1}{T}\int_0^T v(t)dt$$

and similarly for the other sums. Thus, if we write

$$I = \frac{1}{T}\int_0^T ln(\frac{dV(\omega,t)}{dt})$$

$$J = \frac{1}{T}\int_0^T ln(\lambda_n(X(\omega,t)))$$

$$K = \frac{1}{T}\int_0^T ln(\lambda_n(X(\omega,t)))ln(\frac{dV(\omega,t)}{dt})$$

$$L = \frac{1}{T}\int_0^T ln(\lambda_n(X(\omega,t))),$$

the least squares equations can then be written

$$\frac{1}{2}I = lnC + DJ$$

$$\frac{1}{2}K = lnCJ + DI$$

The solutions are

$$D = \frac{K - IJ}{2(L - J^2)}$$

and

$$C = exp(\frac{IL - KJ}{2(L - J^2)})$$

It should be noted that, provided the boundaries are unattainable [23], since in practice n is large, one could, for convenience, forget that the paths should be truncated at n and $1/n$ before the integrals are computed. Otherwise one must integrate only until the boundaries are attained. Now A and B can be estimated from the likelihood ratio, C and D having the values just estimated. Indeed let

$$I_1 = \int_0^T \frac{dX}{\hat{\sigma}_n^2(X)} \text{ and } I_2 = \int_0^T \frac{XdX}{\hat{\sigma}_n^2(X)},$$

$$J_1 = \int_0^T \frac{dt}{\hat{\sigma}_n^2(X)},\ J_2 = \int_0^T \frac{Xdt}{\hat{\sigma}_n^2(X)} \text{ and } J_3 = \int_0^T \frac{X^2dt}{\hat{\sigma}_n^2(X)}.$$

Then the maximum likelihood equation is

$$lnL(A, B) = AI_1 + BI_2 - \frac{A^2}{2}J_1 - ABJ_2 - \frac{B^2}{2}J_3$$

with the solutions

$$A = \frac{I_2J_2 - I_1J_3}{J_2^2 - J_1J_3}$$

$$B = \frac{I_1J_2 - I_2J_1}{J_2^2 - J_1J_3}.$$

The usefulness of these estimators and their properties remain to be assessed. Some simulations should prove useful.

It should be noted that quasi-likelihood methods [18] do not seem to apply here because the parameters do not simultaneously appear in the diffusion coefficient.

Finally one could try to obtain the densities for the processes which are investigated, and attempt to use those for the likelihoods [19]. Unfortunately the differential equations of which the solutions are the densities (if one uses separation of variables) present (for $D > (1/2)$) irregular singular points. Even for $D \leq (1/2)$, the problems are hard, since the solutions have infinite series representations [17]. The densities have thus a form which depends on the value of D, and would thus be useful only if one were to have some a priori information on the range of variability of D.

4 ACKNOWLEDGEMENT

I would like to thank Prof. J.P. Danthine (HEC-Lausanne) for his coaching in matters of finance and Mr. R. Ballif (IDHEAP) without whose help in LaTeX matters this paper would not have been delivered on time.

5 REFERENCES

[1] Arató, M., *Linear Stochastic Systems With Constant Coefficients*, Springer-Verlag, Berlin 1982

[2] Baker, C.R., Optimum quadratic detection of a random vector in Gaussian noise, *IEEE Trans. Communications Technology 14, 802-805 (1966)*

[3] Baker, C.R., On the deflection of a quadratic linear test statistic, *IEEE Trans. Inform. Theory IT-15, 16-21 (1969)*

[4] Baker, C.R., Mutual information, strong equivalence, and signal sample path properties for Gaussian processes, *Inform. and Control 41, 156-164 (1979)*

[5] Baker, C.R., Frey, M.R., Gualtierotti, A.F., Some results on nonGaussian signal detection, to be published in *Colloque GRETSI 1989*

[6] Baker, C.R., Gualtierotti, A.F., Detection of non-Gaussian signals in Gaussian noise, *in Statistical Signal Processing (Wegman, E.J., Smith, J.G., eds.), Dekker, New York 1984*

[7] Baker, C.R., Gualtierotti, A.F., Discrimination with respect to a Gaussian process, *Probab. Th. Rel. Fields 71, 159-182 (1986)*

[8] Baker, C.R., Gualtierotti, A.F., Likelihood ratios and signal detection for non-Gaussian processes, *in Stochastic Processes In Underwater Acoustics (Baker, C.R. ed.), Springer-Verlag, New York 1986*

[9] Balakrishnan, A.V., Parameter estimation in stochastic differential equations: theory and application, *in Developments In Statistics, Vol.1 (Krishnaiah, P.R. ed.), Academic Press, New York 1978*

[10] Basawa, I.V., Prakasa Rao, L.S., *Statistical Inference For Stochastic Processes*, Academic Press, London 1980

[11] Bretagnolle, J., Formule de Chernoff pour les lois empiriques de variables à valeurs dans des espaces généraux, *in Grandes Déviations Et Applications Statistiques, Astérisque 68, Société mathématique de France, Paris 1979*

[12] de Brucq, D., Folliot, G, *Théorie Du Signal*, Masson, Paris 1988

[13] Christie, A.A., The stochastic behavior of common stock variances, *Journal of Financial Economics 10, 407-432 (1982)*

[14] Chung,, K.L., Williams, R.J., *Introduction To Stochastic Integration*, Birkhäuser, Boston 1983

[15] Clark, J.M.C., The design of robust approximations to the stochastic differential equations of non-linear filtering, *in Communications Systems And Random Processes Theory (Skwirzynski, J.K., ed.), Sijthoff and Noordhoff, Alphen aan den Rijn 1978*

[16] Cox, D.R., Miller, H.D., *Stochastic Processes*, Methuen, London 1965

[17] Engelbert, H.J., Schmidt, W., On solutions of one-dimensional stochastic differential equations, *Z. Wahrscheinlichkeitstheorie verw. Gebiete 68, 287-314 (1985)*

[18] Feller, W., Two singular diffusion problems, *Annals of Mathematics 54, 173-182 (1951)*

[19] Godambe, V.P., Heyde, C.C., Quasi-likelihood and optimal estimation, *Int. Stat. Rev. 55, 231-244 (1987)*

[20] Grenander, U., *Abstract Inference*, Wiley, New York 1981

[21] Gualtierotti, A.F., On the stability of signal detection, *IEEE Trans. Inf. Th. IT-29 (1983), 426-433*

[22] Karlin, S., Taylor, H.M., *A First Course In Stochastic Processes*, Academic Press, New York 1975

[23] Karlin, S., Taylor, H.M., *A Second Course In Stochastic Processes*, Academic Press, New York 1981

[24] Kolmogorov, A.N., *Foundations Of The Theory Of Probability* (English translation), Chelsea Publishing Company, New York 1950

[25] Kullback, S., *Information Theory And Statistics*, Wiley, New York 1959

[26] Kullback, S., Leibler, R., On information and sufficiency, *Ann. Math. Stat. 22, 79-86 (1951)*

[27] Kutoyants, Yu.A., *Parameter Estimation For Stochastic Processes*, Helderman Verlag, Berlin 1984

[28] Lipster, E., Shiryayev, A.N., *Statistics Of Random Processes*, Springer-Verlag, New-York 1978

[29] Müller, U.A.,Dacorogna, M.M., Olsen, R.B., Pictet, O.V., Schwarz, M., Morgenegg, C., Statistical study of foreign exchange rates, empirical evidence of a price change scaling law, and intraday analysis, *Olsen & Associates, Zürich 1988*

[30] Neveu, J., *Processus Alétoires Gaussiens*, Laboratoire de Calcul des Probabilités, Paris 1972

[31] Orey, S., The absolute continuity of two diffusions, *Trans. Amer. Math. Soc. 193, 413-426 (1974)*

[32] Osteyee, D.B., Good, I.J., *Information, Weight Of Evidence, The Singularity Between Probability Measures And Signal Detection*, Springer-Verlag, Berlin1974

[33] Pinsker, M.S., *Information And Information Stability Of Random Variables And Processes*, Holden-Day, San Francisco 1960

[34] Poor, V., *An Introduction To Signal Detection And Eestimation*, Springer-Verlag, New York 1988

[35] Rao, C.R., Varadarajan, V.S., Discrimination of Gaussian processes, *Sankhyā Ser.A 25, 303-330 (1963)*

[36] Wasserfallen, W., Kyburz, H., The behavior of flexible exchange rates - A systematic investigation, *Weltwirtschaftliches Archiv 121, 646-660 (1985)*

6 APPENDIX

6.1 For D < $\frac{1}{2}$

Bdry : 0 for D<(1/2)			
A/B	<0	0	>0
< 0	REG	REG	REG
$= 0$	REG	REG	REG
> 0	REG	REG	REG

Bdry : ∞ for D<(1/2)			
A/B	<0	0	>0
<0	NAT	NAT	NAT
$= 0$	NAT	NAT	NAT
>0	NAT	NAT	NAT

6.2 For D = $\frac{1}{2}$

Bdry : 0 for D=(1/2)			
A/B	<0	0	>0
< 0	EXT	EXT	EXT
$= 0$	EXT	EXT	EXT
$< C^2/2$	REG	REG	REG
$\geq C^2/2$	ENT	ENT	ENT

Bdry : ∞ for D=(1/2)			
A/B	<0	0	>0
<0	NAT	NAT	NAT
$= 0$	NAT	NAT	NAT
>0	NAT	NAT	NAT

6.3 For $\frac{1}{2}$ < D < 1

Bdry : 0 for (1/2)<D<1			
A/B	<0	0	>0
< 0	EXT	EXT	EXT
$= 0$	EXT	EXT	EXT
> 0	ENT	ENT	ENT

Bdry : ∞ for (1/2)<D<1			
A/B	<0	0	>0
<0	NAT	NAT	NAT
$= 0$	NAT	NAT	NAT
>0	NAT	NAT	NAT

6.4 For D = 1

Bdry : 0 for D=1				
A/B	<0	0	$< C^2/2$	$\geq C^2/2$
< 0	EXT	EXT	EXT	EXT
$= 0$	NAT	NAT	NAT	NAT
> 0	ENT	ENT	ENT	ENT

Bdry : ∞ for D=1				
A/B	<0	0	$< C^2/2$	$\geq C^2/2$
<0	REG	REG	REG	NAT
$= 0$	NAT	NAT	NAT	NAT
>0	NAT	NAT	NAT	NAT

6.5 For D > 1

Bdry : 0 for D>1			
A/B	<0	0	>0
< 0	EXT	EXT	EXT
$= 0$	NAT	NAT	NAT
> 0	ENT	ENT	ENT

Bdry : ∞ for D>1			
A/B	<0	0	>0
<0	ENT	ENT	ENT
$= 0$	ENT	ENT	ENT
>0	ENT	ENT	ENT

STATISTICAL DATA ANALYSIS AND INFERENCE
Y. Dodge (Editor)

STATISTICAL ANALYSIS OF MACROMOLECULAR SEQUENCES PATTERN PROBABILITIES UNDER THE DINUCLEOTIDE PERMUTATION MODEL

J.Kleffe

Central Institute of Molecular Biology
Academy of Sciences of the GDR
1115 Berlin Buch, Robert Rössle Str.10

Improved DNA sequencing technology has allowed larger samples to be taken and proved DNA sequences to have some characteristic statistical properties. The mostly mentioned property is the dinucleotide composition. It is therefore important to know to what extent this composition alone explains certain patterns in such sequences. We describe a method for computing the probability for a given pattern to occur in a random sequence if the sample space is formed by all sequences with the same dinucleotide composition. A method to construct the sample space itself is of independent interest.

1. INTRODUCTION

Since Sanger and Gilbert independently discovered methods for rapid sequencing of DNA in 1970, nucleic acid sequence data have accumulated at a constantly growing rate. These sequences are represented by long chains of characters A,C, G and T which denote the nucleotides Adenine, Cytosine, Guanine and Thymine. They carry all genetic information of a cell. Therefore methods of sequence comparison are helping to answer several important questions in biology. There is a vast body of literature on this subject. The common approach is to search for and interpret similarities between sequences and to study patterns in individual sequences. But as these sequences are all made of a few number of different characters, they all look more or less alike. Statistical methods are required to make decisions. A frequently used argument is that a pattern appears more or less often than expected by chance. What is this argument good for? Very rarely do such patterns really admit any biological interpretation. This failure has many different reasons with the following in the lead. The concept "chance" depends on a probability model for generating random sequences that often generates too large a sample space. Hence, it declares events being rare in comparison with many others which do not occur in nature. Due to combinatorical explosion, the number of theoretically possible different genetic sequences dwarfs the number of atoms in the universe. The set of sequences nature has produced so far forms a tiny subset. Under these conditions we cannot expect valid results from simple statistical analysis based on equal probability sampling of characters to generate a random sequence. Biological sequences show quite distinctive compositions by individual nucleotides, pairs of nucleotides (dinucleotides) or triplets of them called codons. Therefore Markov chain models are increasingly being investigated; they preserve such compositions on the average. Such compositions are exactly preserved by permutation models although the numerical effort rises considerably. Of special biological significance is the dinucleotide composition, especially in coding regions of DNA (see Josse et al 1961, Swartz et al 1962, Lipman et al 1884 and Nussinov 1980, 1981 a,b and 1984).

Fitch(1983) and Altschul and Erickson(1985) considered an interesting permutation model for random sequences that maintains the dinucleotide composition of a given sequence. This model proved to provide better results when used in Monte Carlo investigations. We took interest in direct computation of the probability that a given pattern appears in a random sequence generated in this way. There are two major difficulties with this problem. One is the complexity of the probability model. The other is the problem of overlapping patterns. For instance, if the pattern P="ACATACA" appears in a randomly generated sequence, it also prepares for its reoccurrence by providing the first three nucleotides ACA at the end of P if P is followed by TACA or the first nucleotide A if P is followed by CATACA. Santibanez-Koref(1987) solved this problem partly under two more simple models. We provide a general solution by introducing a pattern transition matrix that describes these effects. Two methods for generating the entire sample space are of independent interest if the sequences are not too long.

2. THE D-PERMUTATION MODEL

Consider a short nucleic acid sequence

$$S = \text{"ACGTGCAGT"} \tag{1}$$

with nine nucleotides. It includes 8 overlapping pairs (dinucleotides) which form a list

$$EL = (AC,CG,GT,TG,GC,CA,AG,GT) \tag{2}$$

The D-permutation model forms all sequences of length 9 which contain exactly the pairs listed in (2). Every such sequence is called D-permutation of S. Fitch(1983) has noted that each D-permutation of S represents an Eulerian walk in a directed multigraph with vertices A,C,G and T and directed edges given in (2). Let us therefore refer to (2) as the edge list associated with the sequence (1). Further we can associate each dinucleotide in (2) with its second nucleotide to construct a simple one to one correspondence between the second, third, ...,etc. nucleotides in S and the dinucleotides listed in (2). There is no dinucleotide that is associated with the first letter of S, and we have one more letter A in S than pairs in (2), which end on A. However, both numbers are the same for C,G and T. This simple observation has a very important consequence. It implies that each D-permutation of S starts with the same nucleotide. Altschul and Erickson(1985) represented each D-permutation by an edge ordering E(S). This is a table that arranges all edges in 4 lists according to their first nucleotides and in the order they appear in S. For sequence (1) follows

edge ordering	sequence construction	
A-list: AC,AG	AC AG	
C-list: CG,CA	CG CA GT	(3)
G-list: GT,GC,GT	GT GC	
T-list: TG	TG	

This edge ordering describes the way how to form sequence (1). Of course we must know which list to start with and then link together the pairs in corresponding lists from the left to the right. The list associated with the last character of S is called terminal list because the last edge used points to this list. Obviously, edge orderings E(S) and E(S') for two different D-permutations S and S' differ by permutations of the edges in at least one list of their edge orderings. But permuting edges in one or more lists of an edge ordering will not necessarily define a D-permutation of the underlying sequence. The most remarkable result in the paper by Altschul and Erickson(1985) is a simple n.s. condition for this to happen. They defined a last edge graph Z to be formed by the last edges of each but the terminal list and proved:

THEOREM 1: A permutation of the edges in each list of E(S) defines a D-permutation of S iff the last edge graph Z is connected to the terminal vertex (last nucleotide in S).

For instance, the last edge graph Z for the edge ordering (3) includes the edges AG, CA, and GT. It is connected to the terminal T, for it admits a walk from G to T, directly, and walks from A to T and from C to T via vertex G and vertices A and G, respectively.

Theorem 1 is useful for generating random D-permutations of a given sequence. It does not immediately help to decide whether a given set of edges can be arranged to form a sequence that uses all of them. To study this question, let us number the nucleotides A,C,G,T by integers i=1,2,3,4 and denote edges from vertex i to vertex j by pairs (i,j). Define an edge frequency table N = (n(i,j)). The elements n(i,j) in the i-th row and j-th column represent the number of edges from vertex i to vertex j that are available. Row and column totals of this table denoted by n(i,0) and n(0,j) count the edges available from vertex i and to vertex j, respectively. Then simple arithmetics on the numbers of edges from and to a given vertex yields:

THEOREM 2: Assume an edge frequency table is derived from a sequence with first nucleotide i and last nucleotide j.

$$\text{If } i=j \text{ then } n(k,0)=n(0,k) \text{ for all } k \quad (4)$$

$$\text{If } i\neq j \text{ then (4) for all } k\neq i \text{ and } k\neq j \text{ and}$$

$$n(i,0)=n(0,i)+1 \; ; \; n(j,0)=n(0,j)-1 \quad (5)$$

Both cases are simple variations of Euler's famous bridges of Königsberg problem.

Let us now consider an edge ordering of which we do not know that it defines a sequence, i.e. a walk from vertex i to vertex j that makes use of all edges available. But we can start at vertex i and go on as far as possible. Such walk has been named long walk and it is easy to show that under the conditions (4) or (5):

Each long walk starting at vertex i
- ends at vertex j (6)
- exhausts all edges from vertex j (7)
- exhausts all edges to vertex j (8)

These are the only properties of a sequence used by Altschul and Erickson (1985) to prove Theorem 1. Hence, we can put the edges from our edge frequency table into an edge ordering and apply Theorem 1 to check if this edge ordering defines a sequence. It will do so iff we are able to put a corresponding last edge graph in place.

THEOREM 3: A frequency table that satisfies Theorem 2 admits a sequence with first nucleotide i and last nucleotide j iff we can pick a set of 3 edges from the 3 rows $k\neq j$ which form a last edge graph Z that is connected to j.

Obviously, the diagonal elements and the actual size of nondiagonal elements in an edge frequency table do not matter for Theorem 3. We can form a reduced edge frequency table that just reports a unity for each nonzero entry and zeroes along the main diagonal. Then several simple algorithms are at hand to verify the above condition based on Boolean algebra.

2. CHARACTERIZING THE SET OF ALL D-PERMUTATIONS

Let D(S) be the set formed by all D-permutations of a sequence S. Each of its members represents a long walk starting at the same vertex and using the same edge list. Hence, from (6), it must end at the same vertex, too. This fact suggests a simple partition of D(S) into disjoint subsets. Assume S has last nucleotide j, say. Then each D-permutation S_* of S can be expressed as $S_* = S'_* +$ "j" and there are at most 4 choices for the last nucleotide in S'_*. If this choice is k, say, the edge list EL' associated with S'_* is EL/(k,j), i.e. differs from the edge list EL associated with S by deletion of one edge (k,j). Introducing notation D(i,j,EL) for D(S) where S has first nucleotide i, last nucleotide j and edge list EL, we can write

$$D(i,j,EL) = \bigcup_{k} (D(i,k,EL')+"j") \tag{9}$$

where

$$(D(i,k,EL')+"j") = \left\{ S = S'+"j" \mid S' \; D(i,k,EL') \right\} \tag{10}$$

If k is in fact the second last nucleotide of S or of any D-permutation of S, then D(i,k,EL') is not empty. However, this may not be true for each choice of k. For instance, EL could not contain any edge (k,j), or D(S) could not have any member with second last nucleotide k. Theorems 2 and 3 provide the n.s. conditions for D(i,k,EL') to be not empty.

THEOREM 4: Let D(i,j,EL) be not empty, (k,j)∈ EL and form EL' = EL/(k,j). Then EL' satisfies the necessary conditions of Theorem 2 for a sequence with first nucleotide i and last nucleotide k.

Proof follows easily by just watching the change of row and column totals in the edge frequency table if edge (k,j) is removed.

The condition of Theorem 3 depends on the reduced edge frequency table which will only change if we remove an edge (k,j) that was the last one left of its kind. Only this case requires a new search for a last edge graph Z being connected to the terminal vertex. No set D(i,k,EL') on the right hand side of (9) will be empty as long as all edge frequencies stay greater than zero. Another important special case is when nucleotide j was the last one of its kind left in the sequence. Then the j-th column and the j-th row of the edge frequency table reduce to zero. From now on we need no longer test for a connection from vertex j to the terminal vertex.

We can continue this way of partitioning with each of the sets D(i,k,EL') in order to develop a complete tree that gives a convenient survey of D(S). Each step reduces the problem to shorter sequences till it becomes trivial. Figure 1 shows such tree for example (1) where each sequence S represents the whole class D(S). Each sequence in this tree is uniquely associated with its edge frequency table N = (n(i,j)), a squared 4x4-matrix of integers. Hence the number of positions in such tree is bounded by the number of raster points of a 16-dimensional cube with measures n(i,j). But in general, this number will be much less. Each sequence at one level (total number of dinucleotides) of the tree branches not into 16 but at most 4 directions. Therefore many raster points of the cube are not covered.

A computer can quickly generate a complete map of such a tree as shown in

Figure 2. There each line is associated with one sequence of the tree. The first 16 columns are grouped into 4 blocks and provide the entries of the edge frequency tables. The next column carries the last character. The last four columns report pointers to the shorter sequences the tree branches to. They are line numbers put in different columns according to the last character left in the reduced sequence. A blank in one of these columns says no branch taken in

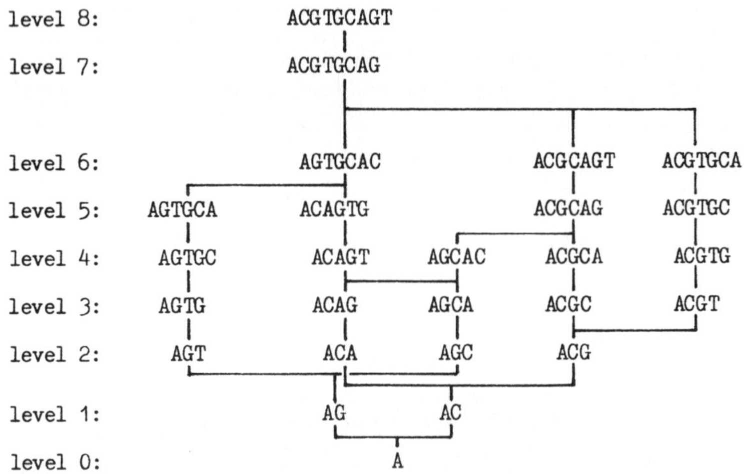

Figure 1: Decomposition of the set D(S) where S = "ACGTGCAGT"

this direction for D(i,k,EL') is empty. We develop this table level by level which are seperated by horizontal lines. The final table is then called structure table of our sequence S and helps to solve a number of problems.

In Figure 2 sequence S has 8 dinucleotides listed in (2). Hence, the structure table starts at level 8 and reports in line 0 the edge frequencies column by column. The first block is formed by the numbers of edges to vertex A. The second block reports those to vertex C etc. T is the last nucleotide in S. Since GT is the only kind of edges to vertex T, 3 out of 4 sets in (9) are empty and we obtain only one class of sequences at level 7 shown in line 1. All sequences in this class have G as the last nucleotide. But now 3 different edges to vertex G are available and verification of the condition of Theorem 3 shows that all of them can be removed. The tree branches into 3 directions represented by 3 lines at level 6. In the next step we process each line at level 6 and obtain 4 new lines at level 5. When processing the lines at level 4 to generate those at level 3, both lines 11 and 13 branch to line 16 since the edge frequency tables become the same. This is not in conflict with the last nucleotide G for it is uniquely defined by the first nucleotide and the frequency table. The fact that each D-permutation of a sequence S has same first and last nucleotide is fundamental for all of the method.

		edge frequency table -A ACGT	-C ACGT	-G ACGT	-T ACGT		branch taken
level 8:	0	0100	1010	1101	0020	T	1
level 7:	1	0100	1010	1101	0010	G	2 3 4
level 6:	2	0100	1010	0101	0010	A	5
	3	0100	1010	1001	0010	C	6 7
	4	0100	1010	1100	0010	T	8
level 5:	5	0000	1010	0101	0010	C	9
	6	0100	0010	1001	0010	A	10
	7	0100	1000	1001	0010	G	11
	8	0100	1010	1100	0000	G	12 13
level 4:	9	0000	1000	0101	0010	G	14
	10	0000	0010	1001	0010	C	15
	11	0100	1000	1000	0010	T	16
	12	0100	1010	0100	0000	A	17
	13	0100	1010	1000	0000	C	18 16
level 3:	14	0000	1000	0100	0010	T	19
	15	0000	0000	1001	0010	G	20
	16	0100	1000	1000	0000	G	21
	17	0000	1010	0100	0000	C	19
	18	0100	0010	1000	0000	A	22
level 2:	19	0000	1000	0100	0000	G	23
	20	0000	0000	1000	0010	T	24
	21	0100	1000	0000	0000	A	23
	22	0000	0010	1000	0000	C	24
level 1:	23	0000	1000	0000	0000	C	25
	24	0000	0000	1000	0000	G	25
level 0:	25	0000	0000	0000	0000	A	

Figure 2: Structure table for S = "ACGTGCAGT"

4. TWO METHODS FOR GENERATING THE SAMPLE SPACE

Altschul and Erickson (1985) described a simple algorithm for equal probability sampling in the class of D-permutations of a given sequence S. It was used in Monte Carlo studies. But there is little evidence on how many sequences have to be sampled for reliable results. This question caused interest in generating the entire population D(S). It is easily done based on the structure table for S. Let us look at Figure 3 which repeats the last characters and the columns "branch taken" of Figure 2. All D-permutations of S begin with the same nucleotide. Hence, at the lowest level 0, the tree given in Figure 1 has only one member formed by the initial nucleotide. It is written in line 25 of the column "sequences" in Figure 3.

		branch taken				sequences
0	T			1		ACGTGCAGT, AGTGCACGT, ACAGTGCGT ACGCAGTGT, AGCACGTGT, ACAGCGTGT
1	G	2	3		4	ACGTGCAG, AGTGCACG, ACAGTGCG ACGCAGTG, AGCACGTG, ACAGCGTG
2	A		5			ACGTGCA
3	C	6		7		AGTGCAC, ACAGTGC
4	T			8		ACGCAGT, AGCACGT, ACAGCGT
5	C			9		ACGTGC
6	A		10			AGTGCA
7	G				11	ACAGTG
8	G	12	13			ACGCAG, AGCACG, ACAGCG
9	G				14	ACGTG
10	C			15		AGTGC
11	T			16		ACAGT
12	A		17			ACGCA
13	C	18		16		AGCAC, ACAGC
14	T			19		ACGT
15	G				20	AGTG
16	G	21				ACAG
17	C			19		ACGC
18	A		22			AGCA
19	G		23			ACG
20	T			24		AGT
21	A		23			ACA
22	C			24		AGC
23	C	25				AC
24	G	25				AG
25	A					A

Figure 3: Generation of the sample space based on the structure table

Now we go through all lines at level 1 and add to sequences referred to the nucleotide shown in column 1. We find AC and AG to be placed in lines 23 and 24, respectively. The same procedure goes on level by level up to the top. In general one line at level q, say, refers to up to four lines at level q-1. Then we have to collect the sequences from all these lines. If the line at level q-1 referred to already carries sets of sequences we must take all of them.

Each such operation corresponds to equation (9). The result of the whole procedure in Figure 3 is a set of 6 sequences at level 8 which form D(S). They are obtained by adding character T to each of the sequences listed in line 1.
This algorithm generates all D-permutations of S simultaneously. But it will collapse on lack of storage if the sample space is large. Let us therefore describe a second algorithm that generates the members of D(S) one by one. The numbers stored in column "branch taken" of the structure table mark as many different passes from the top level down to level 0 as D(S) has members. Therefore, generation of all D-permutations in turn requires to take each pass once and collect the nucleotides left on this pass as last characters. A simple way to do this is to take the first possible branch each time we come to a row with more than one choice. After that we can go back the way we came till we reach a line that offers an alternative choice. Here we stop and go down again to construct the next member. The algorithm stops if no more choice is left at the top level of the structure table.

5. COMPUTATION OF PATTERN PROBABILITIES

Given a sequence S that contains pattern P, biologists are interested in calculating the probability of its occurrence under the D-permutation model. A simple solution to this problem is to generate each member of the sample space and to search for P. But it may be impossible to do so. The number of D-permutations explodes with increasing sequence length. No matter how fast we can find each particular D-permutation, we will quickly be unable to look at all of them one after the other. For instance, if each dinucleotide occures n times in the sequence S, the number of different D-permutations we can form from an edge ordering (3) for given last edge graph Z is

$$N = 64^{-1} q^4 \quad , \quad q = (4n)!(n!)^{-4} \tag{11}$$

Some values of q are given below and call for another way of computing the probability.

n	q	$(n+1)^4$
1	24	16
2	2520	81
3	369600	256
4	63063000	625

(12)

Let P have m characters and let P(i) be the sequence formed by the first i characters of P. P(0) denotes the empty set. We define for each nucleotide j the integers

$$t(i,j) = \left\{ \max_{q=0,\dots,i+1} q \;\middle|\; P(i) + \text{"j"} = R + P(q) \text{ for some } R \right\} \tag{13}$$

for j=1,...,4, i=1,...,m-1 and t(m,j)=m for all j. R stands for a sequence. For instance, if P="ACG", then m=3 and

$$\begin{aligned} P(0) &= \text{""} \\ P(1) &= \text{"A"} \\ P(2) &= \text{"AC"} \\ P(3) &= \text{"ACG"} \end{aligned} \qquad T = (t(i,j)) = \begin{pmatrix} 1 & 0 & 0 & 0 \\ 1 & 2 & 0 & 0 \\ 1 & 0 & 3 & 0 \\ 3 & 3 & 3 & 3 \end{pmatrix} \tag{14}$$

Let us further say that a sequence B belongs to the class C(i) iff

$$i = \left\{ \max_{q=0,\ldots,m-1} q \;\middle|\; B = R + P(q) \text{ for some } R \right\} \tag{15}$$

and P is no subsequence of B. In simpler words, each sequence B is classified according to the longest subpattern P(i), i=0,...,m-1 found at the end of B. If sequence B contains pattern P entirely, we classify it to C(m), independently of the position where P occurs in B. Now, if B gets extended by one additional nucleotide j, it changes membership from class C(i) to class C(t(i,j)). We may therefore consider its membership a state and T a transition matrix that describes the state change. State m is an absorbing state for sequence B stays in this state once it is entered. It remains to remember that the first method described in section 4 develops all sequences in D(S) simultaneously and nucleotide by nucleotide. Therefore, reporting in Figure 3 the counts of sequences being in state i instead of subsequences and updating them from level to level yields these counts for all members of D(S). Figure 4 illustrates the algorithm for P used in (14). P occurs in 4 out of 6 sequences and has probability 2/3. These computations can be performed in time being proportional to the number of lines in the structure table. This number is significantly smaller than the number of D-permutations. For the example considered in (11) and (12), it is less than n+1 to the power 16. Therefore n+1 to the power 4 is reported in (12) for comparison.

The way the transition matrix is defined offers some flexibility of the method. For longer patterns P, biologists often define the relation P⊂S in such a way that S contains all individual nucleotides of P in the same order as in P but with possibly different nucleotides in between. For instance, pattern P="ACG" is contained in sequence S="ACAGCGTGT" since we have the following association (alignment) between the nucleotides in both sequences.

```
ACAGCGTGT                ACAGCGTGT
|| |           or          | ||          (16)
AC G                       A CG
```

The probability for such an event follows if we define

$$t(i,j) = \begin{cases} j & \text{if } i = j-1 \\ i & \text{else} \end{cases} \tag{17}$$

and t(m,j)=m for all j. Figure 4 also shows the computations in this case and yields probability 1.

		branch taken	state counts for T-matrix (14) states 0	1	2	3	(17) states 0	1	2	3	(18) states 0	1	2	3	4	5	6
0	T	1	2			4				6	2						4
1	G	2 3 4	2			4				6	2						4
2	A	5				1				1							1
3	C	6 7	1		1				1	1	1			1			
4	T	8	1			2				3	1						2
5	C	9				1				1							1
6	A	10		1					1			1					
7	G	11	1							1	1						
8	G	12 13	1			2				3	1						2
9	G	14				1				1							1
10	C	15	1						1		1						
11	T	16	1							1	1						
12	A	17				1			1								1
13	C	16	1		1				1	1	1			1			
14	T	19				1				1							1
15	G	20	1					1			1						
16	G	21	1							1					1		
17	C	19				1				1							1
18	A	22		1					1		1						
19	G	23				1				1							1
20	T	24	1					1			1						
21	A	23		1					1			1					
22	C	24	1						1		1						
23	C	25			1				1					1			
24	G	25	1					1							1		
25	A			1				1				1					

Figure 4: State counts for patterns P = "ACG" and P = "A*G" in the Sequence S = "ACGTGCAGT"

Linguistic studies of DNA sequences investigate the frequencies of given patterns (words) in a sequence (text). In such cases we want to know the probability that a given pattern occurs k times. It is obtained by the above method of state counting if we define two-dimensional states as follows. Sequence B is in state (i,j) if it contains pattern P j times and ends on P(i). The required transition matrix is easy to form. Making state (m,k-1) an absorbing state yields the probability of at least k occurrences of pattern P. Withdrawing the upper bound on j results in corresponding counts for any possible k, i.e. the frequency distribution for pattern P as studied by Dayhoff (1984) and Santibanez-Koref (1987) for more simple probability models.

In other cases P is not completely specified. For instance, we could ask for the probability that S contains P = "A*G", where * is a joker or wild card character that stands for A,C,G or T. Such problem can be solved using more than m states. Defining

$$
\begin{aligned}
P(0) &= \text{""} \\
P(1) &= \text{"A"} \\
P(2) &= \text{"AA"} \\
P(3) &= \text{"AC"} \\
P(4) &= \text{"AG"} \\
P(5) &= \text{"AT"} \\
P(6) &= \text{"A*G"}
\end{aligned}
\qquad
T = \begin{pmatrix} 1 & 0 & 0 & 0 \\ 2 & 3 & 4 & 5 \\ 2 & 3 & 6 & 5 \\ 1 & 0 & 6 & 0 \\ 1 & 0 & 6 & 0 \\ 1 & 0 & 6 & 0 \\ 6 & 6 & 6 & 6 \end{pmatrix} \tag{18}
$$

and applying the method of updating state counts yields the probability 2/4 also seen in Figure 4. The joker has not changed the probability that P occurs in S since by the dinucleotide composition of S only C can go between A and G.

6. COMPUTATIONAL LIMITATIONS

The structure table of a sequence defined in section 3 helps to clarify the way of computation. Its size, however, will be the limiting factor for sequence length. Unfortunately we do not yet have a useful estimate for the number of lines in the structure table. The upper bound mentioned in section 3 will quickly yield unacceptably high values. A few examples showed the actual number of lines 4 to 8 times smaller than the upper bound. But still this is a large number. It is therefore a good idea to break the structure table into pieces, one for each level, and process them in batches. The maximum number of lines for a given level turned out to be 5 times smaller than the total number of lines.

	level	edge frequency table -A ACGT	-C ACGT	-G ACGT	-T ACGT	last character string	branch taken			
0	7	0100	1010	1101	0010	GT	1	2		3
1	0	0000	0000	0000	0000	ACGTGCA				
2	6	0100	1010	1001	0010	C	4		5	
3	5	0100	1010	1100	0000	GT	6	7		
4	1	0000	0000	1000	0000	GTGCA	1			
5	2	0100	1000	0000	0000	AGTG		1		
6	3	0000	1010	0100	0000	CA			1	
7	4	0100	1010	1000	0000	C	8		5	
8	2	0000	0010	1000	0000	CA			4	

Figure 5: Reduced structure table for S = "ACGTGCAGT"

Another reduction of size is gained by avoiding lines which only refer to one line at the next lower level. They proved to form the majority of lines in all examples studied so far. Such lines could be replaced by the only lower level line they refer to. Then the column for the last character has to report a list of removed nucleotides. Figure 5 shows one version of a reduced structure table for sequence (1). It was derived from Figure 2 by replacing the above mentioned lines which make first reference to a lower level line. The size of the table came down to 9 lines.

	level	last character string	branch taken	sequences
0	7	GT	1 2 3	ACGTGCAGT,AGTGCACGT,ACAGTGCGT ACGCAGTGT,AGCACGTGT,ACAGCGTGT
1	0	ACGTGCA		ACGTGCA
2	6	C	4 5	AGTGCAC,ACAGTGC
3	5	GT	6 7	ACGCAGT,AGCACGT,ACAGCGT
4	1	GTGCA	1	AGTGCA
5	2	AGTG	1	ACAGTG
6	3	CA	1	ACGCA
7	4	C	8 1	AGCAC,ACAGC
8	2	CA	4	AGCA

Figure 6: Generating sample space using a reduced structure table

The trade off for saved lines is an additional column for the level of each line and the need of storing strings of last characters in place of individual last characters. Further, the computational effort for deriving the reduced structure table rises. But as the structure table represents the entire population of sequences under investigation, this effort is only once for a complete statistical survey. The generation of all D-permutations is shown in Figure 6. We start with the only line at level 0 and copy the character string to the column "sequences". Now the general rule is to go through all lines at a given level in turn and construct sequences in the following way. The last character string given in row q with level r, say, gets added to the r-th position of the sequence (or sequences) already present in the lines referred to. For instance the sequence ACGCA given in line 6 at level 3 was obtained by taking the first 3 characters ACG of the sequence in line 1 plus the last character string "CA" given in line 6.

Computation of probabilities by updating state counts is illustrated in Figure 7. Also here the technique is basically the same as described in section 5. But as we do not use the complete sequences at level s, say, to form the sequences at higher levels q, say, it is convenient to report state counts before updating. We start at line 1 with level 0 and report one sequence in state 0 by definition. Then the general rule is as follows: Assume we are in line r at level q that refers to line t at level $s < q$, say. Then we update the state counts given in line t for the first q–s characters of the last character string in line t and place the result in line r. The example in Figure 7 ends with two sequences in state 0, one sequence in state 2 and 3 sequences in state 3. One more updating for the last characters G and T yields two sequences in state 0 and 4 sequences in state 3. We hope that the reduced structure table allows to handle moderate sequences on main frame computers. Careful and economic coding will be required. Longer sequences can hopefully be treated by approximate methods based on a Markov process.

	level	last character string	branch taken	states 0	1	2	3
0	7	GT	1 2 3	2		1	3
1	0	ACGTGCA		1			
2	6	C	4 5	1	1		
3	5	GT	6 7	1			1
4	1	GTGAC	1		1		
5	2	AGTG	1			1	
6	3	CA	1				1
7	4	C	8 5	1			1
8	2	CA	4	1			

Figure 7: Updating state counts using a reduced structure table

REFERENCES

Altschul,S.F. and Erickson,B.W.(1985), Significance of nucleotide sequence alignments: A method for random sequence permutation that preserves dinucleotide and codon usage. Mol. Biol. Evol.2(6) pp. 526–538

Dayhoff,J.E.(1984), Distinguished words in data sequences: Analysis and applications to neural coding and other fields. Bull. Math. Biol. 46(4) pp. 529–543

Fitch,W.M.(1983), Random sequences. J. Mol. Biol. 163 pp. 171–176

Josse,J.,Kaiser,A.D. and Kronberg,A.(1961), Enzymatic synthesis of deoxyribonucleic acid VIII. Frequencies of nearest neighbor base sequences in deoxyribonucleic acid. J.Biol.Chem. 236 pp. 864–875

Lipman,D.J., Wilbur,W.J., Smith,T.F. and Waterman,M.S.(1984), On the statistical significance of nucleic acid similarities. Nucleic Acids Res. 12 pp.251–226

Nussinov,R.(1980), Some rules in the ordering of nucleotides in DNA. Nucleic Acids Res. 8 pp. 4545–4562

Nussinov,R.(1981a), The universal dinucleotide asymmetry rules in DNA and the amino acid codon choice. J. Mol. Evol. 17 pp. 237–244

Nussinov,R.(1981b), Nearest neighbor nucleotide patterns. J. Biol. Chem. 256 pp. 8458–8462

Nussinov,R.(1984), Doublet frequencies in evolutionary distinct groups. Nucleic Acid Res. 12 pp. 1749–1763

Snatibanez-Koref, M. (1987), Statistische Untersuchungen an DNS-Sequenzen und ein Verfahren zum mehrfachen Sequenzvergleich. Dissertation A, Humboldt Universität zu Berlin

Swartz, M.N., Trautner, T.A. and Kronberg, A. (1962), Enzymatic synthesis of deoxyribonucleic acid XI. Further studies on nearest neighbor base sequences in deoxyribonucleic acid. J. Biol. Chem. 237 pp. 1961-1967

SUMMARY

We describe a method for computing the probability of a given pattern to occure in randomly generated DNA sequences with given dinucleotide composition. These sequences are random permutations of the individual characters in a given sequence such that all overlapping pairs of letters reappear in the same numbers as in which they are found in the original sequence. The number of all so called dinucleotide permutations is significantly smaller than the number of ordinary permutations but still is too large for generating them one after the other. At the other hand, the structure of this set is too complex in order to derive explicit combinatorical expressions for the above mentioned probabilities. But a structure table of order being considerably less than the number of dinucleotide permutations helps to reduce numerical complexity by several orders. Unfortunately, the computations still remain expensive and require main frame computers for even moderate sequence length. We also introduce an interesting technique of pattern transition matrices to deal with the complex problem of overlapping patterns in a general way. This technique also extends to deal with patterns which include jokers or wild card characters or allows to study frequencies of pattern occurrences. It is also not restricted to the dinucleotide permutation model and provides a noncombinatorical alternative to the complex expressions in terms of binomial coefficients that have been published for some simple cases. The calculated probabilities can be used to avaluate the significance of occurrences of patterns in DNA sequences which tends to become overestimated by using ordinary random permutations or other simpler models for the generation of random sequences.

STATISTICAL DATA ANALYSIS AND INFERENCE
Y. Dodge (Editor)

COMPUTING ROBUST COVARIANCES WITH COLLINEAR DATA

Alfio MARAZZI

Institut universitaire de médecine sociale et préventive,
Bugnon 17, 1005 Lausanne, Switzerland.

Several algorithms have been proposed in order to compute a matrix A which solves the equation $\frac{1}{n}\sum u(|Ax_i|)Ax_i(Ax_i)^{\mathrm{T}} = I_p$, where x_i $(i = 1,\ldots n)$ are p-variate observation vectors, u is a given weight function and I_p is the $p \times p$ identity matrix. This problem arises within the affinely invariant approach for the robust estimation of a covariance matrix and in the computation of the weights for bounded influence regression. The algorithms do not work if a high proportion of observations belongs approximatively to a hyperplane of $\mathbb{R}^p$. This paper modifies a fixed point algorithm in order to detect collinearity and find a convenient solution.

1. Introduction

We consider the problem of solving numerically the following equation for a $p \times p$ matrix A:

$$F(A) = I_p, \tag{1.1}$$

where

$$F(A) = \frac{1}{n}\sum_{1}^{n} u(|Ax_i|)Ax_i(Ax_i)^{\mathrm{T}}. \tag{1.2}$$

Here $x_i = (x_{i1}, x_{i2}, \ldots, x_{ip})^{\mathrm{T}}$, $i = 1, \ldots, n$, are n p-vectors of observations and u is a given function. I_p denotes the $p \times p$ identity matrix and $|\cdot|$ the Euclidean norm.

Clearly A is not uniquely defined: it can be multiplied by an arbitrary orthogonal matrix from the left. However, uniqueness can be enforced by suitable side conditions, for example by requiring that it be positive definite symmetric, or that it be lower triangular with a positive diagonal.

This problem arises within the affinely invariant approach for the robust estimation of a covariance matrix, (see for ex. Huber [2], Chapter 8). Indeed, A represents a robust orthogonalization of the matrix $X = (x_{ij})$ and $f_p \cdot (A^{\mathrm{T}}A)^{-1}$ is a robust estimate of the covariance matrix when $x_1, \ldots, x_n$ is a sample of size n from a contaminated p-variate Gaussian distribution centered at the origin. The factor f_p makes the estimate asymptotically consistent. (Although the approach can also provide a robust estimate of an unknown multivariate location, we assume for simplicity that the location is known to be the origin).

Furthermore, solving (1.1) is usually a preliminary step for the computation of *bounded influence regression estimates*, (see for example Hampel et al. [3], Chapter 6) based on the design matrix X. Indeed, in order to reduce the influence of outlying rows of X (or *leverage points*) on the coefficient estimates, one uses a weighted regression procedure, where the weights $w(x_i)$ depend (as a decreasing function) on the distances $(x_i^{\mathrm{T}}A^{\mathrm{T}}Ax_i)^{1/2}$ of the points x_i from the origin.

Some examples of function u are (see Hampel et al. [3]):

$$u(r) \equiv 1, \tag{1.3}$$

$$u(r) = \begin{cases} 1, & 0 \le r^2 \le b^2 \\ b^2/r^2, & b^2 < r^2, \end{cases} \tag{1.4}$$

$$u(r) = \left(\frac{c}{r}\right)^2 + \left(1 - \left(\frac{c}{r}\right)^2\right)\left(2\Phi\left(\frac{c}{r}\right) - 1\right) - 2\left(\frac{c}{r}\right)\varphi\left(\frac{c}{r}\right), \tag{1.5}$$

where $r > 0$, Φ denotes the standard normal cumulative distribution function with density φ, and b, c are given "tuning parameters".

It is shown in Huber [2] that under the following assumptions:

(A1) $u(r)$ is monotonely decreasing and $u(r) > 0$ for $r > 0$,

(A2) There exist a $\sigma_0 > 0$ such that $\frac{1}{n}\sum(\sigma_0|x_i|)^2 u(\sigma_0|x_i|) < 1$.

(A3) $r^2 u(r)$ is bounded and continuous,

(A4) $\mathrm{P}(\mathcal{H}) < 1 - p/K$ for all hyperplanes $\mathcal{H} \subset \mathbb{R}^p$ (where $K := \lim_{r\to\infty} r^2 u(r)$, $\dim(\mathcal{H}) = p - 1$ and $\mathrm{P}(\mathcal{H})$ denotes the fraction of observations lying in $\mathcal{H}$),

the sequence $A^{(0)} = \sigma_0 I_p$, $A^{(1)}$, $A^{(2)}, \ldots$, such that

$$\left(A^{(m+1)\mathrm{T}} A^{(m+1)}\right)^{-1} = \frac{1}{n}\sum u(|A^{(m)}x_i|)x_i x_i^{\mathrm{T}} \tag{1.6}$$

converges to a fixed point A^* which solves (1.1). Indeed, assuming that the matrices $A^{(m)}$ are positive definite and symmetric, it follows from (A1) and (A2) that

$$A^{(0)} = \sigma^0 I_p < A^{(1)} < A^{(2)} < \ldots \tag{1.7}$$

where $A < B$ means that $B - A$ is positive semidefinite. Furthermore, (A4) implies that the sequence $A^{(m)}$ is bounded from above. Therefore, $A^{(m)} \to A^*$ and continuity (A3) implies that A^* satisfies (1.1).

Uniqueness can be proved under somewhat stronger conditions. Convergence of (1.7) is also proved in Maronna [8], Schönholzer [10] and Tyler [11] under slightly different conditions.

The sequence (1.6) provides a straightforward algorithm for solving (1.1). Notice that, in order to improve the numerical properties by avoiding the possibly poorly conditioned matrices $A^{(m)\mathrm{T}} A^{(m)}$, it can be convenient to set $A^{(m+1)} := (I_p - S)A^{(m)}$ and to determine S from:

$$((I_p - S)^{\mathrm{T}}(I_p - S))^{-1} = F(A^{(m)}). \tag{1.8}$$

For example, we can take the Cholesky decomposition BB^{T} of $F(A^{(m)})$ and put $I_p - S := B^{-1}$. (Clearly it is not necessary to compute S explicitly.) Finally, for the numerical computations, it is convenient to work with lower triangular matrices $A^{(m)}$ (and S).

Remark 1. K is obviously related to the values of the "tuning constants" of the function u which are typically chosen in relation to p (see for example Hampel et al. [3], Chapters 5 and 6). Since any set of p points in $\mathbb{R}^p$ is contained in some hyperplane $\mathcal{H}$, (A3) implies that $1 - p/K > \mathrm{P}(\mathcal{H}) \ge p/n$ and therefore $K > np/(n-p) > p$. However, a large value of K may imply a loss of robustness.

Several alternative algorithms for solving (1.1) have been described and compared in Huber [2] and Marazzi [7]. They include Newton-like and conjugate gradient procedures. All the algorithms do not work properly if the columns of the matrix $Y = \text{diag}(\sqrt{u(|Ax_i|)})X$ become nearly collinear during the iterations. In this case, the matrix Y^TY becomes nearly singular and the elements of A inflate. Geometrically this means that the weighted observations y_i are very close to a hyperplane $\mathcal{H} \subset \mathbb{R}^p$. Indeed, it follows from the existence theorem in Huber [2], pp. 216-217 that $A^{(m)} \to \infty$ implies that there is a hyperplane $\mathcal{H} \subset \mathbb{R}^p$ such that $P(\mathcal{H}) \geq 1 - p/K$.

In this paper we propose a simple modification of the fixed point algorithm which determines the pseudorank $k \leq p$ of Y by means of the QR decomposition with column pivoting as described in Lawson and Hanson [4]. A $k \times k$ transformation matrix B_1, which plays the role of A in the subspace of dimension k corresponding to the regular part of Y, is obtained and then imbedded in $\mathbb{R}^p$. A related approach has been proposed in Fries [1].

2. The algorithm

For any $p \times p$ matrix $M = (m_{ij})$ let $||M|| = \sqrt{\sum_{i,j} m_{ij}^2}$ and suppose that $n > p$. For any integer k such that $0 < k \leq p$ let J_k be the $p \times p$ matrix defined by

$$J_k = \begin{pmatrix} I_k & 0 \\ 0 & 0 \end{pmatrix} \begin{matrix} \} & k \\ \} & p-k \end{matrix}$$

where I_k denotes the $k \times k$ identity matrix. Also, let TOL be a given tolerance level for the relative precision of the solution (for ex. $TOL = 0.001$) and $MAXIT$ be a given positive integer number (for ex. $MAXIT = 500$). Finally, let τ be a given tolerance level for the determination of the pseudorank of the weighted data matrix.

Step 0. Set $NIT := 1$, $k^0 := p$, $T := I_p$, $\bar{Z} := X$.
Set the value of the tuning parameter of u.

Step 1. Determine σ_0 such that $\frac{1}{n}\sum(\sigma_0|\bar{z}_i|)^2 u(\sigma_0|\bar{z}_i|) < 1$.
Set $B^0 := \sigma_0 J_{k^0}$.

Step 2. Compute $D := \text{diag}(u(|B^0\bar{z}_i|))$ and set $Y := D^{1/2}\bar{Z}$.

Step 3. *a)* Determine an $n \times n$ orthogonal matrix Q and a $p \times p$ permutation matrix P such that:
- $QYP = R$
- R is upper triangular $(n \times p)$
- the diagonal elements of R are non-increasing in magnitude.

b) Determine the pseudorank k of Y;
this is the largest index j such that the absolute value of the element (j, j) of R is greater than τ.

c) Define

$$R = \begin{pmatrix} R_1 & R_2 \\ 0 & R_3 \end{pmatrix} \begin{matrix} \} & k \\ \} & n-k \end{matrix}$$
$$\underbrace{\quad}_{k} \quad \underbrace{\quad}_{p \quad k}$$

with R_1 $k \times k$ and upper triangular
R_2 $k \times (p - k)$
R_3 $(n - k) \times (p - k)$.

If $k = p$ only the matrix R_1 appears.

Step 4. If $k = k^0$ go to *Step 7.*

Step 5. Determine a $p \times p$ orthogonal matrix V and a $k \times k$ non-singular upper triangular matrix U such that:

$$(R_1 \quad R_2)V = (U \quad 0).$$

Step 6. Set $k^0 := k$, $T := TPV$, $\bar{Z} := \bar{Z}PVJ_{k^0}$.
Adjust the value of the tuning parameter of u and go to *Step 1.*

Step 7. Set $\bar{Z} := \bar{Z}P$, $T := TP$ and

$$B := \sqrt{n}\begin{pmatrix} (R_1{}^{\mathrm{T}})^{-1} & 0 \\ 0 & 0 \end{pmatrix} \begin{matrix} \} \; k^0 \\ \} \; p - k^0 \end{matrix}$$

Step 8. Check convergence:
If $||B - B^0|| < TOL$ or $NIT = MAXIT$ go to *Step 9.*
Otherwise, set $NIT := NIT + 1$, $B^0 := B$ and go to *Step 2.*

Step 9. Exit: set $A := BT^{\mathrm{T}}$.

In step 3, the choice of a pseudorank $k < p$ means that R_3 can be ignored. In other words, the data matrix X is replaced by a matrix $\tilde{X}$ of rank k such that

$$D^{1/2}\tilde{X} = Q^{\mathrm{T}}\begin{pmatrix} R_1 & R_2 \\ 0 & 0 \end{pmatrix} P^{\mathrm{T}} \quad \begin{matrix} \} \; k \\ \} \; n - k \end{matrix} \tag{2.1}$$

and

$$D^{1/2}\tilde{X}PV = Q^{\mathrm{T}}\begin{pmatrix} U & 0 \\ 0 & 0 \end{pmatrix} \quad \begin{matrix} \} \; k \\ \} \; n - k \end{matrix} \tag{2.2}$$

Furthermore,

$$||D^{1/2}\tilde{X} - D^{1/2}X|| = ||D^{1/2}\tilde{X}PV - D^{1/2}XPV|| \leq (p-k)^{1/2}\tau. \tag{2.3}$$

Let $Z = XPV$ and $\tilde{Z} = \tilde{X}PV$. $\tilde{Z}$ can be computed by multyplying the right-hand side of (2.2) by $D^{-1/2}$ and its last $p - k$ columns are null. The original problem of solving (1.1) could now be reformulated for the first k columns of $\tilde{Z}$. Unfortunately, the discrepancy between Z and $\tilde{Z}$, i.e. between X and $\tilde{X}$, is usually non-negligible and quite cumbersome. Therefore we prefer to use an approximation $\bar{X}$ such that $\bar{X}PV = XPVJ_k$. The rows of $\bar{X}$ are the projections of the x_i in the hyperplane of dimension k spanned by the rows of $\tilde{X}$.

Let

$$\bar{F}(A) = \frac{1}{n}A\bar{X}^{\mathrm{T}}\mathrm{diag}(u(|A\bar{x}_i|))\bar{X}A^{\mathrm{T}}. \tag{2.4}$$

Obviously the equation $\bar{F}(A) = I$ has no solution. However, $\bar{F}(A) = G(B)$, where

$$G(B) = \frac{1}{n}B\bar{Z}^{\mathrm{T}}\mathrm{diag}(u(|B\bar{z}_i|))\bar{Z}B^{\mathrm{T}}, \tag{2.5}$$

$\bar{Z} = \bar{X}PV$ and $B = APV$. We restrict attention then to the subspace of dimension k corresponding to the non-null columns of $\bar{Z}$. More precisely, we solve (in step 7) $G(B) = J_k$ by requiring B to be of the form

$$B = \begin{pmatrix} B_1 & 0 \\ 0 & 0 \end{pmatrix} \begin{matrix} \} \; k \\ \} \; p - k \end{matrix}$$

where B_1 is $k \times k$ and lower triangular.

Remark 1. At each iteration the current data matrix of rank k^0 is replaced by a matrix of rank $k \leq k^0$, and the problem is reformulated for the corresponding subspace of dimension k. Eventually no hyperplane $\mathcal{H} \in \mathbb{R}^k$ such that $P(\mathcal{H}) \geq 1 - k/K$ exists, and the algorithm converges (under assumptions (A1), (A2) and (A3)).

Remark 2. The approximate solution of $G(B) = J_k$ given in step 7 takes advantage of the QR decomposition obtained in step 3: observe that $G(B) = \frac{1}{n}(BP)R^{\mathrm{T}}R(BP)^{\mathrm{T}}$ and set BP equal to the pseudoinverse (see for example Rao and Mitra [9]) of $\sqrt{1/n}R$. Several alternative procedures for the computation of a "matrix of improvements" S such that $G((I_p - S)B^0) \approx I_p$ in a non-degenerated problem are available in Marazzi [7]. These procedures can easily be modified and used in step 7. However the convergence of these procedures cannot be ensured without the use of an adaptive step-length algorithm. Furthermore, the QR decomposition can be replaced by the singular value decomposition. Finally, it is possible to improve the starting values defined in step 1.

Remark 3. The pseudoinverse $(A^{\mathrm{T}}A)^+$ of $A^{\mathrm{T}}A$ can easily be computed using the decomposition $A = BT^{\mathrm{T}}$:

$$(A^{\mathrm{T}}A)^+ = PV \begin{pmatrix} (B_1{}^{\mathrm{T}}B_1)^{-1} & 0 \\ 0 & 0 \end{pmatrix} P^{\mathrm{T}}V^{\mathrm{T}} \quad \begin{matrix} \} \; k^0 \\ \} \; p-k^0 \end{matrix} \tag{2.6}$$

Therefore, an estimate of the covariance matrix of the underlying p-variate distribution is given by $f_k \cdot (A^{\mathrm{T}}A)^+$. Here f_k is the scalar factor which makes $f_k \cdot (B_1^{\mathrm{T}}B_1)^{-1}$ an asymptotically consistent estimate for the k^0-dimensional problem (see Huber [2], p.235).

3. Example

Suppose that,

$$x_i = C(z_{i1}, z_{i2}, \ldots, z_{i,p-1}, e_i)^{\mathrm{T}}, \quad i = 1, \ldots n$$

where the $p \times p$ matrix $C = (c_{hk})$ is orthogonal and the e_i are "small" perturbations, which are uncorrelated to the z_{ik} and have a symmetric distribution. In other words, the p-dimensional data points x_i are "close" to a $(p-1)$-dimensional hyperplane in $\mathbb{R}^p$.

In order to choose a value for τ we follow a suggestion in Fries [1]. Write:

$$x_i^{\mathrm{T}} = (x'_{i1} + e'_{i1}, \ldots, x'_{ip} + e'_{ip})$$

where $e'_{ih} = c_{hp}e_i$ and $x'_{ih} = \sum_{j=1}^{p-1} c_{hj}z_{ij}$, and let ξ_1 and ξ_2 be numbers such that $|e_i| \leq \xi_1$ and $|e'_{ih}| \leq \xi_2$, for $i = 1, \ldots n$ and $h = 1, \ldots p$. The p-th diagonal element r_{pp}^2/n of the matrix $R^{\mathrm{T}}R/n$, where R is the upper triangular matrix defined in the algorithm, is of the same order of magnitude as $v = \mathrm{Var}(e_i\sqrt{d_{ii}})$, where $(d_{ij}) = D$. Clearly, if $K^* = \sup_r u(r)$,

$$v \leq \xi_1^2 K^*. \tag{3.1}$$

Moreover, it follows from the orthogonality of C that $e_i = \sum_{j=1}^{p} c_{jp}e'_{ij}$ and hence:

$$v \leq p^2\xi_2^2 K^*. \tag{3.2}$$

As an estimate of ξ_2 we may take, for example, the maximal absolute rounding error of the data x_i. A value of $|r_{pp}^2/n|$ smaller than the right-hand side of (3.2) can then be interpreted as a sign of near collinearity. Therefore we may set

$$\tau = p\xi_2\sqrt{nK^*}. \tag{3.3}$$

The same bound may also be used to detect a collinearity of dimension lower than $p-1$.

Example. Let $p = 3$, $n = 13$,

$$((z_i1, z_i2)) = \begin{pmatrix} +2.0 & +4.0 \\ +4.0 & +2.0 \\ -2.0 & -4.0 \\ -4.0 & -2.0 \\ +2.0 & -4.0 \\ +4.0 & -2.0 \\ -2.0 & +4.0 \\ -4.0 & +2.0 \\ +1.5 & +1.5 \\ +1.5 & -1.5 \\ -1.5 & +1.5 \\ -1.5 & -1.5 \\ z_{13,1} & z_{13,2} \end{pmatrix}, \qquad C_1 = \begin{pmatrix} +\frac{1}{\sqrt{3}} & +\frac{1}{\sqrt{3}} & +\frac{1}{\sqrt{3}} \\ +\frac{1}{\sqrt{2}} & -\frac{1}{\sqrt{2}} & 0 \\ +\frac{1}{\sqrt{6}} & +\frac{1}{\sqrt{6}} & -\frac{2}{\sqrt{6}} \end{pmatrix}.$$

$$(e_1, \ldots, e_{13}) = 10^{-4} \cdot (e_1^*, -1, +2, -2, +3, -3, +4, -4, +5, -5, -5, e_{12}^*, e_{13}^*).$$

We use the function u defined in (1.4) with $b = 1.25\sqrt{k^0}$; we set $\tau = 3 \cdot 10^{-3}\sqrt{13}$, $TOL = 0.001$ and consider the following cases:

a) $C = I_3$, $(z_{13,1}, z_{13,2}) = (0,0)$, $e_1^* = 1$, $e_{12}^* = 5$, $e_{13}^* = 0$
b) $C = C_1$, $(z_{13,1}, z_{13,2}) = (0,0)$, $e_1^* = 1$, $e_{12}^* = 5$, $e_{13}^* = 0$
c) $C = I_3$, $(z_{13,1}, z_{13,2}) = (10,10)$, $e_1^* = 1$, $e_{12}^* = 5$, $e_{13}^* = 0$
d) $C = C_1$, $(z_{13,1}, z_{13,2}) = (10,10)$, $e_1^* = 1$, $e_{12}^* = 5$, $e_{13}^* = 0$
e) $C = I_3$, $(z_{13,1}, z_{13,2}) = (10,10)$, $e_1^* = 10^5$, $e_{12}^* = 10^5$, $e_{13}^* = 10^5$
f) $C = C_1$, $(z_{13,1}, z_{13,2}) = (10,10)$ $e_1^* = 10^5$, $e_{12}^* = 10^5$, $e_{13}^* = 10^5$

The main results are collected in Table 1. Its lines include the elements $(1,1), (2,1), (2,2), (3,1), (3,2), (3,3)$ of the matrices $(B^{\mathrm{T}}B)^+$, $(A^{\mathrm{T}}A)^+$ and $X^{\mathrm{T}}X$ and the values NIT_1 and NIT_2 of NIT at step 5 and step 9.

Computations were programmed in FORTRAN within the framework of the ROBETH library (see Marazzi [5] and Marazzi and Randriamiharisoa [6]) and in double precision. They were performed on a VAX 6210.

Acknowledgement

The author is grateful to Alex Randriamiharisoa for his generous help with programming.

Table 1: results

Case	Matrix (m_{ij})	m_{11}	m_{21}	m_{22}	m_{31}	m_{32}	m_{33}	NIT_1	NIT_2
a)	$(B^T B)^+$	6.846	0.000	6.846	0.000	0.000	0.000	1	2
	$(A^T A)^+$	6.846	0.000	6.846	0.000	0.000	0.000		
	$X^T X/n$	6.846	0.000	6.846	0.000	0.000	0.000		
b)	$(B^T B)^+$	6.846	0.000	6.846	0.000	0.000	0.000	1	2
	$(A^T A)^+$	5.705	-1.141	5.705	2.282	2.282	2.282		
	$X^T X/n$	5.705	-1.141	5.705	2.282	2.282	2.282		
c)	$(B^T B)^+$	7.933	1.087	7.933	0.000	0.000	0.000	1	5
	$(A^T A)^+$	7.933	1.087	7.933	0.000	0.000	0.000		
	$X^T X/n$	14.538	7.692	14.538	0.000	0.000	0.000		
d)	$(B^T B)^+$	8.999	0.217	6.868	0.000	0.000	0.000	1	5
	$(A^T A)^+$	7.499	-1.322	5.723	3.088	2.201	2.644		
	$X^T X/n$	18.396	-2.423	5.834	7.986	1.706	4.846		
e)	$(B^T B)^+$	7.933	1.087	7.933	0.000	0.000	0.000	44	48
	$(A^T A)^+$	7.933	1.087	7.933	0.000	0.000	0.000		
	$X^T X/n$	14.538	7.692	14.538	2.692	1.923	15.384		
f)	$(B^T B)^+$	8.999	0.217	6.868	0.000	0.000	0.000	46	50
	$(A^T A)^+$	7.499	-1.322	5.723	3.088	2.201	2.644		
	$X^T X/n$	23.340	1.410	8.557	1.113	-2.947	12.564		

References

[1] Fries, R., Robuste Kovarianzmatrizen, Diplomarbeit, Seminar für Statistik, Eidgen. Technische Hochschule, Zürich (1980).

[2] Huber P.J., Robust statistic (Wiley, New York, 1981).

[3] Hampel F.R., Ronchetti E.M., Rousseeuw P.J., Stahel W.A., Robust Statistics: the infinitesimal approach (Wiley, New York, 1986).

[4] Lawson C.L. and Hanson R.J., Solving Least Squares Problems (Prentice-Hall, Englewood Cliffs, New Jersey, 1974).

[5] Marazzi A., ROBETH: A subroutine library for robust statistical procedures, in: COMPSTAT 1980 *Proceedings in computational statistics*, (Physika Verlag, Vienna, 1980) 577-583.

[6] Marazzi A., Randriamiharisoa A., ROBETH-ROBSYS: a software for robust statistical computing, ROBETH-85/ROBSYS Document No. 0, Cahiers de recherche et documentation, Institut Universitaire de médecine sociale et préventive, Bugnon 17, CH-1005 Lausanne (1985).

[7] Marazzi A., Algorithms for the computation of weights in bounded influence regression, *Computational Statistics & Data Analysis* **6** (1988) 251-276.

[8] Maronna R., Robust M-estimators of multivariate location and scatter, *Ann. Statist.* **4** (1976) 51-67.

[9] Rao C.R. and Mitra, S.K., Generalized Inverse of Matrices and its Application (Wiley, New York, 1971).

[10] Schönholzer H., Robuste Kovarianz, Ph.D. Thesis. Eidgen. Technische Hochschule, Zürich (1979).

[11] Tyler D.E., A distribution-free M-estimator of multivariate scatter, *Ann. Statist.* **15** (1987) 234-251.

STATISTICAL DATA ANALYSIS AND INFERENCE
Y. Dodge (Editor)
© Elsevier Science Publishers B.V. (North-Holland), 1989

PREPOSTERIOR ANALYSIS IN SOFTWARE TESTING

Nozer D. Singpurwalla

Institute for Reliability and Risk Analysis
The George Washington University, Washington, DC 20052, USA*

Abstract

In this paper we address the important practical problem of how long to test and debug a piece of software before it is released for use. We consider the case of single stage testing and outline two protocols for the case of a double stage testing. Using results from the literature on software reliability, we formulate, for the case of single stage testing, the ensuing optimization problems which can be addressed using numerical techniques.

1. Introduction

The goal of preposterior analysis is to choose an experiment (or experiments) which minimizes overall costs. The overall cost consists of decision losses and the cost of testing. These two costs are in opposition to each other. To lower the decision loss, it will be necessary to run a large experiment – this in turn will increase experimental cost. In undertaking a preposterior analysis we strive to balance the two costs.

In the context of testing software, the experimental design consists of determining an optimal time for which software is to be tested and debugged before it is released. Long periods of testing result in increased probabilities of detecting bugs and decreased probabilities of catastrophic failures. However, longer periods of testing also result in increased costs and the possibility of the software becoming obsolete.

The testing of software could be done in a single stage or in a multiple stages. In this paper, we focus attention on single stage testing, and outline two protocols for the case of double stage testing. Despite its specialized nature, the case of single stage testing poses computationally difficult problems; these we have been able to address here using numerical techniques. The computational problems for

*Research Supported by the Office of Naval Research Contract N00014-85-K-0202 Project NR 042-372 and the U.S. Army Research Office Grant DAAL03-87-K-0056.

the case of multi-stage testing appear to be horrendous, and these we have not been able to deal with; we invite the attention of other researchers to look into such issues. Our review of the literature on software testing suggests that there is a dearth of literature addressing this fundamental and important issue. An exception is the work of Dalal and Mallows (1986) whose results are quite general, but whose model and assumptions are more restrictive than ours.

2. A Model for Software Failures

Any endeavor which attempts to address the question of an optimal time for which a software is to be tested, should have as an essential ingredient a probability model which describes the times between failures of the software or equivalently, the number of bugs that are remaining in the software. Of the several models that are currently being advocated for use, the one by Jelinski and Moranda (1972), given by (2.1) below, appears to have appeal, both because of its simplicity and its sense of reality. In this paper, we base our analysis on the above model but not without remarking that other models, such as the one by Littlewood and Verral (1973) can also be entertained using a methodology analogous to the one given here.

Let $\tilde{T}_1, \tilde{T}_2, \ldots$, be the successive running times between failures of the software; by "times" we mean here CPU times. We assume that $\tilde{T}_i$ can be described by an exponential distribution with a parameter $\Lambda(N-i+1)$, where Λ is an unknown constant of proportionality and N is the unknown number of bugs (faults) in the software. Specifically,

$$P(\tilde{T}_i \geq \tau \mid N, \Lambda) = \exp\{-\Lambda(N-i+1)\tau\}, \qquad \tau \geq 0. \tag{2.1}$$

Following Langberg and Singpurwalla (1985) we shall assume that N has a Poisson distribution with parameter θ and that Λ has a gamma distribution with scale parameter μ and shape parameter α. We shall also suppose that the above distributions for N and Λ are independent. These assumptions are reasonable and provide sufficient flexibility to express a wide range of subjective a priori assumptions about N and Λ [cf. Martz and Waller (1982), p. 338]. Furthermore, as is shown by Langberg and Singpurwalla (1985), these assumptions result in a unification of several other models in software reliability.

Suppose that the software is to be tested for T units of CPU time and that every time a bug is encountered, it is corrected. After testing for T units of time, the software is released for use - this is the case of *single stage testing*. In the case of *two stage testing*, the software is either released after the T units of testing or is tested for an additional T^* units of CPU time and then released. Our decision of whether to test for the additional T^* units of time or not, depends on M(T), the number of bugs encountered in time T. Our aim is to choose an optimal value of T, or optimal values of both T and T^*, depending on whether the testing is single or double stage.

Let us first consider the case of single stage testing with T specified, and suppose that $M(T) = k \geq 0$, with $t_1, t_2, \ldots, t_k$ as the CPU running times between software failures (whenever $k > 1$). Let $\underset{\sim}{t}^{(k)} = \left(t_1, \ldots t_k, T - \sum_{i=1}^{k} t_i\right)$, when $k \geq 1$, and $\underset{\sim}{t}^{(k)} = T$, when $k = 0$.

2.1 Preliminaries.

Given $\underset{\sim}{t}^{(k)}$ and k, the likelihood of N and Λ at q and λ, respectively, is proportional to

$$\frac{q!}{(q-k)!} \lambda^k \exp\left\{- \lambda \sum_{j=1}^{k} (q - j + 1)t_j\right\} \exp\left\{-\lambda(q - k)\left(T - \sum_{j=1}^{k} t_j\right)\right\}$$

$$= \frac{q!}{(q-k)!} \lambda^k \exp\left[-\lambda\left\{\sum_{j=1}^{k} (q - j + 1)t_j + (q - k)\left(T - \sum_{j=1}^{k} t_j\right)\right\}\right]. \tag{2.2}$$

Let $Y_i = \sum_{j=1}^{i} \tilde{T}_j$; then it can be verified [see Joe and Reid (1985)] that when $M(T) \geq 1$, $\underset{\sim}{t}(T) = \frac{1}{T} \sum_{i=1}^{M(T)} Y_i = \frac{1}{T} \sum_{j=1}^{M(T)} (M(T) - j + 1) \tilde{T}_j$ and M(T) are jointly sufficient for N and Λ. Thus by Bayes' Law, the posterior distribution of, N given $\underset{\sim}{t}(T)$ and $M(T) \geq 1$, is, of the form [see Meinhold and Singpurwalla (1983)]

$$P(N=q \geq M(T) \mid (\underset{\sim}{t}(T), M(T)), \theta, \alpha, \mu) = \frac{We^{-\theta}\, \theta^q}{(q-M(T))!}$$

$$\{\mu + \underset{\sim}{t}(T)T + (q - M(T))T\}^{-(\alpha+M(T))}; \tag{2.3a}$$

W is a normalizing constant. When $M(T) = 0$, the above expression becomes

$$P(N = q \geq 0 \mid (T, 0), \theta, \alpha, \mu) = \frac{We^{-\theta}\theta^q}{q!} (\mu + qT)^{-\alpha}. \tag{2.3b}$$

In what follows, we shall suppress θ, α and μ, and write the left hand sides of (2.3a) and (2.3b) as simply $P(N = q \geq M(T) \mid (\underset{\sim}{t}(T), M(T)))$ and $P(N = q \geq 0 \mid (T, 0))$, respectively.

The sampling distributions of M(T) and $\underset{\sim}{t}(T)$ are needed for preposterior analysis. From Theorem 3.3 of Langberg and Singpurwalla (1985), it can be seen that, for $k \geq 0$,

$$P(M(T) = k \mid \theta, \alpha, \mu) \overset{\text{def}}{=} P(M(T) = k)$$

$$= \int_0^\infty \frac{\exp(-\theta(1-e^{-\lambda T}))}{k!} \left[\theta(1-e^{-\lambda T})\right]^k \frac{e^{-\mu\lambda}(\mu\lambda)^{\alpha-1}\mu}{\Gamma(\alpha)} d\lambda. \tag{2.4}$$

From a result of Bain and Weeks (1964), the probability density of $\underset{\sim}{t}(T)$ at any t, given that $M(T) \geq 1$, can be obtained as

$$p(t \mid M(T) = k, T, \bullet) = \int_0^\infty \left[\frac{e^{-\mu\lambda}(\mu\lambda)^{\alpha-1}\mu}{\Gamma(\alpha)}\right] \frac{(b\lambda)^k}{(k-1)!} Te^{-\lambda Tt} \sum_{j=0}^{k_0} (-1)^j \binom{k}{j} (tT - jT)^{k-1} d\lambda, \tag{2.5}$$

for $k_0 < t < (k_0+1)$ and $k_0 = 0, 1, \ldots, (k-1)$; the constant $b = (1-e^{-\lambda T})^{-1}$.

3. The Single Stage Testing Problem

In the single stage testing problem, our goal is to pick an optimum value for T. To do this, we will need to specify a utility function; however, before we do so it is helpful if we conceptualize the testing scenario via the decision tree of Figure 3.1.

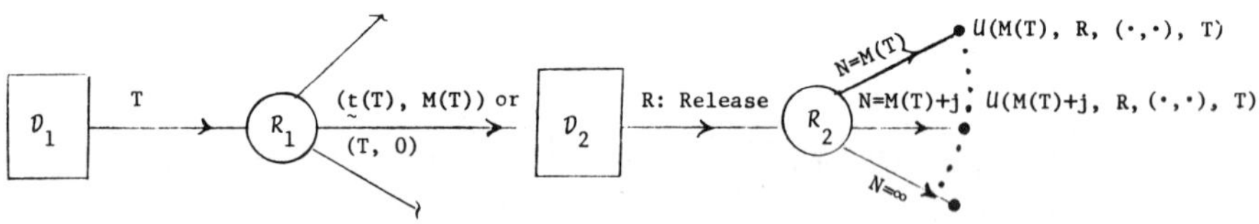

Figure 3.1 The Decision Tree for a Single Stage Testing of Software

The decision tree given above, consists of two decision nodes $\mathfrak{D}_1$ and $\mathfrak{D}_2$ and two random nodes $\mathcal{R}_1$ and $\mathcal{R}_2$. At $\mathfrak{D}_1$ we choose an optimal value for T, say T_0, whereas at $\mathfrak{D}_2$ there is only one action to be taken, namely R: a release of the software. At the random node $\mathcal{R}_1$, we observe the sufficient statistics ($\underset{\sim}{t}(T)$, M(T)), if $M(T) \geq 1$, or (T, 0), if $M(T) = 0$. At the random node $\mathcal{R}_2$, one of the various unknown states of nature $N = M(T) + j$, $j = 0, 1, \ldots$, will be realized, each leading to the utility

$\mathcal{U}[M(T) + j, R, (\underset{\sim}{t}(T), M(T)), T]$, when $M(T) \geq 1$, or
$\mathcal{U}[M(T) + j, R, (T, 0), T]$, when $M(T) = 0$.

A temporal evolution of the decision process from choosing an optimal T to arriving upon an unknown random state of nature and its associated utility is as follows:

Suppose that at $\mathfrak{D}_1$ we were to test the software for T units of CPU time. This brings us to $\mathcal{R}_1$ which results in our observing as data the pair ($\underset{\sim}{t}(T)$, M(T)), or the pair (T, 0); subsequent to this, we arrive at $\mathcal{D}_2$ at which the only action available to us is $\mathcal{R}$: a release of the software. Upon releasing the software, we arrive at $\mathcal{R}_2$ which results in our (eventually) encountering one of the various unknown states of nature $N = M(T) + j$, $j = 0,1, \ldots$, and incurring a utility of $\mathcal{U}[M(T) + j, R, (\cdot\,,\cdot), T]$, where $(\cdot\,,\cdot)$ is $(\underset{\sim}{t}(T), M(T))$ or $(T, 0)$. We would expect the utility function to have the property that it decreases in both T and j since large test times imply increased costs of testing and the risk of obsolescence of the software and large values of j imply an increase in the chance of encountering in-process failures. However, large values of T result in small values of j. Clearly, the choice of a suitable utility function is central for a meaningful implementation of the proposed procedure.

To obtain an optimum value of T, it is usual [see for example Lindley (1985), p. 144] to start from a terminus of the tree, the state $M(T) + j$, and work our way backwards to $\mathcal{D}_1$, following the principle of *maximization of expected utility*. To see this, we note that at $\mathcal{R}_2$ since $M(T) + j$, $j = 0,1, \ldots$, is unknown, the expected utility, $\mathcal{U}[R, (\cdot\,,\cdot), T]$, is of the form

$$\sum_{j=0}^{\infty} \mathcal{U}[M(T) + j, R, (\cdot\,,\cdot), T]\; P(N = M(T) + j \mid R, (\cdot\,,\cdot), T), \tag{3.1}$$

where the last term in the above expression is $P(N = M(T) + j \mid (\cdot\,,\cdot))$ and is evaluated using (2.3a) [(2.3b)] when $(\cdot\,,\cdot)$ is $(\underset{\sim}{t}(T), M(T))$ [(T, 0)].

At $\mathcal{D}_2$ there is only one action to be taken, namely R, and hence maximization of the expected utility at $\mathcal{R}_2$ over all actions at $\mathcal{D}_2$ leads us to the maximum expected utility $\mathcal{U}[(\cdot\,,\cdot), T] = \mathcal{U}[R, (\cdot\,,\cdot), T]$.

At $\mathcal{R}_1$, the data $(\cdot\,,\cdot)$ is unknown, and so $\mathcal{U}(T)$, the expected utility at $\mathcal{R}_1$ (were we to test the software for T units of CPU test time, correct all encountered bugs as they occur and then release the software), is given by

$$\begin{aligned}\mathcal{U}(T) &= \sum_{k=1}^{\infty} \int_t \mathcal{U}[(t, k), T]\; p(t, k \mid T)\, dt + \mathcal{U}[(T, 0), T]\; p(0 \mid T)\\ &= \sum_{k=1}^{\infty} \int_t \mathcal{U}[(t, k), T]\; p(t \mid k, T)\; p(k \mid T)\, dt, + \mathcal{U}[(T, 0), T]\; p(0 \mid T),\end{aligned} \tag{3.2}$$

where $p(k \mid T)$ is the probability mass of M(T) at the point k given T, and $p(t \mid k, T)$ is the probability density of $\underset{\sim}{t}(T)$ at t, given T and $M(T) = k \geq 1$. Note that $p(k \mid T)$ and $p(t \mid k,T)$ can be evaluated using (2.4) and (2.5) respectively.

At $\mathcal{D}_1$ we choose that value of T, say T_0, which maximizes $\mathcal{U}(T)$.

To summarize all of the above, T_0 is that value of T which yields

$$\sup_T \left\{ \sum_{k=1}^{\infty} \int_t \left\{ \left\{ \sum_{j=0}^{\infty} \mathcal{U}\big[k + j, R, (t, k), T\big] \frac{W_1 e^{-\theta}\, \theta^{k+j}}{j!} (\mu + tT + jT)^{-(\alpha+k)} \right\}\right.\right.$$

$$\bullet\left\{\int_0^\infty \frac{e^{-\mu\lambda}(\mu\lambda)^{\alpha-1}\mu}{\Gamma(\alpha)} \frac{(b\lambda)^k}{(k-1)!} T\, e^{-\lambda T t} \sum_{i=0}^{k_0} (-1)^i \binom{k}{i} (tT - iT)^{k-1}\, d\lambda\right\} dt\Bigg]$$

$$\bullet\left[\int_0^\infty \frac{\exp\left[-\theta(1-e^{-\lambda T})\right]}{k!} \left[\theta(1-e^{-\lambda T})\right]^k \frac{e^{-\mu\lambda}(\mu\lambda)^{\alpha-1}}{\Gamma(\alpha)} \mu\, d\lambda\right] +$$

$$\left\{\sum_{j=0}^{\infty} \mathcal{U}[j, R, (T, 0), T]\ \frac{W_2 e^{-\theta}\theta^j}{j!} (\mu+jT)^{-\alpha}\right\}$$

$$\bullet\left\{\int_0^\infty \exp(-\theta(1-e^{-\lambda T}))\ \frac{e^{-\mu\lambda}(\mu\lambda)^{\alpha-1}}{\Gamma(\alpha)}\ \mu\, d\lambda\right\}\Bigg]. \tag{3.3}$$

Recall that $b = (1-e^{-\lambda T})^{-1}$, and the range of integration for t is, $k_0 < t < (k_0+1)$, for each $k_0 = 0, 1, 2, \ldots, (k-1)$; W_1 and W_2 are normalizing constants with

$$W_1 = \left[\sum_{j=0}^{\infty} \frac{e^{-\theta}\theta^{k+j}}{j!\,(\mu+tT+jT)^{\alpha+k}}\right]^{-1} \text{ and}$$

$$W_2 = \left[\sum_{j=0}^{\infty} \frac{e^{-\theta}\,\theta^j}{j!\,(\mu+jT)^{\alpha}}\right]^{-1}.$$

3.1 A Plausible Choice for the Utility Function

From a practical point of view, the choice of a meaningful and realistic utility function would constitute a very important and fundamental step for implementing our procedure. With the above in mind, suppose that the utility of a program that encounters j bugs during its operation is $a_1 + a_2 e^{-a_3 j}$, where a_1, a_2 and a_3 are specified constants. Note that a_1 is the utility of the program with a large (possibly infinite) number of bugs and $a_1 + a_2$ is the utility of the program when it is free of bugs. It is meaningful to suppose that a_1 is a large negative number and that $a_2 > 0$. In writing the above, we are of course assuming that all the bugs are of equal consequence; this may not be reasonable. One way to overcome this limitation would be to group all bugs into different classes and to associate different values of a_2 and a_3 for each class; however such a classification of bugs would require an elaboration of our model. Suppose that C_1 is the cost of fixing a bug which is encountered during the testing phase of the software and suppose that f(T) is a nonnegative monotone increasing function which denotes the cost of testing plus the opportunity cost of not releasing the software until time T [see Dalal and Mallows (1986)]. Possible choices for f(T) are $f(T) = dT^a$, with d, a > 0 as specified constants, or $f(T) = Z\,[Z + \log T]$, for $T < [\geq]\ 1$, and Z a specified constant.

Given the above quite general but flexible specifications, the total utility in testing the software for T units of CPU time, correcting all the $M(t) = k$ bugs encountered during the test, releasing the software and subsequently encountering j bugs during the operation of the software, would be

$$\mathcal{U}[M(T) + j,\ R,\ (\underset{\sim}{t}(T),\ M(T)),\ T] = e^{-bT}[a_1 + a_2\, e^{-a_3 j} - C_1 M(T) - f(T)],$$

where, for some specified constant $b \geq 0$, e^{-bT} is a devaluating factor.

Once the above is done, the rest follows upon plugging in the necessary ingredients in Equations (3.2) and (3.3). The behavior of $\mathcal{U}(T)$ as a function of T for different values of the prior parameters θ and α, and different choices of a_2 is described in Singpurwalla (1989).

3.2 Utility as a Function of the Realized Reliability

Our motivation for the proposed utility function is based on the notion that the realized utility is a function of j, the number of remaining bugs in the program. This point of view ignores Λ, the rate at which the residual bugs are encountered (which in turn may depend on the rate of usage of the software), and also τ, the mission time of the software. As an elaboration of our development, suppose that the utility is a function of $R(\tau)$, the reliability of the software for a mission of duration τ. Suppose that for some specified constants $\alpha > 0$, $a_1 > a_0$, and $\ell = 1, 2, \ldots$, the utility, when regarded as a function of $R(\tau)$ alone, is of the form

$$\mathcal{U}[R(\tau)] = \left[\frac{a_1 - a_0}{1 - e^{-1/\alpha}}\right] e^{-(R(\tau)-1)^{2\ell}/\alpha} + \left[a_1 - \frac{a_1 - a_0}{1 - e^{-1/\alpha}}\right]. \tag{3.4}$$

With the above choice of the utility function, we note that at the random node $\mathcal{R}_2$, given the data $(\underset{\sim}{t}(T), M(T)$ or $(T, 0)$, the test time T and the mission time τ, we will realize a particular value for the reliability, say $(R(\tau)|(\bullet\,,\bullet), T)$ and this in turn will lead us to the total utility $\mathcal{U}[(R(\tau)|(\bullet\,,\bullet), T),\ R,\ (\bullet\,,\bullet), T] = \mathcal{U}[(R(\tau)|(\bullet\,,\bullet), T)] - C_1 M(T) - f(T)$. In writing the above, we have assumed, as before, that C_1 is the cost to fix a bug encountered during the testing phase of the software and f(T) is the cost of testing plus the opportunity cost of not releasing the software for time T.

But $(R(\tau)|(\bullet\,,\bullet), T)$ can be prescribed, via j, the number of bugs remaining in the program and, λ the rate at which they are encountered. Thus extending the conversation to $M(T) + j$, $(R(\tau)|(\bullet\,,\bullet), T) = \sum_{j=0}^{\infty} (R(\tau)|(\bullet\,,\bullet), M(T) + j, T) \bullet$ $P(M(T) + j|(\bullet\,,\bullet), T)$, where the last term in the above expression is evaluated using (2.3a) or (2.3b). Note the analogy between (3.1) and the above expression.

To obtain $(R(\tau)|(\bullet\,,\bullet), M(T)+j, T)$ we extend the conversation to $\lambda \in \Lambda$, and write

$$(R(\tau)|(\bullet\,,\bullet), M(T)+j,T) = \int_0^\infty (R(\tau)|(\bullet\,,\bullet), M(T)+j,T,\lambda)\; p(\lambda|(\bullet\,,\bullet), M(T)+j,T)d\lambda.$$

Since Λ is assumed to have a gamma distribution with scale (shape) $\mu(\alpha)$, the posterior of Λ given $(\bullet\,,\bullet)$, $M(T)+j$, and T, is also a gamma with scale

$\mu + \sum_{i=1}^{M(T)} (M(T)+j-i+1)\,\widetilde{T}_i$ $[\mu+T]$ and shape $\alpha + M(T)$ $[\alpha]$, when $(\bullet, \bullet)$ is $(\underset{\sim}{t}(T), M(T))$ $[T,0]$. Also, $(R(\tau)|(\bullet\,,\bullet),T, M(T)+j,\lambda)$ is $e^{-\lambda j\tau}$, and so, when $M(T) \geq 1$

$$(R(\tau)|(\bullet\,,\bullet), M(T)+j) = \int_0^\infty e^{-\lambda j\tau}\; p(\lambda|(\bullet\,,\bullet), M(T)+j, T)d\lambda$$

$$= \left(\frac{\mu + \sum_{i=1}^{M(T)} (M(T)+j-i+1)\,\widetilde{T}_i}{\mu + \sum_{i=1}^{M(T)} (M(T)+j-i+1)\widetilde{T}_i + j\tau}\right)^{M(T)+\alpha}$$

$$= \left(\frac{\mu + j\sum_{i=1}^{M(T)} \widetilde{T}_i + \sum_{i=1}^{M(T)} (M(T)-i+1)\,\widetilde{T}_i}{\mu + j\sum_{i=1}^{M(T)} \widetilde{T}_i + \sum_{i=1}^{M(T)} (M(T)-i+1)\,\widetilde{T}_i + j\tau}\right)^{M(T)+\alpha}, \text{ and so}$$

$$= \left(\frac{\mu + T\,\underset{\sim}{t}(T) + j\sum_{i=1}^{M(T)} \widetilde{T}_i}{\mu + T\,\underset{\sim}{t}(T) + j\sum_{i=1}^{M(T)} \widetilde{T}_i + j\tau}\right)^{M(T)+\alpha}. \tag{3.5}$$

Similarly, when $M(T) = 0$,

$$(R(\tau)|(\bullet\,,\bullet), M(T)+j) = \left(\frac{\mu+T}{\mu+T+j\tau}\right)^{\alpha}. \tag{3.6}$$

To simplify subsequent calculations, we replace each $\widetilde{T}_i$ in the term $j\sum_{i=1}^{M(T)} \widetilde{T}_i$ of (3.5) by its prior expectation to obtain

$(R(\tau)|(\underset{\sim}{t}(T), M(T)), M(T)+j) \approxeq$

$$\left[\frac{\mu + T\underset{\sim}{t}(T) + \frac{\mu j}{\alpha}\sum_{i=1}^{M(T)} (M(T)+j-i+1)^{-1}}{\mu + T\underset{\sim}{t}(T) + \frac{\mu j}{\alpha}\sum_{i=1}^{M(T)} (M(T)+j-i+1)^{-1}+j\tau}\right]^{M(T)+\alpha} . \tag{3.7}$$

To obtain $\mathcal{U}[R(\tau)|(\cdot\, , \cdot), M(T) + j]$ we plug in the expressions on the right hand side of (3.6) and (3.7) in (3.4), and proceed as in Section 3.1.

To summarize, T_0 is that value of T which yields the result of (3.3) with the following changes:

1. $\mathcal{U}[k+j,R, (t,k),T]$ of (3.3) is replaced by $\mathcal{U}[R(\tau)|(t, k), k+j] - C_1k - f(T)$, and
2. $\mathcal{U}[j, R, (T, 0), T]$ of (3.3) is replaced by $\mathcal{U}[R(\tau)|(T, 0), j] - f(T)$.

The behavior of T_0 as a function of θ, with μ and α fixed, and as a function of α with μ and θ fixed is described in Singpurwalla (1989).

4. The Double Stage Testing Problem

The decision tree for a double stage testing problem is shown in Figure 4.1. The tree has 3 decision nodes and either two or three random nodes depending upon what action is taken at node $\mathfrak{D}_2$.

Suppose that at $\mathfrak{D}_1$ we were to choose T_1, and test the software for T_1 units of CPU time. This brings us to $\mathfrak{R}_1$, which results in our observing as data the pair $(T_1, 0)$ or the pair $(\underset{\sim}{t}(T_1), M(T_1))$; subsequent to this, we arrive at $\mathfrak{D}_2$ at which we may choose one of two possible actions, to release the software, or to test for an additional T_2 units of CPU time. Should we choose to release the software, we arrive at $\mathfrak{R}_2$ which results in our (eventually) encountering one of the various unknown states of nature $N = M(T_1)+j$, $j = 0, 1, \ldots$, each leading to a utility $\mathcal{U}M(T_1) + j, R, (\underset{\sim}{t}(T_1), M(T_1)), T)$, when $M(T_1) \geq 1$ or $\mathcal{U}(M(T_1) + j, R, (T_1,0), T_1)$, when $M(T_1) = 0$. Should we choose to test for an additional T_2 units of CPU time, we would arrive at the random node $\mathfrak{R}_2'$, which results in our observing as additional data, the pair $(\underset{\sim}{t}(T_2), M(T_2))$ or $(T_2, 0)$ where

$$\underset{\sim}{t}(T_2) = \frac{1}{T_2} \sum_{j=M(T_1)+1}^{M(T_1)+M(T_2)} (M(T_1)+M(T_2) - j+1)\tilde{T}_j;$$ subsequent to this, we arrive at $\mathfrak{D}_3$ at which the only action available to us is R: a release of the software. Upon a release of the software at $\mathfrak{D}_3$, we arrive at $\mathfrak{R}_3$ which results in our incurring the utility

$\mathcal{U}(M(T_1) + M(T_2) + j,R, (\underset{\sim}{t}(T_2), M(T_2)), T_2, (\underset{\sim}{t}(T_1), M(T_1)),T_1)$, for $j = 0, 1, \ldots$;
Note that $(\underset{\sim}{t}(T_2), M(T_2))$ could be $(T_2, 0)$ and $(\underset{\sim}{t}(T_1), M(T_1))$ could be $(T_1, 0)$.

Our goal is to pick the optimum values for T_1 and T_2 and also to specify a decision rule which outlines the circumstances under which we either release the software at $\mathfrak{D}_2$ or test for an additional T_2 units of CPU time. The strategy for choosing the optimal values of T_1 and T_2 depends on the testing protocol. Under Protocol 1 we choose the optimal value of T_2, say T_{20}, after testing for T_{10} units of CPU time — where T_{10} is the optimal value of T_1 – and basing our choice of T_{20} upon the observed data $(\underset{\sim}{t}(T_{10}), M(T_{10}))$ or $(T_{10}, 0)$. Under Protocol 2, the optimal value of T_2, say T^*_{20} is to be specified prior to testing for T_{10}

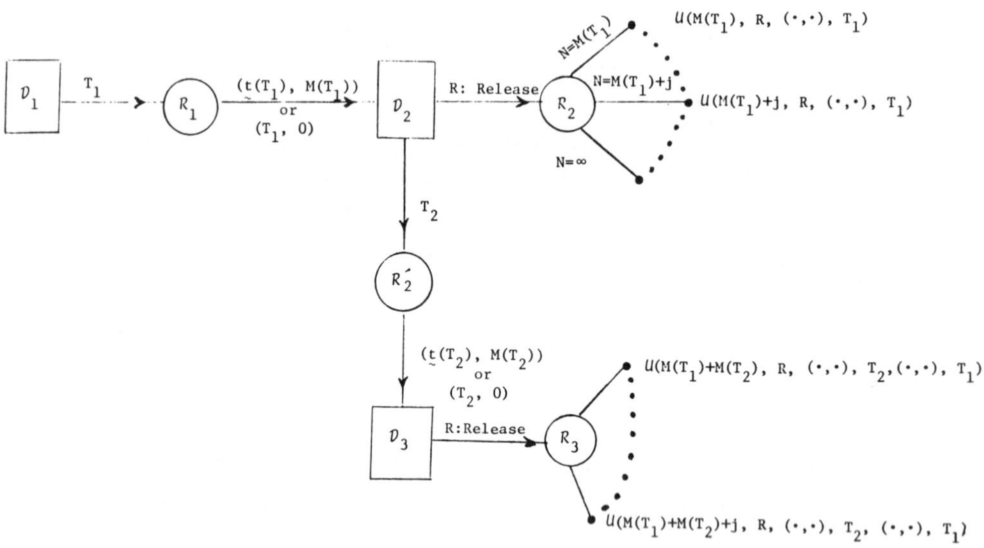

Figure 4.1 Decision Tree for a Double Stage Testing of Software

units of CPU time. Under Protocol 2, the observed data $(\underset{\sim}{t}(T_{10}), M(T_{10}))$ or $(T_{10}, 0)$, does not help us choose T^*_{20} ; it enables us to execute our decision rule at $\mathfrak{D}_2$, namely test for an additional T^*_{20} units of CPU time, or release the software. Whereas Protocol 1 is endowed with a truly sequential characteristic, Protocol 2 may be necessary in situations wherein one finds it desirable to specify, in *advance of experimentation*, the maximum amount of testing that one proposes to undertake. The situation described above can be generalized to the case of testing in more than two stages, for both Protocol 1 and Protocol 2, with Protocol 1 leading us to the case of sequential testing. (cf DeGroot (1970) p. 267 or Berger (1985) p. 442).

4.1 Optimal Testing Under Protocol 1 — The Precursor to Sequential Testing

Using Figure 4.1 as a guide, it is easy to see that $\mathcal{U}(R, (\bullet, \bullet), T_1)$ the expected utility at $\mathcal{R}_2$ is given by

$$\sum_{j=0}^{\infty} \mathcal{U}(M(T_1) + j, R, (\bullet, \bullet), T_1)\, P(N = M(T_1) + j \mid (\bullet, \bullet), T_1), \qquad (4.1)$$

where the last term in the above expression is evaluated via 2.3a[2.3b], when $(\bullet, \bullet)$ is $(\underset{\sim}{t}(T_1), M(T_1))$ $[(T_1, 0)]$. Similarly, $\mathcal{U}(R, (\underset{\sim}{t}(T_2), M(T_2)), T_2, (\underset{\sim}{t}(T_1), M(T_1)), T_1)$, the expected utility at $\mathcal{R}_3$, is given by

$$\sum_{j=0}^{\infty} \mathcal{U}\left[M(T_1) + M(T_2) + j, R, (\underset{\sim}{t}(T_2), M(T_2)), T_2, (\underset{\sim}{t}(T_1), M(T_1)), T_1\right] \bullet$$

$$P(N = M(T_1) + M(T_2) + j|(\underset{\sim}{t}(T_2), M(T_2)), T_2, (\underset{\sim}{t}(T_1), M(T_1)), T_1), \tag{4.2}$$

where the last term of the above expression can be evaluated via an expression analogous to (2.3a). Specifically, when both $M(T_1)$ and $M(T_2)$ are nonzero, then for $q \geq M(T_1) + M(T_2)$,

$$P(N=q \geq M(T_1)+M(T_2)|(\underset{\sim}{t}(T_1),M(T_1)), (\underset{\sim}{t}(T_2), M(T_2)),\theta,\alpha,\mu) = \left(\frac{W\ e^{-\theta}\theta^{q}}{(q-M(T_1)-M(T_2))!}\right) \bullet$$

$$\{\mu + \underset{\sim}{t}(T_1+T_2)\ (T_1+T_2) + (q-M(T_1) - M(T_2))\ (T_1+T_2)\}^{-(\alpha+M(T_1)+M(T_2))}, \tag{4.3}$$

where W is a normalizing constant and

$\underset{\sim}{t}(T_1+T_2) = \frac{1}{T_1+T_2} \sum_{j=1}^{M(T_1)+M(T_2)} (M(T_1) + M(T_2) - j+1)\tilde{T}_j$. In the above expressions $\underset{\sim}{t}(T_i) = T_i$ whenever $M(T_i) = 0$, $i=1, 2$ and $\underset{\sim}{t}(T_1+T_2)=0\{\underset{\sim}{t}(T_1)\}[\underset{\sim}{t}(T_2)]$, whenever $(M(T_1)=M(T_2)=0)\{M(T_2)=0\}[M(T_1) = 0]$; also, it is helpful to observe that $\underset{\sim}{t}(T_1+T_2)(T_1+T_2) = T_1\underset{\sim}{t}(T_1) + T_2\underset{\sim}{t}(T_2) + M(T_2) \sum_{j=1}^{M(T_1)} T_j$.

At $\mathcal{D}_3$ there is only one action to be taken, namely R, and hence $\mathcal{U}((\underset{\sim}{t}(T_2), M(T_2)),T_2, (\underset{\sim}{t}(T_1), M(T_1)), T_1)$, the maximum expected utility at $\mathcal{D}_3$ is indeed $\mathcal{U}(R, (\underset{\sim}{t}(T_2), M(T_2)), T_2, (\underset{\sim}{t}(T_1), M(T_1)),T_1)$.

At $\mathcal{R}_2'$, the data $(\underset{\sim}{t}(T_2), M(T_2))$ is unknown, and so $\mathcal{U}(T_2, (t(T_1), M(T_1)), T_1)$, the expected utility at $\mathcal{R}_2'$ is given by

$$\sum_{k_2=1}^{\infty} \int_{t_2} \mathcal{U}\left[(t_2, k_2), T_2, (\underset{\sim}{t}(T_1), M(T_1)), T_1\right) p(t_2, k_2|T_2, (\underset{\sim}{t}(T_1), M(T_1)), T_1)dt_2$$

$$+ \mathcal{U}((T_2, 0), T_2, (\underset{\sim}{t}(T_1), M(T_1)), T_1)\ p(0|T_2, (\underset{\sim}{t}(T_1), M(T_1)), T_1), \tag{4.4}$$

where $p(t_2, k_2|\bullet) = p(t_2|k_2, T_2, (\underset{\sim}{t}(T_1), M(T_1)), T_1)\ p(k_2|T_2, (\underset{\sim}{t}(T_1), M(T_1)), T_1)$; $p(t_2|\bullet)$ is the probability density of $\underset{\sim}{t}(T_2)$ at t_2 given T_2, $(\underset{\sim}{t}(T_1), M(T_1))$, T_1 and $M(T_2) = k_2 \geq 1$, and $p(k_2|\bullet)$ is the probability mass of $M(T_2)$ at $k_2 \geq 0$, given T_2, $(\underset{\sim}{t}(T_1), M(T_1))$ and T_1. It turns out that an exact evaluation of (4.4) is quite cumbersome. To see why, note that to evaluate $p(k_2|\bullet)$, we may extend the conversation to N and Λ and write

$$p(k_2|\bullet) = \sum_{n \geq M(T_1)}^{\infty} \int_0^{\infty} p(k_2|n, \lambda, T_2, (\underset{\sim}{t}(T_1), M(T_1)), T_1)\, p(n, \lambda|(\underset{\sim}{t}(T_1), M(T_1))d\lambda,$$

where the second term of the above expression is the posterior distribution of N and Λ given $(\underset{\sim}{t}(T_1), M(T_1))$; this is given by Meinhold and Singpurwalla (1983) as their Equation (3.4). The cumbersome nature of the above posterior distribution when put together with the result that the first term of the above integral is $\binom{n}{k_2}(1-e^{-\lambda T_2})^{k_2} e^{-\lambda T_2(n-k_2)}$ – see the proof of Theorem 3.2 of Langberg and Singpurwalla (1985) – calls for the use of approximations. One possibility, is to ignore the effect of $(\underset{\sim}{t}(T_1), M(T_1))$ and use expressions analogous to (2.4) and (2.5) as approximations to $p(k_2|\bullet)$ and $p(t_2|\bullet)$, respectively.

Let T_{20}^* be that value of T_2 which maximizes $\mathcal{U}(T_2, (\underset{\sim}{t}(T_1), M(T_1)), T_1)$. Specifically, T_{20}^* is that value of T_2 which yields

$$\sup_{T_2}\left\{\sum_{k_2=1}^{\infty}\int_{t_2}\left[\left\{\sum_{j=0}^{\infty}\mathcal{U}\left[M(T_1)+k_2+j, R, (t_2, k_2), T_2, (\underset{\sim}{t}(T_1), M(T_1)), T_1\right\}\right.\right.\right. \bullet$$

$$\left[\frac{W_1 e^{-\theta}\,\theta^{M(T_1)+k_2+j}}{j!}\left(\mu+T_1\underset{\sim}{t}(T_1)+T_2t_2+k_2\left[T_1\underset{\sim}{t}(T_1)-\sum_{i=1}^{M(T_1)}\left(M(T_1)-i\right)\right]\right)^{-(\alpha+M(T_1)+k_2)}\right] \bullet$$

$$\left\{\int_0^{\infty}\frac{e^{-\mu\lambda}(\mu\lambda)^{\alpha-1}\mu}{\Gamma(\alpha)}\frac{(b\lambda)^{k_2}}{(k_2-1)!}T_2\,e^{-\lambda T_2 t_2}\sum_{i=0}^{k_0}(-1)^i\binom{k}{i}(t_2T_2-iT_2)^{k_2-1}d\lambda\right\}dt_2\Bigg] \bullet$$

$$\left\{\int_0^{\infty}\frac{\exp\left[-\theta(1-e^{-\lambda T_2})\right]}{k_2!}\left[\theta(1-e^{-\lambda T_2})\right]^{k_2}\frac{e^{-\mu\lambda}(\mu\lambda)^{\alpha-1}}{\Gamma(\alpha)}\mu d\lambda\right\} +$$

$$\left\{\sum_{j=0}^{\infty}\mathcal{U}(M(T_1)+j, R, (T_2, 0), T_2, (\underset{\sim}{t}(T_1), M(T_1)), T_1) \bullet\right.$$

$$\left.\frac{W_2 e^{-\theta}\theta^{M(T_1)+j}}{j!}(\mu+T_1\underset{\sim}{t}(T_1)+jT_2)^{-(\alpha+M(T_1))}\right\}$$

$$\left\{\int_0^{\infty}\exp(-\theta(1-e^{-\lambda T}))\frac{e^{-\mu\lambda}(\mu\lambda)^{\alpha-1}}{\Gamma(\alpha)}\mu d\lambda\right\}\Bigg\}, \tag{4.5}$$

where W_1 and W_2 are normalizing constants, $b = (1-e^{-\lambda T_2})^{-1}$, and the range of integration for t_2 is $k_0 < t_2 < (k_0+1)$, for each $k_0 = 0, 1, \ldots, (k_2-1)$.

At $\mathfrak{D}_2$, we will choose the larger of $\mathcal{U}(R, (\underset{\sim}{t}(T_1), M(T_1)), T_1)$ – see Equation (4.1) – or $\mathcal{U}(T_{20}^*, (\underset{\sim}{t}(T_1), M(T_1)), T_1)$ – see Equation (4.4) – so that $\mathcal{U}((\underset{\sim}{t}(T_1), M(T_1)), T_1)$, the expected utility at $\mathfrak{D}_2$ is max $\{\mathcal{U}(R, (\underset{\sim}{t}(T_1), M(T_1)), T_1), \mathcal{U}(T_{20}^*, (\underset{\sim}{t}(T_1), M(T_1)), T_1\}$.

At $\mathcal{R}_1$, the data $(\underset{\sim}{t}(T_1), M(T_1))$ is unknown, and so $\mathcal{U}(T_1)$ the expected utility at $\mathcal{R}_1$ is given by $\sum_{k_1=1}^{\infty} \int_{t_1} \mathcal{U}((t_1, k_1), T_1)\, p(t_1|k_1, T_1)\, p(k_1|t_1)dt_1$ $+ \mathcal{U}((T_1, 0), T_1)\, p(0|T_1)$, where $p(k_1|T_1)$ is the probability mass of $M(T_1)$ at k_1 – given T_1 – and $p(t_1|k_1, T_1)$ is the probability density of $\underset{\sim}{t}(T_1)$ at t_1, given that $M(T_1) = k_1$. Finally, at $\mathfrak{D}_1$, we choose that value of T_1, say T_{10} which maximizes $\mathcal{U}(T_1)$. Specifically, T_{10} is that value of T_1 which yields

$$\sup_{T_1}\left(\sum_{k_1=1}^{\infty}\int_{t_1}\mathcal{U}((t_1, k_1), T_1)\left\{\int_0^{\infty}\left(\frac{e^{-\mu\lambda}(\mu\lambda)^{\alpha-1}\mu}{\Gamma(\alpha)}\,\frac{(b\lambda)^{k_1}}{(k_1-1)!}\,T_1\, e^{-\lambda T_1 t_1}\right)\bullet\right.\right.$$

$$\left.\left[\sum_{i=0}^{k_0}(-1)^i\binom{k_1}{i}(t_1T_1 - iT_1)^{k_1-1}\right]d\lambda\; dt_1\right\}\bullet$$

$$\left\{\int_0^{\infty}\frac{\exp(-\theta(1-e^{-\lambda T_1}))}{k_1!}\,(\theta(1-e^{-\lambda T_1}))^{k_1}\,\frac{e^{-\mu\lambda}(\mu\lambda)^{\alpha-1}}{\Gamma(\alpha)}\mu\; d\lambda\right\} +$$

$$\left.\left\{\mathcal{U}((T_1, 0), T_1)\int_0^{\infty}\exp(-\theta(1-e^{-\lambda T_1}))\,\frac{e^{-\mu\lambda}(\mu\lambda)^{\alpha-1}}{\Gamma(\alpha)}\,\mu\; d\lambda\right\}\right), \tag{4.6}$$

where $b = (1 - e^{-\lambda T_1})^{-1}$, and the range of integration for t_1 is $k_0 < t_1 < (k_0+1)$, for each $k_0 = 0, 1, \ldots, (k_1 - 1)$.

The computations involved in Equations (4.5) and (4.6) are similar in vein to those of Equation (3.3), albeit with some additional terms.

Thus under Protocol 1, our strategy for testing the software would be as follows:

1. Run the software for T_{10} units of CPU time and record the data $M(T_{10}) = k_1$, say, and $\underset{\sim}{t}(T_{10}) = t_1$, say; correct all the encountered bugs.
2. Replace $\underset{\sim}{t}(T_1)$ and $M(T_1)$ in $\mathcal{U}(T_2, (\underset{\sim}{t}(T_1), M(T_1)), T_1)$ by t_1 and k_1, respectively and let T_{20} be that value of T_2 which maximizes $\mathcal{U}(T_2, (t_1, k_1), T_1)$ – see Equation (4.5).
3. Compute $\mathcal{U}(T_{20}, (t_1, k_1), T_{10})$ and $\mathcal{U}(R, (t_1, k_1), T_{10})$ via Equations (4.1) and (4.4) respectively.

4. Release the software if

$$\mathcal{U}(T_{20}, (t_1, k_1), T_{10}) < \mathcal{U}(R, (t_1, k_1), T_{10}); \qquad (4.7)$$

otherwise, test for an additional T_{20} units of CPU time – correct all encountered errors – and release the software.

Note that the inequality (4.7) provides us with a decision rule at $\mathcal{D}_2$ which is based on the observed data; the latter also enables us to obtain T_{20} the time for which we have to test under Stage 2. Given T_{10} and T_{20}, there exists a region in the (X-Y) plane such that certain pairs of values of (t_1, k_1) lead us to a release of the software.

4.2 Optimal Testing Under Protocol 2

Under Protocol 2 our strategy would be similar to that under Protocol 1 above, except that here Step 2 is not needed and that Step 3 we would compute $\mathcal{U}(T^*_{20}, (t_1, k_1), T_{10})$ and $\mathcal{U}(R, (t_1, k_1), T_{10})$ and that at Step 4, we would release the software if

$$\mathcal{U}(T^*_{20}, (t_1, k_1), T_{10}) < \mathcal{U}(R, (t_1, k_1), T_{10}); \qquad (4.8)$$

otherwise, we would test the software for an additional T^*_{20} units of CPU time – correct all encountered errors – and release the software.

Note that the inequality (4.8) provides us with a decision rule at $\mathcal{D}_2$ which is based on the data (t_1, k_1); however, the latter is not used for determining T^*_{20}, the time for which we have to test under Stage 2.

The key difference between the schemes of Protocols 1 and 2, is that at node $\mathcal{R}'_2$, in the former we average out $(\underset{\sim}{t}(T_2), M(T_2))$ conditioned on (t_1, k_1), whereas in the latter the averaging out of $(\underset{\sim}{t}(T_2), M(T_2))$ is not conditioned on (t_1, k_1).

REFERENCES

[1] Bain, L. J. and Weeks, D. L. , A Note on the Truncated Exponential Distribution. *Annals of Mathematical Statistics, 35, (1964)*, pp. 1366-1367.

[2] Berger, J. O. , Statistical Decision Theory and Bayesian Analysis. Second Edition, (1985), Springer-Verlag, New York.

[3] Dalal, S. R. and Mallows, C. L., When to Stop Testing Software? *Journal of the American Statistical Association, Vol. 83, No. 403*, (1986), pp. 872-879.

[4] DeGroot, M. H., Optimal Statistical Decisions. McGraw-Hill Book Company, Inc. (1970), New York.

[5] Joe, H. and Reid, N., Estimating the Number of Faults in a System. *Journal of the American Statistical Association, Vol. 80, No. 389*, (1985), pp. 222-226.

[6] Jelinski, Z. and Moranda, P. B., Software Reliability Research. Statistical Computer Performance Evaluation (ed. W. Freiberger), (1972), pp. 485-502. Academic Press, New York.

[7] Langberg, N. and Singpurwalla, N. D., A Unification of Some Software Reliability Models. *SIAM Journal of Scientific and Statistical Computing, Vol. 6, No. 3*, (1985), pp. 781-790.

[8] Lindley, D. V., Making Decisions. Second Edition. John Wiley & Sons, Ltd. (1985), New York.

[9] Littlewood, B. and Verrall, J. L., A Bayesian Reliability Growth Model for Computer Software, *Record IEEE Symposium on Computer Software Reliability*, IEEE New York, (1973), pp. 70-77.

[10] Martz, H. F. and Waller, R. A., Bayesian Reliability Analysis. John Wiley and Sons, Inc. (1982), New York.

[11] Meinhold, R. J. and Singpurwalla, N. D., Bayesian Analysis of a Commonly Used Model for Describing Software Failures. *The Statistician* 32 (1&2), (1983), pp. 163-173.

[12] Singpurwalla, N. D., Determination of an Optimal Time Interval for Testing and Debugging Software, Technical Report, GWU/IRRA/Serial TR-89/4, (1989), The George Washington University, Washington, DC 20052.

STATISTICAL DATA ANALYSIS AND INFERENCE
Y. Dodge (Editor)

A VIEW OF HARMONIZABLE PROCESSES

M. M. Rao

Department of Mathematics
University of California, Riverside
Riverside, California 92521

This paper contains an account of harmonizable processes, the structure of their spectral domains in many cases, and certain results on sampling and statistical inference problems. Also some potential avenues for future research in the area are indicated.

1. INTRODUCTION

One of the motivations for the concept of harmonizability is to enlarge the applications of stationary processes while retaining the powerful Fourier analytic methods that play a key role in the latter studies. The flexibility afforded here may be explained directly as follows. If $\{X_n, -\infty < n < \infty\}$ is a sequence of (complex) random variables with means zero and covariances $r(m,n) = E(X_m \bar{X}_n)$, then it is (weakly or Khintchine) stationary if $r(m,n) = \tilde{r}(m-n)$. A simple example of such a process is an orthonormal sequence of $L^2(\Omega, \Sigma, P)$. If one considers a truncation of such a family $\{Y_n, -\infty < n < \infty\}$, where $Y_n = X_n$, $n > 0$, and $Y_n = 0$ for $n \leq 0$, then the new sequence is not stationary but turns out to be a harmonizable process. More generally, if $Y_n = TX_n$, $-\infty < n < \infty$, where T is any bounded linear transformation on $L^2(P)$ ($= L^2(\Omega, \Sigma, P)$), then the new process is always harmonizable. Thus it is advantageous to study this enlarged class, characterize and classify its subclasses from a theoretical point of view. It is equally useful to know the applications of the enlarged class. Thus the following is an account of the results on some recent work including a fair amount of some previously unpublished material on the subject. Here is a brief outline.

After introducing various classes of harmonizability in the next section, a detailed account of the spectral domain analysis of some of the classes is given in Section 3. This is useful since the emphasis in harmonizable as well as stationary classes is on the spectral (or frequency) domain so as to use the powerful methods of the classical Fourier analysis in contrast to the general time domain studies. Then Section 4 treats a problem of sampling the process in a way that allows a reconstruction of the process itself. A brief discussion is given in Section 5 on a computation of spectral functions from data, and an asymptotically unbiased estimator of the spectral bimeasure of a harmonizable family is included. The related class(KF) is also discussed there. Some statistical inference questions on harmonizable processes with low multiplicity are discussed in Section 6, and the final section contains a sketch of a few other asymptotic results (laws of large numbers) as well as certain indications of possible directions for future work.

2. HARMONIZABILITY AND RELATED CHARACTERIZATIONS

A first extension of stationarity was given by Loève in the late 1940's, and a readable account is included in his book [16], under the name "harmonizability." This turns out to be a special case of a general concept due to Bochner [3] called "V-boundedness," and also one of Rozanov's [25] which was again termed "harmonizable." The last two concepts can be shown to be equivalent for second order processes, and they extend Loève's concept. For this reason the Loève concept is now called strongly harmonizable, and the more general one, weakly harmonizable. These are stated precisely as follows:

2.1. Definition

A second order process $\{X_t, t \in G\}$ with mean $E(X_t) = 0$, and continuous covariance $r(s,t) = E(X_s\bar{X}_t)$, is *strongly harmonizable* if r can be expressed as: ($G = \mathbb{R}$ or $\mathbb{Z}$)

$$r(s,t) = \int_{\widehat{G}}\int_{\widehat{G}} e^{is\lambda - it\lambda'} \rho(d\lambda, d\lambda'), \qquad s,t \in G, \tag{1}$$

where $\rho(\cdot,\cdot)$ is a positive (semi-)definite function of finite Vitali variation on $\hat{G}\times\hat{G}$ with $\hat{G} = \mathbb{R}$ or $[-\pi,\pi)$ respectively. The process is *weakly harmonizable* if r has the representation (1) but that ρ satisfies the same conditions except that it has finite Fréchet variation instead of Vitali's variation.

To avoid misunderstanding let us recall the last three terms. A function $\rho : G \times G \to \mathbb{C}$ is positive semi-definite if $\rho(s,t) = \overline{\rho(t,s)}$ and for any $a_i \in \mathbb{C}$, $1 \le i \le n$, $n \ge 1$

$$\sum_{i=1}^{n}\sum_{j=1}^{n} a_i\bar{a}_j\rho(s_i,s_j) \ge 0, \qquad s_i \in G, \tag{2}$$

and positive definite if, moreover, $= 0$ holds only when all $a_i = 0$. Letting $(\Delta_{hk}\rho)(s,t) = \rho(s+h,t+k) - \rho(s,t+k) - \rho(s+h,t) + \rho(s,t)$, the increment, ρ is of Vitali ($=$ ordinary) variation finite if

$$\sup_{\pi,\pi'}\sum_{i=1}^{n}\sum_{j=1}^{n'} |(\Delta_{h_ik_j}\rho)(s_i,t_i)| < \infty \tag{3}$$

where $\pi : s_1 < s_2 < \cdots < s_{n+1}$, $\pi' : t_1 < t_2 < \cdots < t_{n'+1}$ from G ($= \mathbb{R}$ or $\mathbb{Z}$), and $0 \le h_i \le s_{i+1} - s_i$, $0 \le k_i \le t_{i+1} - t_i$; and ρ has Fréchet variation finite if (3) is replaced by

$$\sup_{\pi,\pi'} \{|\sum_{i=1}^{n}\sum_{j=1}^{n'} a_i\bar{b}_i(\Delta_{h_ik_j}\rho)(s_i,t_j)| : |a_i| \le 1, |b_j| \le 1\} < \infty. \tag{4}$$

It is clear that Vitali variation is not less than Fréchet's, and the converse implication is however false. Actually it can be shown that for ρ satisfying (2), (4) always holds but (3) does not, so that weak harmonizability is more general than its strong counterpart.

Since for the case of weak harmonizability ρ is not of finite Vitali variation, the integral in (1) cannot be taken in the usual Lebesgue sense, and a different definition is necessary. The appropriate concept is fortunately available from the work of M. Morse and W. Transue of middle 1940's, and thus (2) is interpreted as the MT-integral. But this is not an absolute integral in contrast to the Lebesgue concept, and the dominated convergence theorem is not valid for it. However, a slightly restricted version remedies this defect, and it is detailed in [6]. Even then the Jordan decomposition is not valid. But these problems seem unavoidable if a generalized concept such as (1) is to be employed for realistic applications. One may turn to the last reference for omitted details; but its results will be used below. See also [17].

If the process is stationary, then $r(s,t) = \tilde{r}(s-t)$ and $\rho(\cdot)$ of (1) is supported on the diagonal $s = t$ of $\hat{G} \times \hat{G}$, so that one has

$$r(s,t) = \int_{\widehat{G}} e^{i(s-t)\lambda}\rho(d\lambda), \qquad s,t \in G. \tag{5}$$

Here $\rho(\cdot)$ is a nonnegative, bounded, and nondecreasing function. It is called the spectral function of the process. In the more general harmonizable case also the possibly complex valued positive semi-definite $\rho(\cdot,\cdot)$ is again termed the *spectral (bi)measure* of the process. These functions reflect several properties of the covariance function $r(\cdot,\cdot)$, but have a better potential for applications than r, and a deeper analysis is possible by employing the finer Fourier techniques with them.

A close relation exists between harmonizability and stationarity. This and related properties are given by the following fundamental characterizations which are also used in some applications.

2.2. Theorem

Let $\{X_t, t \in G\} \subset L_0^2(P)$ ($= L_0^2(\Omega,\Sigma,P)$ of centered random variables) be weakly harmonizable. Then there exists a probability space $(\tilde{\Omega},\tilde{\Sigma},\tilde{P})$ [possibly an enlarged one than the original (Ω,Σ,P)], an orthogonal projection Q on $L_0^2(\tilde{P})$ onto $L_0^2(P)$, and a stationary process $\{Y_t, t \in G\} \subset L_0^2(\tilde{P})$ ($\supset L_0^2(P)$) such that $X_t = QY_t, t \in G$.

Conversely, if $\{Y_t, t \in G\} \subset L_0^2(P)$ is any stationary or (weakly) harmonizable process and T is any bounded linear operator on $L_0^2(P)$, then $X_t = TY_t, t \in G$, is weakly (but not necessarily strongly) harmonizable.

By this result one notes that the Y_n-sequence of the introduction with T as the orthogonal projection onto the closed linear span of $\{X_n, n > 0\}$, denoted $\overline{\text{sp}}\{X_n, n > 0\}$, is weakly harmonizable. It can be shown that the Y-process is not strongly harmonizable. This is an important motivation for the general study. Also the above theorem implies that the class of weakly harmonizable processes is closed under (bounded) linear transformations. The proof of this result in the forward direction depends on a relatively deep theorem of Grothendieck while the reverse direction is a consequence of the following representation theorem and a few properties of the Dunford and Schwartz theory of integration relative to vector measures.

2.3. **Theorem**

Let $\{X_t, t \in G\}$ be a second order process whose covariance function is bounded and continuous. Then the process is weakly harmonizable iff there is a σ-additive function Z on the Borel σ-algebra $\mathcal{B}(\hat{G})$ of $\hat{G}$ into $L_0^2(P)$ such that

$$\text{(i)}\quad E(Z(A)\bar{Z}(B)) = \rho(A,B) \quad \Big(= \int_A\int_B \rho(d\lambda, d\lambda')\Big) \tag{6}$$

$$\text{(ii)}\quad X_t = \int_{\hat{G}} e^{it\lambda} Z(d\lambda), \quad t \in G, \tag{7}$$

where the integral in (7) is in the Dunford-Schwartz sense and ρ is as in Definition 1. In particular, the X_t-process is strongly harmonizable iff ρ of (6) is of bounded Vitali variation, and stationary iff $Z(\cdot)$ has orthogonal values (i.e., $E(Z(A)\bar{Z}(B)) = 0$ for $A \cap B = \emptyset$). In all cases the representation (7) holds, and X_t is uniformly continuous on compacts of G.

A detailed proof of the above results can be obtained, for instance, from [21] where related references to the works of others may be found. Let us present an example of a weakly harmonizable process. Its verification can be seen from the same reference.

Example. Let $\mathcal{H}$ be a Hilbert space and $A : \mathcal{H} \to \mathcal{H}$ be a contractive linear operator. If A^* is its adjoint and $x_0 \in \mathcal{H}$ is any element, define

$$Y_n = A^n x_0,\ n \geq 0;\ \text{and } = (A^*)^n x_0 \text{ if } n < 0. \tag{8}$$

Then $\{Y_n, -\infty<n<\infty\}$ is weakly harmonizable.

In case A is unitary in (8), the corresponding Y-process is stationary. This illustration actually shows a deep relation between the harmonizability study and the Hilbert space operator theory. In the continuous parameter case $\{A^n, n \geq 0\}$ is replaced by a weakly continuous contraction semi-group of operators there.

Another useful weak harmonizability characterization will be presented before turning to an analysis of the spectral properties. Abstracting the characterization of continuous positive semi-definite (point) functions on G as Fourier transforms of positive bounded measures, Bochner introduced the fundamental concept of "V-boundedness" which enables a similar characterization of signed (and more generally vector) measures from their Fourier transforms. Its $L_0^2(P)$-version is relevant here. Thus a measurable mapping (or process) $X : G \to L_0^2(P)$ is called *V-bounded* if the following two conditions are met:

(i) $\|X(t)\|_2^2 = E(|X_t|^2) \leq M < \infty,\ t \in G,$

(ii) $\|\int_G X(t)\overline{\varphi(t)}dt\|_2 \leq C \sup_{\lambda\in\widehat{G}} |\hat{\varphi}(\lambda)|,\ \varphi \in L^1(G),$

where $\hat{\varphi}$ is the Fourier transform of the Lebesgue integrable φ on G and the (vector) integral in (ii) is in the Bochner sense, $0 < C < \infty$ being an absolute constant.

With this concept, the desired result is given by:

2.4. **Theorem**

Let $\{X_t, t \in G\} \subset L_0^2(P)$ be a process. Then it is weakly harmonizable iff it is V-bounded and has a continuous covariance.

Such a compact statement is not possible for the strongly harmonizable case. Several related important results on V-bounded and other processes are in [3]. This concept is useful even when the index set G is a noncommutative locally compact group. An extension of the above theorem for unimodular G and related results with references and necessary details can be found in [24]. (See also [30].)

3. SPECTRAL DOMAINS OF A CLASS OF HARMONIZABLE PROCESSES

Let $\{X_t, t \in G\}$ be a harmonizable process whose covariance function r admits the representation (1) with ρ as its spectral measure. Here ρ is taken to be positive definite so that for each pair of bounded Borel functions $f, g : \hat{G} \to \mathbb{C}$, one may define

$$(f,g) = \int_{\widehat{G}}\int_{\widehat{G}} f(\lambda)\overline{g(\lambda')}\rho(d\lambda, d\lambda') \tag{9}$$

and $(\cdot,\cdot)$ is then an inner product on $B(\hat{G}, \mathbb{C})$, the space of bounded complex Borel functions. Here the integral is in the MT-sense when the process is weakly harmonizable and in Lebesgue's sense if the process is strongly harmonizable. Define

$$\mathcal{L}^2(\rho) = \{f \in \mathbb{C}^{\hat{G}} : f\text{-Borel}, (f,f) = \|f\|^2 < \infty\}. \tag{10}$$

Then $\mathcal{L}^2(\rho)$ is called the *spectral domain space* of the process. In what follows the strongly harmonizable case is considered, so that the integral in (9) is in the Lebesgue sense and the standard computations can be used. The more general case needs the MT-integral and a theorem of Grothendieck, already noted; but the special case clarifies the analytical problems involved for $\mathcal{L}^2(\rho)$.

Thus if $\{X_t, t \in G\}$ is strongly harmonizable, with $G = \mathbb{R}$ or $\mathbb{Z}$, let ρ be its spectral measure on $\hat{G}\times\hat{G}$ which is of bounded (Vitali) variation. Then the hermitian property of ρ implies that $|\rho|(A,B) = |\rho|(B,A)$ where $\rho(\cdot,\cdot)$ denotes its variation measure on $\mathcal{B}(\hat{G})\otimes\mathcal{B}(\hat{G})$ the product Borel σ-algebra of $\hat{G}\times\hat{G}$. Define $\beta : \mathcal{B}(\hat{G}) \to \mathbb{R}^+$ as:

$$\beta(A) = \tfrac{1}{2}\{|\rho|(A,\hat{G}) + |\rho|(\hat{G},A)\} = |\rho|(A,\hat{G}),\ A \in \mathcal{B}(\hat{G}). \tag{11}$$

Consider the Lebesgue space $L^2(\rho)$ of ρ-square integrable scalar functions on $\hat{G}\times\hat{G}$, i.e., $f \in L^2(\rho)$ iff

$$\|f\|_2^2 = \int_{\widehat{G}\times\widehat{G}} |f(s,t)|^2 |\rho|(ds,dt) < \infty.$$

Then $L^2(\rho)$ is a complete normed linear space under the norm $\|\cdot\|_2$ above (cf. [12], III.3.1 and IV.3.6). Since $|\rho|(\hat{G}\times\hat{G}) < \infty$, a function $\varphi : \hat{G} \to \mathbb{C}$, $\mathcal{B}(\hat{G})$-measurable, can be identified with $\tilde{\varphi} = \varphi \times 1 : \hat{G}\times\hat{G} \to \mathbb{C}$ and it belongs to $L^2(\rho)$ iff $|\varphi|^2$ is integrable relative to β of (11). In fact,

$$\int_{\widehat{G}\times\widehat{G}} |\tilde{\varphi}|^2 |\rho|(ds,dt) = \int_{\widehat{G}} |\varphi(s)|^2 \beta(ds). \tag{12}$$

If one considers the subclass of $L^2(\rho)$, of elements such as φ above (i.e., of functions of one variable only), then

$$\int_{\widehat{G}}\int_{\widehat{G}} \varphi(s)\overline{\varphi(t)}\rho(ds,dt)$$

exists iff

$$\int_{\widehat{G}}\int_{\widehat{G}} |\varphi(s)\overline{\varphi(t)}||\rho|(ds,dt) < \infty,$$

and this is true iff $\varphi \in L^2(\rho)$. Thus in our case of positive definite ρ, $\mathcal{L}^2(\rho)$ is defined also for $\varphi \in L^2(\rho)$. Then $\mathcal{L}^2(\rho) \subset L^2(\rho)$, since (φ,φ) in (9) is undefined otherwise. But by (12) such functions are a subclass of $L^2(\beta) \subset L^2(\rho)$, so that $\mathcal{L}^2(\rho) \subset L^2(\beta)$.

To use this relation, consider for $f \in L^1(\beta)$ and $B \in \mathcal{B}(\hat{G})$, the identity:

$$\int_B f(\lambda)\beta(d\lambda) = \frac{1}{2}\int_B\int_B f(s)|\rho|(ds,dt) + \frac{1}{2}\int_B\int_B f(t)|\rho|(ds,dt).$$

If $f \in L^2(\beta)$, then one has with (9)

$$
\begin{aligned}
0 \le B(f,f) &= \int_{\widehat{G}}\int_{\widehat{G}} f(s)\bar{f}(t)\rho(ds,dt) \le \int_{\widehat{G}}\int_{\widehat{G}} |f(s)||\bar{f}(t)||\rho|(ds,dt) \\
&\le \frac{1}{2}\int_{\widehat{G}}\int_{\widehat{G}} [|f(s)|^2 + |f(t)|^2]|\rho|(ds,dt) \\
&= \int_{\widehat{G}} |f(s)|^2 \beta(ds). \qquad (13)
\end{aligned}
$$

Thus for f, g of the above type

$$
\begin{aligned}
|B(f,g)| &\le \int_{\widehat{G}} |f(s)\overline{g(t)}|\beta(ds), \quad \text{by (13)} \\
&\le \|f\|_{2,\beta}\|g\|_{2,\beta}\,, \qquad (14)
\end{aligned}
$$

by the CBS inequality. Hence the sesquilinear form $B(\cdot,\cdot)$ on $L^2(\beta) \times L^2(\beta)$ is positive definite, bounded, and so has a unique extension onto the latter space with the same properties. Denoting this by the same symbol, one has by a classical representation theorem (cf. [26], p. 63)

$$B(f,g) = (Af,g) = \int_{\widehat{G}} (Af)(s)\overline{g(s)}\beta(ds), \qquad (15)$$

for a unique $A : L^2(\beta) \to L^2(\beta)$. Further the positive definiteness of $B(\cdot,\cdot)$ is equivalent to that of A. So by (14) and (15), A is a positive definite bounded operator and so has a unique square root $A^{\frac{1}{2}}$ with the same properties. Thus our positive definiteness hypothesis of ρ translates to that of $A^{\frac{1}{2}}$ in the Hilbert space $L^2(\beta)$ on which it acts. But this means (see the classic [26], Def. 2.14, p. 56) $(A^{\frac{1}{2}})^{-1}$ exists and is bounded, and this is the form that is used in the work below. It may be noted that the latter is actually a consequence of the positive definiteness of A on $L^2(\beta)$. This can be "justified" from the following computation.

The positive definiteness of A on $L^2(\beta)$ implies the existence of a constant $C > 0$ such that $(Af,f) \ge C\|f\|^2_{2,\beta}$, $f \in L^2(\beta)$.

For, by the homogeneity of this expression, one can take $\|f\|_2 = 1$. Let $T = \|A\|$, the operator bound. Then $0 < T < \infty$ since the trivial case that $A = 0$ can be excluded. If $\{E(\lambda), 0 \le \lambda \le T\}$ is the resolution of the identity of A, let $\mu_f(\cdot) = (E(\cdot)f,f)$. Then $0 \le \mu_f(\cdot) \le T$ is an increasing left continuous function with $\mu_f(T) = 1$, i.e., a distribution function. By the classical spectral theorem (cf. [26], Thm. 5.9), one has

$$(Af,f) = \int_0^T \lambda d\mu_f(\lambda).$$

Suppose now that no such $C > 0$ in the positive definiteness definition exists. Then $\inf\{(Af, f) : \|f\|_{2,\beta} = 1\} = 0$, so there is a sequence f_n, $\|f_n\|_{2,\beta} = 1$, and $\lim_n (Af_n, f_n) = 0$. If $\{\mu_{f_n}, n \geq 1\}$ is the corresponding set of distribution functions, by the Helly selection principle there exists a subsequence $\mu_{f_{n_k}} \to \mu_0$, a nondecreasing function, at all continuity points of the latter. Since the domain here is a compact interval, it follows that

$$\lim_{k\to\infty} \mu_{f_{n_k}}(0) = 0 = \mu_0(0) \quad \text{and} \quad \lim_{k\to\infty} \mu_{f_{n_k}}(T) = 1 = \mu_0(T).$$

Hence by the Helly-Bray lemma (cf. [16], p. 180)

$$0 = \lim_{k\to\infty} (Af_{n_k}, f_{n_k}) = \lim_{k\to\infty} \int_0^T \lambda d\mu_{f_{n_k}}(\lambda) = \int_0^T \lambda d\mu_0(\lambda). \tag{16}$$

Also using integration by parts appropriately, one has with (16)

$$T\mu_0(T) = \lim_{k\to\infty} T\mu_{f_{n_k}}(T) = \lim_{k\to\infty} \int_0^T \mu_{f_{n_k}}(\lambda)d\lambda = \int_0^T \mu_0(\lambda)d\lambda. \tag{17}$$

It follows from (17) that

$$0 \leq \int_0^T [\mu_0(T) - \mu_0(\lambda)]d\lambda = 0, \tag{18}$$

and since μ_0 is increasing, this implies $\mu_0(\lambda) = \mu_0(T) = 1$, for almost all λ. But then for each $\varepsilon > 0$, a continuity point of μ_0, using (16) one has

$$0 = \int_0^T \lambda d\mu_0(\lambda) \geq \int_\varepsilon^T \lambda d\mu_0(\lambda) > \varepsilon\mu_0(\varepsilon) = \varepsilon > 0.$$

This contradiction shows that the initial assumption is false and there is a $C > 0$ such that $(Af, f) \geq C\|f\|_{2,\beta}^2$.

Thus resuming the analysis of (15), it follows that $(A^{\frac{1}{2}})^{-1}$ exists as a bounded operator on $L^2(\beta)$, and $\|f\|^2 = (Af, f) = \|A^{\frac{1}{2}}f\|_{2,\beta}^2$ defines another norm on $\mathcal{L}^2(\rho)$. If $\{f_n, n \geq 1\}$ in $\mathcal{L}^2(\rho)$ is a Cauchy sequence, then $\|f_n - f_m\| = \|A^{\frac{1}{2}}(f_n - f_m)\|_{2,\beta} \to 0$, as $n, m \to \infty$. Since $L^2(\beta)$ is complete, there is a $g \in L^2(\beta)$ such that $A^{\frac{1}{2}}f_n \to g$ in $L^2(\beta)$. Define $f = (A^{\frac{1}{2}})^{-1}g$. This is unambiguous by the preceding analysis, and

$$\|f\| = \|A^{\frac{1}{2}}f\|_{2,\beta} = \|g\|_{2,\beta} < \infty. \tag{19}$$

Hence $f \in \mathcal{L}^2(\rho)$ and

$$\|f_n - f\| = \|A^{\frac{1}{2}}(f_n - f)\|_{2,\beta} = \|A^{\frac{1}{2}}f_n - g\|_{2,\beta} \to 0,$$

as $n \to \infty$. So f is the limit in $\mathcal{L}^2(\rho)$ of the Cauchy sequence $\{f_n, n \geq 1\}$. This establishes the following since the linearity of $\mathcal{L}^2(\rho)$ is evident:

3.1. **Theorem**

The spectral domain space $\mathcal{L}^2(\rho)$ of a strongly harmonizable process (with a positive definite spectral measure ρ) is a complete inner product space.

A natural question here is this: Can the positive definiteness of ρ be relaxed to positive semi-definiteness? The stronger hypothesis was needed to establish (19) in using the fact that the domain of $(A^{\frac{1}{2}})^{-1}$ contains g so that f is well defined. Suppose that the positive semi-definite A of (15) has a closed range. This was automatic in the previous case since A^{-1} exists and is bounded there, and it will be shown that the theorem is true with essentially the same proof if the range $\mathcal{R}_A$ is closed.

Indeed by ([2], Thm. 2, p. 475), $A = P_{\mathcal{R}_A}\tilde{A}$ where $P_{\mathcal{R}_A}$ is the orthogonal projection onto the closed range $\mathcal{R}_A$ of A, and $\tilde{A}$ is a bounded operator with a bounded inverse on $L^2(\beta)$ (cf. also the discussion in [2] on p. 476 about the dimensional equivalence of $\mathcal{R}_A^{\perp}$ and $\mathcal{N}_A$, the null space of A). If $A^+ = P_{\mathcal{N}_A^{\perp}}\tilde{A}^{-1}$, then A^+ is a generalized inverse of A and $AA^+ = P_{\mathcal{R}_A}$. These facts allow us to modify the above proof to get the same conclusion.

Thus if $\{f_n, n \geq 1\} \subset \mathcal{L}^2(\rho)$ is a Cauchy sequence as before, with $g_n = A^{\frac{1}{2}}f_n$, $\|f_n - f_m\| = \|g_n - g_m\|_{2,\beta} \to 0$, let g be the limit in $L^2(\beta)$. Since by the current hypothesis $\mathcal{R}_{A^{1/2}}$ is also closed, let B^+ be the generalized inverse of $A^{\frac{1}{2}}$, and then $g \in \mathcal{R}_{A^{1/2}}$. So let $f = B^+g$. Since the domain of B^+ is $L^2(\beta)$, this is well defined, and

$$\|f\| = \|A^{\frac{1}{2}}(B^+g)\|_{2,\beta} = \|P_{\mathcal{R}_{A^{1/2}}}g\|_{2,\beta} \leq \|g\|_{2,\beta} < \infty,$$

since $P_{\mathcal{R}_{A^{1/2}}}$, being an orthogonal projection on $L^2(\beta)$, is a contraction. So $f \in \mathcal{L}^2(\rho)$. Also $A^{\frac{1}{2}} = P_{\mathcal{R}_{A^{1/2}}}A^{\frac{1}{2}}$, and hence

$$\begin{aligned}\|f_n - f\| &= \|A^{\frac{1}{2}}(f_n - f)\|_{2,\beta} = \|P_{\mathcal{R}_{A^{1/2}}}A^{\frac{1}{2}}f_n - A^{\frac{1}{2}}B^+g\|_{2,\beta} \\ &= \|P_{\mathcal{R}_{A^{1/2}}}(A^{\frac{1}{2}}f_n - g)\|_{2,\beta}, \text{ since } P_{\mathcal{R}_{A^{1/2}}} = A^{\frac{1}{2}}B^+, \\ &\leq \|g_n - g\|_{2,\beta} \to 0, \text{ as } n \to \infty.\end{aligned}$$

Hence $\mathcal{L}^2(\rho)$ is again complete, and this can be stated as:

3.2. **Proposition**

Let the spectral function ρ of the strongly harmonizable $\{X_t, t \in G\}$ be positive semi-definite and suppose that its induced operator A (of (15)) has a closed range. [Conditions for this to occur are given below.] *Then $\mathcal{L}^2(\rho)$ is again a complete (semi-)*

inner product space, so that if $\mathcal{N} = \{f \in \mathcal{L}^2(\rho) : \|f\| = 0\}$ *and* $\|\tilde{f}\|' = \|f\|$ *for* $\tilde{f} = f + \mathcal{N}$, *then* $(\mathcal{L}^2(\rho)/\mathcal{N}, \|\cdot\|')$ *is a Hilbert space.*

A sufficient condition that insures the closedness of $\mathcal{R}_B$ for a bounded operator B is as follows. If $\{f_n, n \geq 1\} \subset \mathcal{R}_B$ is any weakly convergent sequence with limit f in $L^2(\beta)$ (i.e., $(f_n, h) \to (f, h)$ for all $h \in L^2(\beta)$), then $f \in \mathcal{R}_B$. Thus $\mathcal{R}_B$ is weakly sequentially complete. In fact, since β is a Radon measure, the space $L^2(\beta)$ is separable here and this implies, by ([12], V.7.16), that $\mathcal{R}_B$ is weakly closed. But then by ([12], V.3.13), $\mathcal{R}_B$ is also strongly closed. Whether $\mathcal{R}_A$ in our special case (of semi-definite ρ) is always closed, has not been settled.

Regarding the preceding two propositions the following complements are in order and they should be useful in applications.

A. The spectral domain $\mathcal{L}^2(\rho)$ in (10) is defined as a subset of $L^2(\rho)$. In case ρ has infinite Vitali variation, then ρ does not necessarily determine a signed (or even a semi-bounded) measure. Then one has to resort to the MT-integral, and a deeper analysis shows (with a theorem of Grothendieck) that there is a suitable Radon measure β on $\hat{G}$ and then $\mathcal{L}^2(\rho)$ is a subset of $L^2(\beta)$. From this point on the argument proceeds as above. Suppose now that $G = \mathbb{R}$ and ρ has a finite Vitali variation only on each finite domain of the plane $\mathbb{R}^2$, so that the variation measure $|\rho|(\cdot,\cdot)$ determines a σ-finite Radon measure. In this case the preceding analysis shows only that $\mathcal{L}^2_E(\rho) = \{f \in L^2(\rho) : f\big|_A = 0$ for $A \subset (E \times E)^c\}$ is complete for $E \subset \mathbb{R}$, a finite domain. If $\mathcal{L}^2(\rho)$ is defined as $\cup\{\mathcal{L}^2_E(\rho) : E \subset \mathbb{R}$, a finite domain$\}$, then one cannot conclude that $\mathcal{L}^2(\rho)$ is complete, although it is a (semi-) inner product space. This general case was originally considered by Cramér [7]. Now define the inner product as

$$(f, g) = \int_{\mathbb{R}} \int_{\mathbb{R}} f(s)\overline{g(t)}\rho(ds, dt) \tag{20}$$

for all compactly supported f, g in $B(\mathbb{R}, \mathbb{C})$. [Since $\tilde{f}(s,t) = f(s) \cdot 1$ need not be in $L^2(\rho)$, the preceding analysis leading to Theorem 3.1 and Proposition 3.2 breaks down, although $L^2(\rho)$ is complete by [12], III.3.6.] In this general case, Cramér [7] suggests that $\mathcal{L}^2(\rho)$ be considered as an inner product space with (20), in its own right and then use $\Lambda^2(\rho)$, the completion of $\mathcal{L}^2(\rho)$, as the spectral domain space. Note that since $\Lambda^2(\rho) \subset L^2(\rho)$ this completed space still consists of functions and not some ideal elements. Thus for most of the applications, with $\Lambda^2(\rho)$, the analysis can be carried out with an occasional additional argument on the limit elements. This implies that the possible noncompleteness of $\mathcal{L}^2(\rho)$ is not an impediment for applications, as indicated in [7]. However, one can show that the positive definiteness of ρ in (1) necessarily has finite Fréchet variation. Consequently, with the MT-integration and the Grothendieck result noted before, one has $\mathcal{L}^2(\rho) \subset L^2(\beta)$ and the completeness follows as above, and no additional argument is needed. Thus the usual applications such as prediction, filtering, and multiplicity theories (cf. Sec. 6 on the latter, and also [6]) can be carried out satisfactorily, with either approach.

B. A special case of Theorem 3.1 was established earlier by Truong-van ([27], Thm. 6). In fact this is given for strongly harmonizable processes of the form $X_t = AY_t$, $t \in G$, where A is a bounded invertible operator on $L_0^2(P)$ onto itself, and the Y_t-process is stationary. The result was given in the multidimensional case. However, the latter extension is not too difficult. The basic idea of Proposition 3.2 already appeared in a tentative form in the thesis of Kelsh ([15], pp. 65–69).

C. The multidimensional case with a different proof, for the weakly harmonizable case, was given in [6], under the same hypothesis. When the latter work was completed, unfortunately, the authors had not seen [27] and so no reference was made to it. The discussion on positive (semi-) definiteness given above should be helpful in the general case also, since not everybody seems to follow the classical Stone [26] definition of these concepts. It should also be noted that positive definiteness for point functions (and kernels) takes a somewhat different appearance than that for the bilinear (or sesquilinear) forms. An extended discussion is given here to draw some attention to these distinctions.

4. SAMPLING PROBLEMS

When the index is $\mathbb{R}$, sampling the process at appropriately chosen time points to base inference is a nontrivial matter since observing at equidistant intervals (the so-called periodic sampling) can introduce an "aliasing" effect, i.e., the sample paths of *different* processes may coincide at these points. This problem exists even for the stationary processes. Thus to avoid this difficulty, the time points of observation must be chosen with care. The following result presents a solution of this problem for a class of harmonizable processes.

Theorem 4.1.

Let $X : \mathbb{R} \to L_0^2(P)$ be a weakly harmonizable process with its spectral function vanishing off a set of diameter $\alpha_0 < \infty$, in $\mathbb{R}^2$. Then for each $\alpha > \alpha_0$, the processes $X_{n,\alpha} : \mathbb{R} \to L_0^2(P)$ defined by

$$X_{n,\alpha}(t) = \sum_{k=-n}^{n} X\left(\frac{k\pi}{\alpha}\right) \frac{\sin(\alpha t - k\pi)\sin^q(pt - \frac{kp\pi}{\alpha})}{(\alpha t - k\pi)p^q(t - \frac{k\pi}{\alpha})}, \tag{21}$$

where $q \geq 0$ is an integer and $0 < p < (\alpha - \alpha_0)/q$, satisfy

$$\|X(t) - X_{n,\alpha}(t)\|_2^2 \leq C_0 L_q(t) p^{-q} (\alpha - \alpha_0 - pq)^{-1} (\alpha/n)^{q+1}. \tag{22}$$

Here $0 < C_0 < \infty$ is an absolute constant and $L_q(t)$ is given by

$$L_q(t) = 2(1 - e^{-\pi})^{-1} (\tfrac{2}{\pi})^{q+1} |\sin \alpha t|. \tag{23}$$

This useful form with an error bound on the right side of (22) is obtained with an approximation of $e^{it\lambda}$ by suitable trigonometric polynomials. However the method of proof is different from the stationary or strongly harmonizable cases, even though the approximations given by (21) have the same form in all cases implying a certain "robustness." This is established by a specialization of [5], and the original motivation as well as the error estimation follow the work in [19]. If $q = 0$ in (21)–(23), the result reduces, in the stationary case, to the classical Kotel'nikov-Shanon formula. Further details will be omitted here.

5. ESTIMATION OF SPECTRAL FUNCTIONS

The importance of spectral functions for an analysis of harmonizable processes is clear from Section 3. Consequently it is useful to have a formula for the calculation of the spectral function from the observed data. Using Theorem 2.2 on the dilation and the classical results on stationary processes, this may be obtained.

Thus if r is the covariance function of a harmonizable process $\{X_t, t \in G\}$ with spectral function F, then it can be given by the following:

$$F(A_1, A_2) = \lim_{0 \leq S_1, S_2 \to \infty} \int_{-S_1}^{S_1} \int_{-S_2}^{S_2} \frac{e^{-i\lambda_1' u} - e^{-i\lambda_1 u}}{-iu} \cdot \frac{e^{i\lambda_2' v} - e^{i\lambda_2 v}}{iv} r(u, v) du dv, \tag{24}$$

where $A_i = (\lambda_i, \lambda_i')$, $\lambda_i \leq \lambda_i'$, and $F(\lambda_i \pm, \lambda_i' \pm) = F(\lambda_i, \lambda_i')$, $i = 1, 2$. Corresponding formula in the discrete parameter is similarly obtainable.

Since $r(\cdot, \cdot)$ is not generally known to the experimenter, an estimator of F is needed. For this, observe that (24) can be expressed as

$$F(A_1, A_2) = \lim_{0 \leq S_1, S_2 \to \infty} E(Y_{S_1, \lambda_1, \lambda_1'} \bar{Y}_{S_2, \lambda_2, \lambda_2'})$$

where

$$Y_{S, \lambda, \lambda'} = \int_{I_{\lambda,\lambda'}} \left(\int_{D_S} e^{ius} X_s ds \right) du$$

with $D_S = (-S, S)$, $I_{\lambda, \lambda'} = (\lambda, \lambda')$. This motivates the estimator

$$\widehat{F}_{S_1, S_2}(A_1, A_2) = \int_{A_1} \int_{A_2} \tilde{Y}_{S_1}(u) \bar{\tilde{Y}}_{S_2}(v) du dv, \tag{25}$$

where

$$\tilde{Y}_S(u) = \int\limits_{D_S} e^{ius} X_s ds.$$

Thus $\widehat{F}$ is computable from the observed segment of the process on D_S. It can be shown that when A_1, A_2 are continuity intervals of F, as in (24), then $\widehat{F}_{S_1,S_2}(A_1, A_2)$ is asymptotically unbiased. This is an easy consequence of known results, e.g. from ([21], Thm. 8.2). When is $\widehat{F}_{S_1,S_2}$ consistent? Conditions similar to those in [18] on higher moments or a use of convergence factors, called "spectral windows" (cf. [29], p. 259ff), can be used. Details are not yet worked out.

There is also interest in finding, if possible, a nonnegative spectral function associated with harmonizable processes. Since it is known that each harmonizable process is of Karhunen class (cf. e.g. [6], p. 59) and the latter has nonnegative spectral functions, one may follow this route. However, in the representation of the covariance as

$$r(s,t) = \int\limits_{\widehat{G}}\int\limits_{\widehat{G}} e^{is\lambda - it\lambda'} F(d\lambda, d\lambda') = \int\limits_{\widehat{G}} g(s,\lambda)\overline{g(t,\lambda)} H(d\lambda),$$

one does not have enough knowledge on $\{g(s,\cdot), s \in G\} \subset L^2(\widehat{G}, H)$, and this method does not necessarily give the desired solution. The alternative is a subclass of harmonizable processes, isolated by Kampé de Feriet and Frenkiel [13], hereafter called *class*(KF). This class was also studied independently by Rozanov [25], and by Parzen [18] under the name "asymptotic stationarity." This will be briefly discussed here.

Let $\{X_t, t \in \mathbb{R}\}$ be a mean continuous process from $L_0^2(P)$, with covariance r. It is then of class(KF) if

$$\lim_{T\to\infty} \frac{1}{T} \int\limits_0^T r(s, s+h) ds = \tilde{r}(h), \qquad h \in \mathbb{R}, \tag{26}$$

exists. One replaces the integral by a sum in the discrete case. It can be verified that $\tilde{r}(\cdot)$ is a stationary covariance, so that by the Bochner-Riesz theorem (cf. [16], p. 209)

$$\tilde{r}(h) = \int\limits_{\mathbb{R}} e^{ih\lambda} \mu(d\lambda), \tag{27}$$

for almost all (Lebesgue measure) h in $\mathbb{R}$, with μ as a bounded nonnegative measure, called the *associated spectral measure* of the process. Moreover, as Rozanov [25] noted, all strongly harmonizable processes belong to it. Recently Dehay [9] has shown that most weakly harmonizable processes, namely those whose spectral bimeasures ρ determine σ-finite Radon measures, belong to this class. However, this class does not exhaust the weakly harmonizable class since it is known (cf. e.g. [22], p. 297) that there exist discrete parameter weakly harmonizable processes which are not in class(KF). On the other hand, there are nonharmonizable (e.g. periodically correlated [14]) processes in class(KF). Thus it is a useful family to analyze both for its spectral properties and for estimation problems. Some interesting remarks on this class in [25] and an elaboration in [6] may be useful in considering further work on these and related future studies.

6. INFERENCE ON PROCESSES OF LOW MULTIPLICITY

It was already noted that each harmonizable process is of Karhunen class. The latter consists of families $\{X_t, t \in T\} \subset L_0^2(P)$ which are representable as

$$X_t = \int_S g(t,\lambda) Z(d\lambda), \qquad t \in T, \tag{28}$$

where $Z(\cdot)$ is a process of orthogonal increments (or orthogonally scattered vector measure on S into $L_0^2(P)$) and $g(t,\cdot) : S \to \mathbb{C}$ is a measurable function in $L^2(S,F)$, with $F(A) = E(|Z(A)|^2)$, $A \in \mathcal{B}(S)$, a σ-algebra of S (cf. e.g. [4], p. 473). Here $Z(\cdot)$ has desirable properties, but $g(t,\cdot)$ can have a very complicated structure. In fact Cramér has shown that there exist even strongly harmonizable processes of any prescribed multiplicity N, an integer, satisfying $1 \le N \le \infty$ (cf. [8], p. 11). On the other hand, for a harmonizable process, its mean square derivative of order k exists if the spectral measure has k (mixed) moments. This translates to these processes, when represented as (28), as: $g(\cdot,\lambda)$ is k-times differentiable and these derivatives belong to $L^2(S,F)$. To take advantage of this property it is desirable to make a preliminary simplification by decomposing the X_t-process into the purely nondeterministic and deterministic parts (called the Wold decomposition) and concentrate on the former, since the deterministic part does not play any significant role in either the prediction or inference studies. In the purely nondeterministic case there is the Cramér-Hida theory which includes the corresponding weakly harmonizable class of arbitrary multiplicity N. In fact, if $\{X_t, t \in \mathbb{R}\} \subset L_0^2(P)$ is a (mean) left continuous with right limits process at each $t \in \mathbb{R}$, then it can be represented as

$$X_t = \sum_{n=1}^{N} \int_{-\infty}^{t} g_n(t,\lambda) Z_n(d\lambda) \tag{29}$$

where $E(Z_n(A)\overline{Z_m}(B)) = \delta_{mn} F_n(A \cap B)$, $\sum_{n=1}^{N} \int_{-\infty}^{t} |g_n(t,\lambda)|^2 dF_n(\lambda) < \infty$, and $F_1 \succ F_2 \succ \cdots \succ F_n$ (i.e., F_n dominates F_{n+1}). Moreover, N is the minimal integer for which (29) holds; it is called the *multiplicity* of the process. Acomparison of (28), with $T = S = \mathbb{R}$, and

(29) illustrates the complicated nature of g_n's, called the *response functions* corresponding to the *innovation elements* Z_n. (See [8] about proofs of these assertions.) An interesting result in this study is that if each g_n and $\frac{\partial g_n}{\partial t}$ are bounded and continuous for each $\lambda \leq t$, $g_n(t,t) = 1$, and F_n has a density f_n $(= \frac{dF_n}{d\lambda}) \not\equiv 0$, which may have at most a finite number of discontinuities in any finite interval of $\mathbb{R}$, then necessarily the multiplicity $N = 1$. Thus

$$X_t = \int_{-\infty}^{t} g(t,\lambda) Z(d\lambda), \tag{30}$$

and $g(t,\lambda) = 0$ for $\lambda > 0$ (cf. [8], Thm. 5.1). In what follows only these processes will be discussed and a signal detection problem from a noisy output is outlined. Note that although stationary processes always have multiplicity one, there exist many nonstationary processes of multiplicity one, so that even this family is quite general.

Let $Y_t = S_t + X_t$, $t \in T$, be an additive model (cf. [23] on such problems) describing the observation process Y_t consisting of a signal process S_t and a noise process X_t both of which are of type (30), each with multiplicity one. In particular, suppose X_t is given by (30), but S_t is of the form

$$S_t = \int_{-\infty}^{t} g(t,\lambda)\xi_\lambda d\mu(\lambda), \tag{31}$$

where ξ_t is a measurable process all of whose sample paths are μ-integrable. Hence the observed Y_t is representable as (here μ is a Radon measure on $\mathbb{R}$):

$$Y_t = \int^{t} g(t,\lambda)\xi_\lambda d\mu(\lambda) + \int^{t} g(t,\lambda) dZ(\lambda). \tag{32}$$

Then the desired testing or detection problem is the hypothesis $H_0 : S_t = 0$ (no signal) *vs.* the alternative $H_1 : S_t \neq 0$. If the probability measures determined by the observation and noise processes are denoted P_Y and P_X, then the detection problem is nontrivial when $P_Y << P_X$ (the P_Y measure is dominated by P_X or $P_Y \prec P_X$). The likelihood ratio $f(w) = \frac{dP_Y}{dP_X}(w)$ can then be used, with the Neyman-Pearson-Grenander theorem, to distinguish H_1 from H_0 with a prescribed acceptance level of risk (cf. [20], p. 200ff). However, to get concrete results one has to assume more. For instance, interesting work was outlined by Baker and Gualtierotti [1], if the $Z(\cdot)$ process in (32) is Brownian motion and $\xi_t d\mu(t)$ in (31) is replaced by $\eta(t)$ which is a solution of

$$d\eta(t) = g(\eta(t))dt + dZ(t), \tag{33}$$

a special Itô-type stochastic differential equation so that $\eta(t)$ is a diffusion process. The calculation of likelihood ratios present challenging problems even with these simplifying assumptions (cf. e.g. [20], §4.4).

7. SOME COMPLEMENTS AND DIRECTIONS

If $\{X_t, t \in \mathbb{R}\}$ is harmonizable, then from the dilation property (cf. Theorem 2.2) and the classical results on stationary processes one can establish the following:

$$\frac{1}{\tau}\int_0^\tau X_t dt \to Z(0), \text{ in } L_0^2(P), \tau \to \infty, \tag{34}$$

and similarly

$$\frac{1}{2\tau}\int_{-\tau}^{\tau} X_t dt \to Z(0), \text{ in } L_0^2(P). \tag{35}$$

Here $Z(\cdot)$ is the representing measure of the process. Thus the X_t-process obeys the weak law of large numbers (WLLN) iff $Z(\cdot)$ does not change the origin, i.e., $Z(0) = 0$. However, the corresponding strong law needs a different set of techniques. Extending the earlier results for the strongly harmonizable processes by V. F. Gaposhkin and others, Dehay [10] has obtained the following result. For the SLLN not only should $Z(\cdot)$ not charge '0' but it should be "diffuse" at '0' which implies that the spectral measure F associated with $Z(\cdot)$ must satisfy $F(I_n, I_n) \to 0$ as $n \to \infty$ for the symmetric intervals I_n descending to 0, as $n \to \infty$. To state the result precisely let $\sigma(X : \tau) = \frac{1}{2\tau}\int_{-\tau}^{\tau} X_s ds$, $X_s = \int_{\mathbb{R}} e^{is\lambda} Z(d\lambda)$. Then both the SLLN and the speed of convergence obtained in [10] can be stated as:

7.1. Theorem

If $\{X_t, t \in \mathbb{R}\}$ is weakly harmonizable, $E(X_t) = 0$, with $Z(\cdot)$ as its representing measure, then the process obeys the SLLN iff there exists a $p \geq 2$ such that $\lim_{n\to\infty} Z(I_n) = 0$ a.e., where $I_n = (-p^{-n}, p^{-n})$. Moreover, if there is a nondecreasing function $g : \mathbb{R}^+ \to \mathbb{R}^+$ with the following two properties:

(a) for each integer $p \geq 2, q_0 \geq 1, 1 < a < \sqrt{p}$, one has $g^2(p^{q+1}) \leq a g^2(p^s)$, $s = p^q$, $q \geq q_0$ integers, and

(b) if β is the Grothendieck measure on $(\mathbb{R}, \mathcal{B}(\mathbb{R}))$ dominating the spectral measure F of the process [this will be the β of (13) in that case], and β has the property that for some $u_0 > 0$,

$$\int_{0<|u|<u_0} [(\log\log|u|^{-1}) g(|u|^{-1})]^2 \beta(du) < \infty,$$

then $\lim_{\tau\to\infty} g(\tau)[\sigma(X : \tau) - Z(0)] = 0$, a.e.

The details of this result involving several estimates for using the Borel-Cantelli lemma suitably, may be found in [10]. It is also well known that each weakly harmonizable process (or a field, meaning the index set $\mathbb{R}$ is replaced by $\mathbb{R}^k$, for example) can be approximated pointwise, in mean, by a sequence of strongly harmonizable ones, and then the metric approximation property of Hilbert space allows one to conclude that the convergence is uniform on compact subsets of the index set. This result has been refined, with even a more elementary argument, by Dehay and Moché [11]. They have shown that the sequential closure on compacta of the set of all processes $X : t \mapsto X(t) = \int_{\mathbb{R}^k} e^{i\langle t,\lambda\rangle} Y(\lambda) d\lambda$, where $Y : \mathbb{R}^k \to L_0^2(P)$ is (Bochner) integrable, contains the weakly harmonizable family as a proper subset. A natural problem here is to characterize the weakly harmonizable class by restricting the strongly harmonizable processes further. By strengthening the topology in this approximation procedure, Marc Mehlman has obtained such a result for his dissertation, at UCR, dealing with a finer structure of harmonizable classes.

There are several interesting directions along which a study on these processes can progress. Extensions of these results to the harmonizable random fields when the index G is not necessarily abelian are possible. Also from the point of view of inference, it appears useful to study such processes if the index set is a hypergroup (i.e., an object generalizing the double coset space of a locally compact group). For some indications of promising avenues in both these directions and for related references on the subject, one may see [24]. The finer aspects of the subject involving consistency and asymptotic distributions of estimators such as $\widehat{F}$ of (25), as well as the mean and covariance functions, seem to demand a serious effort even when the index is $\mathbb{R}$ or $\mathbb{Z}$. From the prediction and filtering points of view, much work remains to be done. A comprehensive treatment of the available results on these problems is recently given by Yaglom ([29], especially see Chapter IV). Most of this work with many concrete examples given in [29] would serve as guide posts for studying the corresponding problems for harmonizable random fields. In fact, each result on stationary processes, such as those initially pioneered by Wiener [28], motivates a study for the harmonizable case although the latter usually needs a different set of tools even when the final formula has a similar appearance in many cases.

Acknowledgement. This work was supported in part by the ONR Contract No. N00014-84-k-0356 (mod 00006).

REFERENCES

1. C. R. Baker and A. F. Gualtierotti (1984), "Detection of non-Gaussian signals in Gaussian noise," *in: Statistical Signal Processing*, Marcel Dekker Inc., New York, 107–116.

2. F. J. Beutler (1965), "The operator theory of the pseudo-inverse-II: unbounded operators with arbitrary range," *J. Math. Anal. Appl.* **10**, 471–493.

3. S. Bochner (1956), "Stationarity, boundedness, almost periodicity of random valued functions," *Proc. Third Berkeley Symp. Math. Statist. & Prob.* **2**, 7–27.

4. S. Cambanis (1975), "The measurability of a stochastic process of second order and its linear space," *Proc. Amer. Math. Soc.* **47**, 467–475.

5. D. K. Chang and M. M. Rao (1983), "Bimeasures and sampling theorems for weakly harmonizable processes," *Stochastic Anal. Appl.* **1**, 21–55.

6. _______ (1986), "Bimeasures and nonstationary processes," *in: Real and Stochastic Analysis*, Wiley, New York, 7–118.

7. H. Cramér (1951), "A contribution to the theory of stochastic processes," *Proc. Second Berkeley Symp. Math. Statist. & Prob.*, 329–339.

8. _______ (1971), *Structural and Statistical Problems For A Class of Stochastic Processes*, Princeton Univ. Press, New Jersey.

9. D. Dehay (1987), "On a class of asymptotically stationary harmonizable processes," *J. Multivar. Anal.* **22**, 251–257.

10. _______ (1987), "Strong law of large numbers for weakly harmonizable processes," *Stochastic Proc. Appl.* **24**, 259–267.

11. D. Dehay and R. Moché (1986), "Strongly harmonizable approximations of bounded continuous random fields," *Stochastic Proc. Appl.* **23**, 327–331.

12. N. Dunford and J. T. Schwartz (1958), *Linear Operators, Part I: General Theory*, Wiley-Interscience, New York.

13. J. Kampé de Feriet et F. N. Frenkiel (1959) "Estimation de la corrélation d'une fonction aléatoire nonstationaire," *C. R. Acad. Sci., Ser. B*, **249**, 348–351.

14. _______ (1962), "Correlation and spectra for nonstationary random functions," *Math Comp.* **16**, 1–21.

15. J. P. Kelsh (1978), *Linear Analysis of Harmonizable Time Series*," Ph.D Dissertation, Univ. of Calif., Riverside.

16. Loève, M. (1955), *Probability Theory*, Van Nostrand, Princeton, New Jersey.

17. R. Moché (1985), *Introduction aux Processus Harmonisables*, Lecture Notes, UER Math., Université des Sciences et Tech. de Lille.

18. E. Parzen (1962), "Spectral analysis of asymptotically stationary time series," *Bull. Inst. Int'l Statist.* **39**(2), 87–103.

19. Z. A. Piranashvili (1967), "On the problem of interpolation of random processes," *Theor. Prob. Appl.* **12**, 647–659.

20. M. M. Rao (1981), *Foundations of Stochastic Analysis*, Academic Press, New York.

21. _______ (1982), "Harmonizable processes: structure theory," *L'Enseign. Math.* **28**, 295–351.

22. _______ (1985), "Harmonizable, Cramér, and Karhunen classes of processes," *in: Handbook of Statistics, Vol 5*, North-Holland, Amsterdam, 279–310.

23. _______ (1989), "Harmonizable signal extraction, filtering and sampling," *in: Topics in NonGaussian Signal Processes*, Springer-Verlag, New York, 98–117.

24. _______ (1989), "Bimeasures and harmonizable Processes," *in: Probability Theory on Groups–IX*, Lect. Notes in Math., Springer-Verlag, Berlin (to appear).

25. Yu. A. Rozanov (1959), "Spectral analysis of abstract functions," *Theor. Prob. Appl.* **4**, 271–287.

26. M. H. Stone (1932), *Linear Transformations in Hilbert Space*, Amer. Math. Soc., Providence, Rhode Island.

27. B. Truong-van (1981), "Une généralisation du théorème de Kolmogorov-Aronszajn processus V-bornés q-dimensionnels: domaine spectral et dilations stationnaires," *Ann. Inst. Hénri Poincaré* **17**, 31–49.

28. N. Wiener (1949), *Extrapolation, Interpolation, and Smoothing of Stationary Time Series*, Wiley, New York.

29. A. M. Yaglom (1987), *Correlation Theory of Stationary and Related Random Functions*, Vols I & II, Springer-Verlag, New York.

30. K. Ylinen (1986), "Random fields on noncommutative locally compact groups," *in: Probability Measures on Groups—VIII*, Lect. Notes in Math. No. 1210, Springer-Verlag, Berlin, 365–386.

Hakim Khayyam Naishapuri
Persian Astronomer, Mathematician and Poet
(12th Century)

One Moment in Annihilation's Waste
One Moment, of the Well of Life to taste-
The Stars are setting and the Caravan
Starts for the Dawn of Nothing-Oh, make haste!

From the Robaiyat